Edited by
Katja Becker

Apicomplexan Parasites

D1642030

Titles of the Series "Drug Discovery in Infectious Diseases"

Selzer, P. M. (ed.)

Antiparasitic and Antibacterial Drug Discovery

From Molecular Targets to Drug Candidates

2009

ISBN: 978-3-527-32327-2

Forthcoming Topics of the Series

- Helminths
- Kinases in Parasites
- Kinetoplastida

Related Titles

Sansonetti, P. (ed.)

Bacterial Virulence

Basic Principles, Models and Global Approaches

2010

ISBN: 978-3-527-32326-5

Kaufmann, S. H. E., Walker, B. D. (eds.)

AIDS and Tuberculosis

A Deadly Liaison

2010

ISBN: 978-3-527-32270-1

Krämer, R., Jung K. (eds.)

Bacterial Signaling

2010

978-3-527-32365-4

Schaible, U. E., Haas, A. (eds.)

Intracellular Niches of Microbes

A Pathogens Guide Through the Host Cell

2009

ISBN: 978-3-527-32207-7

Edited by
Katja Becker

Apicomplexan Parasites

Molecular Approaches toward Targeted Drug
Development

The Editors

Volume Editor:

Prof. Dr. Katja Becker
University of Giessen
Nutritional Biochemistry
Heinrich-Buff-Ring 26-32
35392 Gießen
Germany
katja.becker@ernaehrung.uni-giessen.de

Series Editor:

Prof. Dr. Paul M. Selzer
Intervet Innovation GmbH
BioChemInformatics
Zur Propstei
55270 Schwabenheim
Germany
Paul.Selzer@intervet.com

Cover
Microscopic image of the malaria parasite
Plasmodium falciparum: with kind permission of the
Institut Pasteur, Paris, France.

Ribbon representation of the plasmodial flavoen-
zyme glutathione reductase - a drug target of
Plasmodium falciparum - with the bound cofactor
flavin adenine dinucleotide in ball and stick repre-
sentation (PDB-Code 1ONF): picture courtesy of
Dr. Richard J. Marhöfer and Prof. Dr. Paul M. Selzer,
Intervet Innovation GmbH, Schwabenheim,
Germany.

The parasite and the protein structure are connected
by protein crystals of *P. falciparum* glutathione
reductase: courtesy of Prof. Dr. Katja Becker, Justus
Liebig University, Giessen, Germany.

Limit of Liability/Disclaimer of Warranty: While the
publisher and authors have used their best efforts in
preparing this book, they make no representations
or warranties with respect to the accuracy or
completeness of the contents of this book and
specifically disclaim any implied warranties of
merchantability or fitness for a particular purpose.
No warranty can be created or extended by sales
representatives or written sales materials. The Advice
and strategies contained herein may not be suitable
for your situation. You should consult with a
professional where appropriate. Neither the
publisher nor authors shall be liable for any loss of
profit or any other commercial damages, including
but not limited to special, incidental, consequential,
or other damages.

Library of Congress Card No.: applied for

British Library Cataloguing-in-Publication Data
A catalogue record for this book is available from the
British Library.

**Bibliographic information published by
the Deutsche Nationalbibliothek**
The Deutsche Nationalbibliothek lists this
publication in the Deutsche Nationalbibliografie;
detailed bibliographic data are available on the
Internet at http://dnb.d-nb.de.

© 2011 WILEY-VCH Verlag & Co. KGaA,
Boschstr. 12, 69469 Weinheim, Germany

Wiley-Blackwell is an imprint of John Wiley & Sons,
formed by the merger of Wiley's global Scientific,
Technical, and Medical business with Blackwell
Publishing.

All rights reserved (including those of translation into
other languages). No part of this book may be
reproduced in any form – by photoprinting,
microfilm, or any other means – nor transmitted or
translated into a machine language without written
permission from the publishers. Registered names,
trademarks, etc. used in this book, even when not
specifically marked as such, are not to be considered
unprotected by law.

Cover Design   Adam-Design, Weinheim
Typesetting   Thomson Digital, Noida, India
Printing and Binding   Fabulous Printers Pte Ltd,
Singapore

Printed in Singapore
Printed on acid-free paper

ISBN: 978-3-527-32731-7

Foreword

The Apicomplexa are obligate intracellular eukaryotic parasites that infect metazoans, and bear a huge impact on both animal and human health worldwide. Since both the parasites and the hosts in which they cause disease belong to the eukaryotic kingdom – and despite the fact that astounding chemotherapeutic success has been achieved, notably with chloroquine prior to the emergence and spread of resistance in malarial parasites which rendered the drug useless in most parts of the world – it has been argued that selective intervention based on small molecules would be more difficult to achieve than in the case of bacterial pathogens. Phylogenetic research conducted during recent decades, however, has revealed that the Apicomplexa diverged very early from the main branch of Eukaryotes, and arose from a complex evolutionary path that involved secondary endosymbiosis. It is now well established that the enormous phylogenetic distance between these parasites and their metazoan hosts is reflected by apicomplexan-specific features in terms of organelle complement, biochemical pathways, and the properties of specific enzymes (even in cases where orthologous molecules are present in the vertebrate host).

The widely encompassing collection of chapters that constitute the present volume provides a comprehensive snapshot of ongoing drug discovery activities aimed at exploiting such divergences. The topics covered range from bioinformatics and chemogenomics approaches for target selection, to the characterization of compounds with established parasiticidal activity. A large part of the book is composed of chapters in which apicomplexan features not found in vertebrates are discussed from the perspective of their potential as targets for chemotherapy. These include the apicoplast, the remnant of a photosynthetic organelle retaining pathways that are absent from mammalian cells, and the calcium-dependent protein kinases, which possess a domain organization found only in plants and Alveolates, the phylum that includes the Apicomplexa and Ciliates – are discussed from the perspective of their potential as targets for chemotherapy. Other chapters are devoted to the exploitation of more subtle differences between orthologous enzymatic systems between host and parasite, for example heat shock proteins or the machineries pertaining to redox metabolism or cell cycle control.

Undoubtedly, much remains to be uncovered. It is natural that drug-discovery efforts are largely focused on those enzymes that can be recognized as potential

Apicomplexan Parasites. Edited by Katja Becker
Copyright © 2011 WILEY-VCH Verlag GmbH & Co. KGaA, Weinheim
ISBN: 978-3-527-32731-7

targets. This, however, leaves behind the large fraction of the parasite proteomes (approximately 60% in the case of *Plasmodium*, though a similar situation applies to other Apicomplexa) currently annotated as "hypothetical proteins," and which may hide prime targets, despite the fact that chemogenomics approaches fill this gap to some extent.

Nevertheless, the contents of this book provides glimmers of hope. Some compounds – for example, ferroquine and bi-cationic inhibitors of phospholipid metabolism – have entered clinical trials against malaria, although whether or not novel drugs based on these compounds will reach the market in the not-too-distant future remains to be seen. It is to be hoped that other projects discussed in this book, many of which are in the exploratory/discovery phase, will be pursued to development. The relatively recent implementation of specific institutions to promote the translation of early drug discovery projects into drug development for previously neglected diseases, such as the Medicines for Malaria Venture (MMV) and the Drugs for Neglected Diseases Initiative (DNDi) Public-Private Partnerships, is playing an important part in this endeavor.

Whilst the apicomplexan parasites remain formidable foes, the variety of approaches and targets addressed in this book demonstrate that the "anti-apicomplexan drug discovery pipeline" is well primed, largely as a result of efforts devoted to parasite genomics during the previous decades, and testifies to the vitality of ongoing research in the "arms race" against these pathogens. May this book stimulate interest and dedication of the research community, and thus contribute to successes in this race.

Lausanne
September 2010

Christian Doerig

Contents

Apicomplexan Parasites. Edited by Katja Becker
Copyright © 2011 WILEY-VCH Verlag GmbH & Co. KGaA, Weinheim
ISBN: 978-3-527-32731-7

Preface

Ever since the discovery of apicomplexan parasites by Anthony von Leeuwenhoek, who first detected the oocysts of *Eimeria* in a rabbit in 1674, the term Apicomplexa has comprised a growing number of protists causing severe and often neglected diseases in humans and animals worldwide. The Apicomplexa possess a unique plastid-like organelle, the apicoplast, and an apical complex structure that is involved in host cell invasion. Many of these parasites undergo complex developmental cycles, and exist in both intracellular and extracellular forms in various hosts. Diseases caused by apicomplexan parasites include, among others, malaria, toxoplasmosis, cryptosporidiosis, coccidiosis, babesiosis, and theileriosis. Worldwide, *Plasmodium* accounts for up to 500 million clinical cases of malaria each year, with 1–2 million deaths, mainly in young children, while *Toxoplasma gondii* infects approximately one-third of the human population. Although most *Toxoplasma* infections are benign, severe opportunistic diseases affect immunodeficient or immunosuppressed individuals or children infected *in utero*. Likewise, *Cryptosporidium* causes infections that are especially harmful for immunocompromised individuals. Coccidiosis, a disease mainly of chicken and other poultry, caused by *Eimeria* spp., is a major problem for the global food industry, not least in developed countries. In this case, the annual global loss has been estimated to be in the range of US$0.3–3 billion, and improved strategies for the effective control of eimerian parasites are required. Other parasites in this family include *Babesia* and *Theileria*, which cause a hemolytic disease resembling malaria in cattle and sheep, and a severe and often fatal lymphoproliferative disease in cattle, respectively.

For many parasitologists it is difficult to believe that it has not yet been possible to combat apicomplexan parasites – in spite of intense drug discovery and vaccine development efforts, of numerous available target molecules, novel highly active drug candidates, and approved drugs for which antiparasitic action could be an additional indication. Whilst it has always been difficult for scientists to influence political and industrial priorities and decision making, it is nonetheless possible for the scientific community to make continuous joint efforts and suggestions, with the hope of improving the situation for patients suffering from infections. Over the past years, these efforts have been increasingly supported by initiatives not only to

Apicomplexan Parasites. Edited by Katja Becker
Copyright © 2011 WILEY-VCH Verlag GmbH & Co. KGaA, Weinheim
ISBN: 978-3-527-32731-7

commonly share genomic, transcriptomic, proteomic, and structural data, but also to share and exchange materials and methods as well as chemical compound libraries, data acquired from inhibitor screens, and the results of chemical genetics studies.

This book represents the second volume of a series dealing with drug discovery and development approaches against infectious diseases. Whereas, the first volume "Antiparasitic and Antibacterial Drug Discovery: From Molecular Targets to Drug Candidates" dealt with general drug discovery approaches towards diseases caused by protozoans, multicellular parasites, and bacteria, the present volume is solely dedicated to the Apicomplexa. This decision was made not only on the basis of the worldwide importance of the respective organisms and the diseases they cause, but also because the apicoplast harbors unique enzymes and metabolic pathways that represent highly attractive drug targets. Both aspects strongly suggest the coordination of drug discovery and development strategies against different Apicomplexa.

We are grateful to all authors who contributed articles to this book, for their excellent work and their constructive cooperation. We furthermore are indebted to the Justus Liebig University Giessen, Germany and the Intervet Innovation GmbH, Schwabenheim, Germany for their support and inspiration over the past months. In particular, we wish to thank Timothy Bostick for his excellent editorial assistance.

Giessen and Schwabenheim *Katja Becker and*
September 2010 *Paul M. Selzer*

List of Contributors

Heike Adler
University of Heidelberg
Biochemistry Center
Im Neuenheimer Feld 328
69120 Heidelberg
Germany

Uday Bandyopadhyay[*]
Indian Institute of Chemical Biology
Division of Infectious Diseases and
Immunology
4, Raja S. C. Mullick Road, Jadavpur
Kolkata 700032
India
E-mail: ubandyo_1964@yahoo.com

Sailen Barik[*]
University of South Alabama
College of Medicine
Department of Biochemistry and
Molecular Biology, MSB 2370
307 University Blvd
Mobile, AL 36688-0002
USA

Present address:
Cleveland State University
Center for Gene Regulation in Health
and Disease
Room SR351, 2121 Euclid Avenue
Cleveland, OH 44115
USA
E-mail: s.barik@csuohio.edu

Stefan Baumeister
University of Marburg
FB Biology/Parasitology
35043 Marburg
Germany

Katja Becker[*]
Justus-Liebig-University
Interdisciplinary Research Center
Chair of Nutritional Biochemistry
Heinrich-Buff-Ring 26-32
35392 Giessen
Germany
E-mail: katja.becker@ernaehrung.uni-
giessen.de

Christophe Biot[*]
Université de Lille 1
Unité de Catalyse et Chimie du Solide
UMR CNRS 8181, ENSCL
Bâtiment C7, B.P. 90108
59652 Villeneuve d'Ascq Cedex
France
E-mail: christophe.biot@univ-lille1.fr

Apicomplexan Parasites. Edited by Katja Becker
Copyright © 2011 WILEY-VCH Verlag GmbH & Co. KGaA, Weinheim
ISBN: 978-3-527-32731-7

and

Université de Lille1
Unité de Glycobiologie Structurale et
Fonctionnelle
CNRS UMR 8576, IFR 147
59650 Villeneuve d'Ascq Cedex
France

Subir Biswas
Central Drug Research Institute
Division of Molecular and Structural
Biology M.G. Road
226001 Lucknow
India

Sabine Bork-Mimm*
Bavarian Health and Food Safety
Authority (Bayerisches Landesamt für
Gesundheit und
Lebensmittelsicherheit, LGL)
Department Oberschleissheim
(Dienststelle Oberschleissheim)
Veterinaerstrasse 2
85764 Oberschleissheim
Germany
E-mail: sabine.bork-mimm@lgl.bayern.
de

Conor R. Caffrey
University of California San Francisco
Department of Pathology and the
Sandler Center for Drug Discovery
Byers Hall, CA 94158
USA

Sergio Caldarelli
Université Montpellier 2
Institut des Biomolécules Max
Mousseron
UMR 5247 CNRS-UM1&2
CC1705, Place E. Bataillon
34095 Montpellier Cedex 5
France

Nicholas D.P. Cosford
Sanford Burnham Medical Research
Institute
10901 North Torrey Pines Road
La Jolla, CA 92037
USA

Thomas Dandekar*
University of Würzburg
Department of Bioinformatics
Biocenter
Am Hubland
97074 Würzburg
Germany
E-mail: dandekar@biozentrum.uni-
wuerzburg.de

Elisabeth Davioud-Charvet*
Biochemistry Center of the University
of Heidelberg
Im Neuenheimer Feld 328
69120 Heidelberg
Germany

and

University of Strasbourg
European School of Chemistry,
Polymers and Materials (ECPM)
UMR7509 CNRS
25 rue Becquerel
67087 Strasbourg Cedex 2
France
E-mail: elisabeth.davioud@unistra.fr

Bianca Derrer
University Hospital Heidelberg
Department of Infectious Diseases,
Parasitology
Im Neuenheimer Feld 326
69120 Heidelberg
Germany

Sumanta Dey
Indian Institute of Chemical Biology
Division of Infectious Diseases and
Immunology
4, Raja S. C. Mullick Road, Jadavpur
Kolkata 700032
India

Daniel Dive
Université Lille Nord de France
Institut Pasteur de Lille
UMR CNRS 8024, CIIL, Inserm U 1019
1 rue du Pr. Calmette
59019 Lille Cedex
France

Laurent Fraisse
Sanofi-Aventis Recherche et
Developpement
Unité thérapeutique des maladies
infectieuse
195 route d'Espagne
31036 Toulouse
France

Karin Fritz-Wolf
Max Planck Institute for Medical
Research
Department of Biophysics
Jahnstraße 29
69120 Heidelberg
Germany

Gilles Gargala[*]
University of Rouen
Faculty of Medicine & Pharmacy
Department of Parasitology
Rouen
France
E-mail: gilles.gargala@univ-rouen.fr

Suresh Kumar Gorla
Brandeis University
Department of Biology
Waltham, MA 02454
USA

Robin Das Gupta
University of Bern
Institute of Cell Biology
Baltzerstrasse 4
3012 Bern
Switzerland

Saman Habib[*]
Central Drug Research Institute
Division of Molecular and Structural
Biology M.G. Road
226001 Lucknow
India
E-mail: saman.habib@gmail.com

Anja R. Heckeroth
Intervet Innovation GmbH
Profiling Antiparasitic
Zur Propstei
55270 Schwabenheim
Germany

Christian Hedberg
Department of Chemical Biology
Max Planck Institute of Molecular
Physiology
Otto-Hahn-Strasse 11
44227 Dortmund
Germany

Lizbeth Hedstrom[*]
Brandeis University
Departments of Biology and Chemistry
Waltham, MA 02454
USA
E-mail: hedstrom@brandeis.edu

Andrew Hemphill*
University of Berne
Institute of Parasitology
Länggass-Strasse 122
3012 Berne
Switzerland
E-mail: andrew.hemphill@ipa.unibe.ch

Jan A. Hiss*
Eidgenössische Technische Hochschule
(ETH)
Department of Chemistry and Applied
Biosciences
Zürich
Switzerland
E-mail: jan.hiss@pharma.ethz.ch

Ikuo Igarashi*
Obihiro University of Agriculture and
Veterinary Medicine
National Research Center for Protozoan
Diseases
Obihiro, Hokkaido 080-8555
Japan
E-mail: igarcpmi@obihiro.ac.jp

Corey Johnson
Brandeis University
Graduate Program in Chemistry
Waltham, MA 02454
USA

Esther Jortzik
Justus-Liebig-University
Interdisciplinary Research Center
Heinrich-Buff-Ring 26-32
35392 Giessen
Germany

Barbara Kappes*
University Hospital Heidelberg
Department of Infectious Diseases,
Parasitology
Im Neuenheimer Feld 326
69120 Heidelberg
Germany
E-mail: Barbara.kappes@urz.uni-
heidelberg.de

Denis Matovu Kasozi
University of Gießen (Justus-Liebig-
University)
Interdisciplinary Research Center
Chair of Nutritional Biochemistry
Heinrich-Buff-Ring 26-32
35392 Giessen
Germany

Jihan Khan
Brandeis University
Graduate Program in Chemistry
Waltham, MA 02454
USA

Dominik Kugelstadt
University Hospital Heidelberg
Department of Infectious Diseases
Parasitology
Im Neuenheimer Feld 326
69120 Heidelberg
Germany

and

Genzyme Virotech GmbH
Löwenplatz 5
65428 Rüsselsheim
Germany

Don Antoine Lanfranchi
University of Strasbourg
European School of Chemistry,
Polymers and Materials (ECPM)
UMR7509 CNRS
25 rue Becquerel
67087 Strasbourg Cedex 2
France

Kai Lüersen
Westfalian Wilhelms University
Institute for Zoophysiology
Hindenburgplatz 55
48143 Münster
Germany

Richard J. Marhöfer
Intervet Innovation GmbH
BioChemInformatics
Zur Propstei
55270 Schwabenheim
Germany

Peter Meissner
University of Ulm
Department of Pediatrics
89075 Ulm
Germany

Thorsten Meyer
Intervet Innovation GmbH
Lead Optimization
Zur Propstei
55270 Schwabenheim
Germany

Christian Miculka
Intervet Innovation GmbH
Zur Propstei
55270 Schwabenheim
Germany

Present address:
Merial Limited
3239 Satellite Boulevard
Duluth, GA 30096
USA

Snober S. Mir
Central Drug Research Institute
Division of Molecular and Structural
Biology M.G. Road
226001 Lucknow
India

Ingrid B. Müller[*]
Bernhard-Nocht-Institute for Tropical
Medicine
Biochemical Parasitology
Bernhard-Nocht-Str. 74
20359 Hamburg
Germany
E-mail: ibmueller@bni-hamburg.de

Joachim Müller
University of Berne
Institute of Parasitology
Länggass-Strasse 122
3012 Berne
Switzerland

Norbert Müller
University of Berne
Institute of Parasitology
Länggass-Strasse 122
3012 Berne
Switzerland

*Sylke Müller**
University of Glasgow
Institute of Infection, Immunity and
Inflammation
College of Medical, Veterinary and Life
Sciences
120 University Place
Glasgow G12 8TA
UK
E-mail: s.muller@bio.gla.ac.uk

Eva-Maria Patzewitz
University of Glasgow
Institute of Infection, Immunity and
Inflammation
College of Medical, Veterinary and Life
Sciences
120 University Place
Glasgow G12 8TA
UK

Diana Penarete
Université Montpellier 2
Dynamique des Interactions
Membranaires Normales et
Pathologiques
UMR 5235 CNRS-UM2
CC107, Place E. Bataillon
34095 Montpellier Cedex 5
France

Suzanne Peyrottes
Université Montpellier 2
Institut des Biomolécules Max
Mousseron
UMR 5247 CNRS-UM1&2
CC1705, Place E. Bataillon
34095 Montpellier Cedex 5
France

Bruno Pradines
Institut de Recherche Biomédicale des
Armées, Antenne de Marseille
Unité de Recherche en Biologie et
Epidémiologie Parasitaires
URMITE-UMR 6236
Allée du Médecin Colonel Jamot, Parc le
Pharo, BP 60109
13262 Marseille Cedex
France

G. Sridhar Prasad
CalAsia Pharmaceuticals, Inc.
6330 Nancy Ridge Drive, Suite 102
San Diego, CA 92121
USA

Stefan Rahlfs
Justus-Liebig-University
Interdisciplinary Research Center
Chair of Nutritional Biochemistry
Heinrich-Buff-Ring 26-32
35392 Giessen
Germany

Andreas Rohwer
Intervet Innovation GmbH
BioChemInformatics
Zur Propstei
55270 Schwabenheim
Germany

Jean-François Rossignol
Stanford University School of Medicine
Division of Gastroenterology &
Hepatology
Department of Medicine
Stanford, CA
USA

and

University of Oxford
Exeter College
Glycobiology Institute
Department of Biochemistry
Oxford
UK

Joana M. Santos
University of Geneva
CMU
Department of Microbiology and
Molecular Medicine
1 Rue Michel-Servet
1211 Geneva 4
Switzerland

Theo P.M. Schetters[*]
Microbiology R&D Department
Intervet/Schering-Plough Animal
Health
Wim de Korverstraat 35
5831 Boxmeer
The Netherlands
E-mail: theo.schetters@intervet.com

and

University Montpellier I
LBCM, UFR Pharmacy
Av. Charles Flahault 15
34093 Montpellier Cedex 5
France

R. Heiner Schirmer[*]
University of Heidelberg
Biochemistry Center
Im Neuenheimer Feld 328
69120 Heidelberg
Germany
E-mail: Heiner.Schirmer@bzh.uni-
heidelberg.de

Gisbert Schneider
Eidgenössische Technische Hochschule
(ETH)
Department of Chemistry and Applied
Biosciences
Zürich
Switzerland

Frank Seeber
Robert-Koch-Institut
Nordufer 20
13353 Berlin
Germany

Paul M. Selzer[*]
Intervet Innovation GmbH
BioChemInformatics
Zur Propstei
55270 Schwabenheim
Germany
E-mail: paul.selzer@intervet.com

J. Edward Semple
University of Oxford
Exeter College
Glycobiology Institute
Department of Biochemistry
Oxford
UK

Lisa Sharling
University of Georgia
Center for Tropical and Emerging
Global Diseases
500 D.W. Brooks Drive
Athens, GA 30602
USA

Irmgard Sinning
Biochemiezentrum der Universität
Heidelberg BZH
Im Neuenheimer Feld 328
69120 Heidelberg
Germany

Dominique Soldati-Favre[*]
University of Geneva
CMU
Department of Microbiology and
Molecular Medicine
1 Rue Michel-Servet
1211 Geneva 4
Switzerland
E-mail: dominique.soldati-
favre@unige.ch

Andrew V. Stachulski
University of Oxford
Exeter College
Glycobiology Institute
Department of Biochemistry
Oxford
UK

Janet Storm
University of Glasgow
Institute of Infection, Immunity and
Inflammation
College of Medical, Veterinary and Life
Sciences
120 University Place
Glasgow G12 8TA
UK

Boris Striepen
University of Georgia
Center for Tropical and Emerging
Global Diseases and the Department of
Cellular Biology
500 D.W. Brooks Drive
Athens, GA 30602
USA

Xin Sun
Brandeis University
Graduate Program in Biochemistry
Waltham, MA 02454
USA

Mohamad Alaa Terkawi
Obihiro University of Agriculture and
Veterinary Medicine
National Research Center for Protozoan
Diseases
Obihiro, Hokkaido 080-8555
Japan

Ivo Tews[*]
University of Southampton
School of Biological Sciences
Institute for Life Sciences (IfLS)
B85, Highfield Campus
Southampton SO17 1BJ
UK
E-mail: ivo.tews@soton.ac.uk

Hon Q. Tran
Intervet Innovation GmbH
BioChemInformatics
Zur Propstei
55270 Schwabenheim
Germany

Henri J. Vial[*]
Université Montpellier 2
Dynamique des Interactions
Membranaires Normales et
Pathologiques
UMR 5235 CNRS-UM2
CC107, Place E. Bataillon
34095 Montpellier Cedex 5
France
E-mail: vial@univ-montp2.fr

Rolf D. Walter
Bernhard-Nocht-Institute for Tropical
Medicine
Biochemical Parasitology
Bernhard-Nocht-Str. 74
20359 Hamburg
Germany

Sharon Wein
Université Montpellier 2
Dynamique des Interactions
Membranaires Normales et
Pathologiques
UMR 5235 CNRS-UM2
CC107, Place E. Bataillon
34095 Montpellier Cedex 5
France

Carsten Wrenger[*]
Bernhard-Nocht-Institute for Tropical
Medicine
Biochemical Parasitology
Bernhard-Nocht-Str. 74
20359 Hamburg
Germany
E-mail: wrenger@bni-hamburg.de

Ke Xiao
University of Würzburg
Department of Bioinformatics,
Biocenter
Am Hubland
97074 Würzburg
Germany

Kathleen Zocher
Justus-Liebig-University
Interdisciplinary Research Center
Chair of Nutritional Biochemistry
Heinrich-Buff-Ring 26-32
35392 Giessen
Germany

Part One
Screening, Bioinformatics, Chemoinformatics, and Drug Design

Apicomplexan Parasites. Edited by Katja Becker
Copyright © 2011 WILEY-VCH Verlag GmbH & Co. KGaA, Weinheim
ISBN: 978-3-527-32731-7

1
Drug Discovery Approaches Toward Anti-Parasitic Agents

Andreas Rohwer, Richard J. Marhöfer, Conor R. Caffrey, and Paul M. Selzer[*]

Abstract

Parasitic diseases afflict hundreds of millions of people worldwide, and are a major issue in animal health. Because most drugs available today are old and have many limitations, novel drugs for the treatment of human and animal parasitic diseases are urgently needed. Modern research disciplines such as genomics, proteomics, metabolomics, chemogenomics, and other "-omics" technologies improve the quality of the drug discovery process and influence the design of novel anti-parasitic agents. These include the application of high-throughput technologies such as DNA/RNA sequencing, microarrays, mass spectrometry, high-throughput screening, and bio/chemoinformatics. Here, an overview is provided of the drug discovery workflow, and the steps employed to generate novel drug candidates with anti-parasitic activity are briefly described.

Drug Discovery Initiatives to Accelerate the Development of Novel Anti-Parasitic Drugs for Humans and Animals

Infectious diseases, including those caused or transmitted by parasites, are responsible for substantial morbidity and mortality worldwide and affect several billion people globally, particularly in developing countries [1]. Until recently, infectious diseases were viewed as a problem of the past; however, the emergence of drug-resistant organisms makes the need for new drugs or vaccines more important than ever before. Moreover, as these diseases predominantly afflict inhabitants of poor countries, drug discovery efforts are minimal due to the lack of returns on investment. Accordingly, diseases such as malaria, leishmaniasis, Chagas disease, elephantiasis, or schistosomiasis are often called "neglected diseases" [2].

More recently, the growing realization of the humanitarian and economic consequences of neglected diseases in poor countries has spurred the establishment of new organizations specifically focused on novel anti-parasitic drug development

[*]Corresponding author

Apicomplexan Parasites. Edited by Katja Becker
Copyright © 2011 WILEY-VCH Verlag GmbH & Co. KGaA, Weinheim
ISBN: 978-3-527-32731-7

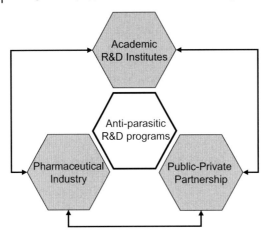

Figure 1.1 Research and development activities to fight neglected diseases. Anti-parasitic research and development programs were initiated by academic R&D centers, public–private partnerships, and the pharmaceutical industry. Intensive collaboration is key to optimizing R&D output.

[3–5]. Collaborations between the pharmaceutical industry, specialized academic drug discovery centers, and public–private partnerships (PPPs) have been initiated to support anti-parasitic drug discovery and development programs (Figure 1.1).

Public–private partnerships focus to combine the skills and resources of academia, the pharmaceutical industry, and contract research teams, with the goal of generating independent research and development (R&D) consortia. Well-known initiatives are the World Health Organization's Special Program for Research and Training in Tropical Diseases (WHO/TDR; http://www.who.int/tdr), the Drugs for Neglected Diseases *initiative* (DND*i*; http://www.dndi.org), the Institute for One World Health (iOWH; http://www.oneworldhealth.org), and the Bill & Melinda Gates Foundation (B&MGF; http://www.gatesfoundation.org). For instance, the DND*i* has built a virtual, not-for profit R&D organization for developing new drugs against kinetoplastid diseases, which include human African trypanosomiasis, visceral leishmaniasis, and Chagas disease. The partners of DND*i* are Doctors Without Borders, the Oswaldo Cruz Foundation of Brazil, the Indian Council for Medical Research, the Kenya Medical Research Institute, the Ministry of Health in Malaysia, the Pasteur Institute in France, and the WHO/TDR. DND*i* has already registered two products; in 2007 and 2008, respectively, the antimalarial drugs fixed-dose artesunate-amodiaquine (AS/AQ) and fixed-dose artesunate-mefloquine (AS/MQ) were launched [2].

Development projects organized by PPPs are often supported by the pharmaceutical industry itself. One particular project in the public eye is the Accelerating Access Initiative (AAI; http://www.ifpma.org/health/hiv/health_aai_hiv.aspx), a global initiative to broaden access to and ensure affordable and safe use of drugs for HIV/AIDS-related illnesses. Related programs, such as the Global Alliance to Eliminate Lymphatic Filariasis (GAELF; http://www.filariasis.org) and the Medicines for Malaria Venture (MMV; http://www.mmv.org), exist for many parasitic diseases. In addition, the pharmaceutical industry also invests directly in anti-parasitic research activities, with many companies having established state-of-the-art research facilities that concentrate exclusively on the development of drugs and vaccines for neglected diseases. Prominent research centers include the Novartis Institute for Tropical Diseases (NITD; http://www.novartis.com/research/nitd), the GlaxoSmithKline Drug Discovery Center for Diseases of the Developing World (DDW; http://www.gsk.com), or the MSD Wellcome Trust Hilleman Laboratories (http://www.hillemanlaboratories.in).

Needless to say, another major source of novel drugs in anti-parasitics stems from the extensive research activities in academic facilities. A relatively recent trend has been the foundation of academic drug discovery centers that focus exclusively on R&D in the field of neglected diseases. These aim to translate basic biomedical research into candidate medicines for neglected diseases. Examples are the Drug Discovery Unit of the University of Dundee (DDU; http://www.drugdiscovery.dundee.ac.uk), the Sandler Center for Drug Discovery (formerly Sandler Center for Basic Research in Parasitic Diseases) at the University of California San Francisco (http://www.sandler.ucsf.edu), and the Seattle Biomedical Research Institute (SBRI; http://www.sbri.org).

It is worth noting that anti-parasitic drug R&D programs are not confined to human medicine. Indeed, the animal health industry also performs intensive research on novel anti-parasitics [6–8]. This is important, as the situation of people in developing countries suffering from neglected diseases is aggravated by drastic economic losses in agriculture due to parasitic infections in farm animals. In this context, some animal health companies support developing countries in the framework of corporate social responsibility activities. For example, Intervet/Schering-Plough Animal Health (ISPAH; http://www.intervet.com) has an ongoing cooperation with the Indian non-governmental organization *Bharatiya Agro Industries Foundation* (BAIF; http://www.baif.org.in), whereby poor farmers in India have access to ISPAH's range of livestock products, including vaccines and anti-parasitic agents. It is expected that more than two million rural families could benefit from this project.

Innovation from Anti-Parasitic Drug Discovery Approaches

A *parasite* is defined as an animal that lives completely at the expense of plants, other animals, or humans [9]. In general, parasites are much smaller than their hosts, show a high degree of specialization for their mode of life, and reproduce more quickly and in greater numbers than their hosts. Parasites belong to a wide range of biologically diverse organisms and, based on their interactions with their hosts, are often classified into three categories: parasitic protozoa; endo-parasites; and ecto-parasites [7, 9].

1) Parasitic protozoa are unicellular microorganisms that infect humans or animals and either live extra- or intracellularly. Representatives include *Plasmodium falciparum*, *Trypanosoma cruzi*, and *Leishmania donovani*; causing malaria, Chagas disease, and leishmaniasis, respectively [9].

2) Endo-parasites are mainly multicellular helminthes that have adapted to live in the host's gastrointestinal tract, or systemically. Well-known endo-parasites are the nematode *Brugia malayi*, the causative agent of lymphatic filariasis, and the blood-fluke *Schistosoma mansoni*, which causes schistosomiasis [9, 10].

3) Ecto-parasites are parasitic organisms that live on the surface of their hosts. In most cases, ecto-parasites do not cause fatal maladies by themselves but affect the health of their hosts by transmitting pathogenic viruses, bacteria, or protozoa [9, 11]. Taxonomically, the majority of ecto-parasites belong to the phylum Arthropoda, and include important organisms such as fleas, flies, and ticks.

Historically, all marketed anti-parasitic products have been discovered by screening synthetic and natural compounds against intact parasites, either in culture or in animal models [12]. Such physiology-based assays, bioassays, or phenotypic assays involve parasites cultured *in vitro*, and exist for many different protozoa, endo-parasites, and ecto-parasites [13–16]. The main benefit of testing candidate compounds directly on whole organisms is that compounds with anti-parasitic activity are immediately apparent, suggesting that they possess the physico-chemical properties that allow them to penetrate the membrane barriers of the parasites in order to reach their molecular targets [17]. Since the simultaneous optimization of lead compounds for optimal anti-parasitic activity and bioavailability is one of the major hurdles in the lead optimization process, the advantage of bioassays should not be underestimated. Bioassays continue to play an important role in today's drug discovery process, particularly during the identification and optimization of novel anti-parasitic compounds [7].

On the other hand, the use of bioassays as the sole screening platform has a disadvantage. Those potential drugs with a high activity against attractive anti-parasitic target molecules, but no activity in bioassays (e.g., due to disadvantageous physico-chemical properties), are discarded. For this reason, alternative target- or mechanism-based drug screening strategies have been developed [8].

The Process of Target-Based Drug Discovery

In contrast to physiology-based drug discovery screens, the target-based approach starts with the identification of a protein that is used to search for new active compounds in *in vitro* screens [18, 19]. The goal of the target-based approach, as with every drug discovery workflow, is to provide drug candidates for the downstream development process, finally ending with a newly registered drug (Figure 1.2). In principle, the target-based approach consists of four major steps: (i) target identification; (ii) target validation; (iii) lead discovery; and (iv) lead optimization, including the *in vitro* profiling of the optimized lead structures (Figure 1.2). Target-based drug discovery is a highly technology-driven process, which particularly benefits from advances in modern "omics" research areas. Omics is a neologism which refers to a broad area of study in biology of fields ending in the suffix "-omics," such as genomics, transcriptomics, proteomics, or metabolomics (http://omics.org). Omics

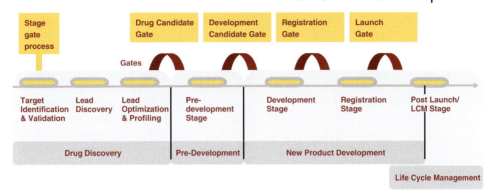

Figure 1.2 Drug discovery and development workflow. The workflow is organized as a stage-gate model; that is, a product development process is divided into stages separated by gates. At each gate, the continuation of the development process is decided by the organization. The drug discovery stage typically consists of target identification and validation, lead discovery and optimization, and profiling.

sciences apply large-scale experiments in order to analyze complete biological entities such as genomes, proteomes, metabolomes, and so on. They are enabled by major advances in modern high-throughput technologies such as DNA/RNA sequencing, microarrays, mass spectrometry, high-throughput screening, or combinatorial and medicinal chemistry – technologies that are increasingly common and affordable [20]. These technologies have already started to improve the quality and quantity of the drug discovery process [21].

Target Identification

Target identification starts with the discovery of a relevant drug target believed to be essential for the survival of a parasitic organism [22]. In order to avoid or minimize potential toxicity effects prior to the development of a new anti-parasitic drug, an optimal drug target would be absent from the host [23]. However, experience has shown that many of the existing anti-parasitic drugs act on target molecules which also exist in the host organisms [10, 24, 25].

Common methods for the selection of potential drug targets are classical biochemistry and molecular biology techniques. For example, the reverse transcriptase-polymerase chain reaction (RT-PCR) can be used to verify the expression of a potential target protein in the critical life stages of a parasite [19, 26]. Alternatively, information that can be "mined" in genome and drug target databases such as EuPathDB (http://w1.eupathdb.org/eupathdb) [27], GeneDB (http://www.genedb.org) [28], and the TDR target database (http://tdrtargets.org) [29], enables the identification of new drug targets. Such databases contain a wealth of data relating to parasite genes, proteins, homologs, transcript expression, single nucleotide polymorphisms (SNPs), cellular localization, and putative functions. It is expected that the data content in these databases will continue to increase due to advances in high-throughput technologies, and their ever-greater data output. For example, the area

Table 1.1 Published genomes of apicomplexan organisms.

Parasite	Taxonomy	Link
Babesia bovis T2Bo	Aconoidasida	Genbank accession AAXT00000000
Theileria annulata str. Ankara	Aconoidasida	http://www.sanger.ac.uk/Projects/
Theileria parva str. Muguga	Aconoidasida	Genbank accession AAGK00000000
Cryptosporidium hominis TU502	Coccidia	http://cryptodb.org/cryptodb/
Cryptosporidium parvum Iowa	Coccidia	http://cryptodb.org/cryptodb/
Plasmodium yoelii str. 17XNL	Aconoidasida	http://plasmodb.org/plasmo/
Plasmodium falciparum 3D7	Aconoidasida	http://plasmodb.org/plasmo/
Toxoplasma gondii ME49	Coccidia	http://toxodb.org/toxo/
Toxoplasma gondii TgCkUg2	Coccidia	http://toxodb.org/toxo/

of functional genomics already enables the determination of complete genomic protein functions by utilizing high-throughput experiments such as microarrays, serial analysis of gene expression (SAGE), ChIP-on-chip experiments, or proteomics [30]. Moreover, the emergence of competitive second- or next-generation DNA-sequencing techniques [31] and further advances in single-molecule DNA-sequencing technologies [32] will lead to the sequencing of additional genomes, including those of parasites [33]. Currently, over 1000 bacterial and 120 eukaryotic genomes have been reported as completely sequenced, and many more are ongoing (http://www.genomesonline.org) [34]. In examining the phylum *Apicomplexa*, approximately 50 genome sequencing projects have now been initiated, of which nine have already been published (Table 1.1). The availability of genome datasets for parasites, their vectors, and hosts provides the basis for another highly effective target identification method, the bioinformatic comparison of genomes [10, 35–38]. Such comparative genomics strategies aim to compare simultaneously two or more genomes in order to identify similarities and differences, and hence identify potential drug targets [18, 19].

Target Validation

When a particular protein has been identified as a potential drug target, the validation of its function is mandatory (Box 1.1) [39]. This involves the demonstration that

Box 1.1: Features of Optimal Anti-Parasitic Targets

A validated target:

- has a clear biological function
- has an essential role for the growth or survival of the parasite
- is expressed during the relevant life stages
- is druggable
- can be screened in a biochemical or cellular assay.

affecting the target will be sufficient for obtaining a significant anti-parasitic effect, and this can be accomplished by genetic studies that include the generation of loss-of-function (Knock-out) and gain-of-function (Knock-in) mutants in animal models [40]. Further common target validation methods are RNA interference, antisense RNA, and antibody-mediated inhibition experiments. Alternatively, the validation of a drug target is performed using chemical compounds [26, 41]. In such cases, experimental compounds with well-understood modes of action are tested directly on parasites and screened for anti-parasitic phenotypes. With positive results, it is inferred that the phenotypic effect is due to the interaction of the chemical compound with its known target. *Chemical validation* is a reliable form of target validation, although it cannot be excluded that the phenotypes resulting from the chemical validation experiment are in fact due to an interaction of the compounds with secondary, unknown, or multiple targets. One benefit in employing chemical validation is that, simultaneously, the druggability of a molecular target –that is, the ability of a target to interact with a small compound that modulates its function – is analyzed [42]. Experience has shown that this is a key prerequisite to successful drug discovery [42]. Since both genetic and chemical validation approaches have their benefits and drawbacks, a drug target is best validated using a combination of the two.

A potential drug target should fulfill additional criteria, including the ease of recombinant expression and purification, and "assayability" in automated biochemical or cellular assays (including their miniaturization) (Box 1.1) [26]. Drug target validation is a complex process that often produces ambiguous results. Accordingly, target validation is a risk-adjusted decision on the overall value of the target protein [43]. However, since the downstream steps in target-based drug discovery include very expensive and time-consuming processes, it is important that potential target molecules are validated in as large a quantity as possible.

Lead Discovery

The next step after the validation of a drug target is to identify compounds that interfere with its function [44]. A general lead discovery workflow is shown in Figure 1.3; this consists of a series of steps of hit identification, hit exploration, hit-to-lead, and lead selection [45]. The compounds should be amenable to chemical optimization, finally leading to a drug candidate. Such compounds are called "leads," and the process is called "lead discovery" [17].

Hit Identification

In order to identify small chemical compounds that interact with the target molecule, a screening campaign (often high-throughput screening, HTS) is performed [46]. Before an HTS can be started, biochemical or cellular assays must first be developed and miniaturized for optimal robotics and throughput [18]. The HTS assays are usually validated for their suitability and robustness by screening a small subset of compounds before the actual high-throughput screen is started. Depending on the type of assay, the HTS can typically involve the examination of more than one million

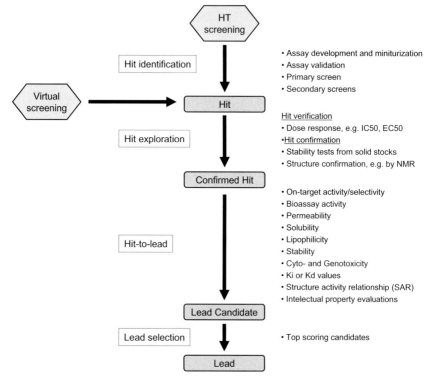

Figure 1.3 A typical lead discovery workflow.

compounds within a few weeks [47, 48]. The primary screen during the hit identification process is carried out on the initial validated target molecule, while secondary screens may be performed on further targets. For example, orthologous targets from additional parasites may be screened if the goal of the drug discovery project is ultimately to produce a compound that exhibits broad anti-parasitic activity, as is often the case in the animal health industry. Another possibility is to consider orthologs from host organisms, and to include only those compounds that show a high activity on the target molecules of the parasite. Thus, potentially toxic compounds may be filtered out in a very early stage [7]. A complementary approach is to involve chemoinformatics to identify hit compounds (Figure 1.3) [49]. Such *in silico* approaches – which are also known as "virtual screening" – deal with the automatic evaluation of large virtual compound libraries in order to prioritize compound subsets [50]. Compared to HTS, virtual screening has two major advantages: (i) the speed and throughput of *in silico* screens are much greater than in experimental set-ups; and (ii) more importantly, virtual screening is not limited to existing in-house compound collections, and provides a fast and cheap way to explore unknown parts of the chemical space [49]. Accordingly, virtual screening can generate target-focused and activity-enriched datasets, which can eventually be tested in experimental HTS assays [49, 51].

Hit Exploration

Hit exploration can be divided into "hit verification" and "hit confirmation," and mainly involves filtering processes to separate appropriate from inappropriate molecules (Figure 1.3) [17]. Hit verification concentrates on the experimental validation of the effectiveness of a compound by measuring the half-maximal inhibitory concentration (IC_{50}) in the case of antagonists, and the half-maximal effective concentration (EC_{50}) for agonists [46]. During hit confirmation, the stability of compounds is proofed. For example, compound solutions are freshly prepared from solid stocks and measured again in order to exclude artifacts from compound degradation in liquid stocks. Hit confirmation also includes the verification of the compound structures using techniques such as mass spectrometry (MS) and nucleic magnetic resonance (NMR) [19, 46]. The result of the hit exploration process is termed a "confirmed hit" (Figure 1.3).

Hit-To-Lead

The next phase in a lead discovery project is the "hit-to-lead" (H2L) process [52]. This is characterized by less filtering and a broader knowledge of the hits for a subsequent prioritization [17]. During the H2L process, data regarding toxicity, bioactivity, and intellectual property are assembled (Figure 1.3). One of the first actions in the H2L process is usually to purchase or synthesize structurally related compounds which are then tested together with the confirmed hits for their target activity and, by screening in bioassays against intact parasites in culture, for their anti-parasitic activity. In this way, data related to the on-target activity and initial structure–activity relationships (SARs) of the compound classes are generated. The experiments also provide the first hints regarding the bioactivity of the compound classes. If the compounds are inactive in the bioassays, then data arising from solubility, lipophilicity, or permeability experiments may explain the lack of bioactivity. Other typical H2L activities are the evaluation of toxicity data from cytotoxicity and genotoxicity tests [53], and an understanding of the patent literature for the corresponding compound classes (Figure 1.3) [54]. The H2L process concludes with a list of lead candidates that fulfill clearly defined criteria (Box 1.2), and from which a lead is selected for chemical optimization.

Box 1.2: Definition of a Lead

A lead:

- possesses specific activity in functional target assays
- exhibits a particular SAR
- shows no indication for genotoxicity
- has adjustable physico-chemical properties
- already features some bioactivity in parasitic bioassays, or at least offers favorable physico-chemical properties needed for bioactivity.

Lead Optimization and Profiling

Leads display certain effects and properties of active drugs, but not with all of the necessary attributes. The missing properties are subsequently introduced in the lead optimization phase, a process that aims to transform an active lead compound into a drug candidate [55]. Lead optimization is a multi-parametric process aimed at simultaneously optimizing several features, such as on-target activity, bioactivity, and stability (Figure 1.4) [7, 56]. Therefore, lead optimization is highly complex, and generally takes the most time in drug discovery projects, largely because the necessary medicinal chemistry is a major bottleneck in the process [57].

The lead optimization workflow can be divided into different phases connected with decision points, any of which might bring an end to the project (Figure 1.4). During the first phase, mainly *in vitro* target assays are used to control the optimization progress, and the emphasis is on improving on-target activity; this is fundamental for achieving biological activity [7]. During this phase, several thousand derivatives might be synthesized, a situation made possible by the major advances in medicinal and combinatorial chemistry that have helped to increase both diversity and yields [58]. With such large-scale synthesis, a clear SAR for a compound class can be determined. In SAR studies, the lead compounds are typically divided into specific regions, after which each in turn is chemically modified. Figure 1.5 shows the results of a SAR experiment with anthelmintic

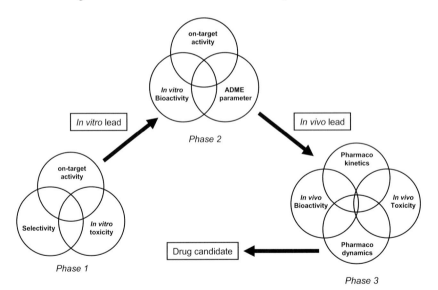

Figure 1.4 Multiparametric lead optimization. Starting with an *in vitro* lead, a chemical compound is optimized until it meets specific criteria defined for drug candidates. The process is multiparametric, and several sometimes conflicting requirements (e.g., high on-target activity, anti-parasite efficacy, low toxicity) must be accomplished simultaneously during the chemical optimization.

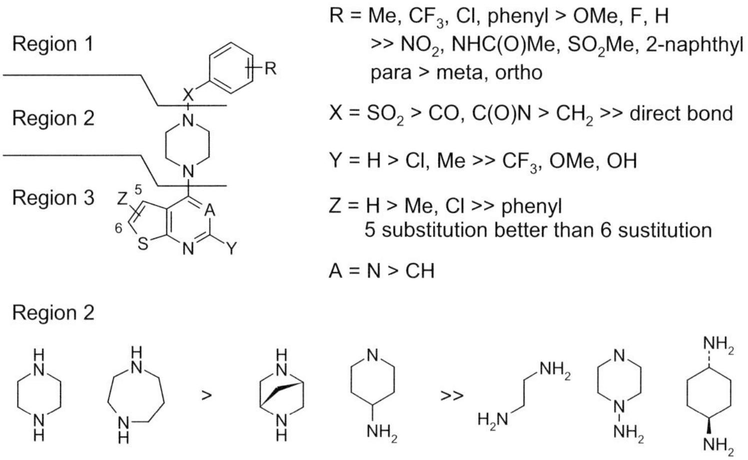

Region 1

Region 2

Region 3

Region 2

R = Me, CF$_3$, Cl, phenyl > OMe, F, H
>> NO$_2$, NHC(O)Me, SO$_2$Me, 2-naphthyl
para > meta, ortho

X = SO$_2$ > CO, C(O)N > CH$_2$ >> direct bond

Y = H > Cl, Me >> CF$_3$, OMe, OH

Z = H > Me, Cl >> phenyl
5 substitution better than 6 sustitution

A = N > CH

Region 2

Figure 1.5 Structure–activity relationship (SAR) of thienopyrimidine analogs. The lead structure was divided into three regions, with all regions being chemically modified sequentially, varying only one motif at a time: first the aromatic head (region 1), followed by the central diamine linker (region 2), and finally the thienopyrimidine core (region 3). The anthelmintic potencies were determined in bioassays. This figure summarizes the nematocidal SARs of compound analogs. For example in region 1, substitution of the para position on the aromatic motif led to analogs with superior activity. Of these substituents, methyl, trifluoromethyl, chloride, and phenyl were observed to be the best. Reproduced with permission from Ref. [59]; © 2009, Wiley-VCH, Weinheim.

thienopyrimidine analogs, and demonstrates how different substitutions can affect nematocidal activity [59]. It is important to note that the lead optimization steps can be efficiently supported by chemoinformatic tools [60]; this is especially true if the protein structure of a target complex is available. Then, modern structure-based drug design methods can be applied to support rational lead optimization [61, 62]. Finally, it is essential to check the toxicity potential of interesting compounds in this first phase of lead optimization, usually by performing *in vitro* toxicity tests [63].

Yet, because on-target activity alone is not sufficient to achieve anti-parasitic activity, it is also essential to consider the biological activity and to ensure that the physico-chemical properties are within the desired range. This is achieved during the second phase of lead optimization, which focuses on monitoring biological activity using *in vitro* parasite models [13–15]. Critical to the biological activity, and thus to the success of any potential drugs, are the ADME parameters (this is an acronym for absorption, distribution, metabolism, and excretion) [64]. ADME deals with the disposition of a drug within organisms, with each of the four criteria having an important influence on the efficiency and pharmacological activity of a compound as a drug [65].

The third phase of lead optimization goes a step further, and includes animal models to assist in making the transition from *in vitro* assays to *in vivo* conditions [7]. For this, the compounds are profiled in model organisms such as mice or rats [66]. In any animal model it is necessary to generate pharmacokinetic, pharmacodynamic, and toxicity profiles. Pharmacokinetics describes the time course of the drug in the body (namely, its ADME behavior [67]) while, in contrast, pharmacodynamics specifies the effect versus concentration relationship. In simple words, pharmacodynamics explores what a drug does to the body over time, whereas pharmacokinetics is the study of what the body does to the drug [68]. Often, the intricacies and expense of animal models limits the numbers of compounds that can be tested annually [7]. Following the successful completion of a lead optimization project, a compound is then considered to be a suitable drug candidate and transferred into a drug development program. In human health, a drug candidate has already been profiled in model organisms, whereas animal health research can go a step further and prepare a clinical profile in target animals.

Examples of Successful Target-Based, Anti-Parasitic Drug Discovery Programs

Proteases are validated as targets for therapy of a number of parasitic diseases, including malaria, leishmaniasis, African trypanosomiasis, and schistosomiasis [5, 25, 69, 70]. Several chemical structures have already been identified as protease inhibitor leads [1]; for example, promising inhibitor leads targeting the falcipain protease for the treatment of *P. falciparum* infection have been discovered by a collaboration between the pharmaceutical company GlaxoSmithKline plc and the University of California San Francisco, supported by the Medicines for Malaria Venture. Although these compounds are far along in the drug development process, their structures remain proprietary [69]. A related example of a target-based drug discovery workflow is the identification of the cysteine protease inhibitor, *N*-methyl-piperazine-phenylalanyl-homophenylalanyl-vinylsulfone-phenyl (K11777 or K777) as a small-molecule therapy of Chagas disease by targeting the parasite's cathepsin L-like cysteine protease, cruzain [24, 71]. The vinyl sulfone class of molecules was originally identified in the mid-1990s in a curtailed industrial drug discovery program (at Khepri Pharmaceuticals) to target bone loss, but the parent molecule of K777 (K11002 or K02) was subsequently transferred (including the intellectual property rights) to an anti-parasitic drug discovery program conducted at the University of California, San Francisco. Following the modification of K02 to improve its bioavailability, K777 was put through a standard development workflow which included on-target mechanism of action studies incorporating crystallography, both *in vitro* and *in vivo* anti-parasitic activity profiling (the latter in mice and dogs), and a suite of ADME and (acute and chronic) toxicity (studies in rodents and dogs). As of early 2010, a dossier is being prepared for filing K777 as an Investigational New Drug (IND) at the US Food and Drug Administration in advance of clinical trials in humans. A structure-guided medicinal chemistry program is also ongoing to identify "back-up" compounds (Figure 1.6) [72].

Another recent success story in the development of novel anti-parasitic drugs is the identification of a new class of anthelmintics, the amino-acetonitrile derivatives (AADs) [73]. In veterinary medicine, there is an urgent need for novel drugs against

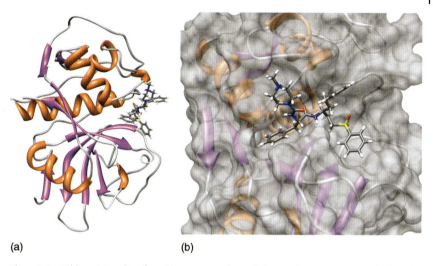

(a)   (b)

Figure 1.6 Ribbon (a) and surface (b) representations of the cysteine protease cruzain from *T. cruzei*, complexed with the inhibitor K11777 [72].

parasitic worms, as some nematodes have developed drug resistance against all available anthelmintic drugs; even worse, some multidrug-resistant worms have appeared [74, 75]. Thus, the development of the AADs, which were discovered in a physiological-based screen, proved to be most welcome [76]. Consequently, an extensive lead optimization program is ongoing which, to date, has resulted in over 600 compounds with different anthelmintic activities, both *in vitro* and *in vivo*, in different hosts [77]. Moreover, the compounds are effective against a wide range of livestock helminths, including several drug-resistant parasites [78]. This indicates a new mode of action for the AADs and, indeed, genetic experiments have shown that they act on unique, nematode-specific subunits of the acetylcholine receptors [79]. If the excellent pharmacokinetic properties and tolerability of the AADs in ruminants can be extended to humans, the class may offer an alternative anthelmintic for human medical practice [73].

Conclusion

The discovery of novel drugs for parasitic diseases is a high-risk, expensive, and lengthy process [22]. The past few years have seen increased financial and infrastructural support for drug discovery and development for parasitic diseases by academic institutes, PPPs, and the pharmaceutical industry [3, 5]. Already this has led to imaginative, comprehensive and dynamic drug discovery and development pipelines (even in the face of sometimes modest financial backing) when compared to those diseases directly impacting developed countries [1]. A closer look, however, at the portfolios of some of the PPPs reveals a plethora of early discovery projects for parasitic diseases that have yet to translate into late development leads. The enormous expense and considerable expertise required to develop late leads and drug candidates (involving medicinal and combinatorial chemistry, and in-life animal studies such as A

DME and toxicology) remain major bottlenecks. This is especially true for academic institutions which, with their relatively finite resources, concentrate either on the biology or chemistry of the drug discovery workflow. Here, a closer collaboration between biology and chemistry groups may lead to an increased efficiency, and indeed a number of specialized academic centers have already arisen specifically focused on R&D for neglected diseases. Most importantly, the pharmaceutical industry, through a variety of internal and external R&D programs, has re-entered the business of drug development for infectious diseases. This is vital, given its decades-long know-how on furthering compounds through to the market. In summary, the continued collaboration between academic groups, PPPs, and the pharmaceutical industry is the key to optimizing R&D output and, eventually, the registration of novel and badly needed anti-parasitic drugs.

References

1 Renslo, A.R. and McKerrow, J.H. (2006) Drug discovery and development for neglected parasitic diseases. *Nat. Chem. Biol.*, **2**, 701–710.

2 Don, R. and Chatelain, E. (2009) Drug discovery for neglected diseases: view of a public-private partnership, in *Antiparasitic and Antibacterial Drug Discovery: From Molecular Targets to Drug Candidates* (ed. P. Selzer), Wiley-VCH, Weinheim, Germany, pp. 33–43.

3 Moran, M.A. (2005) Breakthrough in R&D for neglected diseases: new ways to get the drugs we need. *PLoS Med.*, **2**, e309.

4 Croft, L.C. (2005) Public-private partnership: from there to here. *Trans. R. Soc. Trop. Med. Hyg.*, **99** (Suppl. 1), S9–S14.

5 Caffrey, C.R. and Steverding, D. (2008) Recent initiatives and strategies to developing new drugs for tropical parasitic diseases. *Curr. Opin. Drug Discov.*, **3**, 173–186.

6 Selzer, P.M. (2009) *Preface*, in *Antiparasitic and Antibacterial Drug Discovery: From Molecular Targets to Drug Candidates* (ed. P. Selzer), Wiley-VCH, Weinheim, Germany.

7 Chassaing, C. and Sekljic, H. (2009) Approaches towards anti-parasitic drug

candidates for veterinary use, in *Antiparasitic and Antibacterial Drug Discovery: From Molecular Targets to Drug Candidates* (ed. P. Selzer), Wiley-VCH, Weinheim, Germany, pp. 117–133.

8 Geary, T.G., Woods, D.J., Williams, T., and Nwaka, S. (2009) Target identification and mechanism-based screening for anthelmintics: application of veterinary anti-parasitic research programs to search for new anti-parasitic drugs for human indications, in *Anti-parasitic and Antibacterial Drug Discovery: From Molecular Targets to Drug Candidates* (ed. P. Selzer), Wiley-VCH, Weinheim, Germany, pp. 3–15.

9 Mehlhorn, H. (2008) *Encyclopedia of Parasitology*, Springer Verlag, Heidelberg, Germany.

10 Caffrey, C.R., Rohwer, A., Oellien, F., Marhöfer, R.J., Braschi, S. *et al.* (2009) A comparative chemogenomics strategy to predict potential drug targets in the metazoan pathogen, *Schistosoma mansoni. PLoS One*, **4**, e4413.

11 Krasky, A., Rohwer, A., Marhöfer, R., and Selzer, P.M. (2009) Bioinformatics and chemoinformatics: key technologies in the drug discovery process, in *Anti-parasitic and*

Antibacterial Drug Discovery: From Molecular Targets to Drug Candidates (ed. P. Selzer), Wiley-VCH, Weinheim, Germany, pp. 45–57.

12 Woods, D.J. and Williams, T.M. (2007) The challenges of developing novel anti-parasitic drugs. *Invert. Neurosci.*, **7**, 245–250.

13 Noedl, H., Wongsrichanalai, C., and Wernsdorfer, W.H. (2003) Malaria drug-sensitivity testing: new assays, new perspectives. *Trends Parasitol.*, **19**, 175–181.

14 Coles, G.C. (1990) Recent advances in laboratory models for evaluation of helminth chemotherapy. *Br. Vet. J.*, **146**, 113–119.

15 Kröber, T. and Guerin, P.M. (2007) In vitro feeding assays for hard ticks. *Trends Parasitol.*, **23**, 445–449.

16 Abdulla, M.H., Ruelas, D.S., Wolff, B., Snedecor, J., Lim, K.C., Xu, F. *et al.* (2009) Drug discovery for schistosomiasis: hit and lead compounds identified in a library of known drugs by medium-throughput phenotypic screening. *PLoS Negl. Trop. Dis.*, **3**, e478.

17 Gassel, M., Cramer, J., Kern, C., Noack, S., and Streber, W. (2009) Lessons learned from target-based lead discovery, in *Anti-parasitic and Antibacterial Drug Discovery: From Molecular Targets to Drug Candidates* (ed. P. Selzer), Wiley-VCH, Weinheim, Germany, pp. 99–115.

18 Selzer, P.M., Brutsche, S., Wiesner, P., Schmid, P., and Mullner, H. (2000) Target-based drug discovery for the development of novel antiinfectives. *Int. J. Med. Microbiol.*, **290**, 191–201.

19 Selzer, P.M., Marhöfer, J.R., and Rohwer, A. (2008) *Applied Bioinformatics*, Springer Verlag, Heidelberg, Germany.

20 Kandpal, R., Saviola, B., and Felton, J. (2009) The era of "omics unlimited". *Biotechniques*, **46**, 351–352, 354–355.

21 Cho, C.R., Labow, M., Reinhardt, M., van Oostrum, J., and Peitsch, M.C. (2006) The application of systems biology to drug discovery. *Curr. Opin. Chem. Biol.*, **10**, 294–302.

22 Frearson, J.A., Wyatt, P.G., Gilbert, I.H., and Fairlamb, A.H. (2007) Target assessment for anti-parasitic drug discovery. *Trends Parasitol.*, **23**, 589–595.

23 Köhler, P. and Marhöfer, R. (2009) Selective drug targets in parasites, in *Anti-parasitic and Antibacterial Drug Discovery: From Molecular Targets to Drug Candidates* (ed. P. Selzer), Wiley-VCH, Weinheim, Germany, pp. 75–98.

24 Engel, J.C., Doyle, P.S., Hsieh, I., and McKerrow, J.H. (1998) Cysteine protease inhibitors cure an experimental *Trypanosoma cruzi* infection. *J. Exp. Med.*, **188**, 725–34.

25 Selzer, P.M., Pingel, S., Hsieh, I., Ugele, B., Chan, V.J. *et al.* (1999) Cysteine protease inhibitors as chemotherapy: lessons from a parasite target. *Proc. Natl Acad. Sci. USA*, **96**, 11015–11022.

26 Wolf, C. and Gunkel, N. (2009) Target identification and validation in anti-parasitic drug discovery, in *Anti-parasitic and Antibacterial Drug Discovery: From Molecular Targets to Drug Candidates* (ed. P. Selzer), Wiley-VCH, Weinheim, Germany, pp. 59–73.

27 Aurrecoechea, C., Brestelli, J., Brunk, B.P., Fischer, S., Gajria, B. *et al.* (2010) EuPathDB: a portal to eukaryotic pathogen databases. *Nucleic Acids Res.*, **38** (Database issue), D415–D419.

28 Hertz-Fowler, C., Peacock, C.S., Wood, V., Aslett, M., Kerhornou, A. *et al.* (2004) GeneDB: a resource for prokaryotic and eukaryotic organisms. *Nucleic Acids Res.*, **32** (Database issue), D339–D343.

29 Agüero, F., Al-Lazikani, B., Aslett, M., Berriman, M., Buckner, F.S. *et al.* (2008) Genomic-scale prioritization of drug targets: the TDR Targets database. *Nat. Rev. Drug Discov.*, **7**, 900–907.

30 Chen, X., Jorgenson, E., and Cheung, S.T. (2009) New tools for functional genomic analysis. *Drug Discov. Today*, **14**, 754–760.

31 Ansorge, W.J. (2009) Next-generation DNA sequencing techniques. *Nat. Biotechnol.*, **25**, 195–203.

32 Gupta, P.K. (2008) Single-molecule DNA sequencing technologies for future genomics research. *Trends Biotechnol.*, **26**, 602–611.

33 Winzeler, E.A. (2009) Advances in parasite genomics: from sequences to

regulatory networks. *PLoS Pathog.*, **5**, e1000649.

34 Liolios, K., Mavromatis, K., Tavernarakis, N., and Kyrpides, N.C. (2008) The Genomes OnLine Database (GOLD) in 2007: status of genomic and metagenomic projects and their associated metadata. *Nucleic Acids Res.*, **36** (Database issue), D475–D479.

35 Beckstette, M., Mailänder, J., Marhöfer, J.R., Sczyrba, A., Ohlebusch, E. *et al.* (2004) Genlight: an interactive system for high-throughput sequence analysis and comparative genomics. *J. Integ. Bioinform.*, **8**, 1.

36 Gerber, S., Krasky, A., Rohwer, A., Lindauer, S., Closs, E. *et al.* (2006) Identification and characterisation of the dopamine receptor II from the cat flea *Ctenocephalides felis* (CfDopRII). *Insect Biochem. Mol. Biol.*, **36**, 749–758.

37 Krasky, A., Rohwer, A., Schroeder, J., and Selzer, P.M. (2007) A combined bioinformatics and chemoinformatics approach for the development of new anti-parasitic drugs. *Genomics*, **89**, 36–43.

38 Klotz, C., Marhöfer, R.J., Selzer, P.M., Lucius, R., and Pogonka, T. (2005) *Eimeria tenella*: identification of secretory and surface proteins from expressed sequence tags. *Exp. Parasitol.*, **111**, 14–23.

39 Blake, R.A. (2006) Target validation in drug discovery. *Methods Mol. Biol.*, **356**, 367–378.

40 Hillisch, A. and Hilgenfeld, R. (2003) The role of protein 3D-structures in the drug discovery process, in *Modern Methods of Drug Discovery* (eds A. Hillisch and R. Hilgenfeld), Birkhäuser, Basel, Switzerland, pp. 157–181.

41 Torrie, L.S., Wyllie, S., Spinks, D., Oza, S.L., Thompson, S. *et al.* (2009) Chemical validation of trypanothione synthetase: a potential drug target for human trypanosomiasis. *J. Biol. Chem.*, **284**, 36137–36145.

42 Sugiyama, Y. (2005) Druggability: selecting optimized drug candidates. *Drug Discov. Today*, **10**, 1577–1579.

43 Betz, U.A. (2005) How many genomics targets can a portfolio afford? *Drug Discov. Today*, **10**, 1057–1063.

44 Goodnow, R.A. Jr and Gillespie, P. (2007) Hit and lead identification: efficient practices for drug discovery. *Prog. Med. Chem.*, **45**, 1–61.

45 Ernst, A. and Obrecht, D. (2008) Case studies of parallel synthesis in hit identification, hit exploration, hit-to-lead, and lead-optimization programs, in *High-Throughput Lead Optimization in Drug Discovery* (ed. T. Kshirsagar), CRC Press, Boca Raton, USA, pp. 99–116.

46 Mayr, L.M. and Bojanic, D. (2009) Novel trends in high-throughput screening. *Curr. Opin. Pharmacol.*, **9**, 580–588.

47 Pereira, D.A. and Williams, J.A. (2007) Origin and evolution of high throughput screening. *Br. J. Pharmacol.*, **152**, 53–61.

48 Kenny, B.A., Bushfield, M., Parry-Smith, D.J., Fogarty, S., and Treherne, J.M. (1998) The application of high-throughput screening to novel lead discovery. *Prog. Drug Res.*, **51**, 245–69.

49 Oellien, F., Engels, K., Cramer, J., Marhöfer, J.R., Kern, C., and Selzer, P.M. (2009) Searching new anti-parasitics in virtual space, in *Anti-parasitic and Antibacterial Drug Discovery: From Molecular Targets to Drug Candidates* (ed. P. Selzer), Wiley-VCH, Weinheim, Germany, pp. 323–338.

50 McInnes, C. (2007) Virtual screening strategies in drug discovery. *Curr. Opin. Chem. Biol.*, **11**, 494–502.

51 Köppen, H. (2009) Virtual screening - what does it give us? *Curr. Opin. Drug Discov. Devel.*, **12**, 397–407.

52 Keseru, G.M. and Makara, G.M. (2006) Hit discovery and hit-to-lead approaches. *Drug Discov. Today*, **11**, 741–748.

53 Nogueira, R.C., Oliveira-Costa, J.F., de Sá, M.S., dos Santos, R.R., and Soares, M.B. (2009) Early toxicity screening and selection of lead compounds for parasitic diseases. *Curr. Drug Targets*, **10**, 291–298.

54 Cockbain, J. (2007) Intellectual property rights and patents, in *Comprehensive Medicinal Chemistry II*, vol. 1 (eds J.B. Taylor and D.J. Triggle), Elsevier, Amsterdam, The Netherlands, pp. 779–815.

55 Lindsley, C.W., Weaver, D., Bridges, T.M., and Kennedy, J.P. (2009) Lead

optimization in drug discovery, *Wiley Encyclopedia of Chemical Biology*, vol. **2**, Wiley-VCH, Weinheim, pp. 511–519.

56 Colombo, M. and Peretto, I. (2008) Chemistry strategies in early drug discovery: an overview of recent trends. *Drug Discov. Today*, **13**, 677–684.

57 Wess, G. (2002) How to escape the bottleneck of medicinal chemistry. *Drug Discov. Today*, **7**, 533–535.

58 Di, L. and Kerns, E.H. (2003) Profiling drug-like properties in discovery research. *Curr. Opin. Chem. Biol.*, **7**, 402–408.

59 Meyer, T., Schröder, J., Uphoff, M., Noack, S., Heckeroth, A. *et al.* (2009) Chemical optimization of anthelmintic compounds – a case study, in *Anti-parasitic and Antibacterial Drug Discovery: From Molecular Targets to Drug Candidates* (ed. P. Selzer), Wiley-VCH, Weinheim, Germany, pp. 357–371.

60 Fotouhi N., Gillespie, P., Goodnow, R.A., So, S.S., Han, Y., and Babiss, L.E. (2006) Application and utilization of chemoinformatics tools in lead generation and optimization. *Comb. Chem. High-Throughput Screen.*, **9**, 95–102.

61 Andricopulo, A.D., Salud, L.B., and Abraham, D.J. (2009) Structure-based drug design strategies in medicinal chemistry. *Curr. Top. Med. Chem.*, **9**, 771–790.

62 Guido, R.V. and Oliva, G. (2009) Structure-based drug discovery for tropical diseases. *Curr. Top. Med. Chem.*, **9**, 824–843.

63 Schoonen, W.G., Westerink, W.M., and Horbach, G.J. (2009) High-throughput screening for analysis of in vitro toxicity. *EXS (Basel)*, **99**, 401–452.

64 Wang, J. and Skolnik, S. (2009) Recent advances in physicochemical and ADMET profiling in drug discovery. *Chemistry & Biodiversity*, **6**, 1887–1899.

65 Caldwell, G.W., Yan, Z., Tang, W., Dasgupta, M., and Hasting, B. (2009) ADME optimization and toxicity assessment in early- and late-phase drug discovery. *Curr. Top. Med. Chem.*, **9**, 965–980.

66 Wykes, M.N. and Good, M.F. (2009) What have we learnt from mouse models

for the study of malaria? *Eur. J. Immunol.*, **39**, 2004–2007.

67 Erhardt, P.W. and Proudfoot, J.R. (2007) Drug discovery: historical perspective, current status, and outlook, in *Comprehensive Medicinal Chemistry II*, vol. 1 (eds J.B. Taylor and D.J. Triggle), Elsevier, Amsterdam, The Netherlands, pp. 29–96.

68 Padden, J., Skoner, D., and Hochhaus, G. (2008) Pharmacokinetics and pharmacodynamics of inhaled glucocorticoids. *J. Asthma*, **45**, 13–24.

69 McKerrow, J.H., Rosenthal, P.J., Swenerton, R., and Doyle, P. (2008) Development of protease inhibitors for protozoan infections. *Curr. Opin. Infect. Dis.*, **21**, 668–672.

70 Rosenthal, P.J. (2004) Cysteine proteases of malaria parasites. *Int. J. Parasitol.*, **34**, 1489–1499.

71 Doyle, P.S., Zhou, Y.M., Engel, J.C., and McKerrow, J.H. (2007) A cysteine protease inhibitor cures Chagas' disease in an immunodeficient-mouse model of infection. *Antimicrob. Agents Chemother.*, **51**, 3932–3939.

72 Kerr, I.D., Lee, J.H., Farady, C.J., Marion, R., Rickert, M. *et al.* (2009) Vinyl sulfones as anti-parasitic agents and a structural basis for drug design. *J. Biol. Chem.*, **284**, 25697–25703.

73 Kaminsky, R., Ducray, P., Jung, M., Clover, R., Rufener, L. *et al.* (2008) A new class of anthelmintics effective against drug-resistant nematodes. *Nature*, **452**, 176–180.

74 Wolstenholme, A.J., Fairweather, I., Prichard, R., von Samson-Himmelstjerna, G., and Sangster, N.C. (2004) Drug resistance in veterinary helminths. *Trends Parasitol.*, **20**, 469–476.

75 von Samson-Himmelstjerna, G., Prichard, R.K., and Wolstenholme, A.J. (2009) Anthelmintic resistance as a guide to the discovery of new drugs? in *Anti-parasitic and Antibacterial Drug Discovery: From Molecular Targets to Drug Candidates* (ed. P. Selzer), Wiley-VCH, Weinheim, Germany, pp. 17–32.

76 Prichard R.K. and Geary, T.G. (2008) Drug discovery: fresh hope to can the worms. *Nature*, **452**, 157–158.

77 Ducray, P., Gauvry, N., Pautrat, F., Goebel, T., Fruechtel, J. *et al.* (2008) Discovery of amino-acetonitrile derivatives, a new class of synthetic anthelmintic compounds. *Bioorg. Med. Chem. Lett.*, **18**, 2935–2938.

78 Hosking, B.C., Kaminsky, R., Sager, H., Rolfe, F., and Seewald, W. (2010) A pooled analysis of the efficacy of monepantel, an amino-acetonitrile derivative against gastrointestinal nematodes of sheep. *Parasitol. Res.*, **106**, 529–532.

79 Rufener, L., Mäser, P., Roditi, I., and Kaminsky, R. (2009) *Haemonchus contortus* acetylcholine receptors of the DEG-3 subfamily and their role in sensitivity to monepantel. *PLoS Pathol.*, **5**, e1000380.

2
New Bioinformatic Strategies Against Apicomplexan Parasites

Thomas Dandekar and *Ke Xiao**

Abstract

This chapter introduces basic bioinformatics tools for the analysis of sequences, genomes, and networks, and then discusses apicomplexan-specific challenges: similarity to the host, rapid resistance, and adaptation. Counter-strategies rely on specific databases and genome information. Some of these modern approaches, including gene-drive strategies, refined targeting of structures, modeling resistance, and drug-combination strategies, are explained.

Introduction

Following the provision of basic bioinformatics strategies, which are valuable when identifying new targeting strategies in parasites, an explanation will be provided as to when and where useful apicomplexan-specific modifications are possible. New approaches include striking strategies, such as targeting not the catalytic center but rather the other parts of the protein after extensive host–parasite genome comparisons. In this way, it is possible to successfully identify target substructures in shared protein families that are different between the host and parasite. Examples are provided through recent data relating to cathepsins and other CA/C1 peptidases in plasmodia [1]. Another important strategy is to use insights from genomics and structure target comparisons, not for a pharmacological strategy but rather for a population engineering or gene-drive strategy, such that either apicomplexa or their transmission vectors are targeted [2]. In the latter case, beneficial targeting may involve the transmission of resistance, for example, against plasmodia. Further bioinformatical support strategies typical for apicomplexan antibiotic drug development include multiple hit strategies (i.e., to hit multiple pathways so as to prevent resistance development) as well as dual drugs [3]. In all cases, pathway and structure analyses are combined with further data (expression, etc.) to maximize the differ-

*Corresponding author.

Apicomplexan Parasites. Edited by Katja Becker
Copyright © 2011 WILEY-VCH Verlag GmbH & Co. KGaA, Weinheim
ISBN: 978-3-527-32731-7

ences in drug response between a human or eukaryotic host and an apicomplexan parasite, and to prevent resistance by broad strategies [4, 5].

Databases and Methods

The typical bioinformatics databases that have proven valuable in obtaining access to full genome information or to other large-scale data for the organism of choice, include the NCBI database [6], transcription factors [7, 8], promoter databases [9] and protein and pathway databases [10]; these are summarized in Table 2.1. Besides Name and Web-link we give literature references and public library of medicine article identifiers (PMID).

The last four pointers of Table 2.1 consider furthermore metabolic modeling (metatool, [11]; promoter analysis (GenomatixSuite; [7]; TESS, [8]), and microarray handling (SMD software, [13]). Useful standard tools and links to investigate potential drug targets after screening a genome sequence, in particular with regards to sequence analysis, are listed in Table 2.2.

Although, the basic pointers given are very general and broadly useful as a general primer, for target searches and drug discovery against Apicomplexa these are often insufficient. In particular, there is currently a large body of apicomplexan genome information available as well as target databases (Table 2.3). With such assistance, it is much easier to specifically target either individual Apicomplexa (e.g., PlasmoDB, [14]), their vectors (generally insects), as well as to rank a putative new target against a string of known targets [15]. This is according to the literature and to molecular evidence, since for many drug and target properties the structural data available are limited. As the life cycle of human apicomplexan pathogens involves several (often three) organisms, a differential genome analysis [1] is typically required. Yet, is a potential target in the apicomplexan or the vector organism sufficiently different from the human host as to allow a drug strategy? Furthermore, this also implies two options to hit the target – either on the vector or the apicomplexan level.

Following a basic genome analysis (Tables 2.1 and 2.2), specific knowledge and database material (Table 2.3) are added for the detailed analysis of: (i) apicomplexan genomes; (ii) transmission vectors; and (iii) the human host. Notably, the apicomplexan biology reveals interesting basic differences from the human host, including different organelles and membranes. These represent clear advantages for drug development, such as the specific trans-splicing that occurs in trypanosomes. Moreover, some organelles do not occur in the human host, such as the glycosome [16]. On the other hand, the apicomplexans have also evolved specific strategies for host immune escape, an example being the well-known surface antigen turncoat strategy in trypanosomes. Although this active immune evasion by apicomplexans "comes in different flavors," it provides real challenges for a number of therapeutic strategies, such as vaccine development.

Several bioinformatics databanks, such as KEGG [10], permit "zooming in" on parasite-specific enzymes; indeed, some databanks, such as COG [17], even define

Table 2.1 General overview of databases.

Name	Link	PMID
NCBI databases		
Sequence annotation		
GenBank: sequence annotation for nucleotides, protein sequence, and whole genome sequence	http://www.ncbi.nlm.nih.gov/Genbank/	18940867 [27]
GEO: Gene expression datasets	http://www.ncbi.nlm.nih.gov/gds	12519941 [28]
Gene annotation		
Entrez Gene: a searchable database of genes	http://www.ncbi.nlm.nih.gov/gene	15608257 [6]
UniGene: Gene-oriented clusters of transcript sequences	http://www.ncbi.nlm.nih.gov/unigene	12519941 [28]
OMIM: catalog of human genes and genetic disorders, with links to literature references, sequence records, maps, and related databases	http://www.ncbi.nlm.nih.gov/omim	15608251 [29]
Sanger Institute gene annotation Transcript/translation information, location, SNPs, ortholog prediction, disease matches, related web sites for genes	http://www.ensembl.org/	19033362 [30]
Transfac database Transcription factors databases with their experimentally proven binding sites, and regulated genes	http://www.gene-regulation.com/pub/databases.html	12520026 [9]
Genomatix Promoter Database and annotation	http://www.genomatix.de/products/index.html	11222983 [7]
Stanford Microarray Database(SMD) Integrated microarray databases allow storing, annotating, and analyzing data generated by microarray technology	http://smd.stanford.edu	17182626 [13]
Swiss-Prot A curated protein sequence database with a high level of annotation	http://www.expasy.ch/sprot/	15153305 [31]
KEGG Integrated pathway database resource	http://www.genome.jp/kegg/	19880382 [10]

SNP: single nucleotide polymorphism.

Table 2.2 Useful tools and links.

Name	Link	PMID
Blast Basic Local Alignment Search Tool	http://www.ebi.ac.uk/Tools/blast/	2231712 [32]
ClustalW Multiple sequence alignment program	http://www.ch.embnet.org/software/ClustalW.html/	7984417 [33]
Entrez Cross-database search page	http://www.ncbi.nlm.nih.gov/Entrez/	17148475 [6]
SRS Primary gateway to major databases	http://srs.ebi.ac.uk	11847095 [34]
HMMER Pattern search and discovery tool	http://hmmer.janelia.org/	9918945 [35]
BlockMaker Finding conserved blocks in a set of protein sequences	http://blocks.fhcrc.org/blocks/blockmkr/make_blocks.html	7590261 [36]
SMART Simple Modular Architecture Research Tool	http://smart.embl-heidelberg.de/	18978020 [37]
Metatool Biochemical reaction networks properties computing program	http://pinguin.biologie.uni-jena.de/bioinformatik/networks	16731697 [11]
TESS Transcription Element Search System	http://www.cbil.upenn.edu/cgi-bin/tess/tess	18428685 [8]
GenomatixSuite Integrated bioinformatics software bundle to analyze gene regulation	http://www.genomatix.de/products/GenomatixSuite/index.html	11222983 [7]
YANA Metabolic networks analyzing software package	http://yana.bioapps.biozentrum.uni-wuerzburg.de/	15929789 [12]
S.O.U.R.C.E Unification tool collecting data from many scientific databases	http://smd.stanford.edu/cgi-bin/source/sourceSearch	12519986 [38]

Table 2.3 Apicomplexan databases and tools.

Name	Link	PMID
Genome database and annotation *EuPathDB* (ApiDB): integrated database covering the eukaryotic pathogens in the genera *Cryptosporidium, Giardia, Leishmania, Neospora, Plasmodium, Toxoplasma, Trichomonas,* and *Trypanosoma*	http://eupathdb.org/eupathdb/	19914931 [16]
NCBI Entrez Genome Project: a searchable collection of complete and incomplete large-scale sequencing, assembly, annotation, and mapping projects for cellular organisms	http://www.ncbi.nlm.nih.gov/sites/entrez?cmd=Search&db=genomeprj&term=%22Apicomplexa%22 [Organism]	12519941 [39]
Apicomplexan metabolism pathway Tool *Apicyc:* apicomplexan metabolism pathways prediction and comparison tool	http://apicyc.apidb.org/	17098930 [40]
Genus-specific databases *PlasmoDB:* integrated database for the genus *Plasmodium*	http://plasmodb.org/plasmo/	18957442 [14]
TriTrypDB: integrated database for pathogens of the family Trypanosomatidae (including *Leishmania* and *Trypanosoma* genera	http://tritrypdb.org/tritrypdb/	19914931 [16]
Expression level database		
Transcriptome database *Malaria IDC Database:* transcriptome data characterizing transcriptional differences between strains of *Plasmodium falciparum*	http://malaria.ucsf.edu/comparison/index.php	16493140 [41]
c-DNA database *Full-parasites:* a database for a full-length cDNAs from various parasites	http://fullmal.hgc.jp/	18987005 [42]
Disease-related target database *The TDR Targets database:* identification and ranking of targets against tropical diseases	http://tdrtargets.org/	18927591 [15]
Additional data *Wolbachia Genome Project:* Wolbachia genome data including blast search	http://tools.neb.com/wolbachia/search.html	15780005 [43]

(*Continued*)

Table 2.3 (*Continued*)

Name	Gene Identifier	PMID
GWAS: genome-wide association studies of *Plasmodium falciparum* to identify *seven* genes associated with antimalarial drug responses	PFB0675w, PFD0135w, PFE1445c, MAL7P1.27, MAL7P1.34, PFI0175w, MAL13P1.285	20101240 [26]

this in terms of the individual proteins. Nevertheless, a detailed and strain-specific genome analysis is required to yield optimal targets. Typically, a phylogenetic analysis allows a better description of strain variation and development, for example, with regards to malarial disease genetic variations in plasmodia vectors and hosts in sub-Saharan Africa [18, 19]. For this, different statistical measures allow vector strain differences to be identified simply by their microsatellite variation, and without knowledge of their complete genome sequence. Similarly, recent genome projects have provided unprecedented genomic information, for example, on *Plasmodium falciparum* and *Plasmodium vivax*. Furthermore, whilst the evolutionary history of apicomplexan parasites is complex [20], it suggests numerous plasmodia-specific pathways which, again, may provide valuable pharmacological targets.

Example: Gene-Drive Strategies

Increasing resistance against pharmacological interventions is widespread among apicomplexans, particularly with regards to malaria. An alternative to typical pharmacological approaches, which generally attempt to modify enzyme function by small molecules, are gene-drive strategies (Figure 2.1). These are genetic-modification strategies that, in general, have a super-Mendelian inheritance. The idea is that, in this way, the genetic modification is spread among a population and changes either the vector or parasite properties in a favorable way so as to reduce the illness burden. For example, a vector with the modification would become resistant against plasmodia, whereas with the modification the plasmodia would become less virulent. The primary challenge here is to identify a suitable target gene for the intervention; an example might be the genes that are important for plasmodia transmission in the vector, or plasmodia sexual stage-specific kinases [21]. If the latter were to be modified, the plasmodia carrying a gene-driven construct would only be impaired or killed after having undergone sexual transformation. Before this stage is reached, however, the gene-driven construct would already have multiplied in the asexual cycle, which is a further multiplication factor for the construct (see Figure 2.1).

For optimal gene-driven construct design, the transmission is simulated by suitable programs and calculation [23]; furthermore, a number of gene-driven vectors is already available. Gene-drive strategies function especially well for holoendemic areas, where both targeting and transmission would have optimal chances [2].

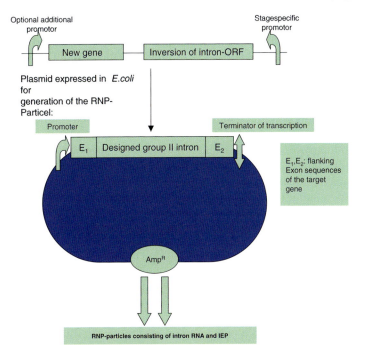

Figure 2.1 Gene-drive strategies. Genetic modification of the organism (green boxes) including a gene conversion vector (blue) is critical. The example shown features gene conversion by a designed group II intron (middle; intron jumps and modifies both copies of the diploid genome leading to super-Mendelian inheritance). In the advanced vector shown, two different promoters (top) separate the gene conversion step (right, the stage-specific promoter allows jumping only at an optimal time point in the life cycle) from the induction of the modified gene ("new gene," promoter at top left).

Although, in the previous discussion only vector-based gene-drive strategies were outlined, interestingly, endo-symbionts and parasites also utilize their full genome in different gene-drive strategies. For instance, the bacterium *Wollbachia* is very effective at changing the gender and thus creating infertility in various insects [23]. A new concept here would be to use such endosymbiotic organisms against malaria disease [24]. In this case, bioinformatics may help in the choice and design for example, by a detailed genome analysis, to identify the optimum *Wollbachia* strain, in an effort to control the spread of the malarial vector, *Anopheles gambiae*.

Combination Strategies

Combination strategies are clearly required as there is, unfortunately, a high-resistance development potential in apicomplexans. Interestingly, specific bioinformatics strategies are available for the creation of such drug combinations and tools, to calculate the combination effects of different drugs. Tools to calculate the effect of drug combinations on metabolism include different algorithms for flux balance analysis; examples include metatool, YANA, or YANAsquare for elementary mode analyses [25]. Such an elementary mode analysis investigates the different enzyme combinations of a network that balance all of the internal metabolites in which a specific chain of enzymes is involved [12]. The list of elementary modes obtained would be an enumeration of all feasible pathways in a metabolic network [25]. The topological analysis of a pathway network, as calculated by elementary modes, can be used to readily identify which key enzymes must be hit in order to block many routes in the network. Furthermore, given target knowledge, the drop-out of different pathways after administering a combination of drugs is also easily calculated [25]. Thus, current investigations involve the effect of drug combinations in redox networks to impair resistance development in malaria. Yet, the opposite is also possible, with "dirty" drugs (such as methylene blue) hitting several pathways and targets at the same time. A pathway analysis with the above-mentioned tools would allow a better description of the effects of such dirty drugs, and their multiple-hit

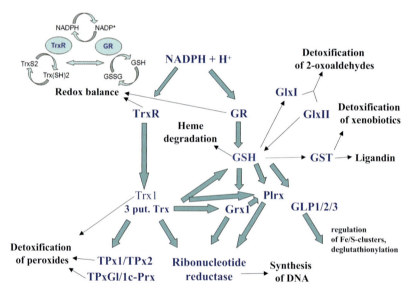

Figure 2.2 Multiple-hit strategies. The example [44] shown is the central redox network around glutathione reductase (GR) and reduced glutathione (GSH). The flow (light blue = major metabolic flow, black arrows = further connected processes, simplified) through these networks can be calculated (key enzymes and substrates are shown in blue), and thus also complex and broad effects of different drugs appropriately modeled; for instance, methylene blue would affect several of the involved enzymes and centrally target GR (middle). This also occurs in the host (human GR) and by methylene blue acting as a subversive substrate, incapacitating the enzymes that metabolize it. However, the multiple side effects improve the drug effect here and prevent resistance, this effect can be calculated in detail. Abbreviations: GLP, glutaredoxin-like protein; Glx, glyoxalase; Grx, glutaredoxin; GSSG, glutathione disulfide; GST, glutathione S-transferase; Plrx, plasmoredoxin; TPx, thioredoxin-dependent peroxidase; TPxGl, glutathione peroxidase-like thioredoxin peroxidase; Trx, thioredoxin; TrxR, thioredoxin reductase.

strategy (Figure 2.2). In fact, this has been heralded as a key to success against apicomplexan parasites, and their spread of resistance [5].

Pro-drugs represent another effort to improve treatment of apicomplexan diseases. In particular, the take-up of a pro-drug may be very low or nontoxic for the host, but then kill the parasite when converted into the drug. This can be easily and appropriately modeled applying YANA and YANAsquare [12, 25], and by including separate pro-drug conversion reactions. *Dual drugs*, instead, try to attack several targets by their different head groups [3]. A further variation of the pro-drug theme is a drug which acts as a subversive substrate. In this case, the drug closely resembles the physiological substrate, but is converted into a toxic compound. In order to model this correctly, flux balance models would need to include an irreversible reaction with the impaired enzyme as substrate. To appropriately model the strengths of this removal from the active pool under actual conditions, YANAsquare calculates the removal rates from experimental data or activity data. Whilst, in each of these three cases, a network analysis would be advantageous (see Chapters 5 to 14 for further detail), it must be stressed that the bioinformatics and specific algorithms detailed here allow the rational design of combination or refined drug strategies, and these will become ever-more important in the fight against apicomplexans, to prevent resistance (Figure 2.3).

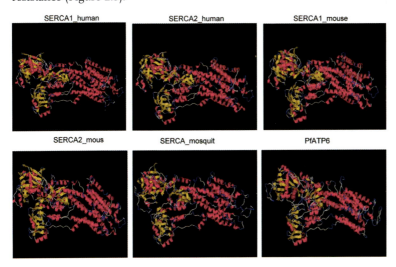

Figure 2.3 Analyzing minor changes leading to resistance. The example shown concerns the artemisin target PfATP6, a sarcoplasmic/endoplasmic reticulum Ca^{2+}-ATPase ortholog (SERCA). Colored homology models (helices: red; strands: brown; loops: blue) compare the structures of the SERCAs from human, mouse, *Anopheles* mosquito, and *P. falciparum* (PfATP6). The minor differences in structure regarding PfATP6 are targeted by artemisin; for example, position 263, which is suggested to be involved in artemisinin binding to PfATP6, or position 769, which is proposed to be related to a decreased sensitivity to artemether *in vitro*, shows that resistance could involve only minor structural changes that, in this case, can be investigated by homology modeling. In fact, after a strong initial success, an initial resistance and spread of resistance against artemisin are now being observed, for example in southeast Asia.

Resistance Modeling in Apicomplexa and Insights from a Gene-Drive Strategy

The rapid development of resistance occurs much more easily than was previously considered in the key artemisin target, PfATP6 (Figure 2.3). Homology modeling has shown that a few point mutations at the domain boundaries of PfATP6 are sufficient, in accordance with recently reported data [26]. In general, structural modeling allows the prediction of resistance development to a certain extent, provided that some of the mutations are known. Similarly, pathway modeling and genome comparisons chart in a broader way (i.e., key proteins required, key pathways required) the routes and potential for resistance development. For pathway predictions, large repositories such as KEGG [10] allow first predictions on the involved enzymes and pathways, after which more accurate results again involve the direct modeling of enzyme and pathway chains, for example using the tools discussed above. By genetic modification of apicomplexans or their vectors a gene drive strategy can target (Figure 2.1) any pathway of interest including resistance pathways by suitable constructs. However, even an escape from gene-driven vectors and agents is possible, and again involves point mutations (e.g., in gene conversion or at the recombination site). Furthermore, this also allows (if desired) the design of "call-back" constructs, leading to resistant individuals.

Missing the Catalytic Target and Identifying Other Important Parts in the Protein

Since the apicomplexans are complex, higher eukaryotic cells, the fact that the human or mammalian host is a complex eukaryote means that it will be especially difficult to identify suitable differences for pharmacological target strategies. The modification of standard target strategies is helpful under these conditions, however. In particular

Falcipain2(pf)   **Falcipain2'(pf)**

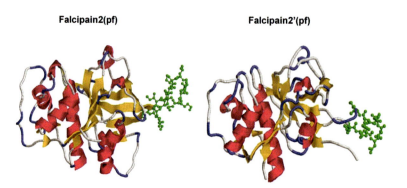

Figure 2.4 Noncatalytic target site strategies. Another strategy for targeting eukaryotic parasites with their relatedness to the host is not to target the putatively conserved catalytic center of enzymes such as papain-like proteases, but rather to seek other, more specific structural differences. The figure shows Falcipain2 (PDB structure: 1yvb, left) and Falcipain2' (model, right), but in a view stressing the hemoglobin-specific substrate binding part at the C terminus (homology models, color coded as in Figure 2.2; the hemoglobin binding motif is shown in green with ball and stick representation).

for enzymes, where the basic catalytic mechanisms are rather similar in both the parasite and host, it is useful to attack other parts of the structure or, alternatively, the host enzyme and to create a pharmacological effect in this way. In this case, glutathione reductase may serve as an excellent example [12]; since an impairment of the host enzyme will help to limit replication of the parasite, a break-down of redox-protection in the infected erythrocyte will also block any further replication of the parasite. Further structure and sequence analyses of glutathione reductase (GR), which is classified as a primary target in malaria [15], would be worthwhile as a number of drugs (including dual drugs [3]) currently exploit different hit strategies against GR.

Further examples include the cathepsins in apicomplexans and the host. In this case, it is very interesting that Falcipain 2 and Falcipain 2′, from *P. falciparum*, are specific hemoglobin-digesting proteases, with the sequence-specific differences being situated in the C-terminal hemoglobin adaption sequence (Figure 2.4). Again, these are small differences, with tools such as sequence, domain, structure, and gene expression analysis allowing an improved design of drugs targeting the C terminus of the cysteine proteases. Furthermore, the exploitation of a gene expression analysis would improve the targeting of a specific stage, for example, the gametocyte stage [1], where only a subset of falcipains is expressed.

Conclusion

In this chapter, it has been shown that drug design and the fight against apicomplexans is particular challenging, as these are complex eukaryotic cells that, in certain aspects, are quite similar to the cells of their human hosts. Nevertheless, a number of different and complementary bioinformatics strategies will allow new and improved strategies to be devised to create drugs against the apicomplexans, although these must then be implemented on a practical basis (e.g., see Chapters 1, 3, 7,9, 16, and 20).

References

1 Xiao, K., Jehle, F., Peters, C., Reinheckel, T., Schirmer, R.H., and Dandekar, T. (2009) CA/C1 peptidases of the malaria parasites *Plasmodium falciparum* and *P. berghei* and their mammalian hosts – a bioinformatical analysis. *Biol. Chem.*, **390**, 1185–1197.

2 Sinkins, S.P. and Gould, F. (2006) Gene drive systems for insect disease vectors. *Nat. Rev. Genet.*, **7**, 427–435.

3 Friebolin, W., Jannack, B., Wenzel, N., Furrer, J., Oeser, T. *et al.* (2008) Antimalarial dual drugs based on potent inhibitors of glutathione reductase from *Plasmodium falciparum*. *J. Med. Chem.*, **51**, 1260–1277.

4 Russo, I., Babbitt, S., Muralidharan, V., Butler, T., Oksman, A., and Goldberg, D.E. (2010) Plasmepsin V licenses *Plasmodium* proteins for export into the host erythrocyte. *Nature*, **463**, 632–636.

5 Kelly, J.X., Smilkstein, M.J., Brun, R., Wittlin, S., Cooper, R.A., Lane, K.D. *et al.* (2009) Discovery of dual function acridones as a new antimalarial chemotype. *Nature*, **459**, 270–273.

6 Maglott, D., Ostell, J., Pruitt, K.D., and Tatusova, T. (2005) Entrez Gene: gene-centered information at NCBI. *Nucleic Acids Res.*, **33**, D54–58.

7 Werner, T. (2001) Target gene identification from expression array data by promoter analysis. *Biomol. Eng.*, **17**, 87–94.

8 Schug, J. (2008) Using TESS to predict transcription factor binding sites in DNA sequence. *Curr. Protoc. Bioinformatics*, **21**, 2.6.1–2.6.15.

9 Matys, V., Fricke, E., Geffers, R., Gossling, E., Haubrock, M., Hehl, R. *et al.* (2003) TRANSFAC: transcriptional regulation, from patterns to profiles. *Nucleic Acids Res.*, **31**, 374–378.

10 Kanehisa, M., Goto, S., Furumichi, M., Tanabe, M., and Hirakawa, M. (2010) KEGG for representation and analysis of molecular networks involving diseases and drugs. *Nucleic Acids Res.*, **38**, D355–D360.

11 Von Kamp, A. and Schuster, S. (2006) Metatool 5.0: fast and flexible elementary modes analysis. *Bioinformatics*, **22**, 1930–1931.

12 Schwarz, R., Musch, P., Von Kamp, A., Engels, B., Schirmer, H. *et al.* (2005) YANA - a software tool for analyzing flux modes, gene-expression and enzyme activities. *BMC Bioinformatics*, **6**, 135.

13 Demeter, J., Beauheim, C., Gollub, J., Hernandez-Boussard, T., Jin, H. *et al.* (2007) The Stanford microarray database: implementation of new analysis tools and open source release of software. *Nucleic Acids Res.*, **35**, D766–D770.

14 Aurrecoechea, C., Brestelli, J., Brunk, B.P., Dommer, J., Fischer, S., Gajria, B. *et al.* (2009) PlasmoDB: a functional genomic database for malaria parasites. *Nucleic Acids Res.*, **37**, D539–D543.

15 Agüero, F., Al-Lazikani, B., Aslett, M., Berriman, M., Buckner, F.S., Campbell, R.K. *et al.* (2008) Genomic-scale prioritization of drug targets: the TDR Targets database. *Nat. Rev. Drug Discov.*, **7**, 900–907.

16 Aurrecoechea, C., Brestelli, J., Brunk, B.P., Fischer, S., Gajria, B., Gao, X. *et al.* (2010) EuPathDB: a portal to eukaryotic pathogen databases. *Nucleic Acids Res.*, **38**, D415–D419.

17 Tatusov, R.L., Fedorova, N.D., Jackson, J.D., Jacobs, A.R., Kiryutin, B. *et al.* (2003) The COG database: an updated version includes eukaryotes. *BMC Bioinformatics*, **4**, 41.

18 Michalakis, Y. and Renaud, F. (2009) Malaria: Evolution in vector control. *Nature*, **462**, 298–300.

19 Wang, R., Zheng, L., Toure, Y.T., Dandekar, T., and Kafatos, F.C. (2001) When genetic distance matters: measuring genetic differentiation at microsatellite loci in whole-genome scans of recent and incipient mosquito species. *Proc. Natl Acad. Sci. USA*, **98**, 10769–10774.

20 Tyagi, N., Swapna, L.S., Mohanty, S., Agarwal, G., Gowri, V.S., Anamika, K. *et al.* (2009) Evolutionary divergence of *Plasmodium falciparum*: sequences, protein-protein interactions, pathways and processes. *Infect. Disord. Drug Targets*, **9**, 257–271.

21 Kuehn, A. and Pradel, G. (2010) The coming-out of malaria gametocytes. *J. Biomed. Biotechnol.*, **2010**, 976827.

22 Burt, A. (2003) Site-specific selfish genes as tools for the control and genetic engineering of natural populations. *Proc. Biol. Sci.*, **270**, 921–928.

23 Marshall, J.M. (2009) The effect of gene drive on containment of transgenic mosquitoes. *J. Theor. Biol.*, **258**, 250–265.

24 Read, A.F. and Thomas, M.B. (2009) Microbiology. Mosquitoes cut short. *Science*, **323**, 51–52.

25 Schwarz, R., Liang, C., Kaleta, C., Kuhnel, M., Hoffmann, E., Kuznetsov, S. *et al.* (2007) Integrated network reconstruction, visualization and analysis using YANAsquare. *BMC Bioinformatics*, **8**, 313.

26 Mu, J., Myers, R.A., Jiang, H., Liu, S., Ricklefs, S., Waisberg, M. *et al.* (2010) Plasmodium *falciparum* genome-wide scans for positive selection, recombination hot spots and resistance to antimalarial drugs. *Nat. Genet.*, **42**, 268–271.

27 Benson, D.A., Karsch-Mizrachi, I., Lipman, D.J., Ostell, J., and Sayers, E.W. (2009) GenBank. *Nucleic Acids Res.*, **37**, D26–D31.

28 Wheeler, D.L., Church, D.M., Federhen, S., Lash, A.E., Madden, T.L. *et al.* (2003) Database resources of the National Center for Biotechnology. *Nucleic Acids Res.*, **31**, 28–33.

29 Hamosh, A., Scott, A.F., Amberger, J.S., Bocchini, C.A., and McKusick, V.A. (2005) Online Mendelian Inheritance in Man (OMIM), a knowledgebase of human genes and genetic disorders. *Nucleic Acids Res.*, **33**, D514–D517.

30 Hubbard, T.J., Aken, B.L., Ayling, S., Ballester, B., Beal, K., Bragin, E. *et al.* (2009) Ensemble 2009. *Nucleic Acids Res.*, **37**, D690–D697.

31 Bairoch, A., Boeckmann, B., Ferro, S., and Gasteiger, E. (2004) Swiss-Prot: juggling between evolution and stability. *Brief. Bioinform.*, **5**, 39–55.

32 Altschul, S.F., Gish, W., Miller, W., Myers, E.W., and Lipman, D.J. (1990) Basic local alignment search tool. *J. Mol. Biol.*, **215**, 403–410.

33 Thompson, J.D., Higgins, D.G., and Gibson, T.J. (1994) CLUSTAL W: improving the sensitivity of progressive multiple sequence alignment through sequence weighting, position-specific gap penalties and weight matrix choice. *Nucleic Acids Res.*, **22**, 4673–4680.

34 Zdobnov, E.M., Lopez, R., Apweiler, R., and Etzold, T. (2002) The EBI SRS server – recent developments. *Bioinformatics*, **18**, 368–373.

35 Eddy, S.R. (1998) Profile hidden Markov models. *Bioinformatics*, **14**, 755–763.

36 Henikoff, S., Henikoff, J.G., Alford, W.J., and Pietrokovski, S. (1995) Automated construction and graphical presentation of protein blocks from unaligned sequences. *Gene*, **163**, GC17–GC26.

37 Letunic, I., Doerks, T., and Bork, P. (2009) SMART 6: recent updates and new developments. *Nucleic Acids Res.*, **37**, D229–D232.

38 Diehn, M., Sherlock, G., Binkley, G., Jin, H., Matese, J.C., Hernandez-Boussard, T. *et al.* (2003) SOURCE: a unified genomic resource of functional annotations, ontologies, and gene expression data. *Nucleic Acids Res.*, **31**, 219–223.

39 Wheeler, D.L., Church, D.M., Federhen, S., Lash, A.E., Madden, T.L., Pontius, J.U. *et al.* (2003) Database resources of the national center for biotechnology. *Nucleic Acids Res.*, **31**, 28–33.

40 Aurrecoechea, C., Heiges, M., Wang, H., Wang, Z., Fischer, S., Rhodes, P. *et al.* (2006) ApiDB: integrated resources for the apicomplexan bioinformatics resource center. *Nucleic Acids Res.*, **35**, D427–D430.

41 Llinás, M., Bozdech, Z., Wong, E.D., Adai, A.T., and Derisi, J.L. (2006) Comparative whole genome transcriptome analysis of three *Plasmodium falciparum* strains. *Nucleic Acids Res.*, **34**, 1166–1173.

42 Wakaguri, H., Suzuki, Y., Katayama, T., Kawashima, S., Kibukawa, E., Hiranuka, K. *et al.* (2009) Full-Malaria/Parasites and Full-Arthropods: databases of full-length cDNAs of parasites and arthropods, update 2009. *Nucleic Acids Res.*, **37**, D520–D525.

43 Foster, J., Ganatra, M., Kamal, I., Ware, J., Makarova, K., Ivanova, N. *et al.* (2005) The *Wolbachia* genome of *Brugia malayi*: endosymbiont evolution within a human pathogenic nematode. *PLoS Biol.*, **3**, e121.

44 Becker, K., Rahlfs, S., Nickel, C., and Schirmer, R.H. (2003) Glutathione-functions and metabolism in the malarial parasite *Plasmodium falciparum*. *Biol. Chem.*, **384**, 551–566.

3
Sorting Potential Therapeutic Targets in Apicomplexa

Jan A. Hiss and *Gisbert Schneider**

Abstract

Malaria has a special place in the list of tropical diseases, due not only to its high death toll but also to the characteristics of its causative agent, the Plasmodia, which belong to the phylum Apicomplexa, and are obligate intracellular parasites of eukaryotic origin. Their intracellular lifestyle is a challenge not only for the parasite itself but also for the drug designer. Antimalarial drugs must reach an extracellular state of the parasite or overcome additional barriers, for example multiple membranes, in order to reach the parasite inside its host cell. This makes the intracellular sorting of target proteins vital for the development and application of antimalarial agents. The parasite possesses unique transport mechanisms that differ from its host cell, such as the sorting of pathogenic factors into the host cell and the retrograde transport of nutrients to the parasite. A further specialty of the malaria-causing Plasmodia (and other Apicomplexa) in rodents and humans is the presence of an additional organelle, the *apicoplast*. This complex plastid, which is essential for the parasite, is of prokaryotic origin and not present in the host cells. It is surrounded by additional membranes comparable to plastids in plants (such as chloroplasts), and originates from a secondary endosymbiotic event. The apicoplast, with its subcompartments, requires special protein-sorting mechanisms. As the apicoplast is absent in the host cell, both this organelle and its proteins are candidates for antimalarial drug targeting. Proteins and sorting pathways as potential targets in antimalarial drug discovery are discussed.

Introduction

The Relevance of Protein Sorting for Drug Delivery to Parasitic Organisms

The delivery of a drug to its target faces different situations *in vitro* and *in vivo*. While the drug may be potent in *in vitro* assays, it may be undeliverable *in vivo*, perhaps due to an inability to cross the cell membrane and reach an intracellular target.

*Corresponding author

Apicomplexan Parasites. Edited by Katja Becker
Copyright © 2011 WILEY-VCH Verlag GmbH & Co. KGaA, Weinheim
ISBN: 978-3-527-32731-7

Consequently, in order for the drug-delivery process to succeed, it is of key importance to recognize:

- the subcellular location of the therapeutic target (accessibility);
- the target's expression pattern (e.g., constitutive or during certain life stages, half-life, antigenic diversity [1]); and
- the mechanism for transport of the therapeutic target to its place of function (compartment).

In the Apicomplexa, the sorting aspect of therapeutic targets becomes even more important and complex, as most Apicomplexa are obligate intracellular parasites [2]. A drug in the bloodstream needs to pass multiple barriers, in addition to the cell membrane, before reaching an intracellular parasitic target [3]. Yet, these additional barriers also represent a challenge for the parasite itself, as it must deliver its proteins to the host cell cytosol or membrane, and thereby cut itself off from the supply of nutrients in the bloodstream [4]. The Apicomplexa live in mammalian hepatocytes or erythrocytes, depending on their life cycle stage. Within the erythrocytes, the parasite faces the problem that its host cell lacks typical nutrient transporters, and that it is living inside a parasitophorous vacuole, which separates it from the erythrocyte cytosol [4]. The erythrocytic stage, or more precisely *parasite replication* inside the erythrocyte, and the eventual release of new merozoites into the bloodstream, causes the clinical symptoms of malaria [5]. Replication takes approximately 48–72 h, depending on the species [6].

At this point, attention will be focused on the Plasmodia that cause human and rodent malaria, and their erythrocytic stage, as representatives of the Apicomplexa. The erythrocyte stage of *Plasmodium* is responsible for the clinical symptoms and host mortality [6]. It should be borne in mind that the intracellular life stage within hepatocytes is a prerequisite for the erythrocytic stage, and this has only recently received additional attention [6, 7]. In this chapter, information is provided concerning the sorting of potential therapeutic targets in the Apicomplexa, and its relevance to the drug-delivery process, while the sorting pathways are themselves proposed as a therapeutic target. Hence, attention is focused on: (i) intracellular parasitic development and survival (ontogenesis); and (ii) its evolutionary history (phylogenesis).

Protein Sorting in Apicomplexa: Ontogenesis and Phylogenesis

The majority of species which belonging to the phylum Apicomplexa are obligate endoparasites [2]. Both, the completion of their life cycle and their survival depend on the host cells providing nutrients [8, 9]. The parasite remodels the erythrocyte for this purpose by inducing the formation of a tubovesicular membrane system [10]. Endoparasitism may be viewed as a strategy to evade detection by the adaptive immune system of the mammalian host [11]. It should be borne in mind that proteins on the pathogen's cellular surface are, therefore, visible to the host immune system only during parasitic states outside the host cell, namely as sporozoites and merozoites [12]. The potential therapeutic points of interference might be, but are not limited to, the following:

- The proteins required for *host cell detection* are blocked – the pathogen cannot complete the life cycle, eventually leading to detection by the immune system.
- Proteins required for *host cell invasion* are blocked by antibodies [13] – the host membrane cannot be invaginated, invasion-associated proteins cannot be secreted by the parasite, and the invasion apparatus cannot be assembled.
- *Invasion/detection-associated proteins* can be recognized by antibodies triggering the adaptive immune response before invasion takes place (vaccination, [14]). Igonet and coworkers have reported that the apical membrane antigen 1 (AMA1), and specifically its ectoplasmic region, can induce immunity against malaria in animal models. A co-crystallization of the antibody and the ectoplasmic domain is shown in Figure 3.1 (PDB: 2J5L).

During the intracellular stages, namely the exoerythrocytic schizogony (hepatocytes) or the trophozoite ("ring form" erythrocytes), proteins on the pathogen's surface are inaccessible to the immune system, and intracellular location thereby helps the parasite to prevent a response by the immune system [12]. Under such circumstances, pathogenic proteins that are actively transported to the host's cell membrane by the parasite become essential. These surface proteins are potentially accessible to detection by antibodies or interaction with drugs. As a prerequisite, these proteins must be secreted into the parasitophorous vacuole and the cytosol of the erythrocyte before they eventually reach the cell membrane of the erythrocyte. Protein export into the host cell has been reported for *Plasmodium, Theileria*, and *Toxoplasma*, but has not

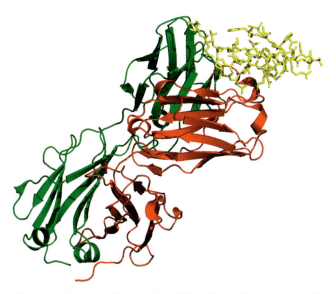

Figure 3.1 Apical membrane antigen 1 from *Plasmodium vivax* in conjunction with monoclonal antibodies taken and adapted from PDB entry 2j5l [14]. Yellow: apical membrane antigen 1 fragment; green: light chain of mouse monoclonal antibody; red: light chain of mouse monoclonal antibody.

yet been proven for other Apicomplexa such as *Eimeria*, *Babesia*, and *Cryptosporidium* [8].

It remains a challenge to identify and predict exported proteins based on their amino acid sequence. However, a major step in this direction was made in 2004, when an amino acid sequence pattern that was sufficient – but not strictly required – for the export of proteins from *P. falciparum* was identified by Hiller *et al.* [15] and Marti *et al.* [16]. This motif, specifically its core amino acids (RxLx[E,Q]), was proposed as the "*Plasmodium* export element" (PEXEL) [16] and "host targeting signal" [15], and its simplistic pattern was used to predict exported proteins. Further analysis revealed that the physico-chemical properties of the amino acids flanking the motif, as well as their position relative to the N terminus, or rather to an N-terminal signal peptide, play a role [17]. Notably, it has been shown that proteins lacking a PEXEL motif (PEXEL-negative proteins, PNEPs) are also exported to the erythrocyte [18, 19]. An example of such a protein is REX2 which, although lacking a PEXEL motif, has 10 N-terminal amino acids that are required for export into the erythrocyte. Furthermore, these amino acids are functionally replaceable by the N-terminal region of another PNEP, SBP1 [20]. This may also hint at additional transport pathways or a differential use of the existing ones, which may also be important with regards to knowledge transfer between species. Van Ooij and coworkers have proposed that 11 proteins are exported and conserved between human and rodent malarial parasites [21].

With regards to drug design, all proteins that reach the erythrocyte cytosol might represent potential therapeutic targets during the intracellular stages. This includes proteins already residing in the cytosol, or those that eventually reach the erythrocyte cell membrane. For intracellular protein transport to the plasma membrane in *P. falciparum*, special parasite-induced organelles ("Maurer's clefts") appear in the erythrocyte cytosol after infection [22, 23]. Evidence is accumulating that Maurer's clefts are secretory organelles [4, 24–27].

Proteins in the erythrocyte cytosol also represent potential drug targets, because drugs could presumably interfere with parasite nutrient uptake, replication inside the host cell, and release from the host cell after replication. Those parasitic proteins that reach the erythrocyte membrane might be available for a drug–antibody interaction. Thus, the therapeutic aspect could derive from immune system recruitment, or by blocking the original function of the protein.

In order to complete their life cycle, the parasites face several challenges while interacting with their host. First, they must detect and attach to appropriate host cells, after which they must invade the host, survive inside (i.e., solve the nutrition dilemma), and eventually replicate and leave the host cells. Each of these tasks makes the parasite potentially vulnerable to therapeutic intervention (Figure 3.2).

Detection and Attachment

The phylum Apicomplexa comprises mostly obligate endoparasites [2]. The parasite needs to detect appropriate host cells in a timely manner; otherwise it will perish due to starvation, detection by the immune system, or failed replication. As a first step,

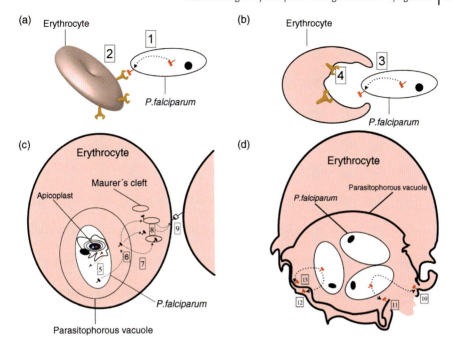

Figure 3.2 Erythrocyte-associated stages of *P. falciparum*. The numbers indicate potential points for therapeutic interference; dotted lines denote protein sorting/transportation pathways. (a) Erythrocyte detection. (1) Interference with the sorting of erythrocyte "detection" protein(s) to the surface of *P. falciparum*. (2) Blocking the interaction between *P. falciparum* "detection" proteins and erythrocyte surface proteins; (b) Erythrocyte invasion. (3) Interference with the sorting of erythrocyte "invasion" protein(s) to the surface of *P. falciparum*. (4) Blocking the interaction between *P. falciparum* "invasion" surface proteins and erythrocyte surface proteins (receptors); (c) Life inside the erythrocyte; interference with the sorting of vital proteins to the (5) apicoplast (dark blue), possibly inside the ER, (6) parasitophorous vacuole, (7) erythrocyte cytosol, (8) Maurer's clefts, (9) erythrocyte membrane; (d) Merozoite release from the erythrocyte, either by membrane rupture (10, 11) or by fusion between the parasitophorous vacuole and the erythrocyte membrane (12, 13).

a successful attachment to the host cell membrane is required to enable invasion. Although some of the relevant proteins are already known, the complete invasion apparatus and all associated proteins remain to be identified. It has been shown by Morahan and coworkers that thrombospondin-related anonymous protein (TRAP) is required for the invasion of hepatocytes [28]. The pathogen expresses erythrocyte-binding antigens (EBAs) of the Duffy-binding-like family for/during the invasion process [29]; for example, BAEBL (also known as EBA-140) mediates invasion and interacts with glycophorin C. Recently, Kobayashi *et al.* demonstrated that binding is mediated primarily by sialic acid [30]. Some of the proteins required for erythrocyte invasion are not continuously expressed by the pathogen, but rather are expressed in a regulated fashion and underlie epigenetic control [31].

Invasion

Host cell invasion requires the assembly of a complex apparatus at the apical membrane of the parasite [2, 32]. These apical complexes give the phylum its name, and their assembly requires a precise, regulated transport in time and space. They are referred to as micronemes, rhoptries, dense granules, and exonemes [2, 33]. The time of expression and location of proteins involved in the invasion are crucial for parasite survival. Tufet-Bayona and coworkers have shown that, for the rodent malaria parasite *Plasmodium berghei*, the RON2 protein is essential for the invasion of erythrocytes. The proteins RON2, RON4, and AMA1 are expressed in merozoites as well as in sporozoites, which suggests their importance for the invasion of erythrocytes and hepatocytes [34]. An interference with the sorting of these proteins could therefore affect the parasite during its different life stages.

Ravindran *et al.* (2009) [35] gave an example where protein sorting influences the invasion of *Toxoplasma gondii*, by showing that 4-bromophenacyl bromide would block the secretion of rhoptry, which in turn influences the motility and invasion of *Toxoplasma* but not the attachment or egress of the parasite. Another protein known to be important for the invasion of *T. gondii* is cathepsin L. This is an example where an inhibitor was first found, and only later was the actual target identified. In 2007, Teo and coworkers identified morpholinurea-leucylhomophenyl-vinyl sulfone phenyl as a potent inhibitor of parasite invasion [36] and subsequently, in 2009, demonstrated that its primary target was indeed cathepsin L [37]. The exact process of how the inhibition of cathepsin L leads to a reduction of invasiveness remains to be clarified, however.

Nutrition

During its intracellular state, the parasite shuts itself off from nutrients in the extracellular medium [4]. Additionally, noninfected erythrocytes are neither able to synthesize proteins nor import soluble substances required by the parasite [2]. Yet, apparently the parasite has found a solution to this dilemma, namely that a tubulovesicular network appears in the erythrocyte after infection, which seems important for nutrient uptake by the parasite [38]. Baumeister and coworkers provided evidence for parasite-encoded proteins involved in inducing "new permeability pathways" in the parasite membrane. The process is accompanied by an increased secretion of proteins from the parasite to the erythrocyte, remodeling the host cell functions [39].

An example of a substance which is essential for parasite survival, but is not present in the erythrocyte, is pantothenic acid (a precursor of coenzyme A). Saliba and coworkers showed that uninfected erythrocytes are impermeable to the precursor, whereas infected cells took it up rapidly [40]. Consequently, analogs of pantothenic acid were actually used as antimicrobial agents [41]. The parasite depends on additional substances such as glucose and choline, which require transporters or specialized organelles for their transport into the parasite. This makes them potential drug targets or, as Kirk and Saliba stated in 2007, "Proteins that mediate the uptake, intracellular trafficking and metabolism of essential nutrients in the *Plasmodium*-infected erythrocyte are potential antimalarial drug targets" [42].

Müller *et al.* have recently reported that protein-anchoring glideosome-associated protein 50 (GAP50) acts as a phosphatase, and is important for the dephosphorylation of host cell proteins and, consequently, their uptake by the parasite [43]. As the degradation of hemoglobin is essential for parasite survival, the *Plasmodium* proteases (plasmepsins [44]) involved in the digestion of hemoglobin become potential targets for chemotherapeutic strategies [45, 46]. As the uptake of additional nutrients from the medium is strictly essential for the parasite [45], Kirk and Saliba have suggested that this phenomenon might enable a "Trojan Horse" strategy for selective drug delivery into infected erythrocytes [42].

Replication/Release

Following the replication of *P. falciparum* into merozoites inside the parasitophorous vacuole, three possible routes of merozoite exit from the erythrocyte have been proposed [47, 48]: (i) rupture of the parasitophorous vacuole and the erythrocyte membrane in a coordinated manner [49, 50]; (ii) the fusion of both membranes [51–53]; and (iii) the parasite leaves the erythrocyte inside an intact parasitophorous vacuole, followed by its break-up outside the erythrocyte [54].

A decade ago, Winograd and coworkers questioned whether the release of merozoites into the bloodstream could be mediated by a nonexplosive event [52]. Despite Clavijo and coworkers having proposed a corresponding model a year earlier [51], Winograd *et al.* corroborated this possibility in failing to observe the diffusion of erythrocyte cytoplasm during merozoite release. A recent model by Kafsack *et al.* (2009) in *Toxoplasma* [55] included a secreted protein named *Toxoplasma* perforin-like protein 1 (TgPLP1), which plays a role in the pore formation that is required for the parasite to exit the parasitophorous vacuole and eventually to leave the host cell. For therapeutic appliances, the interruption of proper TgPLP1 sorting to the host cell membrane, thereby entrapping the parasite inside the cell, might be imagined.

The exact sequence of merozoite formation and release remains a matter of discussion [45]. This also includes the concept that the parasites leave the erythrocyte in an intact parasitophorous vacuole [56]. For hepatocytes, it has been shown recently by Heussler and coworkers that the parasite induces a coordinated cell death, which can be clearly distinguished from apoptosis and necrosis, and is initiated by the rupture of the parasitophorous vacuole [57].

Erythrocyte *apoptosis* is also a host cell response to prevent further infection [11]. Glushakova and coworkers have proposed a biochemical alteration of the erythrocyte membrane and a combination of folding and subsequent rupture to be the method for merozoite egress [48].

Regardless of the actual mechanism involved, it is beyond dispute that the correct sorting of factors to distinctive places of function within and without the parasite is required. At a certain point in the development of the parasite, additional proteins must be translated and translocated to their place of function (e.g., the membrane of the host cell or the parasitophorous vacuole). Interference with the sorting or activity of these proteins might be suited for trapping the parasite inside the erythrocyte.

Cell-Surface Proteins

Within the host cell, the parasite communicates not only with the intra-erythrocytic world, but also with the extra-erythrocytic. It does so by altering its host cell after successful invasion via the export of proteins into the erythrocyte cytosol [58]. The largest reported exported protein is Pf332, with a molecular mass exceeding 1 MDa, and which seems to mediate erythrocyte rigidity and the agglomeration of red blood cells [58]. After invasion, additional organelles termed Maurer's Clefts appear in the erythrocyte cytosol [22, 23]; these are associated with protein secretion from *P. falciparum* [4, 25] (Figure 3.2c).

These fundamental changes in the host cell require that the parasite not only secretes proteins to the parasitophorous vacuole (PV) in which it resides, but also crosses the PV membrane in order to reach the erythrocyte cytosol or become integrated into its membrane [2]. This task is complicated specifically for a parasite that has "...chosen to live in a de-nucleated cell lacking a secretory apparatus" [59, 60]. As a consequence, the parasite must code for its own transport machinery, or use proteins already present in the erythrocyte cytosol; the manner in which this is achieved remains unknown, however. For therapeutic sorting, this fact is most relevant because those proteins that reach the membrane of the erythrocyte play two crucial roles in drug design:

- They may act as pathogenic factors required for survival of the parasite (e.g., leading to erythrocyte agglomeration). Interfering with the function of these proteins could affect the parasite inside the host cell, or interfere with its progression to the next life stage. For example, when the parasite reproduces and leaves the infected cell, such an intervention could hinder new host cells from being readily available, due to inhibited agglomeration factors. As a consequence, the parasite would be exposed for a longer time to the immune system in the bloodstream, thereby increasing the chances of its detection.
- The proteins of pathogenic origin located on the erythrocyte membrane could be detected by antibodies and attract the immune system specifically to those infected erythrocytes.

The most important family of proteins under this aspect is the PfEMP1 (*P. falciparum* erythrocyte membrane proteins), which serve as a primary target for antibodies [61]. Recently, Kunrae *et al.* detailed the expression of a full-length member of the PfEMP1 protein family, and described antibodies raised against the full-length version of VAR2CSA that would inhibit parasite binding to the chondroitin sulfate proteogly-can, which is important for placental malarial infection. [The structure can be found in the Protein Data Bank [62, 63] (PDB ID: 2wau [64])]. Whilst this second aspect places the parasite under selective pressure because it requires pathogenic factors (e.g., agglomeration factors), exposure to the immune system for too long may lead to immunization.

Based on an elegant mathematical analysis, Recker *et al.* [65] introduced a model of how the parasite might circumvent the detection of parasitic proteins on the erythrocyte cell membrane. In this case, the expression of *P. falciparum var* genes that encode proteins from the PfEMP1 family, was analyzed. In acting as virulence factors, these proteins mediate adhesion of the erythrocyte to epithelial cells, which serves as a major trigger for the adaptive immune system [2]. Recker *et al.* suggested that the *var* genes might be expressed in a coordinated manner over time, rather than simultaneously. It would appear that the parasite possesses some form of "molecular clock" that switches gene expression over time. The idea of antigenic variation coordinated with a molecular clock as a response to selective pressure on large *Plasmodium* populations had already been proposed by Rich *et al.* [66]. As a consequence of this timing, the cell surface factors are not presented to the immune system for long enough to trigger an immune response, and so may vary from infection to infection, even within the actual population [65]. It may be assumed that such a mechanism not only requires the specific sorting of relevant factors to the erythrocyte membrane at certain time intervals, but also regulates the elimination or internalization of "obsolete" factors. These could prove to be important drug targets, with different therapeutic scenarios that might include:

- interference with their transport to the erythrocyte membrane;
- detection by antibodies at the cell surface; or
- hindrance of *var* gene product degradation by interfering with the respective degradation system.

Recently, Bergmann-Leitner *et al.* demonstrated that antibodies binding to surface proteins actually affected the maturing schizonts [67], thereby supporting this hypothesis.

Phylogenesis of the Apicomplexa

The Apicomplexa are single-cell eukaryotic parasites [2] that possess typical eukaryotic organelles, such as a nucleus, a Golgi apparatus, and mitochondria, albeit with minor differences; for example, the mitochondria encode only three genes [68]. A more prominent difference in the organelle configuration of the Apicomplexa originates in their evolutionary heritage, the apicoplast [69]. Because this organelle contains metabolic pathways (potentially of plant or bacterial origin) that are absent from the host cell, it represents a potentially attractive target for antimalarial drug development (for a review, see Ref. [70]). It should be borne in mind that the name of the phylum "Apicomplexa" originates from the *apical complexes* formed during invasion [2].

The apicoplast is unique to the Apicomplexa, and is thought to stem from a secondary endosymbiotic event [71, 72]. During the first event a cyanobacterium-like organism, as a predecessor of the apicoplast, was ingested by a eukaryote [73]. The resultant eukaryotic photosynthetic alga, containing the cyanobacterium-like first endosymbiont, was then again engulfed by a nonphotosynthetic eukaryote [72]. As a relic of this second event, additional membranes surround the apicoplast [74].

The exact origins of the membranes remain a matter of debate [75], but in some Apicomplexa the apicoplast is thought to be located "inside" or to be in continuum with the endoplasmic reticulum. The exact location and membrane configuration remains a matter of debate, and has been the subject of recent reviews [73, 76, 77].

An additional and fascinating relic of the endosymbiotic event is the *nucleomorph* in the cryptophytes and chlorarachniophytes [78]. The nucleomorph represents the remnant nucleus of the eukaryotic algae ingested during the endosymbiotic event, and remains present in cryptophytes and chlorarachniophytes. Notably, the nucleomorph still contains genes that become transcribed and translated into functional proteins. Up to four membranes envelop the apicoplast, each of which represents a subcompartment that must be addressed specifically by protein sorting procees [79].

The apicoplast is strictly required for parasite survival [80], as it harbors essential metabolic pathways such as fatty acid synthesis [81]. Thus, the apicoplast – or, more specifically, the proteins targeted toward the apicoplast – represents a potential target for therapeutic intervention by "apicodrugs" [73, 81], as it performs essential metabolic functions such as lipid or heme synthesis [71].

Today, there is increasing evidence that the endosymbiont-derived endoplasmic reticulum-associated protein degradation system (ERAD) is involved in protein sorting to the apicoplast [82–84]. In addition, protein targeting to the apicoplast in *P. falciparum* requires a bipartite N-terminal targeting sequence which consists of a signal peptide that guides the protein to the secretory pathway, followed by a transit peptide which resembles plant signals for apicolast import [74]. For therapeutic applications, a number of different scenarios might be imagined where the sorting of drugs – or, rather, of their targets – would be important that included:

- interfering with apicoplast metabolism;
- sorting proteins required for apicoplast metabolic pathways; and
- interfering with the division of the apicoplast.

Because the apicoplast is vital for *P. falciparum*, if any of these processes is affected it might prove to be harmful to the parasite [80]. The apicoplast has attracted attention in the treatment of other apicomplexan infections, besides *Plasmodium* and *Eimeria tenella* [85]; the organelle is also vital for other members of the Apicomplexa, such as *Toxoplasma gondii* [86]. The interference caused by sorting therapeutic targets to the apicoplast might also trigger the development of vaccines, as the apicoplast is present in all live stages of the parasite [69, 87].

In conclusion, the following aspects make the sorting of therapeutic agents to the apicoplast worthy of investigation:

- Apicoplast functionality is vital for the parasite.
- The apicoplast and its protein-sorting mechanisms are not present in the host cell.
- The apicoplast is present in all life stages of the parasite.

Since the identification of potential drug targets means, in turn, the identification of relevant apicoplast proteins, it is essential that the cellular localization of proteins is known. Bioinformatics methods can assist target hunting by following two concepts:

1) An *indirect approach*, where the proteins or their homologs are known to form part of the metabolic pathways that occur only in the apicoplast; homologous proteins are known to be localized in the apicoplast.
2) A *direct approach*, which involves the prediction of apicoplast proteins based on their amino acid sequences.

Both concepts have advantages and drawbacks. The indirect approach allows for fast prediction and analysis based on existing database knowledge but, due to inaccurate database annotation or differences between species, it is prone to false predictions. A prediction based on the primary sequence depends on experimental validated examples and robust algorithms. From the PlasmoDB [88] database, a set of prediction tools is available based on various machine learning methods:

- PlasmoAP (apicoplast targeting signal in *P. falciparum*) [89]
- PATS (apicoplast targeting signals in *P. falciparum*) [90]
- PlasMit (mitochondrial transit peptides in *P. falciparum*) [91]
- PlasmoCyc (Apicomplexa metabolic pathways) [92]

The prediction tool SignalP 3.0 [93] is also linked and used to predict automatically the signal peptides in plasmodial proteins. Search results using these tools can be combined with each other using logical operations, which allows filtering for certain combinations of protein properties. A search for proteins with different predicted characteristics or motifs across different species is provided in Table 3.1.

Notably, a larger number of proteins seem to be sorted to the apicoplast of *P. falciparum* ($N = 388$) than to the erythrocyte ($N = 292$ PEXEL, $N = 194$ HT). This corroborates the importance of protein sorting to the apicoplast, and its sorting mechanism as a potent target for antimalarial drugs.

In Table 3.1, the zeros arise because PlasmoAP [89], PEXEL [16], and HT analysis [15] are able to analyze *P. falciparum* proteins only, for the following reasons:

Table 3.1 PlasmoDB prediction results.

	P. falciparum	*P. vivax*	*P. yoelii*	*P. berghei*	*P. chabaudi*	*P. knowlesi*
Exported proteins	189	27	16	9	10	28
SP predicted	1043	821	983	1606	913	826
Apicoplast predicted	388	0	0	0	0	0
PEXEL	292	0	0	0	0	0
HT motif	194	0	0	0	0	0
Total no. of annotated proteins	5479	5435	7724	12 235	5098	5110

Exported proteins: ExportPred score > 10 [94]; Signal peptide (SP) predicted: SignalP [93] (SignalP-NN conclusion score > 3, SignalP-NN D-Score > 0.5, SignalP-HMM Signal Probability > 0.5); Apicoplast predicted: PlasmoAP prediction [89]; PEXEL: proteins containing the PEXEL motif [16]; HT: proteins containing the HT motif [15].

- The algorithms were based on and specifically designed for *P. falciparum* proteins. While the other species also possess an apicoplast, the sorting mechanism may be different.
- While the PEXEL or HT motif is functional in *P. falciparum* and artificially transferable to other species, it is absent from the proteome of the others. This leaves room for speculation about alternative protein sorting mechanisms.

Alternative protein-sorting mechanisms in other species – such as the rodent malarial parasite – still need to be integrated into prediction tools. This may also prove valuable for the analysis of *P. falciparum* proteins, as PEXEL alone is insufficient to explain the export of all proteins [18, 19]. Further general prediction tools not specifically trained on the Apicomplexa might prove useful, in particular TargetP (mitochondrial transit peptides) [95], and NtraC (Analysis of exceptional signal peptides > 40 residues) [96].

Conclusion

Protein sorting is vital for the parasite, and includes parasite-specific cellular localizations such as the apicoplast. Because certain aspects of protein sorting in *P. falciparum* are distinct from the host cell, not only the sorted proteins but also the sorting mechanisms might become viable targets. As a consequence, knowledge of both the cellular localization of a potential drug target and its respective sorting process is essential for hit and lead structure finding, and this might provide innovative strategies for combating Apicomplexa infections. Target identification and selection in future antimalarial drug discovery projects could combine various readily available prediction algorithms for cellular localization in combination with database information in the concept of a semantic cloud [97].

In 2009, the translocon of exported proteins (PTEX) was discovered [98]. This multiprotein complex is located in the parasitophorous vacuole membrane, and enables protein transport to the erythrocyte cytosol, and drugs capable of inhibiting the function of the translocon would undoubtedly prove to be effective against the parasite. Yet, translocon assembly could also be hindered, for example by sorting of its components to the PV membrane. Thus, a new and exciting field may emerge for both innovative bioinformatic prediction algorithms and novel tool compounds affecting protein-sorting processes in the Apicomplexa.

References

1 Takala, S.L. and Plowe, C.V. (2009) Genetic diversity and malaria vaccine design, testing and efficacy: preventing and overcoming 'vaccine resistant malaria'. *Parasite Immunol.*, **31**, 560–573.

2 Ravindran, S. and Boothroyd, J.C. (2008) Secretion of proteins into host cells by Apicomplexan parasites. *Traffic*, **9**, 647–656.

3 Samuel, B.U., Hearn, B., Mack, D., Wender, P., Rothbard, J. *et al.* (2003) Delivery of antimicrobials into parasites. *Proc. Natl Acad. Sci. USA*, **100**, 14281–14286.

4 Lanzer, M., Wickert, H., Krohne, G., Vincensini, L., and Braun-Breton, C. (2006) Maurer's clefts: a novel multi-functional organelle in the cytoplasm of *Plasmodium falciparum*-infected erythrocytes. *Int. J. Parasitol.*, **36**, 23–36.

5 Collins, W.E. and Jeffery, G.M. (2007) *Plasmodium* malariae: parasite and disease. *Clin. Microbiol. Rev.*, **20**, 579–592.

6 Mazier, D., Rénia, L., and Snounou, G. (2009) A pre-emptive strike against malaria's stealthy hepatic forms. *Nat. Rev. Drug Discov.*, **8**, 854–864.

7 Wells, T.N., Alonso, P.L., and Gutteridge, W.E. (2009) New medicines to improve control and contribute to the eradication of malaria. *Nat. Rev. Drug Discov.*, **8**, 879–891.

8 Deitsch, K.W. and Wellems, T.E. (1996) Membrane modifications in erythrocytes parasitized by *Plasmodium falciparum*. *Mol. Biochem. Parasitol.*, **76**, 1–10.

9 Gero, A.M. and Kirk, K. (1994) Nutrient transport pathways in *Plasmodium*-infected erythrocytes: what and where are they? *Parasitol. Today*, **10**, 395–399.

10 Lauer, S.A., Rathod, P.K., Ghori, N., and Haldar, K. (1997) A membrane network for nutrient import in red cells infected with the malaria parasite. *Science*, **276**, 1122–1125.

11 Föller, M., Bobbala, D., Koka, S., Huber, S.M., Gulbins, E., and Lang, F. (2009) Suicide for survival–death of infected erythrocytes as a host mechanism to survive malaria. *Cell Physiol. Biochem.*, **24**, 133–140.

12 Smith, J.D. and Craig, A.G. (2005) The surface of the *Plasmodium falciparum*-infected erythrocyte. *Curr. Issues Mol. Biol.*, **7**, 81–93.

13 Collins, C.R., Withers-Martinez, C., Hackett, F., and Blackman, M.J. (2009) An inhibitory antibody blocks interactions between components of the malarial invasion machinery. *PLoS Pathog.*, **5**, e1000273.

14 Igonet, S., Vulliez-Le Normand, B., Faure, G., Riottot, M.M., Kocken, C.H. *et al.* (2007) Cross-reactivity studies of an anti-*Plasmodium vivax* apical membrane antigen 1 monoclonal antibody: binding and structural characterisation. *J. Mol. Biol.*, **366**, 1523–1537.

15 Hiller, N.L., Bhattacharjee, S., van Ooij, C., Liolios, K., Harrison, T. *et al.* (2004) A host-targeting signal in virulence proteins reveals a secretome in malarial infection. *Science*, **306**, 1934–1937.

16 Marti, M., Good, R.T., Rug, M., Knuepfer, E., and Cowman, A.F. (2004) Targeting malaria virulence and remodeling proteins to the host erythrocyte. *Science*, **306**, 1930–1933.

17 Hiss, J.A., Przyborski, J.M., Schwarte, F., Lingelbach, K., and Schneider, G. (2008) The *Plasmodium* export element revisited. *PLoS One*, **3**, e1560.

18 Spielmann, T., Hawthorne, P.L., Dixon, M.W., Hannemann, M., Klotz, K. *et al.* (2006) A cluster of ring stage-specific genes linked to a locus implicated in cytoadherence in *Plasmodium falciparum* codes for PEXEL-negative and PEXEL-positive proteins exported into the host cell. *Mol. Biol. Cell*, **17**, 3613–3624.

19 Spielmann, T. and Gilberger, T.W. (2010) Protein export in malaria parasites: do multiple export motifs add up to multiple export pathways? *Trends Parasitol.*, **26**, 6–10.

20 Haase, S., Herrmann, S., Grüring, C., Heiber, A., Jansen, P.W. *et al.* (2009) Sequence requirements for the export of the *Plasmodium falciparum* Maurer's clefts protein REX2. *Mol. Microbiol.*, **71**, 1003–1017.

21 van Ooij, C., Tamez, P., Bhattacharjee, S., Hiller, N.L., Harrison, T. *et al.* (2008) The malaria secretome: from algorithms to essential function in blood stage infection. *PLoS Pathog.*, **4**, e1000084.

22 Maurer, G. (1900) Die tuepfelung der wirtszelle des tertianaparasiten. *Centralbl. f. Bakt. Abt. I. Orig.*, **28**, 114–125.

23 Maurer, G. (1902) Die malaria perniciosa. *Centralbl. f. Bakt. Abt. I. Orig.*, **32**, 695–719.

24 Hinterberg, K., Scherf, A., Gysin, J., Toyoshima, T., Aikawa, M. *et al.* (1994) Plasmodium *falciparum*: the Pf332 antigen is secreted from the parasite by a brefeldin A-dependent pathway and is translocated to the erythrocyte membrane via the Maurer's clefts. *Exp. Parasitol.*, **79**, 279–291.

25 Przyborski, J.M., Wickert, H., Krohne, G., and Lanzer, M. (2003) Maurer's clefts – a novel secretory organelle? *Mol. Biochem. Parasitol.*, **132**, 17–26.

26 Bhattacharjee, S., van Ooij, C., Balu, B., Adams, J.H., and Haldar, K. (2008) Maurer's clefts of *Plasmodium falciparum* are secretory organelles that concentrate virulence protein reporters for delivery to the host erythrocyte. *Blood*, **111**, 2418–2426.

27 Przyborski, J.M. (2008) The Maurer's clefts of *Plasmodium falciparum*: parasite-induced islands within an intracellular ocean. *Trends Parasitol.*, **24**, 285–288.

28 Morahan, B.J., Wang, L., and Coppel, R.L. (2009) No TRAP, no invasion. *Trends Parasitol.*, **25**, 77–84.

29 Gomez-Escobar, N., Amambua-Ngwa, A., Walther, M., Okebe, J., Ebonyi, A., and Conway, D.J. (2010) Erythrocyte invasion and merozoite ligand gene expression in severe and mild *Plasmodium falciparum* malaria. *J. Infect. Dis.*, **201**, 444–452.

30 Kobayashi, K., Kato, K., Sugi, T., Takemae, H., Pandey, K. *et al.* (2010) *Plasmodium falciparum* BAEBL binds to heparan sulfate proteoglycans on the human erythrocyte surface. *J. Biol. Chem.*, **285**, 1716–1725.

31 Jiang, L., López-Barragán, M.J., Jiang, H., Mu, J., Gaur, D. *et al.* (2010) Epigenetic control of the variable expression of a *Plasmodium falciparum* receptor protein for erythrocyte invasion. *Proc. Natl Acad. Sci. USA*, **107**, 2224–2229.

32 Preiser, P., Kaviratne, M., Khan, S., Bannister, L., and Jarra, W. (2000) The apical organelles of malaria merozoites: host cell selection, invasion, host immunity and immune evasion. *Microbes Infect.*, **2**, 1461–1477.

33 Baum, J., Gilberger, T.W., Frischknecht, F., and Meissner, M. (2008) Host-cell invasion by malaria parasites: insights from *Plasmodium* and *Toxoplasma*. *Trends Parasitol.*, **24**, 557–563.

34 Tufet-Bayona, M., Janse, C.J., Khan, S.M., Waters, A.P., Sinden, R.E., and Franke-Fayard, B. (2009) Localisation and timing of expression of putative *Plasmodium berghei* rhoptry proteins in merozoites and sporozoites. *Mol. Biochem. Parasitol.*, **166**, 22–31.

35 Ravindran, S., Lodoen, M.B., Verhelst, S.H., Bogyo, M., and Boothroyd, J.C. (2009) 4-Bromophenacyl bromide specifically inhibits rhoptry secretion during *Toxoplasma* invasion. *PLoS One*, **4**, e8143.

36 Teo, C.F., Zhou, X.W., Bogyo, M., and Carruthers, V.B. (2007) Cysteine protease inhibitors block *Toxoplasma gondii* microneme secretion and cell invasion. *Antimicrob. Agents Chemother.*, **51**, 679–688.

37 Larson, E.T., Parussini, F., Huynh, M.H., Giebel, J.D., Kelley, A.M. *et al.* (2009) Toxoplasma *gondii* cathepsin L is the primary target of the invasion-inhibitory compound morpholinurea-leucyl-homophenyl-vinyl sulfone phenyl. *J. Biol. Chem.*, **284**, 26839–26850.

38 Baumeister, S., Winterberg, M., Duranton, C., Huber, S.M., Lang, F. *et al.* (2006) Evidence for the involvement of *Plasmodium falciparum* proteins in the formation of new permeability pathways in the erythrocyte membrane. *Mol. Microbiol.*, **60**, 493–504.

39 Crabb, B.S., de Koning-Ward, T.F., and Gilson, P.R. (2010) Protein export in *Plasmodium* parasites: from the endoplasmic reticulum to the vacuolar export machine. *Int. J. Parasitol.*, **40**, 509–513.

40 Saliba, K.J., Horner, H.A., and Kirk, K. (1998) Transport and metabolism of the essential vitamin pantothenic acid in human erythrocytes infected with the malaria parasite *Plasmodium falciparum*. *J. Biol. Chem.*, **273**, 10190–10195.

41 Spry, C., Kirk, K., and Saliba, K.J. (2008) Coenzyme A biosynthesis: an

antimicrobial drug target. *FEMS Microbiol. Rev.*, **32**, 56–106.

42 Kirk, K. and Saliba, K.J. (2007) Targeting nutrient uptake mechanisms in *Plasmodium. Curr. Drug Targets*, **8**, 75–88.

43 Müller, I.B., Knöckel, J., Eschbach, M.L., Bergmann, B., Walter, R.D., and Wrenger, C. (2010) Secretion of an acid phosphatase provides a possible mechanism to acquire host nutrients by *Plasmodium falciparum. Cell Microbiol.*, **12**, 677–691.

44 Coombs, G.H., Goldberg, D.E., Klemba, M., Berry, C., Kay, J., and Mottram, J.C. (2001) Aspartic proteases of *Plasmodium falciparum* and other parasitic protozoa as drug targets. *Trends Parasitol.*, **17**, 532–537.

45 Baumeister, S., Winterberg, M., Przyborski, J.M., and Lingelbach, K. (2009) The malaria parasite *Plasmodium falciparum*: cell biological peculiarities and nutritional consequences. *Protoplasma*, **240**, 3–12.

46 Klemba, M. and Goldberg, D.E. (2002) Biological roles of proteases in parasitic protozoa. *Annu. Rev. Biochem.*, **71**, 275–305.

47 Bannister, L.H. (2001) Looking for the exit: How do malaria parasites escape from red blood cells? *Proc. Natl Acad. Sci. USA*, **98**, 383–384.

48 Glushakova, S., Yin, D., Li, T., and Zimmerberg, J. (2005) Membrane transformation during malaria parasite release from human red blood cells. *Curr. Biol.*, **15**, 1645–1650.

49 Dvorak, J.A., Miller, L.H., Whitehouse, W.C., and Shiroishi, T. (1975) Invasion of erythrocytes by malaria merozoites. *Science*, **187**, 748–750.

50 Wickham, M.E., Culvenor, J.G., and Cowman, A.F. (2003) Selective inhibition of a two-step egress of malaria parasites from the host erythrocyte. *J. Biol. Chem.*, **278**, 37658–37663.

51 Clavijo, C.A., Mora, C.A., and Winograd, E. (1998) Identification of novel membrane structures in *Plasmodium falciparum* infected erythrocytes. *Mem. Inst.*, **93**, 115–120.

52 Winograd, E., Clavijo, C.A., Bustamante, L.Y., and Jaramillo, M. (1999) Release of merozoites from *Plasmodium falciparum* infected erythrocytes could be mediated by a non-explosive event. *Parasitology*, **85**, 621–624.

53 Sherman, I.W., Eda, S., and Winograd, E. (2004) Erythrocyte aging and malaria. *Cell. Mol. Biol.*, **50**, 159–169.

54 Salmon, B.L., Oksman, A., and Goldberg, D.E. (2001) Malaria parasite exit from the host erythrocyte: a two-step process requiring extraerythrocytic proteolysis. *Proc. Natl Acad. Sci. USA*, **98**, 271–276.

55 Kafsack, B.F.C., Pena, J.D.O., Coppens, I., Ravindran, S., Boothroyd, J.C., and Carruthers, V.B. (2009) Rapid membrane disruption by a perforin-like protein facilitates parasite exit from host cells. *Science*, **323**, 530–533.

56 Blackman, M.J. (2008) Malarial proteases and host cell egress: an 'emerging' cascade. *Cell. Microbiol.*, **10**, 1925–1934.

57 Heussler, V., Rennenberg, A., and Stanway, R. (2010) Host cell death induced by the egress of intracellular *Plasmodium* parasites. *Apoptosis*, **15**, 376–385.

58 Glenister, F.K., Fernandez, K.M., Kats, L.M., Hanssen, E., Mohandas, N. *et al.* (2009) Functional alteration of red blood cells by a megadalton protein of *Plasmodium falciparum. Blood*, **113**, 919–928.

59 Przyborski, J.M. and Lanzer, M. (2004) Parasitology. The malarial secretome. *Science*, **306**, 1897–1898.

60 Przyborski, J.M. and Lanzer, M. (2005) Protein transport and trafficking in *Plasmodium falciparum*-infected erythrocytes. *Parasitology*, **130**, 373–388.

61 Khunrae, P., Dahlbäck, M., Nielsen, M.A., Andersen, G., Ditlev, S.B. *et al.* (2010) Full-length recombinant *Plasmodium falciparum* VAR2CSA binds specifically to CSPG and induces potent parasite adhesion-blocking antibodies. *J. Mol. Biol.*, **397**, 826–834.

62 Berman, H.M., Westbrook, J., Feng, Z., Gilliland, G., Bhat, T.N. *et al.* (2000) The protein data bank. *Nucleic Acids Res.*, **28**, 235–242.

63 Berman, H.M., Henrick, K., and Nakamura, H. (2003) Announcing the worldwide Protein Data Bank. *Nat. Struct. Biol.*, **10**, 980.

64 Khunrae, P., Philip, J.M., Bull, D.R., and Higgins, M.K. (2009) Structural comparison of two CSPG-binding DBL domains from the VAR2CSA protein important in malaria during pregnancy. *J. Mol. Biol.*, **393**, 202–213.

65 Recker, M., Arinaminpathy, N., and Buckee, C.O. (2008) The effects of a partitioned *var* gene repertoire of *Plasmodium falciparum* on antigenic diversity and the acquisition of clinical immunity. *Malar. J.*, **7**, 18.

66 Rich, S.M., Hudson, R.R., and Ayala, F.J. (1997) Plasmodium falciparum antigenic diversity: evidence of clonal population structure. *Proc. Natl Acad. Sci. USA*, **94**, 13040–13045.

67 Bergmann-Leitner, E.S., Duncan, E.H., and Angov, E. (2009) MSP-1p42-specific antibodies affect growth and development of intra-erythrocytic parasites of *Plasmodium falciparum*. *Malar. J.*, **8**, 183.

68 Gray, M.W., Burger, G., and Lang, F. (1999) Mitochondrial evolution. *Science*, **283**, 1476–1481.

69 McFadden, G.I., Reith, M.E., Munholland, J., and Lang-Unnasch, N. (1996) Plastid in human parasites. *Nature*, **381**, 482.

70 Ralph, S.A., van Dooren, G.G., Waller, R.F., Crawford, M.J., Fraunholz, M.J. *et al.* (2004) Tropical infectious diseases: metabolic maps and functions of the *Plasmodium falciparum* apicoplast. *Nat. Rev. Microbiol.*, **2**, 203–216.

71 Waller, R.F. and McFadden, G.I. (2005) The apicoplast: a review of the derived plastid of apicomplexan parasites. *Curr. Issues Mol. Biol.*, **7**, 57–79.

72 Dyall, S.D., Brown, M.T., and Johnson, P.J. (2004) Ancient invasions: from endosymbionts to organelles. *Science*, **304**, 253–257.

73 Tonkin, C.J., Foth, B.J., Ralph, S.A., Struck, N., Cowman, A.F., and McFadden, G.I. (2008) Evolution of malaria parasite plastid targeting sequences. *Proc. Natl Acad. Sci. USA*, **105**, 4781–4785.

74 Waller, R.F., Reed, M.B., Cowman, A.F., and McFadden, G.I. (2000) Protein trafficking to the plastid of *Plasmodium falciparum* is via the secretory pathway. *EMBO J.*, **19**, 1794–1802.

75 Cavalier-Smith, T. (2003) Genomic reduction and evolution of novel genetic membranes and protein-targeting machinery in eukaryote-eukaryote chimaeras (meta-algae). *Philos. Trans. R. Soc. Lond. B Biol. Sci.*, **358**, 109–133.

76 Gibbs, S.P. (1979) The route of entry of cytoplasmically synthesized proteins into chloroplasts of algae possessing chloroplast ER. *J. Cell Sci.*, **35**, 253–266.

77 Tonkin, C.J., Struck, N.S., Mullin, K.A., Stimmler, L.M., and McFadden, G.I. (2006) Evidence for Golgi-independent transport from the early secretory pathway to the plastid in malaria parasites. *Mol. Microbiol.*, **61**, 614–630.

78 Moore, C.E. and Archibald, J.M. (2009) Nucleomorph genomes. *Annu. Rev. Genet.*, **43**, 251–264.

79 Sommer, M.S., Gould, S.B., Lehmann, P., Gruber, A., Przyborski, J.M., and Maier, U.G. (2007) Der1-mediated preprotein import into the periplastid compartment of chromalveolates? *Mol. Biol. Evol.*, **24**, 918–928.

80 He, C.Y., Shaw, M.K., Pletcher, C.H., Striepen, B., Tilney, L.G., and Roos, D.S. (2001) A plastid segregation defect in the protozoan parasite *Toxoplasma gondii*. *EMBO J.*, **20**, 330–339.

81 Ralph, S.A., D'Ombrain, M.C., and McFadden, G.I. (2001) The apicoplast as an antimalarial drug target. *Drug Resist. Update*, **4**, 145–151.

82 Agrawal, S., van Dooren, G.G., Beatty, W.L., and Striepen, B. (2009) Genetic evidence that an endosymbiont-derived endoplasmic reticulum-associated protein degradation (ERAD) system functions in import of apicoplast proteins. *J. Biol. Chem.*, **284**, 33683–33691.

83 Spork, S., Hiss, J.A., Mandel, K., Sommer, M., Kooij, T.W., Chu, T., Schneider, G., Maier, U.G., and Przyborski, J.M. (2009) An unusual ERAD-like complex is targeted to the apicoplast of *Plasmodium falciparum*. *Eukaryot. Cell*, **8**, 1134–1145.

84 Kalanon, M., Tonkin, C.J., and McFadden, G.I. (2009) Characterization of two putative protein translocation components in the apicoplast of *Plasmodium falciparum. Eukaryot. Cell*, **8**, 1146–1154.

85 Cai, X., Fuller, A.L., McDougald, L.R., and Zhu, G. (2003) Apicoplast genome of the coccidian *Eimeria tenella. Gene*, **321**, 39–46.

86 Mazumdar, J., Wilson, E.H., Masek, K., Hunter, C.A., and Striepen, B. (2006) Apicoplast fatty acid synthesis is essential for organelle biogenesis and parasite survival in *Toxoplasma gondii. Proc. Natl Acad. Sci. USA*, **103**, 13192–13197.

87 Bannister, L.H., Hopkins, J.M., Fowler, R.E., Krishna, S., and Mitchell, G.H. (2000) A brief illustrated guide to the ultrastructure of *Plasmodium falciparum* asexual blood stages. *Parasitol. Today*, **16**, 427–433.

88 Aurrecoechea, C., Brestelli, J., Brunk, B.P., Dommer, J., Fischer, S. *et al.* (2009) PlasmoDB: a functional genomic database for malaria parasites. *Nucleic Acids Res.*, **37**, D539–D543.

89 Foth, B.J., Ralph, S.A., Tonkin, C.J., Struck, N.S., Fraunholz, M. *et al.* (2003) Dissecting apicoplast targeting in the malaria parasite *Plasmodium falciparum. Science*, **299**, 705–708.

90 Zuegge, J., Ralph, S., Schmuker, M., McFadden, G.I., and Schneider, G. (2001) Deciphering apicoplast targeting signals - feature extraction from nuclear-encoded precursors of *Plasmodium falciparum* apicoplast proteins. *Gene*, **280**, 19–26.

91 Bender, A., van Dooren, G.G., Ralph, S.A., McFadden, G.I., and Schneider, G. (2003) Properties and prediction of mitochondrial transit peptides from *Plasmodium falciparum. Mol. Biochem. Parasitol.*, **132**, 59–66.

92 Caspi, R., Foerster, H., Fulcher, C.A., Kaipa, P., Krummenacker, M. *et al.* (2008) The MetaCyc Database of metabolic pathways and enzymes and the BioCyc collection of Pathway/Genome Databases. *Nucleic Acids Res.*, **36**, D623–D631.

93 Bendtsen, J.D., Nielsen, H., von Heijne, G., and Brunak, S. (2004) Improved prediction of signal peptides: SignalP 3.0. *J. Mol. Biol.*, **340**, 783–795.

94 Sargeant, T.J., Marti, M., Caler, E., Carlton, J.M., Simpson, K. *et al.* (2006) Lineage-specific expansion of proteins exported to erythrocytes in malaria parasites. *Genome. Biol.*, **7**, R12.

95 Emanuelsson, O., Nielsen, H., Brunak, S., and von Heijne, G. (2000) Predicting subcellular localization of proteins based on their N-terminal amino acid sequence. *J. Mol. Biol.*, **300**, 1005–1016.

96 Hiss, J.A., Resch, E., Schreiner, A., Meissner, M., Starzinski-Powitz, A., and Schneider, G. (2008) Domain organization of long signal peptides of single-pass integral membrane proteins reveals multiple functional capacity. *PLoS One*, **3**, e2767.

97 Kim, H.-L., Passant, A., Breslin, J., Scerri, S., and Decker, S. (2008) Review and alignment of tag ontologies for semantically-linked data in collaborative tagging spaces. IEEE International Conference on Semantic Computing, pp. 315–322.

98 de Koning-Ward, T.F., Gilson, P.R., Boddey, J.A., Rug, M., Smith, B.J. *et al.* (2009) A newly discovered protein export machine in malaria parasites. *Nature*, **459**, 945–949.

4
Alternatives to Drug Development in the Apicomplexa

Theo P.M. Schetters

Abstract

A series of measures is being used to control infectious diseases that comprise management practices such as hygiene, sanitation, pasture management, vector control, the use of chemotherapeutics, and vaccines. Especially in the case of apicomplexan parasites, chemotherapeutics aiding the control of vector and parasite infestations are being used. Vaccines in general have received little attention, despite coccidiosis vaccines having been available since the 1950s. Classically, vaccines have attempted to provoke immune reactions that limited parasite proliferation (anti-parasite), which was best obtained with live (sometimes attenuated) vaccine strains, for example, those against *Eimeria*, *Toxoplasma*, *Theileria*, and *Babesia* spp. More detailed research has revealed that, in some cases, it was not the presence of the parasite *per se* but rather the inflammatory responses that they triggered which led to clinical signs. This was also suggested by the fact that hosts that have developed immunity often carry the parasite without overt clinical signs. This led to the development of vaccines to help the host control the disease rather than the parasite (anti-disease vaccines against *Plasmodium* and *Babesia* spp.). To improve the immunogenicity of a non-live vaccine, and to trigger the proper immune response, adjuvants were required. With increasing knowledge of the immune system, a more rational approach for the development of adjuvants ensued. Currently, tailor-made adjuvants that may even contain specific regulatory molecules from the immune system (cytokines) are being developed. It is envisaged that these new strategies will provide impetus to the development of new and effective vaccines against apicomplexan parasites in the near future.

Introduction

Vaccination is an alternative to drugs for controlling infectious diseases. Whilst most vaccines aim to reduce the proliferation of the pathogen in the host, some vaccines merely limit the development of disease, allowing time for the generation of additional protective immune responses upon infection. Additionally, part of vaccine

Apicomplexan Parasites. Edited by Katja Becker
Copyright © 2011 WILEY-VCH Verlag GmbH & Co. KGaA, Weinheim
ISBN: 978-3-527-32731-7

research has been dedicated to the development of safe adjuvants that must stimulate protective immune responses. In order to understand the different strategies used to develop safe and effective vaccines, it is important to understand the basic principles of disease and immunity. In this chapter, the different interactions between parasites and hosts will be discussed; this will be followed by an overview as to how these interactions have been used to develop vaccines against species that belong to the Apicomplexa.

Infection and Immunity

The time period between the moment of infection and the occurrence of clinical signs is called the *incubation period*. This is followed by a period of disease which, depending on the immune status of the host, may lead to recovery or death (Figure 4.1). Immediately after infection with a parasite, a series of reactions is triggered. Some of these may be stimulated by the parasite itself or be the result of parasite activity, as is the case with migrating parasites that destroy host tissue and parasites which proliferate in the host cells [1, 2]. Although an appropriate stimulation of these systems may help to control the invading organism, in many cases the continued activation of these systems is involved in the development of clinical disease.

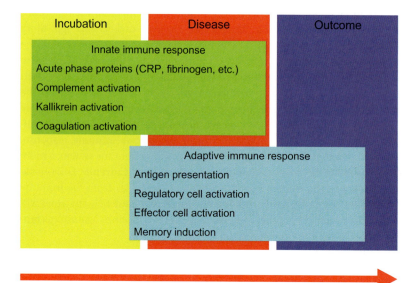

Figure 4.1 Different stages of infection and the activation of innate and adaptive immune responses over time. Initially, effector systems that play a role in innate immunity are activated, followed by effector systems that belong to the adaptive immune response. There is a clear overlap in time, and the balance between the systems determines the outcome of infection.

The earliest reactions involve humoral systems that are constitutively present, including the complement-, kinin-, and coagulation systems. All of the requisite components are present in inactive form, and each of these becomes activated by consecutive enzymatic processes (the "domino effect"). Additionally, an activation of the host cells takes place, which leads to responses that usually take more time before this is translated into an effector response. The activation of these systems can be either specific, and defined as a receptor–ligand interaction, or nonspecific, with no specific ligand involved (this is described below in more detail). The specific (via antigens) and aspecific (via adjuvants) stimulation of the different systems by vaccination is used to tilt the balance from disease to immunity. However, this procedure is not without risk, and in a number of cases the activation of these systems may cause adverse reactions upon vaccination, sometimes due to the antigen and often due to the adjuvant used.

Humoral Systems

Parasite infections trigger a number of humoral systems that comprise bioactive molecules in body fluids, which are produced and secreted by cells. Some of these molecules appear as inactive precursor molecules (zymogens) that can become activated by enzymatic cleavage. This is especially the case with the complement, coagulation, and kallikrein/kinin systems, which are constitutively present in the plasma (Table 4.1). Other molecules are secreted in an active form, but because these are only produced upon stimulation of a cell, they are discussed in the section below (see "Cellular Systems"). The concerted action of the plasma enzyme systems and leukocytes, which leads to the almost instantaneous production/release of acute-phase proteins (mainly by the liver), is called the *acute-phase response* [3]. These proteins display a number of different functions, from the opsonization of foreign particles to stimulating additional effector mechanisms (both innate and adaptive immune systems).

Nonspecific Activation of Humoral Systems

Complement System   Complement activation plays a central role in inflammation, not only because of the direct attack of invading parasites, but also because of the generation of molecules that affect the blood circulation and extravasation of humoral and cellular factors at the sites of infection.

Many parasites sustain the activation of the alternative complement pathway, which is different from direct activation. The current view is that, in the vertebrate host, there is always a background hydrolysis of fluid phase C3, leading to an initial C3-convertase which can generate C3b. Although this molecule is not very stable in solution because of soluble inhibitors and inactivators, when it adheres to surrounding particles (such as microbes) it is prevented from such inhibition and will generate the very active alternative pathway convertase, C3Bb [4]. This leads to the generation of bioactive molecules such as C3a, C4a, and C5a, and the formation of a circular complex composed of the late-stage complement factors C5b–C9. This complex,

Table 4.1 Humoral systems that play a role in innate immunity, the way of activation, and the biological effect.

System	Method of activation[a]	Bioactive compounds	Effect
Complement system	C3b deposition on parasite (-infected cell) surface prevents inactivation of C3b leading to further activation of complement cascade (alternative pathway) C1q activation by antigen–antibody complex (classical pathway)	C3a, C5a, membrane attack complex (MAC; C5b-C9)	Increased vascular permeability, leukocyte attraction, activation of polymorphonuclear leukocytes, pore formation in cell membranes by MAC and lysis of cells, macrophage activation
Coagulation system	Factor VII comes into contact with tissue factor in damaged blood vessels for example, migrating parasites (Tissue factor pathway) Hageman factor is activated when it comes into contact with surfaces of damaged or altered cells for example, parasite-infected cells (contact activation pathway)	Prothrombinase complex that converts prothrombin in active thrombin	Fibrin formation and blood clotting. Platelet activation
Kallikrein–kinin system	Activated Hageman factor (see "Coagulation system") Parasite-derived esterases directly convert kallikreinogen (prekallikrein) into kallikrein	Kinins resulting from proteolytic cleavage of kininogen by kallikrein	Endothelial cell activation leading to arterial dilatation and hypotension. Leukocyte attraction, macrophage activation

a) Parasite activity and parasite-derived compounds play a role in the activation of these systems.

which is referred to as the membrane attack complex (MAC), can lodge itself in the lipid bilayers of the cell membranes, causing pore formation and the lysis of cells such as infected red blood cells and parasites.

Coagulation System   Activation of the coagulation system is another example of activation that does not involve specific ligand–receptor interaction [5]. Central to this system is the destruction of host tissue due to the pathogen (e.g., lysis and alteration of host cells and tissues infected with pathogens). Two activation pathways can be distinguished: (i) the tissue factor pathway; and (ii) the contact activation pathway (these were formerly known as the extrinsic and intrinsic pathways, respectively). The tissue factor pathway is mainly involved in maintaining the integrity of the blood vessels, following damage to which, the coagulation factor VII (FVII) leaves the circulation and comes into contact with the tissue factor (TF); this is expressed on tissue-factor-bearing cells (stromal fibroblasts and leukocytes), and forms an activated complex (TF–FVIIa). The contact activation pathway is activated when the Hageman factor (FXII) comes into contact with the surfaces of altered or damaged cells. Both activation pathways lead to the formation of a prothrombinase complex, which converts prothrombin into thrombin, the enzyme that leads to the formation of fibrin. It should be noted that only host molecules are involved in these cascades, and that activation results because the parasite's activity has provoked the interaction of these host molecules.

Kallikrein–Kinin System   Although the kallikrein–kinin system has been described as being separate from the other systems, it is closely involved in the coagulation system. The central factor of the kallikrein–kinin system, *kallikrein*, is activated by the coagulation system's Hageman factor [6]. The end result of kallikrein activation is the formation of bioactive kinins with a very short half-life. It is suggested that, as with complement activation, there is always a background production of active plasma kallikreins (from neutrophils), and that this might find a more favorable environment for staying in the active form at a site of inflammation. Kinins may induce changes in smooth muscle tone (in blood vessels), or promote the release of hormones, neurotransmitters, and autacoids, or stimulate ion transport in excitable tissues, including the epithelia. A consequence of these activities is, among others, a change in the blood flow through the vertebrate host, fever, and the induction of pain.

Specific Activation of Humoral Systems

Apart from the nonspecific activation described above, some parasites activate these systems directly. Examples are activation of the coagulation system by altered erythrocyte surfaces upon infection with *Babesia* parasites [7], and activation of the kallikrein–kinin system by a *Babesia bovis* esterase or bacterial thermolysin [8].

A special case of specific activation of a plasma system is the activation of the complement system by antigen–antibody complexes. When hosts have been in prior contact with the antigen, specific antibodies may form a complex with it, that can activate C1q, the trigger of the classical complement pathway [9]. It may be made clear from the above that there is not a single functionally defined system involved here, and that infection causes a plethora of reactions comprising all humoral plasma systems. Some of the bioactive molecules affect the activity of cellular components of the innate and adaptive immune system, thus adding to the complexity of the inflammatory response.

Cellular Systems

Parasites also stimulate cellular systems through receptor–ligand interactions. The result of such stimulation can be either nonspecific (the production of bioactive molecules with a broad spectrum of activity, such as cytokines and nitric oxide; NO), or specific (the production of bioactive molecules that interact with parasite-derived molecules such as antibodies). Both of these cellular systems cooperate during infection, and lead to inflammation and immunity (Figure 4.2) [2].

Activation of Nonspecific Cellular Systems

During primary infection, when there has been no previous contact with the parasite, there is a limited stimulation of cells that express pathogen-specific receptors. However, pathogens carry structures that are recognizable by nonspecific receptors on host cells (Signal 0, Figure 4.2); these structures are collectively termed pathogen-associated molecular patterns (PAMPs [10]), and the host receptors that are involved are termed pattern recognition receptors (PRRs; see Ref. [11] for a review). Currently, different classes of PAMPs and PRRs have been described, most of which are not pathogen–species-specific, which means that the same receptor can be triggered by PAMPs of different pathogens (Table 4.2). Some PRRs are stimulated by self

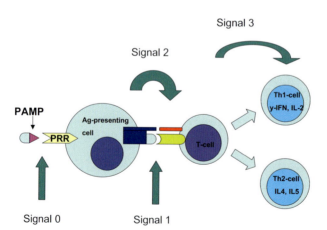

Figure 4.2 Regulation of the immune response. The parasite-derived antigen (pathogen-associated molecular pattern [PAMP]) activates an antigen (Ag)-presenting cell through interaction with a pattern recognition receptor (PRR). The antigen is processed and presented to the T-cell (Signal 1) that, upon co-stimulation (Signal 2), becomes activated. The cytokine context (Signal 3) determines the type of T-cell response. γ-IFN, gamma-interferon; IL, interleukin.

Table 4.2 Receptors on myeloid cells that are involved in the initial recognition of parasite molecules, the biochemical entity, and the cellular localization of the receptor.

Pattern recognition receptor (PRR)	Pathogen-associated molecular pattern (PAMP)	Distribution
TLR Toll-like receptors		
Toll-like receptor 1/2	Triacyl lipoprotein	Surface
Toll-like receptor 2	Lipoprotein; self	Surface
Toll-like receptor 3	Double-stranded RNA	Endolysosome (ER-membrane)
Toll-like receptor 4	Lipopolysaccharides; self	Surface
Toll-like receptor 5	Flagellin	Surface
Toll-like receptor 6/2	Diacyl lipoprotein	Surface
Toll-like receptor 7 (human TLR8)	Single-stranded RNA; self	Endolysosome (ER-membrane)
Toll-like receptor 9	Pathogen CpG-DNA; hemozoin (*Plasmodium falciparum*); *P. falciparum* extract; self	Endolysosome (ER-membrane)
Toll-like receptor 10	Unknown	Endolysosome
Toll-like receptor 11	Profilin-like molecule (*T. gondii*)	Surface
Mannose receptor	Repeated mannose units on pathogen surfaces (also phospholipids, nucleic acids, nonglycosylated proteins)	Surface Macrophages and dendritic cells
NLR NOD-like receptors	Endogenous or microbial molecules	Cytoplasm
NOD1	Peptidoglycan on Gram-negative bacteria (meso-DAP)	Cytoplasm
NOD2	Muramyl dipeptide on Gram + and Gram − bacteria	Cytoplasm
RLR Retinoic acid-inducible gene-1 receptors		
RIG-1	Short ds-RNA	Cytoplasm
MDA5	Long ds-RNA	Cytoplasm
LGP2	Unknown	Cytoplasm
CLR C-type lectin receptors		
Dectin-1	β-Glucan	Surface
Dectin-2	β-Glucan	Surface
MINCLE	SAP 130	Surface

molecules, which ensue after host tissue damage, as occurs when parasites proliferate in and escape from the host cells. These molecules are termed danger-associated molecular patterns (DAMPs [12]). The PRRs are expressed on different myeloid cells, mainly macrophages and dendritic cells, and also on certain endothelial cells, fibroblasts, and epithelial cells of the gut, that are generally considered to be part of the innate immune system. The result of PRR activation is primarily the production of *inflammatory cytokines*; these are regulatory proteins that control the activity of many cells playing a role in inflammation and immunity (see "Regulation of the Immune Response" below). Some cytokines, such as tumor necrosis factor α (TNFα) and γ-interferon (γ-IFN) can be directly cytotoxic to apicomplexan parasites [13]. Additionally, some PRRs are secreted by the cells that express them (e.g., complement receptors, collectins, pentraxins [11]); these molecules can then react with parasites in the circulation, which may lead to a complement-mediated lysis or facilitate phagocytosis by macrophages and granulocytes through opsonization [3]. Furthermore, dendritic cells especially can take up the antigen with PRRs and process it for presentation to lymphoid cells, which are subsequently activated (see "Activation of Specific Cellular Systems"; Figure 4.2).

Activation of Specific Cellular Systems

Immediately after infection, cells with specific antigen receptors are also triggered and activated if costimulated through accessory molecules (Signal 1 and Signal 2 respectively; Figure 4.2). These cells belong to the adaptive immune system. Two major types of cell can be distinguished, based on the primary lymphoid organ where they are generated: (i) the B lymphocytes, derived from bone marrow and, in birds, from the bursa of Fabricius; and (ii) T lymphocytes, derived from the thymus. Although both cell types carry antigen-specific receptors on their surface, only the B lymphocytes can secrete antigen-specific molecules (antibodies). These lymphocytes act in concert with other leukocytes and the cytokines produced by them. The general view here is that parasite-derived antigens are taken up and processed by antigen-presenting cells (APCs), such as dendritic cells [2]. The latter then travel to specific areas in the second lymphoid organs (e.g., follicles in the lymph nodes and spleen), where they come into contact with antigen-specific B and T lymphocytes. The result of this activation is the generation of effector T cells and the maturation of B cells to antibody-forming plasma cells.

Regulation of the Immune Response

Importantly, the inflammatory response, which comprises the onset of the adaptive immune response, should be controlled and phased out at a certain time point after infection. In a number of diseases this is not the case, and a syndrome known as systemic inflammatory response syndrome (SIRS) may result [14]. SIRS has been described in patients infected with *Plasmodium falciparum*, and also in dogs with *Babesia* infection [15–17]. It is suspected also to occur in a number of other blood infections, such as trypanosomiasis (sleeping sickness) and theileriosis infections (East Coast Fever in cattle; [18]). Vaccines that help to control inflammatory responses, without directly affecting proliferation of the parasite, have been developed for the treatment of some of these diseases (see "Anti-Disease Vaccines").

The type of immune response varies, depending on the infectious agent. Some immune responses are skewed toward antibody production, whereas in other cases cell-mediated effector mechanisms such as cytotoxic T cells dominate. This is regulated by the cytokine context of the immune responses; for example, immune responses that are dominated by T helper 1 (T_h1) cells are characterized by the production of γ-IFN and interleukin 2 (IL-2), whereas responses dominated by T_h2 cells are characterized by the production of IL-4 and IL-5 [19]. The balance between T_h1 and T_h2 depends on the cytokine context in the microenvironment in which the APC plays a crucial role (referred to as Signal 3 [20]; Figure 4.2). This is particularly evident in the *Leishmania major*-mouse model. In this case, mice that develop a T_h1 response are immune to challenge infection, whereas mice that develop a T_h2 response succumb upon challenge infection [21]. The regulation and balancing of the innate and adaptive immune responses to control the disease is central to modern vaccine development.

Vaccination

When considering immunity, it is important to realize that there is often a difference between natural immunity that arises after repeated infections with the parasite, and vaccine-induced immunity in which the immune system of the host is trained to produce a protective immunological response that may not occur with natural infection. In the past, vaccinologists have used the principles of natural immunity for the development of live vaccines (see "Anti-Parasite Vaccines" below); however, with a better understanding of immunology, it appears possible to more specifically direct the immune system to generate the protective response by using antigens and adjuvants (Signals 0 and 1; Figure 4.2). Recently, more sophisticated strategies have been used (see "Adjuvants" below), in which the immune response is guided by certain cytokines and other regulatory molecules (Signals 2 and 3; Figure 4.2). Based on the protective principle, it is possible to recognize three different groups of vaccine, namely anti-parasite vaccines, anti-disease vaccines, and therapeutic vaccines.

Anti-Parasite Vaccines

The success of vaccination against viruses and bacteria that induced sterile immunity (i.e., the protection is explained by an anti-microbial effect) led scientists to search for parasite vaccines that would induce anti-parasite immunity. A number of conventional live vaccines were developed against coccidiosis, babesiosis, theileriosis, and toxoplasmosis, that showed a clear anti-parasite effect (e.g., the reduction in oocyst output in chickens vaccinated with live *Eimeria* parasites was typically >90%; [22]). However, not all live vaccines were considered safe. Against a number of parasites, it appeared impossible to design an effective vaccine, because these parasites showed antigenic diversity (differences between different clones of parasites) and antigenic variation (differences within a cloned parasite population), that allowed them to escape the established immune response [23]. It appeared very difficult to replace effective live vaccines with non-live vaccines for a variety of reasons, including protective antigen selection, the choice of adjuvant, and formulation, and subsequently killed vaccines were successfully developed for *Babesia divergens* and *Neospora caninum* (see "Killed Vaccines" below). Although subunit vaccines against the Apicomplexa are scarce, a combination of *P. falciparum*-derived antigens has shown promise as a first vaccine against malaria. Moreover, in the case of *B. divergens* a vaccine based on a single recombinant antigen was developed (see "Subunit Vaccines" below).

Anti-Disease Vaccines

With an increasing knowledge concerning the pathogenesis of a parasitic infection, the way was opened to develop vaccines that specifically aimed to prevent clinical signs, without affecting parasite proliferation *per se*. The existence of anti-disease immunity is already evident in malaria; children first develop anti-disease immunity that does not protect them from subsequent reinfection; however, during adolescence an additional protective immunity is developed with a clear anti-parasite characteristic [24]. Subsequently, it was discovered that *P. falciparum* produces molecules that are pathogenic in themselves and behave as toxins by stimulating PRRs on myeloid cells that trigger the production of inflammatory cytokines through TLR1/2 (see "Activation of Nonspecific Cellular Systems" above). This finding culminated in the identification and synthetic production of the active moiety, glycosylphosphatidyl inositol (GPI) anchors, that function to bind the surface proteins to the parasite membrane [25]. In experimental vaccination-challenge studies, this vaccine was shown to protect mice against the development of clinical and cerebral signs after malaria infection [26]. Moreover, this principle could be included in a multivalent malaria vaccine. It has been suggested that the control of GPI-induced inflammation might also represent a strategy to reduce pathogenicity and mortality due to *Toxoplasma gondii* [27].

A comparable situation exists in *Babesia* infection. Typically, cattle infected with *Babesia bovis* develop a SIRS that is characterized by circulatory shock due to hypotension. It was shown subsequently that the parasite molecules activated the kallikrein and coagulation systems, leading to hypotension and sludging of the infected red blood cells and causing disturbance of the blood circulation ([7]; see "Nonspecific Activation of Humoral Systems" above). Subsequently, *Babesia* antigens obtained either from the plasma of infected animals or from the supernatants of *in vitro* cultures, were shown to protect against the development of this syndrome [28]. In the case of canine babesiosis, this principle has been used to develop commercial vaccines against *Babesia canis* in Europe. The importance of correct adjuvant selection is also exemplified by these vaccines; saponins were shown to be effective, in contrast to oil-based adjuvants ([29]; see "Subunit Vaccines" below).

Therapeutic Vaccines

Similar to the therapeutic use of drugs for infected subjects, it is possible to use vaccines as therapeutics. The earliest example of therapeutic vaccination was in 1885, when Louis Pasteur treated a boy who had been bitten by a rabid dog with an attenuated rabies vaccine [30]. In *Leishmania* infection in dogs, the vaccination of parasite-positive dogs was shown to reduce the parasite burden; moreover, when combined with a chemotherapeutic drug, the effect was greater [31]. In particular, for apicomplexan parasites that survive in the immune host (carrier status), a therapeutic vaccination could be used to reduce the parasite density in certain environments. At present, however, no such vaccines have been developed for the Apicomplexa.

Available Vaccines

Live Vaccines

Vaccination with live organisms has a long tradition, and was practiced as a traditional Chinese medicine long before the discovery of vaccination against smallpox by Jenner [32]. This strategy does not aim to specifically orchestrate the innate and adaptive immune response; rather, it relies on the natural sequence of host-responses that follows infection with a low or controlled dose of pathogen. These vaccines protect against disease by inhibiting parasite proliferation in the vaccinated host when it becomes infected. Several approaches have been taken to ensure that, upon vaccination, the infection does not induce unacceptable clinical symptoms; however, it should be noted that, depending on secondary factors, there is an inherent risk that an overt clinical disease may occur. As a consequence, although live vaccines are still being used, their replacement by inactivated vaccines is a continuous pursuit.

Chemotherapeutically Controlled Infection

In humans, malarial infection is caused by *Plasmodium* parasites. Although four *Plasmodium* species can infect humans, the main problem is caused by *P. falciparum*, which causes severe and cerebral malaria. Currently, the control of malaria infection is achieved by the chemotherapeutic treatment of infected subjects, by the prophylactic treatment of people that travel into endemic areas, and by using bed nets and repellents to prevent mosquito bites. In the endemic situations, immunity develops after the repeated exposure and cure of clinical cases. The prophylactic treatment of visitors does not sustain the development of protective immunity, and these subjects remain fully susceptible to malarial infection [24]. However, it has been shown recently that volunteers receiving prophylactic treatment with chloroquine and infectious mosquito bites once monthly for a period of three months, developed solid immunity against a homologous challenge infection [44]. Whether such a live vaccine would be a realistic treatment option is uncertain, however.

A similar situation exists in some veterinary diseases; for example, babesiosis in cattle is caused by *Babesia* parasites of which four species infect cattle, with *B. bovis* and *Babesia bigemina* (which is prevalent in tropical and subtropical areas) being the most virulent. Clinical cases can be effectively treated with chemotherapeutics such as imidocarb dipropionate (Carbesia®) and diminazene aceturate (Berenil®), with such animals developing immunity after one or more infectious periods [45]. It was soon realized that deliberate infection and subsequent cure would present a more pragmatic and feasible approach, and this is currently practiced with *Theileria* species. The parasite strains used for such treatment are usually virulent when given without a chemotherapeutic drug. Vaccination against East Coast Fever is achieved by infection with tick-derived *Theileria parva* sporozoites, while the drug of choice is presently a long-acting tetracycline [34].

Immunity against *Eimeria* infection in broiler flocks may result from chemotherapeutically controlled infection, especially when ionophoric drugs are used as a prophylactic control measure. Initially, when ionophoric drugs were screened for coccidiostatic activity in chickens housed in wired cages, they were considered mediocre [46]. However, when the chickens were treated with the same regimen but housed in floor pens (which increased accidental reinfection with shed oocysts), the coccidiostatic effects were much more dramatic. It appeared that the low number of oocysts that survived despite ionophore treatment evoked a solid immunity in the flock, which added to the reduction of parasite proliferation. This principle was later exploited commercially in the development of a live coccidiosis vaccine ([33], see below).

Low-Dose Infection

In many diseases, a correlation exists between the infectious dose and the clinical signs that occur. This means, in turn, that an infection with controlled low doses of the pathogen could be used to vaccinate subjects. The best example of this is coccidiosis in chickens and, indeed, a number of live vaccines against coccidiosis in chickens are currently available commercially. All of these are based on a similar principle, namely the administration of a low number of sporulated *Eimeria* oocysts (the infective stage of the parasite) at an early age to chickens placed in a broiler house (for a review, see Ref. [33]). When the vaccine oocysts are taken up by the animals, the progeny that are shed during the following week ensures a second round of infection that stimulates a further protective immunity. In this case, it is essential that *all* of the chickens encounter the oocysts early in life; otherwise they will pick up high numbers of oocysts shed by flock mates when they are still naïve, which would lead to clinical signs.

In some cases, low-dose infections ensue, even when the strain used is considered virulent in other situations. For instance, calves are less susceptible to *B. bovis* infections when compared to adult cattle, possibly due to the presence of a babesiacidal factor in the serum [47]. A similar dose of vaccine given to adult cattle may lead to virulent infection. The current practice is to vaccinate calves with *B. bovis* and *B. bigemina* parasites during the first six months of life. Maternal immunity represents an additional factor that may help an immunizing infection to develop a mild course, although a strong maternal immunity may interfere with the subsequent induction of immunity [35].

Infection with Attenuated Strains

In the case where the wild-type parasite line is (also in low doses) too virulent to be used as a vaccine strain, attenuation can be used to select less-virulent variants of the parasite. Whereas in virology, attenuation is often achieved by a prolonged

Table 4.3 Vaccination strategies in the control of parasitic infections.

Strategy	Parasite	Host	Mechanism	Reference
Low-dose infection	*Eimeria* spp.	Poultry	Antiparasite	[33]
Infection and simultaneous treatment	*Theileria parva*	Cattle	Antiparasite	[34]
Live attenuated				
	Babesia bovis	Cattle	Antiparasite	[35]
	Babesia bigemina	Cattle	Antiparasite	[35]
	Theileria annulata	Cattle	Antiparasite	[36]
	Theileria hirci	Sheep/ goats	Antiparasite	[37]
	Toxoplasma gondii	Sheep	Antiparasite	[38]
	Eimeria species	Poultry	Antiparasite	[32]
Killed				
	Neospora caninum	Cattle	Antiparasite	[39]
	Babesia divergens	Cattle	Antiparasite	[40]
Subunit	*Babesia canis*	Dog	Anti-disease	[29]
	Babesia rossi	Dog	Antiparasite/ anti-disease	[41]
	Babesia divergens	Cattle	Antiparasite	[42]
	Eimeria maxima	Poultry	Maternal antiparasite	[43]

culture of the virus, in parasitology a variety of techniques has been used to obtain less-virulent vaccine strains; these may vary from repeated passage (*in vivo* or *in vitro*), treatment with chemicals, or gamma irradiation. As this also includes serial passages in other than the target animal (e.g., the *Toxoplasma gondii* S48 vaccine that protects against ovine abortion; Table 4.3), it can be argued to what extent cross-species vaccination would differ from that using attenuated strains. The most effective vaccines developed by serial passage in the target animal are those against tropical babesiosis in cattle, caused by *B. bovis* [48]. In this case, it was shown that the serial passage of virulent *B. bovis* through splenectomized calves would select for the parasite with a reduced virulence. The rationale behind this was that, in splenectomized animals, parasites that normally would be removed by the spleen were selected which, when used to infect a eusplenic host, would be more effectively removed from the circulation, thus leading to less-virulent infections that provoked protective immunity. However, the biological stability of such vaccine strains must be verified in order to reduce the chance that the vaccine parasites would regain their virulent character. At present, vaccines containing live *B. bovis* and *B. bigemina* parasites are produced and distributed by governmental organizations in Australia and South Africa.

The selection of less-virulent strains has been applied to the development of live coccidiosis vaccines. Here, it was shown that in the natural population there are less proliferative strains, because they have a shorter developmental cycle. These so-called "precocious" strains have been selected for all seven *Eimeria* species that infect chickens [49], and at least two commercial products are available containing such precocious vaccine strains.

In cases where it is not feasible to attenuate a parasite by repeated passage, it may be possible to grow the *virulent* organism and to attenuate it with a subsequent treatment. For example, gamma-irradiation has been used successfully to irradiate the sporozoites of *P. falciparum*; these sporozoites then invade liver cells but do not develop further into the asexual blood-stage merozoites, thereby allowing protective immunity to develop. Currently, the development of a malaria vaccine based on this principle is at an advanced stage [50].

With the advent of molecular biological techniques, an entire new array of possibilities has been introduced, including a *T. gondii* "drug-addiction-gene" that allowed propagation of the vaccine strain in the presence of the drug required for vaccine production, but prevented the vaccine strain from further proliferation in the mammalian host, while inducing protective immunity [51]. Although, at present, no such vaccine is available but there are clear opportunities for this type of methodology.

Non-Live Vaccines

In order to improve the safety of vaccines, recent investigations have been focused on the use of killed vaccines. Although some inactivated bacterial vaccines may be administered without an adjuvant (most likely due to the fact that some of the bacterial components have inherent adjuvant activity), inactivated parasite vaccines require an adjuvant to induce protective immunity. During the early twentieth century, when bacterial vaccine cocktails were first developed, oils were used as an adjuvant, the rationale being that the oil would prevent the instantaneous release of bacterial toxins into the host, which would lead to serious adverse effects [52]. Although, in this case, the adjuvant was used to increase the safety of the vaccine, it later became apparent that these compounds would also affect the innate and adaptive immune system [53]. Although the mode of action of most adjuvants is unknown (see "Mode of Action" below), it is generally assumed that they affect the cytokine context that influences the type of immune response (Signal 2; Figure 4.2; [54]). In fact, the type of immune response is crucial, as it may determine whether animals are protected or, in some cases, may become more vulnerable to infection (see "Regulation of the Immune Response" above). Hence, the selection of a correct adjuvant is equally as important as the identification of a putative protective antigen.

Killed Vaccines
Different methods of inactivation are used, of which chemical treatment is the preferred approach; this might involve treatment with formaldehyde, beta-propiolactone (BPL), and/or bromoethylamine (BEA). Alternative methods include inactivation by ultrasonic treatment, heating, ultraviolet-light irradiation, and gamma-irradiation. Some examples exist of parasitic vaccines produced along these principles, notably a vaccine against *Neospora caninum*-induced abortion in cattle,

and against *B. divergens* in Austria. The *N. caninum* vaccine contains tachyzoites derived from *in-vitro* cultures that are inactivated with BEA [39], while the adjuvant is a depot-forming preparation containing a polymer of acrylic acid crosslinked with polyallylsucrose. The *B. divergens* vaccine contains formalinized, infected red blood cells, adjuvanted with aluminum hydroxide and saponin [40].

Subunit Vaccines

A vaccine that does not contain whole organisms is referred to as a "subunit vaccine." In general, subunit vaccines contain (partially) purified antigens from whole, killed organisms. Purification may be necessary to remove any putative toxic antigens that might compromise the safety of the vaccine, or to remove immunodominant antigens that might jeopardize the immunogenicity of the protective antigen. Currently, a subunit vaccine against coccidiosis in broilers is available commercially that contains antigens from the gametocytes (sexual forms) of *Eimeria maxima* in an oil adjuvant. The vaccine functions indirectly by inducing a maternal immunity in broiler breeders, in order to protect the progeny against coccidiosis. The effector mechanism is anti-parasitic [43].

Some subunit vaccines do not contain antigen from the parasite's cell itself, but rather contain the excreted/secreted antigens of the parasite. Examples of this include the vaccines against *Babesia canis* and *Babesia rossi* in dogs [29, 41], both of which contain soluble parasite antigens (SPA) derived from the supernatant of *in vitro* parasitic cultures, while saponin is used as the adjuvant. It was shown that in canine babesiosis, depending on the infectious dose, dogs would develop life-threatening pro-inflammatory responses with severe clinical signs. However, when the dogs were vaccinated with SPA, they showed limited clinical signs after a challenge infection, which was associated with reduced levels of SPA in plasma, and without an effect on parasite proliferation. This was a clear example of an anti-disease vaccine.

Special forms of subunit vaccines are those that are produced by recombinant DNA technology. In this case, the isolated gene that encodes for an immunoprotective pathogen-encoded protein is inserted into an appropriate expression system, allowing an overproduction of the relevant antigen. An example of this is the recombinant-merozoite antigen (Bd37) of *B. divergens* [42], for which it was shown that both active vaccination with the recombinant antigen (using saponin as adjuvant), and passive vaccination with a monoclonal antibody directed against Bd37, would provide protection against a challenge infection (J. Kleuskens, *et al.*, personal communication). A vaccine that comprises the partial gene sequences of a number of *P. falciparum* antigens is currently undergoing Phase III clinical trials, after some efficacy had been demonstrated in Phase II trials (see "Mode of Action" below).

DNA Vaccines

Vaccines that do not contain the protective antigen, but instead contain the DNA that encodes for the vaccine antigen, are termed "DNA vaccines" [55]. These may be live viral vectors, but non-live DNA fragments may also be injected into the host under circumstances that maximize the chance of DNA entering a nucleated cell. In this way, the parasite protein is produced and expressed by the host cell, and can be recognized by the cells of the immune system. The feasibility of messenger RNA to induce protective immunity has also been evidenced [56]. Whilst these approaches appear to show much promise in the field of virology, such vaccines have not yet become available in the areas of bacteriology and parasitology. Improved responses can be obtained when subjects are primed with a DNA vaccine and boosted with the (recombinant) protein or viral vectors encoding the vaccine antigen [57, 58]. However, whilst this technique holds much promise, such combinations are presently not commercially feasible.

Adjuvants

When administered in isolation, antigens do not generally induce a protective immune response. As noted above, adjuvants are necessary for increasing the immunogenicity of parasite antigens, and it should be realized that the earlier use of oils and lipids was not intended to affect the immune response (as very little was then known about the immune system), but merely to conserve the organisms used for vaccination (in the case of the smallpox vaccine). Alternatively, adjuvants were intended to retard the release of bacterial toxins from multivalent vaccines, in order to limit any adverse effects resulting from the induction of endotoxin shock.

With increasing insight into the innate and adaptive immune system, it became clear that adjuvants also play a role in the activation of these systems. Indeed, present-day vaccinologists strategically target specific systems in order to trigger the appropriate response [59]. The activation of humoral inflammatory systems (complement, coagulation, kinin–kallikrein; see "Humoral Systems" above), either directly or indirectly by parasite antigens or adjuvants, is generally avoided because these are implicated in the local and systemic adverse reactions, such as the fever seen upon vaccination. Currently, a multitude of compounds are used as adjuvants, and it is impossible to provide an extensive overview of these at this point (for excellent reviews, see Refs [54, 59]). Nonetheless, the basic principles of the mode of action of several adjuvants are detailed below, with examples.

Mode of Action

This aspect of adjuvants is poorly understood. Some improve the uptake of antigen via APCs (Signal 0, Figure 4.2), an effect that is influenced by the properties of the antigen itself, by the pharmaceutical carrier, or by the addition of a molecule that is itself immunostimulatory (e.g., a PRR-agonist). Examples include bacterial toxins such as pertussis toxin, cholera toxin, and heat-labile *Escherichia coli* toxin, which have a high affinity to specific cells, and thereby facilitate the uptake of vaccine antigen via APCs [60].

It is generally assumed that the prolonged exposure of an antigen to the immune system favors the development of an immune response (Signal 1, Figure 4.2). Many adjuvants help to create a depot from which antigen is slowly released, such as aluminum phosphate and aluminum hydroxide, which form a gel-like matrix in order to trap the antigen [59]. A water-in-oil emulsion, in which the antigen is usually contained in the water phase, creates a similar depot effect in the target animal.

The stimulation of APCs by PAMPs in adjuvant preparations may lead to an enhanced expression of co-stimulatory molecules on the cell surface, facilitating the delivery of signal 2 [61]. A saponin injection in mice will increase the expression of major histocompatibility complex (MHC) class II molecules used experimentally to vaccinate against malaria (for a review, see Ref. [62]).

Yet another class of adjuvants acts on the generation/modulation of the immune response through signal 3. Monophosphoryl lipid A (MPLA) is a nontoxic material derived from the lipopolysaccharides of Gram-negative bacteria. This compound stimulates the production of certain cytokines (e.g., γ-IFN) that direct the further development of the immune response (Signal 3, Figure 4.2; [63]).

It may be understood that, usually, several of these modes of action may be combined in an antigen-adjuvant presentation. For example, the AS02D adjuvant in the RTS,S malaria vaccine combines a squalene in water emulsion, saponin and MPLA [64]. It has been hypothesized that the oil-in-water emulsion and saponin provide the depot effect that affects the uptake by APCs, MPLA stimulates the production of regulatory cytokines, and the RTS,S antigens (partial *P. falciparum* protein sequences) stimulate the specific lymphocytes.

Specific Immunomodulation

With an increasing knowledge of the pathways of the immune system, the specific modulation of certain pathways has become a reality. At present, cytokines are used mainly to direct the immune response by influencing the cytokine context (Signal 3, Figure 4.2). Because of the pleiotropic effects of many cytokines (a cytokine usually has more than one biological function), many of the studies are empirical in nature, starting with an "educated guess" [61]. There are essentially two approaches to deliver the relevant cytokine: (i) addition of the cytokine to the vaccine formulation; and (ii) to use recombinant organisms (usually viruses) that carry the required cytokine gene, either alone or in combination with the protective antigen.

Conclusions

Although, at present, antiparasitic drugs prevail in the control of parasitic infestations, effective vaccines are currently under development for such purpose, with basic vaccine research being directed at manipulating the innate and adaptive immune system in order to trigger the immunoprotective response. The vaccine adjuvants and immunopotentiators that orchestrate the immune response through Signals 0, 1, 2, and 3 are key to this situation. Today, genomic tools allow for the selection and testing of putative vaccine candidate antigens for use in subunit vaccines. The discovery of such antigens in *Babesia* and *Plasmodium* spp. will continue to provide leads to seek homologs in other apicomplexan parasites. It is envisaged that, in future, vaccines will be used strategically to control infectious diseases. This means that, in addition to management practices which limit the level of contamination, the alternate and/or combined use of drugs and vaccines will be practiced.

Acknowledgments

The author is an invited Professor at the University of Montpellier I, France, and would like to thank the staff of the University for their support.

References

1 Evans, S.W. and Whicher, J.T. (1993) An overview of the inflammatory response, in *Biochemistry of Inflammation* (eds S.W. Evans and J.T. Whicher), Kluwer Academic Publishers, Dordrecht, Boston, London, pp. 1–15.

2 Male, D.K. and Roitt, I.M. (1989) Adaptive and innate immunity, in *Immunology.* (eds I.M. Roitt, J. Brostoff, and D.K. Male), Churchill Livingstone, Gower Medical Publishing, London, New York, pp. 1.1–1.10.

3 Whicher, J.T. and Westacott, C.I. (1993) The acute phase response, in *Biochemistry of Inflammation* (eds J.T. Whicher and S.E. Evans), Kluwer Academic Publishers, Dordrecht, Boston, London, pp. 243–269.

4 McPhaden, A.R. and Whaley, K. (1993) The complement system and inflammation, in: *Biochemistry of Inflammation* (eds J.T. Whicher and S.E. Evans), Kluwer Academic Publishers, Dordrecht, Boston, London, pp. 17–36.

5 Schalm, O.W., Jain, N.C., and Carroll, E.J. (1975) Blood coagulation and fibrinolysis, in *Veterinary Hematology* (eds O.W. Schalm, N.C. Jain, and E.J. Carroll), Lea and Febiger, Philadelphia, pp. 284–300.

6 Regoli, D., Rhaleb, N.-E., Drapeau, G., and Dion, S. (1993) The kinin system: current concepts and perspectives, in *Biochemistry of Inflammation* (eds J.T. Whicher and S.E. Evans), Kluwer Academic Publishers, Dordrecht, Boston, London, pp. 37–55.

7 Wright, I.G. and Goodger, B.V. (1988) Pathogenesis of babesiosis, in *Babesiosis of Domestic Animals and Man* (ed. M. Ristic), CRC Press Inc., Boca Raton, Florida, pp. 100–118.

8 Takada, Y., Skidgel, R.A., and Erdös, E.G. (1985) Purification of human urinary prekallikrein. Identification of the site of activation by the metalloproteinase thermolysin. *Biochem. J.*, **232**, 851–858.

9 Male, D. and Roitt, I. (1989) Complement, in *Immunology* (eds I.M. Roitt, J. Brostoff, and D.K. Male), Churchill Livingstone, Gower Medical Publishing, London, New York, pp. 13.1–13.16.

10 Medzhitov, R. and Janeway, C.A. Jr (2002) Decoding the patterns of self and nonself by the innate immune system. *Science*, **296**, 298–300.

11 Takeuchi, O. and Akira, S. (2010) Pattern recognition receptors and inflammation. *Cell*, **140**, 805–820.

12 Seong, S.-Y. and Matzinger, P. (2004) Hydrophobicity: an ancient damage associated molecular pattern that initiates innate immune responses. *Nat. Rev. Immunol.*, **4**, 469–478.

13 Stevenson, M.M., Tam, M.F., Wolf, S.F., and Sher, A. (1995) IL-12-induced protection against blood-stage *Plasmodium chabaudi* AS requires IFN-γ and TNF-α and occurs via a nitric oxide-dependent mechanism. *J. Immunol.*, **155**, 2545–2556.

14 Bone, R.C., Balk, R.A., Cerra, F.B., Dellinger, R.P., Fein, A.M. *et al.* (1992) Definitions of sepsis and organ failure and guidelines for the use of innovative therapies in sepsis. The ACCP/SCCM Consensus Conference Committee, American College of Chest Physicians/Society of Critical Care Medicine. *Chest*, **1201**, 1644–1655.

15 Clark, I.A., Alleva, L.M., Budd, A.C., and Bowden, W.B. (2008) Understanding the role of inflammatory cytokines in malaria and related diseases. *Travel Med. Infect. Dis.*, **6**, 67–81.

16 Jacobson, L. (2006) The South African form of severe and complicated canine

babesiosis: clinical advances 1994–2004. *Vet. Parasitol.*, **138**, 126–139.

17 Schetters, T.P.M., Kleuskens, J.A.G.M., Van de Crommert, J., De Leeuw, P.W.J., Finizio, A.-L., and Gorenflot, A. (2009) Systemic inflammatory responses in dogs experimentally infected with *Babesia canis*; a haematological study. *Vet. Parasitol.*, **162**, 7–15.

18 Schetters, T.P.M., Arts, G., Niessen, R., and Schaap, D. (2010) Development of a new score to estimate clinical East Coast Fever in experimentally infected cattle. *Vet. Parasitol.*, **167**, 255–259.

19 Mosmann, T.R. and Sad, S. (1996) The expanding universe of T-cell subsets: Th1, Th2 and more. *Immunol. Today*, **17**, 138–146.

20 Kalinski, P., Hilkens, C.M., Wieringa, E.A., and Kapsenberg, M.L. (1999) T-cell priming by type-1 and type-2 polarised dendritic cells: the concept of a third signal. *Immunol. Today*, **20**, 561–567.

21 Scott, P., Pearce, E., Cheever, A.W., Coffman, R.L., and Sher, A. (1989) Role of cytokines and CD4 + T-cell subsets in the regulation of parasite immunity and disease. *Immunol. Rev.*, **112**, 161–182.

22 Meeusen, E.N.T., Walker, J., Peters, A., Pastoret, P.-P., and Jungersen, G. (2007) Current status of veterinary vaccines. *Clin. Microbiol. Rev.*, **20**, 287–305.

23 Barriga, O.O. (1994) A review on vaccination against protozoa and arthropods of veterinary importance. *Vet. Parasitol.*, **55**, 29–55.

24 Doolan, D.L., Dobaño, C., and Baird, J.K. (2009) Acquired immunity to malaria. *Clin. Microbiol. Rev.*, **22**, 13–36.

25 Campos, M.A.S., Almeida, I.C., Takeuchi, O., Akira, S., Valente, E.P. *et al.* (2001) Activation of toll-like receptor-2 by glycosylphosphatidylinositol anchors from a protozoan parasite. *J. Immunol.*, **167**, 416–423.

26 Schofield, L., Hewitt, M.C., Evans, K., Slomos, M.-A., and Seeberger, P.H. (2002) Synthetic GPI as a candidate anti-toxic vaccine in a model of malaria. *Nature*, **418**, 785–789.

27 Debierre-Grockiego, F. (2010) Glycolipids are potential targets for protozoan parasite diseases. *Trends Parasitol.*, **26**, 404–411.

28 Schetters, T.P.M. and Montenegro-James, S. (1995) Vaccines against babesiosis using soluble parasite antigens. *Parasitol. Today*, **11**, 456–462.

29 Schetters, T.P.M. (2005) Vaccination against canine babesiosis. *Trends Parasitol.*, **21**, 179–184.

30 Pasteur, L. (1885) Méthode pour prévenir la rage après morsure. *CR Acad. Sci.*, **51**, 765–773.

31 Miret, J., Nascimento, E., Sampaio, W., França, J.C., Fujiwara, R.T. *et al.* (2008) Evaluation of an immunochemotherapeutic protocol constituted of N-methyl meglumine antimoniate (Glucantime®) and the recombinant Leish-110f® + MPL-SE® vaccine to treat canine visceral leishmaniasis. *Vaccine*, **26**, 1585–1594.

32 Grabar, P. (1980) The historical background of immunology, in *Basic and Clinical Immunology* (eds H.H. Fudenberg, D.P. Stites, J.L. Caldwell, and J.V. Wells), Lange Medical Publications, Los Altos, California, pp. 16–27.

33 Vermeulen, A.N., Schaap, D.C., and Schetters, T.P.M. (2001) Control of coccidiosis in chickens by vaccination. *Vet. Parasitol.*, **100**, 13–20.

34 Morzaria, S., Nene, V., Bishop, R., and Musoke, A. (2000) Vaccines against *Theileria parva*. *Ann. N. Y. Acad. Sci.*, **916**, 464–473.

35 Bock, R.E., de Vos, A.J., Jorgensen, W.K., Dalgliesh, R.J., and Lew, A.E. (1996) Living babesiosis vaccines for cattle - the Australian experience, in Acta Parasitologica Turcica. Proceedings of the VIII International Congress of Parasitology – New Dimensions in Parasitology, Izmir, Turkey (ed. M.A. Özcel), Turkish Society for Parasitology, pp. 517–529.

36 Shkap, V. and Pipano, E. (2000) Culture-derived parasites in vaccination of cattle against tick-borne diseases. *Ann. N. Y. Acad. Sci.*, **916**, 154–171.

37 Lightowlers, M.W. (1994) Vaccination against animal parasites. *Vet. Parasitol.,* **54**, 177–204.

38 Buxton, D. (1993) Toxoplasmosis: the first commercial vaccine. *Parasitol. Today,* **9**, 335–337.

39 Romero, J.J., Pérez, E., and Frankena, K. (2004) Effect of a killed whole *Neospora caninum* tachyzoites vaccine on the crude abortion rate of Costa Rican dairy cows under field conditions. *Vet. Parasitol.,* **123**, 149–159.

40 Edelhofer, R., Kanout, A., Schuh, M., and Kutzer, E. (1998) Improved disease resistance after *Babesia divergens* vaccination. *Parasitol. Res.,* **84**, 181–187.

41 Schetters, T.P.M., Strydom, T., Crafford, D., Kleuskens, J., Van de Crommert, J., and Vermeulen, A.N. (2007) Immunity against *Babesia rossi* infection in dogs vaccinated with antigens from culture supernatants. *Vet. Parasitol.,* **144**, 10–19.

42 Hadj-Kaddour, K., Carcy, B., Vallet, A., Randozzo, S., Delbecq, S. *et al.* (2007) Recombinant protein Bd37 protected gerbils against heterologous challenges with isolates of *Babesia divergens* polymorphic for the *bd37* gene. *Parasitol.,* **134**, 187–196.

43 Wallach, M.G., Ashash, U., Michael, A., and Smith, N.C. (2008) Field application of a subunit vaccine against an enteric protozoan disease. *Plos ONE,* **3**, article no. e3948.

44 Roestenberg, M., McCall, M., Hopman, J., Wiersma, J., Luty, A.J.F. *et al.* (2009) Protection against a malaria challenge by sporozoite inoculation. *N. Engl. J. Med.,* **361**, 468–477.

45 James, M. (1988) Immunology of babesiosis, in *Babesiosis of Domestic Animals and Man* (ed. M. Ristic), CRC Press, Boca Raton, Florida, pp. 119–130.

46 McDougald, L.R. (1990) Control of coccidiosis in chickens: chemotherapy, in *Coccidiosis of Man and Domestic Animals* (ed. P.L. Long), CRC Press, Boca Raton, Florida, pp. 307–320.

47 Levy, M.G., Clabaugh, G., and Ristic, M. (1982) Age resistance in bovine babesiosis: role of blood factors in resistance to *Babesia bovis. Infect. Immun.,* **37**, 1127–1131.

48 De Waal, D.T. and Combrink, M.P. (2006) Live vaccines against bovine babesiosis. *Vet. Parasitol.,* **138**, 88–96.

49 Chapman, H.D., Cherry, T.E., Danforth, H.D., Richards, G., Shirley, M.W., and Williams, R.B. (2002) Sustainable coccidiosis control in poultry production: the role of live vaccines. *Int. J. Parasitol.,* **32**, 617–629.

50 Hoffman, S.L., Billingsley, P.F., James, E., Richman, A., Loyevsky, M. *et al.* (2010) Development of a metabolically active, non-replicating sporozoite vaccine to prevent *Plasmodium falciparum* malaria. *Hum. Vaccines,* **6**, 97–106.

51 Poppel, N.F.J., Welagen, J., Duisters, R.F.J.J., Vermeulen, A.N., and Schaap, D. (2006) Tight control of transcription in *Toxoplasma gondii* using an alternative tet repressor. *Int. J. Parasitol.,* **36**, 443–452.

52 Whitmore, E.R. (1918) Lipovaccines, with special reference to public health work. *Am. J. Public Health,* **9**, 504–507.

53 Freund, J., Casals, J., and Hosmer, E.P. (1937) Sensitization and antibody formation after injection of tubercle bacilli and paraffin oil. *Proc. Soc. Exp. Biol. Med.,* **37**, 509–513.

54 Kwissa, M., Kastun, S.P., and Pulendran, B. (2007) Science of adjuvants. *Expert Rev. Vaccines,* **6**, 673–684.

55 Gurunathan, S., Wu, C.-Y., Freidag, B.L., and Seder, R.A. (2000) DNA vaccines: a key for inducing long-term cellular immunity. *Curr. Opin. Immunol.,* **12**, 442–447.

56 Pascolo, S. (2008) Vaccination with messenger RNA (mRNA). *Handb. Exp. Pharmacol.,* **183**, 221–235.

57 Scheerlinck, J.-P.Y., Casey, G., McWaters, P., Kelly, J., Woollard, D. *et al.* (2001) The immune response to a DNA vaccine can be modulated by co-delivery of cytokine genes using a DNA prime-protein boost strategy. *Vaccine,* **19**, 4053–4060.

58 Ramshaw, I.A. and Ramsay, A.J. (2000) The prime-boost strategy: exciting prospects for improved vaccination. *Immunol. Today,* **21**, 163–165.

59 Wilson-Welder, J.H., Torres, M.P., Kipper, M.J., Mallapragada, S.K., Wannemuehler, M.J., and Narasimhan, B.

(2009) Vaccine adjuvants: Current challenges and future approaches. *J. Pharm. Sci.*, **98**, 1278–1316.

60 Liang, S. and Hajishengallis, G. (2010) Heat labile enterotoxins as adjuvants or anti-inflammatory agents. *Immunol. Invest.*, **39**, 186–204.

61 Schijns, V.E.J.C. and Degen, W.G.J. (2007) Vaccine immunopotentiators of the future. *Clin. Pharmacol. Ther.*, **82**, 750–755.

62 Read Kensil, C. (1996) Saponins as vaccine adjuvants. *Crit. Rev. Ther. Drug*, **13**, 1–55.

63 Mata-Haro, V., Cekic, C., Martin, M., Chilton, P.M., Casella, C.R., and Mitchell, T.C. (2007) The vaccine adjuvant monophosphoryl lipid A as a TRIF-biased agonist of TLR4. *Science*, **316**, 1628–1632.

64 Casares, S., Brumeanu, T.-D., and Richie, T.L. (2010) The RTS,S malaria vaccine. *Vaccine*, **28**, 4880–4894.

Part Two
Metabolic Pathways and Processes Addressed by Current Drug Discovery Approaches

Apicomplexan Parasites. Edited by Katja Becker
Copyright © 2011 WILEY-VCH Verlag GmbH & Co. KGaA, Weinheim
ISBN: 978-3-527-32731-7

5
Energy Metabolism as an Antimalarial Drug Target

Esther Jortzik and Katja Becker[*]

Abstract

Malaria is a leading cause of mortality and morbidity globally, accounting for 250 million cases with almost one million deaths per year, mainly in young children. The causative agent of malaria in humans, *Plasmodium falciparum*, undergoes a complex life cycle, both in the human host and in the *Anopheles* mosquito vector. In the human host, the merozoite life stage of *P. falciparum* is responsible for the characteristic symptoms. This life stage requires glycolysis for its energy generation; indeed, selective inhibitors of *P. falciparum* glycolytic enzymes have been shown to halt proliferation of the parasite. The inhibition of lactate dehydrogenase by gossypol derivatives, as well as by azole-based compounds, also exhibits antimalarial activity. In agreement with these observations, human pyruvate kinase deficiency exerts a protective effect against malaria infection and replication. This phenomenon is comparable to glucose 6-phosphate dehydrogenase deficiency, and indicates that an inhibition of the glycolytic pathway of the host–parasite cell unit is a valuable target for antimalarial strategies. Several enzymes of the glycolytic pathway that convert glucose to lactate have been extensively studied in the context of structure-based inhibitor design. This chapter summarizes the current knowledge of the function, structure, and regulation of important glycolytic enzymes of *P. falciparum* in comparison to their human counterparts, and elucidates their potential as drug targets.

Abbreviations

Pf	*Plasmodium falciparum*
HK	hexokinase
2-DG	2-deoxy-D-glucose
2-FG	2-fluoro-2-deoxy-D-glucose

[*]Corresponding author

Apicomplexan Parasites. Edited by Katja Becker
Copyright © 2011 WILEY-VCH Verlag GmbH & Co. KGaA, Weinheim
ISBN: 978-3-527-32731-7

PP$_i$	pyrophosphate
ALDO	fructose 1,6-bisphosphate aldolase
FBP	fructose 1,6-bisphosphate
GAP	glyceraldehyde 3-phosphate
F1P	fructose 1-phosphate
TRAP	thrombospondin-related anonymous protein
TIM	triosephosphate isomerase
DHAP	dihydroxyacetone phosphate
GAPDH	glyceraldehyde 3-phosphate dehydrogenase
NO	nitric oxide
PK	pyruvate kinase
PEP	phosphoenolpyruvate
LDH	lactate dehydrogenase
MOCB	methyl-2-oxamoto-5-chlorobenzoate

Introduction

Malaria is one of the oldest burdens of mankind, and even today almost 50% of the world's population is at risk of the disease. The most severe form of malaria is caused by the apicomplexan parasite *Plasmodium falciparum*, which is transmitted to the human host via the bite of the female *Anopheles* mosquito. An initial hepatic stage is followed by an erythrocytic stage, where the parasites live and multiply in the red blood cells. Merozoites exit from erythrocytes into the bloodstream every two days, causing the fever bursts that are characteristic of malaria (for reviews on the biology of *Plasmodium*, see Ref. [1]). Many years of effort to eradicate malaria has resulted in a widespread resistance to effective antimalarial drugs such as chloroquine and sulfadoxine-pyrimethamine, as well as to insecticides. As a consequence, malaria-related morbidity and mortality have each worsened over the past decades in regions where malaria is endemic [2]. Currently, between 10% and 30% of patients with severe malaria die, despite receiving treatment with the best available antimalarial drugs [3]. Although, today, malaria therapy is based on artemisinin-based combination therapies [4], the resistance to current drugs continues to increase. Hence, despite the ongoing but extremely time-consuming development of effective vaccines, the death rate is expected to double in the next 20 years, unless new control methods for malaria are implemented [5]. This, in turn, triggers the quest for new targets, and further substantiates the urgent need for novel therapeutics [2, 6].

Glycolytic Enzymes as Antimalarial Drug Targets

During the asexual intra-erythrocytic sector of their life cycle, malarial parasites reside in a parasitophorous vacuole within the erythrocytes of the human host. A major challenge for the parasites is the acquisition of nutrients, as the red blood cells are highly differentiated with only modest energy requirements, and also lack a nucleus, protein synthesis machinery, and protein trafficking. For its energy supply, the malarial parasite requires ATP to be generated via the fermentation of glucose; this has been confirmed by ATP levels being drastically reduced following treatment of the parasite with glucose-transport inhibitors, whereas mitochondrial inhibitors exhibited only minor effects on ATP levels [7]. *Plasmodium falciparum* requires glucose for growth, and utilizes up to 100-fold more glucose compared to unparasitized erythrocytes [8]. Taken together, these data suggest that the intra-erythrocytic stages of *Plasmodium* depend heavily on ATP generation via the fermentation of glucose as a primary energy source. The major product of glucose metabolism is lactate [9, 10]. Thus, as the malarial parasite does not store energy reserves, a constant supply and metabolism of glucose is critical in order for the parasite to survive, and for the ATP levels to be maintained [11, 12]. This obligatory dependence of the parasite on glycolysis for ATP production means that the glycolytic enzymes represent attractive targets for antimalarial drug development (Table 5.1). In this chapter, attention is focused on the current approaches toward identifying selected glycolytic enzymes for this role.

Table 5.1 Summary of the current knowledge of selected glycolytic enzymes from *P. falciparum* as drug targets.

Enzyme (EC number)	Reaction	Structure	Inhibitor	Reference(s)
Hexokinase (EC 2.7.1.1)	D-hexose + ATP → D-hexose 6-phosphate + ADP	nd	2-Substituted glucose analogs	[18]
Fructose 1,6-bisphosphate aldolase (EC 4.2.1.13)	Aldol cleavage of D-fructose 1,6-bisphosphate to dihydroxyacetone phosphate and glyceraldehyde 3-phosphate	1, 3	Hydroxynaphthaldehyde phosphates	[22, 30, 31]
Triosephosphate isomerase (EC 5.3.1.1)	Isomerization of glyceraldehyde 3-phosphate to dihydroacetone phosphate	1	None	[41, 42]
Glyceraldehyde 3-phosphate dehydrogenase (EC 1.2.1.12)	D-glyceraldehyde 3-phosphate + phosphate + NAD^+ → 1,3-bisphosphoglycerate + $NADH + H^+$	1	None	[64, 65]
Pyruvate kinase I (EC 2.7.1.40)	Phosphoenolpyruvate + ADP → pyruvate + ATP	1	None	[66]
Lactate dehydrogenase (EC 1.1.1.27)	Pyruvate + NADH ↔ lactate + NAD^+	1, 2	Azole-based compounds Gossypol derivatives	[78, 81, 82, 87, 88, 91]

nd: not determined; 1: unbound structure; 2: inhibitor-bound structure; 3: structure with interaction partner.

Hexokinase

The first enzyme of the glycolytic pathway, hexokinase (HK; EC 2.7.1.1), phosphory-lates hexoses in the initial glycolysis reaction. In *Plasmodium*, the catalytic properties of HK seem to be optimized to cope with an increased glucose turnover, as the enzyme's activity is seen to increase proportionally with parasitemia in *P. falciparum* (Pf) cultures [13]. The overall amino acid identity between human and PfHK is only 26.2%, with the highest homology scores in functional domains such as the putative ATP and glucose binding sites. PfHK contains a cluster of hydrophobic residues at the C-terminus, which could function as a membrane anchor sequence, whereas human HK possesses a potential mitochondrial membrane anchor sequence at the N-terminus [14]. Investigations using immune electron microscopy have revealed that PfHK is indeed associated with the parasite membrane structures, which might increase the efficiency of glucose phosphorylation at the site of uptake [15].

When, previously, various glucose analogs were investigated as potential inhibitors of glucose transport into the parasite, one of the most promising candidates was compound 3361, which had been shown to be a selective inhibitor of the hexose transporter and to inhibit the growth of *Plasmodium*, both *in vitro* and *in vivo* [16, 17]. A series of 2-substituted glucose analogs has been shown to inhibit parasite proliferation with varying efficiency, most likely by inhibiting glycosylation in the parasite [18]. Two of these analogs, 2-deoxy-D-glucose (2-DG) and 2-fluoro-2-deoxy-D-glucose (2-FG) had been shown previously to possess antiplasmodial activity linked to the inhibition of glycosylation [19], with 2-FG being the most effective analog [18]. Both, 2-DG and 2-FG inhibit glucose accumulation at least in part by inhibiting glucose phosphorylation by HK, although the exact mechanism of HK inhibition by 2-DG and 2-FG remains to be determined. Whilst glucose analogs may not neces-sarily be suitable lead compounds for antimalarial drug design, these compounds have shown that inhibition of the first step of glycolysis results in a reduced parasite growth [18]. The selective inhibition of *Trypanosoma cruzi* HK by bisphosphonates (which are metabolically stable analogs of PP_i) effectively reduces proliferation of the clinically relevant intracellular amastigote form of the parasite *in vitro* [20, 21]. Consequently, HK may be regarded as a feasible drug target.

Fructose 1,6-Bisphosphate Aldolase

In the glycolytic pathway, fructose 1,6-bisphosphate aldolase (ALDO; EC 4.1.2.13) catalyzes the aldol cleavage of fructose 1,6-bisphosphate (FBP) to dihydroxyacetone phosphate (DHAP) and glyceraldehyde 3-phosphate (GAP). Apart from FBP, fruc-tose 1-phosphate (F1P) can also function as a substrate for the aldol cleavage reaction [22]. As with all other glycolytic enzymes, *P. falciparum* ALDO is encoded by a single-copy gene, which suggests the absence of isoenzymes [23]. In contrast to *Plasmodium*, mammals possess three isoenzymes of ALDO (class A, B, and C), all of which have distinct substrate specificities with regards to the cleavage of FBP and F1P, and are found in all tissues, including erythrocytes [24, 25]. Hence, the design of selective PfALDO inhibitors becomes a major challenge. Typically, the human aldolase isoenzymes have a 69–82% sequence similarity among themselves, while PfALDO has a 50% sequence similarity with all three of the human isoenzymes. Nevertheless, the crystal structure of *Plasmodium* ALDO, when compared to human aldolase, reveals structural differences that can be targeted with structure-based inhibitors [22].

The region of greatest similarity in human and PfALDO is the active site. Typically, the catalytic mechanism of class I aldolase proceeds through a Schiff base intermediate–a covalent adduct formed by the nucleophilic attack of a critical lysine on the C2 carbonyl of FBP. All residues involved in the catalytic mechanism are conserved in *P. falciparum* ALDO.

The C-terminal tail shows a high diversity between aldolases from different species, although the C-terminal Tyr residue is highly conserved and required for the catalytic mechanism [22]. Removal of the C-terminal Tyr results in a decreased aldolase activity [26]. The C-terminal tail of PfALDO represents a promising region to be targeted by selective inhibitors. In human (h) ALDO-A, the C-terminal tail burrows into the active site, while the C-terminal tail in PfALDO is not well defined but seems to be located over the active site cleft. Some nonconservative amino acid changes have been made in the C-terminal tails of both human and plasmodial ALDO; notably, Glu355 and Glu363 of PfALDO have been replaced by Ala and Val in hALDO-A. In addition, PfALDO contains a Lys364–Lys365 tandem, which is a unique characteristic of malarial aldolases [22, 27]. This Lys–Lys tandem may be suitable for designing specific ligands, that might interfere with the proposed motion of the tail during catalysis [22].

A second promising difference occurs in the 290s loop, which has an unusual amino acid sequence and a conformational change when compared to human aldolases. The differences in PfALDO include the deletion of a conserved Leu and the replacement of Lys295 and Lys299 by Ala and His. These changes result in a binding pocket that is significantly different from the corresponding pocket in hALDO-A. The topological and electrostatic differences between the 290s loop regions of PfALDO and hALDO-A suggest the design of a ligand that binds selectively to the pocket of PfALDO [22].

Besides the glycolytic function, PfALDO contributes to the invasion of parasites into the human host cells. The process of host cell invasion is unique, being mediated by the gliding motility motor and requiring a transmembrane link between the parasite cytoskeleton and the host cell. PfALDO mediates the interaction between actin filaments of the actomyosin motor, which mediates gliding motility and host-cell invasion, and the transmembrane adhesion proteins of the thrombospondin-related anonymous protein (TRAP) family, which are responsible and essential for contacting the host-cell receptors [28, 29]. The four residues that define the TRAP-binding pocket in PfALDO are conserved in all human aldolases. Compared to human aldolases, the active-site region of PfALDO exerts an increased flexibility, thereby enabling small molecules to selectively inhibit the binding of TRAP to PfALDO, without impairing human aldolases (Figure 5.1) [30].

The selective inhibition of protozoan aldolases, including that of *P. falciparum*, has been reported. A derivative of hydroxynaphthaldehyde phosphate inhibited the

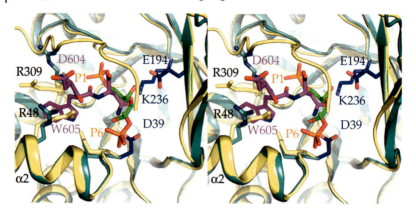

Figure 5.1 Crystal structure of *P. falciparum* aldolase (ALDO) with a bound C-terminal hexapeptide of the thrombonspondin-related anonymous protein (TRAP). The C-terminal tail of TRAP partially overlaps with the substrate-binding site. The PfALDO–TRAP-complex is superimposed with the structure of human ALDO-A with the bound substrate fructose 1,6-phosphate. The TRAP-tail is shown in magenta, residues involved in substrate binding are blue, human ALDO-A in yellow, and fructose 1,6-phosphate is shown as sticks in green with the two phosphate groups in orange. Reproduced with permission from Ref. [30]; © 2007.

aldolases from *Trypanosoma*, *Leishmania*, and *Plasmodium* irreversibly, but was ineffective against human aldolase [31].

Recently, co-immobilized glycolytic enzymes have been shown to be more efficient than solubilized enzymes, which suggests that the clustering of such enzymes might improve their efficiency if the metabolites were to be channeled [32]. As the malarial parasite causes a major rise in the rate of glycolysis, the glycolytic enzymes may be grouped together so as to increase their enzymatic efficiency [33]. Consequently, a compound capable of interfering with the oligomerization of PfALDO would reduce the thermostability and catalytic activity, and might also disrupt the multi-protein organization in the parasite's glycolytic pathway [22]. Thus, the selective targeting of PfALDO might represent an effective antimalarial strategy, not only by disrupting the parasites' energy supply but also by interfering with the host-cell invasion process of the parasite.

Triosephosphate Isomerase

Triosephosphate isomerase (TIM; EC 5.3.1.1) catalyzes the isomerization of GAP to DHAP via an enediolate intermediate, which can either isomerize to GAP or eliminate the phosphate to yield the toxic metabolite, methylglyoxal [34]. Furthermore, TIM is also involved in gluconeogenesis, the hexose monophosphate shunt, and fatty acid biosynthesis. TIM has a high degree of structural homology in all species studied; for example, TIMs of other organisms have been co-crystallized with inhibitors such as phosphoglycolohydroxamate and 2-phosphoglycolate, for both of which the reaction mechanism and catalytic residues have been extensively studied [35, 36].

The high enzymatic efficiency of TIM led to its description as a perfect catalyst [37], the enzymatic reaction being controlled by the movement of a loop comprising 11 residues (residues 166–176), which move around flanking hinge residues and close over the active site. This yields two distinct conformational states– a closed and an opened state [38, 39]. This active site loop motion is the rate-limiting step of the reaction, although the control of its conformation remains unclear. The loop moves over the active site, independent of the presence of a ligand, yet favors the closed state upon ligand binding. Thus, the bound ligands will affect only the ratios of the open and closed states [40–42].

Plasmodial TIM possesses several unique target residues around the active site that concern the conformation of the active site loop and differences in the dimer interface, when compared to the human enzyme [43]. The propensity of the loop to adopt an open or closed state is distinct in PfTIM when compared to other TIMs; notably, the loop of plasmodial TIM occurs predominantly in the open state, but is closed in other TIMs on ligand binding [41, 42]. One unique feature of *P. falciparum* TIM is Phe96, which is replaced by a highly conserved Ser in almost all other known TIM sequences. Ser96 is proximal to the active site, and its function is unknown, despite its high level of conservation [44]. Gayathri *et al.* created three mutants of PfTIM, in which Phe96 was replaced by Ser, His, or Trp. Ser is the residue found in almost all other species, whereas His and Trp provide cyclic side chains, as in Phe. The Phe → Ser and Phe → His mutants showed a significantly reduced ligand binding affinity and a decreased catalytic activity, while the Phe → Trp mutant was substantially active. Both, kinetic and structural analyses demonstrated that the *Plasmodium*-specific Phe96 plays a crucial role in ligand binding and loop movement. The presence of Phe96 and Leu167 facilitates an occurrence of the open state of the loop in PfTIM, while retention of the ligand with an increased probability of open-loop conformation is controlled by Ser73 and Asn233. The residues Phe96, Leu167, Ser73, and Asn233 are found uniquely in *Plasmodium* TIM, and are responsible for the catalytic efficiency of PfTIM [44]. The presence of Phe96 in PfTIM has inspired the design of small-molecule inhibitors, which could specifically interact with Phe96 via hydrophobic or stacking interactions [43].

The dimer interface of TIM consists mainly of four loops. On investigation, engineered, monomeric TIM showed a decreased activity, which suggested that such dimerization might be important for stability and function [45]. A striking difference in the dimer interface of *Plasmodium* and the human enzyme is the residue at position 13; this is a cysteine in TIM from pathogens such as *Plasmodium*, *Leishmania*, and *Trypanosoma*, but a methionine in the human counterpart. Subsequent studies with trypanosomal and leishmanial TIM have shown that to derive an interface cysteine residue by using sulfhydryl reagents induces progressive structural alterations and interrupts the catalytic activity [46–48]. The same phenomenon occurs in PfTIM; for example, labeling the interface residue Cys13 of PfTIM with sulfhydryl-modifying agents resulted in dimer dissociation, with a concomitant loss of enzymatic activity [49]. This might represent a useful approach for developing *Plasmodium*-selective inhibitors for TIM, that specifically inhibit subunit assembly by targeting Cys13 at the dimer interface. Interestingly, dithiodianiline inhibits the TIM from *Trypanosoma cruzi* by disrupting the dimer interface, thereby effectively inhibiting the growth of *T. cruzi* epimastigotes [50].

Another distinct feature of PfTIM is the hydrophobic Leu at position 183 (this is a Glu in the TIM of humans and most other species), which is thought to be involved in membrane attachment in the erythrocyte. Glycolytic enzymes appear to interact with the erythrocyte cytoskeleton, which may be particularly important for the malarial parasite. The specific inhibition of membrane attachment might provide an alternative strategy to target the parasite in the erythrocyte host [43].

Glyceraldehyde 3-Phosphate Dehydrogenase

Glyceraldehyde 3-phosphate dehydrogenase (GAPDH; EC 1.2.1.12) is a key enzyme in glycolysis, and catalyzes the reversible oxidative phosphorylation of GAP into 1,3-bisphosphoglycerate, in the presence of NAD^+ and inorganic phosphate. The interruption of glycolysis by inhibiting GAPDH would lead to a block in ATP production, as well cause the consumption of ATP during generation of the substrate. The GAPDH-catalyzed reaction involves an initial formation of a covalent hemithioacetal intermediate between GAP and the active site cysteine residue; in this case, the hemithioacetal is oxidized to a thioester, with a concomitant reduction of NAD^+ to NADH [51]. In addition to its role in glycolysis, GAPDH displays several nonglycolytic activities related to apoptosis [52], membrane transport and fusion, microtubule bundling, and nuclear RNA export, as well as DNA replication via the transcriptional control of histone gene expression and DNA repair [53]. Interestingly, mammalian GAPDH is also involved in the effects of nitric oxide (NO). When, under oxidative stress, glycolysis is blocked and cellular ATP levels are decreased, the effect is due mainly to the inhibition of GAPDH, which is a major target of oxidative stress and is inactivated following modification of the active site cysteine. Subsequently, NO oxidatively induces the covalent binding of NAD^+ to GAPDH, and inhibits the activity of the latter by *S*-nitrosylating the active site cysteine residue [54–56]. The active site cysteine has also been reported to be nitroalkylated by nitroalkene derivatives [57], and *S*-thiolated by hydrogen peroxide [58]. Furthermore, the *S*-glutathionylation of the active site cysteines by *S*-nitrosoglutathione reversibly inhibits GAPDH [59]. During *P. falciparum* infection in humans, NO has been shown to inactivate not only the liver-invading sporozoites but also the blood-stage gametocytes [60]. The mosquito *Anopheles stephensi* may also limit parasite development by inducing the synthesis of NO [61].

In *P. falciparum*, GAPDH is thought to be involved in the vesicular transport and biogenesis of the apical complex [62]. The activity of PfGAPDH is inhibited by low micromolar concentrations of free ferriprotoporphyrin IX, a toxic byproduct of hemoglobin degradation in infected erythrocytes, whereas the human erythrocytic GAPDH is only slightly inhibited. The biological consequences of this *Plasmodium*-specific inhibition remain hypothetical, although it has been suggested that the inhibition of glycolysis by ferriprotoporphyrin IX (or NO) would enhance glucose flux through the hexose monophosphate shunt, thus enhancing the production of NADPH, which is required for the antioxidant pathways [63].

GAPDH, as a tetramer, is composed of four identical polypeptide chains, each of which contains sulfhydryl groups that are essential for catalytic activity. Despite the relatively high sequence identity between human and *Plasmodium* GAPDH, it is sequence differences that cause modification of the NAD^+-binding groove and the oligomerization behavior of plasmodial GAPDH [64]. In PfGAPDH, a so-called S-loop (residues 181–209) separates the NAD^+-binding cavities of the adjacent subunits. In contrast to GAPDH from humans or other species, plasmodial GAPDH possesses an insertion of two residues in this S-loop, which lines the wide cavity containing the NAD^+-binding groove. This forms, together with the highly conserved Lys197, a more constricted opening than in human GAPDH (Figure 5.2). This variation may lead to differences in the functional behavior of the S-loop of human and PfGAPDH, and could be employed in the development of inhibitors [65]. A second distinct feature is an Ile residue in the adenosyl binding pocket of human GAPDH, that is replaced by Met38 in the malarial enzyme [64], although this minor difference might be too difficult to exploit for selective inhibitors [65]. In all four subunits of the PfGAPDH crystal structure, a distinct area of unexpected extra density was found adjacent to the bound NAD^+; this was thought to be 4-(2-aminoethyl) benzenesulfonyl fluoride (AEBSF), a protease inhibitor used during protein purification. The sulfonylfluoride moiety of the AEBSF molecule is located in a position where the phosphate moiety of a substrate or inhibitor in previous GAPDH structures had been found. Therefore, a tightly bound molecule at this position would block substrate entry and inhibit enzymatic activity–features which may assist in the design of further inhibitors that specifically target *P. falciparum* GAPDH [65].

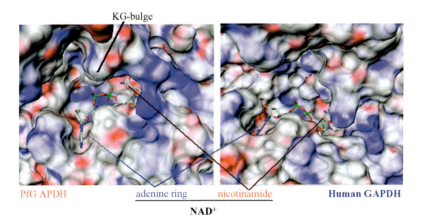

Figure 5.2 Electrostatic environment of the nicotinamide binding pocket of *P. falciparum* and human glyceraldehyde 3-phosphate dehydrogenase. Due to the KG-insertion in PfGAPDH, the nicotinamide binding cavity is more restricted when compared to the human GAPDH. Reproduced with permission from Ref. [65]; © 2006.

Pyruvate Kinases

Pyruvate kinase (PK; EC 2.7.1.40) catalyzes the irreversible transphosphorylation of phosphoenolpyruvate (PEP) and ADP to pyruvate and ATP. *P. falciparum* possesses two isoforms of pyruvate kinase: type I, and the recently discovered type II enzyme. The type I PK shows typical glycolytic properties [66], whereas the type II PK is thought to be involved in fatty acid type II biosynthesis, and is located in the apicoplast [67].

Interestingly, a genetically inherited PK deficiency manifests as a protective effect against malarial parasite replication in human erythrocytes [68]. Currently, PK deficiency is the second most common erythrocytic enzyme disorder, with a prevalence estimated at one case per 20 000 persons [69]. The PK deficiency protects erythrocytes against *P. falciparum* by disrupting the parasite invasion into erythrocytes (in individuals with a homozygous mutation), and enhancing the preferential macrophage-mediated clearance of ring-stage-infected erythrocytes (both in homozygotes and heterozygotes) [68, 70]. This may lead not only to an overall reduction in the parasite burden, but also to a reduction in the number of infected erythrocytes with parasites in the trophozoite or schizont stages. Thus, a heterozygosity for partial or complete loss-of-function alleles provides a modest, but significant, protective effect against clinical malaria, comparable to hemoglobinopathies and glucose 6-phosphate-dehydrogenase deficiency [68]. This indicates the importance of this pathway for the parasite's survival, and PK should be considered to be a potential target when designing new drugs to combat malaria. In many protozoan parasites (e.g., *T. gondii, Trypanosoma brucei, Leishmania mexicana*), PK is highly regulated [71–73] and its activity increases drastically upon infection [74]; this suggests that PK plays an important regulatory role in glycolysis in these parasites. *P. falciparum* PK is also thought to be involved in reducing the levels of 2,3-diphosphoglycerate in red blood cells, thereby contributing to the osmotic fragility of the host cells and hypoxia during malaria [75]. PK has been demonstrated as being highly susceptible to oxidative stress, and is deactivated by reactive oxygen species (ROS). Interestingly, glutathione (GSH) is able not only to protect the activity of PK against deactivation by ROS, but also to recover the activity of the oxidized enzyme [76].

Although the three-dimensional crystal structure of PK-I from *P. falciparum* was recently solved (PDB ID: 3KHD), the details have not yet been published. Likewise, the structure of PK-II from *Plasmodium* has not yet been solved, which makes the development of structure-based inhibitors difficult. Nevertheless, the characteristics of several other PK variants are known. PfPK-I, for example, is not affected by FBP, which serves as an allosteric regulator of most mammalian and bacterial PKs. Likewise, glucose 6-phosphate is an activator of *T. gondii* PK, but does not influence the activity of PfPK-I [66]. These characteristics indicate several *Plasmodium*-specific features, which could be exploited for the development of selective inhibitors.

Previously, the PKs have been considered to be involved exclusively in glycolysis within the cytosol, with non-glycolytic PKs having been found only in the apicomplexan parasites, such as *Plasmodium, Theileria*, and *T. gondii* [67]. Surprisingly, the type II PK of *Plasmodium* is localized in the apicoplast, where it might dephosphory-late PEP imported into the apicoplast via a PEP transporter, thus providing pyruvate for fatty acid synthesis and the non-mevalonate 1-deoxy-D-xylulose-5-phosphate pathway in the apicoplast [67].

Characterization of the non-glycolytic PK of *P. falciparum* would increase the present knowledge of unique metabolic pathways in protozoan parasites, and provide new starting points for selective inhibitor development.

Lactate Dehydrogenase

Lactate dehydrogenase (LDH; EC 1.1.1.27) was one of the first enzymes to be purified from *P. falciparum* [77], and is a well-known target for antimalarial compounds with unique kinetic and structural properties. LDH is the terminal enzyme of anaerobic glycolysis, and converts pyruvate to lactate with concomitant interconversion of NADH and NAD$^+$. LDH would be expected to be essential, as regenerated NAD$^+$ is required for glycolysis. The inhibition of LDH supposedly interrupts ATP production and results in *P. falciparum* cell death, suggesting that LDH might serve as a target for the development of antimalarial drugs [78–81].

LDH from *P. falciparum* has many amino acid residues that are not conserved in LDHs from other species, including those residues that define the substrate and cofactor binding sites. *P. falciparum* and human LDH differ in two structural characteristics [11, 81, 82]. First, the substrate specificity loop region is extended by an insertion of five amino acids, which creates a distinctive cavity adjacent to the catalytic region. Part of the floor of this cleft is created by the nicotinamide end of the cofactor binding pocket. Second, the position of the cofactor NADH is significantly altered in the plasmodial LDH compared to the human counterpart. These changes occur mainly at the nicotinamide end, which differs by about 1.2 Å between the human and malarial LDHs. Three *Plasmodium*-specific residues in the catalytic cavity are likely to be responsible for the altered cofactor association, namely Ser163, Pro250, and Pro246. In addition, the antigenic loop of PfLDH adopts a different conformation when compared to human LDH, which results in antigenic discrimination [81].

The alterations in cofactor association are responsible for differences in the kinetics and the lack of substrate inhibition in PfLDH in comparison to human LDH [78, 83]. Malarial LDH is able to utilize the NADH analog 3-acetylpyridine adenine dinucleotide (APAD$^+$) as a cofactor [78], and is used to monitor *in vivo* parasitemia via the enzyme-based detection of malarial parasites, as well as drug resistance [84]. Another kinetic property of PfLDH is the insensitivity to inhibition by pyruvate or a pyruvate–NAD$^+$ complex [77, 83]. The inhibition of human LDH by pyruvate is mediated by the formation of a covalent adduct between NAD$^+$ and pyruvate, due to slow release of the cofactor. In plasmodial LDH, the responsible residue Ser163 is replaced by leucine, which leads to a reduced association between the enzyme and the cofactor and, subsequently, to a reduced substrate inhibition [85]. The structural and kinetic discrepancies between the human and malarial LDHs provide an excellent starting point for the identification or design of inhibitors that specifically and effectively target the plasmodial enzyme.

Several compounds have been shown to target Pf LDH. Gossypol, a polyphenolic, binaphthyl disesquiterpene found in cottonseed, exhibits antimalarial activity at submicromolar levels, but has been shown to be toxic due to its aldehyde functional groups [78]. Subsequently obtained derivatives that lacked the toxic aldehyde groups retained their antimalarial activity [86]. Derivatives of the sesquiterpene 8-deoxyhemigossylic acid were designed to achieve a high specificity to plasmodial LDH compared to human LDH. Some of these substituted dihydronaphthoic acids have been shown to be highly selective for either plasmodial or human LDH, with selectivities ranging from five- to 190-fold. These compounds inhibit LDH over a low micromolar range, and are competitive inhibitors of cofactor binding [78, 87]. Another gossypol derivative, 7-*p*-trifluoromethylbenzyl-8-deoxyhemigossylic acid, has been shown to inhibit PfLDH with a K_i of 0.2 μM [78]. The co-crystallization of PfLDH with napththalene-based compounds (the core of gossypol) has led to the identification of the gossypol binding site (Figure 5.3). In this case, methyl-2-oxamoto-5-chlorobenzoate (MOCB) was shown to bind to the adenine binding pocket of the cofactor site, while the chlorine atom of MOCB was localized in the same position as the chlorine atom of chloroquine, in a complex of chloroquine and Pf LDH [88]. Chloroquine has been shown also to be a competitive inhibitor of Pf LDH, by binding to the NADH binding pocket [89]. Oxamate inhibits LDH by competing for the binding of pyruvate to LDH, and several oxamic acid derivatives have been developed as selective PfLDH inhibitors, among which two compounds have demonstrated some potential as lead compounds for drug-resistant malaria [90].

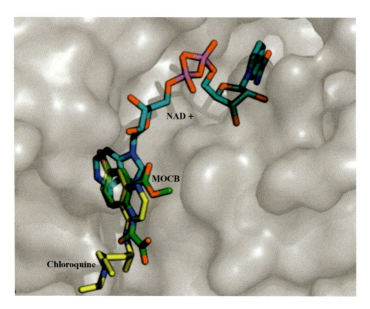

Figure 5.3 Crystal structure of *P. falciparum* lactate dehydrogenase with the naphthalene-based compound methyl-3-oxamoto-5-chlorobenzoate (MOCB). Overlay of the adenine-binding pocket from crystal structures of PfLDH with NADH (cyan), MOCB (green), and chloroquine (yellow). Reproduced with permission from Ref. [88]; © 2005.

Several heterocyclic, azole-based compounds inhibit *P. falciparum* LDH preferentially at submicromolar concentrations, by interacting with the active side residues alongside the NAD$^+$ cofactor. Interestingly, these compounds have also displayed antimalarial activity, confirming that the inhibition of LDH is fatal to the parasite [91]. Overall, these results have supported the concept that the cofactor binding site of PfLDH represents a promising site for the structure-based design of antimalarial drugs.

Conclusion

The discovery of new drug targets and the development of new drugs represents one of the most important gateways in the battle against malaria. Ideally, proteomic and genomic approaches may be complemented by crystallographic studies that characterize the target proteins and protein–inhibitor or protein–substrate complexes, thereby allowing the visualization and improvement of protein–inhibitor contacts. The primary energy source of the asexual erythrocytic-stage *P. falciparum* is ATP, generated via the fermentation of glucose during glycolysis. This unusual route for energy (ATP) generation in *Plasmodium* relative to its human host indicates that an interference with the glycolytic pathway could serve as a valuable source of compounds with antimalarial activities. Indeed, a genetically inherited, human PK deficiency provides protection against clinical malaria, and substantiates the feasibility of glycolytic enzymes as drug targets. Besides glycolysis, several glycolytic enzymes are involved in various cellular processes such as host cell invasion (ALDO), apoptosis, DNA replication (GAPDH), and fatty acid synthesis (TIM, PKII). In this respect, several potent and selective inhibitors have been developed. Notably, the inhibition of LDH by several compounds has resulted in a reduced growth of the parasite, while selective inhibitors of protozoan aldolases have been designed. The crystal structures of almost all glycolytic enzymes have been solved, revealing distinct features between the human and *Plasmodium* enzymes, and allowing the design of structure-based inhibitors. The data summarized in this chapter support the long-held promise that various stages in glycolysis might serve as viable drug targets.

Acknowledgments

These studies were supported by the Deutsche Forschungsgemeinschaft (BE1540/15-1).

References

1 Sherman, I.W. (2005) *Molecular Approaches to Malaria*, 1st edn, ASM Press, Washington, DC.

2 Baird, J.K. (2005) Effectiveness of antimalarial drugs. *N. Engl. J. Med.*, **352**, 1565–1577.

3 Dondorp, A., Nosten, F., Stepniewska, K., Day, N., and White, N. (2005) Artesunate versus quinine for treatment of severe falciparum malaria: a randomised trial. *Lancet*, **366**, 717–725.

4 German, P.I. and Aweeka, F.T. (2008) Clinical pharmacology of artemisinin-based combination therapies. *Clin. Pharmacokinet.*, **47**, 91–102.

5 Breman, J.G. (2001) The ears of the hippopotamus: manifestations, determinants, and estimates of the malaria burden. *Am. J. Trop. Med. Hyg.*, **64**, 1–11.

6 Bustamante, C., Batista, C.N., and Zalis, M. (2009) Molecular and biological aspects of antimalarial resistance in *Plasmodium falciparum* and *Plasmodium vivax*. *Curr. Drug Targets*, **10**, 279–290.

7 Fry, M., Webb, E., and Pudney, M. (1990) Effect of mitochondrial inhibitors on adenosine triphosphate levels in *Plasmodium falciparum*. *Comp. Biochem. Physiol. B*, **96**, 775–782.

8 Roth, E.F., Jr, Raventos-Suarez, C., Perkins, M., and Nagel, R.L. (1982) Glutathione stability and oxidative stress in *P. falciparum* infection *in vitro*: responses of normal and G6PD-deficient cells. *Biochem. Biophys. Res. Commun.*, **109**, 355–362.

9 Scheibel, L.W., Adler, A., and Trager, W. (1979) Tetraethylthiuram disulfide (Antabuse) inhibits the human malaria parasite *Plasmodium falciparum*. *Proc. Natl Acad. Sci. USA*, **76**, 5303–5307.

10 Homewood, C.A. and Neame, K.D. (1983) Conversion of glucose of lactate by intraerythrocytic *Plasmodium berghei*. *Ann. Trop. Med. Parasitol.*, **77**, 127–129.

11 Kirk, K., Horner, H.A., and Kirk, J. (1996) Glucose uptake in *Plasmodium falciparum*-infected erythrocytes is an equilibrative not an active process. *Mol. Biochem. Parasitol.*, **82**, 195–205.

12 Saliba, K.J., Krishna, S., and Kirk, K. (2004) Inhibition of hexose transport and abrogation of pH homeostasis in the intraerythrocytic malaria parasite by an O-3-hexose derivative. *FEBS Lett.*, **570**, 93–96.

13 Roth, E.F. Jr (1987) Malarial parasite hexokinase and hexokinase-dependent glutathione reduction in the *Plasmodium falciparum*-infected human erythrocyte. *J. Biol. Chem.*, **262**, 15678–15682.

14 Olafsson, P., Matile, H., and Certa, U. (1992) Molecular analysis of *Plasmodium falciparum* hexokinase. *Mol. Biochem. Parasitol.*, **56**, 89–101.

15 Olafsson, P. and Certa, U. (1994) Expression and cellular localisation of hexokinase during the bloodstage development of *Plasmodium falciparum*. *Mol. Biochem. Parasitol.*, **63**, 171–174.

16 Joet, T., Eckstein-Ludwig, U., Morin, C., and Krishna, S. (2003) Validation of the hexose transporter of *Plasmodium falciparum* as a novel drug target. *Proc. Natl Acad. Sci. USA*, **100**, 7476–7479.

17 Joet, T., Holterman, L., Stedman, T.T., Kocken, C.H., Van Der Wel, A., Thomas, A.W., and Krishna, S. (2002) Comparative characterization of hexose transporters of *Plasmodium knowlesi*, *Plasmodium yoelii* and *Toxoplasma gondii* highlights functional differences within the apicomplexan family. *Biochem. J.*, **368**, 923–929.

18 van Schalkwyk, D.A., Priebe, W., and Saliba, K.J. (2008) The inhibitory effect of 2-halo derivatives of D-glucose on glycolysis and on the proliferation of the human malaria parasite *Plasmodium falciparum*. *J. Pharmacol. Exp. Ther.*, **327**, 511–517.

19 Udeinya, I.J. and Van Dyke, K. (1981) 2-Deoxyglucose: inhibition of parasitemia and of glucosamine incorporation into glycosylated macromolecules, in malarial parasites (*Plasmodium falciparum*). *Pharmacology*, **23**, 171–175.

20 Hudock, M.P., Sanz-Rodriguez, C.E., Song, Y., Chan, J.M., Zhang, Y., Odeh, S., Kosztowski, T., Leon-Rossell, A., Concepcion, J.L., Yardley, V., Croft, S.L., Urbina, J.A., and Oldfield, E. (2006) Inhibition of *Trypanosoma cruzi* hexokinase by bisphosphonates. *J. Med. Chem.*, **49**, 215–223.

21 Sanz-Rodriguez, C.E., Concepcion, J.L., Pekerar, S., Oldfield, E., and Urbina, J.A. (2007) Bisphosphonates as inhibitors of *Trypanosoma cruzi* hexokinase: kinetic and

metabolic studies. *J. Biol. Chem.*, **282**, 12377–12387.

22 Kim, H., Certa, U., Dobeli, H., Jakob, P., and Hol, W.G. (1998) Crystal structure of fructose-1,6-bisphosphate aldolase from the human malaria parasite *Plasmodium falciparum*. *Biochemistry*, **37**, 4388–4396.

23 Certa, U., Ghersa, P., Dobeli, H., Matile, H., Kocher, H.P., Shrivastava, I.K., Shaw, A.R., and Perrin, L.H. (1988) Aldolase activity of a *Plasmodium falciparum* protein with protective properties. *Science*, **240**, 1036–1038.

24 Penhoet, E.E., Kochman, M., and Rutter, W.J. (1969) Molecular and catalytic properties of aldolase C. *Biochemistry*, **8**, 4396–4402.

25 Penhoet, E.E., Kochman, M., and Rutter, W.J. (1969) Isolation of fructose diphosphate aldolases A, B, and C. *Biochemistry*, **8**, 4391–4395.

26 Drechsler, E.R., Boyer, P.D., and Kowalsky, A.G. (1959) The catalytic activity of carboxypeptidase-degraded aldolase. *J. Biol. Chem.*, **234**, 2627–2634.

27 Meier, B., Dobeli, H., and Certa, U. (1992) Stage-specific expression of aldolase isoenzymes in the rodent malaria parasite *Plasmodium berghei. Mol. Biochem. Parasitol.*, **52**, 15–27.

28 Jewett, T.J. and Sibley, L.D. (2003) Aldolase forms a bridge between cell surface adhesins and the actin cytoskeleton in apicomplexan parasites. *Mol. Cell*, **11**, 885–894.

29 Morahan, B.J., Wang, L., and Coppel, R.L. (2009) No TRAP, no invasion. *Trends Parasitol.*, **25**, 77–84.

30 Bosch, J., Buscaglia, C.A., Krumm, B., Ingason, B.P., Lucas, R., Roach, C., Cardozo, T., Nussenzweig, V., and Hol, W.G. (2007) Aldolase provides an unusual binding site for thrombospondin-related anonymous protein in the invasion machinery of the malaria parasite. *Proceedings of the National Academy of Sciences USA (PNAS)*, **104**, 7015–7020.

31 Dax, C., Duffieux, F., Chabot, N., Coincon, M., Sygusch, J., Michels, P.A., and Blonski, C. (2006) Selective irreversible inhibition of fructose 1,6-bisphosphate aldolase from *Trypanosoma brucei. J. Med. Chem.*, **49**, 1499–1502.

32 Masters, C.J., Reid, S., and Don, M. (1987) Glycolysis – new concepts in an old pathway. *Mol. Cell. Biochem.*, **76**, 3–14.

33 Dobeli, H., Trzeciak, A., Gillessen, D., Matile, H., Srivastava, I.K., Perrin, L.H., Jakob, P.E., and Certa, U. (1990) Expression, purification, biochemical characterization and inhibition of recombinant *Plasmodium falciparum* aldolase. *Mol. Biochem. Parasitol.*, **41**, 259–268.

34 Richard, J.P. (1991) Kinetic parameters for the elimination reaction catalyzed by triosephosphate isomerase and an estimation of the reaction's physiological significance. *Biochemistry*, **30**, 4581–4585.

35 Lolis, E. and Petsko, G.A. (1990) Crystallographic analysis of the complex between triosephosphate isomerase and 2-phosphoglycolate at 2.5-A resolution: implications for catalysis. *Biochemistry*, **29**, 6619–6625.

36 Zhang, Z., Sugio, S., Komives, E.A., Liu, K.D., Knowles, J.R., Petsko, G.A., and Ringe, D. (1994) Crystal structure of recombinant chicken triosephosphate isomerase-phosphoglycolohydroxamate complex at 1.8-A resolution. *Biochemistry*, **33**, 2830–2837.

37 Albery, W.J. and Knowles, J.R. (1976) Evolution of enzyme function and the development of catalytic efficiency. *Biochemistry*, **15**, 5631–5640.

38 Joseph, D., Petsko, G.A., and Karplus, M. (1990) Anatomy of a conformational change: hinged "lid" motion of the triosephosphate isomerase loop. *Science*, **249**, 1425–1428.

39 Desamero, R., Rozovsky, S., Zhadin, N., McDermott, A., and Callender, R. (2003) Active site loop motion in triosephosphate isomerase: T-jump relaxation spectroscopy of thermal activation. *Biochemistry*, **42**, 2941–2951.

40 Rozovsky, S., Jogl, G., Tong, L., and McDermott, A.E. (2001) Solution-state NMR investigations of triosephosphate isomerase active site loop motion: ligand release in relation to active site loop dynamics. *J. Mol. Biol.*, **310**, 271–280.

41 Parthasarathy, S., Balaram, H., Balaram, P., and Murthy, M.R. (2002) Structures of

Plasmodium falciparum triosephosphate isomerase complexed to substrate analogues: observation of the catalytic loop in the open conformation in the ligand-bound state. *Acta Crystallogr. D- Biol. Crystallogr.*, **58**, 1992–2000.

42 Parthasarathy, S., Ravindra, G., Balaram, H., Balaram, P., and Murthy, M.R. (2002) Structure of the *Plasmodium falciparum* triosephosphate isomerase-phosphoglycolate complex in two crystal forms: characterization of catalytic loop open and closed conformations in the ligand-bound state. *Biochemistry*, **41**, 13178–13188.

43 Velanker, S.S., Ray, S.S., Gokhale, R.S., Suma, S., Balaram, H., Balaram, P., and Murthy, M.R. (1997) Triosephosphate isomerase from *Plasmodium falciparum*: the crystal structure provides insights into antimalarial drug design. *Structure*, **5**, 751–761.

44 Gayathri, P., Banerjee, M., Vijayalakshmi, A., Balaram, H., Balaram, P., and Murthy, M.R. (2009) Biochemical and structural characterization of residue 96 mutants of *Plasmodium falciparum* triosephosphate isomerase: active-site loop conformation, hydration and identification of a dimer-interface ligand-binding site. *Acta Crystallogr. D - Biol. Crystallogr.*, **65**, 847–857.

45 Borchert, T.V., Abagyan, R., Kishan, K.V., Zeelen, J.P., and Wierenga, R.K. (1993) The crystal structure of an engineered monomeric triosephosphate isomerase, monoTIM: the correct modelling of an eight-residue loop. *Structure*, **1**, 205–213.

46 Gomez-Puyou, A., Saavedra-Lira, E., Becker, I., Zubillaga, R.A., Rojo-Dominguez, A., and Perez-Montfort, R. (1995) Using evolutionary changes to achieve species-specific inhibition of enzyme action – studies with triosephosphate isomerase. *Chem. Biol.*, **2**, 847–855.

47 Maldonado, E., Soriano-Garcia, M., Moreno, A., Cabrera, N., Garza-Ramos, G., de Gomez-Puyou, M., Gomez-Puyou, A., and Perez-Montfort, R. (1998) Differences in the intersubunit contacts in triosephosphate isomerase from two

closely related pathogenic trypanosomes. *J. Mol. Biol.*, **283**, 193–203.

48 Perez-Montfort, R., Garza-Ramos, G., Alcantara, G.H., Reyes-Vivas, H., Gao, X.G., Maldonado, E., de Gomez-Puyou, M.T., and Gomez-Puyou, A. (1999) Derivatization of the interface cysteine of triosephosphate isomerase from *Trypanosoma brucei* and *Trypanosoma cruzi* as probe of the interrelationship between the catalytic sites and the dimer interface. *Biochemistry*, **38**, 4114–4120.

49 Maithal, K., Ravindra, G., Balaram, H., and Balaram, P. (2002) Inhibition of *Plasmodium falciparum* triose-phosphate isomerase by chemical modification of an interface cysteine. Electrospray ionization mass spectrometric analysis of differential cysteine reactivities. *J. Biol. Chem.*, **277**, 25106–25114.

50 Olivares-Illana, V., Rodriguez-Romero, A., Becker, I., Berzunza, M., Garcia, J., Perez-Montfort, R., Cabrera, N., Lopez-Calahorra, F., de Gomez-Puyou, M., and Gomez-Puyou, A. (2007) Perturbation of the dimer interface of triosephosphate isomerase and its effect on *Trypanosoma cruzi*. *PLoS Negl. Trop. Dis.*, **1**, e1.

51 Daubenberger, C.A., Poltl-Frank, F., Jiang, G., Lipp, J., Certa, U., and Pluschke, G. (2000) Identification and recombinant expression of glyceraldehyde-3-phosphate dehydrogenase of *Plasmodium falciparum*. *Gene*, **246**, 255–264.

52 Ishitani, R. and Chuang, D.M. (1996) Glyceraldehyde-3-phosphate dehydrogenase antisense oligodeoxynucleotides protect against cytosine arabinonucleoside-induced apoptosis in cultured cerebellar neurons. *Proc. Natl Acad. Sci. USA*, **93**, 9937–9941.

53 Sirover, M.A. (2005) New nuclear functions of the glycolytic protein, glyceraldehyde-3-phosphate dehydrogenase, in mammalian cells. *J. Cell Biochem.*, **95**, 45–52.

54 Dimmeler, S. and Brune, B. (1992) Characterization of a nitric-oxide-catalysed ADP-ribosylation of glyceraldehyde-3-phosphate

dehydrogenase. *Eur. J. Biochem.*, **210**, 305–310.

55 Molina y Vedia, L., McDonald, B., Reep, B., Brune, B., Di Silvio, M., Billiar, T.R., and Lapetina, E.G. (1992) Nitric oxide-induced S-nitrosylation of glyceraldehyde-3-phosphate dehydrogenase inhibits enzymatic activity and increases endogenous ADP-ribosylation. *J. Biol. Chem.*, **267**, 24929–24932.

56 Hara, M.R., Cascio, M.B., and Sawa, A. (2006) GAPDH as a sensor of NO stress. *Biochim. Biophys. Acta*, **1762**, 502–509.

57 Batthyany, C., Schopfer, F.J., Baker, P.R., Duran, R., Baker, L.M., Huang, Y., Cervenansky, C., Branchaud, B.P., and Freeman, B.A. (2006) Reversible post-translational modification of proteins by nitrated fatty acids *in vivo*. *J. Biol. Chem.*, **281**, 20450–20463.

58 Schuppe-Koistinen, I., Moldeus, P., Bergman, T., and Cotgreave, I.A. (1994) S-thiolation of human endothelial cell glyceraldehyde-3-phosphate dehydrogenase after hydrogen peroxide treatment. *Eur. J. Biochem.*, **221**, 1033–1037.

59 Mohr, S., Hallak, H., de Boitte, A., Lapetina, E.G., and Brune, B. (1999) Nitric oxide-induced S-glutathionylation and inactivation of glyceraldehyde-3-phosphate dehydrogenase. *J. Biol. Chem.*, **274**, 9427–9430.

60 Good, M.F. and Doolan, D.L. (1999) Immune effector mechanisms in malaria. *Curr. Opin. Immunol.*, **11**, 412–419.

61 Luckhart, S., Vodovotz, Y., Cui, L., and Rosenberg, R. (1998) The mosquito *Anopheles stephensi* limits malaria parasite development with inducible synthesis of nitric oxide. *Proc. Natl Acad. Sci. USA*, **95**, 5700–5705.

62 Daubenberger, C.A., Tisdale, E.J., Curcic, M., Diaz, D., Silvie, O., Mazier, D., Eling, W., Bohrmann, B., Matile, H., and Pluschke, G. (2003) The N′-terminal domain of glyceraldehyde-3-phosphate dehydrogenase of the apicomplexan *Plasmodium falciparum* mediates GTPase Rab2-dependent recruitment to membranes. *Biol. Chem.*, **384**, 1227–1237.

63 Campanale, N., Nickel, C., Daubenberger, C.A., Wehlan, D.A., Gorman, J.J.,

Klonis, N., Becker, K., and Tilley, L. (2003) Identification and characterization of heme-interacting proteins in the malaria parasite, *Plasmodium falciparum*. *J. Biol. Chem.*, **278**, 27354–27361.

64 Satchell, J.F., Malby, R.L., Luo, C.S., Adisa, A., Alpyurek, A.E., Klonis, N., Smith, B.J., Tilley, L., and Colman, P.M. (2005) Structure of glyceraldehyde-3-phosphate dehydrogenase from *Plasmodium falciparum*. *Acta Crystallogr. D - Biol. Crystallogr.*, **61**, 1213–1221.

65 Robien, M.A., Bosch, J., Buckner, F.S., Van Voorhis, W.C., Worthey, E.A., Myler, P., Mehlin, C., Boni, E.E., Kalyuzhniy, O., Anderson, L., Lauricella, A., Gulde, S., Luft, J.R., DeTitta, G., Caruthers, J.M., Hodgson, K.O., Soltis, M., Zucker, F., Verlinde, C.L., Merritt, E.A., Schoenfeld, L.W., and Hol, W.G. (2006) Crystal structure of glyceraldehyde-3-phosphate dehydrogenase from *Plasmodium falciparum* at 2.25 A resolution reveals intriguing extra electron density in the active site. *Proteins*, Wiley-Interscience, **62**, 570–577.

66 Chan, M. and Sim, T.S. (2005) Functional analysis, overexpression, and kinetic characterization of pyruvate kinase from *Plasmodium falciparum*. *Biochem. Biophys. Res. Commun.*, **326**, 188–196.

67 Maeda, T., Saito, T., Harb, O.S., Roos, D.S., Takeo, S., Suzuki, H., Tsuboi, T., Takeuchi, T., and Asai, T. (2009) Pyruvate kinase type-II isozyme in *Plasmodium falciparum* localizes to the apicoplast. *Parasitol. Int.*, **58**, 101–105.

68 Ayi, K., Min-Oo, G., Serghides, L., Crockett, M., Kirby-Allen, M., Quirt, I., Gros, P., and Kain, K.C. (2008) Pyruvate kinase deficiency and malaria. *N. Engl. J. Med.*, **358**, 1805–1810.

69 Zanella, A., Fermo, E., Bianchi, P., and Valentini, G. (2005) Red cell pyruvate kinase deficiency: molecular and clinical aspects. *Br. J. Haematol.*, **130**, 11–25.

70 Min-Oo, G., Fortin, A., Tam, M.F., Nantel, A., Stevenson, M.M., and Gros, P. (2003) Pyruvate kinase deficiency in mice protects against malaria. *Nat. Genet.*, **35**, 357–362.

71 Maeda, T., Saito, T., Oguchi, Y., Nakazawa, M., Takeuchi, T., and Asai, T. (2003)

Expression and characterization of recombinant pyruvate kinase from *Toxoplasma gondii* tachyzoites. *Parasitol Res.*, **89**, 259–265.

72 Ernest, I., Callens, M., Opperdoes, F.R., and Michels, P.A. (1994) Pyruvate kinase of *Leishmania mexicana mexicana*. Cloning and analysis of the gene, overexpression in *Escherichia coli* and characterization of the enzyme. *Mol. Biochem. Parasitol.*, **64**, 43–54.

73 Ernest, I., Callens, M., Uttaro, A.D., Chevalier, N., Opperdoes, F.R., Muirhead, H., and Michels, P.A. (1998) Pyruvate kinase of *Trypanosoma brucei*: overexpression, purification, and functional characterization of wild-type and mutated enzyme. *Protein Expr. Purif.*, **13**, 373–382.

74 Roth, E.F. Jr, Calvin, M.C., Max-Audit, I., Rosa, J., and Rosa, R. (1988) The enzymes of the glycolytic pathway in erythrocytes infected with *Plasmodium falciparum* malaria parasites. *Blood*, **72**, 1922–1925.

75 Dubey, M.L., Hegde, R., Ganguly, N.K., and Mahajan, R.C. (2003) Decreased level of 2,3-diphosphoglycerate and alteration of structural integrity in erythrocytes infected with *Plasmodium falciparum in vitro*. *Mol. Cell. Biochem.*, **246**, 137–141.

76 Ogasawara, Y., Funakoshi, M., and Ishii, K. (2008) Pyruvate kinase is protected by glutathione-dependent redox balance in human red blood cells exposed to reactive oxygen species. *Biol. Pharm. Bull.*, **31**, 1875–1881.

77 Van der Jagt, D.L., Hunsaker, L.A., and Heidrich, J.E. (1981) Partial purification and characterization of lactate dehydrogenase from *Plasmodium falciparum*. *Mol. Biochem. Parasitol.*, **4**, 255–264.

78 Gomez, M.S., Piper, R.C., Hunsaker, L.A., Royer, R.E., Deck, L.M., Makler, M.T., and Van der Jagt, D.L. (1997) Substrate and cofactor specificity and selective inhibition of lactate dehydrogenase from the malarial parasite *P. falciparum*. *Mol. Biochem. Parasitol.*, **90**, 235–246.

79 Menting, J.G., Tilley, L., Deady, L.W., Ng, K., Simpson, R.J., Cowman, A.F., and Foley, M. (1997) The antimalarial drug, chloroquine, interacts with lactate dehydrogenase from *Plasmodium falciparum*. *Mol. Biochem. Parasitol.*, **88**, 215–224.

80 Ridley, R.G. (1997) Plasmodium: drug discovery and development – an industrial perspective. *Exp. Parasitol.*, **87**, 293–304.

81 Dunn, C.R., Banfield, M.J., Barker, J.J., Higham, C.W., Moreton, K.M., Turgut-Balik, D., Brady, R.L., and Holbrook, J.J. (1996) The structure of lactate dehydrogenase from *Plasmodium falciparum* reveals a new target for anti-malarial design. *Nat. Struct. Biol.*, **3**, 912–915.

82 Read, J.A., Winter, V.J., Eszes, C.M., Sessions, R.B., and Brady, R.L. (2001) Structural basis for altered activity of M- and H-isozyme forms of human lactate dehydrogenase. *Proteins*, **43**, 175–185.

83 Berwal, R., Gopalan, N., Chandel, K., Prasad, G.B., and Prakash, S. (2008) *Plasmodium falciparum*: enhanced soluble expression, purification and biochemical characterization of lactate dehydrogenase. *Exp. Parasitol.*, **120**, 135–141.

84 Makler, M.T., Ries, J.M., Williams, J.A., Bancroft, J.E., Piper, R.C., Gibbins, B.L., and Hinrichs, D.J. (1993) Parasite lactate dehydrogenase as an assay for *Plasmodium falciparum* drug sensitivity. *Am. J. Trop. Med. Hyg.*, **48**, 739–741.

85 Sessions, R.B., Dewar, V., Clarke, A.R., and Holbrook, J.J. (1997) A model of *Plasmodium falciparum* lactate dehydrogenase and its implications for the design of improved antimalarials and the enhanced detection of parasitaemia. *Protein Eng.*, **10**, 301–306.

86 Royer, R.E., Deck, L.M., Campos, N.M., Hunsaker, L.A., and Van der Jagt, D.L. (1986) Biologically active derivatives of gossypol: synthesis and antimalarial activities of peri-acylated gossylic nitriles. *J. Med. Chem.*, **29**, 1799–1801.

87 Deck, L.M., Royer, R.E., Chamblee, B.B., Hernandez, V.M., Malone, R.R., Torres, J.E., Hunsaker, L.A., Piper, R.C., Makler, M.T., and Van der Jagt, D.L. (1998) Selective inhibitors of human lactate dehydrogenases and lactate dehydrogenase from the malarial parasite *Plasmodium falciparum*. *J. Med. Chem.*, Elsevier, **41**, 3879–3887.

88 Conners, R., Schambach, F., Read, J.,
Cameron, A., Sessions, R.B., Vivas, L.,
Easton, A., Croft, S.L., and Brady, R.L. (2005)
Mapping the binding site for gossypol-like
inhibitors of *Plasmodium falciparum* lactate
dehydrogenase. *Mol. Biochem. Parasitol.*,
Elsevier, **142**, 137–148.

89 Read, J.A., Wilkinson, K.W., Tranter, R.,
Sessions, R.B., and Brady, R.L. (1999)
Chloroquine binds in the cofactor binding
site of *Plasmodium falciparum* lactate
dehydrogenase. *J. Biol. Chem.*, **274**,
10213–10218.

90 Choi, S.R., Beeler, A.B., Pradhan, A.,
Watkins, E.B., Rimoldi, J.M., Tekwani, B.,
and Avery, M.A. (2007) Generation of

oxamic acid libraries: antimalarials and
inhibitors of *Plasmodium falciparum*
lactate dehydrogenase. *J. Comb. Chem.*, **9**,
292–300.

91 Cameron, A., Read, J., Tranter, R.,
Winter, V.J., Sessions, R.B., Brady, R.L.,
Vivas, L., Easton, A., Kendrick, H.,
Croft, S.L., Barros, D., Lavandera, J.L.,
Martin, J.J., Risco, F., Garcia-Ochoa, S.,
Gamo, F.J., Sanz, L., Leon, L., Ruiz, J.R.,
Gabarro, R., Mallo, A., and Gomez de las
Heras, F. (2004) Identification and activity
of a series of azole-based compounds
with lactate dehydrogenase-directed
anti-malarial activity. *J. Biol. Chem.*, **279**,
31429–31439.

6
Polyamines in Apicomplexan Parasites

Ingrid B. Müller, *Robin Das Gupta, Kai Lüersen, Carsten Wrenger,*
and Rolf D. Walter

Abstract

The polyamines putrescine, spermidine, and spermine are simple-structured cationic molecules which fulfill essential functions in cell growth and differentiation. Consequently, the obstruction of synthesis, uptake, or interconversion of polyamines has been proposed as a potent drug target in cancer therapy, as well as against various infections caused by protozoan parasites. Although polyamines are found throughout the Apicomplexa, members of this group differ widely in their polyamine supply, which either derives via their own biosynthesis or from the host. In this chapter, attention is focused on changes in, and differences between, the parasite and host polyamine metabolism. The use of specific enzyme inhibitors may serve as an excellent tool for the validation of their potential in chemotherapeutic interventions.

Introduction

Polyamines are polycations with flexible aliphatic carbon chains and are well known as essential components for cell growth, proliferation and differentiation. The diamine putrescine as well as the polyamines spermidine and spermine (which bear three and four amine/imine groups, respectively) are ubiquitously present in prokaryotic and eukaryotic cells. To date, only two representatives of the Archaea have been shown to lack polyamines completely [1]. Although, under physiological conditions, the polyamines carry a positive charge on each nitrogen atom, the versatile functions of polyamines are in general all based on their reversible ionic interactions with negatively charged macromolecules such as DNA, RNA, membrane proteins and phospholipids, as well as ion channels. Whilst a discussion on the distinguished polyamine functions based mainly on charge interactions is beyond the scope of this chapter, detailed information on this subject is summarized in Ref. [2].

*Corresponding author

Apicomplexan Parasites. Edited by Katja Becker
Copyright © 2011 WILEY-VCH Verlag GmbH & Co. KGaA, Weinheim
ISBN: 978-3-527-32731-7

From Cancer Research to Anti-Apicomplexa

Members of the phylum Apicomplexa account for some of the most debilitating infections of humans and livestock. Malaria, caused by *Plasmodium* sp., is undoubtedly the most important infectious disease and a major cause of death in developing countries. Other infections by apicomplexan parasites such as *Toxoplasma gondii*, *Cryptosporidium parvum* and *Babesia* sp. are usually asymptomatic in humans but become life-threatening diseases in immune-compromised patients. The latter species, as well as *Theileria* and *Eimeria* sp., are the pathogenic agents of fatal infections in livestock. Although effective treatments are available against these human and animal diseases, the problem has become serious in recent years due to the spread of drug resistance. Since *Plasmodium* and the other Apicomplexa are highly diverse and genetically polymorphic, they are provided with a variety of mechanisms to escape drug treatment which leads to an urgent and pressing need for new drugs that can attack the different metabolic pathways of the parasites in order to avoid cross-resistance with the commonly used drugs. Consequently, the control of apicomplexan infections in humans and animals depends greatly on the identification of novel drug targets, and on the design of their inhibitors. Collected data relating to the polyamine metabolism of the malarial parasite and other parasitic protozoa has been reviewed previously [3–5]. In this chapter, a summary is provided of polyamine metabolism in Apicomplexa, with respect to its potential for chemotherapeutic intervention.

Since polyamines are known to be essential for cell proliferation and differentiation, numerous approaches have been taken to interfere with their metabolism, not only for tumor therapy but also in a preventive role [6–9]. The inhibition of polyamine biosynthesis, with a resultant depletion of polyamines, was initially proposed as a very promising antiproliferative strategy. Unfortunately, however, attempts to target the synthesis by using enzyme inhibitors and polyamine analogs have not yet proved to be as successful in cancer treatment as anticipated. Nonetheless, the use of these compounds either as preventive agents or in combination with other drugs has provided some benefits in multiple cancer trials. For example, alpha-difluoromethylornithine (DFMO), an irreversible inhibitor of the enzyme ornithine decarboxylase (ODC), which catalyzes the initial step in polyamine biosynthesis, did not provide the expected curative effect in clinical trials, but did show promise as a chemopreventive agent [7]. The reasons for the modest success of DFMO in anticancer therapy are based on the complex regulation of the polyamine pool, which is supplied not only by biosynthesis and exogenous polyamines but also by a sophisticated fine-tuning via interconversion. DFMO, as such, is a safe and well-tolerated drug, which is today frequently used in cosmetics for hair removal [10, 11].

As the rapid proliferation of protozoan parasites is thought to be associated with a high demand for polyamines, the blockade of polyamine metabolism by using inhibitors and analogs – both of which have been created in abundance in cancer research – was monitored against infections caused by parasites. Notably, this strategy has been applied successfully against trypanosomes, with eflornithine (i.e., DFMO) being shown as highly effective in the fight against West African human sleeping sickness [12–14]. Such pioneering studies with trypanosomal infections encouraged the use of DFMO to validate the blockade of polyamine synthesis in other protozoan parasites for drug intervention, although unfortunately the effect of DFMO was far less pronounced. Trypanosomes depend on a unique trypanothione system that consists of a spermidine–glutathione conjugate which is essential for maintenance of the intracellular thiol redox state (similar to the glutathione system in other organisms) [15, 16]. Nevertheless, some curative results have been reported for DFMO in the treatment of patients suffering from *Pneumocystis carinii* and *C. parvum* infections [17, 18], and it also prevented infections by *Eimeria tenella* [19]. Furthermore, DFMO showed a potent therapeutic effect on experimental *Leishmania donovani* infections [20], and also inhibited the growth of *Giardia lamblia* as well as of *Acanthamoeba castellani* [21, 22]. The fact that the antiparasitic effects of DFMO treatment could be reversed by the addition of physiological polyamines indicated clearly that the drug functions by blocking the biosynthesis of these compounds. However, the results of experiments with *Anopheles* and rodent model infections with *P. berghei*, as well as with cultured *P. falciparum* erythrocytic stages, varied from promising to modest. In this case, DFMO was shown to block the development of trophozoites in the erythrocytic schizogony, an inhibition which was reversible and cytostatic rather than cytotoxic. However, when compared to the impact on sporogony and liver schizogony, the effect on the erythrocytic schizogony was less pronounced, due either to an inefficient uptake of DFMO or to a lower multiplication rate of the erythrocytic stages [3, 4]. In the other stages, the application of DFMO blocked the sporogonous cycle of *P. berghei* in *Anopheles stephensi* [23], and also protected mice against infection with sporozoites, thus providing evidence for the exoerythrocytic schizogony as a potent target [24–26]. The treatment of mice infected with blood-stage forms of *P. berghei* with DFMO led to controversial results, however, when the drug was seen to block the development of erythrocytic schizonts in cultured *P. falciparum-* and *P. berghei*-infected mice [27, 28], but did not show any convincing effect on their survival [24, 25, 29, 30].

Polyamines in Mammals

The common metabolism of polyamines is outlined in Figure 6.1. In addition to the *de novo* synthetic pathway, both animals and yeast convert polyamines back via an interconversion pathway, with the flux of both pathways being well controlled not only by enzymes that are highly regulated on the transcriptional, translational, and protein levels, but also by mechanisms for polyamine uptake and excretion. In an initial step, arginine is first converted to ornithine by the enzyme arginase (ARG; EC 3.5.3.1), and subsequently decarboxylated to putrescine by ornithine decarboxylase (ODC; EC 4.1.1.17). An alternative putrescine supply follows the reactions of arginine decarboxylase (ADC; EC 4.1.1.19) and of agmatinase (AGM; EC 3.5.3.11). Spermidine and spermine are formed by the enzymes spermidine synthase (SPDS; EC 2.5.1.16) and spermine synthase (SPMS; EC 2.5.1.22), respectively. For this formation, the aminopropyl donor, *S*-adenosylmethionine (AdoMet), is synthesized from methionine and ATP by *S*-adenosylmethionine synthetase (AdoMetSyn; EC 2.5.1.6)

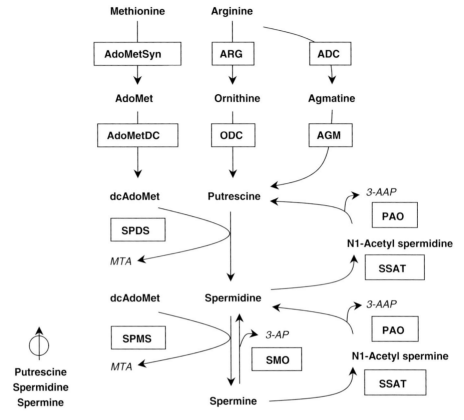

Figure 6.1 Polyamine metabolism in mammalian cells. The uptake, synthesis, and interconversion paths for putrescine, spermidine and spermine are summarized. Arginine is hydrolyzed by arginase (ARG), leading to ornithine, which is subsequently decarboxylated to putrescine by ornithine decarboxylase (ODC). The alternative putrescine formation starts with the reaction of arginine decarboxylase (ADC), followed by the hydrolysis of agmatine by agmatinase (AGM). Methionine and ATP are required to synthesize *S*-adenosyl methionine (AdoMet) by AdoMet synthase (AdoMetSyn). AdoMet is decarboxylated by AdoMet decarboxylase (AdoMetDC) to deliver decarboxylated (dc) AdoMet, the donor of the aminopropyl group for the formation of spermidine, exhibited by the spermidine synthase (SPDS), and of spermine, performed by the spermine synthase (SPMS), both reactions leading to the side product methylthioadenosine (MTA). The interconversion of spermine to spermidine, and further to putrescine, occurs via acetylation by spermidine/spermine N^1-acetyl transferase (SSAT) and oxidation by polyamine oxidase (PAO). The direct oxidation of spermine to spermidine is catalyzed by spermine oxidase (SMO).

and then activated by AdoMet decarboxylase (AdoMetDC; EC 4.1.1.50). The biosynthetic pathway is controlled by the key enzymes ODC and AdoMetDC. The reconversion of spermine via spermidine to putrescine is catalyzed by spermidine/spermine N^1-acetyltransferase (SSAT; EC 2.3.1.57) and polyamine oxidase (PAO,

EC 1.5.3.13); the flux of this interconversion pathway is controlled by the highly inducible SSAT. Additionally, spermine is converted directly to spermidine by spermine oxidase (SMO; EC 1.5.3.16) [8, 31–34].

Comparative Polyamine Metabolism in the Apicomplexa

To date, information regarding polyamine synthesis and metabolism in apicomplexan parasites remains limited and mainly restricted to *Plasmodium* spp., *T. gondii*, *C. parvum*, and *E. tenella*. Like other eukaryotes, the Apicomplexa contain the polyamines putrescine, spermidine, and spermine, although most of the verified information has been derived from studies with *P. falciparum*. Das Gupta *et al.* [35] analyzed the polyamine levels in infected erythrocytes at different developmental stages of *P. falciparum*. From rings to schizonts, the polyamine concentrations were increased up to 15-fold (Figure 6.2a), thus confirming previously reported data [27]. In all developmental stages within the erythrocyte, spermidine represents the major polyamine, followed by putrescine and minor amounts of spermine [35]. Recently, Teng *et al.* [36] analyzed the polyamine pattern in *P. falciparum*, isolated from its host cell. Here, the concentrations in the parasites were determined as 2.9, 5.7, and 0.5 mM for putrescine, spermidine, and spermine, respectively, which differed slightly from former results [35] where the respective concentrations were 1.4, 3.2, and 0.12 mM, taking into consideration the relative volume ratios of parasite (1/3) and host cell (2/3), as calculated according to Ref. [37]. An analysis of the infected erythrocytes led to the conclusion that the polyamines were predominantly ascribed to the parasite compartment (Figure 6.2b). The increased levels of putrescine and spermidine in the host compartment, when compared to noninfected erythrocytes, were derived most likely from the parasite, as the erythrocyte lacks a polyamine biosynthetic process [35].

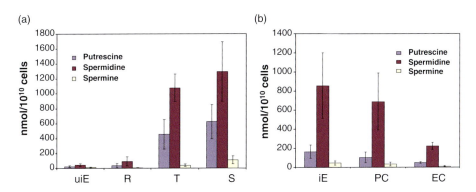

Figure 6.2 Distribution of polyamines in *P. falciparum*-infected erythrocytes. (a) Stage-specific accumulation of putrescine, spermidine, and spermine in ring (R), trophozoite (T), and schizont (S), compared to uninfected erythrocytes (uiE).;(b) Putrescine, spermidine, and spermine distribution in infected erythrocytes (iE). PC, parasite compartment; EC, erythrocyte compartment.

In contrast to the polyamine pattern in *P. falciparum*, where only traces of spermine were identified, an analysis of *T. gondii* tachyzoites revealed spermine to be the main polyamine component, followed by spermidine and putrescine [38]. This situation is undoubtedly caused by the parasite's auxotrophic nature, and by its divergent polyamine metabolism. In both *C. parvum* and *E. tenella* sporozoites, spermidine represented the main polyamine, but was closely followed by spermine and putrescine. Thus, in contrast to the trace amounts of spermine present in *Plasmodium*, both coccidian parasites showed spermine contents comparable to that of spermidine, although the putrescine level was quite low in *C. parvum* [39].

Plasmodial ODC activity was detected more than two decades ago, and later characterized on the enzyme level; this was followed by a description of the activity of AdoMetDC [40–42] and, only recently, also of spermidine synthase [43]. Consequently, polyamine synthesis in *Plasmodium* follows a common route from arginine to spermidine, comparable to the biosynthetic pathway in mammals. Both, ODC and ADC activities were detected in the coccidian parasites *E. tenella* and *C. parvum*, and characterized by specific sensitivities toward DFMO and alpha-difluoromethylarginine (DFMA), respectively [39]. The ODC activity in *E. tenella* extracts indicated a polyamine synthesis similar to that of the plasmodia [38]. Instead of the ODC reaction, the alternative pathway from arginine to putrescine (which is well known from plants and bacteria) is present in *C. parvum*, and is thus unique for apicomplexan parasites [38, 39, 44]. In detail, arginine is decarboxylated by the ADC to agmatine, which is subsequently hydrolyzed to putrescine. Although there is no evidence for ADC activity in *Plasmodium*, an additional bacterial-type ODC (which is upregulated in *P. berghei* sporozoites) was previously predicted by Matuschewski *et al.* [45]. Inhibition studies of the plasmodial AdoMetDC/ODC led to a transcriptional feedback response [46] and to an increased transcription of the bacterial ODC-like *P. falciparum* homolog, suggesting an involvement of this protein in polyamine metabolism [47]. However, the recombinantly expressed enzyme exhibited substrate specificity for lysine, and was hence termed lysine decarboxylase (LDC; EC 4.1.1.18) [3]. Nevertheless, the function of the LDC in the metabolism of *Plasmodium* remains to be elucidated.

In addition to the *de novo* synthesis of putrescine via agmatine, there is strong evidence that *C. parvum* is able to scavenge and interconvert host polyamines by the combined action of SSAT and APAO [48]. To demonstrate that the scavenging of host polyamines occurs via uptake and conversion as the major pathway of polyamine supply, Yarlett *et al.* [48] identified, cloned, and recombinantly expressed the *Cp*SSAT gene for molecular and biochemical characterization. Although the sequence similarity to the human SSAT was quite low, the enzyme showed a substrate preference for spermine before spermidine, and provided the N-acetylated polyamines for oxidation by the PAO with specificity for acetylated polyamines. Hence, *Cp*SSAT was proposed as a target for chemotherapy, and polyamine *cis*-analogues having unsaturated central carbons were demonstrated as potent enzyme inhibitors that could prevent and cure experimental *C. parvum* infections in immunocompromised mice [48]. In *T. gondii*, which is reported as a polyamine auxotroph, neither ODC nor ADC activities were detectable, and gene homologs were not found in the genome database. Moreover, there was strong evidence for the uptake of spermine and its subsequent enzymatic conversion to spermidine and further to putrescine by SSAT and PAO, a polyamine metabolism similar to that found in *Cryptosporidium* [38]. Beyond that, *T. gondii* was also shown to possess a high-affinity transporter for putrescine, indicating an ability to scavenge host-derived putrescine [49]. Sequence homologies to PAO or SSAT were not identified in the *P. falciparum* genome data base [3], indicating an absence of recycling facilities as known from the apicomplexan relatives *T. gondii* and *C. parvum* as well as from mammals [38, 48]. As opposed to *C. parvum* and *T. gondii*, where the interconversion of scavenged spermine to spermidine is the major source of polyamine supply, *P. falciparum* depends on the biosynthesis from arginine via ornithine to putrescine and spermidine [43, 50, 51]. Nevertheless, in order to meet the polyamine demands, transport systems into the cell are important devices. Singh *et al.* [52] showed that *P. knowlesi*-infected erythrocytes acquire polyamines via a putrescine transport system, which leads to an increased uptake in a saturable manner. Ramya *et al.* [53] confirmed the uptake of putrescine and spermidine into *P. falciparum*-infected erythrocytes.

For human erythrocytes, a multicomponent putrescine uptake system involving simple diffusion processes has been described [54]. These data also emphasized the possibility of multiple spermidine uptake mechanisms that were not mediated by simple diffusion. Uptake assays with *P. falciparum* showed the influx of putrescine in infected erythrocytes to be about 60% higher than in uninfected erythrocytes. In both infected and noninfected red blood cells, the addition of spermidine or spermine showed no significant influence on putrescine uptake, indicating its specificity for putrescine [3]. In other experiments, the uptake and metabolism of polyamines in infected erythrocytes were clearly demonstrated for putrescine, but largely denied for spermidine and spermine. It was also shown that exogenous putrescine abolishes the blockade of putrescine synthesis caused by various specific ODC inhibitors in cultured *P. falciparum*, whereas neither spermidine nor spermine were found to compensate for the growth inhibition at considerable rates [35]. Although it is generally accepted that supplementation with putrescine completely compensates for the growth inhibition of *P. falciparum* cultures, the reversal by spermidine and spermine remains controversial. Assaraf *et al.* [55] identified a reversal effect of the DFMO inhibition by adding exogenous spermidine, but they denied such an effect by spermine. In contrast thereto, Bitonti *et al.* [29] as well as Wright *et al.* [56] reported that spermine also reverses the antiproliferative effect. Thus, it remains open to discussion whether or not these strikingly variable results are due to artificial *in vitro* conditions. Nevertheless, all of the data acquired seem to confirm the idea of polyamine uptake and interconversion being an important alternative to the *de novo* synthesis in various apicomplexan parasites.

Only in the case of *P. falciparum* the total set of enzymes involved in the polyamine synthesis pathway was cloned, recombinantly expressed, and biochemically characterized. Consequently, special attention should be drawn to the discovery of certain peculiarities in the plasmodial metabolism. The bifunctional protein AdoMetDC/ODC is a unique feature [51, 57] (Figure 6.3), while the somewhat promiscuous role

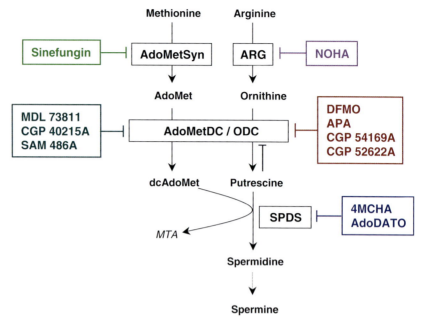

Figure 6.3 Inhibitors attacking polyamine synthesis enzymes in *P. falciparum*, (enzyme abbreviations as in Figure 6.1). Sinefungine inhibits AdoMetSyn, and *N*-hydroxy-L-arginine (NOHA) acts as an ARG inhibitor. The unique bifunctional organization of the AdoMetDC and the ODC as AdoMetDC/ODC is depicted. The AdoMetDC is blocked by compounds MDL 73811, CGP 40215A, and SAM 486A. The ODC inhibitors are alpha-difluoromethylornithine (DFMO), 3-aminooxy-1-propane (APA), and its derivatives CGP 54169A and CGP 52622A. Spermidine (and spermine) synthesis by SPDS is blocked by *trans*-1,4-methyl-cyclohexylamine (4MCHA) and *S*-adenosyl-1,8-diamino-3-thio octane (AdoDATO).

of the plasmodial SPDS, in accepting not only putrescine but also spermidine as substrate, is unusual [43].

Targets and Inhibitors

The Route from Arginine via Ornithine or Agmatine to Putrescine

Arginase The diamine putrescine is synthesized in two steps from arginine by the action of ARG and ODC, or alternatively by that of ADC and AGM (Figure 6.1). Both ARG and AGM belong to the family of ureo hydrolases, which are metal-dependent, share conserved motifs, and thus have a high degree of sequence homology. Only a single gene with similarity to this enzyme family of other organisms was found in the genome of *P. falciparum*, and biochemical characterization clearly ascribed the recombinant enzyme as an arginase [50]. Accordingly, the production of urea and ornithine emphasized the absence of an alternative pathway from arginine via agmatine to putrescine [3]. The main function of the plasmodial ARG is proposed to be not only the production of ornithine for the synthesis of polyamines, but also the regulation of cellular arginine and ornithine levels, as reported for the human ARG in order to control the production of nitric oxide (NO). The origin of ornithine for the putrescine synthesis in *Plasmodium* was shown also to derive from the action of the ornithine aminotransferase, which primarily controls the catabolism of excess ornithine to proline and glutamate, but also catalyzes the reverse reaction from glutamate 5-semialdehyde to ornithine [58]. However, as the small amount of ornithine produced via this route would not explain the high production rate of polyamines, other sources of ornithine supply were proposed. The subsequent demonstration and biochemical characterization of ARG in *P. falciparum* solved this question [50]. Moreover, in a recent report the conversion of intracellular and extracellular arginine to ornithine by the plasmodial parasite was reported, while the systemic depletion of arginine levels was discussed as a factor in human malarial hypoargininemia associated with cerebral malarial pathogenesis. The crucial role of ARG in maintaining the balance of NO and ornithine levels emphasizes its potential for therapeutic intervention; however, disruption of the **PbARG** gene in the rodent malarial parasite *P. berghei* did not affect viability [59]. Furthermore, whilst *N*-hydroxy-L-arginine (NOHA; an intermediate of NO synthesis) inhibited the **PfARG** *in vitro*, the effect on cultured parasites was only marginal. The plasmodial enzyme was shown to differ from its homologs mainly in terms of its high dependency on the metal cofactor for structure; this activity was recognized as being a feature that rendered the enzyme potentially vulnerable [50, 60].

The Reaction Leading to S-Adenosylmethionine (AdoMet)

AdoMet Synthetase (AdoMetSyn)   The second precursor for the polyamine synthesis besides ornithine is generated from methionine and ATP by the action of AdoMet-Syn. Although AdoMet is a donor for methylation reactions, it also participates in the trans-sulfuration pathway and is particularly important for polyamine synthesis, because it provides the aminopropyl moiety for the synthesis of spermidine and spermine. To date, the gene and protein of the plasmodial AdoMetSyn have hardly been characterized, although Chiang *et al.* [61] have modeled its structure using the crystal structure of the dimeric *Escherichia coli* enzyme. More recently, when the recombinantly expressed plasmodial AdoMetSyn was characterized, the data obtained did not support the predicted dimeric structure, as the enzyme appeared to be a monomer. Moreover, the plasmodial enzyme differed from the mammalian form, in that it was not allosterically regulated by AdoMet [3]; the latter situation was also demonstrated for the leishmanial and trypanosomal enzymes [62]. The potential of plasmodial AdoMetSyn as a drug target has not yet been evaluated, although the antibiotic sinefungin, which is structurally related to AdoMet [63], is well known for its plasmodicidal activity [64]. Subsequently, Bachrach and coworkers discussed the mode of action of sinefungin in the depletion of polyamines in *P. falciparum*, followed by the blockade of DNA synthesis and growth at the trophozoite stage [65].

Synthesis of Spermidine and Spermine

AdoMetDC and ODC The bifunctional organization of AdoMetDC and ODC, located on a single polypeptide and organized as N- and C-terminal domains connected by a hinge region, was primarily shown for *P. falciparum* [51] and later confirmed for other *Plasmodium* species [66]. According to the respective dimeric and heterotetrameric structures of the mammalian ODC and AdoMetDC, *Plasmodium* holds a heterotetrameric enzyme complex consisting of two bifunctional polypeptides, which are post-translationally cleaved in the AdoMetDC domain for the formation of the essential pyruvoyl prosthetic group [51, 57]. Bifunctional proteins are not unusual in *Plasmodium*, and are thought to coordinate consecutive reactions, as shown in biosynthesis and metabolism of folate [67]. The biological advantage of the bifunctional organization of AdoMetDC and ODC cannot be explained by the metabolic channeling of substrates or domain–domain interactions, since experiments employing mutagenesis have revealed that each domain operates independently, without influencing the other [57]. This result was further supported by the enzymatic activity of the separately expressed AdoMetDC and ODC domains [57, 66, 68]. It appears advantageous to regulate the abundance of one transcript, and thus one protein of AdoMetDC/ODC, in order to synchronize the substrate supply of putrescine and AdoMet for spermidine formation. This finding is in agreement with the transcription peak of the AdoMetDC/ODC at the trophozoite stage, when the requirement for polyamines is high due to rapid parasite growth and the onset of DNA synthesis [51]. Striking differences have been identified between the parasite and host with regards to the regulation of AdoMetDC/ODC on the protein level. In contrast to the mammalian enzyme, the plasmodial AdoMetDC activity is not stimulated by putrescine and therefore lacks the regulatory mechanism that has been proposed for mammalian cells. Moreover, in contrast to the behavior of the mammalian ODC, the activity of the plasmodial enzyme is strongly inhibited by putrescine; indeed, such feedback permits the control of putrescine synthesis. However, when considering the low intracellular concentrations of decarboxylated AdoMet, the aminopropyl group donor was discussed as a possible limiting factor. It is also notable that the bifunctional AdoMetDC/ODC has a relatively long half-life, at least in comparison to the mammalian monofunctional homologs [57].

For many years, the targeting of ODC and AdoMetDC was recognized as the preferred strategy to inhibit polyamine biosynthesis in tumor cells and protozoan parasites. As the irreversible ODC inhibitor DFMO failed to cure experimental rodent malaria successfully, more powerful ODC inhibitors – including 3-aminooxy-1-propane (APA) and its analogs [69, 70] – were tested on the plasmodial enzyme and on cultured *P. falciparum*. Compared to the irreversible inhibitor DFMO, these competitive inhibitors revealed a 1000-fold stronger effect on cultured *P. falciparum*, which was reflected by a higher affinity toward the plasmodial ODC. Although APA and its analogs consistently depleted the polyamine levels in cultured parasites [35], they had a minimal effect on parasitemia in mice infected with *P. berghei* [3], which was a disappointing outcome in consideration of the drug's short half-life in the host.

Subsequently, Wright *et al.* [56] demonstrated a successful growth inhibition of cultured *P. falciparum* by MDL 73811, an irreversible inhibitor of AdoMetDC. By contrast, when the treatment of *P. berghei*-infected mice with MDL 73811 failed to reduce the parasite load, this was explained by the rapid clearance of the drug in mice. More encouraging results were achieved with CGP 40215A and CGP 48664A (also known as SAM 486A), which were designed within a set of competitive AdoMetDC inhibitors for tumor therapy. Both agents were found to be effective against African trypanosomiasis *in vitro* and *in vivo* [71, 72] and, when analyzed for antimalarial activity, both strongly inhibited the plasmodial AdoMetDC activity and the proliferation of cultured *P. falciparum*, with respective K_i- and IC_{50}-values in the low nanomolar and micromolar ranges [35].

When CGP 40215A was applied to *P. berghei*-infected mice, parasitemia was reduced and a significant curative effect observed [3]. Unfortunately, as the drug did not deplete spermidine levels in cultured *P. falciparum*, the primary target of this drug candidate remained unclear [35]. The second candidate, SAM 486A, underwent various clinical trials in multiple cancers, both as a single drug and in a combination therapy, but met with only limited success [7]. Nonetheless, SAM 486A was recently reported to have efficiently depleted spermidine levels. In a novel therapeutic strategy, HIV-1 replication was suppressed by a reduction in spermidine levels following SAM 486A treatment. This resulted in less hypusine modification and, thereby, less activation of the eukaryotic initiation factor 5A [73], an approach which has now also been proposed for malaria therapy [74, 75].

In conclusion, the structure and organization of the bifunctional protein, as well as its regulation, clearly distinguish the plasmodial AdoMetDC/ODC from the mammalian homologs, thus offering options for parasite-specific inhibition. Both domains, ODC and AdoMetDC, have been modeled and have provided insights into some novel features [76, 77], although the design of parasite-specific inhibitors still awaits a crystal structure, with neither the bifunctional protein nor one of its domains having been crystallized to date.

Spermidine Synthase (SPDS) In *Plasmodium*, SPDS appears to be a single-copy gene, and additional searches in the genome database, using known SPMS amino acid sequences as a query, have failed to identify a second aminopropyl transferase [43]. As with the AdoMetDC/ODC, SPDS is expressed at the late trophozoite stage, which enables an enhanced polyamine synthesis as a prerequisite for the process of schizogony. It should be noted that the plasmodial SPDS possesses an N-terminal extension that is found solely in the SPDS of certain plants, and which predicts a closer relationship to plants than to animal orthologs. Recombinant expression was achieved only after omitting the N terminus, thereby yielding proteins for biochemical studies [43] and crystal analyses [78]. Whilst SPDS preferably catalyzes the formation of spermidine, it can also accept spermidine as a substrate, leading to the formation of spermine (albeit to a lower extent of 15% compared to putrescine). This capacity to catalyze the formation of spermine was suggested as being responsible for the small, but significant, amounts of spermine identified in the erythrocytic stages of *Plasmodium*. Beyond that, the synthesis of spermine is obviously enhanced in cultured *P. falciparum* under conditions when putrescine levels are decreased by blocked ODC activity, a factor which supports the hypothesis of spermidine being an alternative substrate of SPDS [35, 43]. The targeting of SPDS appears to represent a very promising strategy for intervention in the polyamine metabolism of *Plasmodium*, as this apicomplexan parasite clearly lacks an efficient uptake and interconversion system to rescue polyamine depletion. Whereas its apicomplexan relatives *Cryptosporidium* and *Toxoplasma* scavenge host polyamines (thereby circumventing critically low spermidine levels), the malarial parasite depends on its own spermidine biosynthesis for survival and growth. SPDS is absolutely required, as shown by gene-deletion experiments in *Leishmania* and other lower eukaryotes [79–81], and by RNA interference (RNAi) downregulation of the trypanosomal SPDS [82]. Although, a final statement on the druggability of SPDS awaits knockdown, the crucial function of the plasmodial SPDS with regards to the parasite's supply of spermidine has clearly been validated via inhibition studies. For example, when *S*-adenosyl-1,8-diamino-3-thio-octane (AdoDato) impeded the plasmodial SPDS activity, the IC_{50} value obtained (8.5 μM; [78]) was surpassed by *trans*-1,4-methylcyclohexylamine (4MCHA), which inhibited the plasmodial SPDS activity sixfold more efficiently and provided a distinct blockade of the parasite's growth [43]. Remarkably, spermidine supplementation of the culture medium had no significant rescue effect on the parasite growth [43], which might be indicative of an inefficient uptake of spermidine. Previously, 4MCHA has been reported to drastically lower spermidine concentrations in treated hepatoma tissue culture (HTC) cells and rat tissues, but did not affect the growth rates to any marked degree [83–85]. The crystal structures of the plasmodial SPDS, apoenzyme and enzyme complexed with substrates, and also of the inhibitors AdoDATO and 4MCHA, have recently been resolved, and should ease the design of new inhibitors with an improved selectivity toward the parasite's SPDS [78, 86–88].

Conclusion

As polyamines are an absolute requirement for cell growth and proliferation, enzyme inhibitors have been used to analyze the "druggability" of polyamine metabolism in infectious diseases caused by apicomplexan parasites. Although a high potency at both the enzyme and cellular levels has been achieved with some of the compounds used, a general strategy to combat apicomplexan polyamine homeostasis cannot yet be proposed. Members of the phylum Apicomplexa rely on distinct sources of polyamines, whether via their own biosynthesis, via polyamine scavenging from the host, or a combination of both. In general, the polyamine metabolism of apicomplexan parasites is far less complex than that of the human host, a situation that might be advantageous for chemotherapeutic interventions, considering the restricted possibilities of overcoming shortages in parasite polyamine resources. In the case of the malarial parasite, a blockade of biosynthesis is highly warranted, since it appears that *Plasmodium* cannot fully overcome polyamine depletion by increasing its uptake. Nevertheless, the treatment of experimental rodent malarial infections with inhibitors of polyamine synthesis has not yet proved convincing, due most likely to a low bioavailability or a short plasma half-life of the drug candidates. Clearly, both of these drawbacks must be addressed in future studies.

References

1 Hamana, K. and Matsuzaki, S. (1992) Polyamines as a chemotaxonomic marker in bacterial systematics. *Crit. Rev. Microbiol.*, **18**, 261–283.

2 Cohen, S.S. (1998) *A Guide to the Polyamines*, Oxford University Press.

3 Müller, I.B., Das Gupta, R., Lüersen, K., Wrenger, C., and Walter, R.D. (2008) Assessing the polyamine metabolism of *Plasmodium falciparum* as chemotherapeutic target. *Mol. Biochem. Parasitol.*, **160**, 1–7.

4 Müller, S., Coombs, G.H., and Walter, R.D. (2001) Targeting polyamines of parasitic protozoa in chemotherapy. *Trends Parasitol.*, **17**, 242–249.

5 Clark, K., Niemand, J., Reeksting, S., Smit, S., van Brummelen, A.C., Williams, M., Louw, A.I., and Birkholtz, L. (2010) Functional consequences of perturbing polyamine metabolism in the malaria parasite, *Plasmodium falciparum*. *Amino Acids*, **38**, 633–644.

6 Seiler, N. (2003) Thirty years of polyamine-related approaches to cancer therapy. Retrospect and prospect. Part 1. Selective enzyme inhibitors. *Curr. Drug Targets*, **4**, 537–564.

7 Casero, R.A. Jr and Marton, L.J. (2007) Targeting polyamine metabolism and function in cancer and other hyperproliferative diseases. *Nat. Rev. Drug Discov.*, **6**, 373–390.

8 Casero, R.A., and Pegg, A.E. (2009) Polyamine catabolism and disease. *Biochem. J.*, **421**, 323–338.

9 Marton, L.J. and Pegg, A.E. (1995) Polyamines as targets for therapeutic intervention. *Annu. Rev. Pharmacol. Toxicol.*, **35**, 55–91.

10 Friedman, J.H. (2001) Thank God for rich women with mustaches. *Med. Health R. I.*, **84**, 222–223.

11 Hickman, J.G., Huber, F., and Palmisano, M. (2001) Human dermal safety studies with eflornithine HCl 13.9% cream (Vaniqa), a novel treatment for excessive facial hair. *Curr. Med. Res. Opin.*, **16**, 235–244.

12 Bacchi, C.J., Nathan, H.C., Hutner, S.H., McCann, P.P., and Sjoerdsma, A. (1980) Polyamine metabolism: a potential therapeutic target in trypanosomes. *Science*, **210**, 332–334.

13 Burri, C. and Brun, R. (2003) Eflornithine for the treatment of human African trypanosomiasis. *Parasitol. Res.*, **90** (Supp 1), S49–S52.

14 Sjoerdsma, A. and Schechter, P.J. (1999) Eflornithine for African sleeping sickness. *Lancet*, **354**, 254.

15 Fairlamb, A.H., Blackburn, P., Ulrich, P., Chait, B.T., and Cerami, A. (1985) Trypanothione: a novel bis(glutathionyl) spermidine cofactor for glutathione reductase in trypanosomatids. *Science*, **227**, 1485–1487.

16 Müller, S., Liebau, E., Walter, R.D., and Krauth-Siegel, R.L. (2003) Thiol-based redox metabolism of protozoan parasites. *Trends Parasitol.*, **19**, 320–328.

17 McCann, P.P., Bacchi, C.J., Clarkson, A.B. Jr, Bey, P., Sjoerdsma, A., Schecter, P.J., Walzer, P.D., and Barlow, J.L. (1986) Inhibition of polyamine biosynthesis by alpha-difluoromethylornithine in African trypanosomes and *Pneumocystis carinii* as a basis of chemotherapy: biochemical and clinical aspects. *Am. J. Trop. Med. Hyg.*, **35**, 1153–1156.

18 McCann, P.P. and Pegg, A.E. (1992) Ornithine decarboxylase as an enzyme target for therapy. *Pharmacol. Ther.*, **54**, 195–215.

19 Hanson, W.L., Bradford, M.M., Chapman, W.L. Jr, Waits, V.B., McCann, P.P., and Sjoerdsma, A. (1982) alpha-Difluoromethylornithine: a promising lead for preventive

chemotherapy for coccidiosis. *Am. J. Vet. Res.*, **43**, 1651–1653.

20 Mukhopadhyay, R. and Madhubala, R. (1993) Effect of a bis(benzyl)polyamine analogue, and DL-alpha-difluoromethylornithine on parasite suppression and cellular polyamine levels in golden hamster during *Leishmania donovani* infection. *Pharmacol. Res.*, **28**, 359–365.

21 Gillin, F.D., Reiner, D.S., and McCann, P.P. (1984) Inhibition of growth of *Giardia lamblia* by difluoromethylornithine, a specific inhibitor of polyamine biosynthesis. *J. Protozool.*, **31**, 161–163.

22 Kim, B.G., McCann, P.P., and Byers, T.J. (1987) Inhibition of multiplication in *Acanthamoeba castellanii* by specific inhibitors of ornithine decarboxylase. *J. Protozool.*, **34**, 264–266.

23 Gillet, J.M., Charlier, J., Bone, G., and Mulamba, P.L. (1983) *Plasmodium berghei*: inhibition of the sporogonous cycle by alpha-difluoromethylornithine. *Exp. Parasitol.*, **56**, 190–193.

24 Gillet, J., Bone, G., Lowa, P., Charlier, J., Rona, A.M., and Schechter, P.J. (1986) Alpha-Difluoromethylornithine induces protective immunity in mice inoculated with *Plasmodium berghei* sporozoites. *Trans. R. Soc. Trop. Med. Hyg.*, **80**, 236–239.

25 Gillet, J.M., Bone, G., and Herman, F. (1982) Inhibitory action of alpha-difluoromethylornithine on rodent malaria (*Plasmodium berghei*). *Trans. R. Soc. Trop. Med. Hyg.*, **76**, 776–777.

26 Lowa, P.M., Gillet, J., Bone, G., and Schechter, P.J. (1986) alpha-Difluoromethylornithine inhibits the first part of exoerythrocytic schizogony of *Plasmodium berghei* in rodents. *Ann. Soc. Belg. Med. Trop.*, **66**, 301–308.

27 Assaraf, Y.G., Golenser, J., Spira, D.T., and Bachrach, U. (1984) Polyamine levels and the activity of their biosynthetic enzymes in human erythrocytes infected with the malarial parasite, *Plasmodium falciparum*. *Biochem. J.*, **222**, 815–819.

28 Whaun, J.M. and Brown, N.D. (1985) Ornithine decarboxylase inhibition and the malaria-infected red cell:

a model for polyamine metabolism and growth. *J. Pharmacol. Exp. Ther.*, **233**, 507–511.

29 Bitonti, A.J., McCann, P.P., and Sjoerdsma, A. (1987) Plasmodium *falciparum* and *Plasmodium berghei*: effects of ornithine decarboxylase inhibitors on erythrocytic schizogony. *Exp. Parasitol.*, **64**, 237–243.

30 Hollingdale, M.R., McCann, P.P., and Sjoerdsma, A. (1985) Plasmodium *berghei*: inhibitors of ornithine decarboxylase block exoerythrocytic schizogony. *Exp. Parasitol.*, **60**, 111–117.

31 Heby, O. and Persson, L. (1990) Molecular genetics of polyamine synthesis in eukaryotic cells. *Trends Biochem. Sci.*, **15**, 153–158.

32 Pegg, A.E. and McCann, P.P. (1982) Polyamine metabolism and function. *Am. J. Physiol.*, **243**, C212–C221.

33 Seiler, N. and Heby, O. (1988) Regulation of cellular polyamines in mammals. *Acta Biochim. Biophys. Hung.*, **23**, 1–35.

34 Tabor, H. and Tabor, C.W. (1964) Spermidine, spermine, and related amines. *Pharmacol. Rev.*, **16**, 245–300.

35 Das Gupta, R., Krause-Ihle, T., Bergmann, B., Müller, I.B., Khomutov, A.R., Müller, S., Walter, R.D., and Lüersen, K. (2005) 3-Aminooxy-1-aminopropane and derivatives have an antiproliferative effect on cultured *Plasmodium falciparum* by decreasing intracellular polyamine concentrations. *Antimicrob. Agents Chemother.*, **49**, 2857–2864.

36 Teng, R., Junankar, P.R., Bubb, W.A., Rae, C., Mercier, P., and Kirk, K. (2009) Metabolite profiling of the intraerythrocytic malaria parasite *Plasmodium falciparum* by (1)H NMR spectroscopy. *NMR Biomed.*, **22**, 292–302.

37 Yayon, A., Van de Waa, J.A., Yayon, M., Geary, T.G., and Jensen, J.B. (1983) Stage-dependent effects of chloroquine on *Plasmodium falciparum in vitro*. *J. Protozool.*, **30**, 642–647.

38 Cook, T., Roos, D., Morada, M., Zhu, G., Keithly, J.S., Feagin, J.E., Wu, G., and Yarlett, N. (2007) Divergent polyamine metabolism in the Apicomplexa. *Microbiology*, **153**, 1123–1130.

39 Keithly, J.S., Zhu, G., Upton, S.J., Woods, K.M., Martinez, M.P., and Yarlett, N. (1997) Polyamine biosynthesis in *Cryptosporidium parvum* and its implications for chemotherapy. *Mol. Biochem. Parasitol.*, **88**, 35–42.

40 Assaraf, Y.G., Kahana, C., Spira, D.T., and Bachrach, U. (1988) Plasmodium *falciparum*: purification, properties, and immunochemical study of ornithine decarboxylase, the key enzyme in polyamine biosynthesis. *Exp. Parasitol.*, **67**, 20–30.

41 Konigk, E. and Putfarken, B. (1985) Ornithine decarboxylase of *Plasmodium falciparum*: a peak-function enzyme and its inhibition by chloroquine. *Trop. Med. Parasitol.*, **36**, 81–84.

42 Rathaur, S. and Walter, R.D. (1987) Plasmodium *falciparum*: S-adenosyl-L-methionine decarboxylase. *Exp. Parasitol.*, **63**, 227–232.

43 Haider, N., Eschbach, M.L., Dias Sde, S., Gilberger, T.W., Walter, R.D., and Luersen, K. (2005) The spermidine synthase of the malaria parasite *Plasmodium falciparum*: molecular and biochemical characterisation of the polyamine synthesis enzyme. *Mol. Biochem. Parasitol.*, **142**, 224–236.

44 Yarlett, N., Martinez, M.P., Zhu, G., Keithly, J.S., Woods, K., and Upton, S.J. (1996) Cryptosporidium *parvum*: polyamine biosynthesis from agmatine. *J. Eukaryot. Microbiol.*, **43**, 73S.

45 Matuschewski, K., Ross, J., Brown, S.M., Kaiser, K., Nussenzweig, V., and Kappe, S.H. (2002) Infectivity-associated changes in the transcriptional repertoire of the malaria parasite sporozoite stage. *J. Biol. Chem.*, **277**, 41948–41953.

46 Clark, K., Dhoogra, M., Louw, A.I., and Birkholtz, L.M. (2008) Transcriptional responses of *Plasmodium falciparum* to alpha-difluoromethylornithine-induced polyamine depletion. *Biol. Chem.*, **389**, 111–125.

47 van Brummelen, A.C., Olszewski, K.L., Wilinski, D., Llinas, M., Louw, A.I., and Birkholtz, L.M. (2009) Co-inhibition of *Plasmodium falciparum* S-adenosylmethionine decarboxylase/ornithine decarboxylase reveals

perturbation-specific compensatory mechanisms by transcriptome, proteome, and metabolome analyses. *J. Biol. Chem.*, **284**, 4635–4646.

48 Yarlett, N., Wu, G., Waters, W.R., Harp, J.A., Wannemuehler, M.J., Morada, M., Athanasopoulos, D., Martinez, M.P., Upton, S.J., Marton, L.J., and Frydman, B.J. (2007) Cryptosporidium *parvum* spermidine/spermine N1-acetyltransferase exhibits different characteristics from the host enzyme. *Mol. Biochem. Parasitol.*, **152**, 170–180.

49 Seabra, S.H., DaMatta, R.A., de Mello, F.G., and de Souza, W. (2004) Endogenous polyamine levels in macrophages is sufficient to support growth of *Toxoplasma gondii*. *J. Parasitol.*, **90**, 455–460.

50 Müller, I.B., Walter, R.D., and Wrenger, C. (2005) Structural metal dependency of the arginase from the human malaria parasite *Plasmodium falciparum*. *Biol. Chem.*, **386**, 117–126.

51 Müller, S., Da'dara, A., Lüersen, K., Wrenger, C., Das Gupta, R., Madhubala, R., and Walter, R.D. (2000) In the human malaria parasite *Plasmodium falciparum*, polyamines are synthesized by a bifunctional ornithine decarboxylase, S-adenosylmethionine decarboxylase. *J. Biol. Chem.*, **275**, 8097–8102.

52 Singh, S., Puri, S.K., Singh, S.K., Srivastava, R., Gupta, R.C., and Pandey, V.C. (1997) Characterization of simian malarial parasite (*Plasmodium knowlesi*)-induced putrescine transport in rhesus monkey erythrocytes. A novel putrescine conjugate arrests *in vitro* growth of simian malarial parasite (*Plasmodium knowlesi*) and cures multidrug resistant murine malaria (*Plasmodium yoelii*) infection *in vivo*. *J. Biol. Chem.*, **272**, 13506–13511.

53 Ramya, T.N., Surolia, N., and Surolia, A. (2006) Polyamine synthesis and salvage pathways in the malaria parasite *Plasmodium falciparum*. *Biochem. Biophys. Res. Commun.*, **348**, 579–584.

54 Fukumoto, G.H. and Byus, C.V. (1996) A kinetic characterization of putrescine and spermidine uptake and export in human erythrocytes. *Biochim. Biophys. Acta*, **1282**, 48–56.

55 Assaraf, Y.G., Golenser, J., Spira, D.T., Messer, G., and Bachrach, U. (1987) Cytostatic effect of DL-alpha-difluoromethylornithine against *Plasmodium falciparum* and its reversal by diamines and spermidine. *Parasitol. Res.*, **73**, 313–318.

56 Wright, P.S., Byers, T.L., Cross-Doersen, D.E., McCann, P.P., and Bitonti, A.J. (1991) Irreversible inhibition of S-adenosylmethionine decarboxylase in *Plasmodium falciparum*-infected erythrocytes: growth inhibition *in vitro*. *Biochem. Pharmacol.*, **41**, 1713–1718.

57 Wrenger, C., Lüersen, K., Krause, T., Müller, S., and Walter, R.D. (2001) The *Plasmodium falciparum* bifunctional ornithine decarboxylase, S-adenosyl-L-methionine decarboxylase, enables a well balanced polyamine synthesis without domain-domain interaction. *J. Biol. Chem.*, **276**, 29651–29656.

58 Gafan, C., Wilson, J., Berger, L.C., and Berger, B.J. (2001) Characterization of the ornithine aminotransferase from *Plasmodium falciparum. Mol. Biochem. Parasitol.*, **118**, 1–10.

59 Olszewski, K.L., Morrisey, J.M., Wilinski, D., Burns, J.M., Vaidya, A.B., Rabinowitz, J.D., and Llinas, M. (2009) Host-parasite interactions revealed by *Plasmodium falciparum* metabolomics. *Cell Host Microbe*, **5**, 191–199.

60 Wells, G.A., Muller, I.B., Wrenger, C., and Louw, A.I. (2009) The activity of *Plasmodium falciparum* arginase is mediated by a novel inter-monomer salt-bridge between Glu295-Arg404. *FEBS J.*, **276**, 3517–3530.

61 Chiang, P.K., Chamberlin, M.E., Nicholson, D., Soubes, S., Su, X., Subramanian, G., Lanar, D.E., Prigge, S.T., Scovill, J.P., Miller, L.H., and Chou, J.Y. (1999) Molecular characterization of *Plasmodium falciparum* S-adenosylmethionine synthetase. *Biochem. J.*, **344** (Pt 2), 571–576.

62 Reguera, R.M., Balana-Fouce, R., Perez-Pertejo, Y., Fernandez, F.J., Garcia-Estrada, C., Cubria, J.C., Ordonez, C., and Ordonez, D. (2002) Cloning expression and characterization of methionine adenosyltransferase in *Leishmania infantum* promastigotes. *J. Biol. Chem.*, **277**, 3158–3167.

63 Perez-Pertejo, Y., Reguera, R.M., Ordonez, D., and Balana-Fouce, R. (2006) Characterization of a methionine adenosyltransferase over-expressing strain in the trypanosomatid *Leishmania donovani. Biochim. Biophys. Acta*, **1760**, 10–19.

64 Trager, W., Tershakovec, M., Chiang, P.K., and Cantoni, G.L. (1980) Plasmodium *falciparum*: antimalarial activity in culture of sinefungin and other methylation inhibitors. *Exp. Parasitol.*, **50**, 83–89.

65 Messika, E., Golenser, J., Abu-Elheiga, L., Robert-Gero, M., Lederer, E., and Bachrach, U. (1990) Effect of sinefungin on macromolecular biosynthesis and cell cycle of *Plasmodium falciparum. Trop. Med. Parasitol.*, **41**, 273–278.

66 Birkholtz, L.M., Wrenger, C., Joubert, F., Wells, G.A., Walter, R.D., and Louw, A.I. (2004) Parasite-specific inserts in the bifunctional S-adenosylmethionine decarboxylase/ornithine decarboxylase of *Plasmodium falciparum* modulate catalytic activities and domain interactions. *Biochem. J.*, **377**, 439–448.

67 Müller, I.B., Hyde, J.E., and Wrenger, C. (2010) Vitamin B metabolism in *Plasmodium falciparum* as a source of drug targets. *Trends Parasitol.*, **26**, 35–43.

68 Krause, T., Luersen, K., Wrenger, C., Gilberger, T.W., Muller, S., and Walter, R.D. (2000) The ornithine decarboxylase domain of the bifunctional ornithine decarboxylase/S-adenosylmethionine decarboxylase of *Plasmodium falciparum*: recombinant expression and catalytic properties of two different constructs. *Biochem. J.*, **352** (Pt 2), 287–292.

69 Khomutov, R.M., Hyvonen, T., Karvonen, E., Kauppinen, L., Paalanen, T., Paulin, L., Eloranta, T., Pajula, R.L., Andersson, L.C., and Poso, H. (1985) 1-Aminooxy-3-aminopropane, a new and potent inhibitor of polyamine biosynthesis that inhibits ornithine decarboxylase, adenosylmethionine decarboxylase and spermidine synthase. *Biochem. Biophys. Res. Commun.*, **130**, 596–602.

70 Stanek, J., Frei, J., Mett, H., Schneider, P., and Regenass, U. (1992) 2-substituted 3-(aminooxy)propanamines as inhibitors of ornithine decarboxylase: synthesis and biological activity. *J. Med. Chem.*, **35**, 1339–1344.

71 Bacchi, C.J., Brun, R., Croft, S.L., Alicea, K., and Buhler, Y. (1996) In vivo trypanocidal activities of new S-adenosylmethionine decarboxylase inhibitors. *Antimicrob. Agents Chemother.*, **40**, 1448–1453.

72 Brun, R., Buhler, Y., Sandmeier, U., Kaminsky, R., Bacchi, C.J., Rattendi, D., Lane, S., Croft, S.L., Snowdon, D., Yardley, V., Caravatti, G., Frei, J., Stanek, J., and Mett, H. (1996) In vitro trypanocidal activities of new S-adenosylmethionine decarboxylase inhibitors. *Antimicrob. Agents Chemother.*, **40**, 1442–1447.

73 Schäfer, B., Hauber, I., Bunk, A., Heukeshoven, J., Dusedau, A., Bevec, D., and Hauber, J. (2006) Inhibition of multidrug-resistant HIV-1 by interference with cellular S-adenosylmethionine decarboxylase activity. *J. Infect. Dis.*, **194**, 740–750.

74 Blavid, R., Kusch, P., Hauber, J., Eschweiler, U., Sarite, S.R., Specht, S., Deininger, S., Hoerauf, A., and Kaiser, A. (2010) Down-regulation of hypusine biosynthesis in plasmodium by inhibition of S-adenosyl-methionine-decarboxylase. *Amino Acids*, **38**, 461–469.

75 Kaiser, A., Gottwald, A., Wiersch, C., Lindenthal, B., Maier, W., and Seitz, H.M. (2001) Effect of drugs inhibiting spermidine biosynthesis and metabolism on the in vitro development of *Plasmodium falciparum*. *Parasitol. Res.*, **87**, 963–972.

76 Birkholtz, L., Joubert, F., Neitz, A.W., and Louw, A.I. (2003) Comparative properties of a three-dimensional model of *Plasmodium falciparum* ornithine decarboxylase. *Proteins*, **50**, 464–473.

77 Wells, G.A., Birkholtz, L.M., Joubert, F., Walter, R.D., and Louw, A.I. (2006) Novel properties of malarial S-adenosylmethionine decarboxylase as revealed by structural modelling. *J. Mol. Graph. Model.*, **24**, 307–318.

78 Dufe, V.T., Qiu, W., Müller, I.B., Hui, R., Walter, R.D., and Al-Karadaghi, S. (2007) Crystal structure of *Plasmodium falciparum* spermidine synthase in complex with the substrate decarboxylated S-adenosylmethionine and the potent inhibitors 4MCHA and AdoDATO. *J. Mol. Biol.*, **373**, 167–177.

79 Roberts, S.C., Jiang, Y., Jardim, A., Carter, N.S., Heby, O., and Ullman, B. (2001) Genetic analysis of spermidine synthase from *Leishmania donovani*. *Mol. Biochem. Parasitol.*, **115**, 217–226.

80 Hamasaki-Katagiri, N., Tabor, C.W., and Tabor, H. (1997) Spermidine biosynthesis in *Saccharomyces cerevisiae*: polyamine requirement of a null mutant of the SPE3 gene (spermidine synthase). *Gene*, **187**, 35–43.

81 Jin, Y., Bok, J.W., Guzman-de-Pena, D., and Keller, N.P. (2002) Requirement of spermidine for developmental transitions in *Aspergillus nidulans*. *Mol. Microbiol.*, **46**, 801–812.

82 Taylor, M.C., Kaur, H., Blessington, B., Kelly, J.M., and Wilkinson, S.R. (2008) Validation of spermidine synthase as a drug target in African trypanosomes. *Biochem. J.*, **409**, 563–569.

83 Beppu, T., Shirahata, A., Takahashi, N., Hosoda, H., and Samejima, K. (1995) Specific depletion of spermidine and spermine in HTC cells treated with inhibitors of aminopropyltransferases. *J. Biochem. (Tokyo)*, **117**, 339–345.

84 Shirahata, A., Morohohi, T., Fukai, M., Akatsu, S., and Samejima, K. (1991) Putrescine or spermidine binding site of aminopropyltransferases and competitive inhibitors. *Biochem. Pharmacol.*, **41**, 205–212.

85 Shirahata, A., Takahashi, N., Beppu, T., Hosoda, H., and Samejima, K. (1993) Effects of inhibitors of spermidine synthase and spermine synthase on polyamine synthesis in rat tissues. *Biochem. Pharmacol.*, **45**, 1897–1903.

86 Burger, P.B., Birkholtz, L.M., Joubert, F., Haider, N., Walter, R.D., and Louw, A.I. (2007) Structural and mechanistic insights into the action of *Plasmodium falciparum* spermidine synthase. *Bioorg. Med. Chem.*, **15**, 1628–1637.

87 Vedadi, M., Lew, J., Artz, J., Amani, M., Zhao, Y., Dong, A., Wasney, G.A., Gao, M., Hills, T., Brokx, S., Qiu, W., Sharma, S., Diassiti, A., Alam, Z., Melone, M., Mulichak, A., Wernimont, A., Bray, J., Loppnau, P., Plotnikova, O., Newberry, K., Sundararajan, E., Houston, S., Walker, J., Tempel, W., Bochkarev, A., Kozieradzki, I., Edwards, A., Arrowsmith, C., Roos, D., Kain, K., and Hui, R. (2007) Genome-scale protein expression and structural biology of *Plasmodium falciparum* and related Apicomplexan organisms. *Mol. Biochem. Parasitol.*, **151**, 100–110.

88 Jacobsson, M., Garedal, M., Schultz, J., and Karlen, A. (2008) Identification of *Plasmodium falciparum* spermidine synthase active site binders through structure-based virtual screening. *J. Med. Chem.*, **51**, 2777–2786.

7

The Reducing Milieu of Parasitized Cells as a Target of Antimalarial Agents: Methylene Blue as an Ethical Drug

*Peter Meissner, Heike Adler, Denis Kasozi, Karin Fritz-Wolf, and R. Heiner Schirmer**

Abstract

There are five species of the apicomplexan genus *Plasmodium* which cause human malaria. The diseases are distinct entities, and may require different drugs for their treatment. For the different intervention goals, specific parasitized cells and tissues must be considered as targets of the antimalarial agents. As suggested by large-scale *experimenta naturae*, such as the effects of pro-oxidant glucose-6-phosphate dehydrogenase (G6PD) alleles and pro-oxidant foodstuffs, the compromised antioxidant capacity of erythrocytes protects against severe malaria. One mechanism underlying this protection is the IgG-mediated, early phagocytosis of parasitized erythrocytes. As illustrated here for glutathione reductase (GR) deficiency, this mechanism cannot be observed in cultures of parasitized erythrocytes. Pro-oxidant agents such as the GR inhibitor and subversive substrate methylene blue (MB), dapsone, or primaquine most likely result in pharmacological phenocopies of natural protection mechanisms.

Introduction

Drugs used to treat malaria – a disease of the poor – should be also developed for ethical reasons. By using methylene blue (MB) as an example, it is possible to illustrate Sir James Black's statement that, [The] "...most fruitful basis for the discovery of a new drug is to start with an old drug." Recently, MB has been shown to be an analog of the bacterial redox pigment and quorum sensor, pyocyanin, which might explain why it behaves like a natural compound shaped for biological effects and biological compatibility in evolution. MB is active against both schizonts and gametocytes, and should be included in combination therapies against malaria. In this chapter, the details are reported of recent efforts to develop an MB-based drug combination to treat pediatric malaria, not only via clinical trials but also through anthropological and pharmacological studies.

*Corresponding author

Apicomplexan Parasites. Edited by Katja Becker
Copyright © 2011 WILEY-VCH Verlag GmbH & Co. KGaA, Weinheim
ISBN: 978-3-527-32731-7

The Redox Milieu of Parasite-Bearing Cells as a Natural Target

Human malaria is caused by a handful of different *Plasmodium* species; simultaneous infections by two or more *Plasmodium* strains, or even species, are not uncommon in tropical countries [1]. Likewise, two other apicomplexan protozoal parasites, *Babesia microti* and *Babesia divergens* [2], also cause a disease in humans that resembles malaria.

Cells and tissues carrying malarial parasites are most sensitive to oxidative stress, a situation confirmed by *experimenta naturae*, which show that persons in whom the erythrocytes are exposed to pro-oxidative conditions are protected from severe forms of malaria. The basis of this pro-oxidative situation may be a genetic disposition, such as glucose-6-phosphate dehydrogenase (G6PD) deficiency, the intake of foodstuffs containing pro-oxidant redox-cycling compounds (e.g., divicine in broad beans, or certain food additives in red suya dishes), or the administration of oxidizing drugs such as primaquine, dapsone, or MB [3, 4]. As in most other cells, the redox milieu of parasitized erythrocytes is a fragile, delicate biological entity. Typically, the overall redox potential of cytosolic spaces is less negative than $-250\,mV$, and thus far below that of the extracellular spaces of aerobically living organisms, where the redox milieu is governed by atmospheric oxygen. One explanation of the need for a reducing intracellular milieu is that living cells originated in an environment where thiol groups are stable, and where such thiol groups are essential for numerous central life processes, such as enzyme catalysis, deoxynucleotide reduction, and the detoxification of xenobiotics. However, following the intrusion of oxygen into the biosphere, the thiophilic environment was – and still is – challenged enormously by oxidative processes. Consequently, in order to survive, all living cells had to protect themselves by employing redoxin-based networks fueled by the reducing equivalents of NADPH and NADH (Figure 7.1). The network of proteins which maintains redox homeostasis *in situ* is the target of natural antimalarial mechanisms. It should be noted that in the parasitized cells a number of pro-oxidant foodstuff ingredients and synthetic drugs act not only as enzyme inhibitors but also as redox-cycling compounds (Figure 7.2). In this way, the enzyme–drug complexes mimic the action of the antimicrobial enzyme NADPH oxidase [3].

The development of drugs targeting the redox-metabolism of *Plasmodium falciparum*, as a representative of the apicomplexan parasites, has been comprehensively described in reviews which have neither their timeliness nor relevance [1, 3, 5, 6]. Hagai Ginsburg's schemes and contribution represent another source of continuous inspiration (see Malaria parasite metabolic pathways; http://sites.huji.ac.il/malaria/). In addition, the discovery and development of naphthoquinone drugs directed against the antioxidative homodimeric flavoenzymes of the glutathione reductase (GR) family has been detailed by Elisabeth Davioud-Charvet and Don Antoine Lanfranchi (see Chapter 20 and Ref. [4]). The biosynthesis of glutathione and dihydrolipoic acid, the major low-molecular-weight thiols of parasites and host cells, is described by Sylke Müller and colleagues in Chapter 10.

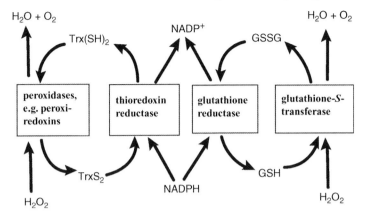

Figure 7.1 Network of antioxidative thiol- and dithiol proteins in *P. falciparum*. The functions of the system include the detoxification of reactive oxygen species and the production of deoxynucleotides from nucleotides. Trx, thioredoxin; GSH, glutathione; GSSG, glutathione disulfide. For *P. falciparum*, five peroxiredoxins and ten Trx- and glutaredoxin-like proteins are known. Individual thioredoxins and glutaredoxins can undergo redox reactions with each other, and with GSH (not shown here). The rate constants of these thiol-disulfide exchanges are in the order of 10 to $100 \, M^{-1} s^{1}$. Glutathione-*S*-transferase catalyzes the formation of glutathione conjugates, can complex heme, and – because of its high intracellular concentration – can act as an efficient glutathione peroxidase. *P. falciparum* does not contain Se-dependent glutathione peroxidase and catalase [1, 3, 5].

Searching for Ethical Drugs Against Malaria

Drugs used to treat malaria, as a disease of the poor, should be also developed as *ethical drugs*, characterized by the acronym *bonaria*, where <u>bon</u> means safe and effective, <u>a</u> affordable for the patients who need it, <u>r</u> already registered for other medical indications, and <u>ia</u> for internationally accessible (K. Becker and R.H. Schirmer, personal communication; this aspect will be further dealt with in the Discussion).

In this chapter, attention is focused on what antimalarial drug discovery, especially in the field of pro-oxidant agents, can learn from clinical malaria and from natural phenomena related to this disease. Furthermore, the redox metabolism offers convincing examples of parasite and host coevolution, as well as evolution of pathogen–host cell interactions [7].

Different Goals Require Different Antimalarial Agents

In all forms of human malaria, the actual disease (characterized *inter alia* by fever bouts) is caused by the rapid multiplication of so-called "schizonts" in erythrocytes. There are, however, other host cells and host tissues essential for maintenance of the parasite lifestyles in humans and in the *Anopheles* mosquito [7, 8]. Consequently, a distinction can be made between three major indications for antimalarial drugs: (i) the prevention and cure of the actual disease caused by blood schizonts; (ii) the prevention of disease transmission by sporozoites, blood schizonts (via transfusion–transmission), or gametocytes; and (iii) the prevention of relapses caused, for example, by liver hypnozoites of *Plasmodium vivax* [9].

Target Cells and Tissues

When an *Anopheles* mosquito attacks the human body, it injects approximately 20 *P. falciparum* sporozoites; the sporozoites subsequently enter 20 hepatocytes, in which they then multiply 30 000-fold within 10–14 days. The liver schizonts are then released and enter the erythrocytes, where they multiply asexually by a factor of up to 16 within 48 h. When the merozoites are subsequently released, it is essential that they find fresh erythrocytes within 1 min; otherwise they will not survive.

Some erythrocytic parasites develop within 14 days into so-called female and male gametocytes which, after being taken up by a female mosquito, can mate to form zygotes that, in the midgut of the mosquito, develop into ookinetes and then oocysts. The latter mature and release sporozoites, which become concentrated in the mosquito's salivary glands from where they are injected into the human skin. The oocysts, which impose as huge tumors in the mosquito, are combated by the insect host via pro-oxidant processes such as encapsulation in melanin-rich cocoons, or destruction with reactive oxygen species (ROS) produced by phagocytes of the hemolymph. Although about 90% of the population of most mosquito species are refractory to *Plasmodium* transmission (i.e., they are successful in killing all of the parasites), the remaining 10% can quite easily support the propagation of *Plasmodia* [7, 10]. Challenging the oocysts, for example with inhibitors of plasmodial glutathione synthesis or GR (K. Buchholz, personal communication), would cause the parasites to be killed in the mosquito, relieve the vector insect from an enormous disease burden, and this would interrupt the transmission pathway from one human host to another. A more practical, and less mosquitophilic, approach is to use insecticides.

Typically, the targeting of *extracellular* forms of the parasite in the human host as a prophylactic measure against malaria is a domain of vaccine research rather than of drug research. Sporozoite motility, however, can also be inhibited effectively by drugs. Recent *in vitro* studies conducted by Frischknecht and coworkers at Heidelberg University have revealed that MB, at submicromolar concentrations, blocks the gliding motility that is characteristic of sporozoites during their long journey through their insect and human hosts, before they settle in their final hepatocyte destination (J. Hellmann and F. Frischknecht, personal communication).

In addition, several mechanisms exist whereby the blood schizonts – and thus malaria as a clinical disease – are transferred directly from person to person by blood transfusion, by organ transplantation, and by therapeutic inoculation (a procedure which was used to cure antibiotic-resistant neurosyphilis by injecting malarial parasites, for which Wagner-Jauregg was awarded the Nobel Prize in 1927). In addition, accidental inoculation can occur among drug addicts who share needles, syringes, or – even worse – contaminated drug solutions for injections. With regards to transfusion–transmission [11], special agents are required for blood sterilization;

typically, the recipient of parasitized blood can be treated with schizonticides and gametocyocidal agents and, in the case of *P. vivax*, later with hypnozoite-eliminating agents such as primaquine.

Another upcoming possibility is pathogen inactivation in suspicious blood units, on the basis of photoactivatable MB or dimethylene blue [11].

Congenital malaria, caused by the transmission of blood schizonts and gametocytes from mother to baby before or during childbirth, is not uncommon in so-called *pregnancy malaria*, where the maternal side of the placenta carries a pool of parasitized erythrocytes adhering to the chorionic villi.

To date, no cases have been reported of malaria as a sexually transmitted disease (STD). It is important to recognize this point as a negative aspect, since STDs attract relatively high public attention and concern on a worldwide basis.

Primaquine and its Redox-Cycling Metabolite, 5-Hydroxyprimaquine, Target Hypnozoites, Gametocytes, and Blood Schizonts

To date, very few clinical studies have been conducted with drugs aimed at combating gametocytes, because it was an accepted principle of ethics commissions that therapeutic measures must exclusively serve the patient, while "altruistic" drugs or vaccines that would prevent a patient from spreading the disease to other persons were considered "unethical." Among the few drugs used against gametocytes the most prominent are primaquine and the artemisinins.

Treatment with the 8-aminoquinoline tissue schizontocide, *primaquine*, has been regarded as a radical cure for relapsing malaria caused by *P. vivax* and other *Plasmodia* that form liver hypnozoites [9, 12]. In addition, primaquine has gametocytocidal activity, thus preventing the transmission of viable parasites from man to mosquito. Primaquine has also been used as one partner of drug combinations used against blood schizonts, the disease-causing form of malarial parasites. Unfortunately, under certain circumstances the clinical utility of primaquine is restricted by its hemolyzing effect in patients with certain types of genetic erythrocytic G6PD deficiency [13]. However, the statement that primaquine should not be administered to any patient with G6PD deficiency is too general, as this would prevent its use in more than 500 million persons worldwide.

When treating malaria patients with primaquine, the parasite-bearing erythrocytes are thought to be recognized by splenic macrophages as the equivalent of senescent red cells. This leads to a selective removal of the parasitized erythrocytes from the circulation [14]. The mechanism of this process is interesting for the present authors' program, which is aimed at developing pro-oxidant agents as antiparasitic drugs (see Figure 7.2 and Table 7.2). 5-Hydroxyprimaquine (5-HPQ) is a human metabolite of primaquine that forms a redox pair with its quinoneimine form; continuous cycling of this redox pair is thought to generate ROS within the erythrocytes. For example, when rat erythrocytes were incubated with 5-HPQ *in vitro*, and subsequently injected into the veins of isologous rats, they were removed rapidly from the circulation as compared to untreated, control erythrocytes [14]. Thus, the ROS attack on the *cytosolic* surface of the membrane generates a signal to remove

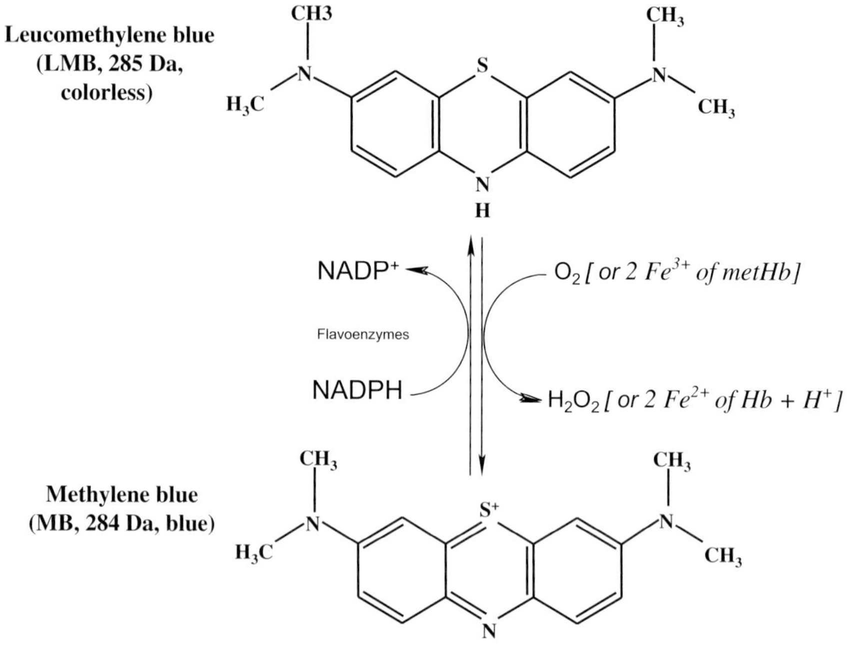

Leucomethylene blue (LMB, 285 Da, colorless)

NADP$^+$ ← O$_2$ [or 2 Fe^{3+} of metHb]

Flavoenzymes

NADPH → H$_2$O$_2$ [or 2 Fe^{2+} of Hb + H$^+$]

Methylene blue (MB, 284 Da, blue)

Figure 7.2 Methylene blue (MB) as a redox-cycling catalyst *in vivo*. Shown are the NAD(P)H-dependent reduction of the blue oxidized form and the rapid reoxidation of the color-free leuco form by auto-oxidation. Pyocyanin undergoes the same redox cycle. In contrast, both divicine and isouramil are reduced by glutathione (GSH) and reoxidized by O$_2$. The other product, glutathione disulfide, is reduced by NADPH in a glutathione reductase-catalyzed reaction. In all cases, the balance equation for one redox cycle is NADPH + H$^+$ + O$_2$ → NADP + H$_2$O$_2$ (see Ref. [15]).

damaged erythrocytes from the circulation. The nature of the crucial target(s), and the mechanism of transfer of this signal across the membrane, are discussed below in the context of MB as a drug.

Pro-Oxidant Mechanisms for Preventing and Curing Malaria as an Acute Disease

Malaria as a clinical disease is caused by blood schizogony – that is, the asexual propagation of the parasite within the erythrocytes.

A number of *experimenta naturae* have suggested that malarial parasites require a highly reducing intra-erythrocytic environment for schizogony [3, 15, 16]. Notably, *P. vivax* may require an even higher antioxidative capacity of the erythrocyte metabolism than other plasmodia, as this parasite is able to propagate only in reticulocytes. Such young red blood cells are biochemically characterized *inter alia* by a very stable redox milieu.

Erythrocytes are normally well equipped with antioxidant cell-biochemical tools, including high activities of catalase, superoxide dismutase, peroxiredoxins and other peroxidases, as well as enzymes that guarantee high concentrations of NADPH and thiol groups in the form of glutathione (2 mM) and protein-bound cysteine residues of hemoglobin (e.g., 10 mM Cys-93). One challenge for the reducing metabolism is that, during maturation of the precursor cells, enormous amounts of free iron ions and heme groups must be handled for hemoglobin production; both, the free iron ions and heme can readily catalyze the formation of ROS, including the OH-radical [17]. Another cause of continuous pro-oxidative challenges is the absence of mitochondria from mature erythrocytes; these organelles maintain physically dissolved O_2 at levels below 10 μM (http://sites.huji.ac.il/malaria/).

In addition, the binding mode of oxygen includes features of O_2^--binding to ferrihemoglobin, which suggests that superoxide or another ROS can be released from the O_2-transporting hemoglobin [1].

In mature erythrocytes, the tetrameric hemoglobin molecules are tightly packed in quasi-crystalline arrays. However, when pro-oxidant conditions prevail in the erythrocytic subregions, disturbances of the hemoglobin structure lead to the formation of hemichrome aggregates that contain intermolecular disulfide bridges and loosened heme groups as products of hemoglobin destabilization and degradation. The hemichromes possess a strong affinity for cytoskeletal proteins [14] and/or the cytoplasmic domain of the plasma membrane anion channel [18]. Following this binding, channel protein oxidation and oligomerization ensue.

The clustering of oxidized anion channels sets off a series of events that lead to changes at the erythrocyte surface which are recognized by specific immunoglobulin G (IgG) molecules. Typically, these normally recognize the band 3 protein clusters of a senescent cell, which accumulate at the end of its lifespan.

The bound IgGs activate complement, and finally trigger the phagocytosis of altered red blood cells. A number of hereditary and acquired pro-oxidant disorders, including the intracellular development of malarial parasites, exacerbates this phenomenon, leading to a more precocious and effective elimination of the diseased erythrocytes [19, 20]. Examples of pro-oxidant protein polymorphisms in persons exposed to malaria, or whose ancestors have been exposed to malaria for numerous generations, include G6PD deficiency, sickle-cell trait, HbSC (a double Hb mutant that contains HbS and HbC in each erythrocyte), beta-thalassemia and, as discussed below, GR deficiency and pyruvate kinase deficiency [20–25]. Pyruvate kinase, as a glycolytic enzyme, contributes to antioxidative defense [24], most likely because it produces the ATP required for the synthesis of glutathione. Some of the above mutant proteins reach polymorphism frequencies; that is, the corresponding alleles occur in more than 1% of the population. In the case of pro-oxidant G6PD mutants, between 10% and 20% of the male population is affected in many countries [13], notably in India, tropical Africa, and Polynesia.

Pro-oxidative food ingredients, such as isouramil and divicine of fava beans, or food additives in certain red suya meals [15], and pro-oxidative drugs such as primaquine, MB, dapsone, and 1,4-naphthoquinones (see Chapter 20) appear to protect the individual from severe malaria in a similar way as do pro-oxidant gene alleles.

Inherited Erythrocyte GR Deficiency as a Model for Malaria Treatment with Human GR Inhibitors

The above discussion on the protective effects of pro-oxidative G6PD alleles led to the following question: Do GR deficiency or drug-induced GR inhibition also lead to the selective removal of red blood cells containing ring stages of *Plasmodium*, by the process of phagocytosis?

In *P. falciparum*-infected red blood cells, the homodimeric flavoenzyme GR regenerates reduced glutathione, which *inter alia* is essential for antioxidant defense. Notably, GR utilizes NADPH which is produced in the pentose phosphate shunt by G6PD and another enzyme. It should be stressed here that NADPH-binding is also essential to maintain the stable, functional form of catalase; thus, low levels of NADPH will lead, in turn, to low levels of this antioxidant enzyme.

A deficiency in GR resulting from an insufficient saturation of the enzyme with its prosthetic group flavin adenine dinucleotide (FAD; a derivative of the vitamin riboflavin) is common. This situation has been studied in the Maremma region of Italy, that has always been notorious for its high incidence of malaria [25, 26]. In some cases, the so-called "nutritional" GR insufficiency may have a hereditary component, with a low affinity of the apoenzyme for the prosthetic group, FAD. In contrast, hereditary apoGR deficiency is rare. In 1976, the group of Dirk Roos described the case of a young woman who presented with favism after a meal of broad beans. She and two of her siblings were found to have normal, stable G6PD levels, but no activity of GR in the red and white blood cells. Subsequent Western blotting of the cell extracts yielded a negative result for GR as a protein. More recently, an investigation was conducted as to whether the GR-deficient erythrocytes would be suitable as host cells for *P. falciparum*, and whether the infected cells were sensitive to IgG-mediated ring-stage phagocytosis [20]. In these studies, the parasite multiplication rate was found to be equal in both GR-deficient and GR-sufficient erythrocytes, while practically no differences were apparent in terms of drug sensitivity (Table 7.1). Hence, GR deficiency may induce changes in the parasite–host unit similar to those described for G6PD deficiency and other pro-oxidant genetic dispositions, and phagocytosis of the ring-stage-infected red blood cells is more pronounced in antioxidation-compromised erythrocytes than in normal red blood cells. In this context, it should be noted that the ring stages do not yet produce hemozoin, a paralyzing poison for phagocytes [19].

Consequently, GR deficiency adds to the paradigm of malaria-protective genetic variations which are based on enhanced IgG-mediated ring-stage phagocytosis, rather than on impaired parasite growth [19].

Erythrocytes pretreated with the disulfide reductase inhibitor bis-chloronitro-sourea (BCNU) either *in vivo* or *in vitro*, have been shown suitable for corroborating and/or extending the findings of the present report. In analogy to the animal model of primaquine treatment [14], the plan is to conduct the following study: to take blood samples from BCNU-treated patients [27], to use the erythrocytes as host cells of *P. falciparum* and, subsequently, to test if the ring stages are more susceptible to phagocytosis than parasitized matched control cells from patients not treated with

Table 7.1 Comparison of the growth and biochemical properties of *P. falciparum* grown in glutathione reductase (GR)-deficient and in control erythrocytes.[a]

	GR-deficient host cells	Normal host cells
Total glutathione in the parasite ($nmol\,mg^{-1}$ protein)	68 ± 2.1	73 ± 1.4
GR activity in parasites ($mU\,mg^{-1}$)	160 ± 6	270 ± 14
Parasite multiplication rate per red blood cell cycle	4.9 ± 0.3	5.4 ± 0.5
IC_{50} of chloroquine (nM)	$8.2 \pm 0.2\ (4.9 \pm 0.2)$	$8.0 \pm 0.2\ (5.9 \pm 0.6)$
IC_{50} of chloroquine for K1 (nM)[b]	160 ± 10	190 ± 11
IC_{50} of methylene blue (nM)	$3.9 \pm 0.2\ (4.2 \pm 0.3)$	$3.8 \pm 0.2\ (4.2 \pm 0.3)$
IC_{50} of methylene blue for K1 (nM)[b]	8.8 ± 0.5	8.1 ± 0.4
IC_{50} of artemisinin (nM)	14 ± 0.5	16 ± 0.8

a)   Before determining the biochemical parameters in the parasites, *P. falciparum* 3D7 was grown over four to five cycles (corresponding to 8–10 days) in the respective red blood cells. Control cells had the same blood group as the patient's. All values given represent mean values of two to three parallel determinations.

b)   The chloroquine-resistant strain K1 is expectedly less sensitive to chloroquine than 3D7, but it is similarly sensitive to MB. All data were taken from Ref. [20].

BCNU. In this context, it would also be of interest to study whether patients under BCNU-treatment have ever contracted malaria [3]. A previous observation that BCNU-pretreated erythrocytes did not serve as host cells of *P. falciparum in vitro* is explained by the additional damage caused by very high BCNU concentrations, for instance, on the glutathione-synthesizing enzymes [20, 28].

Furthermore, the data acquired had an impact on antimalarial drug development strategies, as they indicated that the antimalarial effects of compounds capable of manipulating or inhibiting the activities of human GR are difficult to assess in cell cultures not containing IgG and phagocytes. Thus, the inhibitors of human GR would be expected to be efficient *in vivo*, but not *in vitro*.

The results obtained suggested that human erythrocyte GR should not be neglected as a potential drug target [20, 28]. Another argument in favor of considering a host cell enzyme as a target is the *a priori* prevention of drug resistance.

P. falciparum GR as a Target of Inhibitors and Subversive Substrates

The inhibitors of *Plasmodium falciparum* glutathione reductase (PfGR), together with details of the natural defense mechanisms of the human host (e.g., peroxynitrite production and the fever reaction) that affect PfGR, are listed in Ref. [3].

As a word of warning for drug-design studies, the stable, intensely studied form of PfGR *in vitro*, namely E_{ox}, is but a minor form *in vivo*, whereas the reduced forms of the enzyme, which contain an active site dithiol, dominate in the cell. Only glutathionylated GR, $NADPH\text{-}EH_2$ and EH_4 are likely to occur in the parasite, and these molecular species should be considered as targets of redox state-specific inhibitors [1, 3, 4]. *P. falciparum* thioredoxin reductase (TrxR) is also present *in vivo* in reduced forms. BCNU, for instance, does not affect the E_{ox} form of disulfide reductases, but is an irreversible inhibitor of reduced species, both *in vitro* and *in vivo*.

Fever Bouts Denature the Reduced Forms of PfGR

The conspicuous symptoms of malaria are the fever bouts (also known as "swamp fever," "Roman fever," *the* fever). Such bouts are dangerous for the patient, but much more so also for the parasites; indeed, evidence indicates that a large percentage of the parasites are killed during these fever phases. In culture, the blood schizonts multiply more than 10-fold in four days at 37 °C, but their numbers decrease to 20% when they are grown at 40 °C. These observations suggest that the inactivation of thermolabile enzyme species such as NADPH·GR contributes to the thermosensitivity of the malarial parasites [29]. Labile $EH_2 \cdot NAD(P)H$ and EH_4, rather than stable E_{ox}, are the predominant forms of the disulfide reductases in active cells, and this should be accounted for in both pharmacological and pathophysiological studies of the enzymes. Additionally, redox-cycling agents and inhibitors targeting GR and TrxR at a physiological NADPH concentration of approximately 50 µM should be tested at 40 °C.

Elevated temperatures, even above 55 °C, do not affect human GR. Erythrocytic enzymes, which must remain functional for more than 100 days at 37 °C, have probably been selected for reasons of stability during the course of evolution.

Pharmaceutical Agents as Inhibitors and/or Substrates of Disulfide Reductases

Pharmaceutical agents that interact with GRs and other FAD-containing disulfide reductases can be grouped into two classes, namely enzyme inhibitors and diaphorase substrates. Diaphorase activity is an additional function of several oxidoreductases that are capable of catalyzing the reaction:

$$NAD(P)H + \text{oxidized heteroaromatic compound}$$
$$\rightarrow NAD(P) + \text{reduced heteroaromatic compound}.$$

Many heteroaromatic, redox-active compounds (such as MB) can serve both as an inhibitor (of the disulfide reduction reaction) and as a diaphorase substrate. In this case, there are most likely two binding sites for the two modes of drug action on each enzyme.

The diaphorase substrates, which often are subdivided into turncoat inhibitors, subversive substrates and redox cyclers, are reduced by enzymes at the expense of NADPH, at a site which is not the binding site of the natural disulfide substrate. The reduction of the heteroaromatic compound is the actual diaphorase reaction. Subsequently, the reduced compound undergoes reoxidation or another cell-biochemical reaction. If the reoxidizing agent is O_2, the compound acts catalytically and serves as a redox-cycling agent. In each catalytic cycle NADPH and O_2 are consumed, while NADP and O_2^- or H_2O_2 are formed. Thus, in cooperation with the reductase and O_2, the drug catalyzes the same reaction as NADPH oxidase, a major enzymic defense tool of higher organisms against pathogens. However, other intracellular components can also reoxidize leucoMB, an example being the heme group of methemoglobin. In this way, the effects of MB do not depend exclusively on O_2-driven redox-cycling.

At this point, dihydrolipoamide dehydrogenase (LipDH or DHLD) deserves special mention. This enzyme can serve as a highly efficient diaphorase using NADH and MB as the two substrates, although its physiological reactions are not inhibited by MB and/or other heteroaromatic compounds [3, 30]. In such a case, gene knockout experiments might lead to the conclusions that the enzyme is not essential for the parasite, and therefore not suitable as a target for drug development. However, the second conclusion is not correct because the enzyme – which serves as an NAD(P)H oxidase in the presence of a catalytic drug concentration – plays an essential drug-like role for the production of parasitotoxic compounds (K. Buchholz, personal communication).

The reaction sequence initiated by MB in the presence of molecular oxygen is:

$$NADPH + MB \rightarrow NADP + leucoMB \text{ (}diaphorase\ reaction\ of\ GR\text{)}$$
$$leucoMB + 2\,O_2 \rightarrow MB + H^+ + 2\,O_2^- \text{ (}auto\text{-}oxidation\ in\ vivo\text{)}$$
$$2\,O_2^- + 2\,H^+ \rightarrow H_2O_2 + O_2 \text{ (}enzyme\text{-}catalyzed\ or\ spontaneous\text{)}$$
$$NADPH + O_2 + H^+ \rightarrow NADP + H_2O_2$$

Considering the first two lines, the balance equation is:

$$NADPH + 2\,O_2^- \rightarrow NADP + H^+ + 2O_2^-$$

This corresponds to the reaction catalyzed by the antimicrobial defense enzyme NADPH oxidase.

Is 100% Enzyme Inhibition Necessary?

In the lifestyles of the malarial parasite, where GR activity is essential, PfGR-targeting drugs do not have to lead to 100% inhibition in order to be effective. By using the *in vivo*-relevant form of the Michaelis–Menten equation with a K_M of $100\,\mu M$, $[S] = K_M\,(V_{max}/v - 1)^{-1}$, 93% inhibition is predicted to lead to an increase of the steady-state concentration in GSSG, from $<4\,\mu M$ to $>800\,\mu M$, which is highly parasitotoxic [3]. The situation is similar for TrxR, where enzyme inhibition leads to a low steady-state concentration of reduced thioredoxin. Of course, these estimations are correct only if there is not more than one biological mechanism for the reduction of glutathione disulfide and oxidized thioredoxin, respectively.

Pyocyanin, a Natural Analog of MB, is an Inhibitor and a Diaphoretic Substrate of GR

Recently, the blue pigment pyocyanin (PYO), which is known to act as a signal, quorum sensor, respiratory metabolite, and antimicrobial molecule in *Pseudomonas aeruginosa* and other bacteria [31], has been identified as having antimalarial activity *in vitro* (K. Becker, personal communication). Comparisons of PYO with the phenothiazine derivative MB have suggested that the latter might be regarded as a sulfur analog of the natural phenazine compound PYO (Table 7.2). This working hypothesis explains, at least in part, why synthetic MB behaves like a compound that has been shaped by biological evolution. The similarities of MB and PYO include: (i) their characteristics as positively charged, heteroaromatic compounds; (ii) their reactivity toward cellular reductants such as NADPH or dihydrolipoamide; (iii) the reactivity of their reduced leuco-forms toward triplet O_2; and (iv) their interactions with signaling proteins, flavoenzymes, heme proteins, or transmembrane transporters. Both compounds were shown to have gametocytocidal and schizontocidal effects on malarial parasites in a submicromolar range. However, whereas PYO is too toxic, MB represents a promising agent for MB-based combination therapies against *P. falciparum* malaria in children (see below). Yet, pyocyanin may serve as a valuable tool for studying disease interference. Indeed, retrospective studies are currently under way to determine if patients suffering from *Pseudomonas aeruginosa* infections, which frequently occur in tropical countries, are protected from malaria by the secondary metabolite, PYO.

The binding of PYO to the flavoenzyme GR from human erythrocytes was studied using X-ray crystallography (Figure 7.3). As might be expected from the binding mode of other heterocyclic compounds [3], the structure shows one PYO molecule bound in the cavity located at the interface of the two subunits of the enzyme. The binding pocket is formed by seven amino acids of one subunit, and their counterparts of the other subunit, such that PYO is perfectly sandwiched between two phenylalanine residues (F78 and F78'). Azure B, a demethylated metabolite of MB with antimalarial activity [32], was also found to bind to the crystalline enzyme in a mode very similar to PYO (K. Fritz-Wolf, unpublished results).

Why Reintroduce MB as a Drug Against Pediatric Malaria?

In a famous paper, Sir James W. Black, M.D., the 1988 Nobel Laureate in Medicine [33], defined the basic philosophy of his approach as, "The most fruitful basis for the discovery of a new drug is to start with an old drug." The present authors greatly approve of this approach, and would like to add the term "...from a natural source" because, in that case, natural selection has shaped the compound for appropriate behavior in biological systems [7].

One very promising old drug (in fact, it was the very first synthetic drug) – MB – can serve as a noncompetitive inhibitor of PfGR, and much less of human GR, at therapeutically employed concentrations. This effect was identified by Petra Färber (of the present authors' group) when screening affordable drugs with antimalarial activities as PfGR inhibitors [34]. MB, which has long been used in the treatment of malaria [35], is today the standard medication against inherited and acquired methemoglobinemia, ifosfamid-induced neuropathy, and other pathological conditions (http://www.alzforum.org/new/Schirmer.asp; see also Ref. [36]). However, there is a contraindication for MB. A very recent discovery has been the functional

Table 7.2 Properties of pyocyanin in comparison to methylene blue.

Property	Pyocyanin	Methylene blue
Chemical class	Phenazine	Phenothiazine
Color	Blue	Dark blue
Colorless reduction product	LeucoPYO	LeucoMB (solubility only 25 μM)
$\lambda_{vis.max}$	690 ± 100 nm	663 nm
M_r of the heterocycle	210 Da	284 Da
Milestones	Forbes (1860), Wrede (1924, 1929), Newman (2006)	Caro (1884), Bernthsen (1887), Ehrlich (1891), Wieland (1922), Clark (1925)
Functions *in vivo*	In *P. aeruginosa* respiratory pigment, quorum sensor, transcription factor, antimicrobial agent	Numerous technical, industrial, scientific, and medical applications
Clinical dosage for various diseases	Not applicable, PYO is toxic	2–20 mg kg^{-1} per day
Concentrations *in vivo*	$<100 \mu$M in infected bronchial mucus	5–30μM in whole blood
Midpoint potential at pH 7	-30 mV	$+10$ mV
Intracellular reductants (k-value at pH 7)	NADPH (86 M^{-1} s^{-1}) > NADH > dithiols	Dithiols > NADPH = NADH
Redox cycling catalyst *in vivo*	Reduction by NAD(P)H, reoxidation by O$_2$	Enzyme-catalyzed reduction by NAD(P)H, reoxidation by O$_2$
Reduction by flavoenzymes	LipDH and TrxR $\gg$ GR	LipDH > TrxR > GR
Inhibition of flavoenzymes	GR and TrxR, not LipDH	GR and TrxR, not LipDH
Crystal structure of enzyme-ligand complex	Human GR–PYO complex	Low-resolution data
IC$_{50}$ against *P. falciparum* blood schizonts *in vitro*	86 ± 4.5 nM (3D7) 59 ± 3.3 nM (K1)	3.1 ± 0.4 nM (3D7) 6.6 ± 1.1 nM (K1)
IC$_{50}$ against early (and mature) *P. falciparum* gametocytes *in vitro*	180 nM (560 nM)	34 nM (60 nM)
Tested for biological conduct *in vivo*	In >100 Ma of evolution in bacteria-containing biotopes [31]	In an unpresented variety of uses in medicine and biotechnology

Data on PYO listed from line "Redox-cycling catalyst" onwards were contributed by D. Kasozi, K. Fritz-Wolf and K. Becker (personal communication). This applies also to the IC$_{50}$ values of MB. Most other data on MB were taken from Ref. [30].

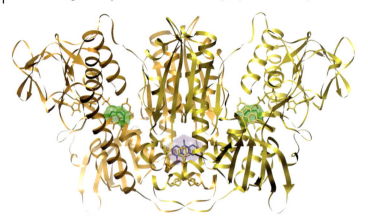

Figure 7.3 Human glutathione reductase homodimer with bound pyocyanin (PYO). The pyocyanin (blue) and FAD (yellow) are represented as ball-and-stick models. Additionally, the surfaces of the catalytic cysteines Cys 58/Cys63 and Cys58'/Cys63' (green) and of PYO (blue) are shown.

interaction that occurs between MB and the serotonin-specific reuptake inhibitors (SSRIs) that are used as antidepressant agents. This undesirable synergy may lead to the potentially fatal "serotonin toxicity syndrome"; consequently, MB must not be administered to patients taking SSRIs [37].

Secondary Drug Design on the Basis of Phenothiazine Drugs?

There are two main reasons for the reluctance to design an optimized derivative of MB on the basis of its binding site in GR. The first reason is that, whilst demethylated MB species and selenoMB are interesting candidates, any derivative would result in major costs for testing the new compound as a drug candidate. The second reason is the fact that antimicrobial drugs acting as ideally fitting ligands of their targets are subject to rapid drug resistance development by the pathogen [38].

Drug resistance is in many cases due to drug tolerance at a molecular level. One reason for the occurrence of drug tolerance is that, in ligand–target interactions, the attractive forces are weak while the repulsive forces are strong. Thus, the introduction of an additional H-atom or a methyl group into the target protein, on the basis of a point mutation, can lead to the complete abolition of any interaction [3].

On a pharmacological level, drug optimization efforts by secondary design may lead to unexpected effects. For example, as shown by Eisenbrand, the carbamoylating agent BCNU was converted to the alkylating drug HeCNU by replacing a chlorine atom with an OH-group [3, 28].

Clinical Trials Using MB-Based Drug Combinations

The effective use of MB in two adult patients (probably with *P. vivax* malaria) was first reported by Guttmann and Ehrlich, in 1891 [39, 40]. Two years later, Ferreira described in detail the first successful oral application of MB to 40 children with malaria in Rio de Janeiro [35]. In this case, oral daily doses of between 25 and 50 mg kg^{-1}, subdivided into several portions, were administered and well tolerated by infants and young children until the malaria symptoms subsided.

In 2003, MB was tested for the first time against uncomplicated *P. falciparum* malaria in Africa in clinical studies. Initially, the safety and pharmacokinetics of MB were studied in healthy and G6PD-deficient adult men [41, 42]. The G6PD-deficiency type G6PDA$^-$ that occurred in this and other study groups did not represent a contraindication for the use of MB, as no hemolysis was observed. It should be noted that about 150 types of G6PD mutation have been identified in populations with roots in malaria-endemic countries, that one in six Africans and one in five Indians carries a G6PD mutation, and that the hemolytic response to drugs is very different among the types and classes of G6PD deficiency [13, 43].

The safety studies were followed by randomized, controlled clinical trials conducted in young children (aged 6–59 months) with malaria [44]. Based on the intention to reverse parasite resistance against chloroquine, MB was combined with chloroquine; this drug combination was termed "Blue CQ." When administered at a dose of 2 mg kg^{-1} twice daily, MB proved to be ineffective, whereas clinical trials using higher doses of MB (6–12 mg kg^{-1} twice daily) showed it to be well tolerated and to have an efficacy of 66% ACPR (adequate clinical and parasitological response) [45]. Unfortunately, the positive effect of MB was overshadowed by parasites rapidly developing resistance to the partner drug, chloroquine. Moreover, the bitter taste of MB solutions necessitated the development of a taste-masked pediatric fluid formulation [35, 46] since, for children aged less than five years, the administration of tablets and capsules is not permitted.

Since 2006, MB has been administered in combination with amodiaquine in several clinical trials in young children in Burkina Faso. The combination proved to be highly effective (95% ACPR on day 28), and even superior to a combination of MB plus artesunate (62% ACPR on day 28), both of which are short-acting drugs [47]. When given as a monotherapy to semi-immune adults with uncomplicated malaria, MB proved to be effective (74% ACPR if given over three days), although several early treatment failures occurred due to the initial slower parasite clearance time of MB compared to the artemisinins [48]. Both experimental (Table 7.2) and clinical [49] evidence has been obtained showing MB to be effective not only against gametocytes (which would lead to MB becoming an asset for ongoing global malaria elimination efforts), but possibly also against hypnozoites (making it an interesting drug candidate for the treatment of *Plasmodium vivax*). Moreover, all of the pharmacokinetic studies with MB showed it to be better tolerated when administered orally than intravenously. Orally administered MB was also shown to provide a very good bioavailability of the drug [50]. In addition, MB demonstrated a much longer plasma half-life (ca. 20 h) than previously assumed [50], such that once-daily dosing should be considered [41, 50].

MB Leads to Blue Coloration of the Urine and, Consequently, of Napkins and Clothes

Although MB has been identified as a promising candidate for the treatment of malaria in the main risk group of young children, there is a major adverse side effect in all cases, in that it leads to a blue coloration of the child's urine and, consequently, also of the mother's clothes. Since the acceptability of such coloration effects was unknown, an anthropological study was undertaken in a malaria-endemic area of rural Burkina Faso to identify the community's perceptions regarding such blue coloration [51]. Perhaps not surprisingly, the results showed that people would accept the drug, even if its color was unusual and it stained the patient's clothing, as long as it is effective against malaria. Moreover, despite modern washing powders proving to be inferior for the purpose, the mothers quickly determined how MB-stained clothing could be cleaned, using traditional washing methods. In conclusion, these studies have not only shed some light on the relationship between color and drug preference in Africa, but also supported the importance of considering community attitudes before commencing public health interventions. An additional benefit was that the blue urine coloration not only improved patient compliance, but also helped to prevent the distribution of counterfeit drugs.

Notably, MB can be converted readily to color-free leucoMB [30]; indeed, REMBER™, a drug that is currently undergoing trials for the treatment of Alzheimer's disease, contains leucoMB [52] (see also http://www.alzforum.org/new/Schirmer.asp), which most likely has the same pharmacological properties as MB as it is readily auto-oxidized. Unfortunately, such instability of leucoMB in the presence of O_2 would render it impractical as an ingredient of antimalarial drug combinations.

Prevention of Resistance Development to Key Antimalarials Using MB as an Additional Partner

Currently, artemisinin-based combination therapy (ACT) represents the first-line treatment of choice for malaria. Recently, however, Samarasekera described the development of artemisinin resistance at the Thai–Cambodia border that may also soon affect sub-Saharan Africa [53]. Hence, the question here is, "What can be done to avoid a repetition of the tragic history of chloroquine resistance?"

As discussed by Müller *et al.* [53], even if fixed-dose ACT were to be widely available, it would be only a question of time until resistance to the partner drug with the longer half-life would develop, leaving the artemisinin component unprotected. Thus, the addition of another antimalarial with a short half-life to existing ACTs might represent an innovative approach. MB, which has a broad activity against *P. falciparum* parasites and acts synergistically with artemisinins [54], might prove to be such a candidate. As noted above, MB has not only been demonstrated as both safe and effective in the malaria patients of Burkina Faso, but also has the potential to greatly reduce the number of *P. falciparum* gametocytes in clinical malaria cases [49].

Perhaps the most obvious question in this context would be whether any resistance of malaria parasites to MB has been observed. This is clearly not the case in animal models, where only a very moderate increase in EC_{50} values can be provoked by administering MB for several months [3], nor in human malaria. The reason for the continuous sensitivity of *P. falciparum* appears to be that (leuco)MB and its metabolites have more than one target. In addition, some targets – such as the growing hemozoin double helix and the GR of the red blood cell – cannot be controlled by the genome of the parasite, which makes the development of rapid resistance against this drug very unlikely.

Discussion

Ethical Drugs: MB as an Example

Although the institutions of post-modern drug research are extremely efficient when developing and marketing drugs for the diseases of affluent societies, as well as "lifestyle" or "performance" drugs, they appear to have lost in part their competence to create drugs for diseases of the poor. Since 1975, less than 1% of all newly approved drugs have been registered for diseases that prevail in the developing countries. One such disease is *P. falciparum* malaria, which affects several hundred million people every year, with the high-risk groups among patients including unprotected tourists, pregnant women and, above all, children aged under five years. Whilst the tourists represent the "happy few" of all nations who have access to adequate prophylaxis or treatment, the main challenges are apparent in the other two high-risk groups. Clearly, the true burden of malaria – from personal, medical, and economical perspectives – is endured by the poor people of tropical countries [39, 55].

The present concept is to complement the modern procedures of drug development [38] by developing ethical drugs that fulfill the *bonaria* criteria. The term "ethical drugs" (*Ethische Präparate*, according to Robert Koch) implies that these drugs are necessary to prevent and cure disease, but that they are unlikely to generate profits. During the classical period of drug research – during the early decades of the twentieth century – physicians and medicinal chemists alike were highly successful at developing and distributing ethical drugs; indeed, at the time it was considered to be a crime, a sin of omission, to withhold treatment. Today, however, in periods of heightened safety concerns, and with the concept of an unlimited liability of both the physicians and the drug companies – the health services would prefer to leave hundreds of millions of people untreated rather than to risk a single case of real or putative toxic effects. This rigor with regards to ethical drugs should be contrasted with the relaxed attitude of today's Western society toward the often serious side effects of "lifestyle" and "performance" drugs.

A related problem to this situation is the "know–do" discrepancy: In other words, we know too much, and we do too little to alleviate the malarial burden [55]. Rather, the development, distribution, and administration of ethical drugs should be handed over to academic public institutions, anthropophilic foundations, and military services. Indeed, for more than 100 years military institutions have been highly successful at developing drugs and vaccines to combat tropical epidemics and other diseases of the poor. Of course, help and support from pharmaceutical industries would be most welcome, but should their role become an essential part of the process?

While developing ethical drugs, it is in everybody's interest to ask challenging questions, including: How are the astronomic costs of developing new drugs calculated by drug providers [55]? Who determines the criteria for registering a drug? Who profits from the achievements of medical and scientific research? And, to whom are such drugs denied [56]?

Today, humankind lives in times of negative ethics, of avoiding mistakes, rather than in times of positive ethics where attempts are made to bring more justice, health, and progress to the world [56]. In his inaugural presidential address in 1937, Franklin D. Roosevelt characterized positive ethics in scientific and medical progress as follows: "*The test of our progress is not whether we add more to the abundance of those who have much; it is whether we provide enough for those who have too little.*"

Conclusion

Both, MB and other pro-oxidant agents can be regarded as pharmacological phenocopies of inborn conditions, such as certain types of pro-oxidant G6PD deficiency or GR deficiency. The pro-oxidant genetic dispositions of red blood cells and pro-oxidant agents often do not affect parasite growth *in vitro*, but rather demonstrate their efficiency *in vivo* by inducing an IgG-mediated phagocytosis of erythrocytes carrying the ring stages of the parasites. The value of natural compounds, such as pyocyanin, as a natural analog of the pro-oxidant drug MB has been highlighted. Indeed, the absorption, distribution, metabolism and excretion (ADME) of administered MB can be better understood by making comparisons with pyocyanin released in *Pseudomonas* infections.

MB has several targets in *P. falciparum*-infected cells. *Inter alia*, it is a diaphorase substrate of *P. falciparum* disulfide reductases, in which case the parasite enzyme is an essential tool for mediating the drug effect. This must be accounted for when developing appropriate inhibitors of disulfide reductases, or when interpreting the results of gene knockout experiments.

The procedure applied to MB appears to show promise for drugs targeted against widespread diseases of the poor. Clearly, when a potential parasite-specific target such as PfGR becomes available for systematic testing, then all affordable registered drugs should be screened as inhibitors of this target. If an active compound were to be identified in this way, then the enormous costs of both preclinical and many clinical trials could be circumvented, and the forbiddingly narrow part of Ridley's "funnel of drug development" [38] could, accordingly, be avoided. Moreover, the desired drug should fulfill most, or all, of the *bonaria* criteria proposed for an ethical drug.

Ongoing studies continue to show that MB is active against not only schizonts but also gametocytes, which are responsible for malarial transmission and thus important in malarial elimination programs. This illustrates a further advantage of the *bonaria* drugs, namely that there is a continuous dynamic interaction between the results of clinical trials and basic research.

Acknowledgments

The authors are indebted to the DFG (SFB 544 "Control of tropical infectious diseases" subprojects B2 and A8) and to the Dream Action Award of DSM-Austria for continuous generous support. Dr. Katja Becker , Giessen, kindly contributed unpublished data.

References

1 Becker, K., Tilley, L., Vennerstrom, J.L., Roberts, D., Rogerson, S., and Ginsburg, H. (2004) Oxidative stress in malaria parasite-infected erythrocytes: host–parasite interactions. *Int. J. Parasitol.*, **34**, 163–189.

2 Vannier, E. and Krause, P.J. (2009) Update on babesiosis. *Interdiscip. Perspect. Infect. Dis.*, **2009**, 984568.

3 Krauth-Siegel, R.L., Bauer, H., and Schirmer, R.H. (2005) Dithiol proteins as guardians of the intracellular redox milieu in parasites: old and new drug targets in trypanosomes and malaria-causing plasmodia. *Angew. Chem. Int. Ed. Engl.*, **44**, 690–715.

4 Wenzel, N.I., Chavain, N., Wang, Y., Friebolin, W., Maes, L., Pradines, B., Lanzer, M. *et al.* (2010) Antimalarial versus cytotoxic properties of dual drugs derived from 4-aminoquinolines and Mannich bases: Interaction with DNA. *J. Med. Chem.*, **53**, 3214–3226.

5 Müller, S. (2004) Redox and antioxidant systems of the malaria parasite *Plasmodium falciparum*. *Mol. Microbiol.*, **53**, 1291–1305.

6 Grellier, P., Maroziene, A., Nivinskas, H., Sarlauskas, J., Aliverti, A., and Cenas, N. (2010) Antiplasmodial activity of quinones: roles of aziridinyl substituents and the inhibition of *Plasmodium falciparum* glutathione reductase. *Arch. Biochem. Biophys.*, **494**, 32–39.

7 Bongfen, S.E., Laroque, A., Berghout, J., and Gros, P. (2009) Genetic and genomic analyses of host-pathogen interactions in malaria. *Trends Parasitol.*, **25**, 417–422.

8 White, N.J. (2008) The role of anti-malarial drugs in eliminating malaria. *Malar. J.*, **7** (Suppl. 1), S8.

9 Wells, T.N., Burrows, J.N., and Baird, J.K. (2010) Targeting the hypnozoite reservoir of *Plasmodium vivax*: the hidden obstacle to malaria elimination. *Trends Parasitol.*, **26**, 145–151.

10 Kumar, S., Christophides, G.K., Cantera, R., Charles, B., Han, Y.S., Meister, S., Dimopoulos, G. *et al.* (2003) The role of reactive oxygen species on Plasmodium melanotic encapsulation in *Anopheles gambiae*. *Proc. Natl Acad. Sci. USA*, **100**, 14139–14144.

11 McCullough, J. (2007) Pathogen inactivation: a new paradigm for preventing transfusion-transmitted infections. *Am. J. Clin. Pathol.*, **128**, 945–955.

12 Vale, N., Moreira, R., and Gomes, P. (2009) Primaquine revisited six decades after its discovery. *Eur. J. Med. Chem.*, **44**, 937–953.

13 Beutler, E. and Duparc, S. (2007) Glucose-6-phosphate dehydrogenase deficiency and antimalarial drug development. *Am. J. Trop. Med. Hyg.*, **77**, 779–789.

14 Bowman, Z.S., Oatis, J.E. Jr, Whelan, J.L., Jollow, D.J., and McMillan, D.C. (2004) Primaquine-induced hemolytic anemia: susceptibility of normal versus

glutathione-depleted rat erythrocytes to 5-hydroxyprimaquine. *J. Pharmacol. Exp. Ther.*, **309**, 79–85.

15 Schirmer, R.H., Schöllhammer, T., Eisenbrand, G., and Krauth-Siegel, R.L. (1987) Oxidative stress as a defense mechanism against parasitic infections. *Free Radic. Res. Commun.*, **3**, 3–12.

16 Becker, K., Koncarevic, S., and Hunt, N.H. (2005) Oxidative stress and antioxidant defense in malarial parasites, in *Molecular Approaches to Malaria* (ed. L. Sherman), American Society for Microbiology, pp. 365–383.

17 Low, F.M., Hampton, M.B., and Winterbourn, C.C. (2008) Peroxiredoxin 2 and peroxide metabolism in the erythrocyte. *Antioxid. Redox Signal.*, **10**, 1621–1630.

18 Cappadoro, M., Giribaldi, G., O'Brien, E., Turrini, F., Mannu, F., Ulliers, D., Simula, G. *et al.* (1998) Early phagocytosis of glucose-6-phosphate dehydrogenase (G6PD)-deficient erythrocytes parasitized by *Plasmodium falciparum* may explain malaria protection in G6PD deficiency. *Blood*, **92**, 2527–2534.

19 Arese, P., Turrini, F., and Schwarzer, E. (2005) Band 3/complement-mediated recognition and removal of normally senescent and pathological human erythrocytes. *Cell Physiol. Biochem.*, **16**, 133–146.

20 Gallo, V., Schwarzer, E., Rahlfs, S., Schirmer, R.H., van Zwieten, R., Roos, D., Arese, P., and Becker, K. (2009) Inherited glutathione reductase deficiency and *Plasmodium falciparum* malaria – a case study. *PLoS One*, **4**, e7303.

21 Mason, P.J., Bautista, J.M., and Gilsanz, F. (2007) G6PD deficiency: the genotype-phenotype association. *Blood Rev.*, **21**, 267–283.

22 Ruwende, C. and Hill, A. (1998) Glucose-6-phosphate dehydrogenase deficiency and malaria. *J. Mol. Med.*, **76**, 581–588.

23 Clark, T.G., Fry, A.E., Auburn, S., Campino, S., Diakite, M., Green, A., Richardson, A. *et al.* (2009) Allelic heterogeneity of G6PD deficiency in West Africa and severe malaria susceptibility. *Eur. J. Hum. Genet.*, **17**, 1080–1085.

24 Ayi, K., Min-Oo, G., Serghides, L., Crockett, M., Kirby-Allen, M., Quirt, I., Gros, P., and Kain, K.C. (2008) Pyruvate kinase deficiency and malaria. *N. Engl. J. Med.*, **358**, 1805–1810.

25 Anderson, B.B., Scattoni, M., Perry, G.M., Galvan, P., Giuberti, M., Buonocore, G., and Vullo, C. (1994) Is the flavin-deficient red blood cell, common in Maremma, Italy, an important defense against malaria in this area? *Am. J. Hum. Genet.*, **55**, 975–980.

26 Dante Alighieri (1320) *Canto V del Purgatorio, Divina Commedia*, verses 130–136.

27 Frischer, H. and Ahmad, T. (1977) Severe generalized glutathione reductase deficiency after antitumor chemotherapy with BCNU [1,3-bis(chloroethyl)-1-nitrosourea]. *J. Lab. Clin. Med.*, **89**, 1080–1091.

28 Zhang, Y., König, I., and Schirmer, R.H. (1988) Glutathione reductase-deficient erythrocytes as host cells of malarial parasites. *Biochem. Pharmacol.*, **37**, 861–865.

29 Schirmer, M., Scheiwein, M., Gromer, S., Becker, K., and Schirmer, R.H. (1999) Disulfide reductases are destabilized by physiologic concentrations of NADPH. *Flavins Flavoproteins*, **13**, 857–862.

30 Buchholz, K., Schirmer, R.H., Eubel, J.K., Akoachere, M.B., Dandekar, T., Becker, K., and Gromer, S. (2008) Interactions of methylene blue with human disulfide reductases and their orthologues from *Plasmodium falciparum*. *Antimicrob. Agents Chemother.*, **52**, 183–191.

31 Dietrich, L.E., Teal, T.K., Price-Whelan, A., and Newman, D.K. (2008) Redox-active antibiotics control gene expression and community behavior in divergent bacteria. *Science* **321** (5893), 1203–1206.

32 Vennerstrom, J.L., Makler, M.T., Angerhofer, C.K., and Williams, J.A. (1995) Antimalarial dyes revisited: xanthenes, azines, oxazines, and thiazines. *Antimicrob. Agents Chemother.*, **39**, 2671–2677.

33 Black, J. (1989) Drugs from emasculated hormones: the principle of syntopic antagonism. *Science*, **245**, 486–493.

34 Färber, P.M., Arscott, L.D., Williams, C.H. Jr, Becker, K., and Schirmer, R.H. (1998) Recombinant *Plasmodium falciparum* glutathione reductase is inhibited by the antimalarial dye methylene blue. *FEBS Lett.*, **422**, 311–314.

35 Ferreira, C. (1893) Therapeutique medicale: sur l'emploi du bleu de méthylène dans la malaria infantile. *Bull. Gen. de Therap. Medicale et Chirurgicale*, **124**, 488–525.

36 Meissner, P. (2010) *Medikamentenresistenz in Afrika: Methylenblau, eine sichere und wirksame Alternative zur Behandlung der Malaria im Kindesalter*, Ruprecht-Karl University, Heidelberg.

37 Ramsay, R.R., Dunford, C., and Gillman, P.K. (2007) Methylene blue and serotonin toxicity: inhibition of monoamine oxidase A (MAO A) confirms a theoretical prediction. *Br. J. Pharmacol.*, **152**, 946–951.

38 Ridley, R.G. (2002) Medical need, scientific opportunity and the drive for antimalarial drugs. *Nature*, **415**, 686–693.

39 Schirmer, H. (2004) Medikamente für die Armen. Zur Biomedizin des 21. Jahrhunderts. *Spektrum der Wissenschaft*, **December**, 110–113.

40 Guttmann, P. and Ehrlich, P. (1891) Ueber die Wirkung des Methylenblau bei Malaria. *Berl. Klin. Wochenschr.*, **28**, 953–956.

41 Rengelshausen, J., Banfield, M., Riedel, K.D., Burhenne, J., Weiss, J., Thomsen, T., Walter-Sack, I. *et al.* (2005) Opposite effects of short-term and long-term St John's wort intake on voriconazole pharmacokinetics. *Clin. Pharmacol. Ther.*, **78**, 25–33.

42 Mandi, G., Witte, S., Meissner, P., Coulibaly, B., Mansmann, U., Rengelshausen, J., Schiek, W. *et al.* (2005) Safety of the combination of chloroquine and methylene blue in healthy adult men with G6PD deficiency from rural Burkina Faso. *Trop. Med. Int. Health*, **10**, 32–38.

43 Wang, J., Luo, E., Hirai, M., Arai, M., Abdul-Manan, E., Mohamed-Isa, Z., Hidayah, N., and Matsuoka, H. (2008) Nine different glucose-6-phosphate dehydrogenase (G6PD) variants in a Malaysian population with Malay, Chinese, Indian and Orang Asli (aboriginal Malaysian) backgrounds. *Acta Med. Okayama*, **62**, 327–332.

44 Meissner, P.E., Mandi, G., Witte, S., Coulibaly, B., Mansmann, U., Rengelshausen, J., Schiek, W. *et al.* (2005) Safety of the methylene blue plus chloroquine combination in the treatment of uncomplicated falciparum malaria in young children of Burkina Faso. *Malar. J.*, **4**, 45.

45 Meissner, P.E., Mandi, G., Coulibaly, B., Witte, S., Tapsoba, T., Mansmann, U., Rengelshausen, J. *et al.* (2006) Methylene blue for malaria in Africa: results from a dose-finding study in combination with chloroquine. *Malar. J.*, **5**, 84.

46 Gut, F., Schiek, W., Haefeli, W.E., Walter-Sack, I., and Burhenne, J. (2008) Cation exchange resins as pharmaceutical carriers for methylene blue: binding and release. *Eur. J. Pharm. Biopharm.*, **69**, 582–587.

47 Zoungrana, A., Coulibaly, B., Sié, A., Walter-Sack, I., Mockenhaupt, F.P., Kouyate, B., Schirmer, R.H. *et al.* (2008) Safety and efficacy of methylene blue combined with artesunate or amodiaquine for uncomplicated falciparum malaria: a randomized controlled trial from Burkina Faso. *PLoS One*, **3**, e1630.

48 Bountogo, M., Mockenhaupt, F.P., Zoungrana, A., Coulibaly, B., Klose, C., Mansmann, U., Burhenne, J. *et al.* (2010) Efficacy of methylene monotherapy in semi-immune adults with uncomplicated falciparum malaria: a controlled trial. *Trop. Med. Int. Health*, **15**, 713–717.

49 Coulibaly, B., Zoungrana, A., Mockenhaupt, F.P., Schirmer, R.H., Klose, C., Mansmann, U., Meissner, P.E., and Müller, O. (2009) Strong gametocytocidal effect of methylene blue-based combination therapy against falciparum malaria: a randomised controlled trial. *PLoS One*, **4**, e5318.

50 Walter-Sack, I., Rengelshausen, J., Oberwittler, H., Burhenne, J., Müller, O.,

Meissner, P., and Mikus, G. (2009) High absolute bioavailability of methylene blue given as an aqueous oral formulation. *Eur. J. Clin. Pharmacol.*, **65**, 179–189.

51 Sanon, M., Mandi, G., De Allegri, M., Schirmer, H., Meissner, P., and Müller, O. (2007) Attitudes towards coloration effects associated with methylene blue malaria treatment in rural Burkina Faso. *Curare*, **30**, 27–34.

52 Oz, M., Lorke, D.E., and Petroianu, G.A. (2009) Methylene blue and Alzheimer's disease. *Biochem. Pharmacol.*, **78**, 927–932.

53 Müller, O., Sie, A., Meissner, P., Schirmer, R.H., and Kouyate, B. (2009)

Artemisinin resistance on the Thai-Cambodian border. *Lancet*, **374**, 1419.

54 Akoachere, M., Buchholz, K., Fischer, E., Burhenne, J., Haefeli, W.E., Schirmer, R.H., and Becker, K. (2005) *In vitro* assessment of methylene blue on chloroquine-sensitive and -resistant *Plasmodium falciparum* strains reveals synergistic action with artemisinins. *Antimicrob. Agents Chemother.*, **49**, 4592–4597.

55 Hubbard, T. and Love, J. (2004) A new trade framework for global healthcare R&D. *PLoS Biol.*, **2**, E52.

56 Hill, A.V. (1946) Scientific ethics. *Chem. Eng. News*, **24**, 1343–1346.

8
Lipids as Drug Targets for Malaria Therapy

Henri J. Vial, Diana Penarete, Sharon Wein, Sergio Caldarelli,
Laurent Fraisse, and Suzanne Peyrottes*

Abstract

Glycerophospholipids, the main *Plasmodium* membrane constituents, mostly originate from the parasite enzymatic machinery, which relies on the scavenging and
downstream metabolism of polar heads and fatty acids. At the blood stage,
P. falciparum combines metabolic pathways found in bacteria, yeasts, and plants.
Coordinated regulations between these pathways appear limited, and the vital
importance of individual metabolic pathways remains to be clarified. The problem
is even more complex as rodent and nonrodent malarial parasites differ in their
phospholipid metabolic pathways. Extensive research on intraerythrocytic *Plasmodium* glycerophospholipid metabolism has revealed potential targets for chemotherapeutic interference; this involves the use of false precursors to deceive the parasite or
inhibitors. The most advanced pharmacological approach is based on the use of
choline analogs, the primary interference of which has been associated with locking
the choline carrier. Such molecules inhibit *P. falciparum* asexual blood stages at
single-digit nanomolar concentrations, and cure *Plasmodium vinckei* malaria infection in mice at doses lower than $1 \, \text{mg} \, \text{kg}^{-1}$. The potency and specificity of these
molecules are most likely due to their unique ability to accumulate, nonreversibly,
inside the intraerythrocytic parasite. The potent antimalarial activity of these derivatives may also be attributed to their compartmentalization within the parasite's
food vacuole, where they bind to ferriprotoporphyrin IX. This exciting new class of
compound is currently under development, while human Phase II clinical trials of
T3/SAR97276 are ongoing for parenteral cures of severe malaria. This clinical
candidate is structurally unrelated to existing antimalarial agents, and acts through
new, independent mechanisms of action. The drug's unique properties are of
tremendous interest in the role of an anti-infectious agent, and indicate that the
targeting of lipid metabolism is a valuable strategy in the development of new
antiparasitic drugs.

*Corresponding author

Apicomplexan Parasites. Edited by Katja Becker
Copyright © 2011 WILEY-VCH Verlag GmbH & Co. KGaA, Weinheim
ISBN: 978-3-527-32731-7

Introduction

Malaria, caused by protozoan parasites of the genus *Plasmodium*, is one of the most prevalent diseases in the world, with approximately 50% of the world's population at risk in 109 countries. In 2006, there were an estimated 250 million cases of malaria, causing one million deaths, mostly in children aged less than five years [1]. In many African countries, malaria ranks as one of the most serious public health problems, with an estimated loss of 40 million disability-adjusted life years (DALYs) per annum. Moreover, developed countries are not shielded from this situation, as the number of imported cases has increased due to the expansion of international transport. Currently, the major factors of recrudescence are the adaptation of both the parasite and the vector to human interventions, that is, antimalarial drugs and insecticides. In most cases, *Plasmodium falciparum* is responsible for malaria (80%), and causes the more severe form of the disease, which is very often fatal. Other causative parasites in humans are *Plasmodium vivax, Plasmodium ovale, Plasmodium malariae,* and the recently reported *Plasmodium knowlesi* [1, 2].

Nowadays, mortality due to malaria is mainly related to the evolution of parasite resistance to the armamentarium of antimalarial drugs. The affordable and widely available antimalarials chloroquine and sulfadoxine–pyrimethamine, which used to from the mainstay of malaria control, are today ineffective in most *P. falciparum* malaria endemic areas. The discovery of artemisinins has provided a new class of highly effective and rapidly acting drugs, and has transformed malaria chemotherapy. Although artemisinin-based combination therapy (ACT) is generally considered the best current treatment for uncomplicated *falciparum* malaria, its widespread use raises the issue of emerging drug resistance. This is of considerable concern in the light of recent reports on decreased sensitivities to artemisinin drugs in southeast Asia [3, 4]. Consequently, the need for new antimalarial strategies involving novel targets appears as crucial as ever. It is hoped that new drugs with independent modes of action will increase the arsenal of drug combinations, thus delaying the development of resistance to the individual components.

Novel antimalarial chemotherapies have already been described from a general standpoint [5–8], and also with respect to specific targets such as fatty acids [9, 10] and phospholipids (PLs) [11, 12]. In this chapter, attention is focused on the various steps of malarial lipids that have been shown to be relevant as antimalarial pharmacological targets. The results presented herein have paved the way to new classes of antimalarial drugs that target membrane biogenesis of the *P. falciparum* erythrocytic stage.

Malarial Lipids: Why are Glycerophospholipids Not Dispensable?

A complex life cycle that requires both invertebrate (mosquito) and vertebrate hosts is common to all *Plasmodium* species. In the human host, the parasite remains mainly intracellular, initially within the hepatocytes, and this is followed by a cyclic development within red blood cells, which causes the clinical disease. Within one week in the liver, a single infected hepatocyte produces some 20 000 short-lived extracellular merozoites, which rapidly invade the red blood cells. There, they multiply rapidly, with each merozoite producing 16–32 offspring every 48 h, resulting in a rapid invasion of further erythrocytes. This rate of production requires a high parasite metabolic activity, including high levels of membrane formation. Whilst the malarial lipid composition has been substantially documented for *P. falciparum* and *P. knowlesi*, albeit only at their blood stages [13–16], very little is known about the parasite-induced changes in red blood cells infected with *P. vivax*. This situation is partially due to the lack of a robust culture and easy-to-study host laboratory animals.

The membranes of *Plasmodium* spp. parasites are mainly composed of glycerophospholipids, and typically lack complex lipids or sterols/ergosterols, in contrast to prokaryotic and yeast/mammalian cells. In uninfected erythrocytes, the main PLs are phosphatidylcholine (PC) (35–40%), phosphatidylethanolamine (PE) (30–35%), sphingomyelin (SM) (15%), and phosphatidylserine (PS) (10%) [17]. Other PLs, such as phosphatidylinositol (PI), phosphatidic acid (PA), phosphatidylglycerol (PG), and cardiolipid (CL), account for less than 3% of the total PLs, while neutral lipids are barely detectable. Upon *Plasmodium* infection, a sixfold increase in erythrocytic PL content is observed at the trophozoite stage (Figure 8.1), along with especially large increases in the PC, PE, and PI contents [18]. In purified parasites, the main PLs are PC (40–50%), PE (35–45%), PI (4–11%), and PS (<5%). The PE content is unusually high compared to its level in other eukaryotes; together, the PC and PE represent 75–85% of the parasite PL composition [13, 15, 18, 19]. A major increase in neutral

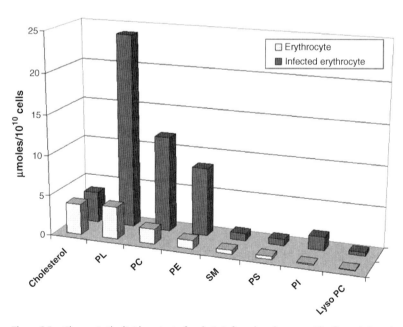

Figure 8.1 Change in the lipid content of malaria-infected erythrocytes. The figure is based on the mature developmental stage, and only the major lipids are shown [14, 18].

lipids is also detected, although the final amounts remain low when compared to PL [20]. Previous studies of *P. berghei* and *P. vinckei* rodent plasmodia have shown similar drastic changes in PL contents and composition upon erythrocyte infection [21, 22]. Compared to the host erythrocytes, the fatty acids grafted onto the glycerol backbone are modified in length and the degree of unsaturation, both of which are decreased [14].

Glycerolipid Acquisition

Lipid metabolism is nonfunctional in uninfected mature erythrocytes [17], and despite exchanges with high- and low-density lipoproteins [23], the import of preformed PLs from the serum does not substantially account for the total PLs found in infected erythrocytes [15, 18]. Actually, the parasite relies on its own robust PL biosynthetic machinery [14, 15].

   Plasmodium has been shown to contain most of the enzymatic activities needed for PL biosynthesis; indeed, the results of biochemical quantification experiments strongly suggest that the parasitic machinery provides the bulk of PLs composing *P. falciparum* membranes at its blood stage [14, 18]. Sequencing of their genome and genomic studies have confirmed that malaria parasites possess a panoply of corresponding genes [15].

Sources of PL Building Units

At the blood stage, the very high biosynthetic capacity of *Plasmodium* operates at the expense of the polar head and fatty acid (FA) building units that mainly originate from the plasma. The entry of choline into the infected red blood cells (IRBC) involves the erythrocytic choline carrier [24] and parasite-induced new permeation pathways (NPP) [25, 26]. Choline is provided to the parasite by a characterized and very efficient organic-cation transporter (OCT) [25], while ethanolamine can be supplied from the poorly available plasmatic ethanolamine, which can cross the membrane by passive diffusion, and from serine. Both, choline and ethanolamine are effectively trapped within the cell by phosphorylation [27–29]. Serine is diverted either from the host red blood cells or from hemoglobin degradation in the food vacuoles [30]. Inositol is scavenged from the host and perhaps synthesized *de novo* via an inositol synthase expressed at the blood stage (PlasmoDB). The glycerol backbone originates from glycerol 3-phosphate (G3P) that is mainly formed from glucose via anaerobic glycolysis reactions [31]. It has also been proposed that G3P might arise from host glycerol recovered by a single aquaglyceroporin (AQP) [32–34] and subsequently phosphorylated by a glycerol kinase [35]. However, both genes are nonessential at the blood stage, even though AQP-null *P. berghei* parasites proliferate more slowly compared to the wild-type parasites [35, 36].

   Plasmatic FAs, but not PLs or neutral lipids, appear to be essential for parasite survival [37, 38], and the amount of FAs derived from the plasma should be sufficient for PL synthesis [19, 39, 40]. Malarial parasites exhibit specificity with regards to both the chain length and level of unsaturation of FAs. The activation of FAs to acyl-CoA thioesters via acyl-CoA synthetase and acyl-CoA binding proteins, is required for most cellular reactions that involve FAs. The FAs are primarily introduced into PLs during the *de novo* synthesis of PA; furthermore, molecular species of PL are strictly controlled by 12 putative acyl-CoA synthetases [10, 15] and choline/ethanolamine-phosphotransferase [41]. Until now, the available biochemical data have suggested that plasmatic FAs were incorporated into malarial lipids without any structural modification during intraerythrocytic development [42], and that the parasite was unable to elongate or desaturate the scavenged FA to any detectable extent [9]. However, the *P. falciparum* genome has been found to contain genes and related biochemical activities of long-chain fatty acid elongases and FA desaturases [10, 15, 43].

An Assembly of Puzzling Metabolic Pathways

Plasmodium falciparum uses a bewildering number of metabolic pathways for PL biosynthesis which, to the present authors' knowledge, have never been identified in the same organism [15]. At this point, a brief description will be provided only of those pathways that have been quantified at the blood stage: (i) the ancestral CDP-DAG-dependent pathway; (ii) eukaryotic *de novo* CDP-choline and CDP-ethanolamine (Kennedy) pathways; and (iii) the serine decarboxylation-phosphoethanolamine methylation (SDPM) route (Figure 8.2). The latter is of great interest as it involves serine decarboxylase (SD) and a novel methyltransferase class that has been characterized in plants and is distributed only sporadically throughout animal genomes. To date, these have been described only in *P. falciparum* and the nematode worm *Caenorhabditis elegans*.

CDP-Choline and CDP-Ethanolamine Pathways

In the so-called Kennedy pathway, choline is phosphorylated into phosphocholine, which is subsequently coupled to CTP, thus generating CDP-choline, which is further converted to PC by a parasitic CDP-diacylglycerol-cholinephosphotransferase (CEPT). A similar *de novo* pathway allows the synthesis of PE from ethanolamine. The final stage of both branches of the Kennedy pathway involves the same CEPT enzyme, catalyzing the formation of PC and PE from CDP-choline and CDP-ethanolamine, respectively [15]. Besides this, most eukaryotic cells, particularly *Saccharomyces cerevisiae*, and liver human cells are able to methylate PE to PC through the PE *N*-methyltransferase (PEMT) pathway [44, 45]. The high capacity of *P. knowlesi*-infected erythrocytes to methylate PE into PC unambiguously indicates the presence of PEMT activity [46], and similar labeling of PE and PC from radioactive ethanolamine and serine indicate that this PEMT pathway is also present in *P. falciparum* [47]. However, the corresponding genes coding for PEMT have not yet been found in *Plasmodium* species. Intriguingly, the PfPMT protein (recombinant or expressed in mutant yeast cells lacks the ability to catalyze the transmethylation of PE [48, 49], and deletion of the *PfPmt* gene in *P. falciparum* parasites failed to reveal the PEMT pathway [50]. Thus, the "enzymatic hole" of the PEMT pathway remains to be clarified.

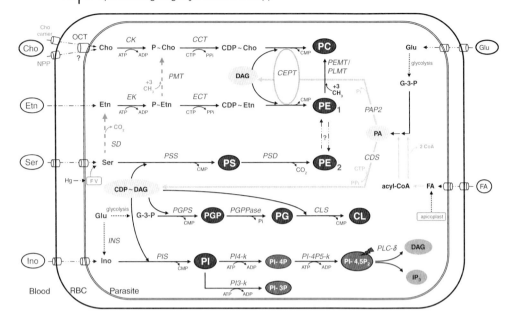

Figure 8.2 Schematic view of *P. falciparum* phospholipid biosynthesis pathways. The SDPM pathway is differentiated by arrows with broken line. PMT and PEMT (unknown gene) activities were only detected in *P. falciparum*; SD activity was only detected in *P. falciparum* and weakly in *P. berghei* (see Ref. [52] for details). Lipids in different shades of grey; enzymes are shown in italics. Abbreviations of metabolites: CL, cardiolipin; DAG, diacylglycerol; CDP~, cytidine-diphospho-; FA, fatty acids; Glu, glucose; G-3-P, glycerol 3-phosphate; Ino, myoinositol; P~, phospho; PA, phosphatidic acid; PC, phosphatidylcholine; PE, phosphatidylethanolamine; PG, phosphatidylglycerol; PGP, phosphatidylglycerolphosphate; PI, phosphatidylinositol; PI-3P, PI 3-phosphate; PI-4P, PI 4-phosphate; PI-4-5P₂, PI 4,5-bisphosphate; PS, phosphatidylserine. Enzymes (with PlasmoDB identifier for putative corresponding gene when identified): CDS, Cytidine diphosphate-DAG synthase (PF14_0097); CK, Cho kinase (PF14_0020); CCT, CTP: P~Cho cytidylyltransferase (PF13_0092); CLS, CL Synthase (MAL6P1.97); EK, Etn kinase (PF11_0257); ECT, CTP: P~Etn cytidylyltransferase (PF13_0253); CEPT, Cho/Etn-phosphotransferase (MAL6P1.145); INS, Ino 1-phosphate synthase (PFE0585c); PAP2, PA phosphatase (PFF1210w); PGPS, PGP synthase; PGPPase, PGP phosphatase; PEMT, PE N-methyltransferase; PIS, PI synthase (MAL13P1.82); PI3-k, PI 3-kinase (PFE0765w); PI4-k, PI 4-kinase (PFE0485w, PFD0965w); PI-4P5-k, PI 4-P 5-kinase (PFA0515w); PLC–δ, phospholipase Cδ (PF10_0132); PLMT, phosphatidyl-N-methylethanolamine N-methyltransferase; PMME, phosphatidyl-N-methylethanolamine; PMT, P~Etn N-methyltransferase (MAL13P1.214); PSD, PS decarboxylase (PFI1370c); PSS, PS synthase CDP-DAG-dependent (MAL8P1.58); SD, Serine decarboxylase.

CDP-DAG-Dependent Ancestral Pathway

Additionally, *Plasmodium* possesses the CDP-DAG-dependent pathway, which provides the anionic phospholipids PS, PI and, eventually, also PG and cardiolipid. The biosynthesized PS is intensively converted into PE via the activity of PS decarboxylase [53].

SDPM Plant-Like Horizontal Pathway

Plasmodium falciparum also has a plant-like pathway that relies on serine to provide additional PC and PE, which is named the serine decarboxylase-phosphoethanolamine methyltransferase (SDPM) pathway. Hereby, serine is first decarboxylated to form ethanolamine, which is then phosphorylated to lead to phosphoethanolamine. The SD enzymatic activity was first described by the present authors' group in *P. knowlesi* and *P. falciparum* [47], while the gene and related SD catalytic activities were subsequently identified in plants [54]. The resulting phosphoethanolamine is either incorporated into PE via the CDP-ethanolamine pathway, or converted into phosphocholine by SAM-dependent triple methylation, which is carried out by a plant-like phosphoethanolamine *N*-methyltransferase (PfPMT; EC 2.1.1.103) [48, 49]. In the Apicomplexa phylum, the SDPM pathway is only conserved in the plasmodia, and is deleted in rodent parasites. The lack of a PMT homologous gene at the expected locus in rodent *Plasmodium* species is likely due to an early deletion, correlated with the separation of rodent from nonrodent *Plasmodium* branches [52]. This highlights a crucial difference in PL metabolism between *Plasmodium* species, and should be borne in mind when developing rational approaches to identify and evaluate new targets for antimalarial therapy.

How Crucial are these Metabolic Pathways?

Pharmacology implies finding the Achilles' heel and appropriate effectors to affect the parasite's growth and prevent its survival. Intuitively – and since PLs represent the major and indispensable lipid components of parasite membrane bilayer structures – any process that hampers PL acquisition should also prevent parasite membrane neogenesis, and would therefore be detrimental to parasite survival. However, the problem remains complex because *P. falciparum* possesses an astonishing variety of biosynthetic pathways for its structural phospholipids, PC and PE. Consequently, the individual pathways are critical, inasmuch as their final products are needed for the construction of cell membranes. Hence, the question arises: How essential and/or how complementary are these metabolic pathways? Essentiality may occur if the different metabolic pathways are attributed to individual subcellular compartments, serve distinct metabolic functions, and/or if specific metabolic pathways are required at the various stages of parasite development. The involvement of each pathway in the formation of one given PL, and its eventual cross-regulations, remain to be clarified. Recently, the metabolic pathways were modified by incubating *P. falciparum*-infected erythrocytes with different concentrations of polar head precursors. Subsequent metabolic flux measurements failed to reveal any significant interactions between the CDP-DAG-dependent and *de novo* PC, and PE pathways (H.J. Vial *et al.* unpublished results), as were demonstrated in yeast [44, 55]. The only interactions observed occurred at very high and nonphysiological precursor concentrations (i.e., millimolar range), and may be related to a direct alteration of the transport or enzymatic steps. The absence of any cross-pathway regulatory mechanisms between the CDP-DAG and Kennedy pathways might indicate that they have distinct roles or specialized functions in the parasite. Regulation of the SDPM pathway involved in PC synthesis by its precursors has recently been established in a comparative study using wild-type and transgenic *P. falciparum* parasites. In this case, exogenous choline produced a dose-dependent repression of PfPMT transcription, a decrease in the corresponding protein, and also induced its proteasomal degradation [56].

Until now, very few genetic studies have demonstrated how crucial each metabolic pathway is to the parasite's biology. PfPMT could be disrupted at the *P. falciparum* blood stage, leading to a substantially reduced multiplication rate and increased cell death; however, this is dependent on the presence of external choline [50]. Since the SDPM and *de novo* CDP-choline pathways use the same initial precursor (phosphocholine) for the ultimate synthesis of PC, their functions are likely not completely redundant. Recent genetic experiments in the rodent malaria parasite *P. berghei* have indicated that the genes coding for PbCK, PbCCT, PbECT, and PbCEPT (see Figure 8.1) are essential for the parasite's survival [51]. However, as this rodent model lacks PMT methyltransferase [52], the extrapolation of a similar crucial role in the human malarial parasite is only tentative. Information is rather scant concerning the lipid function and composition in the hepatic, sexual, or sporozoite stages.

Membrane biogenesis participates in the production of the various subcellular compartments needed by *Plasmodium* throughout its development. In eukaryotic cells, lipid biosynthesis devoted to membrane functions involves multiple cellular compartments, including the cytosol for early water-soluble precursor-dependent reactions, the endoplasmic reticulum (ER), the Golgi apparatus, and mitochondria for the biosynthesis of the ultimate nonpolar lipids. The latter are predominantly microsomal, with some activity located in the mitochondrial inner membrane [55]. The subcellular localization of most malaria enzymes involved in the PL synthetic pathways remains to be determined, although the cytosolic localization of choline kinase (CK) and ethanolamine kinase (EK) [57, 58] is not a surprise. In contrast, *P. falciparum* PS decarboxylase appears to be an ER-resident protein that differs fundamentally from the typical location of type I phosphatidylserine decarboxylase (PSD), which is usually found exclusively in the inner mitochondrial membrane [53]. The PfPMT localization within the parasite Golgi apparatus possibly indicates that the transmethylation reaction occurs therein [59]. Future goals would be to identify these compartments, along with their specific proteins and PLs. Recent fluxomic biochemical studies have highlighted the quantitative importance of individual metabolic pathways [18, 47, 60]. It has been shown that several distinct PE pools in *P. falciparum*-infected erythrocytes are present, and that these depend on their mode of synthesis (from exogenous ethanolamine, serine, or serine-derived ethanolamine), thus suggesting a cellular compartmentalization of these metabolic pathways [47].

Lipid-Based Antimalarial-Chemotherapy: Pharmacological Validation

Because of their major critical role in *Plasmodium* development and survival, the PL biosynthetic pathways represent ideal pharmacological targets, provided that they are

Table 8.1 *In vitro* and *in vivo* antimalarial activity of selected compounds.

Drug classes	Target/mechanisms	Drug/lead compound	*In vitro*	*In vivo*	Reference(s)
Fatty Acids biosynthesis					
	Fab I inhibitors[a]	Triclosan,	60 µM	40	[68, 96]
	Fab B/H inhibitor[a]	Thiolactamin	>10 µM	nd	[68, 96]
	Δ9 fatty acid desaturase inhibitors	Sterculic acid	80 µM	nd	[68, 96]
		Methylsterculate	80 µM		
Fatty acids					
	FA incorporation	Natural or Un-natural FA	>µM	Yes	[73, 74]
False polar head derivatives					
	Membrane alteration (composition, False lipids)	D-2-amino-1-butanol	50 µM	nd	[71]
Choline analogs					
	Phosphatidylcholine synthesis affected	Decamethonium	1 µM	nd	[75]
		Trimethyl dodecyl ammonium, bromide	0.5 µM	nd	[80]
		Tripropyl dodecyl ammonium, bromide	33 nM	nd	[80]
		T1	70 nM	nd	[11]
Bis-cations	Phosphatidylcholine synthesis affected	G25	0.65 nM	0.22	[82]
		T3/SAR97276	3 nM	0.2	[84]
		T4	0.65 nM	0.14	[84]

a) These targets have been recently questioned, since FAS-II does not appear to be essential at the blood stage.
In vitro activity corresponds to EC_{50} against *P. falciparum* growth assessed after 48 h contact with the drug. *In vivo* activity reflects ED_{50} ($mg\,kg^{-1}$) against *P. vinckei* determined after intraperitoneal administration of the compound once daily for four days in infected mice.
nd: not determined.

crucial (see above). Although the pharmacology of lipids is still at the budding stage, a number of concepts have begun to emerge, and many steps within PL metabolism have appeared as potential targets (Table 8.1). The metabolic pathways usually include regulatory steps that control the entire pathway and are of the utmost importance for chemotherapeutic attack. The two enzymes that lead off from PA – that is, CDP-diacylglycerol synthase and PA phosphohydrolase – are of interest because of their assumed roles in controlling the flux into the two branches representing ancestral CDP-DAG-dependent and *de novo* PL biosynthesis. The synthesis of CDP-diacylglycerol by CDS is expected to be a regulated step, as it is a branch point in glycerophospholipid synthesis. The cytidylyltransferases CCT and ECT are clearly the regulatory steps of *de novo* Kennedy pathways in yeast and most mammalian cells [55, 61–63]. In *Plasmodium*, it has been shown that *de novo* PC biosynthesis is controlled by at least two steps, namely CCT and choline transport [25, 64].

Inhibitors of the Fas-II Pathway

Plasmodium ssp. possess the prokaryotic type II fatty acid biosynthesis pathway (FAS-II), with all components of the pathway being localized to the apicoplast [65]. This biosynthetic pathway incorporates several targetable enzymes [66]. Based on the use of triclosan, a well-known inhibitor of bacterial FabI, it was hypothesized that the FAS-II system was essential for the blood stage [67]. Other interesting drugs include isoniazid (inhibition of FabI), thiolactamin, and derivatives (inhibition of FabB and FabH) [68]. Recently, the successful disruption of *FabI* genes in *P. falciparum* and *P. berghei* parasites [69], the deletion of *FabI* in *P. falciparum*, and the deletion of *FabB/F* and *FabZ* in *P. yoelli* [70] have shown that the FAS-II system is needed neither for blood and mosquito stage development, nor for the initial liver infection. On the other hand, genetic tools associated with cell biology studies have indicated that the FAS-II system is crucial only within a short time-window during the late asymptomatic liver development stage [69, 70]. Yet, why the FAS-II system is specifically required for this short period, and why the merozoites are unable to divide and thus initiate blood infection, remain to be clarified. However, two proposed explanations have included a burst of FA requirements for tremendous membrane biogenesis, or a requirement for specific lipid types at this period.

False Precursors to Deceive the Parasite

The first option to prevent PL synthesis that occurs at the expense of exogenous precursors (see Table 8.1) is to use false precursors, such as small analogs of polar heads or unnatural FAs, that can be added to deceive the parasite. These may block the synthesis of the ultimate lipid products, or they may lead to the production of false metabolites that alter the lipid composition of parasite membranes and their physicochemical properties, thus causing death of the parasite (see Ref. [12] and references therein). Such small molecules, which are analogs of choline or ethanolamine and which retain a free hydroxyl group, affect the *in vitro* growth of *P. falciparum*, with an EC_{50} of 50 µM at best (the EC_{50} is the efficient concentration needed to clear parasitemia by 50%). The distance between the nitrogen atom and the hydroxyl group appears to be crucial, and substitutions at the α position (referring to the nitrogen atom) are far more effective than at the β position [71]. These analogs are used by the parasite to build up unnatural PLs, which accumulate at the expense of the natural PE form. Nevertheless, the initial concern was that this might not be a good strategy, due to the potential toxicity of long-term accumulated false metabolites. Similarly, within the FA family, polyunsaturated [72] or naturally occurring [73] FAs were shown to alter parasite growth, most likely by inducing lipid peroxidation or by altering the membrane homeostasis. Non-natural FAs exert an effective inhibition of *P. falciparum* growth, with EC_{50} values of 7 to 90 µM, and this was clearly correlated with the inhibition of acyl-CoA synthetase, the first enzyme to metabolize the FAs [74].

Inhibitors of Glycerophospholipid-Synthesizing Pathways

An alternative was to use effectors that may interfere with the supply of plasmatic precursors, or with the biosynthetic enzymatic process, thereby inhibiting the provision of essential lipid molecules. This may concern the *de novo* biosynthesis of PC or PE, the methylation step of the phosphoethanolamine unit, or the PS biosynthesis pathway.

Locking the choline carrier (a limiting step which provides the parasite with the precursors required for PC biosynthesis) by using commercially available choline analogs led to an EC_{50} of between 0.7 and 10 μM, and did not induce the formation of any false metabolites [75, 76]. Parasite growth arrest using hexadecyltrimethylammonium bromide was attributed to the inhibition of PfCK, the first enzyme in the PC *de novo* pathway [77], but this requires further investigation.

Rationale Based on the Use of Choline Analogs to Block PC Biosynthesis

This led to the rational design of various choline analogs (Figure 8.3) that consisted of mono- and bis-quaternary ammonium salts and, in turn, to optimize their ability to kill *P. falciparum* at the asexual blood stages.

It was first observed within the mono-quaternary ammonium salt series, that an increase in chain lipophilia in the vicinity of the nitrogen atom was beneficial for antimalarial activity, as the EC_{50} was decreased by one order of magnitude from the trimethyl to tripropyl substituents. Regardless of the polar head substitution (methyl, ethyl, hydroxyethyl, and pyrrolidinium), the increase in alkyl chain length from six to 12 methylene groups always improved the activity (Figure 8.4a). The highest activity was observed for the *N,N,N*-tripropyl-*N*-dodecyl ammonium salt, which exhibited an EC_{50} value of 33 nM [78]. Alkylphosphocholine analogs (miltefosine or hexadecylphosphocholine) were also shown to inhibit growth of the erythrocytic parasite, and this phenomenon was correlated with *Plasmodium* PfPmt inhibition [49].

An important and decisive step was reached when duplication of the cationic head groups ("twin-drug") was achieved, which led to a considerable increase in antimalarial activity, with EC_{50} values at single-digit nanomolar concentrations. Bis-ammonium salts were generally 100-fold more active than the mono-ammonium salts [79–81]. Subsequent structure–activity relationship (SAR) studies have highlighted the importance of the spacer that separates the two cationic heads, and also the effect of steric hindrance and lipophilia of the nitrogen atom substituents. An increase in the lipophilic-spacer length between the two nitrogen atoms (from 5 to 21 methylene groups) led to a constant improvement in the activity (Figure 8.4b). For most of these compounds the EC_{50} values were about 1 nM, while the more lipophilic compound (with a 21 methylene-spacer) exhibited an EC_{50} as low as 3 pM.

When the outstanding antimalarial activity of the first-generation lead compound G25 was documented (Figure 8.3), it showed an EC_{50} value of 0.65 nM, and an ability to cure *Aotus/Rhesus* monkeys that had been highly infected by *P. falciparum* (at a very

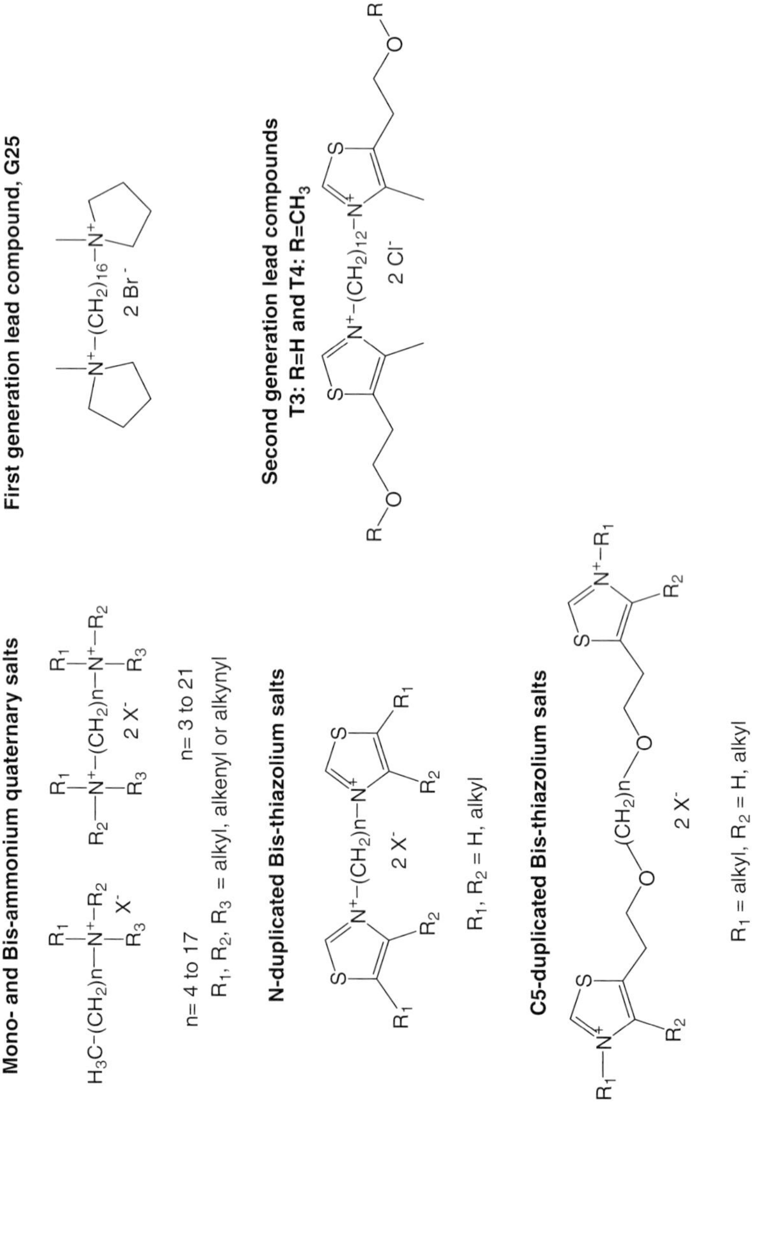

First generation lead compound, G25

Mono- and Bis-ammonium quaternary salts

n= 4 to 17

n= 3 to 21

R₁, R₂, R₃ = alkyl, alkenyl or alkynyl

**Second generation lead compounds
T3: R=H and T4: R=CH₃**

N-duplicated Bis-thiazolium salts

R₁, R₂ = H, alkyl

C5-duplicated Bis-thiazolium salts

R₁ = alkyl, R₂ = H, alkyl

Figure 8.3 General structures of the various series of choline analogs.

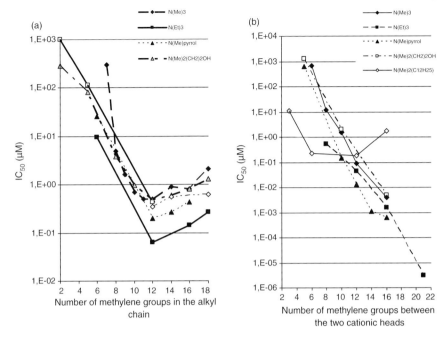

Figure 8.4 Antimalarial activity (*P. falciparum*) of mono-quaternary ammonium (a) and bis-quaternary ammonium salts (b), incorporating various polar heads as a function of the alkyl chain length [80].

low dose of $0.03\,mg\,kg^{-1}$) or *Plasmodium cynomologi*. Complete cures were also observed at a high initial parasitemia, and without recrudescence [82].

In a second-generation lead compound, the pyrolidinium moiety of G25 was replaced by a less-toxic thiazolium ring already present in vitamin B1 [83, 84]. Subsequently, an additional series of mono- and bis-thiazolium salts was synthesized in which the polar heads were linked either with the nitrogen atom, or at the C-5 position of the thiazolium moiety (see Figure 8.3). Among these derivatives, the two most promising (namely T3/SAR97276 and T4) displayed ED_{50} values of $0.2\,mg\,kg^{-1}$ versus *P. vinckei in vivo* (ED_{50} is the efficient dose needed to clear parasitemia by 50%). These results indicated that the development of a pharmacological model for antimalarial agents through the design of choline analogs represents a promising strategy.

Pharmacological Activity Performances of Choline Analogs

The properties of these derivatives may be gleaned from the literature [11, 12]. Herein, attention will be focused on thiazolium-based choline analogs, one of which is currently undergoing clinical development (see T3/SAR97276; see Figure 8.3).

In-Vitro Antimalarial Activity

Bisthiazolium salts with the optimal linker (12 methylene groups) have been shown to exert potent antimalarial activity against the Nigerian, 3D7, and chloroquino-resistant strains of *P. falciparum*, with an EC_{50} below 3 nM [83, 84]. They appear to be equally potent as the first-generation G25 compound (EC_{50} 0.6 nM) [80, 81], and also have powerful toxic properties against *Babesia divergens* (EC_{50} 20–80 nM) [85] (H.J. Vial *et al.*, unpublished results), a malarial-like parasite that proliferates inside mammalian erythrocytes.

Of the utmost interest, *in vitro* experiments have shown that the bis-thiazolium salts T3/SAR97276 and T4 are highly toxic throughout the *P. falciparum* blood stage, exerting a very rapid (i.e., nonreversible) cytotoxic effect after only few hours of contact with the parasites (Figure 8.5). This antimalarial activity against all blood

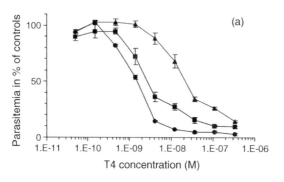

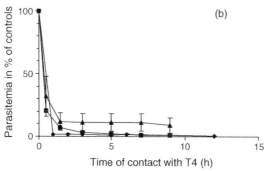

Figure 8.5 Inhibition of *in vitro*-cultured *P. falciparum* (3D7) [84]. (a) Stage-dependent susceptibility. T4 was added at the indicated concentrations to synchronized parasites in culture at ring (●), trophozoite (22 h, ■), or schizont (33 h, ▲) stages. After 4 h, cells were washed twice and resuspended in fresh complete medium; (b), Time course of *P. falciparum* growth inhibition. T4 was added at 40 nM to synchronized cultures at ring, trophozoite, and schizont stages. After selected incubation times, cells were washed twice and resuspended in fresh complete medium. For both experiments, $[^{3}H]$hypoxanthine (0.7 µCi per well) was added at time 52–76 h to monitor the parasite viability. The results are expressed as mean $\pm$ SEM ($n \geq 3$).

stages is most likely due to the rapid and nonreversible accumulation of the compounds within the parasites at all stages. These two points are crucial for understanding the *in vivo* potency of these classes of compounds, even after a single injection, or their capacity to kill the parasites without recrudescence at two- to fourfold the ED_{50} value [84] (H.J. Vial *et al.*, unpublished results).

In-Vivo Antimalarial Activity Against Rodent Malaria

Plasmodium vinckei infection in mice appears to be a suitable model for *P. falciparum* malaria, because of its synchronized cycle and preference for mature erythrocytes. The compounds were evaluated under very severe conditions, such as at high parasitemia or in short-course treatments, with a single injection [84]. Following the daily administration of T3/SAR97276 for four days, the mice were completely cured, with an ED_{50} below $0.3\,\mathrm{mg\,kg^{-1}}$ for the intraperitoneal, intra-muscular and intravenous routes of administration. Remarkably, the ED_{50} of T3/SAR97276 was in the same range at low and high initial parasitemia, namely $0.3\,\mathrm{mg\,kg^{-1}}$ and $\leq 0.5\,\mathrm{mg\,kg^{-1}}$, respectively. The ED_{90} and a complete cure without recrudescence were obtained at two- to fourfold the ED_{50} value. A single injection of T3/SAR97276 at a dosage of $6\,\mathrm{mg\,kg^{-1}}$ led to a complete cure of mice infected with *P. vinckei* at low or moderate parasitemia, thus confirming the potency of the compound.

Comparison with Artesunate and Drug Combinations

T3/SAR97276/was evaluated comparatively to artesunate, which has a very remark-able antimalarial activity due to its high parasite killing rate [86]. T3/SAR97276 showed a more potent activity (ED_{50} $0.5\,\mathrm{mg\,kg^{-1}}$) than artesunate ($ED_{50}$ $>2.5\,\mathrm{mg\,kg^{-1}}$); moreover, whilst a total cure was obtained with T3/SAR97276 at a daily dose of $0.5\,\mathrm{mg\,kg^{-1}}$ for four days, this was not achieved with artesunate dosed even at $10\,\mathrm{mg\,kg^{-1}}$.

Mechanism by Which Choline Analogs Exert Their Antimalarial Activity

Bis-Cationic Choline Analogs are Targeted to the Parasite Where They Accumulate

The structural requirements associated with all choline carriers indicate that the endogenous choline carrier of erythrocytes would not be able to transport drugs inside the infected erythrocyte [87]. This was further confirmed using T16, a radiolabeled bis-thiazolium salt which is not transported by the dedicated choline transporter of yeast [88]. However, bis-cation derivatives [89] and T16 [25] are substrates of the NPP pathway [90, 91] located on the erythrocytic plasma membrane, which may at least partially mediate their entry into infected erythrocytes. Their major route for entering the intracellular parasite is a poly-specific OCT, as distinct from known dedicated eukaryotic choline carriers [25].

One prominent feature of bis-quaternary ammonium derivatives is their ability to accumulate heavily in *Plasmodium*-[82, 84, 92] and *Babesia* [85]-infected erythrocytes.

Choline Analogs Accumulate Within the Parasite and Exert a "Trojan Horse" Effect

Although various series of compounds (e.g., G25, T3/SAR97276 or T4) have been shown to exert a rapid, cytotoxic, nonreversible effect (Figure 8.5), the parasite morphology is not immediately affected. This is most likely due to a swift drug uptake into the parasites, leading to the parasite's death as soon as the drug has been accumulated. This "Trojan horse" effect was confirmed in a study of parasite viability after one or two injections of T3/SAR97276 into infected mice. Parasite infectiousness, as assessed by the injection of blood from treated mice into naive mice, was immediately and drastically stopped by more than 99.9% within 2 h after drug injection, even though the parasites appear to be morphologically normal (S. Wein and H.J. Vial, unpublished results).

Choline Analogs Prevent Malaria Phosphatidylcholine Biosynthesis

There is extensive evidence to indicate that mono-[78], bis-[79] quaternary ammonium, and bis-thiazolium [84, 85] salts exert an early and specific alteration of PC biosynthesis, thus preventing synthesis of the final PC products. Choline strongly antagonized the antimalarial activity of the analogs prepared, while a high correlation between PC and antimalarial inhibition was noted [12].

The mechanisms by which the *de novo* pathway is inhibited have yet to be fully elucidated. Alkyltrimethyl ammonium salts (with a 10-to 16-methylene alkyl chain length) competitively inhibited choline transport into *P. knowlesi*-infected erythrocyte suspensions. The K_i values were found to be in the very low micromolar range (3–8 µM), and the findings closely agreed with previously reported EC_{50}-values and also PC_{50}-values (the concentration required to inhibit PC synthesis by 50%) [78]. Indeed, T16 was seen to be a potent inhibitor of parasite choline uptake, with an EC_{50} of 140 nM. Thus, bis-cationic compounds most likely interfere with the *in vitro* growth of *P. falciparum* by affecting its essential, *de novo* PC biosynthesis. Considering that the antimalarial activity of the drug relies primarily on a high accumulation in infected red blood cells, and that the compound substantially accumulates in the host cell compartment of infected cells, the drugs can likely reach a significantly higher concentration, which is critical for choline entry into the parasite. Interestingly, raising the choline concentration in the medium caused a dose-dependent inhibition of T16 uptake, which was directly proportional to the inhibition of antimalarial activity [25]. Finally, Ben Mamoun and colleagues, in Connecticut, USA, used the yeast *S. cerevisiae* as a surrogate system to identify the targets of these antimalarial compounds. It was striking that G25 relied on a functional PC synthetic machinery for its activity, as well as on a functional PS decarboxylase, PSD1 [88]. However, it cannot be claimed that the antimalarial activity is due solely to the inhibition of choline entry into *Plasmodium*-infected erythrocytes. Rather, at this stage the possibility should be considered that these compounds may induce additional inhibition at later steps of the CDP-choline pathway.

Subsequently, a systematic approach was used to gain insight into the mechanism of action of the bis-thiazolium compound T4 against *P. falciparum*, and to determine how it would specifically inhibit plasmodial PC biosynthesis. Transcriptome profiling demonstrated cell cycle arrest and the general induction of genes involved in gametocytogenesis, but no apparent transcriptional changes in genes were involved in the PC biosynthetic pathways. In contrast, proteomic analysis revealed a significant decrease in the level of the *P. falciparum* Cho/Eth-phosphotransferase (PfCEPT) involved in the final step of PC synthesis. This effect was further supported by the findings of metabolic studies, which suggested that the inhibition of choline transport into the parasite was most likely the primary step. In addition, these compounds might target more than one protein within the PC biosynthesis pathway (choline transporter, cytidylyltransferase and possibly the PC synthesis via the SDPM pathway), and provide evidence of a post-transcriptional regulation of the parasite metabolism in response to external stimuli [60].

Choline Analogs Versus Cholinergic Effects

A crucial issue for drug development is the specificity required to achieve an acceptable therapeutic index. One potential drawback of choline analogs concerns the involvement of choline as a precursor of the acetylcholine neurotransmitter. This may be responsible for the relatively high toxicity of the first-generation compounds, based on their quaternary ammonium structure. Choline entry into infected erythrocytes was further characterized to identify some specific features [24], notably in comparison with the nervous system where it is also a substrate for various enzymatic activities [93]. Choline entry into *Plasmodium* parasites is mediated by carriers that differ from those involved in acetylcholine biosynthesis. High-affinity choline transport (HACT) was demonstrated to be stereospecific for choline analogs, such as α- or β-methylcholine, in contrast to normal and infected erythrocytes [93].

The absence of a plateau for antimalarial activity as a function of the methylene chain length (up to 16) in the bis-ammonium series contrasts strongly with the patterns observed for nicotinic receptors or HACT in synaptosomes, for which inhibition was maximal for 8–12 and 16–18 methylenes, respectively [94, 95]. This indicates that, in both cases – the nicotinic receptor and HACT – the spacer is much shorter than in infected erythrocytes. Besides this, the cholinergic compound hemicholinium (HC3) showed a 75-fold lower EC_{50} in synaptosomes than did G3 (decamethonium bromide, a first-generation bis-quaternary derivative) [78], while their activities were very close against *P. falciparum*.

The marked differences with respect to other cholinergic models (e.g., HACT, nicotinic or muscarinic cholinergic receptor, cholinesterase) in terms of pattern and relative affinity as a function of chain length, steric fit related to the bulk of the nitrogen atom substituents, and antimalarial effects, indicated that the targets differed considerably. This strongly suggests a possible pharmacological discrimination between the antimalarial activity of the choline analogs and their toxic effects, which is mainly mediated via their cholinergic effect on the central nervous system.

One Drug, Many Targets

Pharmacology usually refers to the association of a single drug with a single target. In principle, the interference of a single pharmacological target with a highly potent compound is desirable, because it minimizes the side effects derived from nonspecific drugs. Based on the common use of combination therapies in the field of anti-infectious chemotherapies, it is clear that success may also be achieved via the interaction of one drug with at least two targets.

Compounds also Interact with Plasmodial Hemoglobin Degradation Metabolites in Food Vacuoles

Bis-cationic analogs accumulate to a high extent within the parasite, and are in part recovered in the *P. falciparum* food vacuole, where T16 was found to associate with heme (Figure 8.6); this binding was seen to be partly critical for drug accumulation [85, 92]. It should be noted that the use of a protease inhibitor which prevented hemoglobin degradation and free-heme formation revealed that binding to ferri-protoporphyrin IX or to hemozoin in the parasite contributed significantly to both the accumulation and antimalarial activity of these dual drugs [92]. Thus, in addition to

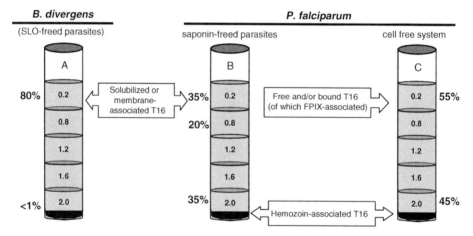

Figure 8.6 Subcellular distribution of a radiolabeled bis-thiazolium salt (T16) in hematozoan parasites and hemozoin interaction. At 50 nM, T16 accumulates inside the *P. falciparum*- or *B. divergens*-infected erythrocytes with a cellular accumulation ratio of 60 and 360, respectively, after 2 h [85]. In the case of *P. falciparum*, a substantial part of the T16 associates with heme and hemozoin inside the food vacuole [92]. Columns A and B: Subcellular distributions of T16 in *B. divergens* and *P. falciparum* parasites. Streptolysin O (SLO)- [85] and saponin-freed [92] parasites were sonicated and overlaid onto discontinuous sucrose gradients. Much of the radioactivity was recovered in the upper fractions in a soluble-free or protein-associated form, or associated with light membranes floating over 0.8 M sucrose [85]. Column C: T16 interacts in a cell-free system with hemozoin crystals. T16 was incubated at 50 nM with FPIX-loaded ghost erythrocyte membrane in conditions that permitted hemozoin formation. The samples were pelleted by centrifugation and applied to a sucrose gradient [92].

their selective inhibition of *de novo* PC biosynthesis, the potent antimalarial activity of bis-cationic derivatives may also involve their compartmentalization within the parasite food vacuole, where they bind to ferriprotoporphyrin IX and decrease the nucleation and growth of β hematin crystals (H.J. Vial and D. Penarete, unpublished results).

The food vacuole has also been shown to accumulate chloroquine [30] and bis-amidine derivatives, such as pentamidine [89]. These lipophilic and weak-base compounds are readily membrane-permeable, and are assumed to accumulate in

Table 8.2 Features of choline analogs, a pharmacologically oriented program based on a hitherto unused therapeutic targets for antimalarial chemotherapy.

Feature	Property
Efficacy of products	• Potent *in vitro* activity against *P. falciparum* (EC50 <5 nM)
	• Active against multiresistant *P. falciparum* malaria (strains and field isolates)
	• Active throughout the blood stage, and rapidly cytotoxic
	• Potent efficacy *in vivo* against *P. vinckei* in mice at dose <1 mg kg^{-1} i.p.
	• Advantageous pharmacokinetic properties in *P. vinckei*-infected mice
	• Curative *in vivo* against *P. falciparum* malaria in *Aotus* monkeys, without recrudescence
	• Curative *in vivo* against *P. cynomolgi* in *Rhesus* monkeys, without recrudescence
	• Attempts to induce drug resistance have failed after 9 months under drug pressure
	• *In vitro* and *in vivo* activity against *B. divergens* and *B. canis*
Mechanism of action	• High accumulation in intracellular parasites
	• An original mode of action (phosphatidylcholine synthesis inhibition)
	• Likely dual molecules, also interacting with hemozoin crystallization process in the malaria food vacuole
Safety	• High *in vitro* selectivity against hematozoan parasites
	• Leading molecules have a convenient therapeutic index
	• Successful preclinical trials achieved
Chemistry	• Compounds are stable and water-soluble
	• Industrial-scale synthesis should be easy and inexpensive
	• Bioprecursors designed to improve oral absorption
Indication (objectives)	• Treating severe malaria by parenteral route (clinical Phase II trials under way)
	• Controlling chemoresistant uncomplicated malaria
Molecule status	• Patented by CNRS/University Montpellier 2; licensed to Sanofi-Aventis
	• The original strategy can be used for other infectious agents (e.g., *Babesia*)
	• Oral absorption needs to be maximized

the parasite acidic food vacuole where they are trapped upon protonation. Conversely, bis-quaternary ammonium salts cannot intrinsically permeate biological membranes without an appropriate carrier or channel. In the light of their high accumulation inside the parasite food vacuole, a suitable transporter is likely involved to facilitate their entry into this *P. falciparum* organelle.

Current Status of Development

Currently, choline analogs are being developed as antimalarial agents by Sanofi-Aventis, through licensing and a collaborative project with CNRS and the University of Montpellier (Table 8.2). On the basis of their *in-vitro* and *in-vivo* antimalarial activities, toxicology, therapeutic index, absorption, solubility and formula, ease of synthesis, and price, the bis-thiazolium salt T3/SAR97276 has been selected as a clinical candidate. Sanofi-Aventis successfully carried out regulatory preclinical and Phase I clinical studies of T3/SAR97276, and in August 2008, Phase II studies were initiated for the treatment of severe *P. falciparum* malaria via the parenteral route in ten research centers in Africa. This open-label, nonrandomized, non-comparative Phase II study was aimed at assessing the antimalarial activity, safety, and pharmacokinetic profile of SAR97276/T3/SAR97276, following single and repeated administrations via the intravenous and intramuscular routes. The results of the study are expected in late 2010.

In parallel, the present authors are pursuing a global research plan in collaboration with Sanofi-Aventis, to produce an orally available formulation of T3/SAR97276 (e.g., a bioprecursor of T3/SAR97276) or a related molecule with a similar mechanism of action that could be developed for the treatment of uncomplicated malaria. Indeed, although these compounds have already achieved a complete cure after oral administration, their absorption remains to be maximized.

Conclusion

The metabolic pathways employed by *Plasmodium* to obtain its lipids, which are the major membrane components and crucial for membrane biogenesis, have been reviewed. Likewise, the various tools and progress in lipid-based antimalarial chemotherapy have been discussed. These metabolic pathways represent attractive targets for drug development, and offer considerable potential for medium-term pharmacological interventions.

Clearly, the most advanced approach targeting lipid metabolism relates to an inhibition of the *de novo* PC biosynthesis, through the use of large molecules possessing one or two cationic heads, which have been designed as choline analogs. Proof of concept has been obtained in rodent and in nonhuman primate models under very severe conditions, and outstanding activities have been observed. The compounds have been shown to accumulate inside *P. falciparum*, leading to a very rapid toxic effect and, most likely, exerting also a "Trojan horse" effect. The clinical development of this powerful class of compounds and Phase II studies in humans are currently under way. The clinical candidate is structurally unrelated to existing antimalarial agents, and acts through new independent mechanisms of action, while its unique properties are of tremendous interest as an anti-infectious agent. This multiple mode of action, which differs from that of current antimalarial agents, represents a major asset as it could help in delaying the development of resistance.

Acknowledgments

These studies were supported by the European Union (FP7, Network of Excellence), the European Virtual Institute dedicated to Malaria Research (Evimalar), the integrated project Antimal (no. IP-018834), an InnoMad/EuroBiomed Grant and Sanofi-Aventis. The authors thank Sandrine Dechamps for the illustration of the malarial phospholipid metabolism, and past and present members of the Vial and Peyrottes laboratories for their contributions to the studies described in this chapter. They are also grateful to major collaborators: Michèle Calas (CNRS, University Montpelier II), Clemens Kocken and Alan Thomas (BPRC, Rijswijk, NL), Steve Ward and Pat Bray, (Liverpool School of Tropical Medicine, UK), Francoise Bressolle (Clinical Pharmacology University Montpellier I), and Socrates Herrera (Fundación Centro de Primates, Universidad del Valle, Cali, Colombia).

References

1 WHO (2008) World malaria report 2008. World Health Organization, Geneva, Switzerland http://www.who.int/malaria/wmr2008/.

2 WHO (2008) Roll Back Malaria, The Global Malaria Action Plan for a malaria-free world. WHO, Geneva, Switzerland http://www.rollbackmalaria.org/gmap/index.html Geneva, p. 274.

3 Noedl, H., Socheat, D., and Satimai, W. (2009) Artemisinin-resistant malaria in Asia. *N. Engl. J. Med.*, **361**, 540–541.

4 Dondorp, A.M., Nosten, F., Yi, P., Das, D., Phyo, A.P. *et al.* (2009) Artemisinin resistance in *Plasmodium falciparum* malaria. *N. Engl. J. Med.*, **361**, 455–467.

5 Biagini, G.A., O'Neill, P.M., Nzila, A., Ward, S.A., and Bray, P.G. (2003) Antimalarial chemotherapy: young guns or back to the future? *Trends Parasitol.*, **19**, 479–487.

6 Biot, C. and Chibale, K. (2006) Novel approaches to antimalarial drug discovery. *Infect. Disord. Drug Targets*, **6**, 173–204.

7 Schlitzer, M. (2007) Malaria chemotherapeutics part I: History of antimalarial drug development, currently used therapeutics, and drugs in clinical development. *ChemMedChem*, **2**, 944–986.

8 Schlitzer, M. (2008) Antimalarial drugs – what is in use and what is in the pipeline. *Arch. Pharm. (Weinheim)*, **341**, 149–163.

9 Mazumdar, J. and Striepen, B. (2007) Make it or take it: fatty acid metabolism of apicomplexan parasites. *Eukaryot. Cell*, **6**, 1727–1735.

10 Tarun, A.S., Vaughan, A.M., and Kappe, S.H. (2009) Redefining the role of de novo fatty acid synthesis in *Plasmodium* parasites. *Trends Parasitol.*, **25**, 545–550.

11 Salom-Roig, X.J., Hamze, A., Calas, M., and Vial, H.J. (2005) Dual molecules as

new antimalarials. *Comb. Chem. High Throughput Screen*, **8**, 49–62.

12 Vial, H.J. and Calas, M. (2001) Inhibitors of phospholipid metabolism, in *Antimalarial Chemotherapy, Mechanisms of Action, Modes of Resistance, and New Directions in Drug Development* (ed. P. Rosenthal), Humana Press, Totowa, NJ, pp. 347–365.

13 Holz, G.G. (1977) Lipids and the malaria parasite. *Bull. World Health Org.*, **55**, 237–248.

14 Vial, H. and Ancelin, M. (1998) Malarial lipids, in *Malaria: Parasite Biology, Biogenesis, Protection* (ed. I. Sherman), ASM Press, Washington DC, pp. 159–175.

15 Vial, H. and Mamoun, C. (2005) Plasmodium lipids: metabolism and function, in *Molecular Approach to Malaria* (ed. I. Sherman), ASM Press, Washington DC, pp. 327–352.

16 Vial, H.J., Eldin, P., Tielens, A.G., and van Hellemond, J.J. (2003) Phospholipids in parasitic protozoa. *Mol. Biochem. Parasitol.*, **126**, 143–154.

17 Van Deenen, L.L.M. and De Gier, J. (1975) Lipids of the red cell membrane, in *The Red Blood Cell* (ed. G. Surgenor), Academic Press, New York, pp. 147–211.

18 Vial, H.J. and Ancelin, M.L. (1992) Malarial lipids. An overview. *Subcell. Biochem.*, **18**, 259–306.

19 Vial, H.J., Ancelin, M.L., Philippot, J.R., and Thuet, M.J. (1990) Biosynthesis and dynamics of lipids in *Plasmodium*-infected mature mammalian erythrocytes. *Blood Cells*, **16**, 531–555.

20 Coppens, I. and Vielemeyer, O. (2005) Insights into unique physiological features of neutral lipids in Apicomplexa: from storage to potential mediation in parasite metabolic activities. *Int. J. Parasitol.*, **35**, 597–615.

21 Lawrence, C.W. and Cenedella, R.J. (1969) Lipid content of *Plasmodium berghei*-infected rat red blood cells. *Exp. Parasitol.*, **26**, 181–186.

22 Stocker, R. *et al.* (1987) Lipids from *Plasmodium vinckei*-infected erythrocytes and their susceptibility to oxidative damage. *Lipids*, **22**, 51–57.

23 Grellier, P., Rigomier, D., Clavey, V., Fruchart, J.C., and Schrével, J. (1991) Lipid

traffic between high density lipoproteins and *Plasmodium falciparum*-infected red blood cells. *J. Cell Biol.*, **112**, 267–277.

24 Ancelin, M.L., Parant, M., Thuet, M.J., Philippot, J.R., and Vial, H.J. (1991) Increased permeability to choline in simian erythrocytes after *Plasmodium knowlesi* infection. *Biochem. J.*, **273** (Pt 3), 701–709.

25 Biagini, G.A., Pasini, E.M., Hughes, R., De Koning, H.P., and Vial, H.J. (2004) *et al.* Characterization of the choline carrier of *Plasmodium falciparum*: a route for the selective delivery of novel antimalarial drugs. *Blood*, **104**, 3372–3377.

26 Kirk, K., Wong, H.Y., Elford, B.C., Newbold, C.I., and Ellory, J.C. (1991) Enhanced choline and Rb + transport in human erythrocytes infected with the malaria parasite *Plasmodium falciparum*. *Biochem. J.*, **278** Pt (2), 521–525.

27 Ancelin, M.L. and Vial, H.J. (1986) Several lines of evidence demonstrating that *Plasmodium falciparum*, a parasitic organism, has distinct enzymes for the phosphorylation of choline and ethanolamine. *FEBS Lett.*, **202**, 217–223.

28 Ancelin, M.L. and Vial, H.J. (1986) Choline kinase activity in Plasmodium-infected erythrocytes: characterization and utilization as a parasite-specific marker in malarial fractionation studies. *Biochim. Biophys. Acta*, **875**, 52–58.

29 Ancelin, M.L., Vial, H.J., and Philippot, J.R. (1986) Characterization of choline and ethanolamine kinase activities in *Plasmodium*-infected erythrocytes, in *Enzymes of Lipid Metabolism II* (eds L. Freysz, H. Dreyfus, R. Massarelli, and S. Gatt), Plenum Press, New York.

30 Goldberg, D.E. (2005) Hemoglobin degradation. *Curr. Top. Microbiol. Immunol.*, **295**, 275–291.

31 Woodrow, C. and Krishna, S. (2005) *Molecular Approaches to Malaria: Glycolysis in Asexual-Stage Parasites* (ed. I.W. Sherman), ASM Press.

32 Beitz, E., Pavlovic-Djuranovic, S., Yasui, M., Agre, P., and Schultz, J.E. (2004) Molecular dissection of water and glycerol permeability of the aquaglyceroporin from *Plasmodium falciparum* by mutational

analysis. *Proc. Natl Acad. Sci. USA*, **101**, 1153–1158.

33 Hansen, M., Kun, J.F., Schultz, J.E., and Beitz, E. (2002) A single, bi-functional aquaglyceroporin in blood-stage *Plasmodium falciparum* malaria parasites. *J. Biol. Chem.*, **277**, 4874–4882.

34 Newby, Z.E., O'Connell, J. 3rd, Robles-Colmenares, Y., Khademi, S., Miercke, L.J., and Stroud, R.M. (2008) Crystal structure of the aquaglyceroporin PfAQP from the malarial parasite *Plasmodium falciparum. Nat. Struct. Mol. Biol.*, **15**, 619–625.

35 Schnick, C., Polley, S.D., Fivelman, Q.L., Ranford-Cartwright, L.C., Wilkinson, S.R., Brannigan, J.A., Wilkinson, A.J., and Baker, D.A. (2009) Structure and non-essential function of glycerol kinase in *Plasmodium falciparum* blood stages. *Mol. Microbiol.*, **71**, 533–545.

36 Promeneur, D., Liu, Y., Maciel, J., Agre, P., King, L.S., and Kumar, N. (2007) Aquaglyceroporin PbAQP during intraerythrocytic development of the malaria parasite *Plasmodium berghei. Proc. Natl Acad. Sci. USA*, **104**, 2211–2216.

37 Mitamura, T., Hanada, K., Ko-Mitamura, E.P., Nishijima, M., and Horii, T. (2000) Serum factors governing intraerythrocytic development and cell cycle progression of *Plasmodium falciparum. Parasitol. Int.*, **49**, 219–229.

38 Vielemeyer, O., McIntosh, M.T., Joiner, K.A., and Coppens, I. (2004) Neutral lipid synthesis and storage in the intraerythrocytic stages of *Plasmodium falciparum. Mol. Biochem. Parasitol.*, **135**, 197–209.

39 Vial, H.J., Thuet, M.J., Broussal, J.L., and Philippot, J.R. (1982) Phospholipid biosynthesis by *Plasmodium knowlesi*-infected erythrocytes: the incorporation of phospholipid precursors and the identification of previously undetected metabolic pathways. *J. Parasitol.*, **68**, 379–391.

40 Vial, H.J., Thuet, M.J., and Philippot, J.R. (1982) Phospholipid biosynthesis in synchronous *Plasmodium falciparum* cultures. *J. Protozool.*, **29**, 258–263.

41 Vial, H.J., Thuet, M.J., and Philippot, J.R. (1984) Cholinephosphotransferase and

ethanolaminephosphotransferase activities in *Plasmodium knowlesi*-infected erythrocytes. Their use as parasite-specific markers. *Biochim. Biophys. Acta*, **795**, 372–383.

42 Krishnegowda, G. and Gowda, D.C. (2003) Intraerythrocytic *Plasmodium falciparum* incorporates extraneous fatty acids to its lipids without any structural modification. *Mol. Biochem. Parasitol.*, **132**, 55–58.

43 Gratraud, P., Huws, E., Falkard, B., Adjalley, S., Fidock, D.A. *et al.* (2009) Oleic acid biosynthesis in *Plasmodium falciparum*: characterization of the stearoyl-CoA desaturase and investigation as a potential therapeutic target. *PLoS One*, **4**, e6889.

44 Carman, G.M. and Henry, S.A. (1989) Phospholipid biosynthesis in yeast. *Annu. Rev. Biochem.*, **58**, 635–669.

45 Li, Z. and Vance, D.E. (2008) Phosphatidylcholine and choline homeostasis. *J. Lipid Res.*, **49**, 1187–1194.

46 Moll, G.N., Vial, H.J., Ancelin, M.L., Op den Kamp, J.A., Roelofsen, B., and van Deenen, L.L. (1988) Phospholipid uptake by *Plasmodium knowlesi* infected erythrocytes. *FEBS Lett.*, **232**, 341–346.

47 Elabbadi, N., Ancelin, M.L., and Vial, H.J. (1997) Phospholipid metabolism of serine in *Plasmodium*-infected erythrocytes involves phosphatidylserine and direct serine decarboxylation. *Biochem. J.*, **324** Pt (2), 435–445.

48 Pessi, G., Choi, J.Y., Reynolds, J.M., Voelker, D.R., and Mamoun, C.B. (2005) *In vivo* evidence for the specificity of *Plasmodium falciparum* phosphoethanolamine methyltransferase and its coupling to the Kennedy pathway. *J. Biol. Chem.*, **280**, 12461–12466.

49 Pessi, G., Kociubinski, G., and Mamoun, C.B. (2004) A pathway for phosphatidylcholine biosynthesis in *Plasmodium falciparum* involving phosphoethanolamine methylation. *Proc. Natl Acad. Sci. USA*, **101**, 6206–6211.

50 Witola, W.H., El Bissati, K., Pessi, G., Xie, C., Roepe, P.D., and Mamoun, C.B. (2008) Disruption of the *Plasmodium falciparum* PfPMT gene results in a complete loss of

phosphatidylcholine biosynthesis via the serine-decarboxylase-phosphoethanolamine-methyltransferase pathway and severe growth and survival defects. *J. Biol. Chem.*, **283**, 27636–27643.

51 Déchamps, S., Wengelnik, K., Berry-Sterkers, L., Cerdan, R., Vial, H.J., and Gannoun-Zaki, L. (2010) The Kennedy phospholipid biosynthesis pathways are refractory to genetic disruption in *Plasmodium berghei* and therefore appear essential in blood stages. *Mol. Biochem. Parasitol.*, **173** (2), 69–80.

52 Dechamps, S., Maynadier, M., Wein, S., Gannoun-Zaki, L., Marechal, E., and Vial, H.J. (2009) Rodent and non-rodent malaria parasites differ in their phospholipid metabolic pathways. *J. Lipid Res.*, **51** (1), 81–96.

53 Baunaure, F., Eldin, P., Cathiard, A.M., and Vial, H. (2004) Characterization of a non-mitochondrial type I phosphatidylserine decarboxylase in *Plasmodium falciparum. Mol. Microbiol.*, **51**, 33–46.

54 Rontein, D., Nishida, I., Tashiro, G., Yoshioka, K., Wu, W.I. *et al.* (2001) Plants synthesize ethanolamine by direct decarboxylation of serine using a pyridoxal phosphate enzyme. *J. Biol. Chem.*, **276**, 35523–35529.

55 Kent, C. (1995) Eukaryotic phospholipid biosynthesis. *Annu. Rev. Biochem.*, **64**, 315–343.

56 Witola, W.H. and Ben Mamoun, C. (2007) Choline induces transcriptional repression and proteasomal degradation of the malarial phosphoethanolamine methyltransferase. *Eukaryot. Cell*, **6**, 1618–1624.

57 Alberge, B., Gannoun-Zaki, L., Bascunana, C., Tran van Ba, C., Vial, H., and Cerdan, R. (2010) Comparison of the cellular and biochemical properties of *Plasmodium falciparum* choline and ethanolamine kinases. *Biochem. J.*, **425**, 149–158.

58 Choubey, V., Guha, M., Maity, P., Kumar, S., and Raghunandan, R. (2006) *et al.* Molecular characterization and localization of *Plasmodium falciparum* choline kinase. *Biochim. Biophys. Acta*, **1760**, 1027–1038.

59 Witola, W.H., Pessi, G., El Bissati, K., Reynolds, J.M., and Mamoun, C.B. (2006) Localization of the phosphoethanolamine methyltransferase of the human malaria parasite *Plasmodium falciparum* to the Golgi apparatus. *J. Biol. Chem.*, **281**, 21305–21311.

60 Le Roch, K.G., Johnson, J.R., Ahiboh, H., Chung, D.W., Prudhomme, J. *et al.* (2008) A systematic approach to understand the mechanism of action of the bisthiazolium compound T4 on the human malaria parasite, *Plasmodium falciparum. BMC Genomics*, **9**, 513.

61 Bakovic, M., Fullerton, M.D., and Michel, V. (2007) Metabolic and molecular aspects of ethanolamine phospholipid biosynthesis: the role of CTP: phosphoethanolamine cytidylyltransferase (Pcyt2). *Biochem. Cell Biol.*, **85**, 283–300.

62 Kent, C. (2005) Regulatory enzymes of phosphatidylcholine biosynthesis: a personal perspective. *Biochim. Biophys. Acta*, **1733**, 53–66.

63 Vance, J.E. (1998) Eukaryotic lipid-biosynthetic enzymes: the same but not the same. *Trends Biochem. Sci.*, **23**, 423–428.

64 Ancelin, M.L. and Vial, H.J. (1989) Regulation of phosphatidylcholine biosynthesis in *Plasmodium*-infected erythrocytes. *Biochim. Biophys. Acta*, **1001**, 82–89.

65 Gardner, M.J., Hall, N., Fung, E., White, O., Berriman, M. *et al.* (2002) Genome sequence of the human malaria parasite *Plasmodium falciparum. Nature*, **419**, 498–511.

66 McFadden, G.I. and Roos, D.S. (1999) Apicomplexan plastids as drug targets. *Trends Microbiol.*, **7**, 328–333.

67 Surolia, N. and Surolia, A. (2001) Triclosan offers protection against blood stages of malaria by inhibiting enoyl-ACP reductase of *Plasmodium falciparum. Nat. Med.*, **7**, 167–173.

68 Carballeira, N.M. (2008) New advances in fatty acids as antimalarial, antimycobacterial and antifungal agents. *Prog. Lipid Res.*, **47**, 50–61.

69 Yu, M., Kumar, T.R., Nkrumah, L.J., Coppi, A., Retzlaff, S. *et al.* (2008) The fatty

acid biosynthesis enzyme FabI plays a key role in the development of liver-stage malarial parasites. *Cell Host Microbe*, **4**, 567–578.

70 Vaughan, A.M., O'Neill, M.T., Tarun, A.S., Camargo, N., Phuong, T.M. *et al.* (2009) Type II fatty acid synthesis is essential only for malaria parasite late liver stage development. *Cell Microbiol.*, **11**, 506–520.

71 Vial, H.J., Thuet, M.J., Ancelin, M.L., Philippot, J.R., and Chavis, C. (1984) Phospholipid metabolism as a new target for malaria chemotherapy. Mechanism of action of D-2-amino-1-butanol. *Biochem. Pharmacol.*, **33**, 2761–2770.

72 Kumaratilake, L.M., Robinson, B.S., Ferrante, A., and Poulos, A. (1992) Antimalarial properties of n-3 and n-6 polyunsaturated fatty acids: *in vitro* effects on *Plasmodium falciparum* and *in vivo* effects on *P. berghei. J. Clin. Invest.*, **89**, 961–967.

73 Krugliak, M., Deharo, E., Shalmiev, G., Sauvain, M., Moretti, C., and Ginsburg, H. (1995) Antimalarial effects of C18 fatty acids on *Plasmodium falciparum* in culture and on *Plasmodium* vinckei petteri and *Plasmodium yoelii nigeriensis in vivo. Exp. Parasitol.*, **81**, 97–105.

74 Beaumelle, B.D. and Vial, H.J. (1988) Correlation of the efficiency of fatty acid derivatives in suppressing *Plasmodium falciparum* growth in culture with their inhibitory effect on acyl-CoA synthetase activity. *Mol. Biochem. Parasitol.*, **28**, 39–42.

75 Ancelin, M.L. and Vial, H.J. (1986) Quaternary ammonium compounds efficiently inhibit *Plasmodium falciparum* growth *in vitro* by impairment of choline transport. *Antimicrob. Agents Chemother.*, **29**, 814–820.

76 Ancelin, M.L., Vial, H.J., and Philippot, J.R. (1985) Inhibitors of choline transport into *Plasmodium*-infected erythrocytes are effective antiplasmodial compounds *in vitro. Biochem. Pharmacol.*, **34**, 4068–4071.

77 Choubey, V., Maity, P., Guha, M., Kumar, S., Srivastava, K., Puri, S.K., and Bandyopadhyay, U. (2007) Inhibition of *Plasmodium falciparum* choline kinase by

hexadecyltrimethylammonium bromide: a possible antimalarial mechanism. *Antimicrob. Agents Chemother.*, **51**, 696–706.

78 Ancelin, M.L., Calas, M., Bompart, J., Cordina, G., Martin, D. *et al.* (1998) Antimalarial activity of 77 phospholipid polar head analogs: close correlation between inhibition of phospholipid metabolism and *in vitro Plasmodium falciparum* growth. *Blood*, **91**, 1426–1437.

79 Ancelin, M.L., Calas, M., Bonhoure, A., Herbute, S., and Vial, H.J. (2003) *In vivo* antimalarial activities of mono- and bis quaternary ammonium salts interfering with *Plasmodium* phospholipid metabolism. *Antimicrob. Agents Chemother.*, **47**, 2598–2605.

80 Ancelin, M.L., Calas, M., Vidal-Sailhan, V., Herbute, S., Ringwald, P., and Vial, H.J. (2003) Potent inhibitors of *Plasmodium* phospholipid metabolism with a broad spectrum of *in vitro* antimalarial activities. *Antimicrob. Agents Chemother.*, **47**, 2590–2597.

81 Calas, M., Ancelin, M.L., Cordina, G., Portefaix, P., Piquet, G., Vidal-Sailhan, V., and Vial, H. (2000) Antimalarial activity of compounds interfering with *Plasmodium falciparum* phospholipid metabolism: comparison between mono- and bisquaternary ammonium salts. *J. Med. Chem.*, **43**, 505–516.

82 Wengelnik, K., Vidal, V., Ancelin, M.L., Cathiard, A.M., Morgat, J.L. *et al.* (2002) A class of potent antimalarials and their specific accumulation in infected erythrocytes. *Science*, **295**, 1311–1314.

83 Hamze, A., Rubi, E., Arnal, P., Boisbrun, M., Carcel, C. *et al.* (2005) Mono- and bis-thiazolium salts have potent antimalarial activity. *J. Med. Chem.*, **48**, 3639–3643.

84 Vial, H.J., Wein, S., Farenc, C., Kocken, C., Nicolas, O. *et al.* (2004) Prodrugs of bisthiazolium salts are orally potent antimalarials. *Proc. Natl Acad. Sci. USA*, **101**, 15458–15463.

85 Richier, E., Biagini, G.A., Wein, S., Boudou, F., Bray, P.G. *et al.* (2006) Potent antihematozoan activity of novel bisthiazolium drug T16: evidence for

inhibition of phosphatidylcholine metabolism in erythrocytes infected with *Babesia* and *Plasmodium* spp. *Antimicrob. Agents Chemother.*, **50**, 3381–3388.

86 Eastman, R.T. and Fidock, D.A. (2009) Artemisinin-based combination therapies: a vital tool in efforts to eliminate malaria. *Nat. Rev. Microbiol.*, **7**, 864–874.

87 Deves, R. and Krupka, R.M. (1979) The binding and translocation steps in transport as related to substrate structure. A study of the choline carrier of erythrocytes. *Biochim. Biophys. Acta*, **557**, 469–485.

88 Roggero, R., Zufferey, R., Minca, M., Richier, E., Calas, M., Vial, H., and Ben Mamoun, C. (2004) Unraveling the mode of action of the antimalarial choline analog G25 in *Plasmodium falciparum* and *Saccharomyces cerevisiae*. *Antimicrob. Agents Chemother.*, **48**, 2816–2824.

89 Stead, A.M., Bray, P.G., Edwards, I.G., DeKoning, H.P., Elford, B.C., Stocks, P.A., and Ward, S.A. (2001) Diamidine compounds: selective uptake and targeting in *Plasmodium falciparum*. *Mol. Pharmacol.*, **59**, 1298–1306.

90 Ginsburg, H. and Stein, W.D. (2005) How many functional transport pathways does *Plasmodium falciparum* induce in the membrane of its host erythrocyte? *Trends Parasitol.*, **21**, 118–121.

91 Kirk, K. (2001) Membrane transport in the malaria-infected erythrocyte. *Physiol. Rev.*, **81**, 495–537.

92 Biagini, G.A., Richier, E., Bray, P.G., Calas, M., Vial, H., and Ward, S.A. (2003) Heme binding contributes to antimalarial activity of bis-quaternary ammoniums. *Antimicrob. Agents Chemother.*, **47**, 2584–2589.

93 Vial, H.J., Eldin, P., Martin, D., Gannoun, L., Calas, M., and Ancelin, M.L. (1999) Transport of phospholipid synthesis precursors and lipid trafficking into malaria-infected erythrocytes. *Novartis Found. Symp.*, **226**, 74–83; discussion 82–78.

94 Barrow, R.B. and Ing, H.R. (1948) Curare-like action of polymethylene bis-quaternary ammonium salts. *Br. J. Pharmacol.*, **3**, 298–304.

95 Roberts, E. (1992) The ligand binding site of the synaptosomal choline transporter: a provisional model based on inhibition studies. *Neurochem. Res.*, **17**, 509–528.

96 Surolia, A., Ramya, T., Ramya, V., and Surolia, N. (2004) FAS't inhibition of malaria. *Biochem. J.*, **383** (3), 401–412.

9
Targeting Apicoplast Pathways in *Plasmodium*

Snober S. Mir, Subir Biswas, and Saman Habib[*]

Abstract

The single apicoplast of *Plasmodium* and related Apicomplexans is essential for survival of the parasite, and is the focus of interest for identifying novel drug targets in order to develop future antimalarial agents. In this chapter, investigations into the apicoplast pathways conducted during the past decade are described, with particular emphasis on apicoplast housekeeping and protein import processes. Progress in molecular characterization of apicoplast-encoded or -targeted proteins is also described. The identification and evaluation of drugs targeting each pathway are addressed in the corresponding section.

Introduction

The apicoplast, a relict plastid which is found in most Apicomplexa, is a potential Achilles' heel for the malaria parasite. Although, today, information continues to be discovered regarding the characteristics of this organelle's four membranes, an even deeper understanding is required in order to identify putative drug targets, in addition to validation of their function(s) and the determination of structure– activity relationships. The post-genomic era has opened up avenues for assessing the potential of newly discovered proteins as inhibitor targets. Notably, the *Plasmodium falciparum* genome project has revealed new targets in the apicoplast, primarily from proteins that are nuclear-encoded but targeted to the organelle. Phylogenetic evidence has indicated that the apicoplast originated from a secondary endosymbiotic event involving an ancient protist engulfing a photosynthetic (probably red) alga; the primary endosymbiotic event would have involved a cyanobacterium and an early eukaryote. The apicoplast is indispensable for the parasite, as its loss or the inhibition of its function leads to immediate or delayed death [1, 2], while mutant progeny with apicoplast segregation defects fail to grow [3]. Although they have undergone substantial modification during the course of evolution, apicomplexan plastids

---

[*]Corresponding author

Apicomplexan Parasites. Edited by Katja Becker
Copyright © 2011 WILEY-VCH Verlag GmbH & Co. KGaA, Weinheim
ISBN: 978-3-527-32731-7

remain fundamentally bacterial in nature, distinguishing themselves from their eukaryotic hosts. The cyanobacterial ancestry of the apicoplast is reflected in much of its metabolic machinery; typically, it harbors metabolic pathways that are not found in the host cell, and is also dependent on its prokaryotic transcriptional and translational machinery – all of which make the apicoplast an attractive target for the identification and design of potential antimalarial agents.

The *P. falciparum* apicoplast contains its own genome (plDNA), a 35 kb double-stranded DNA with a high (86%) A + T content. PlDNA has a limited coding capacity of less than 50 proteins, and most of the genes encoded by it are involved in the housekeeping functions of transcription and translation. In addition, there is a putative protein import machinery component ClpC, a protein involved in [Fe–S] cluster biosynthesis, and seven other open reading frames (ORFs) of unknown function [4]. Whereas, the apicoplast maintains a small genome, the bulk of its proteins are nuclear-encoded and imported into the organelle. Analysis of the *P. falciparum* genome sequence has revealed that approximately 500 proteins may be targeted to the apicoplast [5, 6]; hence, more than 10% of the nuclear genome may encode proteins that function within the organelle. Nuclear-encoded apicoplast-targeted proteins generally contain a bipartite sequence at their N terminus [7]. This bipartite element consists of a signal peptide (SP), which helps in co-translational endoplasmic reticulum (ER) insertion, followed by a transit peptide (TP) that is needed to import and translocate proteins across the apicoplast membranes [8, 9]. Detailed characterization of the TP reveals that positive charges are essential for faithful apicoplast targeting, although how these are recognized is unclear. The SP is removed during co-translational import into the ER, and a protease homologous to the stromal processing peptidase (SPP) of plant chloroplasts is proposed to remove the TP within the apicoplast stroma [10, 11]. The function of the majority of these imported proteins is currently being deciphered.

The localization and functional characterization of nuclear-encoded apicoplast targeted proteins has led to the identification of several biochemical pathways localized in the apicoplast. The organelle has retained a variety of prokaryotic metabolic pathways, such as the fatty acid biosynthesis type II [12], non-mevalonate pathway of isoprenoid biosynthesis [13], iron–sulfur cluster assembly [14], and heme biosynthesis [15, 16], all of which are potential drug targets. In this chapter, attention is focused on the indispensability of the apicoplast to the parasite, and issues are addressed pertaining to its housekeeping functions and protein import machinery. The elucidation of these functions presents unique opportunities for future interventions against *Plasmodium* infection.

Housekeeping Functions

DNA Replication

Replication of the circular genome of the *Plasmodium* apicoplast can be considered the first metabolic process of the organelle. The apicoplast genome has been estimated to be present in 1–15 copies per cell [17, 18] and, at ∼35 kb, it is among the smallest-known plastid genomes. *P. falciparum* plDNA replicates predominantly via the D-loop/bidirectional *ori* mode, with over 95% of the genome comprising circular molecules with commonly visible theta forms [19]. In contrast, *Toxoplasma gondii* plDNA, which shares a high sequence similarity with *P. falciparum* plDNA, exists as a precise oligomeric series of linear tandem arrays of the basic 35 kb genome [20, 21]. A rolling circle mode of DNA replication with the origin of replication at the center of the inverted repeats of rRNA genes is thus envisaged for *T. gondii*. Bidirectional *ori*/D-loop replication in *P. falciparum* plDNA initiates within the inverted repeat (IR) region, and multiple *ori* within the IR are differentially activated during replication [22–24].

The apicoplast genome lacks genes encoding DNA replication enzymes. Thus, all major proteins involved in plDNA replication and organization must be nuclear-encoded and transported to the apicoplast. There is mounting evidence pertaining to the presence of components of the apicoplast replication machinery. Seow *et al.* [25] described a ∼220 kDa multidomain polypeptide (PfPREX) that contains polymerase as well as DNA primase, DNA helicase, and 3′-5′-exonuclease activities. The N-terminal sequence of the nuclear-encoded PfPREX is capable of directing a green fluorescent protein (GFP) reporter to the apicoplast, thus identifying it as a key enzyme for plDNA replication. Chloroquines and sumarin are known DNA polymerase inhibitors [25], and the absence of a human homolog makes this complex a plausible drug target.

The gyrase inhibitors ciprofloxacin and novobiocin specifically inhibit the replication of *Plasmodium* plDNA and also reduce parasite growth in culture, thus validating malarial apicoplast DNA replication as a drug target [26, 27]. A distinct prokaryote-type topoisomerase-II (gyrase) activity within the apicoplast was initially inferred from the differential sensitivity of nuclear and apicoplast DNAs to the topoisomerase II inhibitors VP-16 and ciprofloxacin [28]; the fluoroquinolone ciprofloxacin only affects the 35 kb plastid, whereas the etoposide VP-16, an inhibitor of eukaryotic topoisomerase II, affects both the 35 kb plastid and the nuclear chromosomes of *P. falciparum*. The apicoplast genome does not encode a gyrase, which suggests that the enzyme is nuclear-encoded and targeted into the plastid. Pf*gyrA* and Pf*gyrB* sequences were identified on chromosome 12 of *P. falciparum*, and their products have been localized to the *P. falciparum* apicoplast [27, 29]. A 45 kDa GyrB domain of *Plasmodium vivax* and GyrB of *P. falciparum* have subsequently been characterized for their ATPase and DNA-binding activities [27, 29, 30].

The fact that a single plDNA molecule is approximately 12 μm in circumference, and that several molecules must be packed into an organelle with diameter of only ∼0.3 μm, as well as replicate and divide into daughter molecules without becoming tangled, is indicative of the involvement that a DNA-compacting protein has in plDNA organization. Recently, Ram *et al.* [31] reported that the nuclear genome of *P. falciparum* encodes a HU homolog (PfHU) involved in DNA organization inside the apicoplast. In contrast to bacterial HUs, which bend DNA, the PfHU promotes the concatenation of linear DNA and inhibits DNA circularization. PfHU–DNA complexes exhibit protein concentration-dependent DNA stiffening, intermolecular bundling, and the formation of DNA bridges, followed by the assembly of condensed DNA networks [31]. Attempts at generating PfHU knockouts in *Plasmodium berghei* have been unsuccessful, which suggests that the PfHU gene is indispensable for the survival of *Plasmodium* species [32].

Drugs Targeting DNA Replication

The fluoroquinolone ciprofloxacin is a selective inhibitor of DNA gyrase A, and has also been shown to inhibit apicoplast DNA replication in *P. falciparum*, without affecting the replication of parasite nuclear DNA [26]. Ciprofloxacin stabilizes gyrase binding to DNA, thus increasing the number of double-strand breaks, and an increase in ciprofloxacin-induced specific cleavage of the 35 kb apicoplast DNA has been observed [28]. The drug has been reported to cause parasite death both in the first infection cycle as well as in a delayed-death manner [1, 33]. Chemical derivatives of ciprofloxacin that enhance its antimalarial activity have been recently reported [34]. The coumarin drugs, coumermycin and novobiocin, which inhibit the ATPase activity of GyrB by binding to the monomeric form of the protein, also exhibit antiplasmodial activity. The intrinsic ATPase activity of the 45 kDa GyrB domain of *P. vivax* was shown to be sensitive to coumermycin, although at a much higher K_i than was reported for *Escherichia coli* GyrB [35]. Novobiocin inhibited the ATPase activity of PfGyrB with a K_i value of 1.24 µM; this was approximately 3.5-fold higher than the K_i of novobiocin reported for full-length *E. coli* GyrB (0.35 µM) [27]. Novobiocin also caused a reduction in parasitemia in *P. falciparum* cultures, beginning at about 40 h after treatment at a 10 µM drug concentration, with a higher concentration of the drug causing reduction in parasitemia in the first cycle itself. A specific reduction in the apicoplast/nuclear DNA ratios, as compared to mitochondrial/nuclear DNA ratios, was observed upon novobiocin treatment, indicating an inhibitory effect of the drug on apicoplast DNA replication [27].

Sequence alignment and molecular modeling of PfGyrB on the *E. coli* crystal structure have revealed the presence of two insertions of 50 and 77 residues, respectively, which lack homology with the *E. coli* sequence and appear as loops in the tertiary structure model. The first loop begins immediately after the conserved stretch of residues determined to be essential for the ATPase activity of GyrB. This Asn-rich loop is highly disordered, and likely to influence protein function. The proximity of GyrB unstructured domains to the ATPase domain and the coumarin-binding region may alter the binding affinities of specific inhibitors [27]. *P. falciparum* GyrA, which lacks such unstructured domains, models well on the *E. coli* crystal structure. This has a conserved quinolone resistance-determining region, and is likely to have a fluoroquinolone inhibition profile similar to that of *E. coli*. However, difficulties in the expression of recombinant PfGyrA [27, 29] have made the evaluation of its activity and inhibition profile problematic. A concise summary of drugs inhibiting apicoplast pathways is provided in Table 9.1.

Table 9.1 Molecular targets in the apicoplast and inhibitory activity of drug molecules.

Pathway	Enzyme	Drug	IC_{50}[a]	Reference(s)
DNA replication	Apicoplast DNA gyrase	Ciprofloxacin	*Pf* 8–38 µM	[26, 28, 36]
		Trovofloxacin	*Tg* 30 µM	[37]
		Novobiocin	*Pf* 210 µM	[27, 29]
		Coumermycin A1	*Pf* 60 µM	[27, 29]
Transcription	Apicoplast RNA polymerase β-subunit	Rifampicin	*Pf* 3 µM, *Tg* 3 µM	[38–42]
		Rifabutin	*Tg* 30 µM	[38, 40]
Protein translation	Apicoplast 23S rRNA	Clindamycin	*Pf* 20 nM	[26, 43, 44]
		Azithromycin	*Pf* 2 µM, *Tg* 2 µM	[43–45]
		Thiostrepton	*Pf* 2 µM	[1, 46–50]
		Micrococcin	*Pf* 35 nM	[26, 48, 51, 52]
		Chloramphenicol	*Pf* 10 µM, *Tg* 5 µM	[1, 26, 51, 52]
	Apicoplast 16S rRNA	Doxycycline	*Pf* 11.3 µM	[52, 53]
		Tetracycline	*Pf* 10 µM, *Tg* 20 µM	[39, 52, 53]
	Apicoplast elongation factor-Tu	Amythiamycin	*Pf* 10 nM	[54]
		Kirromycin	*Pf* 50 µM	[54]
		GE2270A	*Pf* 0.3 µM	
	Apicoplast elongation factor-G	Fusidic Acid	*Pf* 60 µM	[55]
Fatty acid biosynthesis	Apicoplast-targeted β-Ketoacyl-ACP synthase III (FabH)	Thiolactomycin	*Pf* 50 µM, *Tg* 100 µM	[8, 56] [56]
	Apicoplast-targeted enoyl-ACP reductase (FabI)	Triclosan	*Pf* 1 µM	[57–59]
	Apicoplast-targeted β-ketoacyl-ACP synthase II (FabF)	Cerulenin	11 µM	[56]
	Apicoplast-targeted acetyl-CoA carboxylase (ACC)	Clodinafop	*Tg* 10 µM	[56, 60, 61]
		Quizlofop	*Tg* 100 µM	[61, 62]
		Haloxypop	*Tg* 100 µM	[61, 62]
		Fenoxaprop	*Pf* 144 µM	[56]
		Diclofop	*Pf* 210 µM	[56]
		Tralkoxydim	*Pf* 181 µM	[56]
Isoprenoid biosynthesis	Apicoplast-targeted DOXP reductoisomerase	Fosmidomycin	*Pf* 290–370 nM	[63]
		FR-900098	*Pf* 90–170 nM	
Import machinery	Hsp70	2-Deoxyspergualin	*Pf* 148 nM	[64]

a) *Pf: Plasmodium falciparum; Tg: Toxoplasma gondii.*

Transcription and RNA Processing

Plastid transcription utilizes a eubacterial ($\alpha_2\beta\beta'$) system of DNA-dependent RNA polymerases, which was inherited with the cyanobacterial endosymbiont [65]. The apicoplast genome encodes part of this system (β subunit by *rpo*B, and β' subunit by *rpo*C1 and *rpo*C2), whereas the remainder (α subunit, as well as sigma factor) are apparently encoded in the nucleus and targeted back to the apicoplast. Despite the eubacterial origin of mitochondria, these endosymbiotic organelles seem to have adopted a single protein phage-like RNA polymerase [65]. It is currently unknown whether the *P. falciparum* apicoplast also uses phage-like RNA polymerase or only the eubacterial form encoded mainly by its genome.

Northern blot analyses of *P. falciparum* have reported the transcription of several genes encoded by the apicoplast genome [4, 66]. The abundance of transcripts of some genes (*ssu* and *lsu* rRNA genes, *rpo*B, and *rpo*C) has been shown to change throughout the cell cycle, and provides evidence of the regulated transcription of apicoplast genes [67]. At present, little is known about RNA processing in the apicoplast, and therefore its potential utility as a drug target is unclear. Many genes appear to be transcribed as operons, although how (or if) these are processed is not known. RNA editing does not seem to occur, but the tRNA-Leu gene contains a group I intron. As both neomycin B and chlortetracycline inhibit group I intron excision, these classes of antibiotic might be worth testing against apicomplexan parasites.

Drugs Targeting Transcription

The RNA polymerase of bacteria and plastids is highly sensitive to rifampicin, and the *in vitro* and *in vivo* antimalarial activity of rifampicin suggests that the drug blocks apicoplast transcription [41, 42]. Rifampicin has been shown to selectively diminish transcripts of apicoplast-encoded genes [4, 47]. One of its derivatives, rifabutin, also shows *in vitro* and *in vivo* activity against *T. gondii* [38, 40]. Trials with rifampicin in patients infected with *P. vivax* showed that although the drug, when given alone, cleared fever and reduced parasitemia initially, it was not curative. All patients treated with the rifampin–primaquine combination cleared both fever and parasitemia, but the therapeutic responses were slower than those following treatment with chloroquine–primaquine [41]. Initial studies using a fixed combination of cotrimoxazole, rifampicin, and isoniazid (Cotrifazid®) had shown efficacy against resistant strains of *P. falciparum* in animal models and in small-scale human studies. However, a multicenter trial to test the safety and efficacy of Cotrifazid concluded that the combination, in its current formulation and regimen, was a poor alternative combination therapy for malaria [68].

Apicoplast Translation

The presence of active translation machinery in the apicoplast has paved the way for investigating prokaryotic translation inhibitors as potential drugs to treat malaria. A significant part of the translational machinery is encoded by the apicoplast genome, which contains genes coding ribosomal proteins, ribosomal RNAs, and tRNAs, together with the elongation factor Tu (EF-Tu). Other components must be imported by the apicoplast, and there is evidence that the nuclear-encoded ribosomal subunits rps9 and rpl28 are targeted to the organelle [8]. Prokaryotic ribosome-like particles have been observed using electron microscopy [69], and polysomes carrying mRNAs and rRNAs specific for the apicoplast have been identified. The apicoplast *tufA* gene product (EF-Tu) has also been detected in the plastid by creating specific antibodies against the protein [50]. A functional prokaryotic translation machinery requires translation initiation factors (IF-1 to -3), elongation factors (EF-Tu, EF-Ts, EF-G), release factors (RFs), and ribosome recycling factor (RRF). The *Plasmodium* nuclear genome contains a set of these factors predicted to be apicoplast-targeted, although the signal sequence element of the bipartite signal-transit targeting sequence is not predicted for some of these. A putative apicoplast, RF-3, has not been identified in the current PlasmoDB annotation. A nuclear-encoded EF-G is predicted to be apicoplast-targeted, and nuclear-encoded EF-Ts has been localized to the organelle (S. Biswas *et al.*, unpublished results). Recombinant EF-Tu and EF-Ts expressed in the present authors' laboratory have been shown to interact *in vitro*, while EF-Ts mediates guanosine diphosphate (GDP) release from EF-Tu in order to recycle the active EF-Tu·GTP form after GTP hydrolysis (S. Biswas *et al.*, unpublished results).

In addition to ribosomal components and translation factors, investigations are now being conducted to determine whether amino-acyl tRNA synthetases (ARS)s might be targeted to the apicoplast as putative drug targets. Recently, genome analysis has identified a set of 37 ARS genes encoded by the nuclear genome [70]. Of these genes, approximately 20 have apicoplast-targeting signals, though only about 12 ARSs may be exclusively targeted to the apicoplast; the remaining ARSs are predicted to be shared between the nucleus, apicoplast, and mitochondrion. The domain architectures of several PfARSs have been reported to be very unusual [70], and may be exploited for inhibitor design and discovery. Of particular interest as a drug target is lysyl-tRNA synthetase (KRS), which exists as a single protein in humans but is present in two forms in *Plasmodium*. One of these forms is predicted to be transported to the apicoplast, and is currently being investigated as a target for specific inhibitors (L. Ribas de Pouplana, personal communication). In addition, unique sequence and structural features of the entire apicoplast tRNA population designate them as putative drug targets [71]. In particular, the single initiator tRNA^{Met} of the apicoplast has a long variable region that is unique for Met tRNAs, and this has recently been suggested as a potential target [71].

Drugs Against the Translation Machinery

Antibiotics with antimalarial activity offer tremendous promise as additives for combination therapy. Geary and Jensen [72] had initially observed that antibiotics inhibited parasite growth, although their effect was slow. Despite this slow action, clinical studies have demonstrated good malaria cure rates with the prokaryotic translation inhibitors tetracycline and clindamycin, when used in combination with a faster-acting drug [73–75]. The antimalarial properties of these drugs have been well validated, and doxycycline is currently recommended by the World Health Organization (WHO) as a prophylactic for travelers to endemic regions [76, 77]. However, the identification of additional targets in the organelle's translation apparatus is warranted for increasing the drug repertoire against resistant strains.

Studies carried out in *T. gondii* have facilitated the present understanding of the antimalarial activities of these antibiotics, and led to an observation of the "delayed-death" phenomenon, wherein the progeny of the parasite treated with drug were more severely hit than the parasite itself [26]. Similar phenotypes were observed in *P. falciparum* treated with tetracycline, clindamycin, and azithromycin [1, 33, 39, 45, 78]. These antibiotics were shown to affect the formation of functional merozoites in the second cycle, such that the abnormal merozoites were incapable of rupturing the host cell in order to establish a fresh infection [39]. The morphology of both the mitochondrion and apicoplast also appeared normal during drug treatment during the first infection cycle, demonstrating that the delayed-death effect was not due to blocking segregation of either of the two organelles [1, 33, 39, 78].

Antibiotics that inhibit prokaryotic translation act on different components of the translation machinery. Both, lincosamides (lincomycin and clindamycin) and macrolides (erythromycin and azithromycin) block protein synthesis by interacting with the peptidyl transferase domain of bacterial 23S rRNA. Clindamycin and chloramphenicol both act by blocking the translocation of peptidyl-tRNAs, and thus interfere with peptidyl transferase activity [79]. Azithromycin has been shown to interact with domains IV and V of the 23S rRNA and the ribosomal proteins L4 and L22 [80]. Tetracyclins interrupt translation by blocking the binding of the peptidyl-tRNA to the acceptor on the ribosomal small subunit [81]. Thiostrepton and micrococcin are also potent inhibitors of parasite growth *in vitro* and *in vivo*. Thiostrepton binds to 23S rRNA and inhibits the activity of EF-G by blocking Pi release after GTP hydrolysis. Thiostrepton has also been reported to inhibit total *P. falciparum* protein synthesis at an IC_{50} greater than that needed to inhibit parasite growth; this suggests that the principal target of the drug is not cytoplasmic protein synthesis [46, 48]. Indeed, a subsequent study comparing the inhibitory effects of thiostrepton on apicoplast versus cytoplasmic translation showed that translation within the apicoplast was indeed the site of the drug's action [50]. This complemented the observation that the GTPase domain of the *P. falciparum* apicoplast LSU rRNA contains A residues at two positions, and is thus the likely target molecule for thiostrepton [47]. Both, mitochondrial and nuclear LSU rRNAs have altered residues at one or both of these positions, and are thus unlikely to interact with the drug. Antibiotics that interact with EF-Tu (e.g., amythiamicin, GE2270A, kirromycin) also exhibit antimalarial activity in blood culture [54], and kirromycin affects apicoplast EF-Tu activity *in vitro* (S. Biswas *et al.*, unpublished results). Fusidic acid, a steroid-based antibiotic, is known to target EF-G, and experiments in the present authors' laboratory have demonstrated effects of this drug on both apicoplast translation and the interaction of apicoplast EF-G with the ribosome (A. Gupta *et al.*, unpublished results). The ribosome-blocking antibiotics telithromycin and quinupristin–dalfopristin, but not linezolid, have also been shown to inhibit the growth of *P. falciparum* [82]. Both of these drugs induced delayed death in the parasite, which suggested that their effect involved an impairment of the apicoplast translation processes.

Currently, there is mounting evidence that apicoplast translation components are the actual targets of antibiotics that exhibit antimalarial activity. A single point mutation in the LSU rRNA gene of the *T. gondii* apicoplast has been shown to confer resistance to clindamycin *in vitro* [79], while azithromycin resistance generated in parasite lines has been attributed to a point mutation in the *P. falciparum* LSU rRNA, as well as in the apicoplast-encoded ribosomal protein subunit Rpl4 [45]. In addition, apicoplast-specific effects such as the disruption of protein import into the organelle by clindamycin and tetracycline [1], as well as a doxycycline-mediated block in expression of the apicoplast genome resulting in the distribution of nonfunctional apicoplasts during erythrocytic schizogony [39], have been implicated in the anti-malarial action of these drugs. A recent study [83] using live microscopy on *P. berghei* has shown that microbial translation inhibitors also block development of the apicoplast during exo-erythrocytic schizogony, thus leading to impaired parasite maturation.

While current knowledge of translation processes in the *Plasmodium* mitochondrion is negligible, the possibility of antimicrobials also targeting mitochondrial translation needs to be addressed. This will be possible when reliable methods and reagents to assess translation in this organelle have been generated.

Protein Import and Targeting

A number of destinations are known for protein targeting in malaria parasites, of which apicoplast targeting is the best studied. The apicoplast depends heavily upon the nucleocytoplasm for its needs, as it has transferred the bulk of its genes to the nucleus [84]. The compartmentalization of this organelle within four membranes complicates the process of protein translocation.

The evolutionary process of the secondary endosymbiotic origin of the apicoplast has given rise to four membranes surrounding the organelle [17, 85]. The primary endosymbiotic event gave rise to the inner membrane (IM) and outer membrane (OM), corresponding to chloroplast inner and outer membranes. The secondary event gave rise to the outermost membrane, which derives as an extension of the endomembrane system and the periplastid membrane (PPM), which is a remnant of the plasma membrane of the secondary endosymbiont [86, 87]. It has been hypothesized that the import machinery of each membrane reflects the origin of that membrane, and specific import machinery should exist at each membrane to facilitate the transport of proteins to the apicoplast [88].

The signal sequence of the bipartite targeting element of apicoplast proteins is comprised of an N-terminal extension, a classic von Heijne-type peptide [89] that acts as a signal for entry into the endomembrane system and helps in the insertion of these proteins into the ER. The transit peptide is a region of variable length, and is enriched in basic amino acids such as lysine and asparagine [8, 90, 91]. The bipartite nature of the presequence has been examined by expressing the N-terminal sequences as GFP fusion proteins. These proteins were localized to the apicoplast lumen [92], but when only the signal peptide was used was the GFP secreted, which indicated that the transit peptide domain played an important role in sorting proteins to the apicoplast, with the sorting machinery lying within the endomembrane system [93, 94]. Furthermore, signal peptides from other proteins could be substituted without affecting targeting to the apicoplast [92]. Mutagenesis experiments with the transit peptide have indicated that an overall positive charge on the N-terminal side is necessary for translocation to the organelle [6]. The absence of acidic amino acids is critical for the fidelity of the transit peptide. Although, a net positive charge is a prerequisite for apicoplast targeting, the exact position of the basic residues does not influence transit peptide fidelity [95]. A median length of 78 residues has been assigned as the transit peptide length [9, 91]. Transit peptides also contain two Hsp70 binding sites, approximately 26 residues apart [6], which would help in the traversal of proteins through the translocon. A point mutation in the Hsp70-binding site of the acyl carrier protein (ACP) transit peptide site led to the disruption of traffic to the apicoplast, indicating that the presence of a molecular chaperone is important for apicoplast targeting [9].

Until recently, it was common belief that a predicted bipartite sequence is sufficient for routing a protein to the apicoplast. However, this view has been challenged by observations made in the case of superoxide dismutase-2, where the bipartite sequence routed the reporter to the mitochondrion [96], and only the full length protein–GFP fusion was localized to the apicoplast [97]. Likewise, thioredoxin-dependent peroxidase, aconitase, and pyruvate kinase are also dually targeted to the apicoplast and mitochondria [98, 99]. The recent characterization of proteins residing in the OM of the apicoplast has revealed a new targeting system, whereby a putative apicoplast transporter (APT1) has been shown to localize to the outermost apicoplast membrane, but to be apparently leaderless [100, 101]. It has also been shown that the transmembrane domain present in the protein acts as an internal signal peptide, leading to its correct localization [100, 101]. Another integral membrane protein seems to follow the same cue; TgFtsH1 lacks the typical bipartite targeting sequence, undergoes processing at the N and C termini, and has only one transmembrane domain that acts as an internal signal sequence, with its deletion leading to cytosolic localization of the protein [102].

Following the passage of proteins through the membranes, the transit peptide is cleaved to yield the mature protein. In this case, SPP, which cleaves the transit peptide into two subfragments, has been identified in *P. falciparum* [10, 11] and bears both a signal and a transit sequence, indicating its residence in the apicoplast lumen. The protease falcilysin then degrades the transit peptide [103]. Finally, the processed proteins are folded into their correct conformations by molecular chaperones, such as Cpn60/Cpn20 and BiP [104].

Traffic Through the Four Membranes

Currently, the molecular mechanisms of translocation across the apicoplast membranes are undergoing extensive investigation, with several major components having been identified. The trafficking route proposed for proteins destined for the apicoplast places the apicoplast between the ER and the Golgi apparatus. Fluorescence imaging of the parasite has revealed an intimate association of the ER and the apicoplast [105], while protein targeting has been shown to be insensitive to Brefeldin A, which blocks the ER to Golgi transport [106]. The apicoplast is likely to be located between the rough ER and the Golgi apparatus [87], such that the proteins cross the outermost membrane when the signal peptide has been cleaved off cotranslationally. The apicoplast-targeted proteins retain the transit peptide for traversal through the remaining three membranes, the PPM, OM, and IM. In this case, the transit peptide recognition function has most likely been transferred to the PPM. The movement of proteins from the PPM to the OM has been shown to be facilitated by the endoplasmic reticulum-associated degradation (ERAD) pathway, which transports misfolded proteins from the ER back to the cytosol for proteosomal degradation. Recently, two versions of the ERAD system have been shown to exist in Apicomplexa, with the second system apparently located in the plastid compartment. Several homologs of ERAD components, such as Der1, Cdc48 (protein translocation motor), and Ufd1 (a cofactor), have been identified and shown to be apicoplast-targeted [104]. Although the plastid ERAD complex contains only minimal components and lacks the complete set of ubiquitination enzymes, its main function appears to be the translocation of proteins across the membranes, away from the ER lumen [107]. Although the ERAD system may hold potential as a drug target, it needs to be better understood in terms of the flexibility exercised by this system for substrate recognition.

Transport across the apicoplast OM and IM is hugely dependent upon the translocon (Tic-Toc) complexes in the membrane [88, 108–110]. The Toc complex has not been discovered as yet, and a main component Toc75 seems to be missing from the *P. falciparum* genome [111]. It has been proposed that a Der1 homolog localized in the OM serves as a replacement for Toc75 [104]. Translocation across the IM is facilitated by a Tic translocon, homologous to the Tic complex in plant chloroplasts [112, 113]. Both, Tic20 and Tic22 have been identified, and the role of Tic20 in the import of proteins to the apicoplast lumen has been elucidated in *T. gondii* [110]. Orthologs of ClpC, an AAA ATPase chaperone implicated as the motor for primary plastid translocation [114], have been localized to the apicoplast. These may function in addition to the ClpC homolog encoded by the apicoplast genome [4].

In addition to protein transport across its four membranes, the apicoplast also exchanges reduced carbon compounds – typically triose phosphates and phosphoenolpyruvate – with the cytosol in order to fuel its activity. This exchange is mediated through the phosphate translocators Pf*o*TPT and Pf*i*TPT, which localize to the outermost and innermost membrane of the apicoplast, respectively [101]. The dependence of the malaria parasite on these transporters makes them attractive targets for drug design. A recently described cell-free assay has helped in the functional characterization of pPTs, and has opened the possibility of testing potential inhibitors against these translocators [115].

Drugs Targeting Import

15-Deoxyspergualin (DSG), which exhibits antimalarial activity *in vitro* and *in vivo*, has been shown to target the transport of proteins to the apicoplast [64]. It has been hypothesized that DSG interferes with the binding of Hsp70 to the transit peptide by interacting with the EEVD motif in the C-terminal domain of Hsp70. This would prevent the transit peptide from remaining in the unfolded conformation that is essential for targeting to the apicoplast. An alternative mode could be competition of the positively charged DSG with the positively charged transit peptides for interaction with the negatively charged pores of the apicoplast translocation complexes such as Toc [116].

Metabolic Pathways

Fatty Acid Biosynthesis

Although *Plasmodium* has long been believed to scavenge fatty acids from its host for sustenance, this notion was challenged by the discovery of nucleus-encoded fatty acid synthesis (FAS) genes and the demonstration of acetate incorporation into *P. falciparum* fatty acids, supporting the presence of a prokaryotic FAS pathway. Early experiments carried out in the blood stages of the parasite indicated that the FASII pathway functions in the apicoplast as a *de novo* pathway for FAS [58, 92]. This type of synthesis is a cyclical process by which two-carbon precursors are iteratively assembled into fatty acids, typically 16 to 18 carbon atoms long. FASII involves multiple chemical reactions executed by several individual enzymes. In *Plasmodium*, following completion of the first committed step of the pathway, which involves the conversion of acetyl-CoA to malonyl-CoA by the enzyme acetyl CoA carboxylase, fatty acid elongation is initiated by the ACP and β-ketoacyl-ACP synthase III (FabH). Subsequent steps involving the activities of β-ketoacyl ACP reductase (FabG), β-hydroxyacyl-ACP dehydratase (FabZ) and enoyl-ACP reductase (FabI) finally produce fatty acyl ACP. Successive cycles through the elongation pathway (involving condensation, reduction, dehydration, and reduction reactions) lead to the formation of fatty acyl ACPs with lengths between 10 and 14 carbons, and with two carbons being added in each round of elongation. The genes corresponding to all enzymes of FAS have been identified, except for a thioesterase required for acyl chain termination.

Biochemical support for FAS in *P. falciparum* came with the functional characterization of the apicoplast-targeted enzymes of FASII [5, 8, 58, 92]. Additionally, acetyl-CoA carboxylase (ACC) is also represented in the *P. falciparum* genome data [61], and has been shown to be apicoplast-targeted in *T. gondii* [60]. The apicoplast multidomain ACC is one of the enzymes warranting full exploration with respect to its potential as a drug target. Both, the ketoacyl-ACP synthase (FabB/F) and enoyl-ACP reductase (ENR/FabI) activities of the FASII pathway are inhibited by thiolactomycin and triclosan [58], with the latter having been proposed as a potent inhibitor of parasite growth in the asexual blood stages [58]. However, the notion that the FASII pathway is active in the blood stages of the parasite has been challenged by the results of recent studies demonstrating that the deletion of FabI in *P. falciparum* [59], and of FabB/FabF, FabZ [117], E1α, and E3 subunit genes of the pyruvate dehydrogenase (PDH) enzyme complex in *P. yoelii* [118], affects parasite growth only during the late liver stages of infection, and does not affect the mosquito, blood and early liver stages. Thus, malaria parasites seem to depend on the *de novo* FASII pathway only at the liver-to-blood transition stage, with the salvage pathway involving fatty acid uptake from the host cell being sufficient to meet their growth and division requirements in the other stages. Doubt has also been cast on FabI as the target of triclosan in *Plasmodium*, since deletion of the gene in the parasite does not alter its susceptibility to the drug [59]. Binding studies have also indicated that *Plasmodium* and bacterial FabI are different, and that the former is not targeted by triclosan [59]. In addition, triclosan inhibits the growth of *Theileria parva*, the apicoplast of which seems to lack FASII components. The effect of triclosan on the host cells suggests that the drug has off-target, non-parasite-specific effects [119].

It is difficult to reconcile these recent data with earlier studies localizing the FASII enzymes ACP [10, 92, 120] and the PDH complex [121] in the apicoplast of the parasite blood stages [122]. Lim and McFadden [122] have suggested that FASII at the blood stage, possibly with low levels of the constituent enzymes, could simply serve as a source of lipoic acid, a potent antioxidant that may protect the parasite against oxidative stress resulting from hemoglobin ingestion. *De novo*-generated lipoic acid could also be used as a cofactor for the mitochondrial α-keto acid dehydrogenase [122, 123]. Whilst further studies are required to resolve the issue, until then the FASII pathway is likely to be considered primarily as a target for malarial prophylactics.

Drugs Against Fatty Acid Biosynthesis

Apart from triclosan, other inhibitors of FabI that exhibit anti-plasmodial activity have been reported, including cerulenin [56, 58], Genz-10850, and Genz-8575 [124]. To date, no useful inhibitors have been reported against β-ketoacyl-ACP reductase (FabG) and β-hydroxyacyl-ACP dehydratase (FabZ), nor against the enzymatic steps upstream of ACC. Apicomplexa have adopted the eukaryotic ACC, which has become plastid-targeted and supports type II FAS [60, 61, 125]. Indeed, ACC has already been established as a target for two classes of successful herbicides, namely the aryloxy-yphenoxy-propionates (fops) and the cyclohexanediones (dims) [62]. Fops have also been reported to inhibit the growth of *T. gondii* in human fibroblasts [61], and of *P. falciparum* [94].

Isoprenoid Biosynthesis

Isoprenoids form an extremely diverse class of compounds, including sterols, ubiquinones, dolichols, and prenylation moieties [126]. All isoprenoid compounds depend on the precursor isomers isopentenyl diphosphate (IPP) and dimethylallyl diphosphate (DMAPP). Two alternate pathways have also been identified for IPP/DMAPP synthesis: animals and fungi rely on the mevalonic acid (MVA) pathway, while a mevalonate-independent pathway which proceeds via 1-deoxy-D-xylulose 5-phosphate (DOXP) has been described for eubacteria, algae, and the chloroplasts of higher plants [127]. The initial step of the MVA-independent pathway is the formation of 1-deoxy-D-xylulose 5-phosphate (DOXP) by condensing pyruvate and glyceraldehyde 3-phosphate, catalyzed by DOXP synthase. In a second step, the enzyme DOXP reductoisomerase synthesizes 2-C-methyl-D-erythritol 4-phosphate (with 2-C-methyl-D-erythrose 4-phosphate as an intermediate), in a single step, via an intramolecular rearrangement followed by a reduction process. Interestingly, in bacteria this DOXP pathway provides biosynthetic intermediates not only of IPP but also of thiamine (vitamin B1 precursors) and pyridoxol (vitamin B6 precursors) [63]. Jomaa *et al.* [116] identified genes encoding DOXP synthase and DOXP reductoisomerase in the *P. falciparum* genome. The N-terminal presequence (consistent with plastid targeting) of the proteins was sufficient to target a recombinant GFP reporter to the apicoplast. Subsequently, the enzymatic activity of the recombinant enzymes was biochemically verified, and DOXP reductoisomerase activity was shown to be sensitive to inhibition by the antibiotic fosmidomycin, which also cured mice of a virulent rodent malarial infection. A model for DOXP isoprenoid synthesis in the *P. falciparum* apicoplast, which traces the pathway from its primary precursors to the finished products DMAPP/IPP, has also been devised [16]. There is no evidence of genes for the mevalonate pathway in *P. falciparum*, nor of any substantial incorporation of labeled mevalonate into the isoprenoids [128]. Thus, the apicoplast-based non-mevalonate pathway is possibly the sole provider of *de novo*-synthesized IPP/DMAPP in parasites, and represents a strong target for antimalarial agents. IPP may be utilized in the synthesis of ubiquinones for the mitochondrial electron transport system, the prenylation of proteins, and the formation of dolichols for glycosylphosphatidyl inositol (GPI)-anchors on *Plasmodium* membrane proteins [16, 129]. DMAPP may also be used in the isopentenylation of apicoplast tRNAs, and thus contribute to organellar translation [16, 129].

Drugs Against DOXP Isoprenoid Biosynthesis

The demonstration of DOXP reductoisomerase as the target of fosmidomycin and its derivative FR-900098 [63, 130] paved the way for the drug to be developed as an antimalarial agent. The effect of fosmidomycin on the levels of the DOXP pathway intermediates was found to be most prominent in the ring and schizont stages of the intraerythrocytic cycle [131], and the drug was subsequently tested in combination with the prokaryotic translation inhibitor clindamycin. Subsequently, the fosmidomycin–clindamycin combination emerged as a potential antimalarial treatment on the basis of the *in vitro* and *in vivo* synergistic activities [13]. The outcome of subsequent clinical studies in Gabonese children, as well as for Thai patients diagnosed with acute uncomplicated *P. falciparum* malaria, has been encouraging [74, 75, 130, 132]. In fact, a six-hourly, five-day, dosing regimen of the combination has been suggested as the shortest duration of treatment that could result in a cure rate greater than 95% [133].

Iron–Sulfur Cluster Assembly

Iron–sulfur cluster proteins play important roles in electron transfer, as well as in redox and non-redox catalysis. The [Fe–S] prosthetic group is also required by enzymes and proteins that participate in anabolic pathways in the apicoplast. These include ferredoxin (Fd), LipA (FAS pathway), IspH, IspG (isoprenoid pathway), and MiaB (tRNA methylthiotransferase) [14, 16]. The apicoplast genome encodes SufB (orf470), a component of the [Fe–S] biosynthetic enzyme complex, while other component enzymes such as NifU, SufA, SufC, SufD and SufS have been identified on the parasite nuclear genome [134]. These nuclear-encoded Fe–S cluster-containing proteins are imported into the apicoplast in an unfolded state [111]. It has been proposed that a requirement for [Fe–S] cluster assembly or repair might at least be a partial explanation for the retention of the apicomplexan plastid's genome and its intrinsic expression system.

The intracellular location of the products of *suf* genes in malaria remains to be established. It is also uncertain whether the specialized mitochondrial and plastid compartments in *P. falciparum* combine their resources for [Fe–S] assembly, or have independent pathways [135]. Among the complete catalog of known *suf* genes that have been recognized in the genomic database of *P. falciparum*, at least one of them targets its product (SufC) to the apicoplast, and recombinant SufC interacts with the apicoplast SufB (B. Kumar and S. Habib, unpublished results). Confirmation that SufC is targeted to the plastid has also been reported with live, transfected *P. falciparum* parasites [135]. An overexpression of bacterial *suf* genes has been observed in response to oxidative stress [136]. In the plastid, where anabolic processes involved in fatty acid and isoprenoid biosynthesis depend on generating a reducing capacity and the repair of oxidative damage, there appear to be multiple opportunities, as well as a requirement for an active *suf* system [137].

Ferredoxin (Fd) and ferredoxin NADP$^+$ reductase (FNR) [138, 139] are proteins known to be imported into the malarial plastid. ApoFd requires a [2Fe–2S] cluster for maturation to Fd [14], and while FNR acquires its flavin adenine nucleotide cofactor inside the plastid [140]. ATP is required for the conversion of apoFd to the holoprotein in the stroma of plastids of higher plants. The interaction between Fd/FNR provides the necessary redox potential to serve the reductive pathways in the plastid [138, 141, 142]. This organellar redox system may be investigated as a target for intervention [122].

Heme Biosynthesis

Heme is an iron-bound tetrapyrrole that serves as an electron-carrying prosthetic group in parasite cytochromes, and also plays a role in *P. falciparum* protein synthesis [143]. Although *P. falciparum* is inundated with heme derived from erythrocyte hemoglobin, it is still dependent on its own essential heme biosynthesis pathway [2, 143].

Whilst the first committed step of tetrapyrrole synthesis is the formation of δ-aminolevulinic acid (ALA), two possible routes to ALA synthesis have been identified. In animals and fungi, ALA synthase (ALAS) employs glycine and succinyl-CoA to create ALA in the mitochondrion, via the Shemin pathway [15, 143]; following this step, the pathway shifts to the apicoplast. The next enzyme in this pathway, ALA dehydratase (ALAD), is imported into *Plasmodium* from the host erythrocyte cytoplasm [11, 144–146]. The nuclear-encoded ALAD gene contains a bipartite presequence for apicoplast targeting [10]. In addition, two other enzymes – porphobilinogen deaminase (PfPBGD) [11] and uroporphyrinogen III decarboxylase (UROD) [147, 148] – have also been localized to the apicoplast. PfPBGD has also been shown to function as uroporphyrinogen III synthase (UROS, HemD) [123, 149], and therefore may compensate for the missing HemD [123]. The remaining three steps of heme synthesis (catalyzed by coproporphyrinogen oxidase, protoporphyrinogen oxidase, and ferrochelatase) most likely occur in the mitochondrion [5, 148]. The localization of ferrochelatase to the mitochondrion has also been confirmed [123]. However, the mechanism of exchange of heme intermediates between the apicoplast and the mitochondrion remains to be understood.

Conclusion

Problems associated with the recombinant expression of soluble *P. falciparum* proteins [150] have hindered the exploration of structure–function and structure–activity relationships of key putative targets of the apicoplast. The presence of large insertions outside conserved protein functional domains in many *Plasmodial* proteins has also complicated *in-silico* docking analysis evaluation. Thus, activity assays that are amenable to low/medium-throughput compound screening modes have been difficult to develop. Consequently, the usual approach adopted to identify potential inhibitors of apicoplast processes has been first to validate specific targets by using known inhibitors, and then to devise chemical derivatives of the inhibitors in an effort to overcome any problems of bioavailability or toxicity. Clearly, a *de novo* approach is required to design drugs capable of inhibiting the activity of validated protein targets in the apicoplast.

Overlaps in the nature of housekeeping processes of the closely associated apicoplast and mitochondrion of *Plasmodium* mean that it becomes imperative to dissect out the protein components in each compartment, and then to evaluate the effects of drugs (particularly those which act on DNA replication, transcription, and translation) on both organelles. The development of procedures to isolate the apicoplast or mitochondrion, and also to generate specific antibodies against mitochondrial encoded proteins, is required to aid in this effort.

References

1 Goodman, C.D., Su, V., and McFadden, G.I. (2007) The effects of anti-bacterials on the malaria parasite *Plasmodium falciparum*. *Mol. Biochem. Parasitol.*, **152**, 181–191.

2 Ramya, T.N., Mishra, S., Karmodiya, K., Surolia, N., and Surolia, A. (2007) Inhibitors of nonhousekeeping functions of the apicoplast defy delayed death in *Plasmodium falciparum*. *Antimicrob. Agents Chemother.*, **51**, 307–316.

3 He, C.Y., Shaw, M.K., Pletcher, C.H., Striepen, B., Tilney, L.G., and Roos, D.S. (2001) A plastid segregation defect in the protozoan parasite *Toxoplasma gondii*. *EMBO J.*, **20**, 330–339.

4 Wilson, R.J., Denny, P.W., Preiser, P.R., Rangachari, K., Roberts, K. *et al.* (1996) Complete gene map of the plastid-like DNA of the malaria parasite *Plasmodium falciparum*. *J. Mol. Biol.*, **261**, 155–172.

5 Gardner, M.J., Hall, N., Fung, E., White, O., Berriman, M. *et al.* (2002) Genome sequence of the human malaria parasite *Plasmodium falciparum*. *Nature*, **419**, 498–511.

6 Foth, B.J., Ralph, S.A., Tonkin, C.J., Struck, N.S., Fraunholz, M. *et al.* (2003) Dissecting apicoplast targeting in the malaria parasite *Plasmodium falciparum*. *Science*, **299**, 705–708.

7 Tonkin, C.J., Foth, B.J., Ralph, S.A., Struck, N., Cowman, A.F., and McFadden, G.I. (2008) Evolution of malaria parasite plastid targeting sequences. *Proc. Natl Acad. Sci. USA*, **105**, 4781–4785.

8 Waller, R.F., Keeling, P.J., Donald, R.G., Striepen, B., Handman, E. *et al.* (1998) Nuclear-encoded proteins target to the plastid in *Toxoplasma gondii* and *Plasmodium falciparum*. *Proc. Natl Acad. Sci. USA*, **95**, 12352–12357.

9 DeRocher, A., Hagen, C.B., Froehlich, J.E., Feagin, J.E., and Parsons, M. (2000) Analysis of targeting sequences demonstrates that trafficking to the *Toxoplasma gondii* plastid branches off the secretory system. *J. Cell Sci.*, **113** (Pt 22), 3969–3977.

10 van Dooren, G.G., Su, V., D'Ombrain, M.C., and McFadden, G.I. (2002) Processing of an apicoplast leader sequence in *Plasmodium falciparum* and the identification of a putative leader cleavage enzyme. *J. Biol. Chem.*, **277**, 23612–23619.

11 Sato, S., Clough, B., Coates, L., and Wilson, R.J. (2004) Enzymes for heme biosynthesis are found in both the mitochondrion and plastid of the malaria parasite *Plasmodium falciparum*. *Protist*, **155**, 117–125.

12 Goodman, C.D. and McFadden, G.I. (2007) Fatty acid biosynthesis as a drug target in apicomplexan parasites. *Curr. Drug Targets*, **8**, 15–30.

13 Wiesner, J. and Jomaa, H. (2007) Isoprenoid biosynthesis of the apicoplast as drug target. *Curr. Drug Targets*, **8**, 3–13.

14 Seeber, F. (2002) Biogenesis of iron-sulphur clusters in amitochondriate and apicomplexan protists. *Int. J. Parasitol.*, **32**, 1207–1217.

15 Varadharajan, S., Dhanasekaran, S., Bonday, Z.Q., Rangarajan, P.N., and Padmanaban, G. (2002) Involvement of delta-aminolaevulinate synthase encoded by the parasite gene in *de novo* haem synthesis by *Plasmodium falciparum*. *Biochem. J.*, **367**, 321–327.

16 Ralph, S.A., van Dooren, G.G., Waller, R.F., Crawford, M.J., Fraunholz, M.J. *et al.* (2004) Tropical infectious diseases: metabolic maps and functions of the *Plasmodium falciparum* apicoplast. *Nat. Rev. Microbiol.*, **2**, 203–216.

17 Kohler, S., Delwiche, C.F., Denny, P.W., Tilney, L.G., Webster, P. *et al.* (1997) A plastid of probable green algal origin in Apicomplexan parasites. *Science*, **275**, 1485–1489.

18 Matsuzaki, M., Kikuchi, T., Kita, K., Kojima, S., and Kuroiwa, T. (2001) Large amounts of apicoplast nucleoid DNA and its segregation in *Toxoplasma gondii*. *Protoplasma*, **218**, 180–191.

19 Williamson, D.H., Preiser, P.R., Moore, P.W., McCready, S., Strath, M., and Wilson, R.J. (2002) The plastid DNA of the malaria parasite *Plasmodium falciparum* is replicated by two mechanisms. *Mol. Microbiol.*, **45**, 533–542.

20 Williamson, D.H., Preiser, P.R., and Wilson, R.J. (1996) Organelle DNAs: The bit players in malaria parasite DNA replication. *Parasitol. Today*, **12**, 357–362.

21 Williamson, D.H., Denny, P.W., Moore, P.W., Sato, S., McCready, S., and Wilson, R.J. (2001) The in vivo conformation of the plastid DNA of *Toxoplasma gondii*: implications for replication. *J. Mol. Biol.*, **306**, 159–168.

22 Williamson, D.H., Janse, C.J., Moore, P.W., Waters, A.P., and Preiser, P.R. (2002) Topology and replication of a nuclear episomal plasmid in the rodent malaria *Plasmodium berghei*. *Nucleic Acids Res.*, **30**, 726–731.

23 Singh, D., Chaubey, S., and Habib, S. (2003) Replication of the *Plasmodium falciparum* apicoplast DNA initiates within the inverted repeat region. *Mol. Biochem. Parasitol.*, **126**, 9–14.

24 Singh, D., Kumar, A., Raghu Ram, E.V., and Habib, S. (2005) Multiple replication origins within the inverted repeat region of the *Plasmodium falciparum* apicoplast genome are differentially activated. *Mol. Biochem. Parasitol.*, **139**, 99–106.

25 Seow, F., Sato, S., Janssen, C.S., Riehle, M.O., Mukhopadhyay, A. *et al.* (2005) The plastidic DNA replication enzyme complex of *Plasmodium falciparum*. *Mol. Biochem. Parasitol.*, **141**, 145–153.

26 Fichera, M.E. and Roos, D.S. (1997) A plastid organelle as a drug target in apicomplexan parasites. *Nature*, **390**, 407–409.

27 Raghu Ram, E.V., Kumar, A., Biswas, S., Kumar, A., Chaubey, S. *et al.* (2007) Nuclear gyrB encodes a functional subunit of the *Plasmodium falciparum* gyrase that is involved in apicoplast DNA replication. *Mol. Biochem. Parasitol.*, **154**, 30–39.

28 Weissig, V., Vetro-Widenhouse, T.S., and Rowe, T.C. (1997) Topoisomerase II inhibitors induce cleavage of nuclear and 35-kb plastid DNAs in the malarial parasite *Plasmodium falciparum*. *DNA Cell Biol.*, **16**, 1483–1492.

29 Dar, M.A., Sharma, A., Mondal, N., and Dhar, S.K. (2007) Molecular cloning of apicoplast-targeted *Plasmodium falciparum* DNA gyrase genes: unique intrinsic ATPase activity and ATP-independent dimerization of PfGyrB subunit. *Eukaryot. Cell*, **6**, 398–412.

30 Dar, A., Prusty, D., Mondal, N., and Dhar, S.K. (2009) A unique 45-amino-acid region in the toprim domain of *Plasmodium falciparum* gyrase B is essential for its activity. *Eukaryot. Cell*, **8**, 1759–1769.

31 Ram, E.V., Naik, R., Ganguli, M., and Habib, S. (2008) DNA organization by the apicoplast-targeted bacterial histone-like protein of *Plasmodium falciparum*. *Nucleic Acids Res.*, **36**, 5061–5073.

32 Sasaki, N., Hirai, M., Maeda, K., Yui, R., Itoh, K. *et al.* (2009) The *Plasmodium* HU homolog, which binds the plastid DNA sequence-independent manner, is essential for the parasite's survival. *FEBS Lett.*, **583**, 1446–1450.

33 Dahl, E.L. and Rosenthal, P.J. (2007) Multiple antibiotics exert delayed effects against the *Plasmodium falciparum* apicoplast. *Antimicrob. Agents Chemother.*, **51**, 3485–3490.

34 Dubar, F., Anquetin, G., Pradines, B., Dive, D., Khalife, J., and Biot, C. (2009) Enhancement of the antimalarial activity of ciprofloxacin using a double prodrug/bioorganometallic approach. *J. Med. Chem.*, **52**, 7954–7957.

35 Khor, V., Yowell, C., Dame, J.B., and Rowe, T.C. (2005) Expression and characterization of the ATP-binding domain of a malarial *Plasmodium vivax* gene homologous to the B-subunit of the bacterial topoisomerase DNA gyrase. *Mol. Biochem. Parasitol.*, **140**, 107–117.

36 Divo, A.A., Sartorelli, A.C., Patton, C.L., and Bia, F.J. (1988) Activity of fluoroquinolone antibiotics against *Plasmodium falciparum in vitro*. *Antimicrob. Agents Chemother.*, **32**, 1182–1186.

37 Khan, A.A., Slifer, T., Araujo, F.G., and Remington, J.S. (1996) Trovafloxacin is active against *Toxoplasma gondii*. *Antimicrob. Agents Chemother.*, **40**, 1855–1859.

38 Araujo, F.G., Slifer, T., and Remington, J.S. (1994) Rifabutin is active in murine models of toxoplasmosis. *Antimicrob. Agents Chemother.*, **38**, 570–575.

39 Dahl, E.L., Shock, J.L., Shenai, B.R., Gut, J., DeRisi, J.L., and Rosenthal, P.J. (2006) Tetracyclines specifically target the apicoplast of the malaria parasite *Plasmodium falciparum*. *Antimicrob. Agents Chemother.*, **50**, 3124–3131.

40 Olliaro, P., Gorini, G., Jabes, D., Regazzetti, A., Rossi, R. *et al.* (1994) In-vitro and in-vivo activity of rifabutin against *Toxoplasma gondii*. *J. Antimicrob. Chemother.*, **34**, 649–657.

41 Pukrittayakamee, S., Viravan, C., Charoenlarp, P., Yeamput, C., Wilson, R.J., and White, N.J. (1994) Antimalarial effects of rifampin in *Plasmodium vivax* malaria. *Antimicrob. Agents Chemother.*, **38**, 511–514.

42 Strath, M., Scott-Finnigan, T., Gardner, M., Williamson, D., and Wilson, I. (1993) Antimalarial activity of rifampicin *in vitro* and in rodent models. *Trans. R. Soc. Trop. Med. Hyg.*, **87**, 211–216.

43 Fichera, M.E., Bhopale, M.K., and Roos, D.S. (1995) *In vitro* assays elucidate peculiar kinetics of clindamycin action against *Toxoplasma gondii*. *Antimicrob. Agents Chemother.*, **39**, 1530–1537.

44 Pfefferkorn, E.R. and Borotz, S.E. (1994) Comparison of mutants of *Toxoplasma gondii* selected for resistance to azithromycin, spiramycin, or clindamycin. *Antimicrob. Agents Chemother.*, **38**, 31–37.

45 Sidhu, A.B., Sun, Q., Nkrumah, L.J., Dunne, M.W., Sacchettini, J.C., and Fidock, D.A. (2007) In vitro efficacy, resistance selection, and structural modeling studies implicate the malarial parasite apicoplast as the target of azithromycin. *J. Biol. Chem.*, **282**, 2494–2504.

46 Clough, B., Strath, M., Preiser, P., Denny, P., and Wilson, I.R. (1997) Thiostrepton binds to malarial plastid rRNA. *FEBS Lett.*, **406**, 123–125.

47 McConkey, G.A., Rogers, M.J., and McCutchan, T.F. (1997) Inhibition of *Plasmodium falciparum* protein synthesis. Targeting the plastid-like organelle with thiostrepton. *J. Biol. Chem.*, **272**, 2046–2049.

48 Rogers, M.J., Bukhman, Y.V., McCutchan, T.F., and Draper, D.E. (1997) Interaction of thiostrepton with an RNA fragment derived from the plastid-encoded ribosomal RNA of the malaria parasite. *Rna*, **3**, 815–820.

49 Sullivan, M., Li, J., Kumar, S., Rogers, M.J., and McCutchan, T.F. (2000) Effects of interruption of apicoplast function on malaria infection, development, and transmission. *Mol. Biochem. Parasitol.*, **109**, 17–23.

50 Chaubey, S., Kumar, A., Singh, D., and Habib, S. (2005) The apicoplast of *Plasmodium falciparum* is translationally active. *Mol. Microbiol.*, **56**, 81–89.

51 Beckers, C.J., Roos, D.S., Donald, R.G., Luft, B.J., Schwab, J.C. *et al.* (1995) Inhibition of cytoplasmic and organellar protein synthesis in *Toxoplasma gondii*. Implications for the target of macrolide antibiotics. *J. Clin. Invest.*, **95**, 367–376.

52 Budimulja, A.S., Syafruddin, Tapchaisri, P., Wilairat, P., and Marzuki, S. (1997) The sensitivity of *Plasmodium protein* synthesis to prokaryotic ribosomal inhibitors. *Mol. Biochem. Parasitol.*, **84**, 137–141.

53 Pradines, B., Spiegel, A., Rogier, C., Tall, A., Mosnier, J. *et al.* (2000) Antibiotics for prophylaxis of *Plasmodium falciparum* infections: *in vitro* activity of doxycycline against Senegalese isolates. *Am. J. Trop. Med. Hyg.*, **62**, 82–85.

54 Clough, B., Rangachari, K., Strath, M., Preiser, P.R., and Wilson, R.J. (1999) Antibiotic inhibitors of organellar protein synthesis in *Plasmodium falciparum*. *Protist*, **150**, 189–195.

55 Black, F.T., Wildfang, I.L., and Borgbjerg, K. (1985) Activity of fusidic acid against *Plasmodium falciparum in vitro*. *Lancet*, **1**, 578–579.

56 Waller, R.F., Ralph, S.A., Reed, M.B., Su, V., Douglas, J.D. *et al.* (2003) A type II pathway for fatty acid biosynthesis presents drug targets in *Plasmodium falciparum*. *Antimicrob. Agents Chemother.*, **47**, 297–301.

57 McLeod, R., Muench, S.P., Rafferty, J.B., Kyle, D.E., Mui, E.J. *et al.* (2001) Triclosan inhibits the growth of *Plasmodium falciparum* and *Toxoplasma gondii* by inhibition of apicomplexan Fab I. *Int. J. Parasitol.*, **31**, 109–113.

58 Surolia, N. and Surolia, A. (2001) Triclosan offers protection against blood stages of malaria by inhibiting enoyl-ACP reductase of *Plasmodium falciparum*. *Nat. Med.*, **7**, 167–173.

59 Yu, M., Kumar, T.R., Nkrumah, L.J., Coppi, A., Retzlaff, S. *et al.* (2008) The fatty acid biosynthesis enzyme FabI plays a key role in the development of liver-stage malarial parasites. *Cell Host. Microbe.*, **4**, 567–578.

60 Jelenska, J., Crawford, M.J., Harb, O.S., Zuther, E., Haselkorn, R. *et al.* (2001) Subcellular localization of acetyl-CoA carboxylase in the apicomplexan parasite *Toxoplasma gondii*. *Proc. Natl Acad. Sci. USA*, **98**, 2723–2728.

61 Zuther, E., Johnson, J.J., Haselkorn, R., McLeod, R., and Gornicki, P. (1999) Growth of *Toxoplasma gondii* is inhibited by aryloxyphenoxypropionate herbicides targeting acetyl-CoA carboxylase. *Proc. Natl Acad. Sci. USA*, **96**, 13387–13392.

62 Zagnitko, O., Jelenska, J., Tevzadze, G., Haselkorn, R., and Gornicki, P. (2001) An isoleucine/leucine residue in the carboxyltransferase domain of acetyl-CoA carboxylase is critical for interaction with aryloxyphenoxypropionate and cyclohexanedione inhibitors. *Proc. Natl Acad. Sci. USA*, **98**, 6617–6622.

63 Jomaa, H., Wiesner, J., Sanderbrand, S., Altincicek, B., Weidemeyer, C. *et al.* (1999) Inhibitors of the nonmevalonate pathway of isoprenoid biosynthesis as antimalarial drugs. *Science*, **285**, 1573–1576.

64 Ramya, T.N., Surolia, N., and Surolia, A. (2007) 15-deoxyspergualin inhibits eukaryotic protein synthesis through eIF2alpha phosphorylation. *Biochem. J.*, **401**, 411–420.

65 Gray, M.W. and Lang, B.F. (1998) Transcription in chloroplasts and mitochondria: a tale of two polymerases. *Trends Microbiol.*, **6**, 1–3.

66 Gardner, M.J., Feagin, J.E., Moore, D.J., Spencer, D.F., Gray, M.W. *et al.* (1991) Organisation and expression of small subunit ribosomal RNA genes encoded by a 35-kilobase circular DNA in *Plasmodium falciparum*. *Mol. Biochem. Parasitol.*, **48**, 77–88.

67 Feagin, J.E. and Drew, M.E. (1995) *Plasmodium falciparum*: alterations in organelle transcript abundance during the erythrocytic cycle. *Exp. Parasitol.*, **80**, 430–440.

68 Genton, B., Mueller, I., Betuela, I., Casey, G., Ginny, M. *et al.* (2006) Rifampicin/Cotrimoxazole/Isoniazid versus mefloquine or quinine + sulfadoxine-pyrimethamine for malaria: a randomized trial. *PLoS Clin. Trials*, **1**, e38.

69 Hopkins, J., Fowler, R., Krishna, S., Wilson, I., Mitchell, G., and Bannister, L. (1999) The plastid in *Plasmodium falciparum* asexual blood stages: a three-dimensional ultrastructural analysis. *Protist*, **150**, 283–295.

70 Bhatt, T.K., Kapil, C., Khan, S., Jairajpuri, M.A., Sharma, V. *et al.* (2009) A genomic glimpse of aminoacyl-tRNA synthetases in malaria parasite *Plasmodium falciparum*. *BMC Genomics*, **10**, 644.

71 Putz, J., Giege, R., and Florentz, C. (2010) Diversity and similarity in the tRNA world: overall view and case study on malaria-related tRNAs. *FEBS Lett.*, **584**, 350–358.

72 Geary, T.G., Delaney, E.J., Klotz, I.M., and Jensen, J.B. (1983) Inhibition of the growth of *Plasmodium falciparum in vitro* by covalent modification of hemoglobin. *Mol. Biochem. Parasitol.*, **9**, 59–72.

73 Baird, J.K. (2005) Effectiveness of antimalarial drugs. *N. Engl. J. Med.*, **352**, 1565–1577.

74 Borrmann, S., Adegnika, A.A., Matsiegui, P.B., Issifou, S., Schindler, A. *et al.* (2004) Fosmidomycin-clindamycin for *Plasmodium falciparum* infections in African children. *J. Infect. Dis.*, **189**, 901–908.

75 Borrmann, S., Lundgren, I., Oyakhirome, S., Impouma, B., Matsiegui, P.B. *et al.* (2006) Fosmidomycin plus clindamycin for treatment of pediatric patients aged 1 to 14 years with *Plasmodium falciparum* malaria. *Antimicrob. Agents Chemother.*, **50**, 2713–2718.

76 Kain, K.C., Shanks, G.D., and Keystone, J.S. (2001) Malaria chemoprophylaxis in the age of drug resistance. I. Currently recommended drug regimens. *Clin. Infect. Dis.*, **33**, 226–234.

77 Ryan, E.T. and Kain, K.C. (2000) Health advice and immunizations for travelers. *N. Engl. J. Med.*, **342**, 1716–1725.

78 Dahl, E.L. and Rosenthal, P.J. (2008) Apicoplast translation, transcription and genome replication: targets for antimalarial antibiotics. *Trends Parasitol.*, **24**, 279–284.

79 Camps, M., Arrizabalaga, G., and Boothroyd, J. (2002) An rRNA mutation identifies the apicoplast as the target for clindamycin in *Toxoplasma gondii. Mol. Microbiol.*, **43**, 1309–1318.

80 Hansen, L.H., Mauvais, P., and Douthwaite, S. (1999) The macrolide-ketolide antibiotic binding site is formed by structures in domains II and V of 23S ribosomal RNA. *Mol. Microbiol.*, **31**, 623–631.

81 Brodersen, D.E., Clemons, W.M. Jr, Carter, A.P., Morgan-Warren, R.J., Wimberly, B.T., and Ramakrishnan, V. (2000) The structural basis for the action of the antibiotics tetracycline, pactamycin, and hygromycin B on the 30S ribosomal subunit. *Cell*, **103**, 1143–1154.

82 Barthel, D., Schlitzer, M., and Pradel, G. (2008) Telithromycin and quinupristin-dalfopristin induce delayed death in *Plasmodium falciparum. Antimicrob. Agents Chemother.*, **52**, 774–777.

83 Stanway, R.R., Witt, T., Zobiak, B., Aepfelbacher, M., and Heussler, V.T. (2009) GFP-targeting allows visualization of the apicoplast throughout the life cycle of live malaria parasites. *Biol. Cell*, **101**, 415–430.

84 Foth, B.J. and McFadden, G.I. (2003) The apicoplast: a plastid in *Plasmodium falciparum* and other Apicomplexan parasites. *Int. Rev. Cytol.*, **224**, 57–110.

85 McFadden, G.I., Reith, M.E., Munholland, J., and Lang-Unnasch, N. (1996) Plastid in human parasites. *Nature*, **381**, 482.

86 Dyall, S.D., Brown, M.T., and Johnson, P.J. (2004) Ancient invasions: from endosymbionts to organelles. *Science*, **304**, 253–257.

87 McFadden, G.I. (1999) Plastids and protein targeting. *J. Eukaryot. Microbiol.*, **46**, 339–346.

88 van Dooren, G.G., Schwartzbach, S.D., Osafune, T., and McFadden, G.I. (2001) Translocation of proteins across the multiple membranes of complex plastids. *Biochim. Biophys. Acta*, **1541**, 34–53.

89 von Heijne, G., Steppuhn, J., and Herrmann, R.G. (1989) Domain structure of mitochondrial and chloroplast targeting peptides. *Eur. J. Biochem.*, **180**, 535–545.

90 Ralph, S.A., Foth, B.J., Hall, N., and McFadden, G.I. (2004) Evolutionary pressures on apicoplast transit peptides. *Mol. Biol. Evol.*, **21**, 2183–2194.

91 Zuegge, J., Ralph, S., Schmuker, M., McFadden, G.I., and Schneider, G. (2001) Deciphering apicoplast targeting signals – feature extraction from nuclear-encoded precursors of *Plasmodium falciparum* apicoplast proteins. *Gene*, **280**, 19–26.

92 Waller, R.F., Reed, M.B., Cowman, A.F., and McFadden, G.I. (2000) Protein trafficking to the plastid of *Plasmodium falciparum* is via the secretory pathway. *EMBO J.*, **19**, 1794–1802.

93 Roos, D.S., Crawford, M.J., Donald, R.G., Kissinger, J.C., Klimczak, L.J., and Striepen, B. (1999) Origin, targeting, and function of the apicomplexan plastid. *Curr. Opin. Microbiol.*, **2**, 426–432.

94 Waller, R.F. and McFadden, G.I. (2005) The apicoplast: a review of the derived plastid of apicomplexan parasites. *Curr. Issues Mol. Biol.*, **7**, 57–79.

95 Tonkin, C.J., Roos, D.S., and McFadden, G.I. (2006) N-terminal positively charged amino acids, but not their exact position, are important for apicoplast transit peptide fidelity in *Toxoplasma gondii. Mol. Biochem. Parasitol.*, **150**, 192–200.

96 Yung, S. and Lang-Unnasch, N. (1999) Targeting of a nuclear encoded protein to the apicoplast of *Toxoplasma gondii. J. Eukaryot. Microbiol.*, **46**, 79S–80S.

97 Brydges, S.D. and Carruthers, V.B. (2003) Mutation of an unusual mitochondrial targeting sequence of

SODB2 produces multiple targeting fates in *Toxoplasma gondii. J. Cell Sci.*, **116**, 4675–4685.

98 Pino, P., Foth, B.J., Kwok, L.Y., Sheiner, L., Schepers, R. *et al.* (2007) Dual targeting of antioxidant and metabolic enzymes to the mitochondrion and the apicoplast of *Toxoplasma gondii. PLoS Pathog.*, **3**, e115.

99 Saito, T., Nishi, M., Lim, M.I., Wu, B., Maeda, T. *et al.* (2008) A novel GDP-dependent pyruvate kinase isozyme from *Toxoplasma gondii* localizes to both the apicoplast and the mitochondrion. *J. Biol. Chem.*, **283**, 14041–14052.

100 Karnataki, A., Derocher, A.E., Coppens, I., Feagin, J.E., and Parsons, M. (2007) A membrane protease is targeted to the relict plastid of toxoplasma via an internal signal sequence. *Traffic*, **8**, 1543–1553.

101 Mullin, K.A., Lim, L., Ralph, S.A., Spurck, T.P., Handman, E., and McFadden, G.I. (2006) Membrane transporters in the relict plastid of malaria parasites. *Proc. Natl Acad. Sci. USA*, **103**, 9572–9577.

102 Karnataki, A., DeRocher, A.E., Feagin, J.E., and Parsons, M. (2009) Sequential processing of the *Toxoplasma* apicoplast membrane protein FtsH1 in topologically distinct domains during intracellular trafficking. *Mol. Biochem. Parasitol.*, **166**, 126–133.

103 Ponpuak, M., Klemba, M., Park, M., Gluzman, I.Y., Lamppa, G.K., and Goldberg, D.E. (2007) A role for falcilysin in transit peptide degradation in the *Plasmodium falciparum* apicoplast. *Mol. Microbiol.*, **63**, 314–334.

104 Sommer, M.S., Gould, S.B., Lehmann, P., Gruber, A., Przyborski, J.M. and Maier, U.G. (2007) Der1-mediated preprotein import into the periplastid compartment of chromalveolates? *Mol. Biol. Evol.*, **24**, 918–928.

105 Tonkin, C.J., Pearce, J.A., McFadden, G.I., and Cowman, A.F. (2006) Protein targeting to destinations of the secretory pathway in the malaria parasite *Plasmodium falciparum. Curr. Opin. Microbiol.*, **9**, 381–387.

106 DeRocher, A., Gilbert, B., Feagin, J.E., and Parsons, M. (2005) Dissection of brefeldin A-sensitive and -insensitive

steps in apicoplast protein targeting. *J. Cell Sci.*, **118**, 565–574.

107 Spork, S., Hiss, J.A., Mandel, K., Sommer, M., Kooij, T.W. *et al.* (2009) An unusual ERAD-like complex is targeted to the apicoplast of *Plasmodium falciparum. Eukaryot. Cell*, **8**, 1134–1145.

108 Parsons, M., Karnataki, A., and Derocher, A.E. (2009) Evolving insights into protein trafficking to the multiple compartments of the apicomplexan plastid. *J. Eukaryot. Microbiol.*, **56**, 214–220.

109 Parsons, M., Karnataki, A., Feagin, J.E., and DeRocher, A. (2007) Protein trafficking to the apicoplast: deciphering the apicomplexan solution to secondary endosymbiosis. *Eukaryot. Cell*, **6**, 1081–1088.

110 van Dooren, G.G., Tomova, C., Agrawal, S., Humbel, B.M., and Striepen, B. (2008) Toxoplasma *gondii* Tic20 is essential for apicoplast protein import. *Proc. Natl Acad. Sci. USA*, **105**, 13574–13579.

111 Tonkin, C.J., Kalanon, M., and McFadden, G.I. (2008) Protein targeting to the malaria parasite plastid. *Traffic*, **9**, 166–175.

112 Kouranov, A. and Schnell, D.J. (1997) Analysis of the interactions of preproteins with the import machinery over the course of protein import into chloroplasts. *J. Cell Biol.*, **139**, 1677–1685.

113 Kouranov, A., Wang, H., and Schnell, D.J. (1999) Tic22 is targeted to the intermembrane space of chloroplasts by a novel pathway. *J. Biol. Chem.*, **274**, 25181–25186.

114 Vojta, L., Soll, J., and Bolter, B. (2007) Requirements for a conservative protein translocation pathway in chloroplasts. *FEBS Lett.*, **581**, 2621–2624.

115 Lim, L., Kalanon, M., and McFadden, G.I. (2009) New proteins in the apicoplast membranes: time to rethink apicoplast protein targeting. *Trends Parasitol.*, **25**, 197–200.

116 Ramya, T.N., Karmodiya, K., Surolia, A., and Surolia, N. (2007) 15-deoxyspergualin primarily targets the trafficking of apicoplast proteins in

Plasmodium falciparum. J. Biol. Chem., **282**, 6388–6397.

117 Vaughan, A.M., O'Neill, M.T., Tarun, A.S., Camargo, N., Phuong, T.M. *et al.* (2009) Type II fatty acid synthesis is essential only for malaria parasite late liver stage development. *Cell Microbiol.*, **11**, 506–520.

118 Pei, Y., Tarun, A.S., Vaughan, A.M., Herman, R.W., Soliman, J.M. *et al.* (2010) *Plasmodium* pyruvate dehydrogenase activity is only essential for the parasite's progression from liver infection to blood infection. *Mol. Microbiol.*, **75**, 957–971.

119 Lizundia, R., Werling, D., Langsley, G., and Ralph, S.A. (2009) *Theileria* apicoplast as a target for chemotherapy. *Antimicrob. Agents Chemother.*, **53**, 1213–1217.

120 Waller, R.F. and McFadden, G.I. (2000) In situ hybridization for electron microscopy. *Methods Mol. Biol.*, **123**, 259–277.

121 Foth, B.J., Stimmler, L.M., Handman, E., Crabb, B.S., Hodder, A.N., and McFadden, G.I. (2005) The malaria parasite *Plasmodium falciparum* has only one pyruvate dehydrogenase complex, which is located in the apicoplast. *Mol. Microbiol.*, **55**, 39–53.

122 Lim, L. and McFadden, G.I. (2010) The evolution, metabolism and functions of the apicoplast. *Philos. Trans. R. Soc. Lond. B. Biol. Sci.*, **365**, 749–763.

123 van Dooren, G.G., Stimmler, L.M., and McFadden, G.I. (2006) Metabolic maps and functions of the *Plasmodium* mitochondrion. *FEMS Microbiol. Rev.*, **30**, 596–630.

124 Kuo, P.C., Shi, L.S., Damu, A.G., Su, C.R., Huang, C.H. *et al.* (2003) Cytotoxic and antimalarial beta-carboline alkaloids from the roots of *Eurycoma longifolia*. *J. Nat. Prod.*, **66**, 1324–1327.

125 Harwood, J.L. (1996) Recent advances in the biosynthesis of plant fatty acids. *Biochim. Biophys. Acta*, **1301**, 7–56.

126 Sacchettini, J.C. and Poulter, C.D. (1997) Creating isoprenoid diversity. *Science*, **277**, 1788–1789.

127 Lichtenthaler, H.K. (2000) Non-mevalonate isoprenoid biosynthesis: enzymes, genes and inhibitors. *Biochem. Soc. Trans.*, **28**, 785–789.

128 Mbaya, B., Rigomier, D., Edorh, G.G., Karst, F., and Schrevel, J. (1990) Isoprenoid metabolism in *Plasmodium falciparum* during the intraerythrocytic phase of malaria. *Biochem. Biophys. Res. Commun.*, **173**, 849–854.

129 Naik, R.S., Davidson, E.A., and Gowda, D.C. (2000) Developmental stage-specific biosynthesis of glycosylphosphatidylinositol anchors in intraerythrocytic *Plasmodium falciparum* and its inhibition in a novel manner by mannosamine. *J. Biol. Chem.*, **275**, 24506–24511.

130 Wiesner, J., Ortmann, R., Jomaa, H., and Schlitzer, M. (2007) Double ester prodrugs of FR900098 display enhanced in-vitro antimalarial activity. *Arch. Pharm. (Weinheim)*, **340**, 667–669.

131 Cassera, M.B., Gozzo, F.C., D'Alexandri, F.L., Merino, E.F., del Portillo, H.A. *et al.* (2004) The methylerythritol phosphate pathway is functionally active in all intraerythrocytic stages of *Plasmodium falciparum. J. Biol. Chem.*, **279**, 51749–51759.

132 Na-Bangchang, K., Ruengweerayut, R., Karbwang, J., Chauemung, A., and Hutchinson, D. (2007) Pharmacokinetics and pharmacodynamics of fosmidomycin monotherapy and combination therapy with clindamycin in the treatment of multidrug resistant falciparum malaria. *Malar. J.*, **6**, 70.

133 Ruangweerayut, R., Looareesuwan, S., Hutchinson, D., Chauemung, A., Banmairuroi, V., and Na-Bangchang, K. (2008) Assessment of the pharmacokinetics and dynamics of two combination regimens of fosmidomycin-clindamycin in patients with acute uncomplicated falciparum malaria. *Malar. J.*, **7**, 225.

134 Ellis, K.E., Clough, B., Saldanha, J.W., and Wilson, R.J. (2001) Nifs and Sufs in malaria. *Mol. Microbiol.*, **41**, 973–981.

135 Wilson, R.J. (2005) Parasite plastids: approaching the endgame. *Biol. Rev. Camb. Philos. Soc.*, **80**, 129–153.

136 Nachin, L., Loiseau, L., Expert, D., and Barras, F. (2003) SufC: an unorthodox cytoplasmic ABC/ATPase required for [Fe-S] biogenesis under oxidative stress. *EMBO J.*, **22**, 427–437.

137 Wilson, R.J., Rangachari, K., Saldanha, J.W., Rickman, L., Buxton, R.S., and Eccleston, J.F. (2003) Parasite plastids: maintenance and functions. *Philos. Trans. R. Soc. Lond. B Biol. Sci.*, **358**, 155–162.

138 Pandini, V., Caprini, G., Thomsen, N., Aliverti, A., Seeber, F., and Zanetti, G. (2002) Ferredoxin-NADP$^+$ reductase and ferredoxin of the protozoan parasite *Toxoplasma gondii* interact productively *in vitro* and *in vivo*. *J. Biol. Chem.*, **277**, 48463–48471.

139 Vollmer, M., Thomsen, N., Wiek, S., and Seeber, F. (2001) Apicomplexan parasites possess distinct nuclear-encoded, but apicoplast-localized, plant-type ferredoxin-NADP$^+$ reductase and ferredoxin. *J. Biol. Chem.*, **276**, 5483–5490.

140 Onda, Y. and Hase, T. (2004) FAD assembly and thylakoid membrane binding of ferredoxin:NADP$^+$ oxidoreductase in chloroplasts. *FEBS Lett.*, **564**, 116–120.

141 Kimata-Ariga, Y., Kurisu, G., Kusunoki, M., Aoki, S., Sato, D. *et al.* (2007) Cloning and characterization of ferredoxin and ferredoxin-NADP$^+$ reductase from human malaria parasite. *J. Biochem.*, **141**, 421–428.

142 Kimata-Ariga, Y., Saitoh, T., Ikegami, T., Horii, T., and Hase, T. (2007) Molecular interaction of ferredoxin and ferredoxin-NADP$^+$ reductase from human malaria parasite. *J. Biochem.*, **142**, 715–720.

143 Surolia, N. and Padmanaban, G. (1992) De novo biosynthesis of heme offers a new chemotherapeutic target in the human malarial parasite. *Biochem. Biophys. Res. Commun.*, **187**, 744–750.

144 Bonday, Z.Q., Dhanasekaran, S., Rangarajan, P.N., and Padmanaban, G. (2000) Import of host delta-aminolevulinate dehydratase into the malarial parasite: identification of a new drug target. *Nat. Med.*, **6**, 898–903.

145 Padmanaban, G. and Rangarajan, P.N. (2000) Heme metabolism of *Plasmodium* is a major antimalarial target. *Biochem. Biophys. Res. Commun.*, **268**, 665–668.

146 Sato, S. and Wilson, R.J. (2003) Proteobacteria-like ferrochelatase in the malaria parasite. *Curr. Genet.*, **42**, 292–300.

147 Nagaraj, V.A., Arumugam, R., Chandra, N.R., Prasad, D., Rangarajan, P.N., and Padmanaban, G. (2009) Localisation of *Plasmodium falciparum* uroporphyrinogen III decarboxylase of the heme-biosynthetic pathway in the apicoplast and characterisation of its catalytic properties. *Int. J. Parasitol.*, **39**, 559–568.

148 Nagaraj, V.A., Prasad, D., Rangarajan, P.N., and Padmanaban, G. (2009) Mitochondrial localization of functional ferrochelatase from *Plasmodium falciparum*. *Mol. Biochem. Parasitol.*, **168**, 109–112.

149 Nagaraj, V.A., Arumugam, R., Gopalakrishnan, B., Jyothsna, Y.S., Rangarajan, P.N., and Padmanaban, G. (2008) Unique properties of *Plasmodium falciparum* porphobilinogen deaminase. *J. Biol. Chem.*, **283**, 437–444.

150 Mehlin, C., Boni, E., Buckner, F.S., Engel, L., Feist, T. *et al.* (2006) Heterologous expression of proteins from *Plasmodium falciparum*: results from 1000 genes. *Mol. Biochem. Parasitol.*, **148**, 144–160.

10
Lipoic Acid Acquisition and Glutathione Biosynthesis in Apicomplexan Parasites

*Janet Storm, Eva-Maria Patzewitz, and Sylke Müller**

Abstract

The thiols lipoic acid and glutathione (GSH) are cofactors of enzymatic reactions crucial to the metabolism of most organisms, and play a vital role in maintaining redox homeostasis. The dithiol-containing fatty acid derivative lipoic acid can be acquired through *de novo* biosynthesis and salvage. In apicomplexans, salvage is confined to the mitochondrion, while biosynthesis occurs solely in the apicoplast. Lipoic acid biosynthesis depends on the supply of octanoyl-acyl carrier protein, an intermediate of type II fatty acid biosynthesis (FASII). This links the occurrence of lipoic acid biosynthesis directly to that of FASII, suggesting that the biosynthesis of lipoic acid is essential for the exo-erythrocytic development of *Plasmodium* and growth of *Toxoplasma gondii* tachyzoites, as described for FASII. Lipoic acid salvage is important for the intra-erythrocytic and sexual development of *Plasmodium*, and inhibition of the respective lipoic acid protein ligase 1 and 2 of malarial parasites might offer potential for interfering with these stages of parasite development.

The suitability of the glutathione biosynthesis pathway for the development of new antimalarials is also discussed. *Plasmodium* parasites synthesize GSH *de novo*, and it was shown that the deletion of the gene encoding γ-glutamylcysteine synthetase (γGCS), the first enzyme of GSH biosynthesis, only marginally affects the intra-erythrocytic development of *Plasmodium berghei*. This is in contrast to the lethal effects of the specific inhibition of this enzyme by L-buthionine sulfoximine on the human malarial parasite *Plasmodium falciparum in vitro*. In addition, a knockout of the gene in *P. falciparum* has, so far, proved impossible. These observations suggest differences in the requirements for GSH biosynthesis in murine and human malaria species during intra-erythrocytic development, which need to be considered when validating this pathway for future drug development.

*Corresponding author

Apicomplexan Parasites. Edited by Katja Becker
Copyright © 2011 WILEY-VCH Verlag GmbH & Co. KGaA, Weinheim
ISBN: 978-3-527-32731-7

Introduction

The tripeptide glutathione (GSH; γ-glutamylcysteinyl-glycine) and the disulfide-containing lipoic acid (LA) are low-molecular-weight thiols which act as cofactors for enzymatic reactions crucial to the metabolism of almost all living organisms. They exist in both reduced and oxidized forms, a feature which makes them important regulators of the intracellular redox homeostasis. GSH is the most abundant low-molecular-weight thiol (1–10 mM) in most eukaryotes and Gram-negative bacteria [1]. The majority of cellular GSH occurs in its reduced form, which is maintained by the activity of the NADPH-dependent disulfide oxidoreductase, glutathione reductase (GR); thus, the redox state of GSH is directly dependent on the availability of NADPH in the cell. In most organisms, the major source of NADPH is believed to be the pentose phosphate shunt, although other sources of NADPH such as $NADP^+$-dependent glutamate dehydrogenases or $NADP^+$-dependent isocitrate dehydrogenase might also play a role in providing the crucial cofactor [2–7]. GR maintains the ratio of GSH to glutathione disulfide (GSSG) high in order to guarantee maintenance of the intracellular redox environment. The redox potential of the GSH/GSSG redox pair is in the range of −0.26 V, and thus the potential to undergo reducing and oxidizing reactions depends on the respective reaction partner that the tripeptide encounters. Similarly, the free dihydrolipoic acid/lipoic acid (DHLA/LA) redox couple has a low redox potential of −0.32 V, which principally makes it a powerful reductant which can, for instance, reduce GSSG and vitamin C as well as reactive oxygen species (ROS) [8]. However, the majority of intracellular LA is bound covalently to the α-keto acid dehydrogenase complexes (KADH) and the glycine cleavage system, where it acts as an essential cofactor for the oxidative decarboxylation of α-keto acids and glycine [9, 10]. The protein-bound form of LA also exerts antioxidant functions, as described for mammalian cells [11].

Other important redox systems involved in regulating the intracellular reducing environment are various thioredoxin-$(SH)_2$/thioredoxin-(S-S) redox pairs, and the $NAD(P)H/NAD(P)^+$ redox couple (as noted above), with a redox potential of −0.32 V. This is in the same range as that of the DHLA/LA pair, and emphasizes the potential of the latter to act effectively in redox reactions [11–14].

The Apicomplexa possess a variety of redox systems and are able to synthesize GSH and LA *de novo* [15–19]. Functional antioxidant and redox machineries are absolutely required for the parasites' survival, because they encounter the ROS generated by the host's immune system, and generate ROS through their own metabolic activities [15, 17, 19].

In this chapter, attention will be focused on the mechanisms that apicomplexan parasites employ to acquire LA, either through *de novo* biosynthesis or salvage. In addition, the *de novo* biosynthesis of GSH in apicomplexans will be discussed, notably in *Plasmodium*. The potential relative roles of these two pathways in survival of the parasites will be discussed, and their potential for future intervention strategies against malaria analyzed.

The Biological Role of LA

Eukaryotes possess three distinct KADHs, namely pyruvate dehydrogenase (PDH), α-ketoglutarate dehydrogenase (KGDH), and branched chain α-keto acid dehydrogenase (BCKDH). These complexes are involved in amino acid and energy metabolism, and consist of a substrate-specific α-keto acid decarboxylase (E1-subunit), an acyltransferase (E2-subunit), and a dihydrolipoamide dehydrogenase (E3-subunit) [10]. The LA is attached covalently to the lipoyl domain of the E2-subunit via an amide linkage with a specific lysine residue, and the oxidized dithiol accepts the acyl moiety generated by the E1-subunit through decarboxylation of an α-keto acid. The lipoamide arm of the E2-subunit transfers the acyl moiety to coenzyme A (CoA) to form acyl-CoA, which is then released from the complex, while the sulfurs of the lipoamide arm are reduced to dihydrolipoamide. In the final step of the reaction, dihydrolipoamide is oxidized by the E3-subunit and NADH is generated [10]. The glycine cleavage system is involved in folate metabolism, and consists of four protein subunits: P-protein, H-protein, T-protein, and L-protein. The LA is attached covalently to the H-protein, and shuttles the reaction intermediates to and from the other proteins in the complex, in similar fashion as was described above for KADH [9].

Lipoic acid, as a dietary component, is taken up by mammalian cells either via an Na^+-dependent multivitamin transporter [20] or a proton-linked monocarboxylic acid transporter [21], and is ligated to the LA-requiring proteins either via a mammalian-like or a bacteria-like salvage pathway [22–24]. Mammals, bacteria, yeast, plants and parasitic protozoa also synthesize LA *de novo* by the action of two enzymes: octanoyl-acyl carrier protein (ACP): protein N-octanoyltransferase (LipB), and lipoate synthase (LipA) [24–26]. LA biosynthesis in the Apicomplexa is restricted to the plastid-like organelle called the apicoplast [16, 18], while LA salvage is confined to the mitochondrion [18, 27–29]. Most investigations involving LA salvage and biosynthesis in apicomplexans have been focused on *Plasmodium falciparum* and *Toxoplasma gondii* [29].

Lipoic Acid Biosynthesis

Lipoic acid biosynthesis in the apicoplast relies on the supply of octanoyl-ACP (Oct-ACP), an intermediate of type II fatty acid biosynthesis (FASII) operating in this organelle. Oct-ACP is ligated to the apo-E2-subunit of PDH, the only apicoplast-located KADH, by LipB [24, 30–32]. Two sulfurs are then incorporated into positions C6 and C8 of the octanoyl-moiety by LipA, an *S*-adenosylmethionine-dependent, [Fe–S] cluster-containing enzyme [33]. LipB is highly specific for its thioester substrate, and is unable to transfer salvaged, free LA to the apo-proteins [24, 28]. Surprisingly, a disruption of the *lipB* gene (Mal8P1.37) in the erythrocytic stage of *P. falciparum* had no effect on parasite viability; by contrast, progression through their intraerythrocytic cell cycle was moderately accelerated [34]. The main phenotype of the *lipB*-null mutant was a drastic reduction in the total (reduced and oxidized) LA content by approximately 95%, resulting in a reduced lipoylation of the PDH E2-subunit. The remaining lipoylation of PDH is thought to result from the activity of a lipoic acid protein ligase-like protein (LplA 2), which was shown to be dually targeted to the apicoplast and mitochondrion [34]. In *Escherichia coli*, LipB function can be replaced by a lipoic acid protein ligase (LplA), albeit with a greatly reduced efficiency [24]. The low level of PDH lipoylation might be sufficient to sustain FASII which, until very recently, was considered an essential pathway for the erythrocytic development of *Plasmodium*. This proposal was based on investigations conducted by Surolia and Surolia [35], who found that triclosan inhibited FabI, an enzyme of the FASII pathway of *Plasmodium*, and killed the parasites both *in vitro* and *in vivo*. However, the results of recent studies with *P. falciparum*, *P. yoeli* and *P. berghei* have strongly suggested that the FASII pathway is not essential for the blood-stage development of the parasites, but that it has an important/essential role for late liver-stage development [36–38]. Taken together, these data suggest that LipB function, which relies on a supply of Oct-ACP from FASII, is only important at a time when the FAS intermediate can be supplied; this, in turn, implies that LA biosynthesis is essential during the exo-erythrocytic development of the malarial parasites. It also implies that PDH is not essential during blood-stage growth, because without LA (which is generated by LA biosynthesis using Oct-ACP as a precursor) the enzyme complex is nonfunctional, given that external LA is not supplied to the apicoplast.

In the case of *T. gondii*, however, the situation is different. A conditional null mutant of apicoplast ACP resulted in the abrogation of PDH lipoylation, presumably because LA was no longer supplied via the LA biosynthesis pathway. As a consequence, the inhibition of FASII resulted in defects in apicoplast maintenance and biogenesis, which eventually led to the parasites' death [39]. Thus, it was concluded that one of the major roles of apicoplast FASII is to provide Oct-ACP, the precursor for LA biosynthesis and eventually PDH lipoylation, presumably to generate acetyl-CoA to feed back into FAS. This is reminiscent of the situation in plant and mammalian mitochondria [40, 41]. In *Eimeria tenella*, genes encoding the enzymes of the FASII pathway have also been identified, but it is not known if – and during which stage of parasite development – the pathway is of importance [42]. According to Seeber *et al.* [43], *Babesia bovis* and *Theileria annulata* lack the genes encoding the proteins of FASII, PDH, LipB, and LipA [43], while Mogi *et al.* [44] and Lau *et al.* [45] have identified three of the four subunits of PDH in the *Babesia* genome, suggesting that they potentially possess PDH activity. This contradictory information emphasizes that experimental evidence is required to fully elucidate the requirement of FASII and, consequently, of LA biosynthesis and PDH activity for the respective parasites. Overall, the various reports have suggested that the requirement for FASII activity might depend on the parasite's host cell, with both environmental and *in-silico* and experimental data suggesting that the potential to generate and acquire fatty acids differs quite considerably between the apicomplexan parasites [46]. *Plasmodium*, *Babesia*, and *Theileria* are all transmitted by blood-feeding insects, and replicate asexually in the red blood cells or lymphocytes of their mammalian host. It is possible that their demand for fatty acids/lipids is met by taking them up from their host and incorporating them into the required lipid species [47, 48] during their development in the red blood cells, lymphocytes, and the insect vectors. This would certainly explain why *Plasmodium* does not need FASII during its blood stage and insect stage growth, and consequently does not rely on LA biosynthesis or PDH activity during these developmental stages. However, as opposed to *Babesia* and *Theileria*, *Plasmodium* has to undergo an exo-erythrocytic developmental cycle in the liver before entering the erythrocytic cycle. The development in the liver appears to be absolutely dependent on FASII activity, and presumably also requires LA biosynthesis and PDH activity to allow synthesis of acetyl-CoA and reducing equivalents. As outlined above, FASII is essential for *T. gondii*, a parasite that infects almost all nucleated cells, which suggests that its needs for fatty acids are not met solely by uptake from the host cell. A number of studies have provided evidence that *T. gondii* is able to commandeer metabolites (including lipids and their precursors) from its host cell [49–52]. Furthermore, the fact that *T. gondii* genome contains one gene encoding FASI, and possibly two genes encoding polyketide synthase [46], suggests that these parasites require a more elaborate set of fatty acids and lipids than, for instance, the malaria parasite during its erythrocytic and sexual development. FASI also occurs in *Cryptosporidium* and *Eimeria* [42, 53], and might be linked to the formation of oocysts which are shed into the environment. It is possible that *Plasmodium* finds itself in a similar situation as *T. gondii* during the liver-stage development, where supplies of fatty acids from the host are insufficient to support the extensive multiplication that *Plasmodium* undergoes in the liver. This is rather surprising, given that the liver is the predominant site of fatty acid metabolism in mammals, and presumably should provide a sufficiently large supply of building blocks for parasite lipids. It is known that lipoproteins are required for a successful liver-stage infection, and UIS3 has been shown to be involved in lipid or fatty acid uptake across the parasitophorous vacuolar membrane (PVM) [54]. But that being the case, why does *Plasmodium* require active FASII when the host cell is actually providing lipids and fatty acids to the developing schizont? The formation of thousands of individual merozoites during development in the liver requires a vast biosynthesis of membranes within a relatively short period of time. It has been shown that the membrane of the developing parasites in the liver invaginates profoundly during the cytomere stage, supporting the need for substantial amounts of lipids [55]. This also suggests that the requirement for lipids during this parasite stage is far greater than during the other developmental stages, and potentially explains the dependence on FASII biosynthesis and, also the dependence on PDH activity and thus LA biosynthesis. These hypotheses need to be addressed experimentally to obtain a full understanding as to why such pathways are essential for the malaria parasites only during liver-stage development. Another possibility is that the parasites rely on a specific fatty acid that cannot be supplied in sufficiently high concentrations in the liver, and therefore must be synthesized *de novo* by the parasite's FASII. Such a situation was found, for instance, in *Trypanosoma brucei* bloodstream-form parasites, which rely on large amounts of myristic acid to guarantee the correct lipidation of their GPI anchors [56]. Interestingly, the GPI-anchored protein MSP1 is no longer present in liver-stage *Plasmodium* lacking FAS II activity [37], which suggests that the external supply of certain fatty acids required for

GPI anchor biosynthesis is insufficient during liver-stage development, whereas this demand can be covered during the blood-stage development, even if FASII is nonfunctional.

Lipoic Acid Salvage

Both, *P. falciparum* and *T. gondii* are not only able to synthesize LA *de novo*, but they also can scavenge free LA, which is used solely to lipoylate mitochondrial KADH and the glycine cleavage system [18, 27, 28].

Whilst the way in which *Plasmodium* acquires LA from its host is unclear, several suggestions have been made as to how *T. gondii* achieves this. The amount of free, intracellular LA is negligible, as it is consumed by the host cell itself [57]. However, it is possible that *T. gondii* is able to take up lipoylated peptides that result from the degradation of host mitochondrial E2 subunits of KADH, and which might serve as a source for LA, provided that the parasite possesses a lipoamidase activity to release the LA from the peptides. A similar process was reported for the acquisition of LA by the intracellular bacterium *Listeria monocytogenes* [58]. This hypothesis is particularly appealing for *T. gondii*, because the PVM forms a tight association with the host cell mitochondria and ER, and the metabolites are potentially provided from the host organelles to the parasite [28, 59]. In addition, *T. gondii* sequesters host endolysosomes in its PVM via host microtubules [60], which also could be a source of free LA. This is certainly not possible for *P. falciparum*-infected erythrocytes, albeit a similar situation might occur during liver-stage development. It was also shown that, during *Plasmodium* liver-stage development, the PVM becomes porous and allows the passive transfer of molecules up to 855 kDa in size [61], which would allow LA-containing host peptides to be scavenged by the parasites. The erythrocytic stages of *Plasmodium* possibly take up free LA from their host [62] via the new permeation pathway, after which LA can enter the parasite via the pantothenate uptake system [63].

Once LA has entered the parasite mitochondrion a bacterial-like salvage enzyme, LplA, is responsible for the ligation of LA to the apo-E2 subunits of BCKDH, KGDH, and to the H-protein of the glycine cleavage system [18, 27]. LplA catalyzes a two-step reaction: (i) LA is activated by ATP, generating a LA–AMP intermediate that remains attached to the active site of the enzyme while pyrophosphate is released; and (ii) LA is transferred to the lipoyl domain of the apo-target protein, while AMP is released [24, 64]. This two-step reaction, catalyzed by the single LplA protein, requires the activity of two enzymes in mammalian cells, namely a lipoate-activating enzyme and a lipoyltransferase [22, 23]. However, it was found recently in *Thermoplasma acidophilum*, that a second Lpl protein (LplB) is required for ligase activity [65]. *Plasmodium* possesses two LplA-like proteins, LplA1 and LplA2. Of these, LplA1 is located in the mitochondrion, while LplA2 is present in both the mitochondrion and apicoplast [18, 34]. *T. gondii* contains a gene encoding a potential LplA1 (TGGT1_110620; TGME49_071820; TGVEG_019270), and also possesses a gene that has a C-terminal LplA2-like domain, although whether it has LplA2 activity remains the subject of future investigation (accession number: XP_002370259). *B. bovis*, *Theileria parva*, and *T. annulata* also contain genes that potentially encode two distinct LplA proteins (LplA1: BBOV_I003930, *TP01_1192, TA09645; LplA2:* BBOV_I002370, *TP03_0090, TA03985*) while *Cryptosporidium muris* possesses only a gene encoding a potential LplA1 (CMU_008530). Both, LplA1 and LplA2 of *Plasmodium* (PF13_0083 and PFI1160w, respectively) were shown to be enzymatically active [27, 34], although the proteins have not yet been fully biochemically characterized to elucidate their substrate specificities and kinetic parameters. However, the roles of the *Plasmodium* LplA proteins were investigated using a reverse genetics approach in both *P. falciparum* and *P. berghei*. The disruption of both genes in *P. falciparum* proved to be unsuccessful, as the gene loci appeared refractory to recombination and it was impossible to target the loci with knockout as well as knock-in constructs. This is clearly an unsatisfactory situation, and allows few conclusions to be drawn as to the validity of these proteins for future drug development. Therefore, an attempt was made to generate *lplA1* and *lplA2* gene deletions in *P. berghei*, with both gene loci being targeted with knock-in and knock out constructs [29, 66]. Despite replacing the *lplA1* gene with a selectable marker, it was impossible to isolate a stable population of the null mutant parasites [66], which suggested that the protein is important for the growth and survival of the erythrocytic stages of the malarial parasites. This hypothesis was supported by the fact that the inhibition of LplA1 by the LA analog 8-bromo-octanoate inhibited the intra-erythrocytic growth of *P. falciparum* [27]. This, in turn, indicated that LplA1 plays an essential role in the intra-erythrocytic development of *Plasmodium*, and that the second lipoic acid protein ligase LplA2 could not replace the LplA1 function. Transcriptional expression profiles [67] suggested that the two genes were expressed at different times of the developmental cycle of the parasites; notably, *lplA1* appeared to be expressed primarily during the blood-stage development, while *lplA2* showed its highest mRNA expression during sexual development, although the mRNA was also present in the blood stages. This potentially accounted for the finding that the knockout of *lplA2* does not affect the intra-erythrocytic growth of *P. berghei*, but rather leads to a reduction in ookinete numbers and totally abolishes the formation of oocysts in the insect vector (S. Günther *et al.*, unpublished results).

Thus, it seems that mitochondrial lipoylation by LplA1 is critical for the blood stages of *Plasmodium*, suggesting that the activities of the mitochondrial KADH or glycine cleavage system have significant functions during these stages of parasite development. Another interpretation of these results might be that protein-bound LA acts as an antioxidant in the mitochondrion, and maintains the organelle's function by protecting it against oxidative stress. This was suggested previously for *Mycobacteria*, where protein-bound LA was shown to interact with a thioredoxin-like protein and thus proved instrumental for antioxidant defense [68]. Indeed, LA bound to mammalian α-ketoglutarate dehydrogenase reduces thioredoxin [14], and is possibly also involved in the organelle's redox regulation [11]. However, it remains to be elucidated whether protein-bound LA has a similar role in redox control and as an antioxidant in the apicomplexans, in addition to its well-known function as an essential cofactor for intermediary metabolism.

The Biological Roles of GSH

Apart from acting as principal thiol redox buffer, GSH is an important cofactor for detoxifying and antioxidant enzymes, such as glutathione-*S*-transferases (GST), thiol transferases such as glutaredoxins and glyoxalases, and also GSH-dependent peroxidases [69]. The tripeptide is also responsible for maintaining disulfide isomerases in their active state, and thus is crucially involved in the formation of disulfides in secreted proteins [70, 71]. These multiple functions mean that GSH is a vital component of the cell's redox and antioxidant metabolism.

GSH Biosynthesis

While the levels of reduced GSH are primarily maintained by GR, the pool of GSH is replenished via biosynthesis when GSSG or GS–X adducts are lost from the cell through efflux pumps. GSH is synthesized *de novo*, in two consecutive steps, from the amino acids glutamate, cysteine, and glycine. In the first (ATP-dependent) step, the enzyme γ-glutamylcysteine synthetase (γGCS) forms a γ-amino linkage between glutamate and cysteine, thus generating the GSH precursor, γ-glutamylcysteine (γGC). In the second (also ATP-dependent) step, glutathione synthetase (GS) adds glycine to γGC, thus generating GSH. The hydrolysis of GSH by intracellular peptidases is prevented by the presence of the unusual γ-linkage.

Biochemical characterization studies have shown γGCS to be the rate-limiting enzyme for GSH biosynthesis in a number of organisms; moreover, the enzyme is feedback-inhibited by GSH [72–75]. The deduced amino acid sequence of γGCS identified in apicomplexan parasites is closely related to that of γGCS in other non-plant eukaryotes, and shows only a limited sequence similarity to plant and prokaryotic γGCS [76]. Similar to *T. brucei*, but opposed to mammals, *Plasmodium* does not possess a gene encoding a regulatory subunit of γGCS, which suggests that the regulation of enzyme activity of parasite γGCS differs from that of the mammalian enzyme [72–74]. Unfortunately, biochemical characterization of the *Plasmodium* enzyme has so far been hampered by an inability to express the protein recombinantly, and consequently there is no information available regarding the regulation of activity, or its kinetic parameters. All *Plasmodium* species studied to date contain genes encoding the two enzymes comprising the GSH biosynthetic pathway. The genes encoding γGCS and GS were identified in *P. berghei* (γGCS: PB001283.02.0, GS:300407.00.0), *P. chabaudi chabaudi* (γGCS: PCAS_081960, GS: PCAS_111140), *P. falciparum* (γGCS: PF10925w, GS: PFE0606c), *P. knowlesi* (γGCS: PKH_071610, GS: PKH_011060), *P. vivax* (γGCS: PVX_099360, GS: PVX_080630), and *P. yoelli yoelli* (γGCS: PY01606, GS: PY07248).

In the case of *T. gondii*, a putative γGCS gene has been identified in type I (TGGT1_114930), type II (TGME49_026800) and type III strains (TGVEG_023050). A GS gene has also been annotated for a type II strain (TGME49_026800).

In the *Cryptosporidium parvum* genome, no putative *γgcs* or *gs* genes have been annotated, although a putative GSH peroxidase and a GST (both GSH-dependent) were identified [77]. It has been shown experimentally that *C. parvum* oocysts and sporozoites contain GSH; furthermore, it was also shown that the GSH level in *C. parvum* is sensitive to treatment with the specific γGCS inhibitor L-buthionin sulfoximine (L-BSO), implying that these parasites also possess an active GSH biosynthetic pathway [78, 79].

Thus, it appears that all apicomplexans studied to date contain an active GSH biosynthetic pathway, which suggests that they are independent of their host for supplies of the tripeptide.

In *Arabidopsis thaliana*, γGCS was exclusively localized to the plastids [80, 81], while γGCS activity in *Zea mays*, *Nicotiana tabacum*, and *Spinacia oleracea* was found predominantly in the plastids and, to a lesser extent, in the cytoplasm [82, 83]. This was in contrast to GS, which was primarily present in the cytoplasm, with only small amounts in the plastids in *A. thaliana* and *N. tabacum* [80, 84]. The distinct localizations of γGCS and GS in plants imply the presence of a transporter that allows the transfer of γGC from plastids to cytoplasm, in order to allow for GSH biosynthesis. In mammalian cells and yeast, GSH biosynthesis is cytoplasmic, and the tripeptide is imported into all organelles requiring GSH [70, 85–87]; this finding was comparable to by our unpublished data indicating that γGCS and GS are both localized to the *Plasmodium* cytoplasm.

Until recently, the biosynthesis of GSH was believed to be essential for survival of the intra-erythrocytic development of *Plasmodium* parasites. However, it has been shown that a deletion of the *γgcs* gene in the mouse malarial parasite *P. berghei* affected its intra-erythrocytic development only marginally, but inhibited its sexual development in the insect vector [88]. These findings were contrary to previous observations which showed that the irreversible inhibitor of γGCS, L-BSO, has a lethal effect on the human malarial parasite *P. falciparum in vitro*, by depleting its intracellular GSH levels [89–92]. This plasmodicidal effect was partially abolished by supplying (externally) high levels of GSH or GSH-ethyl ester. One reason as to why *P. falciparum*-infected red blood cells might have been particularly affected by the inhibition of GSH biosynthesis was that they appeared to lose GSH rapidly; typically, 50% of the total GSH was lost within 2.5 h [89, 92]. This implied that *P. falciparum* relies on the active biosynthesis of GSH during its intra-erythrocytic growth, although this seemed not to be the case for *P. berghei*. GSH biosynthesis also plays a pivotal role in the survival of other eukaryotes; typically, knockout or knockdown of the *γgcs* gene was lethal for *T. brucei*, *Saccharomyces cerevisiae*, *Candida albicans*, *Dictyostelium discoideum*, *Mus musculus*, and *A. thaliana* [93–97]. The loss of GSH biosynthesis in all of these organisms could, however, be reversed by supplementation with either GSH or γGC. In some cases, other thiols such as *N*-acetylcysteine or dithiothreitol were also able to compensate for the loss of GSH biosynthesis [93, 94].

Interestingly, the *P. berghei γgcs* null mutants displayed only a mild growth defect during intra-erythrocytic development, with a closer investigation of the intra-erythrocytic stages of the mutant parasites revealing that they still contained between 4% and 14% total GSH. This suggested that, during their intra-erythrocytic life, the parasites were able to obtain GSH directly from their host; this apparently allowed them to sustain adequate levels of GSH in order to guarantee an intracellular reducing environment, as well as a suitable supply of GSH to maintain the antioxidant and detoxification reactions. Not only the inhibition data obtained with the specific γGCS inhibitor L-BSO, but also the present authors' reverse genetic approaches on *P. falciparum*, have suggested that the human malarial parasite *P. falciparum* would differ in its requirements for an active GSH biosynthesis. To date, it has not been possible to generate either *γgcs* or *gs* null mutants of *P. falciparum*, despite the fact that both gene loci were successfully targeted by various control constructs and are not refractory to recombination (E.-M. Patzewitz *et al.*, unpublished observation). One possible explanation for this discrepancy might be that the *in vitro* conditions under which *P. falciparum* are maintained do not provide sufficiently high levels of GSH for the parasites to scavenge the tripeptide, in order to compensate for the lost GSH biosynthesis. However, even supplementation of the culture medium with millimolar concentrations of GSH did not allow the deletion of either of the two genes in the human malarial parasite. These data suggest that the ability to take up GSH from the host cell appears to differ between *P. berghei* and *P. falciparum*. GSH does not easily cross the cell membrane; rather, most mammalian cells depend on the action of γ-glutamyl-transpeptidase, an enzyme located on the outside of the plasma membrane, which transfers the γ-glutamate moiety to an acceptor amino acid [98]. The preferred acceptor for glutamate is cysteine or homocysteine, so that γ-Glu-Cys or γ-Glu-Cys-Cys are generated and taken up into the cell. Usually, the peptides are used to re-synthesize GSH via the γ-glutamyl cycle [99], but uninfected red blood cells do not express γ-glutamyl-transpeptidase and thus cannot take up GSH from the plasma. Whilst the *P. falciparum*-infected red blood cells take up GSH, the free parasites appear unable to import the tripeptide [92], which suggests that *P. falciparum* may lack the ability to transport GSH across its plasma membrane. In agreement with these findings, the gene encoding for γ-glutamyl-transpeptidase has not yet been identified in the genome of *P. falciparum*, nor that of other *Plasmodium* species, which suggests that the tripeptide might simply reach the parasite via endocytosis. Nonetheless, the fact that a gene encoding the transpeptidase cannot be found in the parasite genomes does not exclude the possibility that a gene encoding a functionally similar protein, but which differs in its primary structure from those characterized to date, does exist. Exactly *how* GSH enters *P. berghei* is yet to be established, and it cannot be excluded that a similar mechanism might exist in *P. falciparum*. Clearly, this is an area that requires further investigation if the discrepancies between results obtained concerning the essential role of GSH biosynthesis during the blood-stage development of the parasite are to be fully understood.

Taken together, these results suggest that the physiological and metabolic requirements of different *Plasmodium* species vary, and that detailed and species-specific analyses are required to fully establish whether a metabolic pathway is essential for the intra-erythrocytic growth of the human malaria parasite. This situation applies even if the pathway has been dismissed as a suitable target for drug development in a model parasite, such as *P. berghei*.

To date, very few data are available on the role of the second protein involved in GSH biosynthesis, namely glutathione synthase (GS). The only apicomplexan the only apicomplexan GS that is known to be recombinantly expressed and has been biochemically characterized is that of *P. falciparum* [100]. However, as outlined above, all present endeavors to generate GS null mutants have failed so far, presumably because the gene is essential for parasite survival.

The fact that GSH seems to be of vital importance during the insect-stage development of *P. berghei* is interesting, and requires further investigation. It has been suggested that this dependence is due primarily to the fact that the mitochondria of oocysts are irreversibly damaged as a consequence of the loss of function of γGCS. One reason for this phenotype might be that mitochondrial metabolism is enhanced during these developmental stages, compared to the intra-erythrocytic stages of parasite development. This would require a higher level of antioxidant defense against the ROS that would result from elevated levels of electron transport chain activity. Another reason to rely on GSH biosynthesis might involve the insect vector not supplying sufficient GSH to be taken up by the parasites to compensate for the loss of γ-GCS function or, alternatively, that the parasites are unable to take up GSH from their insect host.

Conclusion

New intervention strategies to combat parasitic diseases are urgently required. Antioxidants play vital roles in the survival of parasitic protozoa, and both GSH and LA have been identified as crucial components of a parasite's metabolism and antioxidant defense. Indeed, it has been suggested that the acquisition and biosynthesis of these materials offer potential for the development of future drugs that act specifically against the Apicomplexa. In this chapter, a summary has been provided of the information currently available supporting the notion that LA biosynthesis, similar to FASII, has essential functions for development of the exo-erythrocytic stages of *Plasmodium* (Figure 10.1), although experimental evidence to support this proposal has yet to be generated. In addition, the potential was discussed of LA salvage comprising the two lipoic acid protein ligases, LplA1 and LplA2, for the design of future antimalarial agents, either to interfere with the development of the parasite during the red blood cell cycle (LplA1), or to inhibit its sexual development in the mosquito (LplA2) (see Figure 10.1). However, in order to fully understand the distinct functions of the two proteins at different times of parasite development, there is a need to develop more detailed insights into the biochemical properties, expression pattern and metabolic roles of the proteins during the respective parasite life stages.

The rodent and human malarial species *P. berghei* and *P. falciparum* appear to have distinct requirements for the biosynthesis of GSH during red blood cell-stage development, with the enzyme involved in the first step of biosynthesis being essential in *P. falciparum*, but not in *P. berghei*. Hence, several questions need to be addressed in detail to fully understand the differences observed between the two malarial parasite species with respect to the importance of GSH biosynthesis;

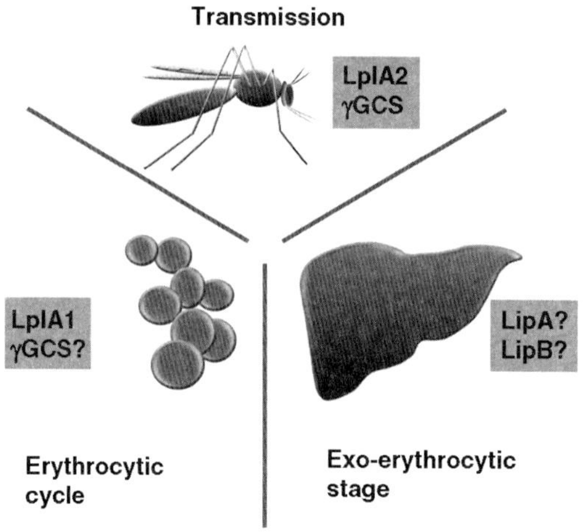

Transmission

LplA2
γGCS

LplA1
γGCS?

LipA?
LipB?

Erythrocytic
cycle

Exo-erythrocytic
stage

Figure 10.1 Potential drug targets involved in lipoic acid (LA) biosynthesis, LA salvage, and glutathione (GSH) biosynthesis throughout the life cycle of *Plasmodium*. Genes involved in LA salvage are essential in the red blood cell stages (LplA1) or in the sexual stages (LplA2), whereas the first step of GSH biosynthesis catalyzed by γGCS is essential only in the mosquito stages of *P. berghei*, but appears to be essential in the erythrocytic stages of *P. falciparum*. LA acid biosynthesis (LipA and LipB) is potentially essential for liver-stage development of the parasite.

- Are they attributable to the different environments in which the parasites are maintained?
- Are they due to the different abilities of the two *Plasmodium* species to efficiently take up and utilize GSH from their host cell?
- Are they the result of different metabolic requirements for GSH in the two *Plasmodium* species?

Yet, before these questions can be satisfactorily answered, it must be judged as to whether GSH biosynthesis represents a suitable target for future malaria intervention against blood-stage parasites, despite the pathway offering great potential for the development of transmission blocking agents (see Figure 10.1).

References

1 Meister, A. and Anderson, M.E. (1983) Glutathione. *Annu. Rev. Biochem.*, **52**, 711–760.
2 Barrett, M.P. (1997) The pentose phosphate pathway and parasitic protozoa. *Parasitology Today*, **13**, 11–16.
3 Atamna, H., Pascarmona, G., and Ginsburg, H. (1994) Hexose-monophosphate shunt activity in intact *Plasmodium falciparum*-infected erythrocytes and in free parasites. *Mol. Biochem. Parasitol.*, **67**, 79–89.
4 Roth, E. Jr (1990) *Plasmodium falciparum* carbohydrate metabolism: a connection between host cell and parasite. *Blood Cells*, **16**, 453–460; discussion 461–466.

5 Bro, C., Regenberg, B., and Nielsen, J. (2004) Genome-wide transcriptional response of a *Saccharomyces cerevisiae* strain with an altered redox metabolism. *Biotechnol. Bioeng.*, **85**, 269–276.

6 Werner, C., Stubbs, M.T., Krauth-Siegel, R.L., and Klebe, G. (2005) The crystal structure of *Plasmodium falciparum* glutamate dehydrogenase, a putative target for novel antimalarial drugs. *J. Mol. Biol.*, **349**, 597–607.

7 Wrenger, C. and Muller, S. (2003) Isocitrate dehydrogenase of *Plasmodium falciparum*. *Eur. J. Biochem.*, **270**, 1775–1783.

8 Bilska, A. and Wlodek, L. (2005) Lipoic acid - the drug of the future? *Pharmacol. Rep.*, **57**, 570–577.

9 Douce, R., Bourguignon, J., Neuburger, M., and Rebeille, F. (2001) The glycine decarboxylase system: a fascinating complex. *Trends Plant. Sci.*, **6**, 167–176.

10 Perham, R.N. (2000) Swinging arms and swinging domains in multifunctional enzymes: catalytic machines for multistep reactions. *Annu. Rev. Biochem.*, **69**, 961–1004.

11 Bunik, V.I. (2003) 2-Oxo acid dehydrogenase complexes in redox regulation. *Eur. J. Biochem.*, **270**, 1036–1042.

12 Ying, W. (2008) NAD$^+$/NADH and NADP$^+$/NADPH in cellular functions and cell death: regulation and biological consequences. *Antioxid. Redox Signal.*, **10**, 179–206.

13 Dietz, K.J. (2003) Redox control, redox signaling, and redox homeostasis in plant cells. *Int. Rev. Cytol.*, **228**, 141–193.

14 Bunik, V. and Follmann, H. (1993) Thioredoxin reduction dependent on alpha-ketoacid oxidation by alpha-ketoacid dehydrogenase complexes. *FEBS Lett.*, **336**, 197–200.

15 Becker, K., Tilley, L., Vennerstrom, J.L., Roberts, D., Rogerson, S., and Ginsburg, H. (2004) Oxidative stress in malaria parasite-infected erythrocytes: host–parasite interactions. *Int. J. Parasitol.*, **34**, 163–189.

16 Thomsen-Zieger, N., Schachtner, J., and Seeber, F. (2003) Apicomplexan parasites contain a single lipoic acid synthase located in the plastid. *FEBS Lett.*, **547**, 80–86.

17 Muller, S. (2004) Redox and antioxidant systems of the malaria parasite *Plasmodium falciparum*. *Mol. Microbiol.*, **53**, 1291–1305.

18 Wrenger, C. and Muller, S. (2004) The human malaria parasite *Plasmodium falciparum* has distinct organelle-specific lipoylation pathways. *Mol. Microbiol.*, **53**, 103–113.

19 Ding, M., Kwok, L.Y., Schluter, D., Clayton, C., and Soldati, D. (2004) The antioxidant systems in *Toxoplasma gondii* and the role of cytosolic catalase in defence against oxidative injury. *Mol. Microbiol.*, **51**, 47–61.

20 Prasad, P.D. and Ganapathy, V. (2000) Structure and function of mammalian sodium-dependent multivitamin transporter. *Curr. Opin. Clin. Nutr. Metab. Care*, **3**, 263–266.

21 Takaishi, N., Yoshida, K., Satsu, H., and Shimizu, M. (2007) Transepithelial transport of alpha-lipoic acid across human intestinal Caco-2 cell monolayers. *J. Agric. Food Chem.*, **55**, 5253–5259.

22 Fujiwara, K., Hosaka, H., Matsuda, M., Okamura-Ikeda, K., Motokawa, Y., Suzuki, M., Nakagawa, A., and Taniguchi, H. (2007) Crystal structure of bovine lipoyltransferase in complex with lipoyl-AMP. *J. Mol. Biol.*, **371**, 222–234.

23 Fujiwara, K., Takeuchi, S., Okamura-Ikeda, K., and Motokawa, Y. (2001) Purification, characterization, and cDNA cloning of lipoate-activating enzyme from bovine liver. *J. Biol. Chem.*, **276**, 28819–28823.

24 Cronan, J.E., Zhao, X., and Jiang, Y. (2005) Function, attachment and synthesis of lipoic acid in *Escherichia coli. Adv. Microb. Physiol.*, **50**, 103–146.

25 Morikawa, T., Yasuno, R., and Wada, H. (2001) Do mammalian cells synthesize lipoic acid? Identification of a mouse cDNA encoding a lipoic acid synthase located in mitochondria. *FEBS Lett.*, **498**, 16–21.

26 Yasuno, R. and Wada, H. (1998) Biosynthesis of lipoic acid in *Arabidopsis*: cloning and characterization of the cDNA for lipoic acid synthase. *Plant Physiol.*, **118**, 935–943.

27 Allary, M., Lu, J.Z., Zhu, L., and Prigge, S.T. (2007) Scavenging of the cofactor lipoate is essential for the survival of the malaria parasite *Plasmodium falciparum. Mol. Microbiol.*, **63**, 1331–1344.

28 Crawford, M.J., Thomsen-Zieger, N., Ray, M., Schachtner, J., Roos, D.S., and Seeber, F. (2006) Toxoplasma *gondii* scavenges host-derived lipoic acid despite its de novo synthesis in the apicoplast. *EMBO J.*, **25**, 3214–3222.

29 Gunther, S., Storm, J., and Muller, S. (2009) *Plasmodium falciparum*: organelle-specific acquisition of lipoic acid. *Int. J. Biochem. Cell Biol.*, **41**, 748–752.

30 Foth, B.J., Stimmler, L.M., Handman, E., Crabb, B.S., Hodder, A.N., and McFadden, G.I. (2005) The malaria parasite *Plasmodium falciparum* has only one pyruvate dehydrogenase complex, which is located in the apicoplast. *Mol. Microbiol.*, **55**, 39–53.

31 Gunther, S., McMillan, P.J., Wallace, L.J., and Muller, S. (2005) Plasmodium *falciparum* possesses organelle-specific alpha-keto acid dehydrogenase complexes and lipoylation pathways. *Biochem. Soc. Trans.*, **33**, 977–980.

32 McMillan, P.J., Stimmler, L.M., Foth, B.J., McFadden, G.I., and Muller, S. (2005) The human malaria parasite *Plasmodium falciparum* possesses two distinct dihydrolipoamide dehydrogenases. *Mol. Microbiol.*, **55**, 27–38.

33 Booker, S.J., Cicchillo, R.M., and Grove, T.L. (2007) Self-sacrifice in radical S-adenosylmethionine proteins. *Curr. Opin. Chem. Biol.*, **11**, 543–552.

34 Gunther, S., Wallace, L., Patzewitz, E.M., McMillan, P.J., Storm, J., Wrenger, C., Bissett, R. *et al.* (2007) Apicoplast lipoic acid protein ligase B is not essential for *Plasmodium falciparum. PLoS Pathog.*, **3**, e189.

35 Surolia, N. and Surolia, A. (2001) Triclosan offers protection against blood stages of malaria by inhibiting enoyl-ACP reductase of *Plasmodium falciparum. Nat. Med.*, **7**, 167–173.

36 Tarun, A.S., Vaughan, A.M., and Kappe, S.H. (2009) Redefining the role of de novo fatty acid synthesis in *Plasmodium* parasites. *Trends Parasitol.*, **25**, 545–550.

37 Vaughan, A.M., O'Neill, M.T., Tarun, A.S., Camargo, N., Phuong, T.M., Aly, A.S., Cowman, A.F., and Kappe, S.H. (2009) Type II fatty acid synthesis is essential only for malaria parasite late liver stage development. *Cell Microbiol.*, **11**, 506–520.

38 Yu, M., Kumar, T.R., Nkrumah, L.J., Coppi, A., Retzlaff, S., Li, C.D., Kelly, B.J. *et al.* (2008) The fatty acid biosynthesis enzyme FabI plays a key role in the development of liver-stage malarial parasites. *Cell Host Microbe*, **4**, 567–578.

39 Mazumdar, J., Wilson, E.H., Masek, K., Hunter, CA., and Striepen, B. (2006) Apicoplast fatty acid synthesis is essential for organelle biogenesis and parasite survival in *Toxoplasma gondii. Proc. Natl Acad. Sci. USA*, **103**, 13192–13197.

40 Gueguen, V., Macherel, D., Jaquinod, M., Douce, R., and Bourguignon, J. (2000) Fatty acid and lipoic acid biosynthesis in higher plant mitochondria. *J. Biol. Chem.*, **275**, 5016–5025.

41 Witkowski, A., Joshi, A.K., and Smith, S. (2007) Coupling of the *de novo* fatty acid biosynthesis and lipoylation pathways in mammalian mitochondria. *J. Biol. Chem.*, **282**, 14178–14185.

42 Lu, J.Z., Muench, S.P., Allary, M., Campbell, S., Roberts, C.W., Mui, E., McLeod, R.L. *et al.* (2007) Type I and type II fatty acid biosynthesis in *Eimeria tenella*: enoyl reductase activity and structure. *Parasitology*, **134**, 1949–1962.

43 Seeber, F., Limenitakis, J., and Soldati-Favre, D. (2008) Apicomplexan mitochondrial metabolism: a story of gains, losses and retentions. *Trends Parasitol.*, **24**, 468–478.

44 Mogi, T. and Kita, K. (2010) Diversity in mitochondrial metabolic pathways in parasitic protists *Plasmodium* and

Cryptosporidium. Parasitol Int., **59**, 305–312.

45 Lau, A.O., McElwain, T.F., Brayton, K.A., Knowles, D.P., and Roalson, E.H. (2009) *Babesia bovis*: a comprehensive phylogenetic analysis of plastid-encoded genes supports green algal origin of apicoplasts. *Exp. Parasitol.*, **123**, 236–243.

46 Mazumdar, J. and Striepen, B. (2007) Make it or take it: fatty acid metabolism of apicomplexan parasites. *Eukaryot. Cell*, **6**, 1727–1735.

47 Palacpac, N.M., Hiramine, Y., Mi-ichi, F., Torii, M., Kita, K., Hiramatsu, R., Horii, T., and Mitamura, T. (2004) Developmental-stage-specific triacylglycerol biosynthesis, degradation and trafficking as lipid bodies in *Plasmodium falciparum*-infected erythrocytes. *J. Cell Sci.*, **117**, 1469–1480.

48 Vial, H.J., Thuet, M.J., and Philippot, J.R. (1982) Phospholipid biosynthesis in synchronous *Plasmodium falciparum* cultures. *J. Protozool.*, **29**, 258–263.

49 Charron, A.J. and Sibley, L.D. (2002) Host cells: mobilizable lipid resources for the intracellular parasite *Toxoplasma gondii*. *J. Cell Sci.*, **115**, 3049–3059.

50 Coppens, I. (2006) Contribution of host lipids to *Toxoplasma* pathogenesis. *Cell Microbiol.*, **8**, 1–9.

51 Coppens, I., Sinai, A.P., and Joiner, K.A. (2000) Toxoplasma *gondii* exploits host low-density lipoprotein receptor-mediated endocytosis for cholesterol acquisition. *J. Cell Biol.*, **149**, 167–180.

52 Gupta, N., Zahn, M.M., Coppens, I., Joiner, K.A., and Voelker, D.R. (2005) Selective disruption of phosphatidylcholine metabolism of the intracellular parasite *Toxoplasma gondii* arrests its growth. *J. Biol. Chem.*, **280**, 16345–16353.

53 Zhu, G., Marchewka, M.J., Woods, K.M., Upton, S.J., and Keithly, J.S. (2000) Molecular analysis of a Type I fatty acid synthase in *Cryptosporidium parvum*. *Mol. Biochem. Parasitol.*, **105**, 253–260.

54 Mikolajczak, S.A., Jacobs-Lorena, V., MacKellar, D.C., Camargo, N., and Kappe, S.H. (2007) L-FABP is a critical host factor for successful malaria liver stage development. *Int. J. Parasitol.*, **37**, 483–489.

55 Sturm, A., Graewe, S., Franke-Fayard, B., Retzlaff, S., Bolte, S., Roppenser, B., Aepfelbacher, M. *et al.* (2009) Alteration of the parasite plasma membrane and the parasitophorous vacuole membrane during exo-erythrocytic development of malaria parasites. *Protist*, **160**, 51–63.

56 Lee, S.H., Stephens, J.L., and Englund, P.T. (2007) A fatty-acid synthesis mechanism specialized for parasitism. *Nat. Rev. Microbiol.*, **5**, 287–297.

57 Akiba, S., Matsugo, S., Packer, L., and Konishi, T. (1998) Assay of protein-bound lipoic acid in tissues by a new enzymatic method. *Anal. Biochem.*, **258**, 299–304.

58 Keeney, K.M., Stuckey, J.A., and O'Riordan, M.X. (2007) LplA1-dependent utilization of host lipoyl peptides enables *Listeria* cytosolic growth and virulence. *Mol. Microbiol.*, **66**, 758–770.

59 Sinai, A.P., Webster, P., and Joiner, K.A. (1997) Association of host cell endoplasmic reticulum and mitochondria with the *Toxoplasma gondii* parasitophorous vacuole membrane: a high-affinity interaction. *J. Cell Sci.*, **110** (Pt 17), 2117–2128.

60 Coppens, I., Dunn, J.D., Romano, J.D., Pypaert, M., Zhang, H., Boothroyd, J.C., and Joiner, K.A. (2006) *Toxoplasma gondii* sequesters lysosomes from mammalian hosts in the vacuolar space. *Cell*, **125**, 261–274.

61 Bano, N., Romano, J.D., Jayabalasingham, B., and Coppens, I. (2007) Cellular interactions of *Plasmodium* liver stage with its host mammalian cell. *Int. J. Parasitol.*, **37**, 1329–1341.

62 May, J.M., Qu, Z.C., and Nelson, D.J. (2007) Uptake and reduction of alpha-lipoic acid by human erythrocytes. *Clin. Biochem.*, **40**, 1135–1142.

63 Kirk, K. and Saliba, K.J. (2007) Targeting nutrient uptake mechanisms in *Plasmodium. Curr. Drug Targets*, **8**, 75–88.

64 Fujiwara, K., Toma, S., Okamura-Ikeda, K., Motokawa, Y., Nakagawa, A., and Taniguchi, H. (2005) Crystal structure of lipoate-protein ligase A from *Escherichia coli*. Determination of the lipoic acid-binding site. *J. Biol. Chem.*, **280**, 33645–33651.

65 Christensen, Q.H. and Cronan, J.E. (2009) The *Thermoplasma acidophilum* LplA-LplB complex defines a new class of bipartite lipoate-protein ligases. *J. Biol. Chem.*, **284**, 21317–21326.

66 Gunther, S., Matuschewski, K., and Muller, S. (2009) Knockout studies reveal an important role of *Plasmodium* lipoic acid protein ligase A1 for asexual blood stage parasite survival. *PLoS One*, **4**, e5510.

67 Le Roch, K.G., Zhou, Y., Blair, P.L., Grainger, M., Moch, J.K., Haynes, J.D., De La Vega, P. *et al.* (2003) Discovery of gene function by expression profiling of the malaria parasite life cycle. *Science*, **301**, 1503–1508.

68 Bryk, R., Lima, C.D., Erdjument-Bromage, H., Tempst, P., and Nathan, C. (2002) Metabolic enzymes of mycobacteria linked to antioxidant defense by a thioredoxin-like protein. *Science*, **295**, 1073–1077.

69 Becker, K., Rahlfs, S., Nickel, C., and Schirmer, R.H. (2003) Glutathione–functions and metabolism in the malarial parasite *Plasmodium falciparum*. *Biol. Chem.*, **384**, 551–566.

70 Chakravarthi, S. and Bulleid, N.J. (2004) Glutathione is required to regulate the formation of native disulfide bonds within proteins entering the secretory pathway. *J. Biol. Chem.*, **279**, 39872–39879.

71 Chakravarthi, S., Jessop, C.E., and Bulleid, N.J. (2006) The role of glutathione in disulphide bond formation and endoplasmic-reticulum-generated oxidative stress. *EMBO Rep.*, **7**, 271–275.

72 Lueder, D.V. and Phillips, M.A. (1996) Characterization of *Trypanosoma brucei* gamma-glutamylcysteine synthetase, an essential enzyme in the biosynthesis of trypanothione (diglutathionylspermidine). *J. Biol. Chem.*, **271**, 17485–17490.

73 Tu, Z. and Anders, M.W. (1998) Expression and characterization of

human glutamate-cysteine ligase. *Arch. Biochem. Biophys.*, **354**, 247–254.

74 Huang, C.S., Chang, L.S., Anderson, M.E., and Meister, A. (1993) Catalytic and regulatory properties of the heavy subunit of rat kidney gamma-glutamylcysteine synthetase. *J. Biol. Chem.*, **268**, 19675–19680.

75 Fraser, J.A., Saunders, R.D., and McLellan, L.I. (2002) *Drosophila melanogaster* glutamate-cysteine ligase activity is regulated by a modifier subunit with a mechanism of action similar to that of the mammalian form. *J. Biol. Chem.*, **277**, 1158–1165.

76 Copley, S.D. and Dhillon, J.K. (2002) Lateral gene transfer and parallel evolution in the history of glutathione biosynthesis genes. *Genome Biol.*, **3**, research0025.

77 Abrahamsen, M.S., Templeton, T.J., Enomoto, S., Abrahante, J.E., Zhu, G., Lancto, C.A., Deng, M. *et al.* (2004) Complete genome sequence of the apicomplexan, *Cryptosporidium parvum*. *Science*, **304**, 441–445.

78 Al-Adhami, B.H., Nichols, R.A., Kusel, J.R., O'Grady, J., and Smith, H.V. (2006) *Cryptosporidium parvum* sporozoites contain glutathione. *Parasitology*, **133**, 555–563.

79 Meister, A. (1983) Selective modification of glutathione metabolism. *Science*, **220**, 472–477.

80 Wachter, A., Wolf, S., Steininger, H., Bogs, J., and Rausch, T. (2005) Differential targeting of GSH1 and GSH2 is achieved by multiple transcription initiation: implications for the compartmentation of glutathione biosynthesis in the Brassicaceae. *Plant. J.*, **41**, 15–30.

81 Pasternak, M., Lim, B., Wirtz, M., Hell, R., Cobbett, C.S., and Meyer, A.J. (2008) Restricting glutathione biosynthesis to the cytosol is sufficient for normal plant development. *Plant J.*, **53**, 999–1012.

82 Ruegsegger, A. and Brunold, C. (1993) Localization of [gamma]-glutamylcysteine synthetase and glutathione synthetase activity in maize seedlings. *Plant Physiol.*, **101**, 561–566.

83 Hell, R. and Bergmann, L. (1990) γ-Glutamylcysteine synthetase in higher plants: catalytic properties and subcellular localization. *Planta*, **180**, 603–612.

84 Hell, R. and Bergmann, L. (1988) Glutathione synthetase in tobacco suspension cultures: catalytic properties and localization. *Physiol. Plant.*, **72**, 70–76.

85 Banhegyi, G., Lusini, L., Puskas, F., Rossi, R., Fulceri, R., Braun, L., Mile, V. *et al.* (1999) Preferential transport of glutathione versus glutathione disulfide in rat liver microsomal vesicles. *J. Biol. Chem.*, **274**, 12213–12216.

86 Hwang, C., Sinskey, A.J., and Lodish, H.F. (1992) Oxidized redox state of glutathione in the endoplasmic reticulum. *Science*, **257**, 1496–1502.

87 Lash, L.H. (2006) Mitochondrial glutathione transport: physiological, pathological and toxicological implications. *Chem. Biol. Interact.*, **163**, 54–67.

88 Vega-Rodriguez, J., Franke-Fayard, B., Dinglasan, R.R., Janse, C.J., Pastrana-Mena, R., Waters, A.P., Coppens, I. *et al.* (2009) The glutathione biosynthetic pathway of *Plasmodium* is essential for mosquito transmission. *PLoS Pathog.*, **5**, e1000302.

89 Luersen, K., Walter, R.D., and Muller, S. (2000) Plasmodium falciparum-infected red blood cells depend on a functional glutathione de novo synthesis attributable to an enhanced loss of glutathione. *Biochem. J.*, **346**Pt (2), 545–552.

90 Meierjohann, S., Walter, R.D., and Muller, S. (2002) Regulation of intracellular glutathione levels in erythrocytes infected with chloroquine-sensitive and chloroquine-resistant *Plasmodium falciparum*. *Biochem. J.*, **368**, 761–768.

91 Ayi, K., Cappadoro, M., Branca, M., Turrini, F., and Arese, P. (1998) *Plasmodium falciparum* glutathione metabolism and growth are independent of glutathione system of host erythrocyte. *FEBS Lett.*, **424**, 257–261.

92 Atamna, H. and Ginsburg, H. (1997) The malaria parasite supplies glutathione to its host cell– investigation of glutathione transport and metabolism in human erythrocytes infected with *Plasmodium falciparum*. *Eur. J. Biochem.*, **250**, 670–679.

93 Grant, C.M., MacIver, F.H., and Dawes, I.W. (1996) Glutathione is an essential metabolite required for resistance to oxidative stress in the yeast *Saccharomyces cerevisiae*. *Curr. Genet.*, **29**, 511–515.

94 Baek, Y.U., Kim, Y.R., Yim, H.S., and Kang, S.O. (2004) Disruption of gamma-glutamylcysteine synthetase results in absolute glutathione auxotrophy and apoptosis in *Candida albicans*. *FEBS Lett.*, **556**, 47–52.

95 Shi, Z.Z., Osei-Frimpong, J., Kala, G., Kala, S.V., Barrios, R.J., Habib, G.M., Lukin, D.J. *et al.* (2000) Glutathione synthesis is essential for mouse development but not for cell growth in culture. *Proc. Natl Acad. Sci. USA*, **97**, 5101–5106.

96 Huynh, T.T., Huynh, V.T., Harmon, M.A., and Phillips, M.A. (2003) Gene knockdown of gamma-glutamylcysteine synthetase by RNAi in the parasitic protozoa *Trypanosoma brucei* demonstrates that it is an essential enzyme. *J. Biol. Chem.*, **278**, 39794–39800.

97 Kim, B.J., Choi, C.H., Lee, C.H., Jeong, S.Y., Kim, J.S., Kim, B.Y., Yim, H.S., and Kang, S.O. (2005) Glutathione is required for growth and prespore cell differentiation in *Dictyostelium*. *Dev. Biol.*, **284**, 387–398.

98 Deneke, S.M. and Fanburg, B.L. (1989) Regulation of cellular glutathione. *Am. J. Physiol.*, **257**, L163–L173.

99 Meister, A. (1988) Glutathione metabolism and its selective modification. *J. Biol. Chem.*, **263**, 17205–17208.

100 Meierjohann, S., Walter, R.D., and Muller, S. (2002) Glutathione synthetase from *Plasmodium falciparum*. *Biochem. J.*, **363**, 833–838.

11
Antimalarial Drugs and Molecules Inhibiting Hemozoin Formation

*Uday Bandyopadhyay** and Sumanta Dey*

Abstract

Blood-stage malarial parasites are responsible for clinical manifestations of the disease. During its proliferation, the parasite digests the host protein hemoglobin within the erythrocytes; this occurs within the digestive vacuole, through a sequential metabolic process involving multiple proteases. A massive degradation of hemoglobin leads to the generation of large amounts of heme, which is extremely toxic to the parasite. In order to protect itself from the self-created toxic substance, the malarial parasite has evolved a distinct mechanism to detoxify free heme, through its conversion into an insoluble crystalline pigment known as hemozoin (Hz). Although the exact mechanism of Hz formation remains obscure, it is considered to be the most validated target for antimalarial development. In this chapter, the chemistry and mechanism of action of antimalarial drugs and molecules inhibiting Hz formation are discussed.

Introduction

Malaria is a major cause of morbidity and mortality in many areas of the world, particularly in sub-Saharan Africa and Southeast Asia. At least 300–500 million clinical cases of malaria occur worldwide each year, resulting in the death of between 0.5 and 2.5 million people [1, 2]. The erythrocyte is the safest place for the malarial parasite to hide from the host's immune system, and indeed the erythrocytic stages of the malarial parasites are responsible for the clinical manifestations in humans [3]. Hemoglobin (Hb) is the major protein inside the erythrocyte and the parasite has evolved a unique metabolic pathway to digest Hb. Hb degradation occurs inside the food vacuole (FV), which involves numerous enzymes (Figure 11.1). Free heme (Fe^{+III}), the degradation product of Hb, is extremely toxic to the parasite in many ways [4]. Heme can inactivate many enzymes of the parasite like plasmepsins,

---

*Corresponding author

Apicomplexan Parasites. Edited by Katja Becker
Copyright © 2011 WILEY-VCH Verlag GmbH & Co. KGaA, Weinheim
ISBN: 978-3-527-32731-7

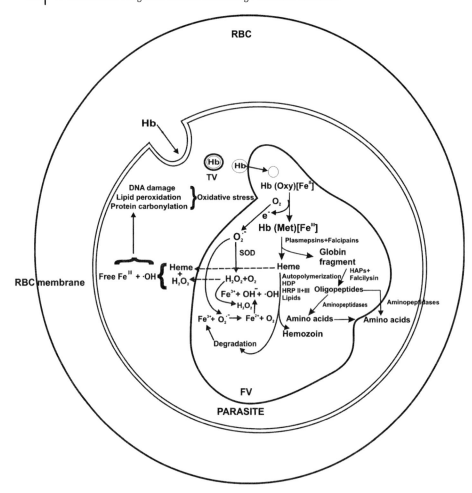

Figure 11.1 Hemoglobin degradation and toxicity caused by free heme (Fe^{III}). Hb is endocytosed from the red blood cell cytoplasm into the parasite. A transport vesicle (TV) loaded with Hb enters the food vacuole (FV) and is lysed, releasing the Hb inside. The oxyhemoglobin Hb [Hb oxy (Fe^{II})] is then oxidized to methemoglobin [Hb Met(Fe^{III})], digested by parasite proteases, and then degrades to release free heme (Fe^{III}) and amino acids. The toxic effect of heme in the parasite is mediated through the generation of reactive oxygen species (ROS). SOD, Superoxide dismutase: HAP, Histo-Aspartic Proteases.

falcipains [5, 6], glyceraldehyde-3-phosphate dehydrogenase (PfGAPDH) [7] and 6-phosphogluconate dehydrogenase of hexose monophosphate shunt [8]. In addition to enzyme inactivation, free heme (Fe^{+III}) causes hemolysis [9] and induces crosslinking of cytoskeletal proteins of RBC such as spectrin and protein 4.1 and makes the RBC membrane unstable [10]. Electrons liberated in the process of oxidation of hemoglobin (Fe^{+II}) to methemoglobin (Fe^{+III}) results in the generation of reactive oxygen species (ROS) in the FV [11] causing oxidative stress [12] (Figure 11.1).

Moreover, heme also induces lipid peroxidation through its interaction with hydrogen peroxide [13]. Heme may generate hydroxyl radical ($^{\bullet}$OH) through Fenton reaction in the FV [14], which develops oxidative stress. Thus, release of heme from Hb poses serious threat to parasite, but the parasite has the ability to detoxify free heme through the formation of a less toxic, highly insoluble and compact crystalline product, hemozoin (Hz) inside the FV for its survival [15]. Heme may undergo a process of degradation instead of polymerization outside the FV with the release of heme iron [16], which may further generate ROS. The current antimalarial drugs mostly act against this pathway of Hz formation, allowing the parasite's own metabolic waste to accumulate inside the erythrocytes. Hz formation is essential for parasite survival and is characteristic of malaria parasite. Any molecule that can interfere or prevent Hz formation by interacting with heme, will be a potential antimalarial drug [17]. Since the process of Hz formation is basically a biomineralization or biocrystalization process, development of direct resistance against drug, which acts through the inhibition of Hz formation, is difficult. Hz formation is considered as a perfectly validated antimalarial drug target; therefore, synthesis or identification of new inhibitors of Hz formation will be helpful to develop novel antimalarial.

Drugs/Molecules Inhibiting Hemozoin Formation

Many antimalarial drugs have the ability to interact with heme, and thereby to inhibit Hz formation. Examples include quinolines, azoles, isonitriles, xanthones, methylene blue, and their derivatives, all of which adopt the aforementioned strategy to kill the parasites. The chemistry of these compounds, together with their structure–activity relationships (SARs) and mechanism of action are discussed in the following subsections.

Quinolines and Their Derivatives

Quinoline-containing antimalarial drugs, such as chloroquine, amodiaquine, amopyroquine, tebuquine, mepacrine, pyronaridine, halofantrine, quinine, epiquinine, quinidine, and bisquinoline are popular chemotherapeutic compounds against malaria (Table 11.1). Quinine has been used for 300 years, ever since extracts of the bark of the *Cinchona* tree were first shown to have antimalarial activity. Investigators in Germany were the first to prepare a synthetic antimalarial called plasmoquine, later named pamaquine, but the use of this drug was discontinued due to its high toxicity. A less toxic drug came in the form of an 8-aminoquinoline, named primaquine, where the basic side chain of this compound was found to be the active moiety. Later, resochin, a 4-aminoquinoline, was also synthesized in Germany, but subsequently was renamed as chloroquine and entered clinical trials in 1943. After World War II, another 4-aminoquinoline, amodiaquine, was introduced in the United States as an alternative to chloroquine [18].

Table 11.1 Antimalarial activity of some quinoline molecules.

Compound	Structure	Antimalarial activites	
		In vitro IC$_{50}$ (nM) (CQ-S)[a]	*In vitro* IC$_{50}$ (nM) (CQ-R)[b]
Chloroquine		14.0	192.1
Amodiaquine		7.8	18.5
Tebuquine		9.5	13.1
Amopyroquine		5.3	11.5
Pyronaridine		5.7	9.1

Table 11.1 (*Continued*)

Compound	Structure	Antimalarial activites	
		In vitro IC$_{50}$ (nM) (CQ-S)[a]	*In vitro* IC$_{50}$ (nM) (CQ-R)[b]
Mepacrine		12.9	43.3
Mefloquine		23.4	9.4
Halofantrine		5.8	2.8
Quinine		34.2	81.2
Quinidine		21.5	50.6
Bisquinoline		123	25

a) CQ-S, chloroquine-sensitive *P. falciparum*.
b) CQ-R, chloroquine-resistant *P. falciparum*.

Chemistry of Quinoline Molecules

Quinoline, a benzopyrodine, is an aromatic heterocyclic compound having nitrogen as a heteroatom. Both, chloroquine and amodiaquine are important 4-aminoquinolines, having a basic chain on position 4 of the quinoline nucleus. Tebuquine and amopyroquine are the important derivatives of amodiaquine. Quinolinemethanol alkaloids are important quinoline derivatives with additional hydroxyl group(s); quinine and mefloquine are the important compounds of this class. There are eight major Cinchona bark alkaloids, which occur as four pairs of enantiomers; the (+) isomers have the $2R,3S$ configuration, while the (−) isomers have the $2S,3R$ configuration. The (+) isomers are cinchonine, quinidine, dihydrocinchonine and dihydroquinidine, while their negative counterparts are cinchonidine, quinine, dihydrocinchonidine and dihydroquinine, respectively. Isoquinoline, another benzopyrodine, is an isomer of quinoline. The bisquinolines, where two quinoline nuclei are combined through an aliphatic or aromatic linker, are also well-known quinoline derivatives. Most of the other quinoline compounds are derivatives of the compounds already described. Another class of compound, the aryl(amino) carbinols, was synthesized by replacing the quinoline moiety of the quinolinemethanols with other aromatic groups. Two compounds belonging to this class a 9-phenanthrenemethanol, halofantrine, and its fluorine analog, lumefantrine, deserve attention in this field. Early synthetic studies in Germany led to the production of the 8-aminoquinolines, primaquine and pamaquine. In further studies, the basic side chain of pamaquine was attached to a number of heterocyclic ring systems, which led to the synthesis of the acridine derivative, quinacrine (also known as atebrine or mepacrine). Pyronaridine, a derivative of mepacrine, has also been synthesized and has shown good antimalarial potency.

Mechanism of Action of Quinoline Molecules

Several mechanisms have been proposed for the action of quinoline drugs. The 4-aminoquinolines, such as chloroquine and amodiaquine, and the quinolinemethanols, such as quinine and mefloquine, act in the same way. Two main factors determine the potency of these antimalarial drugs; the first factor is their accumulation inside the FV of the parasite, and the second factor is their affinity for heme. The mechanism of action of the most widely used quinoline, chloroquine, is discussed here as an example.

Accumulation of Quinoline Inside the FV

Although normal uninfected red blood cells hardly take up chloroquine, its concentration is raised 1000-fold inside the malarial parasite and in isolated FVs. Three major hypotheses have been proposed to account for chloroquine accumulation, although the exact mechanism remains the subject of debate. Earlier studies indicated that the accumulation of chloroquine was due to an "ion trapping" or "weak-base" mechanism [19, 20]. By nature, chloroquine is a diprotic weak base ($pK_{a1} = 8.1$, $pK_{a2} = 10.2$). According to the weak-base model, unprotonated chloroquine can easily traverse the membrane of an infected red blood cell and move down the pH gradient, finally accumulating inside the FV [21–24]. Inside the acidic FV (pH 5.0–5.2), the chloroquine becomes protonated and becomes trapped there. The second hypothesis is based on a carrier-mediated mechanism for chloroquine uptake [25]. In this case, a Na^+/H^+ exchanger located on the parasite plasma membrane seemed to be involved in the transport of chloroquine [26]. The existence of a V-type H^+-ATPase for pH regulation may also have a role in chloroquine accumulation [27]. The third major hypothesis is known as the "ferriprotoporphyrin (FP) IX receptor" hypothesis [28]. According to this hypothesis, free heme (FP IX) molecules in the FV might act as an intravacuolar receptor for chloroquine [29]. To summarize, both ion trapping and receptor binding may be responsible for chloroquine uptake, while the difference in pH between the external and FV compartments may control the total level of chloroquine uptake. The rate of uptake may also be influenced by the activities of different ion channels.

Binding of Quinoline with Free Heme

Chloroquine forms a tight complex with free heme with a $1:2$ stoichiometry [28, 30–32], and binds heme noncovalently [28, 33]. It interacts with the μ-oxo dimer form of oxidized heme and prevents Hz formation [32, 34]; such interaction involves π–π stacking [35]. The binding constant for the FP–chloroquine complex is difficult to measure experimentally, but isothermal titration microcalorimetric studies [36] have shown a relatively low association constant ($K_A = 4.0 \times 10^5 \, M^{-1}$) at pH 6.5 in comparison to previous studies. Mefloquine interacts only weakly with free FP, with a recently reported K_A value of $1.2 \times 10^4 \, M^{-1}$ at pH 6.5 [36]. The FP–chloroquine interaction inhibits Hz formation [37] and leads to toxicity due to free heme or heme–chloroquine complex accumulation [38]. Chloroquine can also initiate a reverse reaction to convert Hz to monomeric heme (FP IX); hence, the reaction was termed "Hz depolymerization" [39].

Structure–Activity Relation of Quinolines

The quinoline methanols such as quinine and its diastereomer 9-epiquine were found to be effective against resistant strains. *Erythro* configurations at the C-8 and C-9 positions of quinine analogs are more active than the *threo* isomers for some, but not all analogs [40]. Modifications on the quinuclidine ring will not affect its activity, but alterations to the stereochemistry have varied effects. An analysis of the arylamino alcohols – that is, the quinolinemethanols, the phenanthrenemethanols, and some pyridinemethanols – clearly suggested that the aromatic ring systems of these compounds were interchangeable to some degree [41]. The emergence of chloroquine -resistant strains led to the synthesis of many new derivatives, using the chloroquine scaffold in order to increase the efficiency of this drug against the resistant strains. Different derivatives of chloroquine have been found useful in predicting the structure–activity relationships of this class of compound. Many studies have shown that the presence of an electron-withdrawing group at the 7-position of the quinoline ring of quinoline antimalarials is responsible for inhibition of Hz formation and its antimalarial activity. Both, chloroquine and amodiaquine have a chloro group on the 7-position. The strong electron-withdrawing groups, such as halogens, nitriles, carbonyls, and nitro groups, cause a considerable decrease in pK_{a1} (the acid dissociation constant of the quinolinium cation), whereas the electron-releasing groups such as amino, alcohol, or alkyl groups, raise pK_{a1} relative to that of the 7-H derivative [42]. The 7-halo substituted compounds are the most active antimalarials among the 4-aminoquinoline series. Quinoline drugs lacking the 7-chloro group do not inhibit Hz formation, despite forming complexes with heme; however, replacement of the 7-chloro group with a bromo or a nitro group, but not 7-amino or 7-chloroderivatives, retains its ability to inhibit Hz formation [43]. The introduction of a CF_3 group at the 7-position reduces antimalarial activity against chloroquine-susceptible and chloroquine-resistant *P. falciparum* [44], but short-chain analogs with a 7-CF_3 group show excellent activity. Modifications of the quinoline ring usually reduce its malarial activity except for the two derivatives, *N*-oxide and 5-aza chloroquine, which seem to be more potent than chloroquine [45]. The antimalarial activity is affected by chemical modifications to the position and the nature of the substituents on the quinoline ring system or alteration of the 4-amino side chain. Various substitutions on the 6′ position on the quinoline ring are tolerated, and some substituents in the 2′, 7′, and 8′ positions may even be beneficial [41].

Chloroquine bears (*N,N*-diethyl)-1,4-diaminopentane on the 4-position of the ring, and the importance of this side chain was explored by synthesizing chloroquine analogs with varying diamino alkane side chains at the same position. Those analogs with diamino alkane side chains shorter than four carbons, or longer than seven carbons, were active against both sensitive and resistant strains of *P. falciparum* [46]. Chloroquine analogs with shortened side chains such as diethylamino-ethyl, diethylaminopropyl, dimethylaminoisopropyl, and diethylamino isopropyl, were effective against the resistant strains of *P. falciparum* [47]. N_1-aralkyl-N_2-quinolin-4-yl-diamine derivatives of chloroquine were also very effective against both sensitive and resistant strains of the parasite [48].

Amodiaquine has been shown to bind to heme and to inhibit heme polymerization *in vitro*, with almost similar efficiency as chloroquine [18]. The better activity of amodiaquine over chloroquine can be attributed to its greater parasite-specific accumulation [22]. However, amodiaquine is extensively metabolized *in vivo* to its primary metabolite, desethyl-amodiaquine, which is markedly less active against chloroquine-resistant isolates [49]. Amodiaquine analogs such as amopyroquine have been shown to retain significant antimalarial activity due to the presence of metabolically stable Mannich side chains [50, 51]. Tebuquine, an amodiaquine derivative, is significantly more active, both *in vitro* and *in vivo*, than amodiaquine and chloroquine [52]. The *in vivo* efficacy of tebuquine is thought to be partially due to the replacement of the diethylamine side chain of amodiaquine with a *N-tert*-butylamine functionality, thereby inhibiting metabolism of the side chain to metabolites that display cross-resistance [53]. Structure–activity studies of quinine analogs such as amodiaquine have indicated that a hydroxyl group on C-9 is necessary for activity [18]. Toxicity is a major drawback of amodiaquine, due to the formation of quinoneimine metabolites [54], therefore to counter this problem of toxicity, isomers of amodiaquine has been synthesized [55]. Isoquine has excellent *in vivo* activity, with ED_{50} of 1.6 and 3.7 mg kg^{-1} against the *P. yoelii* NS strain, compared to 7.9 and 7.4 mg kg^{-1} for amodiaquine [55]. Replacement of the 4′-hydroxy function of amodiaquine with various alkyl, aryl, or heteroaryl substituents also reduces the toxicity [56]. Pyronaridine, in which the usual quinoline heterocycle is replaced by an aza-acridine, displays the aminophenol substructure as in amodiaquine. Pyronaridine has two Mannich-base side chains in contrast to amodiaquine, which has only one chain. Pyronaridine has been shown to be highly efficient against *P. falciparum* and *P. vivax*, and has very few side effects [57, 58]. Some quinoline di-Mannich base compounds have been shown to have higher activity than pyronaridine [59]. Quinoline compounds that are active against chloroquine-resistant malaria strains tend to be lipophilic in character [53]. A minimum of two nitrogens, a quinoline N, and an alkylamino N, are present in the 4-aminoquinolines. In this case, the important parameter is the inter-nitrogen separation, which determines their heme-binding ability [52] and, thus, their antimalarial activity [18]. An aminoalkyl side chain is present in most of the active quinoline antimalarial drugs. The terminal amino group in the aminoalkyl side chain is required for accumulation in the vacuole, and replacement of the aminoalkyl side chain with an alkyl chain or a hydroxyl group usually causes a reduction in antimalarial activity [60]. Tertiary amine compounds are generally more effective than either primary or secondary amines [61]. For example, aryl amino alcohols such as mefloquine (which is structurally related to quinine), halofantrine (9-phenanthrenemethanol), and lumefantrine are effective against chloroquine-resistant malaria. The orientations of the hydroxyl and amine groups of mefloquine have some role in determining its antimalarial activity [62]. However, in addition to being very costly, mefloquine may cause neuropsychiatric side effects [63], while halofantrine has been found to induce heart arrhythmia [64]. Lumefantrine or benflumetol, a halofantrine derivative, though less active than halofantrine, is not associated with any dangerous cardiac side effects [65]. Novel pyrrolizidinyl derivatives of 4-amino-7-chloroquinoline, where the amino group of quinoline is linked to a pyrrolizidinylalkyl ((hexahydro-1*H*-pyrrolizin-7a-yl)alkyl) moiety, were also shown to be active *in vivo* against *P. berghei*. These compounds were all tolerated and have very low toxicity [66]. Some hybrid 4-aminoquinolines with thiourea, triazines, pyrimidines, and oxalamide functionalities in their side chains have been shown active against the chloroquine-resistant and -sensitive strains [67, 68]. Several bisquinolines are active against chloroquine-resistant strains of malaria parasites [69–71], with early examples being the bis(quinolyl)piperazines, such as piperaquine and hydroxypiperaquine, dichloroquinazine and 1,4-*bis*(7-chloro-4-quinolylamino)piperazine). A mixture of dichloroquinazines was shown to be effective against *P. falciparum* [72], with piperaquine and its analog, hydroxypiperaquine, being shown more potent than chloroquine against both chloroquine-sensitive and -resistant strains of malarial parasites [73]. These compounds have also provided interesting results in clinical trials [74]. A series of bisquinolines in which two aminoquinolines are coupled by diamidine linkages have also offered promising antimalarial effects [71, 75], with a correlation between activity and heme polymerization suggesting a similar mode of action to that of chloroquine [76]. Bisquinolines with alkyl ether and piperazine bridges were shown to be more effective than those with alkylamine bridges *in vivo*, and these compounds were able to inhibit Hz formation [36].

Xanthones

Xanthones and their derivatives have for many years been shown to possess antimalarial activity [77]. Xanthones or 9*H*-xanthen-9-ones (dibenzo-γ-pirone) comprise an important class of oxygenated heterocycles, the biological activities of which are associated with their tricyclic scaffold but vary depending on the nature and/or position of the different substituents. Xanthones have been shown to inhibit Hz formation [78] and to interact with heme [79], with their antimalarial activity correlating positively with the number of hydroxyl groups present on the compound [80]. For example, 2,3,4,5,6-pentahydroxyxanthone was found to be active against both the sensitive and resistant strains of *P. falciparum* [78] (Table 11.2), while 1,3,6,8- tetrahydroxyxanthone was shown to be highly potent against *P. berghei in vivo*, and superior to the other oxygenated xanthones in the series [80]. Other compounds in the series which offered *in vivo* activity against *P. berghei* included norlichexanthone, 1,3,8-trihydroxy-6-methylxanthone and di-C-allyl-dihydroxyxanthone; these compounds reduced the number of infected red blood cells by 44.3%, 37.0%, and 33.4%, respectively [80]. Xanthones bearing hydroxyl groups at any peri-position (1 or 8) show a decreased antimalarial activity; such derivatives lose their affinity for heme due to an intramolecular H-bond between the OH group and the carbonyls [81]. The activity of xanthones can be enhanced by substituting the 1,1-dimethylallyl chain, or by the addition of a pyranic ring. The substitution of two isopentenyl chains, or the combination of one isopentenyl chain and a pyranic ring, also improves the activity, but hydroxylation of the prenyl side chain has no beneficial effect [82]. In an effort to increase the heme affinity of xanthones, straight-chain disubstituted series from C2 to C8 (except for C7), each with terminal diethylamino groups, were synthesized. In this case, the protonable nitrogens provided a strong ionic association with heme's propionate groups. Typically, 3,6-bis-ε-(*N,N*-diethylamino) amyloxyxanthone (C5) and 3,6-bis-ζ-*N,N*-(diethylamino)-hexyloxyxanthone (C6) showed the highest potency against the D6 clone of *P. falciparum*, with IC_{50} values of 0.1 and 0.075 μM, respectively [83].

Isonitriles

A series of terpene isonitriles isolated from marine sponges, along with diisocyanoadociane and axisonitrile-3, were each shown to exhibit antimalarial activity [84, 85]. An isonitrile or an isocyanide bears the functional group $R-N\equiv C$. Some of these active isonitriles can inhibit β-hematin formation [85]. Analogs of antimalarial marine isonitrile diterpoids have been synthesized in a completely diastereoselective fashion and, their *in vitro* antimalarial activity having been assayed against *P. falciparum* [86] (Table 11.2). Several easily accessible synthetic isonitriles have also been screened for their antimalarial activity against *P. falciparum* and multidrug-resistant *Plasmodium yoelii* in Swiss mice model [87]. For example, kalihinol C, a bis-isonitrile diterpenoid isolated from *Acanthella* sp., is a highly functionalized tricyclic core containing isonitrile, isothiocyanate, formamide, and chlorine moieties. It inhibits *P. falciparum* growth *in vitro* at nanomolar concentration [88].

Table 11.2 Antimalarial activity of xanthones and isonitriles.

Family	Compounds	Antimalarial activity IC$_{50}$(µM)-Xanthones IC$_{50}$(nM)-Isonitriles D$_6$[a)
Xanthone	2,3,4,5,6-pentahydroxyxanthone	0.7 ± 0.5
	2,3,4,5,- tetrahydroxyxanthone	9.0 ± 1.0
	2-hydroxyxanthone	>50
	3,6-Bis-ζ-(N,N-diethylamino)- hexyloxy xanthone, C6	0.07 ± 0.02
Isonitrile		9.4
		14.48

(Continued)

Table 11.2 (Continued)

Family	Compounds	Antimalarial activity $IC_{50}(\mu M)$-Xanthones $IC_{50}(nM)$-Isonitriles D_6[a]
		47.4
		196.65

a) D_6, chloroquine-sensitive *P. falciparum*.

Acridines

Acridine compounds known to have antimalarial activity are essentially 9-aminoacridine derivatives such as quinacrine and 9-anilinoacridine [89, 90]. These are structurally related to anthracene with three fused rings, the center ring containing a nitrogen as a heteroatom. Acridone is based on the acridine skeleton, with a carbonyl group at the 9 position. These compounds are weakly basic in nature. Although acridines are potent inhibitors of parasite DNA topoisomerase II both *in vitro* and *in situ* [91, 92], their antimalarial activity may also be due to an ability to inhibit Hb digestion or Hz formation in the parasite FV [93]. Antimalarial drugs that contain an acridine nucleus can bind strongly to heme and inhibit the crystallization process. A series of 9-anilinoacridines were also shown to form stable drug–hematin complexes, and to enhance the hematin-induced lysis of red blood cells [94]. Various substitutions were made on the acridine ring and anilino ring of 9-anilinoacridines to increase the antimalarial efficacy of this class of compound. Various substitutions on the acridine ring were made with Cl or NH_2 at positions 3 and 6 [94], and their activities evaluated (Table 11.3). It has been shown that diamino substitution at positions 3 and 6 of the acridine ring of the 9-anilinoacridines increases their antimalarial potency against *P. falciparum* [95], with both 3-Cl and 3,6-diCl showing poor antimalarial activity with respect to their di-NH_2 analogs. The nature of the substituents on the anilino rings also had a strong influence on the antimalarial potency. The presence of a $1'$-$N(CH_3)_2$ substituent in the anilino ring increased the inhibition of β-hematin formation, and all compounds with this group offered a good antimalarial potency, irrespective of the nature of substitution in the acridine nucleus. [94]. Studies have been conducted on two series of acridine derivatives.

Table 11.3 Inhibition of hemozoin formation and antimalarial activities of acridines.

9-anilinoacridine

Compound	Acridine substituent	1'-Anilino substituent	Hemozoin formation inhibition (mM)	IC$_{50}$(μM) K1 [a]	IC$_{50}$ (μM) T9/94 [b]
1	H	N(CH$_3$)$_2$	0.80	0.90	1.25
2	3-NH$_2$	N(CH$_3$)$_2$	0.35	0.25	0.625
3	3,6-DiNH$_2$	N(CH$_3$)$_2$	0.034	0.024	0.125
4	3,6-DiNH$_2$	CH$_2$N(CH$_3$)$_2$	0.040	0.034	>1.25
5	3,6-DiNH$_2$	NHSO$_2$CH$_3$	0.030	0.025	>1.25
6	3,6-DiNH$_2$	OH	0.050	0.047	>1.25
7	3,6-DiNH$_2$	OCH$_3$	0.15	0.12	>1.25
8	3,6-DiNH$_2$	SO$_2$NH$_2$	0.020	0.015	>1.25
9	3-Cl	N(CH$_3$)$_2$	1.1	1.0	0.625
10	3-Cl	NHCH$_3$	1.2	1.1	0.125
11	3-Cl	NHSO$_2$CH$_3$	1.0	1.1	0.625
12	3,6-diCl	CH$_2$N(CH$_3$)$_2$	0.25	0.20	>1.25
13	3,6-diCl	NHSO$_2$CH$_3$	3.5	2.1	0.062
14	3,6-diCl	SO$_2$NH$_2$	6.2	7.0	1.25

a) K1, chloroquine-sensitive *P. falciparum*.
b) T9/94, chloroquine-resistant *P. falciparum*.

One series of 9-amino acridine derivatives had several cationic charges either on the acridine ring or on the chain grafted at the 9-amino position (either a six-carbon ammonium chain or an amine-peptidic chain of different length and nature); this led to an increase in the deriveative's pK_a. The second series of acridyl derivatives was positively charged only on the peptide chain grafted on the acridine ring; in this way, two derivatives were synthesized which differed in the way that the peptidic chain was connected to the acridine (either by NHCO or by CONH) and in the total length of the chain [96] (Table 11.4). According to the results obtained, it was concluded that the presence of a cationic charge on the acridine nucleus was required to provide significant antimalarial activity. Moreover, the length of the chain attached to the acridine ring should be shortened for a better accumulation in the FV. The structure–activity relationship of these compounds showed that not only two positive charges but also 6-chloro and 2-methoxy substituents on the acridine ring were required to exert an effective antimalarial activity. Very recently, a new antimalarial chemotype was synthesized that combined the heme-targeting character of acridones with a chemosensitizing component that counteracted resistance to quinoline antimalarial drugs [97].

Azoles

The antimalarial activities of antifungal azoles, such as clotrimazole, ketoconazole, and miconazole, are well known [98–103]. The azoles are five-membered heterocyclic compounds that contain one or more heteroatoms in their rings, at least one of which is nitrogen. The rings also contain two double bonds that are noncumulative. The simplest form of this class is pyrrole, which contains only one nitrogen and no other heteroatom. The azoles used in medicine are either diazoles or triazoles (with two or three nitrogen atoms in the azole ring, respectively). Ketoconazole, miconazole and clotrimazole are termed imidazoles, a form of diazole that form histidine, whereas fluconazole and itraconazole are classified as triazoles. The azoles have a quite high affinity for heme (Table 11.5) [98]. They also inhibit reduced glutathione-dependent heme catabolism and enhance heme-induced hemolysis [99]. The azoles remove heme both from the histidine-rich peptide–heme complex, and the reduced glutathione–heme complex [100]. These azoles form stable heme–azole complexes with two nitrogenous ligands derived from the imidazole moieties of two azole molecules [100]. The azoles, including clotrimazole (IC$_{50}$ = 12.9 μM), econazole (IC$_{50}$ = 19.7 μM), ketoconazole (IC$_{50}$ = 6.5 μM), and miconazole (IC$_{50}$ = 21.4 μM) reversibly blocked FP crystal growth and induced oxidative stress [98]. Gemma et al. synthesized novel antimalarials based on the clotrimazole scaffold, which has been shown to interact with heme and to inhibit Hz formation in vitro [102, 103].

Methylene Blue (MB)

The phenothiazinium salt methylene blue [3,7-bis (dimethylamino) phenothiazinium chloride; MB] is the oldest known synthetic antimalarial drug, having been first

Table 11.4 Inhibition of hemozoin formation and antimalarial activities of acridines.

Compound	Hemozoin formation inhibition (mM)	IC$_{50}$ (μM) W2[a]	IC$_{50}$ (μM) Bre1[b]
R_1=OCH$_3$, R_2=Cl, R_3=NH$_3{}^{+}$, n=1	4.6	0.25 ± 0.03	0.28 ± 0.05
R_1=OCH$_3$, R_2=Cl, R_3=NH$_3{}^{+}$, n=2	1.0	0.25 ± 0.04	0.30 ± 0.06
R_1=OCH$_3$, R_2=Cl, R_3=NH$_3{}^{+}$, n=3	0.4	0.18 ± 0.05	0.17 ± 0.04
R_1=OCH$_3$, R_2=Cl, R_3=CH$_3$, n=3	1.5	27.2 ± 9.0	16.3 ± 6.7
R_1=OCH$_3$, R_2=Cl, R_3=ACr, n=3	1.3	4.55 ± 0.85	1.11 ± 0.69
R_1=H, R_2=H, R_3= NH$_3{}^{+}$, n=3	0.3	0.60 ± 0.2	0.43 ± 0.10

(Continued)

Table 11.4 (Continued)

Compound	Hemozoin formation inhibition (mM)	IC$_{50}$ (μM) W2[a]	IC$_{50}$ (μM) Bre1[b]
  X= (structure with NH$_3^+$)	1.1	50.0 ± 2.2	26.7 ± 5.7
  X= (structure with NH$_3^+$)	1.8	20.9 ± 4.3	21.6 ± 7.9

a) W2, chloroquine-sensitive *P. falciparum*.
b) Bre1, chloroquine-resistant *P. falciparum*.

identified in this role by Paul Ehrlich in about 1891. Its analogs, azure A (AZA), B (AZB), C (AZC), thion (TH), celestine blue (CB), and phenosaphranin, also show antimalarial activity [104] (Table 11.5). Xanthines, azines, oxazines, and thiazines are potent against three different strains of *P. falciparum*, each having a different drug susceptibility/resistance profile [105]. The thiazine (phenothiazine) dyes were found to be the most active antimalarials while showing low cytotoxicity. MB proved to be among the most potent of this group [105], the antimalarial activity of MB and its analogs being similar to that of the 4-aminoquinolines and based on the prevention of Hz formation. All analogs of MB inhibit the growth of various strains of *P. falciparum* *in vitro*, with IC$_{50}$ values in the range of 2×10^{-9} M to 1×10^{-7} M, with the rank order MB ≈ AZA > AZB > AZC > TH > PS > CB [104]. Their ability to bind heme followed the same rank order. All compounds effectively suppress the growth of *P. vinckei petteri* in vivo at a dose ranging between 1.2 to 5.2 mg kg^{-1}, while MB or AZB also suppress *P. yoelii nigeriensis* growth in the range of 9–11 mg kg^{-1} [104]. Although the toxicity of MB may prevent its use, the parent phenothiazine structure can be a useful scaffold for synthesizing new antimalarial drugs. Recently, it has been shown that chlorpromazine and some novel phenothiazines are able to inhibit the growth of both chloroquine-sensitive and -resistant strains of *P. falciparum* by inhibiting Hz formation [106]. A MB–chloroquine combination is very effective in this role [107], while MB–artensuate and MB–amodiaquine combinations are also presently undergoing clinical trials [108], with both combinations having shown strong gametocy-tocidal effects against *P. falciparum* [109]. MB, which inhibits Hz formation (as well as

Table 11.5 Inhibition of hemozoin formation and antimalarial activity of azoles and methylene blue derivatives.

Family	Compound	Structure	Hemozoin formation inhibition IC$_{50}$ (μM)	Antimalarial activity *in vitro* IC$_{50}$(nM)
	Clotrimazole		12.9 ± 0.6	245.0 (CQ-S)[a] 553.0 (CQ-R)[b]
Azole	Ketoconazole		6.5 ± 0.6	9400 (CQ-S)[a] 1000 (CQ-R)[b]
	Miconazole		21.4 ± 2.2	400 (CQ-R)[b]

(Continued)

Table 11.5 (Continued)

Family	Compound	Structure	Hemozoin formation inhibition IC$_{50}$ (μM)	Antimalarial activity in vitro IC$_{50}$(nM)
Phenothiazinium salt	Methylene Blue		28.9 ± 2.3	3.58 ± 2.22 (CQ-S)[a] 3.99 ± 2.31 (CQ-R)[b]
	Azure A		33.7 ± 1.7	10.2 ± 4.8 (CQ-S)[a] 10.5 ± 4.8 (CQ-R)[b]
	Azure B		–	8.31 ± 4.32 (CQ-S)[a] 12.7 ± 7.36 (CQ-R)[b]

a) CQ-S, chloroquine-sensitive *P. falciparum*.
b) CQ-R, chloroquine-resistant *P. falciparum*.

P. falciparum glutathione reductase) within the parasite's FV and prevents methemoglobinemia, is also undergoing clinical trials [107].

Porphyrins and Metalloporphyrin Complexes

Protoporphyrin is a metal-free porphyrin, which combines with iron to form the heme of iron-containing proteins. Non-iron protoporphyrins (protoporphyrin IX and hematoporphyrin) and metal-substituted protoporphyrins such as tin protoporphyrin IX (SnPP), zinc protoporphyrin IX (ZnPP), and zinc deuteroporphyrin IX, 2,4 bisglycol (ZnBG) have each been shown to possess antimalarial activity. These heme analogs inhibit Hz formation at micromolar concentrations *in vitro* (ZnPP $\ll$ SnPPb $<$ ZnBG) [110]. The mechanism of action of the non-iron porphyrins is same as that of chloroquine. The porphyrin rings of these compounds undergo a π–π interaction with that of hematin at acidic pH levels, resulting in the formation of π–π adducts. The hydroxyl groups present on this class of compounds enhance their binding to the iron of heme. The presence of two α-hydroxyethyl side chains leads to hematoporphyrin being sixfold more potent than protoporphyrin IX [111]. Many metal-substituted protoporphyrin IXs, including (Cr(III) PPIX, Co(III) PPIX, Mn(III) PPIX, Cu(II) PPIX, Mg(II) PPIX, Zn(II), PPIX, and Sn(IV) PPIX, act as inhibitors of Hz formation. In this case, the central metal ion plays a significant role in the efficacy of the metalloporphyrins to inhibit Hz formation. Typically, Mg(II) PPIX, Zn(II) PPIX, and Sn(IV) PPIX are much more effective in preventing Hz formation than the free ligand protoporphyrin IX and chloroquine [112]. When ten metalloporphyrins were tested for their ability to inhibit Hz formation and to kill malarial parasites in culture, the IC_{50} values for inhibiting parasite growth ranged from 15.5 to 190 μM. In a trophozoite lysate-mediated Hz formation assay, SnPPIX, GaPPIX, and GaDPIX each exerted a potent inhibitory activity similar to that of chloroquine [113]. Bilirubin, the FP degradation product, was also shown to inhibit both Hz formation and *P. falciparum* growth [114].

Miscellaneous Antimalarials

Recently, many antimalarials have been reported which prevent Hz formation and enhance free heme toxicity. These include [(aryl)arylsufanylmethyl]pyridines (AASMPs), diamidine compounds, imino complexes and other synthetic compounds, alkaloids derived from *Cryptolepis sanguinolenta* and their related derivatives, and extracts from species of the families Dioncophyllaceae and Ancistrocladaceae.

[(Aryl)arylsufanylmethyl]pyridines (AASMPs)

A series of AASMPs, which represent a new class of trisubstituted methanes, have been synthesized and shown to interact with heme and inhibited Hz formation [115], *in vitro* mediated by whole-cell lysate of *P. yoelii* (MDR) and *P. falciparum* (NF-54) (Table 11.6). Compounds 4a, 4b, and 4e were found to be highly potent to show antimalarial activity *in vitro* ($IC_{50} = 4$–8 μM, P$<$0.001). Further, *in vivo* studies were conducted with these selected compounds in a rodent model infected with *P. yoelli* (MDR) and were found to be highly potent (Table 11.6).

Table 11.6 Inhibition of hemozoin formation and antimalaral activity of AASMPs.

Compound	Structure	Mean IC_{50} for hemozoin formation (μM) $\pm$ SEM		Dose (mg/Kg) daily for 4 days	Mean % suppression of parasitemia $\pm$ SEM on day 4
		P. yoelli (MDR)	*P. falciparum* (NF-54)		
4a		11 ± 4	18 ± 4	10 25 50	20 ± 4 30 ± 5 60 ± 7
4b		16 ± 5	26 ± 5	25 50 100	30 ± 4 50 ± 7 80 ± 6
4e		40 ± 5	20 ± 4	10 25 50	40 ± 3 65 ± 6 85 ± 5

Diamidine Compounds

Diamidines, pentamidine in particular, have long been used as chemotherapeutic agents against infectious diseases [116]. Pentamidine (Figure 11.2), which has also been shown to possess antimalarial activity [117], enters the FV of *P. falciparum* where it binds to heme and inhibits Hz formation [118]. The antimalarial efficacy of some diamidines follows the rank-order: propamidine, stilbamidine, pentamidine, berenil. All of these were potent against both chloroquine-susceptible clones (HB3 and 3D7) and chloroquine-resistant clones (K1 and TM6), with IC_{50}-values for pentamidine of 126 ± 56, 88 ± 56 (HB3 and 3D7), 66 ± 15, and 65 ± 19 (K1 and TM6), respectively. In future, if problems regarding their toxicity and absorption can be overcome, these drugs may be used as lead compounds.

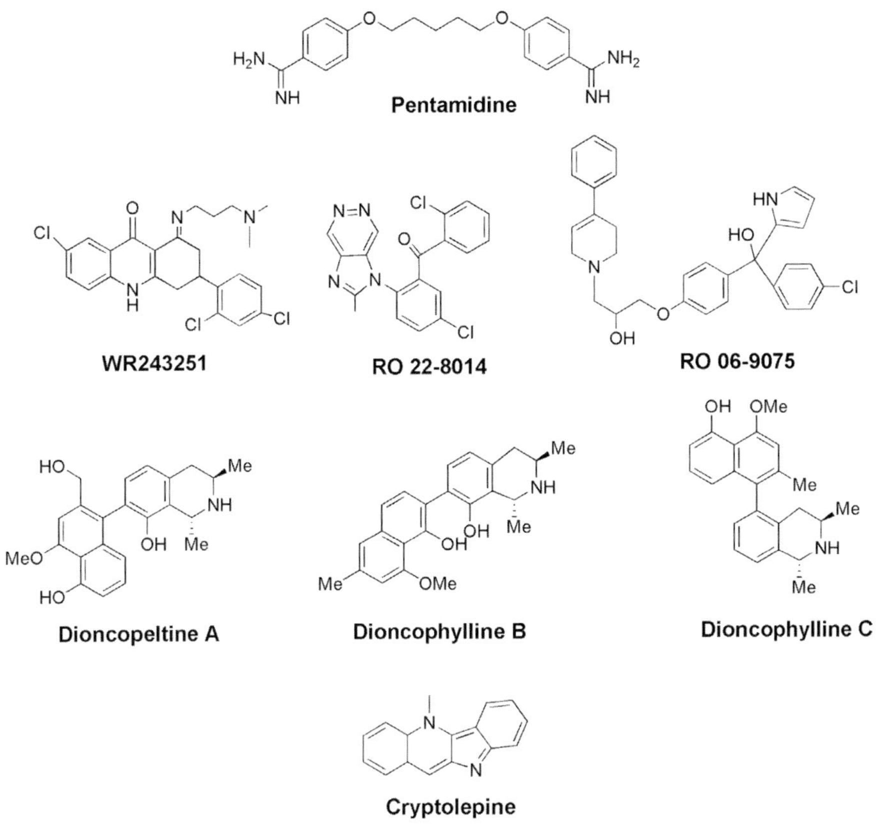

Pentamidine

WR243251   **RO 22-8014**   **RO 06-9075**

Dioncopeltine A   **Dioncophylline B**   **Dioncophylline C**

Cryptolepine

Figure 11.2 Potential antimalarials related to heme metabolism and hemozoin formation.

Imino Complexes and Other Synthetic Compounds

Hexadentate ethylenediamine-N',N'-bis [propyl (2-hydroxy-(R)-benzylimino)] metal (III) complexes [(R)-ENBPI-M(III)] and the corresponding [(R)benzylamino] analog [(R)-ENBPA-M(III)], a group of lipophilic monocations, has potential antimalarial activity. Racemic mixtures of Al(III), Fe(III) or Ga(III) [but not In(III)] (R)-ENBPI metallo-complexes were shown to kill intraerythrocytic malarial parasites in a stage-specific manner, with the R = 4,6-dimethoxy-substituted ENBPI Fe(III) complex being most potent (IC$_{50}$ = 1 μM). These imino complexes prevent parasite growth by inhibiting Hz formation [119]. Dihydroacridine (WR243251), a derivative of floxacrine that has been developed from the lead compound mepacrine [120], as well as triarylmethanol (R006-9075) and benzophenone (R022-8014) (Figure 11.2), each inhibit Hz formation [120, 121]. The Schiff-base coordination complex, an unusual class of antimalarial agent, also has the same target. Cyproheptadine, an antihistaminic drug, has shown significant schizontocidal activity (20–25 mg kg^{-1} for four days) in the blood against a lethal multidrug-resistant (MDR) strain of *P. yoelii nigeriensis*. Cyproheptadine was also shown to inhibit Hz formation, dose-dependently, both *in vivo* and *in vitro* [122].

Plant Extracts and Synthetic Derivatives of the Active Principles

Naturally occurring isomeric indoloquinoline alkaloids such as cryptolepine, iso-cryptolepine and neocryptolepine, and dimeric indoloquinoline alkaloids such as cryptoquindoline and biscryptolepine, are found in the extracts of *Cryptolepis sanguinolenta* (Asclepiadaceae or Periplocaceae), which grows abundantly in the West African subregion. Extracts of this plant have long been used for the treatment of malaria. When the antimalarial activities of these alkaloids were evaluated following their isolation, both cryptolepine (IC$_{50}$ 0.2–0.6 µM) and isocryptolepine (IC$_{50}$ ~0.8 µM) were shown to be active (Figure 11.2) [123]. A series of substituted derivatives of neocryptolepine was also prepared and evaluated, and shown to inhibit β-hematin formation and to be active against chloroquine-resistant *P. falciparum* strains [124]. Some of these derivatives had a higher antiplasmodial activity and a lower cytotoxicity than the original lead, neocryptolepine. Of these compounds, 2-bromoneocryptolepine was seen to be the most selective, with an IC$_{50}$ value against chloroquine-resistant *P. falciparum* of 4.0 µM [125]. Isoneocryptolepine, a new synthetic indolo-quinoline isomer, and a quaternary derivative, *N*-methyl-isocryptolepinium iodide, also showed high antiplasmodial activities against the chloroquine-resistant *P. falciparum* strain K1 (IC$_{50}$ values of 0.23 ± 0.04 and 0.017 ± 0.004 µM, respectively). Isoneocryptolepine was also found to act as an inhibitor of β-hematin formation. Another new class of natural products, known as naphthylisoquinoline alkaloids, has been identified in extracts of species belonging to the families Dioncophyllaceae and Ancistrocladaceae that have long been used in the traditional medicine of several African and Asian countries [126]. The stem bark extract of *Triphyophyllum peltatum* (family Dioncophyllaceae) prepared in $CH_2Cl_2NH_3$ showed a high antimalarial activity, with *in vitro* IC$_{50}$ values of 0.014 and 0.081 µg ml^{-1} against *P. falciparum* and *P. berghei*, respectively. When alkaloids isolated from *T. peltatum*, such as dioncophylline A, dioncophylline B, dioncophylline C, and dioncopeltine A, were tested for their antimalarial activity [127], the IC$_{50}$s of dioncophyllines B and C and dioncopeltine A (Figure 11.2) were far below 1 µg ml^{-1} against both *P. falciparum* and *P. berghei in vitro*, while the corresponding value for dioncophylline A was about 1 µg ml^{-1} in both cases. Dioncophylline C completely cleared the parasites from the blood of OF1 mice infected with *P. berghei*, while both dioncophylline B and dioncopeltine A had suppressed the parasitemia to a significant degree (47% and 99%, respectively) by day 4. Dioncophylline A did not show any effect, however [128]. Whilst the mechanism of action of these alkaloids has not yet been elucidated, dioncophylline C is known to form a complex with heme in solution, which is highly similar to complexes formed by FPIX and antimalarials of the quinoline family [129]. Febrifugine, the active component of a Chinese herb, *Dichroa febrifuga*, has also attracted attention as a potential antimalarial drug [130].

Conclusion

The emergence of resistance against conventional antimalarial drugs has undoubtedly accelerated the development of novel antimalarials, and also the reevaluation of some existing drugs after modification. During the past decade, major contributions by research groups worldwide have helped to provide an understanding of Hb digestion in the FV of the malaria parasite. Indeed, it is now realized that Hb digestion leads to the release of redox-active free heme, the toxicity of which is overcome by the parasite via the application of a unique mechanism. Although many hypotheses have been proposed regarding the process of such detoxification, the mechanistic details require clarification. The less-toxic Hz, which is formed due to the sequestration of free heme, protects the parasite against its self-created toxic byproduct. Moreover, neither β-hematin nor Hz are chemically fixed entities, and are not susceptible to mutations; rather, their unique formation makes them attractive targets for many antimalarial agents. In addition to the 4-aminoquinolines, the xanthones, porphyrins, acridines, azoles, MB derivatives and many natural products have each been shown to be effective against this target. The future antimalarials inhibiting Hz formation should enter and accumulate inside FV and act against resistant parasite with no toxicity in the host. In years to come, although malarial parasite will inevitably develop resistance against most antimalarial drugs, Hz formation will remain the most attractive target for combating malaria.

References

1 Breman, J.G., Alilio, M.S., and Mills, A. (2004) Conquering the intolerable burden of malaria: what's new, what's needed: a summary. *Am. J. Trop. Med. Hyg.*, **71**, 1–15.

2 Ndugwa, R.P., Ramroth, H., Muller, O., Jasseh, M., Sie, A., Kouyate, B., Greenwood, B. *et al.* (2008) Comparison of all-cause and malaria-specific mortality from two West African countries with different malaria transmission patterns. *Malar. J.*, **7**, 15.

3 Miller, L.H., Baruch, D.I., Marsh, K., and Doumbo, O.K. (2002) The pathogenic basis of malaria. *Nature*, **415**, 673–679.

4 Kumar, S. and Bandyopadhyay, U. (2005) Free heme toxicity and its detoxification systems in human. *Toxicol. Lett.*, **157**, 175–188.

5 Gluzman, I.Y., Francis, S.E., Oksman, A., Smith, C.E., Duffin, K.L., and Goldberg, D.E. (1994) Order and specificity of the *Plasmodium falciparum* hemoglobin degradation pathway. *J. Clin. Invest.*, **93**, 1602–1608.

6 Vander Jagt, D.L., Hunsaker, L.A., and Campos, N.M. (1987) Comparison of proteases from chloroquine-sensitive and chloroquine-resistant strains of *Plasmodium falciparum*. *Biochem. Pharmacol.*, **36**, 3285–3291.

7 Campanale, N., Nickel, C., Daubenberger, C.A., Wehlan, D.A., Gorman, J.J., Klonis, N., Becker, K. *et al.* (2003) Identification and characterization of heme-interacting proteins in the malaria parasite, *Plasmodium falciparum. J. Biol. Chem.*, **278**, 27354–27361.

8 Famin, O. and Ginsburg, H. (2003) The treatment of *Plasmodium falciparum*-infected erythrocytes with chloroquine leads to accumulation of ferriprotoporphyrin IX bound to particular parasite proteins and to the inhibition of the parasite's 6-phosphogluconate dehydrogenase. *Parasite*, **10**, 39–50.

9 Chou, A.C. and Fitch, C.D. (1981) Mechanism of hemolysis induced by ferriprotoporphyrin IX. *J. Clin. Invest.*, **68**, 672–677.

10 Solar, I., Dulitzky, J., and Shaklai, N. (1990) Hemin-promoted peroxidation

of red cell cytoskeletal proteins. *Arch. Biochem. Biophys.*, **283**, 81–89.

11 Atamna, H. and Ginsburg, H. (1993) Origin of reactive oxygen species in erythrocytes infected with *Plasmodium falciparum*. *Mol. Biochem. Parasitol.*, **61**, 231–241.

12 Becker, K., Tilley, L., Vennerstrom, J.L., Roberts, D., Rogerson, S., and Ginsburg, H. (2004) Oxidative stress in malaria parasite-infected erythrocytes: host–parasite interactions. *Int. J. Parasitol.*, **34**, 163–189.

13 Klouche, K., Morena, M., Canaud, B., Descomps, B., Beraud, J.J., and Cristol, J.P. (2004) Mechanism of in vitro heme-induced LDL oxidation: effects of antioxidants. *Eur. J. Clin. Invest.*, **34**, 619–625.

14 Golenser, J., Marva, E., Har-El, R., and Chevion, M. (1991) Induction of oxidant stress by iron available in advanced forms of *Plasmodium falciparum*. *Free Radic. Res. Commun.*, **12–13** (Pt 2), 639–643.

15 Egan, T.J. (2008) Haemozoin formation. *Mol. Biochem. Parasitol.*, **157**, 127–136.

16 Ginsburg, H., Famin, O., Zhang, J., and Krugliak, M. (1998) Inhibition of glutathione-dependent degradation of heme by chloroquine and amodiaquine as a possible basis for their antimalarial mode of action. *Biochem. Pharmacol.*, **56**, 1305–1313.

17 Kumar, S., Guha, M., Choubey, V., Maity, P., and Bandyopadhyay, U. (2007) Antimalarial drugs inhibiting hemozoin (beta-hematin) formation: a mechanistic update. *Life Sci.*, **80**, 813–828.

18 Foley, M. and Tilley, L. (1998) Quinoline antimalarials: mechanisms of action and resistance and prospects for new agents. *Pharmacol. Ther.*, **79**, 55–87.

19 Homewood, C.A., Warhurst, D.C., Peters, W., and Baggaley, V.C. (1972) Lysosomes, pH and the anti-malarial action of chloroquine. *Nature*, **235**, 50–52.

20 Ginsburg, H., Nissani, E., and Krugliak, M. (1989) Alkalinization of the food vacuole of malaria parasites by quinoline drugs and alkylamines is not correlated

with their antimalarial activity. *Biochem. Pharmacol.*, **38**, 2645–2654.

21 Fitch, C.D., Yunis, N.G., Chevli, R., and Gonzalez, Y. (1974) High-affinity accumulation of chloroquine by mouse erythrocytes infected with *Plasmodium berghei*. *J. Clin. Invest.*, **54**, 24–33.

22 Hawley, S.R., Bray, P.G., Park, B.K., and Ward, S.A. (1996) Amodiaquine accumulation in *Plasmodium falciparum* as a possible explanation for its superior antimalarial activity over chloroquine. *Mol. Biochem. Parasitol.*, **80**, 15–25.

23 Krogstad, D.J., Schlesinger, P.H., and Gluzman, I.Y. (1985) Antimalarials increase vesicle pH in *Plasmodium falciparum*. *J. Cell Biol.*, **101**, 2302–2309.

24 Krogstad, D.J. and Schlesinger, P.H. (1987) The basis of antimalarial action: non-weak base effects of chloroquine on acid vesicle pH. *Am. J. Trop. Med. Hyg.*, **36**, 213–220.

25 Ferrari, V. and Cutler, D.J. (1991) Simulation of kinetic data on the influx and efflux of chloroquine by erythrocytes infected with *Plasmodium falciparum*. Evidence for a drug-importer in chloroquine-sensitive strains. *Biochem. Pharmacol.*, **42** (Suppl), S167–S179.

26 Wunsch, S., Sanchez, C.P., Gekle, M., Grosse-Wortmann, L., Wiesner, J., and Lanzer, M. (1998) Differential stimulation of the Na^+/H^+ exchanger determines chloroquine uptake in *Plasmodium falciparum*. *J. Cell Biol.*, **140**, 335–345.

27 Saliba, K.J. and Kirk, K. (1999) pH regulation in the intracellular malaria parasite, *Plasmodium falciparum*. H(+) extrusion via a v-type h(+)-atpase. *J. Biol. Chem.*, **274**, 33213–33219.

28 Chou, A.C., Chevli, R., and Fitch, C.D. (1980) Ferriprotoporphyrin IX fulfills the criteria for identification as the chloroquine receptor of malaria parasites. *Biochemistry*, **19**, 1543–1549.

29 Bray, P.G., Janneh, O., Raynes, K.J., Mungthin, M., Ginsburg, H., and Ward, S.A. (1999) Cellular uptake of chloroquine is dependent on binding to ferriprotoporphyrin IX and is independent of NHE activity in

Plasmodium falciparum. J. Cell Biol., **145**, 363–376.

30 Moreau, S., Perly, B., and Biguet, J. (1982) Interaction of chloroquine with ferriprotophorphyrin IX. Nuclear magnetic resonance study. *Biochimie*, **64**, 1015–1025.

31 Adams, P.A., Berman, P.A., Egan, T.J., Marsh, P.J., and Silver, J. (1996) The iron environment in heme and heme-antimalarial complexes of pharmacological interest. *J. Inorg. Biochem.*, **63**, 69–77.

32 Egan, T.J., Mavuso, W.W., Ross, D.C., and Marques, H.M. (1997) Thermodynamic factors controlling the interaction of quinoline antimalarial drugs with ferriprotoporphyrin IX. *J. Inorg. Biochem.*, **68**, 137–145.

33 Cohen, S.N., Phifer, K.O., and Yielding, K.L. (1964) Complex formation between chloroquine and ferrihaemic acid in vitro, and its effect on the antimalarial action of chloroquine. *Nature*, **202**, 805–806.

34 Leed, A., DuBay, K., Ursos, L.M., Sears, D., De Dios, A.C., and Roepe, P.D. (2002) Solution structures of antimalarial drug-heme complexes. *Biochemistry*, **41**, 10245–10255.

35 Webster, G.T., McNaughton, D., and Wood, B.R. (2009) Aggregated enhanced Raman scattering in Fe(III)PPIX solutions: the effects of concentration and chloroquine on excitonic interactions. *J. Phys. Chem. B.*, **113**, 6910–6916.

36 Dorn, A., Vippagunta, S.R., Matile, H., Jaquet, C., Vennerstrom, J.L., and Ridley, R.G. (1998) An assessment of drug-haematin binding as a mechanism for inhibition of haematin polymerisation by quinoline antimalarials. *Biochem. Pharmacol.*, **55**, 727–736.

37 Fitch, C.D. (1986) Antimalarial schizontocides: ferriprotoporphyrin IX interaction hypothesis. *Parasitol. Today*, **2**, 330–331.

38 Macomber, P.B., O'Brien, R.L., and Hahn, F.E. (1966) Chloroquine: physiological basis of drug resistance in *Plasmodium berghei. Science*, **152**, 1374–1375.

39 Pandey, A.V. and Tekwani, B.L. (1997) Depolymerization of malarial hemozoin: a novel reaction initiated by blood schizontocidal antimalarials. *FEBS Lett.*, **402**, 236–240.

40 Hofheinz, W. and Merkli, B. (1984) Quinine and quinine analogues, in *Antimalarial Drugs II* (eds. W. Peters and W.H.G. Richards), Springer Verlag, Berlin.

41 Sweeney, T.R. (1984) Drugs with quinine-like action, *Antimalarial Drugs II: Current Antimalarials and New Drug Developments*, Springer-Verlag, Berlin, Heidelberg, pp. 267–313.

42 Kaschula, C.H., Egan, T.J., Hunter, R., Basilico, N., Parapini, S., Taramelli, D., Pasini, E. *et al.* (2002) Structure-activity relationships in 4-aminoquinoline antiplasmodials. The role of the group at the 7-position. *J. Med. Chem.*, **45**, 3531–3539.

43 Egan, T.J. (2001) Structure-function relationships in chloroquine and related 4-aminoquinoline antimalarials. *Mini Rev. Med. Chem.*, **1**, 113–123.

44 De, D., Krogstad, F.M., Byers, L.D., and Krogstad, D.J. (1998) Structure-activity relationships for antiplasmodial activity among 7-substituted 4-aminoquinolines. *J. Med. Chem.*, **41**, 4918–4926.

45 O'Neill, P.M., Bray, P.G., Hawley, S.R., Ward, S.A., and Park, B.K. (1998) 4-Aminoquinolines–past, present, and future: a chemical perspective. *Pharmacol. Ther.*, **77**, 29–58.

46 De, D., Krogstad, F.M., Cogswell, F.B., and Krogstad, D.J. (1996) Aminoquinolines that circumvent resistance in *Plasmodium falciparum* in vitro. *Am. J. Trop. Med. Hyg.*, **55**, 579–583.

47 Ridley, R.G., Hofheinz, W., Matile, H., Jaquet, C., Dorn, A., Masciadri, R., Jolidon, S. *et al.* (1996) 4-aminoquinoline analogs of chloroquine with shortened side chains retain activity against chloroquine-resistant *Plasmodium falciparum. Antimicrob. Agents Chemother.*, **40**, 1846–1854.

48 Hofheinz, W. and Masciadri, R. (1999) Quinoline derivatives for treating malaria. US Patent 5948791.

49 Churchill, F.C., Patchen, L.C., Campbell, C.C., Schwartz, I.K., Nguyen-Dinh, P., and Dickinson, C.M. (1985) Amodiaquine as a prodrug: importance of metabolite(s) in the antimalarial effect of amodiaquine in humans. *Life Sci.*, **36**, 53–62.

50 Peters, W. and Robinson, B.L. (1992) The chemotherapy of rodent malaria. XLVII. Studies on pyronaridine and other Mannich base antimalarials. *Ann. Trop. Med. Parasitol.*, **86**, 455–465.

51 Hawley, S.R., Bray, P.G., O'Neill, P.M., Naisbitt, D.J., Park, B.K., and Ward, S.A. (1996) Manipulation of the N-alkyl substituent in amodiaquine to overcome the verapamil-sensitive chloroquine resistance component. *Antimicrob. Agents Chemother.*, **40**, 2345–2349.

52 O'Neill, P.M., Willock, D.J., Hawley, S.R., Bray, P.G., Storr, R.C., Ward, S.A., and Park, B.K. (1997) Synthesis, antimalarial activity, and molecular modeling of tebuquine analogues. *J. Med. Chem.*, **40**, 437–448.

53 Bray, P.G., Hawley, S.R., Mungthin, M., and Ward, S.A. (1996) Physicochemical properties correlated with drug resistance and the reversal of drug resistance in *Plasmodium falciparum*. *Mol. Pharmacol.*, **50**, 1559–1566.

54 Neftel, K.A., Woodtly, W., Schmid, M., Frick, P.G., and Fehr, J. (1986) Amodiaquine induced agranulocytosis and liver damage. *Br. Med. J. (Clin. Res. Ed.)*: **292**, 721–723.

55 O'Neill, P.M., Mukhtar, A., Stocks, P.A., Randle, L.E., Hindley, S., Ward, S.A., Storr, R.C. *et al.* (2003) Isoquine and related amodiaquine analogues: a new generation of improved 4-aminoquinoline antimalarials. *J. Med. Chem.*, **46**, 4933–4945.

56 Paunescu, E., Susplugas, S., Boll, E., Varga, R., Mouray, E., Grosu, I., Grellier, P. *et al.* (2009) Replacement of the 4′-hydroxy group of amodiaquine and amopyroquine by aromatic and aliphatic substituents: synthesis and antimalarial activity. *ChemMedChem*, **4**, 549–561.

57 Chang, C., Lin-Hua, T., and Jantanavivat, C. (1992) Studies on a new antimalarial compound: pyronaridine. *Trans. R. Soc. Trop. Med. Hyg.*, **86**, 7–10.

58 Fu, S. and Xiao, S.H. (1991) Pyronaridine: A new antimalarial drug. *Parasitol. Today*, **7**, 310–313.

59 Kotecka, B.M., Barlin, G.B., Edstein, M.D., and Rieckmann, K.H. (1997) New quinoline di-Mannich base compounds with greater antimalarial activity than chloroquine, amodiaquine, or pyronaridine. *Antimicrob. Agents Chemother.*, **41**, 1369–1374.

60 Egan, T.J., Hunter, R., Kaschula, C.H., Marques, H.M., Misplon, A., and Walden, J. (2000) Structure-function relationships in aminoquinolines: effect of amino and chloro groups on quinoline-hematin complex formation, inhibition of beta-hematin formation, and antiplasmodial activity. *J. Med. Chem.*, **43**, 283–291.

61 McChesney, E.W. and Fitch, C.D. (1984) 4-Aminoquinolines, in *Antimalarial Drugs II: Current Antimalarials and New Drug Developments* (eds W. Peters and W.H.G. Richards), Springer-Verlag, Berlin, Heidelberg, pp. 30–60.

62 Karle, J.M. and Karle, I.L. (1991) Crystal structure and molecular structure of mefloquine methylsulfonate monohydrate: implications for a malaria receptor. *Antimicrob. Agents Chemother.*, **35**, 2238–2245.

63 Sowunmi, A., Salako, L.A., Oduola, A.M., Walker, O., Akindele, J.A., and Ogundahunsi, O.A. (1993) Neuropsychiatric side effects of mefloquine in Africans. *Trans. R. Soc. Trop. Med. Hyg.*, **87**, 462–463.

64 Bouchaud, O., Imbert, P., Touze, J.E., Dodoo, A.N., Danis, M., and Legros, F. (2009) Fatal cardiotoxicity related to halofantrine: a review based on a worldwide safety data base. *Malar. J.*, **8**, 289.

65 van Vugt, M., Ezzet, F., Nosten, F., Gathmann, I., Wilairatana, P., Looareesuwan, S., and White, N.J. (1999) No evidence of cardiotoxicity during antimalarial treatment with artemether-lumefantrine. *Am. J. Trop. Med. Hyg.*, **61**, 964–967.

66 Sparatore, A., Basilico, N., Casagrande, M., Parapini, S., Taramelli, D., Brun, R., Wittlin, S. *et al.* (2008) Antimalarial activity of novel pyrrolizidinyl derivatives of 4-aminoquinoline. *Bioorg. Med. Chem. Lett.*, **18**, 3737–3740.

67 Sunduru, N., Srivastava, K., Rajakumar, S., Puri, S.K., Saxena, J.K., and Chauhan, P.M. (2009) Synthesis of novel thiourea, thiazolidinedione and thioparabanic acid derivatives of 4-aminoquinoline as potent antimalarials. *Bioorg. Med. Chem. Lett.*, **19**, 2570–2573.

68 Sunduru, N., Nishi, Palne, S., Chauhan, P.M., and Gupta, S. (2009) Synthesis and antileishmanial activity of novel 2,4,6-trisubstituted pyrimidines and 1,3,5-triazines. *Eur. J. Med. Chem.*, **44**, 2473–2481.

69 Li, Y.T., Hu, Y.G., Huang, H.Z., Zhu, D.Q., Huang, W.J., Wu, D.L., and Qian, Y.L. (1981) Hydroxypiperaquine phosphate in treatment of falciparum malaria. *Chin. Med. J. (Engl.)*, **94**, 301–302.

70 Vennerstrom, J.L., Ellis, W.Y., Ager, A.L. Jr, Andersen, S.L., Gerena, L., and Milhous, W.K. (1992) Bisquinolines. 1. *N,N*-bis(7-chloroquinolin-4-yl) alkanediamines with potential against chloroquine-resistant malaria. *J. Med. Chem.*, **35**, 2129–2134.

71 Raynes, K. (1999) Bisquinoline antimalarials: their role in malaria chemotherapy. *Int. J. Parasitol.*, **29**, 367–379.

72 Le Bras, J., Deloron, P., and Charmot, G. (1983) Dichlorquinazine (alpha 4-aminoquinoline) effective in vitro against chloroquine-resistant *Plasmodium falciparum. Lancet*, **1**, 73–74.

73 Chen, L. (1991) Recent studies on antimalarial efficacy of piperaquine and hydroxypiperaquine. *Chin. Med. J. (Engl.)*, **104**, 161–163.

74 Davis, T.M., Hung, T.Y., Sim, I.K., Karunajeewa, H.A., and Ilett, K.F. (2005) Piperaquine: a resurgent antimalarial drug. *Drugs*, **65**, 75–87.

75 Ridley, R.G., Matile, H., Jaquet, C., Dorn, A., Hofheinz, W., Leupin, W., Masciadri, R. *et al.* (1997) Antimalarial activity of the bisquinoline trans-*N1,N2*-bis (7-chloroquinolin-4-yl)cyclohexane-1,2-diamine: comparison of two stereoisomers and detailed evaluation of the *S,S* enantiomer, Ro 47-7737. *Antimicrob. Agents Chemother.*, **41**, 677–686.

76 Raynes, K., Foley, M., Tilley, L., and Deady, L.W. (1996) Novel bisquinoline antimalarials. Synthesis, antimalarial activity, and inhibition of haem polymerisation. *Biochem. Pharmacol.*, **52**, 551–559.

77 Winter, R.W., Ignatushchenko, M., Ogundahunsi, O.A., Cornell, K.A., Oduola, A.M., Hinrichs, D.J., and Riscoe, M.K. (1997) Potentiation of an antimalarial oxidant drug. *Antimicrob. Agents Chemother.*, **41**, 1449–1454.

78 Ignatushchenko, M.V., Winter, R.W., Bachinger, H.P., Hinrichs, D.J., and Riscoe, M.K. (1997) Xanthones as antimalarial agents; studies of a possible mode of action. *FEBS Lett.*, **409**, 67–73.

79 Xu Kelly, J., Winter, R., Riscoe, M., and Peyton, D.H. (2001) A spectroscopic investigation of the binding interactions between 4,5-dihydroxyxanthone and heme. *J. Inorg. Biochem.*, **86**, 617–625.

80 Fotie, J., Nkengfack, A.E., Rukunga, G., Tolo, F., Peter, M.G., Heydenreich, M., and Fomum, Z.T. (2003) In-vivo antimalarial activity of some oxygenated xanthones. *Ann. Trop. Med. Parasitol.*, **97**, 683–688.

81 Ignatushchenko, M.V., Winter, R.W., and Riscoe, M. (2000) Xanthones as antimalarial agents: stage specificity. *Am. J. Trop. Med. Hyg.*, **62**, 77–81.

82 Hay, A.E., Helesbeux, J.J., Duval, O., Labaied, M., Grellier, P., and Richomme, P. (2004) Antimalarial xanthones from *Calophyllum caledonicum* and *Garcinia vieillardii. Life Sci.*, **75**, 3077–3085.

83 Kelly, J.X., Winter, R., Peyton, D.H., Hinrichs, D.J., and Riscoe, M. (2002) Optimization of xanthones for antimalarial activity: the 3,6-bis-omega-diethylaminoalkoxyxanthone series. *Antimicrob. Agents Chemother.*, **46**, 144–150.

84 Angerhofer, C.K., Pezzuto, J.M., Konig, G.M., Wright, A.D., and Sticher, O. (1992) Antimalarial activity of sesquiterpenes from the marine sponge *Acanthella klethra. J. Nat. Prod.*, **55**, 1787–1789.

85 Wright, A.D., Wang, H., Gurrath, M., Konig, G.M., Kocak, G., Neumann, G., Loria, P. *et al.* (2001) Inhibition of heme detoxification processes underlies the antimalarial activity of terpene isonitrile compounds from marine sponges. *J. Med. Chem.*, **44**, 873–885.

86 Schwarz, O., Brun, R., Bats, J.W., and Schmalz, H.-G. (2002) Synthesis and biological evaluation of new antimalarial isonitriles related to marine diterpenoids. *Tetrahedron Lett.*, **43**, 1009–1013.

87 Singh, C., Srivastav, N.C., and Puri, S.K. (2002) In vivo active antimalarial isonitriles. *Bioorg. Med. Chem. Lett.*, **12**, 2277–2279.

88 White, R.D., Keaney, G.F., Slown, C.D., and Wood, J.L. (2004) Total synthesis of ($+/-$)-kalihinol C. *Org. Lett.*, **6**, 1123–1126.

89 Chavalitshewinkoon, P., Wilairat, P., Gamage, S., Denny, W., Figgitt, D., and Ralph, R. (1993) Structure-activity relationships and modes of action of 9-anilinoacridines against chloroquine-resistant *Plasmodium falciparum* in vitro. *Antimicrob. Agents Chemother.*, **37**, 403–406.

90 Santelli-Rouvier, C., Pradines, B., Berthelot, M., Parzy, D., and Barbe, J. (2004) Arylsulfonyl acridinyl derivatives acting on *Plasmodium falciparum. Eur. J. Med. Chem.*, **39**, 735–744.

91 Auparakkitanon, S. and Wilairat, P. (2000) Cleavage of DNA induced by 9-anilinoacridine inhibitors of topoisomerase II in the malaria parasite *Plasmodium falciparum. Biochem. Biophys. Res. Commun.*, **269**, 406–409.

92 Ciesielska, E., Pastwa, E., and Szmigiero, L. (1997) Inhibition of mammalian topoisomerase I by 1-nitro-9-aminoacridines. Dependence on thiol activation. *Acta Biochim. Pol.*, **44**, 775–780.

93 Egan, T.J., Mavuso, W.W., and Ncokazi, K.K. (2001) The mechanism of beta-hematin formation in acetate solution. Parallels between hemozoin formation and biomineralization processes. *Biochemistry*, **40**, 204–213.

94 Auparakkitanon, S., Noonpakdee, W., Ralph, R.K., Denny, W.A., and Wilairat, P. (2003) Antimalarial 9-anilinoacridine compounds directed at hematin. *Antimicrob. Agents Chemother.*, **47**, 3708–3712.

95 Gamage, S.A., Tepsiri, N., Wilairat, P., Wojcik, S.J., Figgitt, D.P., Ralph, R.K., and Denny, W.A. (1994) Synthesis and *in vitro* evaluation of 9-anilino-3,6-diaminoacridines active against a multidrug-resistant strain of the malaria parasite *Plasmodium falciparum. J. Med. Chem.*, **37**, 1486–1494.

96 Guetzoyan, L., Yu, X.M., Ramiandrasoa, F., Pethe, S., Rogier, C., Pradines, B., Cresteil, T. *et al.* (2009) Antimalarial acridines: synthesis, in vitro activity against *P. falciparum* and interaction with hematin. *Bioorg. Med. Chem.*, **17**, 8032–8039.

97 Kelly, J.X., Smilkstein, M.J., Brun, R., Wittlin, S., Cooper, R.A., Lane, K.D., Janowsky, A. *et al.* (2009) Discovery of dual function acridones as a new antimalarial chemotype. *Nature*, **459**, 270–273.

98 Chong, C.R. and Sullivan, D.J. Jr (2003) Inhibition of heme crystal growth by antimalarials and other compounds: implications for drug discovery. *Biochem. Pharmacol.*, **66**, 2201–2212.

99 Huy, N.T., Kamei, K., Yamamoto, T., Kondo, Y., Kanaori, K., Takano, R., Tajima, K. *et al.* (2002) Clotrimazole binds to heme and enhances heme-dependent hemolysis: proposed antimalarial mechanism of clotrimazole. *J. Biol. Chem.*, **277**, 4152–4158.

100 Huy, N.T., Kamei, K., Kondo, Y., Serada, S., Kanaori, K., Takano, R., Tajima, K. *et al.* (2002) Effect of antifungal azoles on the heme detoxification system of malarial parasite. *J. Biochem.*, **131**, 437–444.

101 Trivedi, V., Chand, P., Srivastava, K., Puri, S.K., Maulik, P.R., and Bandyopadhyay, U. (2005) Clotrimazole inhibits hemoperoxidase of *Plasmodium* *falciparum* and induces oxidative stress. Proposed antimalarial mechanism of clotrimazole. *J. Biol. Chem.*, **280**, 41129–41136.

102 Gemma, S., Campiani, G., Butini, S., Kukreja, G., Joshi, B.P., Persico, M., Catalanotti, B. *et al.* (2007) Design and synthesis of potent antimalarial agents based on clotrimazole scaffold: exploring an innovative pharmacophore. *J. Med. Chem.*, **50**, 595–598.

103 Gemma, S., Campiani, G., Butini, S., Kukreja, G., Coccone, S.S., Joshi, B.P., Persico, M. *et al.* (2008) Clotrimazole scaffold as an innovative pharmacophore towards potent antimalarial agents: design, synthesis, and biological and structure-activity relationship studies. *J. Med. Chem.*, **51**, 1278–1294.

104 Atamna, H., Krugliak, M., Shalmiev, G., Deharo, E., Pescarmona, G., and Ginsburg, H. (1996) Mode of antimalarial effect of methylene blue and some of its analogues on *Plasmodium falciparum* in culture and their inhibition of *P. vinckei petteri* and *P. yoelii nigeriensis* in vivo. *Biochem. Pharmacol.*, **51**, 693–700.

105 Vennerstrom, J.L., Makler, M.T., Angerhofer, C.K., and Williams, J.A. (1995) Antimalarial dyes revisited: xanthenes, azines, oxazines, and thiazines. *Antimicrob. Agents Chemother.*, **39**, 2671–2677.

106 Kalkanidis, M., Klonis, N., Tilley, L., and Deady, L.W. (2002) Novel phenothiazine antimalarials: synthesis, antimalarial activity, and inhibition of the formation of beta-haematin. *Biochem. Pharmacol.*, **63**, 833–842.

107 Schirmer, R.H., Coulibaly, B., Stich, A., Scheiwein, M., Merkle, H., Eubel, J., Becker, K. *et al.* (2003) Methylene blue as an antimalarial agent. *Redox Rep.*, **8**, 272–275.

108 Zoungrana, A., Coulibaly, B., Sie, A., Walter-Sack, I., Mockenhaupt, F.P., Kouyate, B., Schirmer, R.H. *et al.* (2008) Safety and efficacy of methylene blue combined with artesunate or amodiaquine for uncomplicated falciparum malaria: a randomized controlled trial from Burkina Faso. *PLoS One*, **3**, e1630.

109 Coulibaly, B., Zoungrana, A., Mockenhaupt, F.P., Schirmer, R.H., Klose, C., Mansmann, U., Meissner, P.E. *et al.* (2009) Strong gametocytocidal effect of methylene blue-based combination therapy against falciparum malaria: a randomised controlled trial. *PLoS One*, **4**, e5318.

110 Martiney, J.A., Cerami, A., and Slater, A.F. (1996) Inhibition of hemozoin formation in *Plasmodium falciparum* trophozoite extracts by heme analogs: possible implication in the resistance to malaria conferred by the beta-thalassemia trait. *Mol. Med.*, **2**, 236–246.

111 Basilico, N., Monti, D., Olliaro, P., and Taramelli, D. (1997) Non-iron porphyrins inhibit beta-haematin (malaria pigment) polymerisation. *FEBS Lett.*, **409**, 297–299.

112 Cole, K.A., Ziegler, J., Evans, C.A., and Wright, D.W. (2000) Metalloporphyrins inhibit beta-hematin (hemozoin) formation. *J. Inorg. Biochem.*, **78**, 109–115.

113 Begum, K., Kim, H.S., Kumar, V., Stojiljkovic, I., and Wataya, Y. (2003) In vitro antimalarial activity of metalloporphyrins against *Plasmodium falciparum*. *Parasitol. Res.*, **90**, 221–224.

114 Kumar, S., Guha, M., Choubey, V., Maity, P., Srivastava, K., Puri, S.K., and Bandyopadhyay, U. (2008) Bilirubin inhibits *Plasmodium falciparum* growth through the generation of reactive oxygen species. *Free Radic. Biol. Med.*, **44**, 602–613.

115 Kumar, S., Das, S.K., Dey, S., Maity, P., Guha, M., Choubey, V., Panda, G. *et al.* (2008) Antiplasmodial activity of [(aryl) arylsulfanylmethyl]pyridine. *Antimicrob. Agents Chemother.*, **52**, 705–715.

116 Werbovetz, K. (2006) Diamidines as antitrypanosomal, antileishmanial and antimalarial agents. *Curr. Opin. Invest. Drugs*, **7**, 147–157.

117 Bray, P.G., Barrett, M.P., Ward, S.A., and de Koning, H.P. (2003) Pentamidine uptake and resistance in pathogenic protozoa: past, present and future. *Trends Parasitol.*, **19**, 232–239.

118 Stead, A.M., Bray, P.G., Edwards, I.G., DeKoning, H.P., Elford, B.C., Stocks,

P.A., and Ward, S.A. (2001) Diamidine compounds: selective uptake and targeting in *Plasmodium falciparum*. *Mol. Pharmacol.*, **59**, 1298–1306.

119 Goldberg, D.E., Sharma, V., Oksman, A., Gluzman, I.Y., Wellems, T.E., and Piwnica-Worms, D. (1997) Probing the chloroquine resistance locus of *Plasmodium falciparum* with a novel class of multidentate metal(III) coordination complexes. *J. Biol. Chem.*, **272**, 6567–6572.

120 Wiesner, J., Ortmann, R., Jomaa, H., and Schlitzer, M. (2003) New antimalarial drugs. *Angew. Chem. Int. Ed. Engl.*, **42**, 5274–5293.

121 Brinner, K.M., Mi Kim, J., Habashita, H., Gluzman, I.Y., Goldberg, D.E., and Ellman, J.A. (2002) Novel and potent anti-malarial agents. *Bioorg. Med. Chem.*, **10**, 3649–3661.

122 Agrawal, R., Tripathi, R., Tekwani, B.L., Jain, S.K., Dutta, G.P., and Shukla, O.P. (2002) Haem polymerase as a novel target of antimalarial action of cyproheptadine. *Biochem. Pharmacol.*, **64**, 1399–1406.

123 Van Miert, S., Hostyn, S., Maes, B.U., Cimanga, K., Brun, R., Kaiser, M., Matyus, P. *et al.* (2005) Isoneocryptolepine, a synthetic indoloquinoline alkaloid, as an antiplasmodial lead compound. *J. Nat. Prod.*, **68**, 674–677.

124 El Sayed, I., Van der Veken, P., Steert, K., Dhooghe, L., Hostyn, S., Van Baelen, G., Lemiere, G. *et al.* (2009) Synthesis and antiplasmodial activity of aminoalkylamino-substituted neocryptolepine derivatives. *J. Med. Chem.*, **52**, 2979–2988.

125 Jonckers, T.H., van Miert, S., Cimanga, K., Bailly, C., Colson, P., De Pauw-Gillet, M.C., van den Heuvel, H. *et al.* (2002) Synthesis, cytotoxicity, and antiplasmodial and antitrypanosomal activity of new neocryptolepine derivatives. *J. Med. Chem.*, **45**, 3497–3508.

126 Bringmann, G. and Pokorny, F. (1995) The naphthylisoquinoline alkaloids, in *The Alkaloids* (ed. G. Cordell), Academic Press, New York, pp. 127–271.

127 Francois, G., Bringmann, G., Dochez, C., Schneider, C., Timperman, G., and Ake Assi, L. (1995) Activities of extracts and naphthylisoquinoline alkaloids from *Triphyophyllum peltatum, Ancistrocladus abbreviatus* and *Ancistrocladus barteri* against *Plasmodium berghei* (Anka strain) in vitro. *J. Ethnopharmacol.*, **46**, 115–120.

128 Francois, G., Timperman, G., Eling, W., Assi, L.A., Holenz, J., and Bringmann, G. (1997) Naphthylisoquinoline alkaloids against malaria: evaluation of the curative potentials of dioncophylline C and dioncopeltine A against *Plasmodium berghei* in vivo. *Antimicrob. Agents Chemother.*, **41**, 2533–2539.

129 Schwedhelm, K.F., Horstmann, M., Faber, J.H., Reichert, Y., Bringmann, G., and Faber, C. (2007) The novel antimalarial compound dioncophylline C forms a complex with heme in solution. *ChemMedChem*, **2**, 541–548.

130 Jiang, S., Zeng, Q., Gettayacamin, M., Tungtaeng, A., Wannaying, S., Lim, A., Hansukjariya, P. *et al.* (2005) Antimalarial activities and therapeutic properties of febrifugine analogs. *Antimicrob. Agents Chemother.*, **49**, 1169–1176.

12
Exploiting the Vitamin Metabolism of Apicomplexa as Drug Targets

Carsten Wrenger* and Ingrid B. Müller

Abstract

Apicomplexan parasites have a significant impact on health of human and livestock, and chemotherapy remains problematic. The most severe apicomplexan parasite, *Plasmodium* – the pathogenic agent of malaria – causes a devastating and quite often deadly disease, leading to important public health problems in the tropics. Due to the high mutational rate of *P. falciparum*, and its resultant ability to adapt rapidly to environmental changes, drug resistance to standard medications such as chloroquine or antifolates is increasing. Hence, there is a continual need for the development of new chemotherapeutic agents. Previous claims have suggested that antiparasitic compounds should target only the parasite, without harming the human host, and in this sense parasite-specific metabolism represents an ideal drug target. Vitamins, and their metabolism, belong to such a drug target group, as has been established for vitamin B_9. In this chapter, attention is focused on the current knowledge of apicomplexan vitamin metabolism and its validation as a drug target, notably with regards to recently identified vitamin B_1 and B_6 biosyntheses.

Introduction

Vitamin A (retinol) has been reported not only to be involved in normal immune function [1], but also to have a protective role in malaria infection. The latter effect was demonstrated when an increased susceptibility to malaria of vitamin A-deficient ducks was reversed by vitamin A supplementation [2, 3]. Likewise, the appearance of *Plasmodium falciparum* infection among preschool children in Papua New Guinea [4] was reduced following supplementation with vitamin A, with an enhanced clearance of the parasitized erythrocytes [5]. Although, as yet, the precise mechanism of this resistance to malaria remains unknown, β-carotene (the precursor of vitamin A) has been reported to show a strong free-radical scavenging activity, without toxic side effects, even at high concentrations [6]. Such protection against malaria was also

*Corresponding author

Apicomplexan Parasites. Edited by Katja Becker
Copyright © 2011 WILEY-VCH Verlag GmbH & Co. KGaA, Weinheim
ISBN: 978-3-527-32731-7

observed in mice [7]. Interestingly, although the biosynthesis of β-carotene was recently shown to play a role in antioxidant defense within the parasite [8], genes encoding for enzymes that participate in β-carotene biosynthesis have yet not been identified in the *Plasmodium* genome database [9].

Vitamin B2 (riboflavin) forms part of the cofactors flavin adenine dinucleotide (FAD) and flavin mononucleotide (FMN), and thereby is essentially involved in the catalysis of various oxido reductases, such as NADH dehydrogenase or glutathione reductase. Riboflavin influences malaria morbidity, as demonstrated in Papua New Guinea, where riboflavin-deficient children were shown to have reduced levels of malarial infections [10, 11]; similar observations were made in both India [12] and The Gambia [13]. Moreover, vitamin B_2 analogs demonstrated an inhibitory effect on parasite growth *in vitro* and *in vivo* via a reduced glutathione reductase activity [14]. The reduction of vitamin B_2 levels has been proposed to create an oxidative environment, most likely due to a reduced activity of the erythrocytic riboflavin-dependent glutathione peroxidase. Indeed, a higher level of lipid peroxidation was observed in riboflavin-deficient children with malaria infection [15]. Notably, riboflavin-deficient rats were also shown to be more resistant to malaria [16]. In contrast, the incubation of infected erythrocytes with high doses of riboflavin led to an inhibition of hemozoin production and a decrease in the size of the food vacuole, which led in turn to a reduced parasite growth [17]. Clearly, further studies are required to elucidate the role of vitamin B_2 in malarial infections.

Vitamin B5 (pantothenate) is required in the synthesis of coenzyme A (CoA), an essential cofactor for both energy metabolism and fatty acid biosynthesis. Pantothenate cannot be synthesized by humans, due to an absence of enzymes in the first part of the CoA-biosynthetic pathway. A growth analysis of *P. falciparum* in a minimal medium confirmed an absolute requirement for vitamin B_5 in growth of the malarial parasite [18, 19], which implies uptake of the vitamin. In addition, genes encoding for the respective enzymes were not found in *Plasmodium*, nor in the apicomplexan relative *Cryptosporidium*. However, a genome analysis of *Toxoplasma gondii* identified genes encoding for proteins which were apparently involved in the synthesis of vitamin B_5 [20]. An analysis of the vitamin B_5 import revealed that the uptake had occurred via the "new permeability pathways" (NPP) [21], which are not restricted to pantothenate. The NPPs have a broad substrate acceptance, including nutrients or inorganic ions [22–24]; pantothenate is subsequently transported by a H^+: pantothenate symport mechanism into the parasite [25]. Intracellularly, the pantothenate is phosphorylated by pantothenate kinase (PanK) [21], a key enzyme in CoA biosynthesis. Although humans are also able to phosphorylate pantothenate, differences in the kinetic and regulatory properties of PanK make the enzyme an attractive drug target [26]. Indeed, pantothenate analogues that specifically inhibit the plasmodial PanK have been identified [19, 27]. In the next step toward CoA biosynthesis, 4-phosphopantothenate is attached to cysteine, by the action of phosphopantothenoylcysteine synthetase, to form 4-phosphopantothenoylcysteine. Subsequently, phosphopanthenoylcysteine decarboxylase mediates the decarboxylation of 4-phosphopantothenoylcysteine to 4-phosphopantetheine, which is then converted to CoA in consecutive steps by pantetheine phosphate adenylyltransferase and de-phospho-CoA kinase [26].

The *vitamin B₉* (folate) metabolism is an approved drug target exploited for the treatment of various infectious diseases, including those caused by the malaria parasite, and has been the subject of various reviews [28–32]. Vitamin B_9 in its reduced form, tetrahydrofolate, serves as a cofactor for the transfer of one-carbon units, which is essential in DNA synthesis [32]. The malarial parasite *P. falciparum*, and its relative *T. gondii*, each possess the enzymes required for the *de novo* synthesis of 7,8-dihydrofolate. As vitamin B_9 cannot be biosynthesized by humans, this pathway represents an ideal drug target. Indeed, dihydropteroate synthase (DHPS) is targeted by "sulfa" drugs (e.g., sulfonamides and sulfones), and has long been used in the treatment of malaria patients. However, the parasite also imports folate from the host metabolism, which is subsequently reduced by dihydrofolate reductase (DHFR) to tetrahydrofolate. Although the human host also exhibits DHFR activity, susceptibility toward pyrimethamine (a potent inhibitor) is much greater for the plasmodial enzyme. The inhibitory effect of pyrimethamine is not restricted to the plasmodial DHFR; it is also very effective against the DHFR of *Eimeria tenella*, but less so against the chicken counterpart [33]. For synergistic effects, sulfa drugs are commonly used in a combination therapy with pyrimethamine (Fansidar®) to treat malaria, although in recent years resistance to these drugs has become apparent [30, 31].

Both, *vitamin C* (L-ascorbic acid) and *vitamin E* (α-tocopherol) are known to be effective constituents in protection against oxidative stress generated by reactive oxygen species (ROS) [6]. It is well established that the malarial parasite requires a potent anti-oxidant defense system for survival [34] and, indeed, several reports suggested that deficiencies in the vitamin E supply might be protective against malarial infections [35]. The ROS are generated by the parasite itself during hemoglobin degradation, while the host's immune system produces oxygen radicals to combat infectious agents [36]. It has been shown in several studies that vitamin E deficiency leads to malaria protection [37–40], an effect which can be explained by an apparently higher vulnerability to ROS due to the missing antioxidant [41]. A similar role was proposed for vitamin C.

Vitamin B₁ Metabolism: A Novel Drug Target?

The B1 group of vitamers consists of four molecules, namely thiamine, thiamine monophosphate (TMP), thiamine pyrophosphate (TPP), and thiamine triphosphate (TTP). While TTP was proposed to have a neurophysiological role, recent analyses have suggested a more basic function in cellular metabolism, though further experiments are required to elucidate a precise role [42]. The predominant form of vitamin B_1 is TPP; this is the catalytic active form, and thereby the cofactor of various enzymes mainly involved in carbohydrate metabolism, such as 2-oxoglutarate dehydrogenase, pyruvate dehydrogenase (PDH), branched-chain 2-oxo acid dehydrogenase (BCODH), or transketolase [43, 44]. Whilst bacteria, plants, and some fungi are able to synthesis thiamine *de novo*, mammals are dependent on assimilating this nutrient from their diet. Indeed, a reduction in dietary vitamin B_1 uptake leads to Wernicke's disease and beriberi. The biosynthesis of vitamin B_1 has been well elucidated in both *Escherichia coli* and in the fungus, *Saccharomyces cerevisiae* [45, 46]. In *P. falciparum*, the biosynthesis of thiamine proceeds via the combination of two different branches, thiazole and pyrimidine. The thiazole branch utilizes 5-(2-hydroxyethyl)-4-methylthiazole (THZ), which is phosphorylated by THZ kinase (ThiM) to 5-(2-hydroxyethyl)-4-methylthiazole phosphate (THZ-P), with high substrate specificity. The other key intermediate derives from the pyrimidine branch, which provides 4-amino-5-hydroxymethyl-2-methylpyrimidine (HMP). In two steps, the HMP/HMP-P kinase (ThiD) phosphorylates HMP to HMP-P and, subsequently, to HMP-PP (Figure 12.1) [45, 47–49]. In bacteria, HMP is derived via purine biosynthesis by the HMP synthesis enzyme, ThiC [50], whereas in yeast Thi5-p is proposed to combine pyridoxine 5-phosphate and histidine to HMP-P [46, 51, 52]. Subsequently, the respective phosphorylated pyrimidine and thiazole moieties, HMP-PP and THZ-P, are merged by the thiamine phosphate synthase (ThiE) to TMP [53, 54] (Figure 12.1). In some bacteria, TMP is directly phosphorylated by TMP kinase (ThiL) to obtain TPP [45]. However, to date, no homology to a gene encoding ThiL has been found in the genome database of *P. falciparum*; rather, a thiamine pyrophosphokinase (TPK), which accepts only thiamine and not its phosphorylated forms as substrate [55]. Consequently, TMP must be dephosphorylated prior to pyrophosphorylation by TPK, to obtain the active cofactor TPP. As no plausible homologs to known specific phosphatases of this type have been identified in the database, it is possible that this dephosphorylation is catalyzed by a nonspecific phosphatase. A recently identified candidate for this role was a *para*-nitrophenyl phosphatase, which has been shown to exhibit a broad substrate profile that included TMP (Figure 12.1) [56].

In bacteria, vitamin B_1 biosynthesis has been suggested to be exploitable for chemotherapy by using the naturally occurring HMP analog, bacimethrin (MeO-HMP), or the synthetic CF_3-HMP compound [57, 58] (Figure 12.2). Although the former compound was successfully tested as a substrate for the plasmodial ThiD protein, bacimethrin did not affect the proliferation rate of the parasite, possibly due to a limited uptake [49]. However, HMP derivatives as such should be considered to be promising drug candidates. A similar result was obtained by using the thiamine analog, deazathiamine, a potent *in vitro* inhibitor of TPP-dependent enzymes which also failed to show any significant reduction in parasite growth. In contrast, the thiazole precursor of deazathiamine inhibited the proliferation of cultured *P. falciparum* [32].

Culturing *P. falciparum* in a thiamine-free medium for a few days showed no adverse effect on the parasite's growth behavior [18]; however, extending the time period of culture in this minimal medium indicated a significant need for an external supply of thiamine, or its precursor HMP [49]. Consistent with these results, a homolog of the yeast-like Thi5 enzyme was not found in the plasmodial genome database, implying a missing or insufficient correlation to the vitamin B_6 and B_1 biosynthesis pathways, which is in contrast to its yeast counterpart [59–61]. Even further, a homolog encoding for the bacterial ThiC protein, which catalyzes the formation of HMP from the aminoimidazole ribonucleotide, an intermediate of the purine synthesis [45], was also absent [54]. The latter fact was not surprising, however,

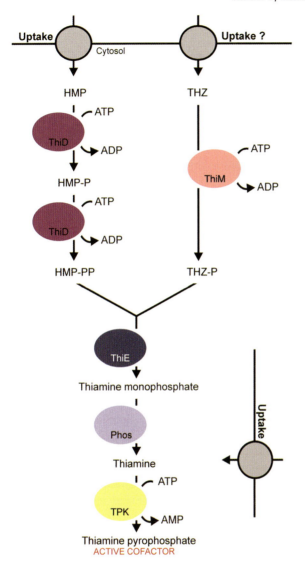

Figure 12.1 Vitamin B₁ biosynthesis of the human malaria parasite *Plasmodium falciparum*. Vitamin B₁ *de novo* synthesis consists of two branches, the pyrimidine branch and the thiazole branch. In the latter case, 5-(2-hydroxyethyl)-4-methylthiazole (THZ) is phosphorylated to 5-(2-hydroxyethyl)-4-methylthiazole phosphate (THZ-P) by THZ kinase (ThiM). In the pyrimidine branch, 4-amino-5-hydroxymethyl-2-methylpyrimidine (HMP), which is imported by the parasite, is phosphorylated in two consecutive steps to HMP-PP by HMP/HMP-P kinase (ThiD). HMP-PP and THZ-P are merged by the thiamine phosphate synthase (ThiE). Thiamine monophosphate is dephosphorylated by a phosphatase (Phos) to create a "thiamine pool." Subsequently, thiamine is diphosphorylated to thiamine pyrophosphate by thiamine pyrophosphokinase (TPK).

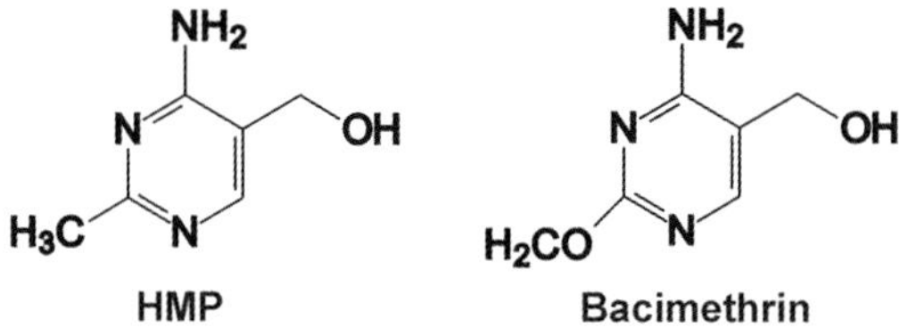

Figure 12.2 Structural comparison of 4-amino-5-hydroxymethyl-2-methylpyrimidine (HMP) and the naturally occurring antimicrobial compound, bacimethrin.

due to the fact that purine biosynthesis is absent from the malarial parasite [62]. Considering that *Plasmodium* can salvage B_1 vitamers for the formation of TPP, this offers several novel chemotherapeutic strategies to import pro-drugs in the form of HMP or thiamine derivatives. As an intracellular parasite, *P. falciparum* needs to import nutrients from the host cell, the erythrocyte, which reveals the highest levels of vitamin B_1 within the bloodstream [43]. In fact, the erythrocyte is capable of diphosphorylating thiamine to TPP via the action of its TPK, thereby trapping vitamin B_1 within the host cell [43]. However, the phosphorylated molecules are characterized by rather limited import capabilities, and in order to obtain the phosphorylated molecules the parasite must dephosphorylate them prior to uptake. Very recently, the proposal was made that a secreted phosphatase with a broad substrate profile – including TPP – was suggested to take over this function [63].

Interestingly, cultivation of the parasite in a THZ-deficient medium led to normal growth, which suggested that *P. falciparum* might be capable of generating THZ *de novo*. Despite using bioinformatic tools, genes encoding for THZ biosynthesis enzymes could not be identified in the plasmodial genome database [9]. For various other organisms, a synthesis of either the pyrimidine or the thiazole moiety was discussed, which required that the other part be salvaged [43, 64, 65]. This might also be the case for the malarial parasite, although a THZ biosynthesis cannot be excluded. Genome analyses of further apicomplexa, such as *Eimeria tenella* and *Toxoplasma gondii*, have indicated an absence of the vitamin B1 biosynthetic enzymes – an astonishing finding, considering that these organisms are close relatives.

Vitamin B_6 Metabolism in Apicomplexa

Pyridoxal 5-phosphate (PLP) acts as the active cofactor for more than 140 distinct enzymatic reactions, all of which are involved in the conversion of amino acids via reactions such as decarboxylation and transamination. In all of these cases, PLP serves as a carbonyl-reactive coenzyme and, as such, is indispensable for enzyme activity [66]. PLP is produced from the B_6 vitamers pyridoxine, pyridoxal, and pyridoxamine. Whereas, almost all bacteria, fungi, and plants possess their own routes of vitamin B_6 biosynthesis, mammals do not synthesize B_6 vitamers *de novo*, and therefore depend entirely on the uptake of this indispensable nutrient from their diet.

Vitamin B_6 biosynthesis has been extensively analyzed in *Escherichia coli*, where two branches must be combined. The first branch starts with erythrose 4-phosphate, which is modified by GapA, PdxB, and PdxF, to finally become 4-phospho-hydroxy-L-threonine (4PHT). The second branch leads to 1-deoxy-D-xylulose 5-phosphate (DOXP) via the intermediates of glycolysis, glyceraldehyde 3-phosphate, and pyruvate. The enzyme responsible for DOXP synthesis is DOXP synthetase (DXPS), while DOXP and 4PHT are merged to form pyridoxine 5-phosphate (PNP) by the bacterial enzymes PdxA and PdxJ. However, PNP is not the active cofactor of vitamin B6-dependent enzymes, and it must first be oxidized by PNP oxidase (PdxH) to the active form pyridoxal 5-phosphate (PLP) (Figure 12.3a) [67]. Salvaged B_6-vitamers can also be activated by a pyridoxine/pyridoxal kinase (PdxK) and deactivated by a phosphatase (Phos), as indicated in Figure 12.3a [68]. Dephosphorylated or salvaged pyridoxal is reduced by the pyridoxal reductase (Plr1) in an NADH-dependent reaction to pyridoxine [69] (Figure 12.3a). Due to the participation of DOXP via biosynthesis, this pathway is referred to as the DOXP-dependent pathway [67, 70, 71]. Recently, a distinctly different synthesis pathway has been identified in plants, fungi, and some bacteria, which was originally assigned to be involved in detoxification of singlet oxygen ($^1O_2$) [72–74].

However, an analysis of fungi mutants deficient in SOR1 (singlet oxygen resistance), and therefore sensitive for singlet oxygen, showed that the product of this gene also participated in pyridoxine biosynthesis [72, 75]. The SOR1 enzyme (also named Pdx1) belongs to the highly conserved enzyme family SNZ in *Saccharomyces cerevisiae*. The SNZ1 protein has been shown to interact with the SNO1 protein [72, 76] (also named Pdx2), which is a member of another preserved family in yeast, consisting of three SNO enzymes [77]. This route of vitamin B_6 biosynthesis has been shown to occur in various other organisms, such as *Bacillus subtilis* and *Arabidopsis thaliana* [75, 78–80]. The substrates of this biosynthesis were identified as ribose 5-phosphate and glyceraldehyde 3-phosphate, while glutamine served as a nitrogen source [79, 81]. While DOXP is not present, this route of vitamin B_6 biosynthesis has been named the DOXP-independent pathway [67]. By using labeling experiments, Cassera *et al.* [82] suggested the presence of a vitamin B_6 metabolite in *P. falciparum*, although in subsequent studies the participating enzymes were analyzed in detail and assigned to the DOXP-independent pathway of *P. falciparum* [83, 84]. Recently, this pathway was also identified in the apicomplexan parasite *T. gondii* [85]. Both apicomplexan pathways consist of the single-copy genes *Pdx1* and *Pdx2* (Figure 12.3b). As has been shown for the yeast proteins SNZ1 and SNO1, the homologs of Pdx1 and Pdx2 in *T. gondii* and *P. falciparum*, the apicomplexan proteins must act together in order to achieve enzyme activity [76, 83, 85]. In contrast to the bacterial-like formation of vitamin B6 (DOXP-dependent pathway) leading to pyridoxine 5-phosphate, synthesis via the DOXP-independent pathway results in PLP, the active form of vitamin B_6. PLP is also synthesized by the apicomplexan enzymes, as shown clearly by activity studies of PLP-dependent enzymes such as plasmodial ornithine decarboxylase [86], which is a key enzyme in polyamine metabolism [85].

The crystal structures of the *B. subtilis* and *Geobacillus stearothermophilus* Pdx1 and Pdx2 homologs have already been solved [87, 88]. Only recently, the structure of the entire Pdx1/Pdx2 complex from *B. subtilis* and *Thermotoga maritima* has been analyzed, identifying the PLP synthase as a multimeric protein complex consisting

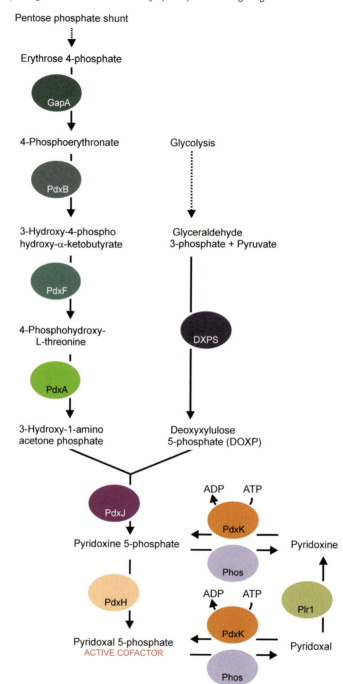

Figure 12.3 Comparison of vitamin B$_6$ biosynthesis in bacteria following the DOXP-dependent pathway (a) and in the apicomplexan parasites *P. falciparum* and *T. gondii* (b), which directly synthesize *de novo* the active cofactor PLP via the DOXP-independent pathway. For abbreviations, please see the text. The dotted arrows are indicative of multiple reaction steps.

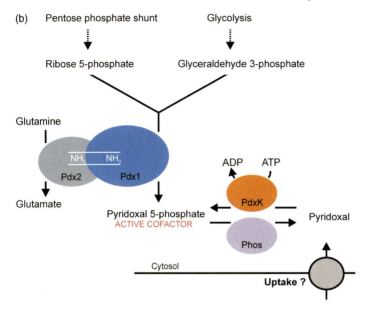

Figure 12.3 *(Continued)*

of 12 Pdx1 proteins decorated by 12 Pdx2 proteins [89, 90]. Interestingly, analyses of the genome data available from *Eimeria* or *Cryptosporidium*, as well as from other protozoan parasites such as *Trypanosoma* and *Leishmania*, do not reveal any indices for genes involved in vitamin B_6 *de novo* synthesis. This clearly indicates that PLP biosynthesis is not generally common in protozoan parasites. Rather, such parasites, which are dependent on external sources of vitamin B_6, must uptake this nutrient in the same manner as the human host. Following import, the B_6 vitamers are immediately phosphorylated and thereby trapped within the cell; this catalysis is carried out by PdxK [91–93]. In addition to the *de novo* synthesis, the malarial parasite also holds a PdxK, which would enable *P. falciparum* to salvage B_6 vitamers [83] (Figure 12.3b). The main carriers of vitamin B_6 in mammals are the erythrocytes [94], which leads to questions regarding the sense of the parasite's PLP synthesis. PLP is bound to serum albumin and hemoglobin [43], and is thereby less accessible to the parasite. It has been shown that cultured parasites do not depend on an external source of pyridoxine [18], and therefore the role of endogenous PdxK has not yet been identified, as to what purpose the enzymes serve in acquiring vitamin B_6, or whether they act solely as interconverting enzymes to suit the needs of the parasite. The crucial role of PfPdxK in blood-stage parasites was recently demonstrated, however [95], when two newly synthesized pyridoxyl-tryptophan methyl esters – PT3 and PT5 – were analyzed for their potential as pro-drugs against *P. falciparum*. Pro-drugs are nontoxic precursors that are converted by endogenous enzymes to their toxic forms, and thereby channeled into the organism's metabolism. Both pro-drugs were shown to be effective substrates of the plasmodial PdxK, and both inhibited (in their phosphorylated form) the plasmodial ornithine decarboxylase (ODC), which is a key regulatory enzyme in polyamine biosynthesis. When these compounds were further applied to infected erythrocytes the growth of parasites was retarded, but the human cells were only marginally affected [95]. Moreover, the antiproliferative effect was most likely directly linked to the action of PfPdxK, which activates the pro-drugs via phosphorylation. In their phosphorylated form, the drugs become trapped in the parasite and interfere with PLP-dependent enzymes by mimicking their cofactor. Based on this specific phosphorylation reaction, a new strategy to combat malaria has been proposed, which has already led to the successful inhibition of PLP-dependent enzymes in *P. falciparum* [32].

Conclusion

As noted above, apicomplexan parasites are characterized by their high mutational rate and, consequently, by their rapid adaptation to environmental changes, which leads in turn to the development of resistance against commonly used drugs. In this sense, a desperate search for novel therapeutic agents is required, where the compounds developed will target only the parasite, without causing harm to the human host. In the case of the Apicomplexa, the modification of vitamin metabolism has a long-standing history with regards to intervening in the salvage and biosynthesis of vitamin B_9. A recent analysis of CoA biosynthesis involving vitamin B_5 showed PanK to be a promising drug target for interference. As with vitamin B_9, the malarial parasite exhibits the *de novo* synthesis of both vitamin B_1 and B_6. Indeed, homologous vitamin B_1 biosynthesis has already been targeted in bacteria, underlying antimicrobial potency. Unfortunately, the antibiotic molecule bacimethrin had marginal effects on the viability of the parasite, most likely due to its limited uptake, although this might be resolved in the future by making structural improvements to the molecule. Promising results have also been obtained by exploiting vitamin B6 metabolism, using pro-drugs (which are metabolized by the parasite to toxic cofactors) as novel antimalarial agents. This approach has highlighted the feasibility of simultaneously targeting a variety of enzymes, and thereby reducing the onset of drug resistance.

Acknowledgments

The relevant studies of the present authors were supported by grants from the Deutsche Forschungsgemeinschaft (WA 395/15 and WR 124/2).

References

1 Semba, R.D. (1998) The role of vitamin A and related retinoids in immune function. *Nutr. Rev.*, **56**, S38–S48.

2 Roos, A., Hegsted, D.M., and Stare, F.J. (1946) Nutritional studies with the duck. IV. The effect of vitamin deficiencies on

the course of *P. lophurae* infection in the duck and the chick. *J. Nutr.*, **32**, 473–484.

3 Krishnan, S., Krishnan, A.D., Mustafa, A.S., Talwar, G.P., and Ramalingaswami, V. (1976) Effect of vitamin A and undernutrition on the susceptibility of rodents to a malarial parasite *Plasmodium berghei*. *J. Nutr.*, **106**, 784–791.

4 Shankar, A.H., Genton, B., Semba, R.D., Baisor, M., Paino, J., Tamja, S., Adiguma, T. *et al.* (1999) Effect of vitamin A supplementation on morbidity due to *Plasmodium falciparum* in young children in Papua New Guinea: a randomised trial. *Lancet*, **354**, 203–209.

5 Serghides, L. and Kain, K.C. (2002) Mechanism of protection induced by vitamin A in *falciparum* malaria. *Lancet*, **359**, 1404–1406.

6 Shankar, A.H. (2000) Nutritional modulation of malaria morbidity and mortality. *J. Infect. Dis.*, **182**, S37–53.

7 Laniyan, T.A., Goodall, H.B., and Dick, H.M. (1989) The effects of dietary oils on murine malaria. *Trans. R. Soc. Trop. Med. Hyg.*, **83**, 863.

8 Tonhosolo, R., D'Alexandri, F.L., de Rosso, V.V., Gazarini, M.L., Matsumura, M.Y., Peres, V.J., Merino, E.F. *et al.* (2009) Carotenoid biosynthesis in intra-erythrocytic stages of *Plasmodium falciparum*. *J. Biol. Chem.*, **284**, 9974–9985.

9 Gardner, M.J., Hall, N., Fung, E., White, O., Berriman, M., Hyman, R.W., Carlton, J.M. *et al.* (2002) Genome sequence of the human malaria parasite *Plasmodium falciparum*. *Nature*, **419**, 498–511.

10 Oppenheimer, S.J., Bull, R., and Thurnham, D.I. (1983) Riboflavin deficiency in Madang infants. *PNG Med. J.*, **26**, 17–20.

11 Thurnham, D.I., Oppenheimer, S.J., and Bull, R. (1983) Riboflavin status and malaria in infants in Papua New Guinea. *Trans. R. Soc. Trop. Med. Hyg.*, **77**, 423–424.

12 Dutta, P., Pinto, J., and Rivlin, R. (1985) Antimalarial effects of riboflavin deficiency. *Lancet*, **2**, 1040–1043.

13 Bates, C.J., Powers, H.J., Lamb, W.H., Anderson, B.B., Perry, G.M., and Vullo, C.

(1986) Antimalarial effects of riboflavin deficiency. *Lancet*, **1**, 329–330.

14 Cowden, W.B. and Clark, I.A. (1987) Antimalarial activity of synthetic riboflavin antagonists. *Trans. R. Soc. Trop. Med. Hyg.*, **81**, 533.

15 Das, B.S., Thurnham, D.I., Patnaik, J.K., Das, D.B., Satpathy, R., and Bose, T.K. (1990) Increased plasma lipid peroxidation in riboflavin-deficient, malaria-infected children. *Am. J. Clin. Nutr.*, **51**, 859–863.

16 Kaikai, P. and Thurnham, D.I. (1983) The influence of riboflavin deficiency on *Plasmodium berghei* infection in rats. *Trans. R. Soc. Trop. Med. Hyg.*, **77**, 680–686.

17 Akompong, T., Ghori, N., and Haldar, K. (2000) *In vitro* activity of riboflavin against the human malaria parasite *Plasmodium falciparum*. *Antimicrob. Agents Chemother.*, **44**, 88–96.

18 Divo, A.A., Geary, T.G., Davis, N.L., and Jensen, J.B. (1985) Nutritional requirements of *Plasmodium falciparum* in culture. I. Exogenously supplied dialyzable components necessary for continuous growth. *J. Protozool.*, **32**, 59–64.

19 Saliba, K.J., Ferru, I., and Kirk, K. (2005) Provitamin B5 (pantothenol) inhibits growth of the intraerythrocytic malaria parasite. *Antimicrob. Agents Chemother.*, **49**, 632–637.

20 Müller, S. and Kappes, B. (2007) Vitamin and cofactor biosynthesis pathways in *Plasmodium* and other apicomplexan parasites. *Trends Parasitol.*, **23**, 112–121.

21 Saliba, K.J., Horner, H.A., and Kirk, K. (1998) Transport and metabolism of the essential vitamin pantothenic acid in human erythrocytes infected with the malaria parasite *Plasmodium falciparum*. *J. Biol. Chem.*, **273**, 10190–10195.

22 Ginsburg, H., Krugliak, M., Eidelman, O., and Cabantchik, Z.I. (1983) New permeability pathways induced in membranes of *Plasmodium falciparum* infected erythrocytes. *Mol. Biochem. Parasitol.*, **8**, 177–190.

23 Ginsburg, H. (1994) Transport pathways in the malaria-infected erythrocyte: characterization and their use as

potential targets for chemotherapy. *Mem. Inst. Oswaldo Cruz*, **89**, 99–109.

24 Kirk, K., Horner, H.A., Elford, B.C., Ellory, J.C., and Newbold, C.I. (1994) Transport of diverse substrates into malaria-infected erythrocytes via a pathway showing functional characteristics of a chloride channel. *J. Biol. Chem.*, **269**, 3339–3347.

25 Saliba, K.J. and Kirk, K. (2001) H⁺-coupled pantothenate transport in the intracellular malaria parasite. *J. Biol. Chem.*, **276**, 18115–18121.

26 Spry, C., Kirk, K., and Saliba, K.J. (2008) Coenzyme A biosynthesis: an antimicrobial drug target. *FEMS Microbiol. Rev.*, **32**, 56–106.

27 Saliba, K.J. and Kirk, K. (2005) CJ-15,801, a fungal natural product, inhibits the intraerythrocytic stage of *Plasmodium falciparum in vitro* via an effect on pantothenic acid utilisation. *Mol. Biochem. Parasitol.*, **141**, 129–131.

28 Nzila, A., Ward, S.A., Marsh, K., Sims, P.F., and Hyde, J.E. (2005) Comparative folate metabolism in humans and malaria parasites (part I): pointers for malaria treatment from cancer chemotherapy. *Trends Parasitol.*, **21**, 292–298.

29 Nzila, A., Ward, S.A., Marsh, K., Sims, P.F., and Hyde, J.E. (2005) Comparative folate metabolism in humans and malaria parasites (part II): activities as yet untargeted or specific to *Plasmodium. Trends Parasitol.*, **21**, 334–339.

30 Hyde, J.E. (2005) Drug-resistant malaria. *Trends Parasitol.*, **21**, 494–498.

31 Hyde, J.E. (2007) Drug-resistant malaria - an insight. *FEBS J.*, **274**, 4688–4698.

32 Müller, I.B., Hyde, J.E., and Wrenger, C. (2010) Vitamin B metabolism in *Plasmodium falciparum* as a source of drug targets. *Trends Parasitol.*, **26**, 35–43.

33 Wang, C.C., Stotish, R.L., and Poe, M. (1975) Dihydrofolate reductase from *Eimeria tenella*: rationalization of chemotherapeutic efficacy of pyrimethamine. *J. Protozool.*, **22**, 564–568.

34 Müller, S., Gilberger, T.W., Krnajski, Z., Lüersen, K., Meierjohann, S., and Walter, R.D. (2001) Thioredoxin and glutathione system of malaria parasite *Plasmodium falciparum. Protoplasma*, **217**, 43–49.

35 Levander, O.A. and Ager, A.L. Jr (1993) Malarial parasites and antioxidant nutrients. *Parasitology*, **107**, 95–106.

36 Becker, K., Tilley, L., Vennerstrom, J.L., Roberts, D., Rogerson, S., and Ginsburg, H. (2004) Oxidative stress in malaria parasite-infected erythrocytes: host-parasite interactions. *Int. J. Parasitol.*, **34**, 163–189.

37 Eckman, J.R., Eaton, J.W., Berger, E., and Jacob, H.S. (1976) Role of vitamin E in regulating malaria expression. *Trans. Assoc. Am. Physicians*, **89**, 105–115.

38 Levander, O.A., Ager, A.L. Jr, Morris, V.C., and May, R.G. (1989) Menhaden-fish oil in a vitamin E-deficient diet: protection against chloroquine-resistant malaria in mice. *Am. J. Clin. Nutr.*, **50**, 1237–1239.

39 Levander, O.A., Ager, A.L. Jr, Morris, V.C., and May, R.G. (1990) *Plasmodium yoelii*: comparative antimalarial activities of dietary fish oils and fish oil concentrates in vitamin E-deficient mice. *Exp. Parasitol.*, **70**, 323–329.

40 Taylor, D.W., Levander, O.A., Krishna, V.R., Evans, C.B., Morris, V.C., and Barta, J.R. (1997) Vitamin E-deficient diets enriched with fish oil suppress lethal *Plasmodium yoelii* infections in athymic and scid/bg mice. *Infect. Immun.*, **65**, 197–202.

41 Stocker, R., Hunt, N.H., Buffinton, G.D., Weidemann, M.J., Lewis-Hughes, P.H., and Clark, I.A. (1985) Oxidative stress and protective mechanisms in erythrocytes in relation to *Plasmodium vinckei* load. *Proc. Natl Acad. Sci. USA*, **82**, 548–551.

42 Makarchikov, A.F., Lakaye, B., Gulyai, I.E., Czerniecki, J., Coumans, B., Wins, P., Grisar, T., and Bettendorff, L. (2003) Thiamine triphosphate and thiamine triphosphatase activities: from bacteria to mammals. *Cell. Mol. Life Sci.*, **60**, 1477–1488.

43 Friedrich, W. (1988) *Vitamins*, Walter de Gruyter, Berlin, NY.

44 Pohl, M., Sprenger, G.A., and Müller, M. (2004) A new perspective on thiamine catalysis. *Curr. Opin. Biotechnol.*, **15**, 335–342.

45 Begley, T.P., Downs, D.M., Ealick, S.E., McLafferty, F.W., Van Loon, A.P., Taylor, S., Campobasso, N. *et al.* (1999) Thiamin biosynthesis in prokaryotes. *Arch. Microbiol.*, **171**, 293–300.

46 Wightman, R. and Meacock, P.A. (2003) The THI5 gene family of *Saccharomyces cerevisiae*: distribution of homologues among the hemiascomycetes and functional redundancy in the aerobic biosynthesis of thiamine from pyridoxine. *Microbiology*, **149**, 1447–1460.

47 Mizote, T., Tsuda, M., Smith, D.D., Nakayama, H., and Nakazawa, T. (1999) Cloning and characterization of the *thiD/J* gene of *Escherichia coli* encoding a thiamin-synthesizing bifunctional enzyme, hydroxymethylpyrimidine kinase/phosphomethylpyrimidine kinase. *Microbiology*, **145**, 495–501.

48 Reddick, J.J., Kinsland, C., Nicewonger, R., Christian, T., Downs, D.M., Winkler, M.E., and Begley, T.P. (1998) Overexpression, purification and characterization of two pyrimidine kinases involved in the biosynthesis of thiamine: 4-amino-5-hydroxymethyl-2-methylpyrimidine kinase and 4-amino-5-hydroxymethyl-2-methy pyrimidine phosphate kinase. *Tetrahedron*, **54**, 15983–15991.

49 Wrenger, C., Eschbach, M.L., Müller, I.B., Laun, N.P., Begley, T.P., and Walter, R.D. (2006) The vitamin B1 *de novo* synthesis of the human malaria parasite *Plasmodium falciparum* depends on external provision of 4-amino-5-hydroxymethyl-2-methylpyrimidine. *Biol. Chem.*, **387**, 41–51.

50 Zhang, Y., Taylor, S.V., Chiu, H.J., and Begley, T.P. (1997) Characterization of the *Bacillus subtilis thiC* operon involved in thiamine biosynthesis. *J. Bacteriol.*, **179**, 030–3035.

51 Hohmann, S. and Meacock, P.A. (1998) Thiamin metabolism and thiamin diphosphate-dependent enzymes in the yeast *Saccharomyces cerevisiae*: genetic regulation. *Biochim. Biophys. Acta*, **1385**, 201–219.

52 Morett, E., Korbel, J.O., Rajan, E., Saab-Rincon, G., Olvera, L., Olvera, M., Schmidt, S. *et al.* (2003) Systematic discovery of analogous enzymes in thiamin biosynthesis. *Nat. Biotechnol.*, **21**, 790–795.

53 Peapus, D.H., Chiu, H.J., Campobasso, N., Reddick, J.J., Begley, T.P., and Ealick, S.E. (2001) Structural characterization of the enzyme–substrate, enzyme–intermediate, and enzyme–product complexes of thiamin phosphate synthase. *Biochemistry*, **40**, 10103–10114.

54 Wrenger, C., Knöckel, J., Walter, R.D., and Müller, I.B. (2008) Vitamin B1 and B6 in the malaria parasite: requisite or dispensable? *Braz. J. Med. Biol. Res.*, **41**, 82–88.

55 Eschbach, M.L., Müller, I.B., Gilberger, T.W., Walter, R.D., and Wrenger, C. (2006) The human malaria parasite *Plasmodium falciparum* expresses an atypical N-terminally extended pyrophosphokinase with specificity for thiamine. *Biol. Chem.*, **387**, 1583–1591.

56 Knöckel, J., Bergmann, B., Müller, I.B., Rathaur, S., Walter, R.D., and Wrenger, C. (2008) Filling the gap of intracellular dephosphorylation in the *Plasmodium falciparum* vitamin B1 biosynthesis. *Mol. Biochem. Parasitol.*, **157**, 241–243.

57 Zilles, J.L., Croal, L.R., and Downs, D.M. (2000) Action of the thiamine antagonist bacimethrin on thiamine biosynthesis. *J. Bacteriol.*, **182**, 5606–5610.

58 Lawhorn, B.G., Gerdes, S.Y., and Begley, T.P. (2004) A genetic screen for the identification of thiamin metabolic genes. *J. Biol. Chem.*, **279**, 43555–43559.

59 Tazuya, K., Adachi, Y., Masuda, K., Yamada, K., and Kumaoka, H. (1995) Origin of the nitrogen atom of pyridoxine in *Saccharomyces cerevisiae*. *Biochim. Biophys. Acta*, **1244**, 113–116.

60 Zeidler, J., Ullah, N., Gupta, R.N., Pauloski, R.M., Sayer, B.G., and Spenser, I.D. (2002) 2′-Hydroxypyridoxol, a biosynthetic precursor of vitamins B(6) and B(1) in yeast. *J. Am. Chem. Soc.*, **124**, 4542–4543.

61 Zeidler, J., Sayer, B.G., and Spenser, I.D. (2003) Biosynthesis of vitamin B1 in yeast. Derivation of the pyrimidine unit from pyridoxine and histidine. Intermediacy of urocanic acid. *J. Am. Chem. Soc.*, **125**, 13094–13105.

62 Sherman, I.W. (1979) Biochemistry of *Plasmodium* (malarial parasites). *Microbiol. Rev.*, **43**, 453–495.

63 Müller, I.B., Knöckel, J., Eschbach, M.L., Bergmann, B., Walter, R.D., and Wrenger, C. (2010) Secretion of an acid phosphatase provides a possible mechanism to acquire host nutrients by *Plasmodium falciparum. Cell. Microbiol.*, **12**, 677–691.

64 Schopfer, W.H. (1949) *Plants and Vitamins*, Chronica Botanica Company, Waltham, Mass., USA.

65 Rodionov, D.A., Vitreschak, A.G., Mironov, A.A., and Gelfand, M.S. (2002) Comparative genomics of thiamin biosynthesis in procaryotes. New genes and regulatory mechanisms. *J. Biol. Chem.*, **277**, 48949–48959.

66 Percudani, R. and Peracchi, A.A. (2003) Genomic overview of pyridoxal-phosphate-dependent enzymes. *EMBO Rep.*, **4**, 850–854.

67 Fitzpatrick, T.B., Amrhein, N., Kappes, B., Macheroux, P., Tews, I., and Raschle, T. (2007) Two independent routes of *de novo* vitamin B6 biosynthesis: not that different after all. *Biochem. J.*, **407**, 1–13.

68 Yang, Y., Tsui, H.C., Man, T.K., and Winkler, M.E. (1998) Identification and function of the *pdx*Y gene, which encodes a novel pyridoxal kinase involved in the salvage pathway of pyridoxal 5′-phosphate biosynthesis in *Escherichia coli* K-12. *J. Bacteriol.*, **180**, 1814–1821.

69 Morita, T., Takegawa, K., and Yagi, T. (2004) Disruption of the *plr*1 + gene encoding pyridoxal reductase of *Schizosaccharomyces pombe. J. Biochem.*, **135**, 225–230.

70 Roa, B.B., Connolly, M., and Winkler, M.E. (1989) Overlap between pdxA and ksgA in the complex pdxA-ksgA-apaG-apaH operon of *Escherichia coli* K-12. *J. Bacteriol.*, **171**, 4767–4777.

71 Lam, H.M., Tancula, E., Dempsey, W.B., and Winkler, M.E. (1992) Suppression of insertions in the complex *pdx*J operon of *Escherichia coli* K-12 by Ion and other mutations. *J. Bacteriol.*, **174**, 1554–1567.

72 Padilla, P.A., Fuge, E.K., Crawford, M.E., Errett, A., and Werner-Washburne, M.

(1998) The highly conserved, coregulated SNO and SNZ gene families in *Saccharomyces cerevisiae* respond to nutrient limitation. *J. Bacteriol.*, **180**, 5718–5726.

73 Osmani, A.H., May, G.S., and Osmani, S.A. (1999) The extremely conserved *pyroA* gene of *Aspergillus nidulans* is required for pyridoxine synthesis and is required indirectly for resistance to photosensitizers. *J. Biol. Chem.*, **274**, 23565–23569.

74 Ehrenshaft, M., Jenns, A.E., Chung, K.R., and Daub, M.E. (1998) SOR1, a gene required for photosensitizer and singlet oxygen resistance in *Cercospora fungi*, is highly conserved in divergent organisms. *Mol. Cell*, **1**, 603–609.

75 Ehrenshaft, M., Bilski, P., Li, M.Y., Chignell, C.F., and Daub, M.E. (1999) A highly conserved sequence is a novel gene involved in *de novo* vitamin B6 biosynthesis. *Proc. Natl Acad. Sci. USA*, **96**, 9374–9378.

76 Dong, Y.X., Sueda, S., Nikawa, J., and Kondo, H. (2004) Characterization of the products of the genes SNO1 and SNZ1 involved in pyridoxine synthesis in *Saccharomyces cerevisiae. Eur. J. Biochem.*, **271**, 745–752.

77 Rodriguez-Navarro, S., Llorente, B., Rodriguez-Manzaneque, M.T., Ramne, A., Uber, G., Marchesan, D., Dujon, B. *et al.* (2002) Functional analysis of yeast gene families involved in metabolism of vitamins B1 and B6. *Yeast*, **19**, 1261–1276.

78 Belitsky, B.R. (2004) Physical and enzymological interaction of *Bacillus subtilis* proteins required for *de novo* pyridoxal 5′-phosphate biosynthesis. *J. Bacteriol.*, **186**, 1191–1196.

79 Raschle, T., Amrhein, N., and Fitzpatrick, T.B. (2005) On the two components of pyridoxal 5′-phosphate synthase from *Bacillus subtilis. J. Biol. Chem.*, **280**, 32291–32300.

80 Tambasco-Studart, M., Titiz, O., Raschle, T., Forster, G., Amrhein, N., and Fitzpatrick, T.B. (2005) Vitamin B6 biosynthesis in higher plants. *Proc. Natl Acad. Sci. USA*, **102**, 13687–13692.

81 Burns, K.E., Xiang, Y., Kinsland, C.L., McLafferty, F.W., and Begley, T.P. (2005)

Reconstitution and biochemical characterization of a new pyridoxal-5′-phosphate biosynthetic pathway. *J. Am. Chem. Soc.*, **127**, 3682–3683.

82 Cassera, M.B., Gozzo, F.C., D'Alexandri, F.L., Merino, E.F., del Portillo, H.A., Peres, V.J., Almeida, I.C. *et al.* (2004) The methylerythritol phosphate pathway is functionally active in all intraerythrocytic stages of *Plasmodium falciparum. J. Biol. Chem.*, **279**, 51749–51759.

83 Wrenger, C., Eschbach, M.L., Müller, I.B., Warnecke, D., and Walter, R.D. (2005) Analysis of the vitamin B6 biosynthesis pathway in the human malaria parasite *Plasmodium falciparum. J. Biol. Chem.*, **280**, 5242–5248.

84 Gengenbacher, M., Fitzpatrick, T.B., Raschle, T., Flicker, K., Sinning, I., Müller, S., Macheroux, P. *et al.* (2006) Vitamin B6 biosynthesis by the malaria parasite *Plasmodium falciparum*: Biochemical and structural insights. *J. Biol. Chem.*, **281**, 3633–3641.

85 Knöckel, J., Müller, I.B., Bergmann, B., Walter, R.D., and Wrenger, C. (2007) The apicomplexan parasite *Toxoplasma gondii* generates pyridoxal phosphate *de novo*. *Mol. Biochem. Parasitol.*, **152**, 108–111.

86 Müller, I.B., Das Gupta, R., Lüersen, K., Wrenger, C., and Walter, R.D. (2008) Assessing the polyamine metabolism of *Plasmodium falciparum* as chemotherapeutic target. *Mol. Biochem. Parasitol.*, **160**, 1–7.

87 Bauer, J.A., Bennett, E.M., Begley, T.P., and Ealick, S.E. (2004) Three-dimensional structure of YaaE from *Bacillus subtilis*, a glutaminase implicated in pyridoxal-5′-phosphate biosynthesis. *J. Biol. Chem.*, **279**, 2704–2711.

88 Zhu, J., Burgner, J.W., Harms, E., Belitsky, B.R., and Smith, J.L. (2005)

A new arrangement of (beta/alpha)8 barrels in the synthase subunit of PLP synthase. *J. Biol. Chem.*, **280**, 27914–27923.

89 Zein, F., Zhang, Y., Kang, Y.N., Burns, K., Begley, T.P., and Ealick, S.E. (2006) Structural insights into the mechanism of the PLP synthase holoenzyme from *Thermotoga maritima. Biochemistry*, **45**, 14609–14620.

90 Strohmeier, M., Raschle, T., Mazurkiewicz, J., Rippe, K., Sinning, I., Fitzpatrick, T.B., and Tews, I. (2006) Structure of a bacterial pyridoxal 5′-phosphate synthase complex. *Proc. Natl Acad. Sci. USA*, **103**, 19284–19289.

91 Yang, Y., Zhao, G., and Winkler, M.E. (1996) Identification of the *pdxK* gene that encodes pyridoxine (vitamin B6) kinase in *Escherichia coli* K-12. *FEMS Microbiol. Lett.*, **141**, 89–95.

92 Kerry, J.A., Rohde, M., and Kwok, F. (1986) Brain pyridoxal kinase. Purification and characterization. *Eur. J. Biochem.*, **158**, 581–585.

93 Scott, T.C. and Phillips, M.A. (1997) Characterization of *Trypanosoma brucei* pyridoxal kinase: purification, gene isolation and expression in *Escherichia coli. Mol. Biochem. Parasitol.*, **88**, 1–11.

94 Fonda, M.L. and Harker, C.W. (1982) Metabolism of pyridoxine and protein binding of the metabolites in human erythrocytes. *Am. J. Clin. Nutr.*, **35**, 1391–1399.

95 Müller, I.B., Wu, F., Bergmann, B., Knöckel, J., Walter, R.D., Gehring, H., and Wrenger, C. (2009) Poisoning pyridoxal 5-phosphate-dependent enzymes: a new strategy to target the malaria parasite *Plasmodium falciparum. PLoS ONE*, **4**, e4406.

13
Vitamin Biosynthetic Pathways, the PLP Synthase Complex, and the Potential for Targeting Protein–Protein Interaction

*Ivo Tews** and *Irmgard Sinning*

Abstract

Vitamin biosynthetic pathways may provide targets for the development of anti-parasitic drugs when absent from mammals or parasitic host organisms. The targeting of a specific biosynthetic route requires that the vitamin in question is essential for the parasite's survival. In this chapter, several vitamin biosynthetic pathways are discussed as malarial drug targets. Recent advances have been made in the area of vitamin B_6 biosynthesis by studying a key multienzyme complex, pyridoxal 5-phosphate (PLP) synthase. Today, complex assembly and ensuing enzyme activation have been very well described, through a combination of X-ray crystallography, calorimetry, homology modeling, and kinetic analyses. The study of protein complexes using these techniques may be valuable for drug design applications, specifically for systems where protein–protein interaction can be targeted.

Introduction

Many vitamin biosynthetic routes were first discovered and described during the 1960s and 1970s, mostly in bacterial systems; as a consequence, by the end of the twentieth century many of the essentials of vitaminology had become firmly established. Nonetheless, knowledge in this area remains far from complete, and with the onset of the genomic era new biosynthetic routes continue to be identified and salvage pathways discovered, some of which include intriguing arrays of chemistry [1].

Microorganisms often establish secondary routes for vitamin biosynthesis, or employ mechanisms for the uptake of vitamins via specific transporters (Figure 13.1). The balance between these pathways is controlled by the nutritional state or environmental cues. For example, while vitamin uptake may be sufficient for a basic supply, during phases of critical development an additional biosynthesis may be necessary. Thus, if a vitamin biosynthetic route is to be considered as a drug target, there will be a need to consider other salvage pathways that might obviate a potential treatment.

---

*Corresponding author

Apicomplexan Parasites. Edited by Katja Becker
Copyright © 2011 WILEY-VCH Verlag GmbH & Co. KGaA, Weinheim
ISBN: 978-3-527-32731-7

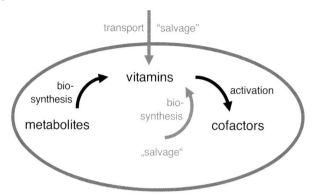

Figure 13.1 A prokaryotic or eukaryotic cell in which vitamin biosynthesis takes place. Salvage pathways or vitamin uptake pathways may also be present.

Several vitamins have the potential to be exploited as drug targets in the Apicomplexa. Recently, details have been provided as to how vitamins B_1 and B_6, as well as folate and pantothenic acid, are important for the development of apicomplexan parasites, such as *Plasmodium* [2, 3] (see also Chapter 12). The following provides a somewhat "patchy" list of malarial infections where either the application of a vitamin (as an enzymatic cofactor) or its biosynthesis have been studied:

- the enzymes dihydropteroate synthase and dihydrofolate reductase in folate biosynthesis [4];
- the biosynthesis of pantothenic acid [5] and uptake of pantothenic acid, vital for parasite survival in the intra-erythrocytic stages [6];
- vitamin B_1 biosynthesis [7] or the uptake of 5-(2-hydroxy-ethyl)-4-methylthiazole for vitamin B_1 biosynthesis [8];
- cobalamins with antimalarial activity through a postulated inhibitory effect on β-hematin formation [9], or the vitamin B_{12}-dependent enzyme methionine synthase [10];
- vitamin B_6/PLP biosynthesis [11–13] or the requirement for a PLP cofactor in amino acid metabolism [14, 15]; and
- the requirement for vitamins A, C, and E to counteract severe oxidative stress during the plasmodial intra-erythrocytic stages [16].

Interestingly, vitamin B_6 also functions as a potent antioxidant [17], and *Plasmodium* is able to synthesize B_6 [12, 13] but not vitamins A, C, and E.

Vitamin B_6 is a vitamer, and collectively refers to three different forms, namely pyridoxal (PL), pyridoxamine (PM), and pyridoxine (PN), all of which can be taken up by membrane transporters. Within the cell, the phosphorylated forms shown in Figure 13.2 are generated. Pyridoxal 5-phosphate (PLP) and pyridoxamine 5-phosphate (PMP) constitute the active cofactor forms.

The vitamin B_6 biosynthetic pathway in *Plasmodium* requires the enzymes Pdx1 and Pdx2 [12, 13], while the kinase PdxK establishes the salvage pathway to convert

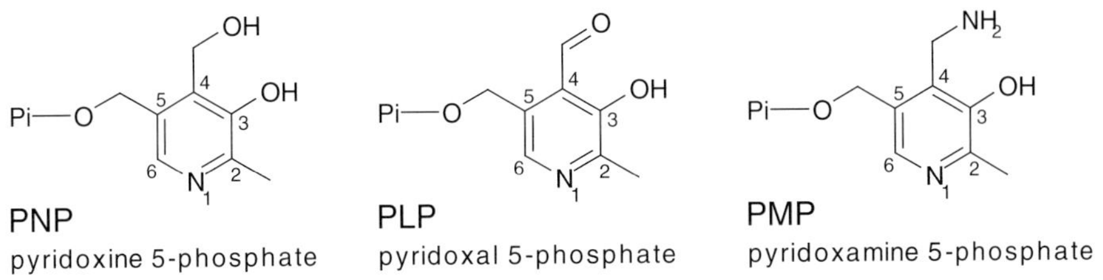

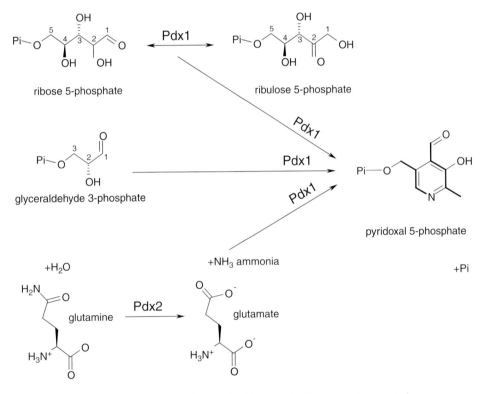

PNP
pyridoxine 5-phosphate

PLP
pyridoxal 5-phosphate

PMP
pyridoxamine 5-phosphate

Figure 13.2 The phosphorylated forms of the vitamer B$_6$. Pi, inorganic phosphate.

the vitamin to the cofactor [12]. Both pathways have been investigated as potential drug targets [2, 3, 15].

The biosynthesis of PLP, as summarized in Figure 13.3, requires the two carbohydrates glyceraldehyde 3-phosphate and ribose 5-phosphate [18, 19], in addition to a nitrogen source for the ring nitrogen in the PLP heterocycle. Hence,

ribose 5-phosphate

ribulose 5-phosphate

glyceraldehyde 3-phosphate

pyridoxal 5-phosphate

+H$_2$O

+NH$_3$ ammonia

+Pi

glutamine

glutamate

Figure 13.3 The main enzymatic route of pyridoxal 5-phosphate biosynthesis uses the synthase enzyme Pdx1 and the glutaminase enzyme Pdx2 in a heteromeric amidotransferase complex. Pdx2 hydrolyses L-glutamine to yield glutamate and ammonia. Pdx1 incorporates the ammonia generated by Pdx2 to yield PLP from substrates ribose 5-phosphate and glyceraldehyde 3-phosphate. A side reaction is the isomerization of ribose 5-phosphate to ribulose 5-phosphate.

PLP synthase is a glutamine amidotransferase (GATase), capable of transferring a nitrogen atom – in the form of ammonia – to an acceptor molecule [20]. The GATases have an elaborate molecular architecture in order to sequester ammonia from the aqueous environment, while in PLP synthase Pdx1 and Pdx2 form an enzyme complex in which Pdx2 hydrolyzes glutamine to yield ammonia, which is transferred through the interior of the Pdx1 protein to the catalytic center where PLP is generated. This universal pathway is found in Plantae [21], yeasts [22], unicellular organisms such as the Apicomplexa [11–13], and in most bacteria [23]. (Note: a less common pathway in the γ-division of proteobacteria, best described for *Escherichia coli*, leads to the generation of PNP [19, 24].)

The Pdx1 and Pdx2 enzymes do not simply heterodimerize, but rather form a 24-subunit complex. The regulation of enzymatic activity during complex assembly offers unique opportunities in drug design, as interference with the ordered process would lead to an inactive enzyme. However, this is an ambitious target when faced with the general criticism of how to achieve affinity and selectivity for an inhibitor targeted at a protein interface. In contrast to active site pockets or allosteric sites, protein interfaces are large, topologically complex, and often devoid of specific pockets in which inhibitors may be anchored [25].

The problem is presented here as a study of protein–protein interaction in the PLP synthase enzyme complex, mainly by using techniques of calorimetry and X-ray crystallography. Although these studies were undertaken to understand the complex assembly and activation of PLP synthase, the way in which the techniques can be used to generate knowledge that might potentially be useful for drug design is also discussed.

Protein–Protein Interaction

Targeting the Interaction

The classic approach to designing drugs that target enzymes is to make use of the enzyme's active site or substrate-binding pocket, or of an allosteric binding site. This requires an understanding of the precise binding mode of the substrate, and also knowledge of the mechanistic details of catalysis in order to identify substrate analogs, transition-state analogs, or other inhibitors that might serve as lead components of the design cycle. Such a strategy can lead to small, bioavailable, and potent compounds. Protein interactions in multiprotein systems, however, pose the problem of how to design specific compounds when no specific, discrete binding pockets can be addressed, because the large interface area often lacks such pronounced features. A structural knowledge of both the autonomous proteins, as well as of the protein interactions, is a crucial advantage in these systems (see Ref. [26] for a roadmap to inhibitor design).

Proteins interact with each other via a specialized subset of amino acids, the most notable (statistically) of which are proline, isoleucine, tyrosine, tryptophan, aspartate, and arginine. This information can be used in the design of peptidomimetic drugs [27]. The same knowledge can also be fed back to create new libraries enriched with compounds geared toward protein–protein interactions [28]. The art of drug design is often to move away from the peptidomimetics in order to identify chemically equivalent compounds that can bind to the protein surfaces [29].

In the past, several reviews have outlined the problem of designing inhibitors of protein–protein interactions [28–33]. In particular, study cases have included protein–protein interactions in signaling cascades disrupted by inhibitor action, whereby the development of commercial drugs has, on occasion, been successful. Examples of this include:

- Cancer: blockade of the interaction of the regulatory protein MDM2 with the tumor suppressor p53 [34, 35]; interference with the interaction of the Bcl [36–38] or X-IAP families [39] of anti-apoptotic signaling molecules; and blocking chaperone action, for example, through Hsp90 [17].
- Conditions such as asthma, arthritis, heart disease and cancer: targeting the integrin cell surface receptor system [40, 41].
- Rheumatic conditions: changes in inflammatory responses via tumor necrosis factor α (TNFα), the functional trimeric state of which can be disrupted by inhibitors [42].
- Immunosuppression: disrupting cytokine action via interleukin-2, for example, in organ transplant rejection [43, 44], or targeting the calcineurin network in inflammatory or autoimmune responses [45].
- Pathogenesis, diabetes, and a variety of other human disease conditions: targeting the JNK pathway [46–48].
- Development of novel antibiotics: targeting the interaction between FtsZ and ZipA required during bacterial cell division [49, 50].

PLP synthase is a multicomponent enzymatic system requiring protein–protein interaction for catalytic function. Two examples involve similar problems in enzymology:

- Non-insulin diabetes: glycogen phosphorylase, characterized by X-ray crystallography with an inhibitor bound neither in the catalytic center nor in the previously characterized allosteric site capable of changing the equilibrium in favor of the low activity T-state over the active R-state [51].
- Inflammatory and autoimmune responses: dimeric nitric oxide synthase (NOS; oxidizes arginine to yield the signaling molecule NO), where the inhibitor binds to the active site pocket in a noncompetitive manner in order to convert the enzyme into an inactive, monomeric species [52].

The Pdx1 and Pdx2 proteins form many well-characterized interactions when they form the active enzyme complex. PLP synthase thus presents itself as a case for the specific design of protein-associated inhibitors.

The Structure of PLP Synthase

Details of the molecular assembly of PLP synthase are known from X-ray crystallographic three-dimensional (3-D) structural determinations (Figure 13.4). The

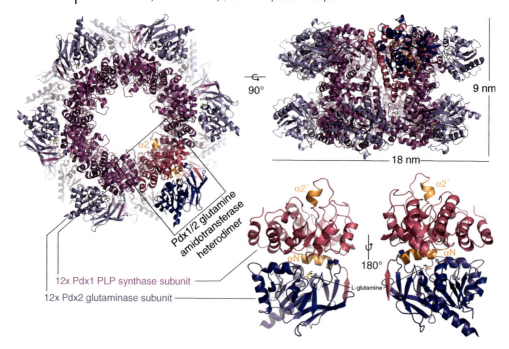

Figure 13.4 Structural model of *Plasmodium falciparum* PLP synthase, where the Pdx1 subunits are based on a homology model after *Bacillus subtilis* Pdx1 (PDB entry 2NV2; [56]), and the Pdx2 subunits were generated by superposition with the experimentally determined Pdx2 structure from *Plasmodium falciparum* (PDB entry 2ABW; [13]); the model has been manually updated and energy-minimized [64]. The Pdx1 subunits form the dodecameric core, consisting of two hexameric rings (magenta and red). Pdx1 obtains a $(\beta\alpha)_8$/barrel or TIM-barrel fold assembling into hexameric rings, and two such rings then form the dodecameric core particle. Up to 12 Pdx2 glutaminase subunits (blue) attach to the Pdx1 synthase core. The Pdx2 subunits only contact Pdx1 subunits, and not each other. The Pdx2 substrate L-glutamine is seen bound in the active site of Pdx2 at the Pdx1/Pdx2 interface. Highlighted in orange are the helix αN of Pdx1, which is responsible for the main interaction between the two proteins, and helix $\alpha2'$, which is required for catalysis. These two helices were seen ordered in the Pdx1/Pdx2 complex, but disordered in autonomous Pdx1 in the *Bacillus subtilis* structures [56].

structures determined from bacteria include: *Geobacillus stearothermophilus* Pdx1$_{12}$ (PDB:1ZNN) [53]; *Thermotoga maritima* Pdx1$_{12}$Pdx2$_{12}$ (PDB:2ISS) [54]; *Bacillus subtilis* Pdx2 (PDB:1R9G and PDB:2NV0), Pdx1$_{12}$ (PDB:2NV1), and Pdx1$_{12}$Pdx2$_{12}$ (PDB:2NV2) [55, 56]; *Methanocaldococcus jannaschii* Pdx1$_{12}$ (PDB:2YZR); and *Thermus thermophilus HB8* Pdx2 (PDB:2YWD) and Pdx1$_{12}$ (PDB:2ZBT). Eukaryotic PLP synthase subunits have also been determined for *Plasmodium falciparum* Pdx2 (PDB:2ABW) [13] and *Saccharomyces* Pdx1$_6$ (PDB:3FEM) [57].

The PLP synthase contains a core of 12 Pdx1 subunits that form a cylinder, while up to 12 Pdx2 subunits attach to the Pdx1 core [54, 56]. Although it is unclear whether all of the Pdx2 binding sites must be occupied, crystallographic structures have demonstrated the presence of fully decorated particles.

Methods for Characterization of Protein–Protein Interaction

In addition to the crystallographic structural determinations, PLP synthase complexes have been studied using gel filtration and size-exclusion chromatography (SEC) [57–59], optionally with static light scattering to minimize the influence of the molecular shape on the determined molecular weight [58, 59], and by analytical ultracentrifugation [53, 56, 57]. An introduction to these techniques is provided in Ref. [60] for protein–ligand interactions. The techniques described are equally valid for the study of protein–protein interactions.

The oligomeric nature of the Pdx1 subunit of PLP synthase has been assessed using analytical ultracentrifugation. The experiments carried out describe a hexamer–dodecamer equilibrium [53] with a fitted K_d of 3.6 µM for the dissociation of *Bacillus subtilis* $Pdx1_{12}$ into hexamers [56]. It is unclear whether the dodecamer is required for catalytic activity, but the results of studies with the Pdx1 protein from *Saccharomyces cerevisiae* (also called Snz1 [57]), have suggested that the hexamers should be catalytically competent. The use of analytical ultracentrifugation also showed the yeast Pdx1 to be mainly hexameric, with a significantly higher K_d for the yeast Pdx1 dodecamer compared to that of the bacterial enzymes [57]. Consistently, the reported 3-D structure is also hexameric [57]. In order to understand the differences between bacterial and yeast proteins with respect to their oligomerization behavior, a structural comparison was carried out. This led to the identification of a single amino acid insertion in the yeast sequences, that was suggested to alter the dodecamer interface. However, when this amino acid was inserted into the bacterial protein, a significant (fivefold) lowering of the interaction occurred.

Analytical ultracentrifugation was also used to study the association of Pdx2 subunits with the Pdx1 oligomer. In this case, whereas the *Bacillus subtilis* Pdx1 dodecamer showed a well-defined molecular weight, the Pdx1/Pdx2 complex showed a distribution over a wide molecular weight range. These data suggested that the association of Pdx2 with the Pdx1 core was noncooperative, with individual Pdx2 subunits being attached independently of each other [56].

Protein Structure and the Early Stages of Drug Design

The techniques described until now (i.e., crystallography, analytical ultracentrifugation, SEC, static light scattering) describe the oligomeric state of a protein. Consequently, if various oligomeric species occur they can be analyzed and, in some cases, also separated. Likewise, these techniques can be used to study the complex formation of two or more interacting proteins.

Proteins are not rigid entities, however, and protein interaction is usually accompanied by some structural change(s) [61]. In fact, there will be an equilibrium of stable states for any given protein, and the conversion between states will be driven by thermal motion. In fact, thermal motion not only affects the displacement of amino acid side chains, but also involves the movement of, or changes in, both secondary structures and entire folded domains.

Conformational change is well-known in catalysis, and there is a particular requirement for it in enzymatic transition-state theory [62], even though it is difficult to predict and may even be ignored in classical *in silico* drug design [63]. For the study of protein–protein interactions, it is appropriate to map the structural changes that occur upon protein complex formation, and better yet to aim to understand what drives the observed structural changes. Several structures of autonomous proteins and protein complexes are required, and a description of the thermodynamics of complex formation would also be helpful. Such knowledge can be acquired using calorimetric methods.

Isothermal Titration Calorimetry

Several studies using isothermal titration calorimetry (ITC) have been used to characterize PLP synthase complexes [64–66]. The advantage is that ITC allows a thermodynamic profile to be obtained of an interaction, provided that this interaction produces a measurable heat signal [67, 68]. This is usually the case for interactions between two proteins, between ligands and receptors, or between inhibitors and enzymes. (Note: The details of recent ITC applications are available in the annual accounts of the *Journal of Molecular Recognition* [69].)

In these experiments, one sample is titrated against the other in a series of individual injections, thereby sampling across the stoichiometric point of the interaction (Figure 13.5). The recorded data are deconvoluted to give least-square fits to ΔS and ΔH of the interaction, as well as the binding constant K_a and the stoichiometry N. The binding constant determined is an association constant K_a that can easily be inverted to give the customary dissociation constant, K_d. The free Gibbs energy ΔG is then calculated from the fitted binding parameters [70].

Not all interactions can be measured with sufficient accuracy, however. For example, very weak interactions will lead to a shallow binding curve that cannot be fitted, whereas in very tight interactions (nearly) all of the titrated ligands will be bound at first (until the stoichiometric ratio is reached), after which significant binding is no longer observed and the resulting rectangular binding curve cannot be fitted. The range between the two extremes is known as the *K*-window [67], and is expressed by the c-value, which relates the total amount of protein to the association constant K_a (Figure 13.5). In practice, measurements for c-values between 1 and 1000 are typically possible; however, when converted to dissociation constants, determination in the K_d range between approximately 10 nM and about 1 mM is usually possible.

A full determination of the thermodynamic profile of an interaction includes an assessment of the temperature dependence of ΔH, which allows the heat capacity change ΔC_P of the interaction to be derived (Figure 13.5). Negative heat capacity

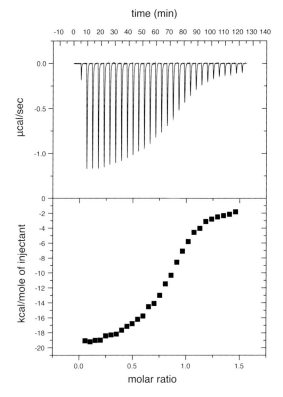

some basic thermodynamics:

$$\Delta G = \Delta H - T \Delta S$$
$$\Delta G = -RT \ln K_a$$
$$= RT \ln K_d$$
$$\text{and} \quad K_d = 1 / K_a$$

the c-value:

$$c = K_a \, M_{tot} \, N$$
where
$$K_a = \text{association constant}$$
$$M_{tot} = \text{total protein}$$
$$N = \text{stoichiometry}$$

heat capacity ΔC_P:

$$\Delta C_P = d(\Delta H) / dT$$
$$= T \, (d(\Delta S) / dT)$$

Figure 13.5 Typical data from an ITC experiment. The top panel shows the binding heats determined from mixing two proteins; the area under the peaks is integrated and plotted, as seen in the lower panel. The 29 injections of Pdx2 into the measuring cell with Pdx1 (using proteins from *Bacillus subtilis*) obtain a sigmoidal binding curve, titrating across the stoichiometric binding point, and would approach zero when fully titrated and corrected for dilution heats. Also given are some thermodynamic relations required for data analysis.

changes indicate hydrophobic/apolar interactions, whereas polar interactions are reflected in positive heat capacity changes [71].

Optimizing Entropic and Enthalpic Binders in Drug Design, Using ITC

The optimization of protein–ligand binding in drug design is not always intuitive, and often lacks suitable knowledge for predicting interaction entropies. Nonetheless, calorimetry may not only fill this gap but also present itself as an ideal technique for rational drug design [72]. Whilst there is also scant understanding of protein plasticity [61], calorimetry in combination with 3-D structural data can compensate for the lack of predictability. For these reasons, ITC has become well established in drug design [73, 74]; notably, it has been instrumental in understanding the interaction between drugs and target DNAs (for a review, see Ref. [75]).

Data acquired from the plasmepsins, which are essential proteases of the *Plasmodium* food vacuole, have allowed an introduction of the enthalpy/entropy funnel concept in drug design [76, 77]. The proposal was made as to how a determination of the thermodynamic parameters ΔH and ΔS would be much more valuable than simply optimizing the binding on the basis of the affinity K_a alone [78]. The model states that low affinity is realized by almost any combination of ΔH and ΔS, but predicts that high binding affinities inevitably fall into a narrowed enthalpy/entropy relationship. This leads to a requirement to balance the thermodynamic parameters, avoiding extremes that cannot be optimized. This concept has proved valid in the enthalpic optimization of HIV-1 protease inhibitors [79].

PLP Synthase and Characterization of its Protein Interfaces by ITC

Isothermal calorimetry is often applied to develop inhibitors that bind in the active sites of enzymes. Whilst the current knowledge of PLP biosynthesis remains fragmentary and is under intense investigation [18, 19], ITC has not yet been employed in the design of specific inhibitors to target the active site of PLP synthase. As a technique, however, ITC is ideally suited to acquire further understanding of protein complex formation, with such knowledge in turn being applied to the design of inhibitors against protein–protein interaction.

Structural Changes Accompany Protein Complex Formation

Several protein structures of the Pdx1 and Pdx2 proteins have formed the basis for an interpretation of the ITC data. In this case, *Bacillus subtilis* PLP synthase is used as a reference, since the isolated Pdx1 and Pdx2 proteins have been studied using X-ray crystallography, while details of the Pdx1/Pdx2 complex have also been investigated (see Figure 13.4). In this example, the complex formation is accompanied by marked structural changes.

The N terminus of Pdx1 is disordered in autonomous Pdx1, but is ordered in the Pdx1/Pdx2 PLP synthase complex (Figure 13.4; see Ref. [56]). This segment is directly involved in complex formation, and forms the small beta-strand βN, and a three-turn alpha helix named αN. Other structural changes in Pdx1 are understood to accompany active site formation, and thus will have a direct impact on catalysis. The active site is covered by helix α2′; as with the N-terminal segment, the helix is disordered in autonomous Pdx1 but ordered in the Pdx1/Pdx2 PLP synthase complex [56]. The C terminus, which is approximately 30 amino acids in length, is usually disordered in PLP synthase structures; however, when partly seen (as in the *Saccharomyces* Pdx1 structure), it contacts the active site of a neighboring Pdx1 subunit in the hexameric ring [57]. Subsequently, with this contact, Pdx1 establishes cooperativity [80]. Both, the α2′ and the C-terminal segment may have a function in secluding the active site from the aqueous environment. Finally, activation of the glutaminase Pdx2 by contact with the Pdx1 PLP synthase subunit has been documented [13, 23].

Since the described structural changes are likely provoked by complex formation between Pdx1 and Pdx2, it is essential to understand the cause and effect. Structural changes can be explained by combining the structural analysis with kinetic studies and ITC.

Setting Up a Test Case: The Problem with PLP Synthase

Purified *Plasmodium* Pdx2 is inactive in solution, but gains activity when the synthase subunit Pdx1 is added [13]. The crystallographic 3-D structure of the *Plasmodium falciparum* Pdx2 glutaminase shows the enzyme in an inactive state (Figure 13.6a; see Ref. [13]). The oxyanion hole – which is a crucial feature of the active site of glutaminase – was not formed, and this leads to the hypothesis that the interaction between the two proteins will result in structural changes that culminate in the formation of an oxyanion hole structure.

During catalysis, the oxyanion hole takes up the partial negative charge of the substrate transition state. The theory of oxyanion hole formation was developed by analyzing proteases [81, 82] which, like glutaminases, have a catalytic triad mechanism and follow a similar catalytic scheme. Knowledge of the transition-state theory for this class of enzymes plays a critical role, not only in understanding catalysis but also in inhibitor design [83].

An analysis of the structure of a bacterial homolog from *Bacillus subtilis* [56] showed, simultaneously, both inactive and active forms of the enzyme in the crystal. This is possible because two Pdx2 proteins comprised the unit forming the crystal (also called the asymmetric unit), and the two proteins were found to differ

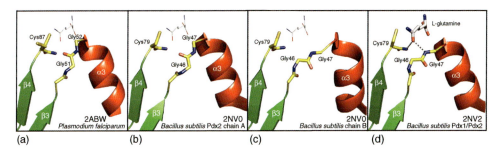

Figure 13.6 Activation of the glutaminase subunit Pdx2 occurs through a peptide-flip mechanism. (a) The structure of *Plasmodium falciparum* Pdx2 in an inactive conformation (PDB entry 2ABW; [13]). The catalytic Cys87 was observed with two alternative side chain conformations, and the carbonyl oxygen of the Gly51–Gly52 peptide is pointed toward the putative oxyanion hole and would clash with the oxygen of the substrate L-glutamine (shown as lines); (b, c) The equivalent region of *Bacillus subtilis* Pdx2 (PDB entry 2NV0; [56]). The crystal asymmetric unit contained two protein chains, which are distinguished by different conformations in the oxyanion hole-forming region. While chain A shows – similar to the *Plasmodium* structure – an obstructed oxyanion hole, this hole is formed in chain B by a reorientation of the Gly46–Gly47 peptide. In the latter conformation, the peptide nitrogen points toward the substrate and allows for its binding; (d) The active conformation is observed in the Pdx1/Pdx2 complex with bound glutamine substrate (PDB entry 2NV2; [56]).

structurally in this critical region. Notably, the oxyanion hole was not formed in one protein (Figure 13.6b), but was formed in the second protein (Figure 13.6c), with only the latter conformation allowing the substrate (L-glutamine) to bind. This critical change of a peptide close to the Pdx2 catalytic center is termed a "peptide-flip" (Figure 13.6b and c, [13, 56]). For the structure of the *Bacillus subtilis* Pdx1–Pdx2 PLP synthase complex, only active Pdx2 protein subunits with the bound substrate glutamine were present [56]. This was similarly the case for the structure of the second bacterial complex from *Thermotoga maritima*, although in this case no substrate glutamine was observed bound to the enzyme [54].

A simplistic model can be generated from the 3-D structures, postulating how protein–protein interaction can lead to a concomitant activation of the glutaminase Pdx2. In the apo-state, the Pdx2 enzyme has an obstructed oxyanion hole, and the loop between α1 and β1 is not well ordered (Figure 13.7a). The main player in protein–protein interaction is the N-terminal helix αN of Pdx1, which provokes

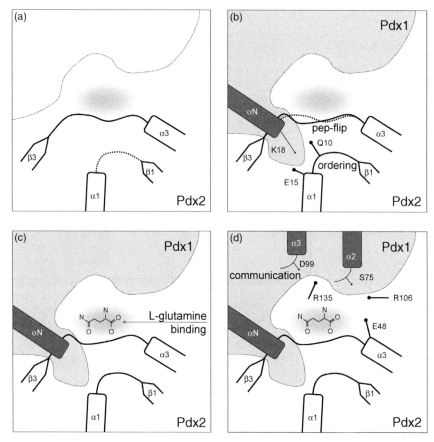

Figure 13.7 Understanding glutaminase activation by PLP synthase complex formation. For details, see the text.

ordering of the loop between α1 and β1 of Pdx2 (Figure 13.7b). Residue Gln10 is now seen to interact with the oxyanion-forming region, and causes the peptide-flip between the two glycine residues to form the oxyanion hole and to allow substrate L-glutamine to bind. The problem is now defined as follows:

- The 3-D structures of PLP synthase suggest, but do not prove, that structural elements are required for enzyme activation and oxyanion hole formation. How can this information be obtained?
- Must the Pdx1–Pdx2 complex form before L-glutamine can bind, or can L-glutamine bind to Pdx2 and the Pdx2–Gln complex subsequently bind to Pdx1?

Thermodynamic Profile of the Pdx1/Pdx2 Protein-Protein Interaction

To address these questions, the binding stoichiometry N, the association constant K_a, the free Gibbs energy ΔG, the entropy ΔS, and the enthalpy ΔH, as well as the heat capacity change ΔC_p were determined by ITC for the Pdx1/Pdx2 interaction. ΔC_p is derived from a set of measurements at different temperatures, and describes the temperature dependence of enthalpies ΔH (see Figure 13.5). All data must be corrected for protonation enthalpies, requiring measurement of the interaction in different buffer systems. Full thermodynamic descriptions were obtained for the interaction of Pdx1 and Pdx2 from *Bacillus subtilis* [65] and *Plasmodium falciparum* PLP synthases [64].

   To give some impression of the amount of material and time required for mapping the interactions, some data are provided here. The calorimeter used in the studies was a VP-ITC instrument (MicroCal, Inc., now GE Healthcare). All experiments were conducted in triplicate, with one additional measurement being necessary to obtain a buffer control value (titrant against buffer only). For the determination of ΔC_p, ITC experiments were carried out at seven different temperatures, ranging from 5 to 37 °C. To determine the buffer protonation enthalpies, measurements at a single temperature (25 °C) were carried out in three buffer systems with different ionization enthalpies. All measurements were carried out with Pdx2 as the titrant; this required about 450 μl of a 250 μM solution of Pdx2, and 2.2 ml of a 30 μM solution of Pdx1. To obtain the full dataset, about 100 mg of Pdx2 and 50 mg of Pdx1 were required. On average, two to three ITC measurements can be carried out per day (the total required net instrument time then calculates as two to three weeks). For PLP synthase, these figures are doubled, as measurements are required in both the absence and presence of the substrate L-glutamine.

   One experimental complication is caused by the fact that Pdx2 is an enzyme, and the binding of substrate L-glutamine is essential to map the interaction of Pdx1 with Pdx2. Therefore, the investigation was carried out with variant Pdx2, where the histidine of the catalytic triad was exchanged with asparagine. This variant could bind substrate, but had no measureable enzymatic activity. With active enzymes, the heat of the interaction would be shadowed by the reaction enthalpy of glutamine hydrolysis.

Similarities and Differences in PLP Synthase Complex Formation

In the first study, the bacterial system from *Bacillus subtilis* was analyzed [65], but later a thermodynamic profile was reported for the *Plasmodium falciparum* system [64]. The two enzyme systems were compared in a subsequent study. Finally, several protein variants were generated in *Bacillus subtilis* to probe for the communication network between the proteins [66]. Since 3-D structural knowledge was available for the bacterial PLP synthase complex [56], the thermodynamic profile of interaction between Pdx1 and Pdx2 proteins could be related to its structure.

Comparison of the thermodynamic parameters [65] for complex formation in the absence and presence of substrate glutamine for the bacterial system showed an approximately 23-fold tighter interaction of Pdx1 with Pdx2 when the substrate was present. PLP synthase has a dissociation constant K_d of $0.3\,\mu M$ in the presence of substrate, but only $6.9\,\mu M$ in its absence. Fitted stoichiometries of $N = 0.5$ indicated that the complex does not fully form under experimental conditions without glutamine. In the presence of glutamine, an ordering of the complex was observed, considering enthalpies and entropies of the interaction. It is intuitive from the positioning of the glutamine binding sites at the protein interface how glutamine "glues" the complex together.

The 3-D structures [54, 56] suggested how helix αN on Pdx1 forms the main interaction of the Pdx1/Pdx2 interface (Figure 13.4 and 13.7b). The importance of the contacts made by helix αN were fully appreciated when studying the thermodynamic profile of *Plasmodium* PLP synthase, and comparing it to the earlier data from *Bacillus* [64]. The differences seen in the ΔC_p values were explained solely by different interactions around αN, with a critical exchange of a salt bridge in the *Bacillus* proteins by a hydrophobic interaction in the *Plasmodium* proteins. This leads to differences in the determined heat capacities in absence of substrate L-glutamine, and these differences vanish in the Michaelis complexes when glutamine was present. Helix αN thus is critical for the formation of the initial encounter complex, but other interactions become important when the complex tightens in the presence of L-glutamine to prepare for catalysis. Since no 3-D structure of the plasmodial PLP synthase complex was available, the study used a homology model (see Figure 13.4), which illustrated how the thermodynamic data experimentally underpinned the homology modeling.

Understanding the Communication of Proteins

Although helix αN is central for complex formation in PLP synthases (Figure 13.7b), this helix does not directly form the active site. Hence, it cannot be explained how the glutaminase Pdx2 is activated by the contact, despite the suggestion that ordering of the loop between $\alpha 1$ and $\beta 1$ and the peptide-flip of the oxyanion hole-forming region occur. To test the influence of αN on complex formation and enzyme activation, the coding sequence for the helix was swapped from the bacterial to the plasmodial gene, so that a chimeric plasmodial Pdx1 protein resulted [64]. This chimera was no longer able to interact with plasmodial Pdx2; instead, it interacted with the bacterial Pdx2 protein. In contrast to wild-type proteins, an absence or presence of substrate glutamine did not alter the catalytic parameters, and thus the swap-variant was no longer able to sense the presence of substrate. The communication network between the two proteins was disturbed in the chimeric complex.

An analysis of individual interactions on the protein interface was then performed by: (i) replacing a particular interacting residue, usually with alanine; (ii) determining the glutaminase catalytic activities; and (iii) comparing the thermodynamic footprint of the variant with that of wild-type protein. This study was carried out with the *Bacillus* system [66], with the probed amino acids at the interface being broadly classified into three categories:

- Exchanges that lead to a loss of interaction, for example, $Pdx2_{R106A}$ or $Pdx2_{R135A}$.
- Exchanges that allow interaction but do not respond to glutamine substrate, for example, $Pdx1_{D99A}$.
- Exchanges that allow interaction but do not respond to glutamine substrate or show glutaminase activity, for example, $Pdx2_{E48A}$.

The updated model of protein–protein interaction in PLP synthase suggests that helix αN is required for interaction of the two proteins, and this interaction prepares the glutaminase for catalytic activity, promoting formation of the oxyanion hole as a prerequisite to glutamine substrate binding (Figure 13.7c). However, this process is not efficient, and an ordered interaction around $Pdx1_{D99}$ is required to attain catalytic activity, also requiring residues $Pdx2_{R106A}$ and $Pdx2_{R135A}$ to support the interaction (Figure 13.7d). The $Pdx2_{E48A}$ exchange demonstrates the requirement for spatial coordination: the residue helps set up a water network around the substrate and residue $Pdx2_{R106}$, and thus organizes the active site and the protein interaction. Without this residue no catalytic activity is measured, despite the fact that Glu48 is not a catalytic residue. Finally, glutamine hydrolysis by Pdx2 is transmitted through $Pdx1_{S75}$ and $Pdx1_{D99}$ to the PLP synthase center (Figure 13.7d).

The PLP synthase example illustrates that crystallographic 3-D structures do not convey the full information required to understand structural change. However, this information is essential in order to understand protein–protein interaction and, in this case, enzyme activation. This is despite the fact that in the case of PLP synthase, the structures of the inactive and active forms were known. Only thermodynamic data, combined with site-directed mutagenesis and kinetic analysis, allowed a proper description of the contribution of individual amino acids to the interaction of the proteins. By using this knowledge, an insightful design of inhibitors that bind at the PLP synthase protein interface might be attempted.

Conclusion: Chances for Drug Design

In this chapter, vitamin biosynthesis, the study of protein–protein interactions, and structure-based drug design have been connected. The malarial PLP synthase multienzyme complex illustrates how a protein–protein interaction can be studied by using a combination of X-ray crystallography, homology modeling, and calorimetry.

PLP synthase is an interesting example of an enzyme where a catalytic center lies on a protein interface, thus opening the opportunity for intervention with specific inhibitors that target protein–protein interaction. The calorimetric characterization is helpful for understanding the forces that drive protein–protein interaction, and for ensuing structural change, both of which are essential prerequisites for a drug design project. As such, the methodology described is widely applicable for systems involving protein–protein interactions.

The goal of designing inhibitors against protein–protein interaction would seem a difficult one, as protein interfaces are usually large and lack suitable high-affinity and selective sites for the binding of small molecules [42]. The general criticism is that the resultant inhibitors tend to be complex with respect to the number of stereocenters and ring types, leading to difficult syntheses [25]. From this, it follows that these inhibitor types are under-represented in corporate screening libraries, and that this lack should be made up for by insightful design [28]. In the context of the design process, this would underline arguments that are in favor of using the full biophysical orchestra of techniques [84, 85]. Furthermore, protein–protein interaction inhibitors always embody the issue of selectivity and thus, of toxicity. In the PLP synthase case, the main Pdx1–Pdx2 interaction motif is helical, and examples of where such an interaction has been successfully abstracted into a small molecule to efficiently replace the interaction have been reported [35, 37]. As such, Pdx1 presents a unique opportunity for continued study in this direction.

Acknowledgments

This work was supported by the Deutsche Forschungsgemeinschaft (IT 368/4).

References

1 Begley, T.P., Chatterjee, A., Hanes, J.W., Hazra, A., and Ealick, S.E. (2008) Cofactor biosynthesis – still yielding fascinating new biological chemistry. *Curr. Opin. Chem. Biol.*, **12**, 118–125.

2 Muller, S. and Kappes, B. (2007) Vitamin and cofactor biosynthesis pathways in *Plasmodium* and other apicomplexan parasites. *Trends Parasitol.*, **23**, 112–121.

3 Muller, I.B., Hyde, J.E., and Wrenger, C. (2010) Vitamin B metabolism in *Plasmodium falciparum* as a source of drug targets. *Trends Parasitol.*, **26**, 35–43.

4 Yuthavong, Y., Kamchonwongpaisan, S., Leartsakulpanich, U., and Chitnumsub, P. (2006) Folate metabolism as a source of molecular targets for antimalarials. *Future Microbiol.*, **1**, 113–125.

5 Spry, C., Kirk, K., and Saliba, K.J. (2008) Coenzyme A biosynthesis: an antimicrobial drug target. *FEMS Microbiol. Rev.*, **32**, 56–106.

6 Kirk, K. and Saliba, K.J. (2007) Targeting nutrient uptake mechanisms in *Plasmodium. Curr. Drug Targets*, **8**, 75–88.

7 Bozdech, Z. and Ginsburg, H. (2005) Data mining of the transcriptome of *Plasmodium falciparum*: the pentose phosphate pathway and ancillary processes. *Malar. J.*, **4**, 17.

8 Wrenger, C., Eschbach, M.L., Muller, I.B., Laun, N.P., Begley, T.P., and Walter, R.D.

(2006) Vitamin B1 de novo synthesis in the human malaria parasite *Plasmodium falciparum* depends on external provision of 4-amino-5-hydroxymethyl-2-methylpyrimidine. *Biol. Chem.*, **387**, 41–51.

9 Chemaly, S.M., Chen, C.T., and van Zyl, R.L. (2007) Naturally occurring cobalamins have antimalarial activity. *J. Inorg. Biochem.*, **101**, 764–773.

10 Krungkrai, J., Webster, H.K., and Yuthavong, Y. (1989) Characterization of cobalamin-dependent methionine synthase purified from the human malarial parasite, *Plasmodium falciparum*. *Parasitol. Res.*, **75**, 512–517.

11 Knockel, J., Muller, I.B., Bergmann, B., Walter, R.D., and Wrenger, C. (2007) The apicomplexan parasite *Toxoplasma gondii* generates pyridoxal phosphate de novo. *Mol. Biochem. Parasitol.*, **152**, 108–111.

12 Wrenger, C., Eschbach, M.L., Muller, I.B., Warnecke, D., and Walter, R.D. (2005) Analysis of the vitamin B6 biosynthesis pathway in the human malaria parasite *Plasmodium falciparum*. *J. Biol. Chem.*, **280**, 5242–5248.

13 Gengenbacher, M., Fitzpatrick, T.B., Raschle, T., Flicker, K., Sinning, I. *et al.* (2006) Vitamin B6 biosynthesis by the malaria parasite *Plasmodium falciparum*: biochemical and structural insights. *J. Biol. Chem.*, **281**, 3633–3641.

14 Amadasi, A., Bertoldi, M., Contestabile, R., Bettati, S., Cellini, B. *et al.* (2007) Pyridoxal 5′-phosphate enzymes as targets for therapeutic agents. *Curr. Med. Chem.*, **14**, 1291–1324.

15 Muller, I.B., Wu, F., Bergmann, B., Knockel, J., Walter, R.D., Gehring, H., and Wrenger, C. (2009) Poisoning pyridoxal 5-phosphate-dependent enzymes: a new strategy to target the malaria parasite *Plasmodium falciparum*. *PLoS One*, **4**, e4406.

16 Muller, S. (2004) Redox and antioxidant systems of the malaria parasite *Plasmodium falciparum*. *Mol. Microbiol.*, **53**, 1291–1305.

17 Mooney, S. and Hellmann, H. (2010) Vitamin B6: Killing two birds with one stone? *Phytochemistry*, **71**, 495–501.

18 Hanes, J.W., Keresztes, I., and Begley, T.P. (2008) 13C NMR snapshots of the complex reaction coordinate of pyridoxal phosphate synthase. *Nat. Chem. Biol.*, **4**, 425–430.

19 Hanes, J.W., Ealick, S.E., Tews, I., and Begley, T.P. (2010) Biosynthesis of pyridoxal phosphate, in *Comprehensive Natural Products Chemistry*, (ed. T.P. Begley), Vol. **7**, Chapter 9, Elsevier, p. 7.

20 Zalkin, H. and Smith, J.L. (1998) Enzymes utilizing glutamine as an amide donor. *Adv. Enzymol. Relat. Areas Mol. Biol.*, **72**, 87–144.

21 Tambasco-Studart, M., Titiz, O., Raschle, T., Forster, G., Amrhein, N., and Fitzpatrick, T.B. (2005) Vitamin B6 biosynthesis in higher plants. *Proc. Natl Acad. Sci. USA*, **102**, 13687–13692.

22 Rodriguez-Navarro, S., Llorente, B., Rodriguez-Manzaneque, M.T., Ramne, A., Uber, G. *et al.* (2002) Functional analysis of yeast gene families involved in metabolism of vitamins B1 and B6. *Yeast*, **19**, 1261–1276.

23 Raschle, T., Amrhein, N., and Fitzpatrick, T.B. (2005) On the two components of pyridoxal 5′-phosphate synthase from *Bacillus subtilis*. *J. Biol. Chem.*, **280**, 32291–32300.

24 Fitzpatrick, T.B., Amrhein, N., Kappes, B., Macheroux, P., Tews, I., and Raschle, T. (2007) Two independent routes of de novo vitamin B6 biosynthesis: not that different after all. *Biochem J.*, **407**, 1–13.

25 Fry, D.C. (2006) Protein-protein interactions as targets for small molecule drug discovery. *Biopolymers*, **84**, 535–552.

26 Arkin, M.R. and Wells, J.A. (2004) Small-molecule inhibitors of protein–protein interactions: progressing towards the dream. *Nat. Rev. Drug Discov.*, **3**, 301–317.

27 Sillerud, L.O. and Larson, R.S. (2005) Design and structure of peptide and peptidomimetic antagonists of protein–protein interaction. *Curr. Protein Pept. Sci.*, **6**, 151–169.

28 Sperandio, O., Reynes, C.H., Camproux, A.C., and Villoutreix, B.O. (2010) Rationalizing the chemical space of protein–protein interaction inhibitors. *Drug Discov. Today*, **15**, 220–229.

29 Fry, D.C. (2008) Drug-like inhibitors of protein-protein interactions: a structural examination of effective protein mimicry. *Curr. Protein Pept. Sci.*, **9**, 240–247.

30 Toogood, P.L. (2002) Inhibition of protein–protein association by small molecules: approaches and progress. *J. Med. Chem.*, **45**, 1543–1558.

31 Berg, T. (2008) Small-molecule inhibitors of protein–protein interactions. *Curr. Opin. Drug Discov. Devel.*, **11**, 666–674.

32 Arkin, M.R. and Whitty, A. (2009) The road less traveled: modulating signal transduction enzymes by inhibiting their protein–protein interactions. *Curr. Opin. Chem. Biol.*, **13**, 284–290.

33 Domling, A. (2008) Small molecular weight protein–protein interaction antagonists: an insurmountable challenge? *Curr. Opin. Chem. Biol.*, **12**, 281–291.

34 Vassilev, L.T., Vu, B.T., Graves, B., Carvajal, D., Podlaski, F. *et al.* (2004) In vivo activation of the p53 pathway by small-molecule antagonists of MDM2. *Science*, **303**, 844–848.

35 Grasberger, B.L., Lu, T., Schubert, C., Parks, D.J., Carver, T.E. *et al.* (2005) Discovery and cocrystal structure of benzodiazepinedione HDM2 antagonists that activate p53 in cells. *J. Med. Chem.*, **48**, 909–912.

36 Petros, A.M., Dinges, J., Augeri, D.J., Baumeister, S.A., Betebenner, D.A. *et al.* (2006) Discovery of a potent inhibitor of the antiapoptotic protein Bcl-xL from NMR and parallel synthesis. *J. Med. Chem.*, **49**, 656–663.

37 Oltersdorf, T., Elmore, S.W., Shoemaker, A.R., Armstrong, R.C., Augeri, D.J. *et al.* (2005) An inhibitor of Bcl-2 family proteins induces regression of solid tumours. *Nature*, **435**, 677–681.

38 Nguyen, M., Marcellus, R.C., Roulston, A., Watson, M., Serfass, L. *et al.* (2007) Small molecule obatoclax (GX15-070) antagonizes MCL-1 and overcomes MCL-1-mediated resistance to apoptosis. *Proc. Natl Acad. Sci. USA*, **104**, 19512–19517.

39 Oost, T.K., Sun, C., Armstrong, R.C., Al-Assaad, A.S., Betz, S.F. *et al.* (2004) Discovery of potent antagonists of the antiapoptotic protein XIAP for the treatment of cancer. *J. Med. Chem.*, **47**, 4417–4426.

40 Xiao, T., Takagi, J., Coller, B.S., Wang, J.H., and Springer, T.A. (2004) Structural basis for allostery in integrins and binding to fibrinogen-mimetic therapeutics. *Nature*, **432**, 59–67.

41 Xiong, J.P., Stehle, T., Zhang, R., Joachimiak, A., Frech, M., Goodman, S.L., and Arnaout, M.A. (2002) Crystal structure of the extracellular segment of integrin alphaVbeta3 in complex with an Arg-Gly-Asp ligand. *Science*, **296**, 151–155.

42 Fry, D.C. and Vassilev, L.T. (2005) Targeting protein–protein interactions for cancer therapy. *J. Mol. Med.*, **83**, 955–963.

43 Rickert, M., Wang, X., Boulanger, M.J., Goriatcheva, N., and Garcia, K.C. (2005) The structure of interleukin-2 complexed with its alpha receptor. *Science*, **308**, 1477–1480.

44 Waal, N.D., Yang, W., Oslob, J.D., Arkin, M.R., Hyde, J. *et al.* (2005) Identification of nonpeptidic small-molecule inhibitors of interleukin-2. *Bioorg. Med. Chem. Lett.*, **15**, 983–987.

45 Roehrl, M.H., Kang, S., Aramburu, J., Wagner, G., Rao, A., and Hogan, P.G. (2004) Selective inhibition of calcineurin-NFAT signaling by blocking protein–protein interaction with small organic molecules. *Proc. Natl Acad. Sci. USA*, **101**, 7554–7559.

46 Manning, A.M. and Davis, R.J. (2003) Targeting JNK for therapeutic benefit: from junk to gold? *Nat. Rev. Drug Discov.*, **2**, 554–565.

47 Stebbins, J.L., De, S.K., Machleidt, T., Becattini, B., Vazquez, J. *et al.* (2008) Identification of a new JNK inhibitor targeting the JNK-JIP interaction site. *Proc. Natl Acad. Sci. USA*, **105**, 16809–16813.

48 Chen, T., Kablaoui, N., Little, J., Timofeevski, S., Tschantz, W.R. *et al.* (2009) Identification of small-molecule inhibitors of the JIP-JNK interaction. *Biochem. J.*, **420**, 283–294.

49 Rush, T.S. 3rd, Grant, J.A., Mosyak, L., and Nicholls, A. (2005) A shape-based 3-D scaffold hopping method and its application to a bacterial protein–protein interaction. *J. Med. Chem.*, **48**, 1489–1495.

50 Jennings, L.D., Foreman, K.W., Rush, T.S. 3rd, Tsao, D.H., Mosyak, L. *et al.* (2004) Design and synthesis of indolo[2,3-a] quinolizin-7-one inhibitors of the ZipA-FtsZ interaction. *Bioorg. Med. Chem. Lett.*, **14**, 1427–1431.

51 Oikonomakos, N.G., Skamnaki, V.T., Tsitsanou, K.E., Gavalas, N.G., and Johnson, L.N. (2000) A new allosteric site in glycogen phosphorylase b as a target for drug interactions. *Structure*, **8**, 575–584.

52 McMillan, K., Adler, M., Auld, D.S., Baldwin, J.J., Blasko, E. *et al.* (2000) Allosteric inhibitors of inducible nitric oxide synthase dimerization discovered via combinatorial chemistry. *Proc. Natl Acad. Sci. USA*, **97**, 1506–1511.

53 Zhu, J., Burgner, J.W., Harms, E., Belitsky, B.R., and Smith, J.L. (2005) A new arrangement of (beta/alpha)8 barrels in the synthase subunit of PLP synthase. *J. Biol. Chem.*, **280**, 27914–27923.

54 Zein, F., Zhang, Y., Kang, Y.N., Burns, K., Begley, T.P., and Ealick, S.E. (2006) Structural insights into the mechanism of the PLP synthase holoenzyme from *Thermotoga maritima. Biochemistry*, **45**, 14609–14620.

55 Bauer, J.A., Bennett, E.M., Begley, T.P., and Ealick, S.E. (2004) Three-dimensional structure of YaaE from *Bacillus subtilis*, a glutaminase implicated in pyridoxal-5′-phosphate biosynthesis. *J. Biol. Chem.*, **279**, 2704–2711.

56 Strohmeier, M., Raschle, T., Mazurkiewicz, J., Rippe, K., Sinning, I., Fitzpatrick, T.B., and Tews, I. (2006) Structure of a bacterial pyridoxal 5′-phosphate synthase complex. *Proc. Natl Acad. Sci. USA*, **103**, 19284–19289.

57 Neuwirth, M., Strohmeier, M., Windeisen, V., Wallner, S., Deller, S. *et al.* (2009) X-ray crystal structure of *Saccharomyces cerevisiae* Pdx1 provides insights into the oligomeric nature of PLP synthases. *FEBS Lett.*, **583**, 2179–2186.

58 Muller, I.B., Knockel, J., Groves, M.R., Jordanova, R., Ealick, S.E., Walter, R.D., and Wrenger, C. (2008) The assembly of the plasmodial PLP synthase complex follows a defined course. *PLoS One*, **3**, e1815.

59 Knockel, J., Jordanova, R., Muller, I.B., Wrenger, C., and Groves, M.R. (2009) Mobility of the conserved glycine 155 is required for formation of the active plasmodial Pdx1 dodecamer. *Biochim. Biophys. Acta*, **1790**, 347–350.

60 Chowdhry, B.Z. and Harding, S.E. (2001) *Protein–Ligand Interactions: Hydrodynamics and Calorimetry: A Practical Approach. The Practical Approach Series*, Oxford University Press, Oxford; New York.

61 Carlson, H.A. (2002) Protein flexibility and drug design: how to hit a moving target. *Curr. Opin. Chem. Biol.*, **6**, 447–452.

62 Falke, J.J. (2002) Enzymology. A moving story. *Science*, **295**, 1480–1481.

63 Teague, S.J. (2003) Implications of protein flexibility for drug discovery. *Nat. Rev. Drug Discov.*, **2**, 527–541.

64 Flicker, K., Neuwirth, M., Strohmeier, M., Kappes, B., Tews, I., and Macheroux, P. (2007) Structural and thermodynamic insights into the assembly of the heteromeric pyridoxal phosphate synthase from *Plasmodium falciparum. J. Mol. Biol.*, **374**, 732–748.

65 Neuwirth, M., Flicker, K., Strohmeier, M., Tews, I., and Macheroux, P. (2007) Thermodynamic characterization of the protein-protein interaction in the heteromeric *Bacillus subtilis* pyridoxal phosphate synthase. *Biochemistry*, **46**, 5131–5139.

66 Wallner, S., Neuwirth, M., Flicker, K., Tews, I., and Macheroux, P. (2009) Dissection of contributions from invariant amino acids to complex formation and catalysis in the heteromeric pyridoxal 5-phosphate synthase complex from *Bacillus subtilis. Biochemistry*, **48**, 1928–1935.

67 Wiseman, T., Williston, S., Brandts, J.F., and Lin, L.N. (1989) Rapid measurement of binding constants and heats of binding using a new titration calorimeter. *Anal. Biochem.*, **179**, 131–137.

68 Harrous, M.E. and Parody-Morreale, A. (1997) Measurement of biochemical affinities with a Gill titration calorimeter. *Anal. Biochem.*, **254**, 96–108.

69 Bjelic, S. and Jelesarov, I. (2008) A survey of the year 2007 literature on applications of isothermal titration calorimetry. *J. Mol. Recognit.*, **21**, 289–312.

70 Leavitt, S. and Freire, E. (2001) Direct measurement of protein binding energetics by isothermal titration calorimetry. *Curr. Opin. Struct. Biol.*, **11**, 560–566.

71 Prabhu, N.V. and Sharp, K.A. (2005) Heat capacity in proteins. *Annu. Rev. Phys. Chem.*, **56**, 521–548.

72 Whitesides, G.M. and Krishnamurthy, V.M. (2005) Designing ligands to bind proteins. *Q. Rev. Biophys.*, **38**, 385–395.

73 Chaires, J.B. (2008) Calorimetry and thermodynamics in drug design. *Annu. Rev. Biophys.*, **37**, 135–151.

74 Holdgate, G.A. and Ward, W.H. (2005) Measurements of binding thermodynamics in drug discovery. *Drug Discov. Today*, **10**, 1543–1550.

75 Haq, I. and Ladbury, J. (2000) Drug-DNA recognition: energetics and implications for design. *J. Mol. Recognit.*, **13**, 188–197.

76 Nezami, A., Kimura, T., Hidaka, K., Kiso, A., Liu, J. *et al.* (2003) High-affinity inhibition of a family of *Plasmodium falciparum* proteases by a designed adaptive inhibitor. *Biochemistry*, **42**, 8459–8464.

77 Nezami, A., Luque, I., Kimura, T., Kiso, Y., and Freire, E. (2002) Identification and characterization of allophenylnorstatine-based inhibitors of plasmepsin II, an antimalarial target. *Biochemistry*, **41**, 2273–2280.

78 Ruben, A.J., Kiso, Y., and Freire, E. (2006) Overcoming roadblocks in lead optimization: a thermodynamic perspective. *Chem. Biol. Drug Des.*, **67**, 2–4.

79 Ohtaka, H., Muzammil, S., Schon, A., Velazquez-Campoy, A., Vega, S., and Freire, E. (2004) Thermodynamic rules for the design of high affinity HIV-1 protease inhibitors with adaptability to mutations and high selectivity towards unwanted targets. *Int. J. Biochem. Cell Biol.*, **36**, 1787–1799.

80 Raschle, T., Speziga, D., Kress, W., Moccand, C., Gehrig, P. *et al.* (2009) Intersubunit cross-talk in pyridoxal 5′-phosphate synthase, coordinated by the C terminus of the synthase subunit. *J. Biol. Chem.*, **284**, 7706–7718.

81 Henderson, R. (1970) Structure of crystalline alpha-chymotrypsin. IV. The structure of indoleacryloyl-alpha-chymotrypsin and its relevance to the hydrolytic mechanism of the enzyme. *J. Mol. Biol.*, **54**, 341–354.

82 Robertus, J.D., Kraut, J., Alden, R.A., and Birktoft, J.J. (1972) Subtilisin; a stereochemical mechanism involving transition-state stabilization. *Biochemistry*, **11**, 4293–4303.

83 Malthouse, J.P. (2007) 13C- and 1H-NMR studies of oxyanion and tetrahedral intermediate stabilization by the serine proteinases: optimizing inhibitor warhead specificity and potency by studying the inhibition of the serine proteinases by peptide-derived chloromethane and glyoxal inhibitors. *Biochem. Soc. Trans.*, **35**, 566–570.

84 Lundqvist, T. (2005) The devil is still in the details – driving early drug discovery forward with biophysical experimental methods. *Curr. Opin. Drug Discov. Devel.*, **8**, 513–519.

85 McCormack, J.G. (2006) Applying science to drug discovery. *Biochem. Soc. Trans.*, **34**, 238–242.

14
Targeting Prokaryotic Enzymes in the Eukaryotic Pathogen *Cryptosporidium*

Suresh Kumar Gorla, Corey Johnson, Jihan Khan, Xin Sun,
Lisa Sharling, Boris Striepen, and Lizbeth Hedstrom*

Abstract

Cryptosporidium parvum is a major cause of diarrhea and malnutrition in the developing world, and a potential bioterrorism agent. *C. parvum* contains very streamlined nucleotide biosynthetic pathways. Several genes encoding these enzymes appear to have been obtained by horizontal gene transfer from bacteria and algae, and thus are highly diverged from the host. Such enzymes represent attractive targets for the development of anticryptosporidial drugs. A program of drug discovery exploiting these unexpectedly diverged enzymes has been undertaken. Efforts to develop anticryptosporidial drugs targeting *C. parvum* IMP dehydrogenase and thymidine kinase are summarized in this chapter.

Introduction

The "vicious cycle" of diarrhea and malnutrition imperils the health of infants and small children in the impoverished regions of the world [1, 2]. While the deaths attributed to acute diarrheal disease are legion, the legacy of chronic diarrhea stunts the physical and intellectual growth of an even larger cohort of children. Both, *Cryptosporidium parvum* and *Cryptosporidium hominis* are increasingly recognized as major players within the complex etiology of diarrheal disease. These apicomplexan parasites cause acute, self-limiting gastrointestinal disease in immunocompetent patients [3]. Cryptosporidiosis can become protracted and life-threatening in immunocompromised patients [4]; chronic cryptosporidiosis is a common complication in AIDS patients, occurring frequently even in the absence of epidemic outbreaks. The Cryptosporidia produce spore-like oocysts that are resistant to most commonly used methods of water treatment, and several large outbreaks have been traced to contaminated water [5]. As oocysts can be obtained with modest effort, and water supplies easily accessed, *Cryptosporidium* also poses a credible bioterrorism

*Corresponding author

Apicomplexan Parasites. Edited by Katja Becker
Copyright © 2011 WILEY-VCH Verlag GmbH & Co. KGaA, Weinheim
ISBN: 978-3-527-32731-7

threat [6–8]. At present, no vaccine is available to treat these conditions, and the currently approved therapy is largely ineffective. Consequently, effective drugs are urgently required to manage cryptosporidiosis in children or adults suffering from AIDS.

Horizontal Gene Transfer Provides Diverged Targets for Drug Development

The selective toxicity of antimicrobial drugs usually results from a disruption of the enzymes and pathways that are crucial for microbial proliferation but absent from the mammalian host. While bacteria typically possess many unique targets, such opportunities for selective toxicity are scarce in eukaryotic pathogens, where essential enzymes and pathways are generally shared with the host. The prospects for selective inhibition are further blocked by the close structural similarity of these essential proteins that arises from their common evolutionary origin. However, some protozoan pathogens have remodeled their genomes via horizontal gene transfer, replacing eukaryotic genes with analogous ones from bacteria or algae. These transfers may be the consequence of an endosymbiotic relationship, and encompass a large collective of genes and pathways (e.g., the algal endosymbiont found in most *Apicomplexa*) or they may represent single gene events. The enzymes encoded by these genes are divergent from their host counterparts, and therefore present attractive opportunities for drug development. In the case of *Cryptosporidium*, several key enzymes in the nucleotide biosynthetic pathways are encoded by such scavenged genes. The "druggability" of these pathways is well established in cancer and antiviral therapy. In this chapter, the efforts made to exploit these unexpectedly divergent enzymes, in order to develop novel anticryptosporidial agents, will be summarized.

The Nucleotide Biosynthetic Pathways of *Cryptosporidium*

Genomic analysis indicates that *C. parvum* is entirely dependent on the host to obtain the purine and pyrimidine nucleotides that are the basic building blocks of DNA and RNA. Purine nucleotide biosynthesis follows a highly streamlined pathway, starting with the uptake of adenosine via a nucleoside transporter (Figure 14.1a). The production of guanine nucleotides requires the sequential conversion of adenosine 5-monophosphate (AMP) to inosine 5-monophosphate (IMP), xanthosine 5-monophosphate (XMP), and finally to guanosine 5-monophosphate (GMP). Importantly, *C. parvum* is resistant to 6-thioxanthine, and no purine phosphoribosyltransferases have been identified in the *Cryptosporidium* genomes, confirming that the parasite cannot salvage purine bases. IMP dehydrogenase (IMPDH; the prefix *Cp* will denote enzymes from *C. parvum*) catalyzes the oxidation of IMP to XMP with the concomitant reduction of NAD^+. Phylogenetic analyses have shown that the *Cp*IMPDH gene groups consistently with eubacteria, suggesting that the IMPDH gene was transferred from an ε-proteobacterium [9]. As the prokaryotic and eukaryotic IMPDHs exhibit differences in their kinetic and inhibition profiles, the divergent

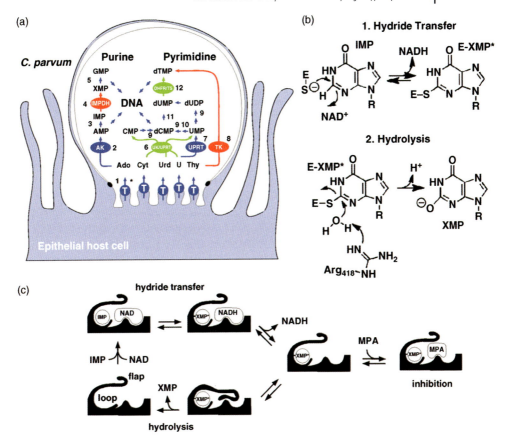

Figure 14.1 The nucleotide biosynthetic pathways of *C. parvum* are a phylogenetic mosaic. (a) The predicted nucleotide salvage pathways. Green denotes enzymes with association to plants or algae, while red denote enzymes with strong associations to eubacteria. Two arrows indicate two or more enzymatic steps. The membrane topology of the feeder organelle is schematized. The localization of the transporter (T) is hypothetical. 1, adenosine transporter; 2, adenosine kinase (AK); 3, adenosine deaminase; 4, IMPDH; 5, GMP synthetase; 6, UKUPRT; 7, UPRT; 8, TK; 9, ribonucleotide reductase; 10, cytidine triphosphate synthetase; 11, deoxycytidine monophosphate deaminase; 12, dihydrofolate reductase-thymidylate synthase (DHFR-TS). Reproduced from Ref. [10]; © 2004, National Academy of Sciences; (b) The mechanism of the IMPDH reaction; (c) The conformational changes during the IMPDH catalytic cycle. Figure modified from Ref. [12]; © 2004, American Society for Biochemistry and Molecular Biology.

phylogenetic origins of the host and parasite enzymes provide a unique opportunity to develop parasite-selective inhibitors.

Unlike other protozoan parasites, *Cryptosporidium* lacks *de novo* pyrimidine nucleotide biosynthesis, and must also salvage pyrimidines from the host. Genomic analyses have suggested that both pyrimidine bases (uracil) and nucleosides (thymidine, uridine) can serve as sources for pyrimidine nucleosides (Figure 14.1a); the incorporation of BrdU into *C. parvum* DNA confirms the presence of deoxyuridine salvage pathways [10]. *C. parvum* appears to have obtained two pyrimidine salvage enzymes via horizontal gene transfer: as with *Cp*IMPDH, *Cryptosporidium* thymidine kinase (CpTK) has bacterial origins, while the bifunctional uridine kinase-uracil phosphoribosyltransferase (CpUK-UPRT) appears to have been obtained from algae [10]. These enzymes may also be uniquely diverged targets for anticryptosporidial therapy.

*Cp*IMPDH is a Target for Anticryptosporidial Drugs

Validation of *Cp*IMPDH as a Drug Target

Several IMPDH inhibitors, most notably mycophenolic acid and ribavirin, are already in clinical use for immunosuppressive, antiviral, and anticancer chemotherapy [11]. Mycophenolic acid and ribavirin will also inhibit *Cp*IMPDH [12]. While the presence of guanosine protects host cells from IMDPH inhibition, mycophenolic acid and ribavirin inhibit *C. parvum* growth even in the presence of guanosine, which not only confirms that the parasite cannot salvage guanosine but also validates *Cp*IMPDH as a potential drug target.

The Mechanism of the IMPDH Reaction

IMPDH catalyzes a two-step reaction (Figure 14.1b) [13]. The first step is a redox reaction, in which IMP and NAD^+ bind to the active site in random order, the active site Cys attacks IMP, and a hydride is transferred to NAD^+, forming the complexes E-XMP* and NADH [11]. NADH is released and a mobile flap moves into the vacant site, forming the closed conformation required for the hydrolysis of E-XMP* to XMP [11]. This complex reaction series offers several potential targets for inhibition; notably, an inhibitor could interfere with the binding of IMP or NAD^+, or with flap closure. The IMP site is highly conserved, and the inhibitors that target this site (e.g., ribavirin monophosphate and mizoribine monophosphate) inhibit all IMPDHs [11]. In contrast, the NAD binding site is significantly diverged, which suggests that it might be possible to identify parasite selective inhibitors that bind to this site. Importantly, the initial redox reaction is fast, and the subsequent hydrolysis reaction is slow, so that E-XMP* is the predominant enzyme form. As a consequence, many inhibitors actually compete with the flap for the vacant NADH site, hence preventing the hydrolysis of E-XMP*. The selectivity of these inhibitors is determined by the structure of the NADH binding site and the properties of the flap. For example, mycophenolic acid (MPA) is a potent inhibitor of eukaryotic IMPDHs and a poor inhibitor of prokaryotic enzymes. MPA traps E-XMP* by binding in the nicotinamide portion of the vacant NADH site (Figure 14.1c). The nicotinamide site of eukaryotic IMPDHs contains Arg322 and Gln441 (human IMPDH2 numbering); these residues favor the binding of MPA. In contrast, prokaryotic IMPDHs contain Lys and Glu at these positions (Figure 14.2). Additionally, the open conformation is favored in

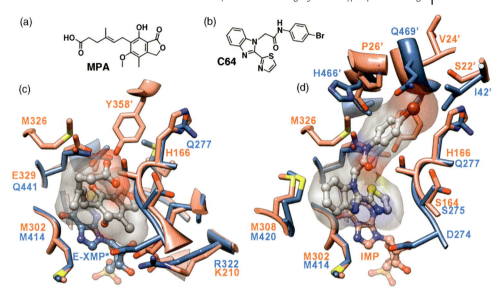

Figure 14.2 Interactions of selective inhibitors with eukaryotic and prokaryotic IMPDHs. (a) Structure of mycophenolic acid, a specific inhibitor of eukaryotic IMPDHs; (b) Structure of *Cp*IMPDH-selective inhibitor **C64**; (c) Structure of the MPA binding site. Residues within 5 Å of MPA are displayed. MPA is shown in gray with a transparent surface; Chinese hamster IMPDH is shown in blue (PDN accession number 1JR1, nearly identical to human IMPDH2 [19]), and *Cp*IMPDH is shown in orange (PDB accession 3KHJ [18]); residues from the adjacent monomer are denoted with a'; (d) Structure of the C64 binding site (color scheme as in panel (c)). Molecular graphics images were produced using the UCSF Chimera package from the Resource for Biocomputing, Visualization, and Informatics at the University of California, San Francisco (supported by NIH P41 RR-01 081) [40].

human IMPDHs, which facilitates MPA binding, whereas the closed conformation dominates in prokaryotic enzymes. Hence, MPA is a potent inhibitor for human IMPDH, but not for microbial IMPDHs [11].

Characterization of CpIMPDH

*Cp*IMPDH was expressed in *Escherichia coli*, and the purified recombinant protein subjected to a series of experiments to elucidate its mechanism, structure, and inhibition [12]. As predicted by phylogenetic analyses, MPA is a poor inhibitor of *Cp*IMPDH, as observed with other prokaryotic IMPDHs. The steady-state kinetic parameters of *Cp*IMPDH are also similar to those of other prokaryotic enzymes. In an attempt to guide this drug development effort, the complete kinetic mechanism of *Cp*IMPDH was delineated [14]. As in all IMPDHs, the hydride transfer step is fast relative to the hydrolysis step in *Cp*IMPDH and, as in other prokaryotic IMPDHs, the values of these rate constants are higher than for human IMPDH2. These results demonstrate significant structural and functional differences between *Cp*IMPDH and the host counterparts that can be exploited to design parasite-selective inhibitors.

The Development of *Cp*IMPDH Selective Inhibitors

A High-Throughput Screen for *Cp*IMPDH-Selective Inhibitors

While the IMP site is highly conserved among IMPDHs, the NAD site is quite variable, and therefore is the most promising binding site for *Cp*IMPDH-selective inhibitors [12]. Hence, the substrate concentrations in a high-throughput screen for *Cp*IMPDH inhibitors were adjusted so that the IMP site was always occupied, but the NAD site was available, thus selecting for inhibitors that bind to the NAD site [15]. The initial high-throughput screen yielded 10 validated hits from 44 000 compounds, with IC_{50} values ranging from 0.13 to 19 µM and selectivity ranging from nine- to 400-fold over human IMPDHs. Several of these compounds displayed antiparasitic activity in a tissue culture model of *C. parvum* infection. Kinetic characterization suggested that these compounds would bind to the NAD site as expected [15]. All of the compounds were shown to bind in the nicotinamide portion of the NAD site, which suggested that they stacked against the purine ring of IMP, as observed with inhibitors of human IMPDHs. A subsequent screen of 85 000 compounds yielded another 50 hits.

Medicinal Chemistry Optimization

A program of medicinal chemistry optimization was commenced to develop the initial hits into therapeutically useful compounds. Intriguingly, all but one of the compounds contained two aromatic systems coupled to an amide or similar linker. These initial hits were sorted into approximately 15 groups of similar structures; each group was given a series name (e.g., series A), where the high-throughput hit was the founding member. Preliminary structure–activity relationships (SARs) were established with commercially available analogs, after which series were chosen for further optimization based on the preliminary SAR, potency, antiparasitic activity, and synthetic accessibility. Details of current progress for two of these hits are summarized below.

Structure–Activity Relationships of A Derivatives

Hit **A1** was identified as a moderately potent ($IC_{50} = 3.0$ µM) and selective inhibitor of *Cp*IMPDH in the first screen for parasite-selective IMPDH inhibitors [15]. The initial optimization focused on potency in the enzyme inhibition assay, while subsequent efforts focused on antiparasitic activity and metabolic stability. The following SARs were established through the substitution of various functional groups of the aniline ring (Figure 14.3a). The (*S*)-stereoisomer of **A1** is active ($IC_{50} = 1.8$ µM), while the (*R*)-stereoisomer is inactive ($IC_{50} > 15$ µM) [16]. In the aniline ring, the substitution of R_2 with *p*-fluorine, *p*-hydroxyl, and *p*-methoxymethyl ether increased potency by up to approximately threefold. The best substitution on the phenoxy ring (Y) was the 2,3-dichloro (**A30**) combination, which exhibited a fivefold increase in potency against *Cp*IMPDH [17]. This discovery motivated further modification with a fused phenyl ring in the 2,3-positions, which resulted in a threefold increase in inhibitory activity (**A50**). Whereas, removal of the methyl

ID	X	Y	R1	R2	IC$_{50}$ (µM)	
					-BSA	+BSA
A1	CH	H	Me	4-Cl	3.0 ± 0.5	3.9 ± 0.5
A30	CH	H	(S)-Me	4-Cl	1.2 ± 0.4	1.2 ± 0.1
A31	CH	H	(R)-Me	4-Cl	> 5000	ND
A36	CH	2,3-di-Cl	Me	4-Cl	0.68 ± 0.09	0.9
A50			Me	4-Cl	1.1 ± 0.1	3.3 ± 0.2
A67			Me	4-Cl	0.7 ± 0.2	0.9 ± 0.2

Cmpd	R$_1$	R$_2$	X	IC$_{50}$ (nM)	
				-BSA	+BSA
A74	Me	4-Cl	CH	130 ± 30	800 ± 200
A89	Me	4-Cl	N	24 ± 8	230 ± 60
A90	Me	3,4-di-Cl	N	20 ± 10	700 ± 200
A110	(R)-Me	4-Cl	N^{+}-O^{-}	13 ± 5	50 ± 20

ID	R$_1$	R$_2$	IC$_{50}$ (nM)	
			-BSA	+BSA
C		4-MeOPh	1200±200	N.D
C10		4-ClPh	120±40	203±21
C14		4-BrPh	60±30	83±15
C86		3,4-di-ClPh	30±10	89±6
C90		2-Naphthyl	7±4	20±11
C61		4-ClPh	30±10	52±8
C64		4-BrPh	28±9	27
C84		3,4-di-ClPh	18±5	48±13
C97		2-Naphthyl	8±3	23±17
C91		2-Naphthyl	8±3	14±5
C96		1-Naphthyl	>5000	N.A

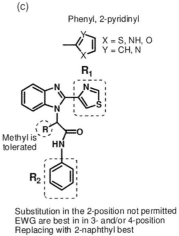

(a)

EWG substitution in para-position best

Small alkyl group best (S) more potent

Fused phenyl ring more potent

Heterocycles more potent

(b)

Small alkyl group still best (R) more potent

Heterocycles more potent N^{+}-O^{-} best

(c)

Phenyl, 2-pyridinyl

X = S, NH, O
Y = CH, N

R_1

Methyl is tolerated

R_2

Substitution in the 2-position not permitted
EWG are best in in 3- and/or 4-position
Replacing with 2-naphthyl best

Figure 14.3 (a) Structure–activity relationships (SARs) for *Cp*IMPDH inhibitors. The tables summarize potency data against *Cp*IMPDH for a selected group of derivatives, and the schemata summarize the overall SAR for each series. EWG, electron-withdrawing groups. (a) SAR for A series amides (data from Refs [16, 17]); (b) SAR for A series triazoles (data from Refs [16, 17]); (c) SAR for C series (data from Refs [18]).

group (**A28**) caused a diminished activity, its replacement with isopropyl or cyclo-propyl groups had no significant effect on the potency, while its replacement with a phenyl group resulted in a loss of activity. Replacement of the C-4 of naphthalene in **A50** with a sp^2-hybridized nitrogen resulted in a twofold increase in potency (**A67**). Despite these improvements, the potency of these amide compounds remained >100 nM.

Concern was expressed regarding the stability of the amide bond, and consequently this linker was replaced with the 1,2,3-triazole (**A74**); such substitution caused a 10-fold increase in potency (Figure 14.3b). Substitutions about the aniline ring were tolerated, but not as potent as the halogen analogs. The 4-chloro (**A89**) and 3,4-di-chloro (**A90**) analogs were by far the best substitutions on the pendent phenyl ring. When both enantiomers of **A90** were evaluated for their inhibitory activity, unlike the **A** amides the (R)-enantiomers of the triazoles proved, somewhat curiously, to be the most potent (e.g., **A105**, $IC_{50} = 9$ nM). The N-oxide derivatives were made to increase polarity, increasing solubility and decreasing nonspecific protein binding. The best compounds had IC_{50} values in the low nanomolar range, and did not inhibit the human IMPDHs ($IC_{50} > 5 \mu M$).

Structure–Activity Relationships of C Derivatives

The benzimidazole derivative **C** ($IC_{50} = 1.2 \mu M$) was also identified in the first high-throughput screen (Figure 14.3c). As above, the SAR of the aniline ring was first probed, whereby the potency was improved 10-fold when the 4-OMe group was replaced with lipophilic ($+ \pi$) and electron-withdrawing ($+ \sigma$) substituents such as chlorine (**C10**). Replacement with bromine (**C14**) improved the potency by an additional 20-fold. Similar trends were observed with other inhibitor series (see above; [17]). Mono-substitutions at the 2- or 3-position and bulky substitutions at the 4-position were detrimental (S. Kirubakaran, *et al.*, unpublished results). When the SAR of the thiazole ring was then explored, inhibitory activity was shown to be lost when the thiazole group was replaced with methyl. Inhibitory reactivity was increased to 28 nM when the benzimidazole connection was moved from 4-thiazole to 2-thiazole (**C64**), or when the 4-thiazole was replaced with a 2-pyridyl ring (**C91**). Importantly, it was possible to obtain a crystal structure of the E · IMP · **C64** complex [18], which has proven invaluable in further efforts to optimize the enzyme inhibition.

The Structure of the E · IMP · C64 Complex Reveals the Structural Basis of Inhibitor Selectivity

The structure of the E · IMP · **C64** complex was solved by molecular replacement to a 2.8 Å resolution (PDB accession number (3KHJ)). Crystals had the symmetry space group of $P2_1$ with two tetramers in the asymmetric unit; the eight monomers were clearly observed. **C64** was visible in three monomers. The presence of the **Br** atom allowed an unambiguous assignment of the aniline ring system. Interestingly, **C64** was shown to bind in a unique and unanticipated manner (see Figure 14.2).

Eukaryotic IMPDH inhibitors such as MPA bind within an IMPDH monomer, stacking against the purine ring of E-XMP* in a parallel fashion and blocking the entrance of the flap [19]. In contrast, the thiazole ring of **C64** stacks against the purine ring of IMP in perpendicular fashion, and the remainder of the molecule extends across the subunit interface into a pocket in the adjacent monomer, where the bromoaniline moiety interacts with Tyr358. This pocket is not found in eukaryotic IMPDHs (Figure 14.2), which explains the selectivity of these compounds. It should be noted that, while the structure is of the E · IMP · **C64** complex, the kinetic characterization indicates that **C64** also interacts with E-XMP*; presumably **C64** binds analogously in this complex, blocking the movement of the flap into the NAD site.

The crystal structure also revealed the presence of a cavity adjacent to the bromoaniline ring of **C64**, which suggested that replacement of the bromoaniline moiety with more bulky substituents would increase the potency. Consequently, when such derivatives were synthesized and tested, the 3,4-dichloroaniline (**C86**) and 2-naphthylamine (**C90**) improved potency by factors of 2 and 4, respectively [18], whereas the 1-naphthyl ring replacement (**C96**) resulted in a complete loss of inhibitory activity. Unfortunately, although some highly potent inhibitors were obtained in this series, the compounds proved to be rapidly metabolized in mouse microsomes – a problem which must clearly be overcome if these compounds are to advance into a mouse model of cryptosporidiosis.

The Screening Pipeline for Antiparasitic Activity

One step in the drug development program that has proved most difficult has been the assaying of compounds for anticryptosporidial activity, in part because *C. parvum* cannot be cultured continuously *in vitro*. Although several assays are available, they are typically labor-intensive, expensive, or not amenable to a higher throughput [20–25]. Additionally, as *Cryptosporidium* is only minimally experimentally tractable, confirming that the inhibitors will target *Cp*IMPDH as designed is not straightforward. In order to overcome these obstacles, a pipeline of novel tools has been developed to facilitate screening for antiparasitic activity. A key element of this pipeline is a novel strain of the related parasite *Toxoplasma gondii*, that has been genetically engineered to mirror the purine biosynthetic pathways of *Cryptosporidium*. In contrast to *Cryptosporidium*, *Toxoplasma* is a robust genetic model organism. In this case, the endogenous *T. gondii* IMPDH was removed by gene targeting, and replaced by the *Cp*IMPDH gene. In addition, the endogenous *HXGPRT* gene was removed by gene targeting, so that this strain would rely on *Cp*IMPDH for the production of guanine nucleotides [16]; this novel strain was designated *T. gondii-Cp*IMPDH. In order to facilitate screening, a fluorescent protein marker was also introduced [26]. Subsequently, the effect of the *Cp*IMPDH inhibitors on the growth of this strain was tested in parallel with two control strains that rely on the native *T. gondii* IMPDH, or are not susceptible to IMPDH inhibition at all, due to heir ability to salvage xanthine. If the inhibitors target *Cp*IMPDH as designed, then the control *T. gondii* would be resistant, but *T. gondii-Cp*IMPDH would be sensitive. This simple and easily conducted assay provides a large signal window and critical target validation that currently is not possible in assays involving *C. parvum*. An automated high content

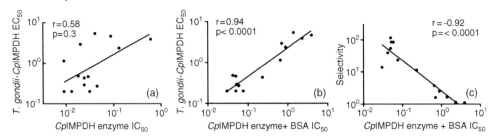

Figure 14.4 (a) A relatively weak ($r = 0.58$) and statistically insignificant (p-value $= 0.3$) correlation between the compound IC_{50} values for the CpIMPDH enzyme and the EC_{50} for proliferation of the *T. gondii-CpIMPDH* parasite; (b) However, a strong, positive correlation exists between the potency of CpIMPDH enzyme inhibition when assayed in the presence of BSA and inhibition of *T. gondii-CpIMPDH* proliferation ($r = -0.94$, $p < 0.0001$); (c) Selectivity in the *T. gondii* model, as determined by the relative inhibition of the *T. gondii-CpIMPDH* parasite over wild-type *T. gondii* clone, also correlates well with the potency of enzyme inhibition in the presence of BSA ($r = -0.92$, $p < 0.0001$). Reproduced from Ref. [16].

imaging assay was established to monitor *C. parvum* growth that enhances throughput [16]. Together, these tools should allow the rapid evaluation of antiparasitic activity.

Curiously, in the initial experiments using the triazole series of **A** compounds, no correlation was observed between the values of IC_{50} in the enzyme assay and the values of EC_{50} for the inhibition of *T. gondii-CpIMPDH* growth. It was reasoned that these inhibitors might bind nonspecifically to serum proteins in the media used in the tissue culture model, and to test this hypothesis the enzyme inhibition was re-evaluated in the presence of bovine serum albumin (BSA). Consequently, the values of IC_{50} were increased 10-fold for some inhibitors, which indicated that these compounds do indeed bind strongly to the BSA. Satisfyingly, a strong correlation was observed between antiparasitic activity and the value of IC_{50} for enzyme inhibition, as determined in the presence of BSA (Figure 14.4). The compounds **A100**, **A102**, **A109**, and **A110** displayed a more than 50-fold selectivity for *T. gondii-CpIMPDH* over the wild-type strain, with submicromolar values of EC_{50} [16]. These compounds also displayed anticryptosporidial activity. Notably, **A110** had sufficient potency and metabolic stability to be advanced into a mouse model of *C. parvum* infection.

Targeting the Pyrimidine Nucleotide Pathways

Protozoan parasites generally must salvage for purines, but are able to synthesize pyrimidines *de novo*. However, genomic analyses have suggested that *C. parvum* relies on salvage for all nucleotides. A closer look at thymidine nucleotide biosynthesis in *C. parvum* reveals that there are redundant pathways for the synthesis of dTMP, via thymidine kinase (TK) and thymidylate-synthase-dihydrofolate reductase [10, 27–29]. This observation was curious, and called into question whether the blockade of dTMP synthesis in *C. parvum* might represent a viable strategy for the development of a chemotherapeutic drug. *Cryptosporidium* is the only member of the Apicomplexa that must salvage pyrimidines, and it is the only Apicomplexa with a TK. The *Cryptosporidium* TK gene was not vertically inherited, but does appear to have been obtained via a horizontal gene transfer from a bacterium [10]. The success achieved in specifically targeting the bacterial-like IMPDH in *C. parvum* encouraged the pursuit of this unique CpTK. It should be noted that TK may prove valuable both as a potential drug target, and for its ability to activate nucleoside prodrugs – a feature that has been exploited with great success in the development of antiviral chemotherapy [30].

CpTK is 23% identical to human TK1 and 62% identical to *E. coli* TK. Bacterial TKs, unlike the promiscuous viral TKs, have similar substrate specificity to mammalian TKs. However, bacterial and mammalian TKs do have different allosteric regulation properties, and the low sequence identity between CpTK and hTK1 further suggests that there might structural differences which could be exploited for specific inhibitor design. Unfortunately, few of these differences are at the active site, and the enzymatic properties of CpTK are very similar to those of the human enzyme [31]. The absence of any striking differences between CpTK and human TK, in addition to the presence of redundant pathways for dTMP biosynthesis, suggest that CpTK is not a good target for anticryptosporidial drugs.

In contrast, CpTK-activated prodrugs may provide an effective strategy against cryptosporidiosis. CpTK phosphorylates only very close analogs of thymidine, such as

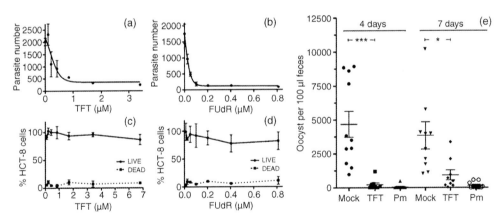

Figure 14.5 Effect of TFT and FUdR on parasite growth in models of *C. parvum* infection. (a,b) *C. parvum* growth was assayed by high-content imaging. (a) TFT; (b) FUdR; (c, d) Host cell cytotoxicity was assessed using the LIVE/DEAD® assay (Invitrogen). Data are representative of two independent experiments. (c) TFT; (d) FUdR; (e) Treatment with TFT in an IL-2 mouse model of *C. parvum* infection. Data shown as numbers of parasite oocysts in mouse feces at 4 and 7 days post infection with treatment with PBS (Mock), 200 mg kg^{-1} TFT, and 2000 mg kg^{-1} paromomycin (Pm). ***, $p < 0.0006$; *, $p < 0.02$. Reproduced from Ref. [31]; © 2010, Public Library of Science.

trifluorothymidine (TFT) and 5-fluorodeoxyuridine (FUdR) [31]. Both, TFT and FUdR are currently being used for the treatment of cancer [32], and both prodrugs are good substrates for mammalian TK. These prodrugs, once phosphorylated, are incorporated into DNA and induce strand breaks, ultimately leading to cell death [33, 34]. Remarkably, TFT and FUdR both inhibit *C. parvum* growth at concentrations much lower than host cell proliferation in a tissue culture model of infection (Figure 14.5). Furthermore, TFT treatment caused a promising reduction of parasite load in a very stringent interleukin-2 (IL-2) knockout mouse model of cryptosporidiosis [31].

The mechanistic basis of the remarkable sensitivity of the parasite to TFT and FUdR is unknown, and clearly warrants further study. Compared to other Apicomplexa, *C. parvum* possesses relatively few biosynthetic enzymes and a plethora of transporters, and so presumably relies on salvage for most of its anabolic metabolism. Nucleoside analogs conveniently circumvent the drug uptake issue by simply "looking" like a metabolite – they are often good substrates for nucleoside transporters [35]. *C. parvum* must have an efficient uptake system in order to obtain nucleosides from the host; indeed, differences in the efficiency of host and parasite uptake could account for the parasite selectivity of TFT and FUdR. Moreover, it may be possible to exploit these systems with other nucleoside prodrugs.

Nucleoside prodrugs typically require activation to the triphosphate via several enzymes before they can be incorporated into nucleic acids and produce cell damage. Although selectivity is usually determined by the initial conversion to the monophosphate, the subsequent phosphorylations can also modulate selectivity. For example, the phosphorylation of AZT-MP by thymidylate kinase (TMPK) is slow in human cells [36]. Likewise, the action of TFT and FUdR is also dependent on production of the di- and triphosphate forms by the *C. parvum* TMPK (CpTMPK). Studies are currently under way in the present authors' laboratory to elucidate the kinetics and specificity of CpTMPK. Importantly, TMPKs are essential enzymes, and the active site of CpTMPK is different from the active site of the host enzyme, despite the fact that both enzymes share a eukaryotic origin. It is believed that CpTMPK offers a promising opportunity for therapeutic intervention in addition to *Cp*IMPDH and CpTK.

Curiously, several other nucleoside prodrugs inhibit *C. parvum* growth *in vitro*, but are not substrates for CpTK [31, 37]. This observation suggests that *C. parvum* must have another pyrimidine nucleoside kinase. One attractive candidate is uridine kinase-uracil phosphoribosyl transferase (CpUK-UPRT), which appears to have been obtained from algae [10]; it is thought that, like other Apicomplexa, *Cryptosporidium* had an algal endosymbiont that was later lost [27, 38, 39]. Unfortunately, little is known at present regarding the substrate specificity of CpUK-UPRT, and further characterization is required not only to elucidate its role in the activation of the prodrugs, but also to gauge more fully its potential value as a specific target for inhibitor design. Likewise, little is presently known regarding the substrate specificity of CpUK-UPRT. Moreover, it is unclear whether any further characterization of the enzyme will elucidate its role in the activation of the prodrugs, nor whether it can be specifically targeted for inhibitor design.

Conclusion

Nucleotide biosynthetic pathways represent a rich source of drug targets, and have been exploited for the treatment of numerous conditions, in particular the inhibition of cell growth in cancer and infections. Current parasite drug discovery programs are built on the sturdy foundation of knowledge provided by this extensive body of work. The presence of enzymes with prokaryotic and plant origins makes these pathways particularly attractive for *Cryptosporidium* drug discovery. These drug discovery programs are well advanced in the cases of *Cp*IMPDH and CpTK, while CpTMPK provides another highly promising opportunity.

References

1 Snelling, W.J., Xiao, L., Ortega-Pierres, G., Lowery, C.J., Moore, J.E. *et al.* (2007) Cryptosporidiosis in developing countries. *J. Infect. Dev. Ctries*, **1**, 242–256.

2 Petri, W.A. Jr, Miller, M., Binder, H.J., Levine, M.M., Dillingham, R., and Guerrant, R.L. (2008) Enteric infections, diarrhea, and their impact on function and development. *J. Clin. Invest.*, **118**, 1277–1290.

3 DuPont, H.L., Chappell, C.L., Sterling, C.R., Okhuysen, P.C., Rose, J.B., and Jakubowski, W. (1995) The infectivity of *Cryptosporidium parvum* in healthy volunteers. *N. Engl. J. Med.*, **332**, 855–859.

4 Hunter, P.R. and Nichols, G. (2002) Epidemiology and clinical features of *Cryptosporidium* infection in immunocompromised patients. *Clin. Microbiol. Rev.*, **15**, 145–154.

5 Carey, C.M., Lee, H., and Trevors, J.T. (2004) Biology, persistence and detection of *Cryptosporidium parvum* and *Cryptosporidium hominis* oocyst. *Water Res.*, **38**, 818–862.

6 Arrowood, M.J. and Sterling, C.R. (1987) Isolation of *Cryptosporidium* oocysts and sporozoites using discontinuous sucrose and isopycnic Percoll gradients. *J. Parasitol.*, **73**, 314–319.

7 Upton, S.J. (1997) In vitro cultivation, in *Cryptosporidium and Cryptosporidiosis* (ed R. Fayer), CRC Press, Boca Raton, pp. 181–207.

8 Fayer, R. (2004) *Cryptosporidium*: a water-borne zoonotic parasite. *Vet. Parasitol.*, **126**, 37–56.

9 Striepen, B., White, M.W., Li, C., Guerini, M.N., Malik, S.B. *et al.* (2002) Genetic complementation in apicomplexan parasites. *Proc. Natl Acad. Sci. USA*, **99**, 6304–6309.

10 Striepen, B., Pruijssers, A.J., Huang, J., Li, C., Gubbels, M.J., Umejiego, N.N. *et al.* (2004) Gene transfer in the evolution of parasite nucleotide biosynthesis. *Proc. Natl Acad. Sci. USA*, **101**, 3154–3159.

11 Hedstrom, L. (2009) IMP dehydrogenase: structure, mechanism and inhibition. *Chem. Rev.*, **109**, 2903–2928.

12 Umejiego, N.N., Li, C., Riera, T., Hedstrom, L., and Striepen, B. (2004) *Cryptosporidium parvum* IMP dehydrogenase: Identification of functional, structural and dynamic properties that can be exploited for drug design. *J. Biol. Chem.*, **279**, 40320–40327.

13 Hedstrom, L. and Gan, L. (2006) IMP dehydrogenase: structural schizophrenia and an unusual base. *Curr. Opin. Chem. Biol.*, **10**, 520–525.

14 Ricra, T.V., Wang, W., Josephine, H.R., and Hedstrom, L. (2008) A kinetic alignment of orthologous inosine-5'-monophosphate dehydrogenases. *Biochemistry*, **47**, 8689–8696.

15 Umejiego, N.N., Gollapalli, D., Sharling, L., Volftsun, A., Lu, J. *et al.* (2008) Targeting a prokaryotic protein in a eukaryotic pathogen: identification of lead compounds against cryptosporidiosis. *Chem. Biol.*, **15**, 70–77.

16 Sharling, L., Liu, X., Gollapalli, D.R., Maurya, S.K., Hedstrom, L., and

Striepen, B. (2010) A screening pipeline for antiparasitic agents targeting *Cryptosporidium* inosine monophosphate dehydrogenase. *PLoS Negl. Trop. Dis.*, **4**, e794.

17 Maurya, S.K., Gollapalli, D.R., Kirubakaran, S., Zhang, M., Johnson, C.R. *et al.* (2009) Triazole inhibitors of *Cryptosporidium parvum* inosine 5'-monophosphate dehydrogenase. *J. Med. Chem.*, **52**, 4623–4630.

18 MacPherson, I.S., Kirubakaran, S., Gorla, S.K., Riera, T.V., D'Aquino, J.A. *et al.* (2010) The structural basis of *Cryptosporidium*-specific IMP dehydrogenase inhibitor selectivity. *J. Am. Chem. Soc.*, **132**, 1230–1231.

19 Sintchak, M.D., Fleming, M.A., Futer, O., Raybuck, S.A., Chambers, S.P. *et al.* (1996) Structure and mechanism of inosine monophosphate dehydrogenase in complex with the immunosuppressant mycophenolic acid. *Cell*, **85**, 921–930.

20 Shahiduzzaman, M., Dyachenko, V., Obwaller, A., Unglaube, S., and Daugschies, A. (2009) Combination of cell culture and quantitative PCR for screening of drugs against *Cryptosporidium parvum*. *Vet. Parasitol.*, **162**, 271–277.

21 MacDonald, L.M., Sargent, K., Armson, A., Thompson, R.C., and Reynoldson, J.A. (2002) The development of a real-time quantitative-PCR method for characterisation of a *Cryptosporidium parvum* in vitro culturing system and assessment of drug efficacy. *Mol. Biochem. Parasitol.*, **121**, 279–282.

22 Cai, X., Woods, K.M., Upton, S.J., and Zhu, G. (2005) Application of quantitative real-time reverse transcription-PCR in assessing drug efficacy against the intracellular pathogen *Cryptosporidium parvum* in vitro. *Antimicrob. Agents Chemother.*, **49**, 4437–4442.

23 Di Giovanni, G.D. and LeChevallier, M.W. (2005) Quantitative-PCR assessment of *Cryptosporidium parvum* cell culture infection. *Appl. Environ. Microbiol.*, **71**, 1495–1500.

24 Fontaine, M. and Guillot, E. (2002) Development of a TaqMan quantitative PCR assay specific for *Cryptosporidium parvum*. *FEMS Microbiol. Lett.*, **214**, 13–17.

25 Godiwala, N.T., Vandewalle, A., Ward, H.D., and Leav, B.A. (2006) Quantification of in vitro and in vivo *Cryptosporidium parvum* infection by using real-time PCR. *Appl. Environ. Microbiol.*, **72**, 4484–4488.

26 Gubbels, M.J. and Striepen, B. (2004) Studying the cell biology of apicomplexan parasites using fluorescent proteins. *Microsc. Microanal.*, **10**, 568–579.

27 Abrahamsen, M.S., Templeton, T.J., Enomoto, S., Abrahante, J.E., Zhu, G. *et al.* (2004) Complete genome sequence of the apicomplexan, *Cryptosporidium parvum*. *Science*, **304**, 441–445.

28 Xu, P., Widmer, G., Wang, Y., Ozaki, L.S., Alves, J.M. *et al.* (2004) The genome of *Cryptosporidium hominis*. *Nature*, **431**, 1107–1112.

29 Vasquez, J.R., Gooze, L., Kim, K., Gut, J., Petersen, C., and Nelson, R.G. (1996) Potential antifolate resistance determinants and genotypic variation in the bifunctional dihydrofolate reductase-thymidylate synthase gene from human and bovine isolates of *Cryptosporidium parvum*. *Mol. Biochem. Parasitol.*, **79**, 153–165.

30 De Clercq, E. (2004) Antiviral drugs in current clinical use. *J. Clin. Virol.*, **30**, 115–133.

31 Sun, X.E., Sharling, L., Muthalagi, M., Mudeppa, D.G., Pankiewicz, K.W. *et al.* (2010) Prodrug activation by *Cryptosporidium* thymidine kinase. *J. Biol. Chem.*, **285**, 15916–15922.

32 Temmink, O.H., Emura, T., de Bruin, M., Fukushima, M., and Peters, G.J. (2007) Therapeutic potential of the dual-targeted TAS-102 formulation in the treatment of gastrointestinal malignancies. *Cancer Sci.*, **98**, 779–789.

33 Emura, T., Suzuki, N., Fujioka, A., Ohshimo, H., and Fukushima, M. (2005) Potentiation of the antitumor activity of alpha, alpha, alpha-trifluorothymidine by the co-administration of an inhibitor of thymidine phosphorylase at a suitable molar ratio in vivo. *Int. J. Oncol.*, **27**, 449–455.

34 Yin, M.B. and Rustum, Y.M. (1991) Comparative DNA strand breakage induced by FUra and FdUrd in human ileocecal adenocarcinoma (HCT-8) cells: relevance to cell growth inhibition. *Cancer Commun.*, **3**, 45–51.

35 Zhang, J., Visser, F., King, K.M., Baldwin, S.A., Young, J.D., and Cass, C.E. (2007) The role of nucleoside transporters in cancer chemotherapy with nucleoside drugs. *Cancer Metastasis Rev.*, **26**, 85–110.

36 Lavie, A. and Konrad, M. (2004) Structural requirements for efficient phosphorylation of nucleotide analogs by human thymidylate kinase. *Mini Rev. Med. Chem.*, **4**, 351–359.

37 Woods, K.M. and Upton, S.J. (1998) Efficacy of select antivirals against *Cryptosporidium parvum* in vitro. *FEMS Microbiol. Lett.*, **168**, 59–63.

38 Xu, P., Widmer, G., Wang, Y., Ozaki, L.S., Alves, J.M. *et al.* (2004) The genome of *Cryptosporidium hominis*. *Nature*, **431**, 1107–1112.

39 Zhu, G., Marchewka, M.J., and Keithly, J.S. (2000) *Cryptosporidium parvum* appears to lack a plastid genome. *Microbiology*, **146**Pt (2), 315–321.

40 Pettersen, E.F., Goddard, T.D., Huang, C.C., Couch, G.S., Greenblatt, D.M. *et al.* (2004) UCSF Chimera – a visualization system for exploratory research and analysis. *J. Comp. Chem.*, **25**, 1605–1612.

Part Three
Drug Targets in Apicomplexan Parasites

Apicomplexan Parasites. Edited by Katja Becker
Copyright © 2011 WILEY-VCH Verlag GmbH & Co. KGaA, Weinheim
ISBN: 978-3-527-32731-7

15
Novel Apicomplexan Phosphatases and Immunophilins as Domain-Specific Drug Targets

Sailen Barik

Abstract

Protozoan parasites of the phylum Apicomplexa are major public health hazards. Malaria, caused by *Plasmodium falciparum*, continues to be a global scourge of humanity, but especially so in the developing nations. Another member of this family, *Toxoplasma gondii*, is also a significant health problem throughout the world, including Europe and the US. Complete genome sequences are now available for a number of Apicomplexa, including these two. Bioinformatic prediction of the sequences and biochemical analysis of the encoded proteins have revealed a number of unique protein phosphatases and immunophilins with unique domains and properties not found in the human host. In this chapter, this select group of divergent and unique parasitic proteins is summarized, and their potential to serve as targets for efficient and specific antiparasitic drug development in the future is discussed.

Introduction

Primitive protozoa of the Apicomplexan family are either facultative or obligatory parasites of higher eukaryotic host cells. Notable members of this family (and the diseases that they cause in humans) include: *Plasmodium falciparum* (malaria), *Toxoplasma gondii* (toxoplasmosis), and *Cryptosporidium parvum* (cryptosporidiosis; diarrhea).

While the first two parasites are transmitted to humans by anopheles mosquitoes and cats, respectively, in which the sexual cycle of these parasites takes place, *C. parvum* is primarily a waterborne parasite, requiring no vector. Malaria is a global health problem, but is particularly endemic in tropical and subtropical regions. According to the World Health Organization estimate, 350–500 million cases of malaria occur each year, leading to the death of about two million people, the majority of whom are young children in developing countries. In sub-Saharan Africa, for example, a child dies from malaria every 30 seconds.

Apicomplexan Parasites. Edited by Katja Becker
Copyright © 2011 WILEY-VCH Verlag GmbH & Co. KGaA, Weinheim
ISBN: 978-3-527-32731-7

Toxoplasmosis is also common globally, transmitted through the ingestion of raw or undercooked meat, and as mentioned before, via domestic cats. In the US alone, almost one-third of the population is estimated to be toxoplasma-positive by immunological criteria. Toxoplasmosis is especially hazardous in pregnancy, often leading to miscarriage or babies born with blindness, encephalitis, or severe brain dysfunction. In immunocompromised adults, such as AIDS victims or those receiving immunosuppressive therapy (e.g., organ recipients), systemic toxoplasmosis can cause blindness and even lethality.

Cryptosporidium, which is ingested through contaminated water, has been increasingly recognized as the cause of diarrhea outbreaks when water supplies become contaminated by bad sewer planning, or in earthquakes, floods, and other disasters. As with toxoplasmosis, cryptosporidiosis is a major problem in immunocompromised individuals, leading to severe diarrheal disease, gallbladder disease (cholecystitis), and inflammation of the pancreas (pancreatitis). Other, less-extensively studied parasites in this family include *Babesia* and *Theileria*, such as *Babesia bovis*, which causes babesiosis, a hemolytic disease resembling malaria in cattle and sheep, and *Theileria parva* and *Theileria annulata*, both of which cause theileriosis, a severe and often fatal lymphoproliferative disease in cattle.

The genomes of these and a number of other Apicomplexa have been fully sequenced, and can be found with various degrees of annotation in databases such as PlasmoDB, ToxoDB, CryptoBB, and ParaDB. Current antiparasitic drug development efforts include the identification of parasitic genes with sequences that are either not found in humans, or are dissimilar from their human counterparts (orthologs). A brief list of such potential candidates, using biochemical and structural considerations, is provided in the following sections.

"Classical" Protein Phosphatases and their Inhibitors

The Apicomplexan genome, as exemplified by *P. falciparum*, encodes a number of phosphatases of both the phosphoprotein phosphatase (PPP) and metal ion-dependent phosphoprotein phosphatase (PPM) classes [1]. A recent search of "protein phosphatase" as a keyword query on PlasmoDB yielded 18 entries in the annotated *P. falciparum* genome, although other phosphatase-like sequences were also identified. Research in the present author's laboratory has led to the unbiased characterization of a number of these Ser/Thr and dual-specificity protein phosphatases, both biochemically and by direct mRNA sequencing [2–9]. In general, the PPP group of enzymes comprises Ser/Thr phosphatases found in all living organisms, and even in some bacteriophages of the lambdoid family [10, 11]. They are characterized by a highly conserved ~250 amino acid catalytic core (Figure 15.1), which contains a number of signature sequence motifs including: GDXHG, GDXXDRG, RGNHE, and SAPNYC (where X is a nonconserved amino acid) [12–15]. The Ser/Thr phosphatases were originally classified into three major classes, namely PP1, PP2A/PP4/PP6, and PP2B, based on their combined profile of substrate specificity and inhibitor sensitivity. In contrast to the PPP class, the PPM class of Ser/Thr

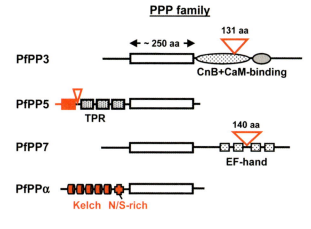

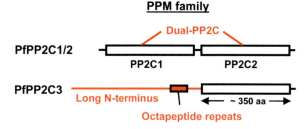

Figure 15.1 Domain diagram of representative unique apicomplexan proteins. Three *P. falciparum* (Pf) and *T. gondii* (Tg) protein families are shown. In the PPP family, the central catalytic core is 250 amino acids long. The most unusual features are shown in red. Selected actual sequences of these domains are presented in Figures 15.2 and 15.3 In PfPP2C1/2 and TgFCBP57, the individual domains are not unique, but their dual occurrence is, as indicated by the red color.

phosphatases (primarily PP2C), requires divalent cations such as Mg^{2+} for activity. The PP2C sequences are more divergent, and show no sequence homology with the PPP class.

The hunt for protein phosphatase inhibitors began in earnest soon after reversible phosphorylation was recognized as the predominant form of post-translational regulation in all living cells. Currently, a shortlist of the commonly used inhibitors (and their sources) includes okadaic acid (produced by marine dionoflagellates), calyculin A, microcystins, nodularins (cyanobacteria), tautomycin, tautomycetin, fostriecin (soil bacteria), and cantharidin (blister beetles) [16]. Traditionally, differential degrees of inhibition by toxins at discriminatory concentrations has served as the principal diagnostic tool of the PPP phosphatases. As a rule, the PP2A class is most sensitive to the toxins; for instance, okadaic acid (OA) inhibits PP2A with an IC_{50} in the low nanomolar range, whereas PP1 is less sensitive, with an IC_{50} of 15–20 nM OA, and PP2B (PP3) is highly resistant. Thus, low concentrations of OA have been used as a selective inhibitor of the PP2A class. Fostreicin inhibits PP2A over PP1 with even more selectivity than OA, whereas calyculin, microcystin, tautomycin, and cantharidin are less discriminatory. Not unexpectedly, PP2C, which is highly dissimilar from PPPs in sequence, is completely resistant to all PPP inhibitors, including OA.

A number of these classical phosphatases have been characterized in *P. falciparum*, including PfPP1 (AF195248), PfPP2A, and PfPP2B [2, 3, 5, 7–9]. The biochemical properties of these plasmodial enzymes were indistinguishable from their human counterparts. Thus, PfPP1 was sensitive to tautomycin, and PfPP2A to OA and microcystin. Both, the present author's group and others have also shown that the erythrocytic growth of *P. falciparum* could be inhibited by some of these inhibitors. In a recent example, such studies led to the discovery of a novel role for PfPP1 [17]. *P. falciparum* transposes a Golgi-like compartment, known as Maurer's clefts, into the cytoplasm of infected erythrocytes and uses it to deliver specific parasite molecules to the erythrocyte surface. The PP1 inhibitor, calyculin A, when added to the late schizont stage of the parasite, caused hyperphosphorylation of the parasitic skeleton-binding protein 1 (PfSBP1), which suggested that PfSBP1 is a substrate of PfPP1. As PfSBP1 is a trans-membrane protein of the Maurer's clefts, this treatment interfered with the release of progeny parasites from the blood cells, effectively creating an antimalarial effect.

In spite of their enormous contributions to laboratory research and to the biochemical classification of the PPP enzymes, the potential toxicity of these established inhibitors remains a matter of concern. OA, in fact, is the infamous agent of "diarrhetic shellfish poisoning," an acute intestinal diarrhea that results from eating shellfish such as oysters contaminated with this marine dinoflagellete toxin [16, 18]. Clearly, the high sequence conservation of the classical phosphatases across the parasitic and host genomes makes them unexciting targets for antimalarial drug development.

Fortunately, the advent of large-scale sequencing technology has revealed many more phosphatases in the parasite genomes [1, 19]. Hence, the most exotic apicomplexan phosphatases that contain substantially unique accessory domains and/or novel sequences within their catalytic subunit, are reviewed in the following subsections.

Unique, Authenticated Phosphatases: PP3, PP5, PP7, PPα, and PP2C

PP3

Previously designated as PP2B, and also referred to as calcineurin (CN), PP3 is a Ca^{2+}-activated phosphatase that requires its accessory subunit, calcineurin B (CNB), and the Ca^{2+}-binding protein, calmodulin (CaM), for optimal activity. In mammals,

PP3 dephosphorylates the transcription factor NFATc (Nuclear Factor of Activated T cell, cytoplasmic), causing it to be activated and to translocate into the nucleus, where it promotes the transcription of interleukins, leading to the stimulation and differentiation of T cells. The PP3 of *P. falciparum* (Pf) (XP_001349486) and *T. gondii* (Tg) (AAM97278) have been characterized. In both parasites, the PP3 and the cognate CNB subunit (AAO33818, AAM97279), and CaM subunits showed significant sequence similarity with the corresponding human orthologs [3]. A diagnostic property of PP3 is its inhibition by cyclosporin A (CsA), a cyclic 11-amino acid-long peptide that is produced naturally by a fungus. CsA does not directly bind to PP3; instead it first binds to a class of immunophilins, called cyclophilin (CyP), and the CsA–CyP complex then jams into the catalytic pocket of PP3. The PfPP3 and TgPP3 are also sensitive to CsA in the presence of the respective cognate cyclophilins, demonstrating their catalytic similarity and the equivalence of the CyP/CsA-binding pocket [3, 8, 20, 21]. Indeed, CsA showed antimalarial activity both *in vitro* and *in vivo* [22, 23], and at least one class of CsA resistance in Pf has been mapped to mutations in the PP3 catalytic subunit [2]. In other studies, TgPP3 activity appeared to be important for parasite egress from the infected human cells [20]. Nonetheless, in complementary studies, the antimalarial activity of various synthetic CsA analogs did not correlate with their potency to inhibit bovine PP3 *in vitro* [24, 25]. Thus, it is postulated that novel antimalarial compounds must be specifically designed against parasitic PP3.

A sequence comparison with the human ortholog [3] revealed that the most unique region of the apicomplexan PP3 is a large insert between the CNB-binding and CaM-binding domains (Figure 15.1). The insert in PfPP3 is approximately 131 amino acids long, and is highly rich in Asn, Asp, and Glu, which often occur in clusters (Figure 15.2a). In contrast, the TgPP3 insert is only 38 amino acids long, and has an unremarkable sequence. It is possible that these inserts loop out in the higher-order structure of both PP3, allowing the remainder of the protein to fold properly, as does their human ortholog. It will be interesting to determine if the inserts in PfPP3 and TgPP3 can be targeted by structure-based small molecules, and what effect this will have on the basic catalytic function and its modulation by CNB or CaM. To the present author's knowledge, PP3 has not been biochemically characterized in any other of the Apicomplexa.

PP5

The PP5 class of PPP phosphatases [25] is characterized by an N-terminal tetra-tricopeptide repeats (TPRs), a 34-amino acid protein–protein interaction motif of degenerate sequence homology, consisting of two helices connected by a flexible loop [26]. In the basal state, the TPRs of PP5 interact with the C terminus, thus altering the structure of the central catalytic core and lowering its activity [28]. In a still unclear mechanism, polyunsaturated fatty acids prevent this interaction and increase the phosphatase activity of PP5 [28, 29].

The PP5 of only one apicomplexan, namely *P. falciparum*, has been characterized in terms of both biochemistry and sequence (AY054983). Combined biochemical

(a) *Unique inserts in Pf and Tg calcineurin (PP3)*

```
Pf insert:383VLPKEVIQILNYIEENNKRINEMNFNNNDDNVQYEDNGPYINQSNNNNNNNNKDNKFDDITYDDHKK
KEKDKRNKISSNGNMQDNNQLYDHSEGHNNYNDEDEFFKNVKKTDTNNNNNNEEEDEEEDEEEEE513
Tg insert: 379DEDVDDVELPPAVLSIMKAHLPSDEASGQRHPPAGDNR416
```

(b) *An extra TPR linked by a unique Asn-rich spacer in PfPP5*

```
                   Extra PfTPR                   Helical spacer      TPR1-
PfPP5   LLKTCDALKNIGNKYFKENNYIISLRYYTEAIDLIKKSEQCPNNVTDIDNENNTNNLHDFDVEVD   130
HsPP5   ------------------------------------------------------------AEELK   32
Pred. Str.  hhhhhhhhhhhhhhcchhhhhhhhhhhhhhhhhhhhhchhhhhhhhhhhhhhhhhhhccc
```

```
              TPR1 (Contd.)                            TPR2
PfPP5   DEDKELFKEYYNKSAISKKSDFISIKETDIHIYYTNRSFCHIKLENYGTAIEDIDEAIKINPYYA   195
HsPP5   TQANDYFKAKDYENAIKFYSQAIELNPSNAIYYGNRSLAYLRTECYGYALGDATRAIELDKKYI   96
```

```
              TPR3 (Contd.)                   Helix----
PfPP5   KAYYRKGCSYLLLSDLKRASECFQKVLKLT-KDKNSELKLKQCKKLIFEQQFQK-   248
HsPP5   KGYYRRAASNMALGKFRAALRDYETVVKVKPHDKDAKMKYQECNKIVKQKAFER-   150
```

(c) *Unique insert in PfPP7*

```
PfPP7   EKFLNTEWKNECFEHLYEALLKADLSLRETLMVEDKNLDGKVSFAEFEQVLRDLNIDLSNEQIRI 685
HsPP7   EAHS---------TLVETLYRYRSDLEIIFNAIDTDHSGLISVEEFRAMWKLFSSHYN------ 603
```

```
PfPP7   LVRLINSNSLCNNTNLQENDKIDVAEFIGKMRVCYRLSINKDYVNNEKIQKLIETIGKHILSDSA 750
HsPP7   ---------------------------------------------------------------- 
```

```
                              Unique insert in Pf
PfPP7   DTANYHYKFYEENNERHNSERRKRSSVIKSVALFQKFKNYDNFGNGYLDYNDFVKAIKNFDMNKI 815
HsPP7   ---------------------------------------------------------------- 
```

```
PfPP7   SKEVEFEVDDDILMELAKSIDITKSSKINFLEFLQAFYVVNKSKY 860
HsPP7   -----VHIDDSQVNKLANIMDLNKDGSIDFNEFLKAFYV--VHRY 641
```

(d) *Octapeptide repeats in PfPP2C*

```
447VDGNKNKNVNGNKYEHVDGNKYEHVDGNKYEHVDGNKYEHVDGNKNKNVDGNKNKNVDGNKYEHVYDNQKNN
```

Figure 15.2 (a) Sequence of two internal insertions in Pf and Tg calcineurins (PP3). The amino acid numbers are indicated; (b) A portion of the N-terminal sequence of PfPP5 is compared with the human (*Hs*) ortholog to indicate the unique features of PfPP5. The structure of the unique TPR and spacer of PfPP5, predicted by HHpred [32], is indicated on the bottom (h = helix, c = coil/loop). Residues important for function in human PP5 [28, 31] are highlighted in gray. However, due to the degeneracy of TPR sequences, no attempt was made to mark the amino acid identities between human and PfPP3; (c) Sequence of the unique insert in PfPP7, absent in *Hs*PP7. Residues similar between the two are highlighted in gray; two EF hands are boxed; (d) Sequence of the octapeptide repeat motifs in PfPP2C. Note the identical repeats as well as the minor variations.

studies from two groups [6, 30] have revealed that, like the human ortholog, PfPP5 is specifically activated by polyunsaturated fatty acids, such as arachidonic and oleic acids, and inhibited by OA with an IC_{50} of ~5 nM. By both criteria, PfPP5 was indistinguishable from human PP5, and thus OA cannot serve as a useful antimalarial agent, based on this target. In terms of sequence, essentially all of the human residues in TPRs 1–3 (highlighted in gray in Figure 15.2b) that are important for interaction with Hsp90, are also conserved in the PfPP5; this includes even those not conserved in some other TPRs, such as Asp84 and Tyr95 (in human numbers), which are respectively changed to Ser and Asn in HsFKBP38 [28, 31].

In sharp contrast, the extra-catalytic, N-terminal TPR domain of PfPP5 presents an interesting diversion. While all other known PP5 sequences – including human PP5 and the predicted TgPP5 (XP_002367424) sequences – contain three TPRs, PfPP5 contains four (Figure 15.1); this extra TPR is the most N-terminal and is attached to the three downstream TPRs by a 22-amino acid spacer (Figure 15.2b). Secondary structure prediction revealed that the extra TPR may indeed consist of two helices, and that the linking spacer is also helical (Figure 15.2b). Based on a structural analysis of *Hs*PP5, it has been suggested that the flexibility of the triple-TPR region is critical to its regulatory role [27, 28, 31]. It has been proposed that the extra TPR and the helical linker of PfPP5 may allow additional degrees of flexibility critical for parasitic functions, and thus offer an atypical but Pf-specific target for drug development.

PP7

With a length of 959 amino acids, the PfPP7 (XP_001348398) is an unusually long member of the PPP family. However, its sequence homology and biochemical properties indicate that the enzyme is similar to the human PP7 and *Drosophila* PP7/RdgC phosphatases, which are respectively 653 and 661 residues long [3, 7, 26, 33]. Sequence comparison also showed that the extra length of PfPP7 is due primarily to a large, 141-residue long insertion between two predicted EF-hand motifs (Figure 15.2c). PfPP7 appears to contain a total of four EF-hand motifs (Figure 15.1), found in a variety of Ca^{2+}-binding proteins. True to the PP7 class, the Pf enzyme was indeed found to require Ca^{2+} for optimal activity [3]. Interestingly, a naturally occurring proteolytic fragment of PfPP7 (amino acids 231–523), which corresponded to the catalytic core and lost the whole insert, also lost Ca^{2+}-dependence [7], thus supporting the role of the EF-hand motifs of the insert in Ca^{2+}-binding. Based on the results of these studies, it appears that the insert in PfPP7 can be inhibited in a novel manner, such that only the Ca^{2+}-mediated physiological regulation is lost, leading to the abrogation of a specific signaling pathway in the malarial parasite. Thus, PfPP7 may be a bona fide antimalarial drug target. The exact status of PP7 orthologs in the other Apicomplexa is currently unclear, although sequences with distant similarity have been identified in *T. gondii*, *T. annulata*, and *B. bovis*.

PP2C

Sequence analysis predicts that the Pf genome contains seven members of the PPM/PP2C family [1]. As implied earlier, there is currently no specific inhibitor for PP2C. At this point, particular attention should be drawn to two PfPP2C paralogs that are extraordinarily unique and which, since there is no standard numbering system for any paralog family, will be called arbitrarily PfPP2C-1/2 and PfPP2C-3.

PP2C1/2 (AAC77359) is a highly unusual 920 residue-long polypeptide containing two heterologous PP2C sequences in tandem (see Figure 15.1) [34]. In essence, it is a chimera of two PP2C sequences, named PP2C-1 and PP2C-2, and may have been generated by a rare event of parasitic genome shuffling that brought the two coding sequences together to create a single continuous open reading frame (ORF). Studies of recombinantly expressed portions of the protein showed that each individual unit is catalytically active, and contains the hallmark drug-resistance of the PP2C class [34]. More interestingly, this PP2C was found to exist as a holoenzyme, which is a dimer of the double – that is $(PP2C1/2)_2$ – essentially making it a tetramer containing four catalytic units in one complex [34]. In a search for its possible function, a Far-Western analysis was conducted which led to an identification of the translation elongation factor PfEF-1β as one of its substrates. PfEF-1β was efficiently phosphorylated by protein kinase C, such that the phosphorylation resulted in a 400% increase in nucleotide exchange activity [35]. PKC-phosphorylated PfEF-1beta was readily and selectively dephosphorylated by PfPP2C1/2, which downregulated the nucleotide exchange activity to its basal level. The identification of a translation elongation factor as substrate for PfPP2C1/2 indicated that this phosphatase serves an important regulatory function, and may be an appropriate target for specific antimalarial drug design due to its unique higher-order structure. It should be noted that this is the only chimeric double PP2C currently known and, in fact, chimeric proteins in general are extremely rare in biology. The details of another chimeric protein, the FCBP chaperone, as first discovered in *T. gondii*, will be presented later in the chapter.

PP2C3 (AY158901) contains 906 amino acid residues and an unremarkable catalytic core. However, it also possesses a ∼570 amino acid-long N-terminal extension (see Figure 15.1) that is composed of a variety of repeats, such as homopolymeric runs of Lys, Asp, and Asn. Most interestingly, it has a 72 residue-long stretch consisting of nine imperfect repeats of the sequence VDGNKNKN (Figure 15.2d). Recent biochemical studies of recombinant full-length PfPP2C3 have revealed that it is indeed a bona fide PP2C-like phosphatase (unpublished results). It is presumed that the N-terminal extension may have an essential role in protein–protein inter-action, and that it may serve as a drug target, although this remains to be seen.

PPα

Perhaps the most unusual of all apicomplexan phosphatases is the one tentatively named PPα in *P. falciparum* [36]. This is a homolog of the plant-specific BSU subfamily [36, 37], and is characterized by the presence of a diagnostic Kelch motif (Figure 15.3a and b). A very similar phosphatase is also found in *T. gondii*

Figure 15.3 (a, b) Two alternative demarcations of the five Kelch repeats (1 through 5) of PPα. In each repeat, only the consensus GG-Y/W-W motif is highlighted; due to the degeneracy of Kelch sequences, no attempt was made to mark the other amino acid identities. However, the high degree of similarity between Pf and Tg is still obvious, as indicated by asterisks and dots below the sequence. The predicted β-strands in each repeat are indicated above the sequence with arrows; (c) Sequence of the Asn/Ser-rich (highlighted) linker following the Kelch repeats in PfPPα. The TgPPα (not shown) also contains an insertion of comparable length in a similar position, but has a very different sequence, rich in Ala, Glu, Pro, and Ser.

(a) *PPα kelch repeats (Pattern 1)*

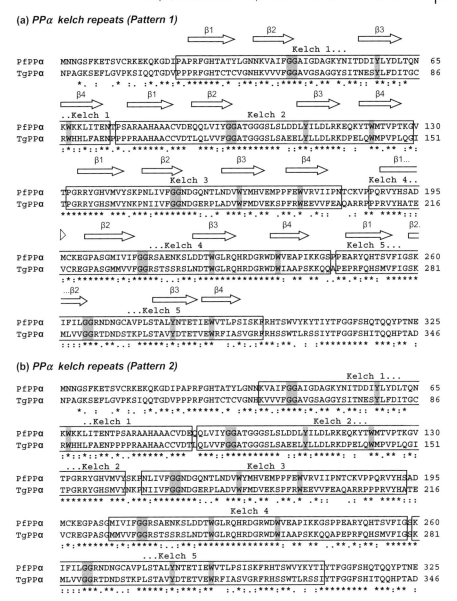

```
                      β1              β2                        β3
                    ──────>        ──────>                  ──────>
                                                   Kelch 1...
PfPPα   MNNGSFKETSVCRKEKQKGDIPAPRFGHTATYLGNNKVAIFGGAIGDAGKYNITDDIYLYDLTQN  65
TgPPα   NPAGKSEFLGVPKSIQQTGDVPPPRFGHTCTCVGNHKVVVFGGAVGSAGGYSITNESYLFDITGC  86
         *.  :   .* :. :*.**:*.******.* :**:**.:****:*.** *.**:: **:*:*

          β4          β1           β2                    β3          β4
        ──────>     ──────>      ──────>              ──────>      ──────>
        ..Kelch 1                          Kelch 2
PfPPα   KWKKLITENTPSARAAHAAACVDEQQLVIYGGATGGGSLSLDDLYILDLRKEQKYTWMTVPTKGV  130
TgPPα   RWHHLFAENPPPPRAAHAACCVDTLQLVVFGGATGGGSLSAEELYLLDLRKDPELQWMPVPLQGI  151
        :*::*::**.*..******.*** ***:::****** ::**:*****:  **.** :*:

          β1            β2           β3           β4              β1...
        ──────>        ──────>      ──────>      ──────>
                                Kelch 3                        Kelch 4..
PfPPα   TPGRRYGHVMVYSKPNLIVFGGNDGQNTLNDVWYMHVEMPPFEWVRVIIPNTCKVPPQRVYHSAD  195
TgPPα   TPGRRYGHSMVYNKPNIIVFGGNDGERPLADVWFMDVEKSPFRWEEVVFEAQARRPPPRVYHATE  216
        ******** ***.***:*********::.* ***:*.** .**.* .*::  .: ** ****:::

              β2              β3          β4           β1      β2.
            ──────>        ──────>      ──────>      ──────>   ─┐
                       ...Kelch 4                   Kelch 5...
PfPPα   MCKEGPASGMIVIFGGRSAENKSLDDTWGLRQHRDGRWDWVEAPIKKGSPPEARYQHTSVFIGSK  260
TgPPα   VCREGPASGMMVVFGGRSTSSRSLNDTWGLRQHRDGRWDWIAAPSKKQQAPEPRFQHSMVFIGSK  281
        :*:********:*:*****::.:**:****************: ** ** ..**.*:** ******

        ...β2              β3      β4
        ──────>        ──────>   ──────>
                          ...Kelch 5
PfPPα   IFILGGRNDNGCAVPLSTALYNTETIEWVTLPSISKFRHTSWVYKYTIYTFGGFSHQTQQYPTNE  325
TgPPα   MLVVGGRTDNDSTKPLSTAVYDTETVEWRFIASVGRFRHSSWTLRSSIYTFGGFSHITQQHPTAD  346
        :::::***.**..:  *****:*:***:**  ..*.::**:**.  :  :*********  ***:** :
```

(b) *PPα kelch repeats (Pattern 2)*

```
                                                           Kelch 1...
PfPPα   MNNGSFKETSVCRKEKQKGDIPAPRFGHTATYLGNNKVAIFGGAIGDAGKYNITDDIYLYDLTQN  65
TgPPα   NPAGKSEFLGVPKSIQQTGDVPPPRFGHTCTCVGNHKVVVFGGAVGSAGGYSITNESYLFDITGC  86
         *.  :   .* :. :*.**:*.******.* :**:**.:****:*.** *.**:: **:*:*

        ..Kelch 1                          Kelch 2...
PfPPα   KWKKLITENTPSARAAHAAACVDEQQLVIYGGATGGGSLSLDDLYILDLRKEQKYTWMTVPTKGV  130
TgPPα   RWHHLFAENPPPPRAAHAACCVDTLQLVVFGGATGGGSLSAEELYLLDLRKDPELQWMPVPLQGI  151
        :*::*::**.*..******.*** ***:::****** ::**:*****:  **.** :*:

        ...Kelch 2                         Kelch 3
PfPPα   TPGRRYGHVMVYSKPNLIVFGGNDGQNTLNDVWYMHVEMPPFEWVRVIIPNTCKVPPQRVYHSAD  195
TgPPα   TPGRRYGHSMVYNKPNIIVFGGNDGERPLADVWFMDVEKSPFRWEEVVFEAQARRPPPRVYHATE  216
        ******** ***.***:*********::.* ***:*.** .**.* .*::  .: ** ****:::

                                 Kelch 4
PfPPα   MCKEGPASGMIVIFGGRSAENKSLDDTWGLRQHRDGRWDWVEAPIKKGSPPEARYQHTSVFIGSK  260
TgPPα   VCREGPASGMMVVFGGRSTSSRSLNDTWGLRQHRDGRWDWIAAPSKKQQAPEPRFQHSMVFIGSK  281
        :*:********:*:*****::.:**:****************: ** ** ..**.*:** ******

        ...Kelch 5
PfPPα   IFILGGRNDNGCAVPLSTALYNTETIEWVTLPSISKFRHTSWVYKYTIYTFGGFSHQTQQYPTNE  325
TgPPα   MLVVGGRTDNDSTKPLSTAVYDTETVEWRFIASVGRFRHSSWTLRSSIYTFGGFSHITQQHPTAD  346
        :::::***.**..: *****:*:***:**  ..*.::**:**.  :  :*********  ***:** :
```

(c) *Asn/Ser-rich domain in PfPPα*

```
361 NDNLKHSSSDLRNVNSYNLNSQDVINTQQHNISTNNQFNVSNELYDLKNNASICNTLANVPIVPNVQNVPNVP
    NTHYRNMFDTSSNSSVFRLSNRPMSNKIRLSAHAHAVQENGSDFAFLVRKISIDKLEEEGRKINNGVLCTPVNYISEF
    KNTVYDKIITTLLLNPNITQFEIQYNHN 541
```

(Figure 15.3a and b), *Cryptosporidium hominis*, and *Theileria parva*, but in no other apicomplexan genome so far, nor in the human genome [1]. Both PfPPα (XP_001348804) and TgPPα (EEE31369) have been recombinantly expressed, and their phosphatase activity confirmed *in vitro* (unpublished data). The phosphatase activity of BSU1 (NP_171844), one of the four BSU paralogs in *Arabidopsis thaliana*, has also been demonstrated [37]. There is another notable similarity, however, in that BSU1 is detected almost exclusively in the rapidly growing *A. thaliana* cells [37], while microarray studies have also demonstrated increased PfPPα mRNA levels in the dividing stages (trophozoite, schizont) of *P. falciparum* [39]. However, the greatest expression of PfPPα is seen in the sexual stage of the parasite [36], which suggests that its role is mainly restricted to the sexual stage, which takes place in the mosquito vector. The co-occurrence of this unique phosphatase in both plants and the Apicomplexa is perhaps not surprising, as many Pf genes are more similar to their homologs in plants (e.g., *A. thaliana*) than in animals. This is consistent with the theory that the Apicomplexa originated by the endosymbiosis of nonphotosynthetic eukaryotes with red algae [40–43]. However, the name Apicomplexan in fact derives from the cytoplasmic organelles in these parasites, known as apicoplasts, that are reminiscent of plant chloroplasts.

Structurally, the Kelch motif is a segment of 44–56 amino acids in length, repeated four to seven times [44]. Like TPR, the Kelch sequences are relatively degenerate, but each Kelch unit contains eight conserved residues, including four hydrophobic ones, immediately followed by a glycine doublet (GG) and then two spaced aromatic residues, Y and/or W (Figure 15.3a and b). The crystal structures of galactose oxidase and other Kelch proteins have revealed that the Kelch repeats form a conserved tertiary structure, in which each unit contributes a four-stranded β-sheet that resembles the blade of a propeller, such that all repeats together form a complete β-propeller (Figure 15.4). The number of Kelch motifs, and their location in the polypeptide vary widely; both PfPPα and TgPPα belong to the N-terminal propeller class (see Figure 15.1).

The function of the β-propeller in the parasitic PPα is currently unknown [1, 38], but in other proteins, the β-propeller domain – like TPR – seems to be involved in protein–protein interactions [44]. The cellular distributions of β-propeller proteins include various intracellular compartments, cell surfaces, as well as the extracellular milieu. Curiously, a substantial number of β-propeller proteins are associated with actin [43], and it will be interesting to see if PPα interacts with the Apicomplexa actin, which is a major constituent of the parasitic cell and plays an important role in its motility and invasion abilities. The Apicomplexa actin also appears to have unique properties and regulations [45, 46], such as short filaments and an absence of profilin, and thus it is a potential drug target. The interaction between a unique phosphatase and a unique microfilament system makes this parasitic pathway worthy of further investigation. It is envisioned that drugs may be designed to abrogate this interaction, leading to the inhibition of a variety of parasitic processes that depend on actin, such as motility, infection, cell division, and egress. Such drugs would also be strongly antiparasitic, due to their multiple target pathways. Of note, the inhibition of cytoskeletal proteins in general is an important area of

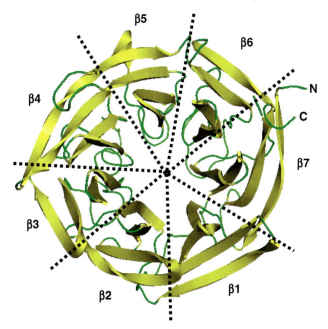

Figure 15.4 A representative β-propeller structure from the galactose oxidase of *Hypomyces rosellus* (PDB 1GOF) [44], consisting of β-strands (yellow) and flexible loops (green). The seven propellers (β1–β7, demarcated by the dotted lines) and the direction of the N- and C-termini are shown. Note the circular propeller shape with a center; the remainder of the protein has been removed for clarity.

antimicrobial and antimetastatic drug development. A prime example of this is taxol or paclitaxel (Bristol-Myers Squibb), which originally was isolated from the bark of the Pacific Yew tree, *Taxus brevifolia*. Taxol interferes with the naturally reversible breakdown of microtubules in animal cells by stabilizing the microtubules, and is currently used to treat many types of cancer and to prevent restenosis. The prokaryotic homolog of tubulin, FtsZ, is essential for bacterial cell division. Indeed, synthetic antimicrobials that bind specifically to the FtsZ pocket, in analogy to the Taxol-binding site of tubulin, exhibit potent and selective anti-staphylococcal activities [47].

Nestled between the Kelch repeat domain and the catalytic core of PPα is a ~180 amino acid-long stretch that is enriched in Asn and Ser, and contains repeats of single residues as well as tripeptides (Figure 15.3c). As seen before, tandem repeats frequently occur in *Plasmodium* proteins, and the results of recent spectroscopic studies, including nuclear magnetic resonance (NMR) and circular dichroism (CD), have indicated that these regions are largely flexible [48], often serving as linkers between domains, and this is likely to be the case also for PPα.

Lastly, the catalytic domain of PPα, despite containing the signature PPP motifs (as noted above), is unique in that it also contains five prominent insertions [36, 38]. Since four of these inserts are also conserved in all four BSU paralogs of

A. thaliana [37, 38], they may have an essential regulatory role which can be inhibited by specifically designed compounds.

The novel domains presented here can be targeted not only by structure-based small molecules, but also possibly with synthetic peptides. For example, the membrane-associated His-rich protein-1 (MAHRP-1) is a Maurer's cleft-resident molecule that has been recently described as being important for the trafficking of PfEMP-1, a major virulence factor, to the Pf-infected erythrocyte membrane [49]. Specific effects of 20-mer-long synthetic peptides spanning the complete MAHRP-1 sequence were studied in *Plasmodium*-infected erythrocytes [49]. A high-activity binding peptide with saturatable binding to EMP was identified, raising hopes about this region's potential as a therapeutic target against malaria.

Classical Immunophilins: CyP and FKBP

The immunophilin superfamily consists of highly conserved proteins with peptidyl prolyl *cis-trans* isomerase (PPIase) activity which act as a chaperone by accelerating the isomerization of X-Pro peptide bonds in nascently synthesized polypeptides [50]. The immunophilins belong to two major families and bind family-specific fungal metabolites, whereas the CyPs bind CsA, and the FK506-binding proteins (FKBPs) bind FK506 (tarcolimus) and rapamycin (sirolimus), both of which are structurally unrelated to CsA. In both families, the drugs bind to and inhibit the respective PPIase domains that are approximately 100 amino acids long. In a parallel mechanism, as described above, the drug–immunophilin complexes bind to PP3 (calcineurin), and this results in an elevated phosphorylation of NFATc and an inhibition of immune activation. Despite their common PPIase activity, there is no significant sequence homology between the CyP and FKBP families.

The apicomplexan CyPs first attracted attention in 1981, when Thommen-Scott discovered the antiparasitic activity of CsA in rodent malaria [22]. Since then, the inhibitory effect of CsA has been documented by many investigators against a variety of Apicomplexa [23], including *P. berghei-* and *P. chaubadi*-mediated malaria in mice, *P. falciparum* and *P. vivax* in human erythrocytic culture, and the growth of *T. gondii*, and *C. parvum*. The CsA effect may result from a direct inhibition of the CyP chaperone activity, and/or the inhibition of parasitic PP3 by the CyP–CsA complex [2, 20, 24, 25].

Multiple immunophilin genes are common in all organisms. The human genome, for example, has at least 16 Cyps and 15 FKBPs which exhibit a range of molecular weights and intracellular locations. Essentially, all human immunophilins have apparent homologs in other higher eukaryotes and plants. Indeed, sequence homology and biochemical analyses have revealed that *P. falciparum* contains three well-characterized CyP paralogs and a single FKBP, whereas other apicomplexan parasites, such as *T. gondii*, also have multiple CyPs that are less characterized. These sequences are highly similar to their human counterparts. In addition to the drug-binding catalytic domain, the larger CyPs and FKBPs tend to contain accessory domains such as RING finger, WD40, TPR, and RNA-recognition motifs and organelle localization signals [50–52].

One predictable side effect of CsA-treatment of parasitic diseases is that of immune suppression, which may result in an aggravation of the clinical course. The development of non-immunosuppressive CsA analogs that retain antiparasitic activity may eliminate this problem, however [24, 25]. The other adverse side effect may derive from an inhibition of the chaperone function of the immunophilin itself. Overall, in view of the multiple human CyPs, the potential toxicity of CsA as an effective antimalarial dosage cannot yet be ruled out. Thus, the goal is to target novel structures in the parasitic chaperones or unique parasitic chaperones.

A Potentially Novel Structural Center in PfFKBP35

Plasmodium falciparum has a single FKBP that is approximately 35 kDa in size, and hence is named FKBP35. An interesting property of this FKBP is that it is able to modestly inhibit PfPP3, without needing FK506 [53, 54]. Recently the X-ray crystal structure of the PfFKBP35–FK506 complex was solved at 2.35 Å resolution, and revealed the following interesting features [55]. Compared to the human FKBP12–FK506 complex reported earlier [PDB 1FKJ], a structural difference was seen in the β5–β6 segment, which lines the FK506 binding site. In PfFKBD35, Cys106 and Ser109 substituted for His87 and Ile90, respectively, of human FKBP12 and were 4–5 Å from the nearest atom of FK506 (Figure 15.5). The authors suggested that the substituted amino acids and the resultant altered structure may allow for the rational design of FK506 analogs with specific antiparasitic activity. Indeed, as the structure shows (Figure 15.5), it may be possible to design a smaller, modified FK506

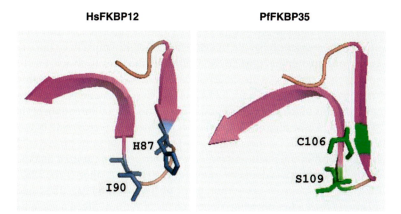

Figure 15.5 A key area of structural distinction between human FKBP12 (PDB 2PPN) and PfFKBP35 (PDB 2VN1) [55]. The two different amino acids are highlighted in blue and green, respectively, and their structures represented in sticks. The β-strands are in magenta, and the loops are orange. Note the clear structural difference between the two orthologs, viewed from roughly the same angle.

which would not only fit into the smaller opening of PfFKBP35 but also capitalize in the presence of Cys and form a covalent thiol adduct, causing strong and irreversible inhibition. Of note, this is akin to the covalent bond formed between microcystin and the Cys residue of the conserved SAPNYC motif in PP1 and PP2A, which is responsible for the exceptional inhibitory effect of this natural toxin.

In yet another hopeful detour, the discovery has been made recently of a founding member of a novel dual-family immunophilin in *T. gondii*, as described in the following section.

Novel Dual-Family Immunophilin, TgFCBP57

In 2005, the details were reported of a novel group of dual-family immunophilins which contained *both* CyP and FKBP domains [56], and for which was proposed the acronym FCBP (FK506- and CsA-binding protein) (see Figure 15.1). The FCBP of *T. gondii*, which is approximately 57.1 kDa in size and hence is named TgFCBP57, contains one N-terminal FKBP and one C-terminal CyP domain, joined by three TPR repeats. Subsequently, essentially similar FCBPs have been identified in the available genome sequences of other Apicomplexa, namely, *Theileria annulata, Theileria parva,* and *Babesia bovis* [51]. A structure–function analysis of the recombinant TgFCBP57 revealed that both domains were functional, and exhibited family-specific drug-sensitivity [56], while the individual domains of TgFCBP57 also inhibited calcineurin (PP3) in the presence of the appropriate drug. *T. gondii* growth was inhibited by CsA and FK506 in a moderately synergistic manner. The knockdown of TgFCBP57 by RNA interference revealed its essentiality for *T. gondii* growth. Clearly, the FCBPs are novel chaperones which can be targeted by existing drugs.

Yet, the question remains as to whether dual-family immunophilins are novel enough for new and specific drug design, and this in turn raises the question: "Why did Nature create?" In other words – what can they do that the separate CyPs and FKBPs cannot? It was reasoned that the fundamental difference must lie in the fact that dual-family immunophilins provide two chaperone modules (CyP and FKBP) in *cis*, by virtue of their residing in the same polypeptide. It was also conjectured that the FCBP may promote a *simultaneous* folding of two nascent subunits of a large multi-subunit complex much more efficiently, as correct folding of the two clients is essential to the *de novo* assembly of the complex. In contrast, the recruitment of two separate chaperone molecules in *trans* to the same place, and at the same time, would be a far more improbable and inefficient event. It was further hypothesized that the flexible peptide linking the two domains in FCBP, such as the TPR in TgFCBP57, would act like a hinge so as to allow movement of the two modular chaperone domains relative to each other, allowing client proteins of various dimensions to be accommodated. Such a folding machine might be compared with a spring-loaded pair of tongs, or a human hand that can tightly hold and then release objects of a range of sizes. If this were indeed true, then a specific drug might be designed to target either the hinge region or parts of the two domains simultaneously, to prevent their movement – and this would be analogous to locking the tongs. Moreover, these drugs would be specific against such bifunctional chaperones, without affecting the monofunctional chaperones of the host.

Conclusion

The collective evidence derived from biochemical studies and sequence comparisons has made it abundantly clear that the apicomplexan parasites possess a number of unique phosphatases and chaperones that are distinct from their human counterparts. Yet, as these two classes of proteins perform essential biological functions in all living cells, it is logical to hypothesize that the novel domains in these orthologs are bona fide targets for highly effective and specific antiparasitic drug development. Whilst such drugs may include rationally designed structure-based small molecules or competing peptides, they will in either case be highly welcome weapons in the battle against parasitic diseases, which have plagued humanity from ancient times to the present day.

References

1 Wilkes, J.M. and Doerig, C. (2008) The protein-phosphatome of the human malaria parasite *Plasmodium falciparum*. *BMC Genomics*, **9**, 412.

2 Kumar, R., Musiyenko, A., and Barik, S. (2005) *Plasmodium falciparum* calcineurin and its association with heat shock protein 90: mechanisms for the antimalarial activity of cyclosporin A and synergism with geldanamycin. *Mol. Biochem. Parasitol.*, **141**, 29–37.

3 Kumar, R., Musiyenko, A., Oldenburg, A., Adams, B., and Barik, S. (2004) Post-translational generation of constitutively active cores from larger phosphatases in the malaria parasite, *Plasmodium falciparum*: implications for proteomics. *BMC Mol. Biol.*, **5**, 6.

4 Kumar, R., Musiyenko, A., Cioffi, E., Oldenburg, A., Adams, B., Bitko, V., Krishna, S.S., and Barik, S. (2004) A zinc-binding dual-specificity YVH1 phosphatase in the malaria parasite, *Plasmodium falciparum*, and its interaction with the nuclear protein, pescadillo. *Mol. Biochem. Parasitol.*, **133**, 297–310.

5 Kumar, R., Adams, B., Oldenburg, A., Musiyenko, A., and Barik, S. (2002) Characterisation and expression of a PP1 serine/threonine protein phosphatase (PfPP1) from the malaria parasite, *Plasmodium falciparum*: demonstration of its essential role using RNA interference. *Malar. J.*, **1**, 5.

6 Dobson, S., Kar, B., Kumar, R., Adams, B., and Barik, S. (2001) A novel tetratricopeptide repeat (TPR) containing PP5 serine/threonine protein phosphatase in the malaria parasite, *Plasmodium falciparum. BMC Microbiol.*, **1**, 31.

7 Dobson, S., Bracchi, V., Chakrabarti, D., and Barik, S. (2001) Characterization of a novel serine/threonine protein phosphatase (PfPPJ) from the malaria parasite, *Plasmodium falciparum. Mol. Biochem. Parasitol.*, **115**, 29–39.

8 Dobson, S., May, T., Berriman, M., Del Vecchio, C., Fairlamb, A.H., Chakrabarti, D., and Barik, S. (1999) Characterization of protein Ser/Thr phosphatases of the malaria parasite, *Plasmodium falciparum*: inhibition of the parasitic calcineurin by cyclophilin-cyclosporin complex. *Mol. Biochem. Parasitol.*, **99**, 167–181.

9 Garcia, A., Cayla, X., Barik, S., and Langsley, G. (1999) A family of PP2

phosphatases in *Plasmodium falciparum* and parasitic protozoa. *Parasitol. Today*, **15**, 90–92.

10 Barik, S. (1993) Expression and biochemical properties of a protein serine/threonine phosphatase encoded by bacteriophage lambda. *Proc. Natl Acad. Sci. USA*, **90**, 10633–10637.

11 Zhuo, S., Clemens, J.C., Hakes, D.J., Barford, D., and Dixon, J.E. (1993) Expression, purification, crystallization, and biochemical characterization of a recombinant protein phosphatase. *J. Biol. Chem.*, **268**, 17754–17761.

12 Ansai, T., Dupuy, L.C., and Barik, S. (1996) Interactions between a minimal protein serine/threonine phosphatase and its phosphopeptide substrate sequence. *J. Biol. Chem.*, **271**, 24401–24407.

13 Zhuo, S., Clemens, J.C., Stone, R.L., and Dixon, J.E. (1994) Mutational analysis of a Ser/Thr phosphatase. Identification of residues important in phosphoesterase substrate binding and catalysis. *J. Biol. Chem.*, **269**, 26234–26238.

14 Wera, S. and Hemmings, B.A. (1995) Serine/threonine protein phosphatases. *Biochem. J.*, **311**, 17–29.

15 Barton, G.J., Cohen, P.T., and Barford, D. (1994) Conservation analysis and structure prediction of the protein serine/threonine phosphatases. Sequence similarity with diadenosine tetraphosphatase from *Escherichia coli* suggests homology to the protein phosphatases. *Eur. J. Biochem.*, **220**, 225–237.

16 Swingle, M., Ni, L., and Honkanen, R.E. (2007) Small-molecule inhibitors of ser/thr protein phosphatases: specificity, use and common forms of abuse. *Methods Mol. Biol.*, **365**, 23–38.

17 Blisnick, T., Vincensini, L., Fall, G., and Braun-Breton, C. (2006) Protein phosphatase 1, a *Plasmodium falciparum* essential enzyme, is exported to the host cell and implicated in the release of infectious merozoites. *Cell. Microbiol.*, **8**, 591–601.

18 Bialojan, C. and Takai, A. (1988) Inhibitory effect of a marine-sponge toxin, okadaic acid, on protein phosphatases. Specificity and kinetics. *Biochem. J.*, **256**, 283–290.

19 Bajsa, J., Duke, S.O., and Tekwani, B.L. (2008) *Plasmodium falciparum* serine/threonine phoshoprotein phosphatases (PPP): from housekeeper to the "holy grail". *Curr. Drug Targets*, **9**, 997–1012.

20 Moudy, R., Manning, T.J., and Beckers, C.J. (2001) The loss of cytoplasmic potassium upon host cell breakdown triggers egress of *Toxoplasma gondii*. *J. Biol. Chem.*, **276**, 41492–41501.

21 High, K.P., Joiner, K.A., and Handschumacher, R.E. (1994) Isolation, cDNA sequences, and biochemical characterization of the major cyclosporin-binding proteins of *Toxoplasma gondii*. *J. Biol. Chem.*, **269**, 9105–9112.

22 Thommen-Scott, K. (1981) Antimalarial activity of cyclosporin A. *Agents Actions*, **11**, 770–773.

23 Bell, A. (2009) Letter to the Editor on effect of cyclosporine on parasitemia and survival of *Plasmodium berghei*-infected mice. *Biochem. Biophys. Res. Commun.*, **378**, 678–679.

24 Bell, A., Wernli, B., and Franklin, R.M. (1994) Roles of peptidyl-prolyl cis-trans isomerase and calcineurin in the mechanisms of antimalarial action of cyclosporin A, FK506, and rapamycin. *Biochem. Pharmacol.*, **48**, 495–503.

25 Bell, A., Monaghan, P., and Page, A.P. (2006) Peptidyl-prolyl cis-trans isomerases (immunophilins) and their roles in parasite biochemistry, host–parasite interaction and antiparasitic drug action. *Int. J. Parasitol.*, **36**, 261–276.

26 Andreeva, A.V. and Kutuzov, M.A. (1999) RdgC/PP5-related phosphatases: novel components in signal transduction. *Cell. Signal.*, **11**, 555–562.

27 Das, A.K., Cohen, P.W., and Barford, D. (1998) The structure of the tetratricopeptide repeats of protein phosphatase 5: implications for TPR-mediated protein-protein interactions. *EMBO J.*, **17**, 1192–1199.

28 Yang, J., Roe, S.M., Cliff, M.J., Williams, M.A., Ladbury, J.E., Cohen, P.T., and Barford, D. (2005) Molecular basis for TPR domain-mediated regulation of protein phosphatase 5. *EMBO J.*, **24**, 1–10.

29 Sinclair, C., Borchers, C., Parker, C., Tomer, K., Charbonneau, H., and Rossie, S. (1999) The tetratricopeptide repeat domain and a C-terminal region control the activity of Ser/Thr protein phosphatase 5. *J. Biol. Chem.*, **274**, 23666–23672.

30 Lindenthal, C. and Klinkert, M.Q. (2002) Identification and biochemical characterisation of a protein phosphatase 5 homologue from *Plasmodium falciparum*. *Mol. Biochem. Parasitol.*, **120**, 257–268.

31 Cliff, M.J., Harris, R., Barford, D., Ladbury, J.E., and Williams, M.A. (2006) Conformational diversity in the TPR domain-mediated interaction of protein phosphatase 5 with Hsp90. *Structure*, **14**, 415–426.

32 Söding, J. (2005) Protein homology detection by HMM-HMM comparison. *Bioinformatics*, **21**, 951–960.

33 Andreeva, A.V. and Kutuzov, M.A. (2009) PPEF/PP7 protein Ser/Thr phosphatases. *Cell. Mol. Life Sci.*, **66**, 3103–3110.

34 Mamoun, C.B., Sullivan, D.J. Jr, Banerjee, R., and Goldberg, D.E. (1998) Identification and characterization of an unusual double serine/threonine protein phosphatase 2C in the malaria parasite *Plasmodium falciparum*. *J. Biol. Chem.*, **273**, 11241–11247.

35 Mamoun, C.B. and Goldberg, D.E. (2001) *Plasmodium* protein phosphatase 2C dephosphorylates translation elongation factor 1β and inhibits its PKC-mediated nucleotide exchange activity in vitro. *Mol. Microbiol.*, **39**, 973–981.

36 Li, J.L. and Baker, D.A. (1998) A putative protein serine/threonine phosphatase from *Plasmodium falciparum* contains a large N-terminal extension and five unique inserts in the catalytic domain. *Mol. Biochem. Parasitol.*, **95**, 287–295.

37 Mora-García, S., Vert, G., Yin, Y., Caño-Delgado, A., Cheong, H., and Chory, J. (2004) Nuclear protein phosphatases with Kelch-repeat domains modulate the response to brassinosteroids in *Arabidopsis*. *Genes Dev.*, **18**, 448–460.

38 Kutuzov, M.A. and Andreeva, A.V. (2002) Protein Ser/Thr phosphatases with kelch-like repeat domains. *Cell. Signal.*, **14**, 745–750.

39 Bozdech, Z., Llinás, M., Pulliam, B.L., Wong, E.D., Zhu, J., and DeRisi, J.L. (2003) The transcriptome of the intraerythrocytic developmental cycle of *Plasmodium falciparum*. *PLoS Biol.*, **1**, E5.

40 Dzierszinski, F., Popescu, O., Toursel, C., Slomianny, C., Yahiaoui, B., and Tomavo, S. (1999) The protozoan parasite *Toxoplasma gondii* expresses two functional plant-like glycolytic enzymes. Implications for evolutionary origin of apicomplexans. *J. Biol. Chem.*, **274**, 24888–24895.

41 Keeling, P.J. and Palmer, J.D. (2001) Lateral transfer at the gene and subgenic levels in the evolution of eukaryotic enolase. *Proc. Natl Acad. Sci. USA*, **98**, 10745–10750.

42 Keeling, P.J. and Palmer, J.D. (2008) Horizontal gene transfer in eukaryotic evolution. *Nat. Rev. Genet.*, **9**, 605–618.

43 Lim, L. and McFadden, G.I. (2010) The evolution, metabolism and functions of the apicoplast. *Philos. Trans. R. Soc. Lond. B. Biol. Sci.*, **365**, 749–763.

44 Adams, J., Kelso, R., and Cooley, L. (2000) The kelch repeat superfamily of proteins: propellers of cell function. *Trends. Cell Biol.*, **10**, 17–24.

45 Morrissette, N.S. and Sibley, L.D. (2002) Cytoskeleton of apicomplexan parasites. *Microbiol. Mol. Biol. Rev.*, **66**, 21–38.

46 Baum, J., Papenfuss, A.T., Baum, B., Speed, T.P., and Cowman, A.F. (2006) Regulation of apicomplexan actin-based motility. *Nat. Rev. Microbiol.*, **4**, 621–628.

47 Haydon, D.J., Stokes, N.R., Ure, R., Galbraith, G., Bennett, J.M., Brown, D.R., Baker, P.J., Barynin, V.V., Rice, D.W., Sedelnikova, S.E., Heal, J.R., Sheridan, J.M., Aiwale, S.T., Chauhan, P.K., Srivastava, A., Taneja, A., Collins, I., Errington, J., and Czaplewski, L.G. (2008) An inhibitor of FtsZ with potent and selective anti-staphylococcal activity. *Science*, **321** 1673–1675.

48 Matsushima, N., Yoshida, H., Kumaki, Y., Kamiya, M., Tanaka, T., Izumi, Y., and Kretsinger, R.H. (2008) Flexible structures and ligand interactions of tandem repeats

consisting of proline, glycine, asparagine, serine, and/or threonine rich oligopeptides in proteins. *Curr. Protein Pept. Sci.*, **9**, 591–610.

49 García, J., Curtidor, H., Gil, O.L., Vanegas, M., and Patarroyo, M.E. (2009) A Maurer's cleft-associated *Plasmodium falciparum* membrane-associated histidine-rich protein peptide specifically interacts with the erythrocyte membrane. *Biochem. Biophys. Res. Commun.*, **380**, 122–126.

50 Barik, S. (2006) Immunophilins: for the love of proteins. *Cell. Mol. Life Sci.*, **63**, 2889–2900.

51 Krücken, J., Greif, G., and von Samson-Himmelstjerna, G. (2009) In silico analysis of the cyclophilin repertoire of apicomplexan parasites. *Parasit. Vectors*, **2**, 27.

52 Adams, B., Musiyenko, A., Kumar, R., and Barik, S. (2005) A novel class of dual-family immunophilins. *J. Biol. Chem.*, **280**, 24308–24314.

53 Monaghan, P. and Bell, A. (2005) A *Plasmodium falciparum* FK506-binding protein (FKBP) with peptidyl-prolyl cis-trans isomerase and chaperone activities. *Mol. Biochem. Parasitol.*, **139**, 185–195.

54 Kumar, R., Adams, B., Musiyenko, A., Shulyayeva, O., and Barik, S. (2005) The FK506-binding protein of the malaria parasite, *Plasmodium falciparum*, is a FK506-sensitive chaperone with FK506-independent calcineurin-inhibitory activity. *Mol. Biochem. Parasitol.*, **141**, 163–173.

55 Kotaka, M., Ye, H., Alag, R., Hu, G., Bozdech, Z., Preiser, P.R., Yoon, H.S., and Lescar, J. (2008) Crystal structure of the FK506 binding domain of *Plasmodium falciparum* FKBP35 in complex with FK506. *Biochemistry*, **47**, 5951–5961.

56 Adams, B., Musiyenko, A., Kumar, R., and Barik, S. (2005) A novel class of dual-family immunophilins. *J. Biol. Chem.*, **280**, 24308–24314.

16

Dehydrogenases and Enzymes of the Mitochondrial Electron Transport Chain as Anti-Apicomplexan Drug Targets

Kathleen Zocher, Stefan Rahlfs, and Katja Becker[*]

Abstract

Dehydrogenases and mitochondrial redox-associated processes are essentially involved in central energy metabolism of the apicomplexan parasites, including *Plasmodium, Toxoplasma, Babesia, Eimeria, Cryptosporidium, Haemoproteus,* and *Neospora* species. Despite the great potential of these enzymes and pathways to serve as target sites for novel antiparasitic control strategies, only a few of them have yet been characterized in detail. In this chapter, a review is provided of the current knowledge of mitochondrial enzymes of the electron transport chain and some dehydrogenases of the tricarboxylic acid (TCA) cycle of apicomplexan parasites, with respect to their function, structure, localization, and potential as drug targets.

Introduction

A constant energy supply is essential to maintain the structure and function of life. In eukaryotic organisms, this is achieved by an oxygen-dependent energy production, coupled to exergonic redox reactions. The electron transport necessary for these reactions is carried out by a reaction chain effected by four multienzyme complexes in the inner mitochondrial membrane, with the electrons that are required for the process originating from several biochemical reactions.

The mitochondrial enzymes of the Apicomplexa have attracted increasing attention during the past decade, since their significance as drug targets was recognized. Recent reviews on this topic [1–6] serve as a résumé of information about mitochondrial processes and their concomitant enzymes in human pathogenic apicomplexan parasites.

Many of the Apicomplexa have an electron transport chain (ETC) which roughly resembles that of the yeast, *Saccharomyces cerevisiae*. This includes the presence of succinate : ubiquinone oxidoreductase, ubiquinol : cytochrome-c oxidoreductase, and cytochrome oxidase activities [2]. Furthermore, the genomes of the Apicomplexa

[*]Corresponding author

Apicomplexan Parasites. Edited by Katja Becker
Copyright © 2011 WILEY-VCH Verlag GmbH & Co. KGaA, Weinheim
ISBN: 978-3-527-32731-7

encode five enzymes, which can supply electrons to the inner-membrane complex III via ubiquinone [1]. First, there is an NADH dehydrogenase that oxidizes NADH to NAD^+ and donates electrons to coenzyme Q. Succinate dehydrogenase (complex II) and malate : quinone oxidoreductase presumably are the only membrane-bound enzymes of the tricarboxylic acid (TCA) cycle which are directly integrated in the mitochondrial ETC by delivering reducing equivalents via $FADH_2$. Due to uncertainties about the TCA cycle in *Plasmodium*, it is not known from where the electrons donated from these enzymes ultimately originate [1]. The best-characterized enzyme that donates electrons to coenzyme Q is dihydroorotate dehydrogenase (DHOHD), although additional electrons can be delivered via the glycerol 3-phosphate shuttle, which consists of an NAD^+- and FAD-linked glycerol 3-phosphate dehydrogenase. The diversity in the parasites' metabolism and the enzymes associated to the ETC – especially the TCA cycle and other mitochondrial redox-associated enzymes – facilitate the development of promising new antimalarial agents.

Additionally, some of these enzymes [e.g., isocitrate dehydrogenase (IDH) and glutamate dehydrogenase (GDH)] supposedly supply an important portion of the NADPH that is required for the reduction of glutathione disulfide (GSSG) and thioredoxin [7], which are essential for antioxidant defense and redox regulation, respectively [8].

Enzymes of the Electron Transport Chain

Instead of the multisubunit NADH : ubiquinone oxidoreductase typically found in eukaryotes, the apicomplexan genes code for a type II alternative NADH dehydrogenase [1]. These oxidoreductases are single-polypeptide, non-proton-pumping enzymes [2, 9], whose main function is related to respiratory chain-linked NADH turnover. The enzymes are characterized by their lack of any transmembrane domain, and their insensitivity to complex I inhibitors such as rotenone and piericidin A [10]. However, the PfNDH2 (PFI0735c) was shown to be involved in the redox reaction of NADH oxidation with subsequent quinol production [11]. Until now, no potent inhibitors of *P. falciparum* type II NADH dehydrogenase have been identified. Indeed, the only high-affinity NDH2 inhibitor described to date is 1-hydroxy-2-alkyl-4(1*H*)quinolone (HDQ) (C12), which possesses a structural similarity to ubiquinone [12]. This compound, and some derivatives of it, were shown also to inhibit the activities of *Yarrowia lipolytica* NDH2 and *T. gondii* NDH2-I [12, 13]. No details of the crystal structure of any apicomplexan NDH2 are available to date.

The second complex of the ETC, also of apicomplexan parasites – the succinate : ubiquinone oxidoreductase (succinate dehydrogenase; SDH) – converts succinate to fumarate by reducing FAD in the process. The two genes of the major catalytic subunits, SDHA (PfSDH1) for the flavoprotein (Fp; PF10_0334) and SDHB (PfSDH2) for the iron–sulfur protein (Ip; PFL0630w), which form the soluble part of *P. falciparum* complex II, have been cloned and characterized by Takeo *et al.* [14]. In contrast to these subunits, PfSDH3 and PfSDH4 (heterodimer; membrane anchor b-type cytochrome) appear to be highly divergent from their mitochondrial orthologs, and are not annotated in the current databases at NCBI, GeneDB, or PlasmoDB [15]. SDH activity in parasite homogenates was determined by Suraveratum *et al.* for *P. falciparum* [16], and by Kawahara *et al.* for *P. yoelii* [17]. The results of both studies showed the kinetic properties to be similar to those of the mammalian enzymes. In the case of *T. gondii*, SDH activity was identified found in cell homogenates of the tachyzoites, but not of the bradyzoites [18]. No updated information is available for the succinate : ubiquinone oxidoreductase of the other Apicomplexa.

The ubiquinone, which is reduced by reducing equivalents of different origin, donates electrons to complex III (cytochrome bc_1 complex; cytochrome c reductase). All three of the core catalytic subunits of complex III have homologs in the *P. falciparum* genome: cytochrome b (encoded on the mitochondrial genome); the Rieske iron–sulfur protein (PF14_0373); and cytochrome c_1 (PF14_0038) [1]. In 1993, Fry and Pudney [19] demonstrated that the parasite mitochondrial cytochrome bc_1 complex (complex III) could be inhibited by atovaquone, a substituted 2-hydroxy-naphthoquinone, at a 1000-fold lower concentration than the complex in mammalian mitochondria. It was further shown that mutations in the coenzyme Q binding site of cytochrome b correlated with the resistance to atovaquone, strongly suggesting that this complex was the major target of this compound [20]. Later, other investigators noted a competitive inhibition mechanism of ubiquinol oxidation that acted via interactions with the Rieske iron-sulfur protein and cytochrome b in the coenzyme Q binding pocket [21, 22]. Additionally, atovaquone was shown to inhibit the cytochrome bc_1 complex of *T. gondii* and *Babesia* spp. [23, 24]. Unfortunately, high-level resistance to atovaquone occurred at a relatively high frequency, and correlated with mutations at position 268 (Tyr) of cytochrome b [4]. On the basis of a structural model, which was generated by Kessl *et al.* [21] and describes the binding mode of atovaquone in the ubiquinol oxidation pocket of cytochrome b, different *in silico* approaches were conducted in order to identify new and more specific inhibitors [25, 26]. For docking analyses, the protein structure of the *S. cerevisiae* enzyme (PDB-ID 1KYO) was used because it showed a similar inhibition behavior by atovaquone (Figure 16.1). Thereby, a 2-hydroxy-naphtoquinone with a branched side chain [25] and (1*H*-pyridin-4-ylidene)amine derivatives [26] were identified as putative drug candidates. Winter *et al.* [27] synthesized haloalkoxy-acridone derivatives, which also appeared to target the bc_1 complex, as well as 4-pyridone analogs of the clopidol class that are presently at an advanced stage of development [28, 29].

Interference with the transfer of electrons via the bc_1 complex in turn prevents enzymes such as DHODH, which are dependent on oxidized CoQ, from being able to carry out their own catalytic reaction via CoQ reduction [30]. DHODH is a flavin-containing protein that catalyzes the conversion of dihydroorotate (DHO) to orotic acid (OA) in the fourth and only redox reaction of the *de novo* pyrimidine biosynthesis [31]. Whereas, malarial parasites depend solely on the *de novo* biosynthesis of pyrimidines [32–34], *Cryptosporidium* is able to scavenge preformed pyrimidines of the host via the salvage pathway [35]. The importance of the balance between a salvage mechanism and *de novo* biosynthesis is well illustrated by the case of *T. gondii*, which falls between *Plasmodium* and *Cryptosporidium* in its abilities to salvage pyrimidines [36]. The dependency of the malarial parasites on *de novo* pyrimidine

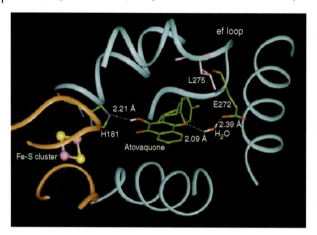

Figure 16.1 A structural model of the binding pocket of yeast bc$_1$ complex with bound atovaquone. Cytochrome b is shown in cyan, and a portion of the Rieske iron-sulfur protein is in orange. The iron–sulfur cluster is at the lower left, and the iron and sulfur atoms are colored purple and yellow, respectively. Atovaquone is hydrogen-bonded to His 181 of the Rieske protein, and to a water molecule that forms a bridge to Glu272 through a second hydrogen bond. Color code: oxygen = red; nitrogen = blue; hydrogen = white. Modified according to Ref. [26].

synthesis makes this pathway an attractive drug target. Indeed, different approaches for combating malaria have used this strategy to identify novel antimalarial agents. Drug interventions to block pyrimidine synthesis have traditionally targeted the folate pathway with antifolates [36], the principal antimalarial antifolate drugs being pyrimethamine (PYR), proguanil (PG), and sulfa drugs, including sulfonamides such as sulfadoxine (SDX) and sulfones such as dapsone [36]. However, many studies have also concentrated on the identification of new chemical compounds capable of inhibiting PfDHODH (PFF0160c). A native DHODH activity in *P. falciparum* was primarily described by Kungrai [37], and the protein crystal structure in complex with A77 1726 as bound inhibitor resolved by Hurt *et al.* [38] (Figure 16.2). By transfecting parasites with a fusion protein of *S. cerevisiae* DHODH gene and green fluoresecent protein (GFP), the dependency of PfDHODH on CoQ could be examined [30]. Thus, the yeast DHODH was shown to be independent of CoQ, when using fumarate as an electron acceptor [39]. In a standard 48 h growth inhibition assay, untransfected parasites were fully susceptible to atovaquone, with an IC$_{50}$ of about 1 nM, whereas the transgenic parasites were not inhibited even by 2.25 μM atovaquone. This suggested that the malarial mitochondrial ETC's main function is the reduction of ubiquinone, which is required as an electron acceptor for PfDHODH [30]. The first reported high-throughput screening (HTS) against malarial DHODH employed a 220 000-compound library, and resulted in the identification of four different chemical classes with antimalarial potential, namely phenylbenzamides, ureas, naphthamides, and triazolopyrimidine-based compounds [40]. When the X-ray structures of the PfDHODH that bound to three inhibitors from the triazolopyr-

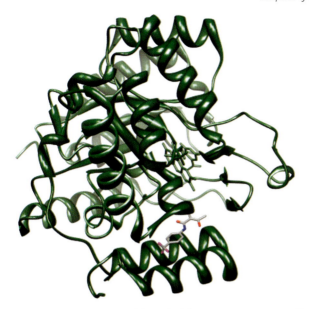

Figure 16.2 PfDHODH with bound inhibitor A771726 (PDB-ID 1TV5). Orotic acid and FMN are represented as green sticks. The inhibitor A771726 is also shown in stick representation, and colored by its atom types (oxygen = red; nitrogen = blue; fluorine = pink).

imidine-based series were resolved [41], they provided not only significant insights into the emerging structure–activity relationship (SAR) but also information regarding how these ligands could be further optimized in order to achieve a highly specific inhibition of only the parasitic enzyme. A second HTS with a library of 208 000 small molecules was conducted by Patel *et al.* [42]. Structure-based methods and the synthesis of brequinar and A771726 analogs were also investigated, but proved to be unsuccessful [43]. To date, no concise information has been made available on the mechanism of DHODH inhibition, although the on-hand results of various research groups have indicated that all inhibitors against PfDHODH share an overlapping binding site. Unfortunately, whether such overlap occurs with the CoQ binding site has not yet been clearly proven, even on the basis of crystallographic data.

The electrons from complex III are donated to cytochrome c, which by itself donates the electrons to complex IV. *Plasmodium falciparum* contains homologs of all three subunits of the catalytic core of complex IV, CoxI, and CoxIII encoded on the mitochondrial genome, as well as the nuclear-encoded CoxIIa/b (PF14_0288; PF13_0327) [1]. Complex IV in other eukaryotes contains several other proteins that function in regulatory roles, but in the *P. falciparum* genome none of these is apparent [1]. The *Plasmodium* complex IV activity has been detected by Kungrai *et al.* [44], and confirmed that known inhibitors such as cyanide would reduce oxygen consumption and depolarize $\Delta\Psi$ across the inner membrane [45, 46].

Cryptosporidium parasites have eliminated nearly all of the electron transport complexes, except for a simple ETC from NADH dehydrogenase and malate : quinone oxidoreductase to an alternative ubiquinol oxidase [2]. The sensitivity of parasites to inhibitors of the mitochondrial ETC underlines the importance of these reactions to the parasites [4], and their potential as drug targets.

Dehydrogenases of the TCA Cycle as Drug Targets

As the pyruvate dehydrogenase of all is localized to the apicoplast [47], the major carbon flow from the cytoplasm to the mitochondrion will be disconnected. For this reason, it remains unclear as to whether the apicomplexan parasites possess a complete TCA cycle, and which function this putative cycle might fulfill. Among the apicomplexan parasites – with the exception of *Cryptosporidium* spp. – all genes relevant to the enzymes of the TCA cycle are present in their genomes [1].

The evidence obtained to date suggests that, during the erythrocytic stages, the TCA cycle of the malarial parasite's mitochondria is unlikely to perform as a cycle [4]. Consequently, in the following subsections, attention will be focused on the dehydrogenases that are primarily involved in this pathway.

In the genome of *P. falciparum*, only one gene encoding an $NADP^+$-dependent IDH (PF13_0242) could be identified [48]; likewise, for *T. gondii* an $NADP^+$-dependent IDH activity was determined [18]. As Vander Jagt *et al.* [7] described an $NADP^+$-dependent PfIDH-activity in the asexual stages of the parasite, it would seem unlikely that the parasite's enzyme would supply electrons to the ETC, unless the plasmodial ETC was able to receive reducing equivalents via NADPH. In the reaction of PfIDH, a further pathway for the generation of NADPH is available to the parasites, providing the enzyme with a potential role in mitochondrial redox metabolism. In fact, upon oxidative stress the PfIDH is upregulated on both the mRNA and protein levels [49]. Inhibition studies on *P. knowlesi* and *P. falciparum* IDH, using 156 μM chloroquine, led to a slightly more than 50% reduction of PkIDH activity and a 30% reduction of PfIDH activity, respectively [48, 50]. While cupric and argentic ions almost completely deactivated the *P. knowlesi* enzyme at 0.1 mM concentration, the *P. falciparum* enzyme activity was inhibited by 68% with $CuSO_4$, but by only 20% with $AgNO_3$. At 5 mM adenosine triphosphate (ATP), the PfIDH activity could be reduced to 42%. Although, to date, the structural characterization of the IDH of the Apicomplexa has not been resolved, this would appear worthwhile if only to accelerate the approach to drug discovery.

Vaidya and Mather [4] hypothesized that the major entry point for the TCA "cycle" in malarial parasites appeared to be α-ketoglutarate produced from glutamate, through the likely reaction of glutamate dehydrogenase (Figure 16.3). *Plasmodium falciparum* possesses three isoenzymes of GDH (GDH 1–3), with PfGDH1 (PF14_0164) being the first to be biochemically and structurally characterized [51, 52]. In addition, PfGDH2 (PF14_0286) has been studied in detail by the present authors' group (K. Zocher *et al.*, unpublished results). PfGDH 1 and 2 are both $NADP^+$-dependent enzymes, which suggests that they are mainly involved in the

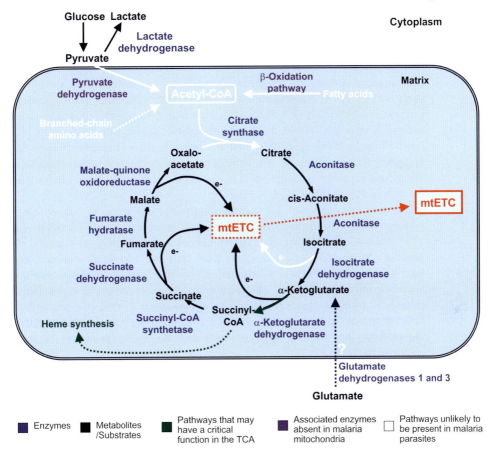

Figure 16.3 Pathways and function of the tricarboxylic acid (TCA) "cycle" in malarial parasites. The enzymes and substrates of the classical TCA cycle are shown. Because pyruvate dehydrogenase is localized to the apicoplast, pyruvate generated through glycolysis is unlikely to be the source of acetyl CoA. Additionally, the enzymes of fatty acid β-oxidation or the catabolism of branched-chain amino acids seem to be absent from the parasite's mitochondria. mtETC, mitochondrial electron transport chain. Modified according to Ref. [4].

anabolism of glutamate and the assimilation of ammonium. Localization studies with PfGDH–GFP fusion proteins (also conducted by the present authors' group) revealed that the NAD$^+$-dependent GDH3 (PF8_0132) and the NADP$^+$-dependent GDH1 were both cytosolic proteins, whereas GDH2 was located to the apicoplast (K. Zocher *et al.*, unpublished results). In other organisms, such as yeast, the need for substrate cycling between the NAD$^+$- and the NADP$^+$-dependent GDHs, both of which were cytosolic, could be demonstrated [53]. The NADP$^+$-GDH is assumed to be primarily responsible for replenishment of the NADP$^+$ pool, whereas the main function of the NAD$^+$-GDH is to metabolize the glutamate generated therein [53].

Alternatively, glutamate can be converted into α-ketoglutarate by aspartate aminotransferase, while oxaloacetate is transaminated to aspartate. However, this enzyme is not predicted to be mitochondrially located. As none of the enzymes mentioned above are localized to the mitochondrion, the hypothesis that α-ketoglutarate produced from glutamate by GDH would serve as a major substrate entry for the TCA in *Plasmodium* must be questioned [4]. Indeed, when approaching this issue it would be helpful to clarify whether the predicted oxoglutarate/malate transporter (PF08_0031; tricarboxylate/dicarboxylate exchanger), as involved in the exchange of TCA cycle intermediates, would also function in the opposite direction. It would also be important to determine whether the above-mentioned aspartate aminotransferase was located to the mitochondrion, despite the lack of any predicted signal sequence.

The enzyme catalyzing the reaction from α-ketoglutarate to succinyl-CoA, α-ketoglutarate dehydrogenase (multienzyme complex; E1: PF08_0054; E2: PF13_0121; E3: PFL1550w), can also be deemed an important drug target, as its product – succinyl-CoA – serves as a substrate for 5-aminolevulinate synthase. The latter enzyme catalyzes the condensation of succinyl-CoA with glycine to produce 5-aminolevulinate, providing the rate-limiting first step of heme biosynthesis. Notably, *P. falciparum* parasites depend on *de novo* heme biosynthesis during their blood stages, as they cannot utilize heme derived from host's hemoglobin [54]. For the parasite, the synthetic heme is required in the cytochromes and in other components of the ETC.

The final step of the TCA cycle is catalyzed by a fifth dehydrogenase, namely malate dehydrogenase (MDH). Previously, Lang-Unnasch *et al.* [55] showed that *P. falciparum* contained a single isoenzyme of MDH, that was thought to localize to the cytosol of the parasite, whilst within the mitochondrion a malate : quinone oxidoreductase (PFF_0815w) was described that supplied electrons directly to the ETC via $FADH_2$ [1]. Consequently, MDH was recognized as an important link between glycolysis, the TCA cycle, and the aspartate malate/oxaloacetate shuttle [56]. Since the cytosolic MDH was considered to belong to the "LDH-like MDH" group, it was regarded as an interesting drug target [57]. In fact, Tripathi *et al.* [58] subsequently showed that PfMDH might complement the PfLDH function of NAD/NADH coupling, such that dual inhibitors of both enzymes would be required to target this pathway in the development of new antimalarial drugs.

Conclusion

During the past decade, the enzymes of the ETC of the Apicomplexa have undergone extensive study. In the case of *P. falciparum*, all four enzyme complexes have been investigated, mainly to elucidate their function and their interactions. Although both PfNDH2 and succinate : ubiquinone oxidoreductase have been functionally characterized, no potent inhibitors have yet been detected for these schemes. Nonetheless, the data reported by Painter *et al.* [30] have suggested that complex III, as well as the CoQ-mediated downstream enzymes and reactions, have emerged as attractive drug targets. As the role of ATP synthase is not yet fully understood, its importance in ATP production or its other functions require further investigation. To date, the main function of the parasitic mitochondrial ETC has been suggested as the regeneration of CoQ, which is required for the DHODH reaction [30]. The absence of pyruvate dehydrogenase, an $NADP^+$-dependent IDH, together with a lack of mitochondrially located glutamate dehydrogenase, indicate that the TCA cycle's enzymatic steps differ notably from those of the host. Consequently, both knockout studies and interaction studies of the TCA cycle enzymes should be conducted in future to confirm the importance of each. An additional essential point to improve the present understanding of these fundamental processes of the TCA cycle would be to determine the means by which intermediates are transported from the cytosol to the mitochondria, and *vice versa*, together with all existent shuttle systems. Only with this information at hand might it be possible to validate the potential drug targets of these pathways, and to control their specific inhibition.

References

1 Van Dooren, G.G., Stimmler, L.M., and McFadden, G.I. (2006) Metabolic maps and functions of the *Plasmodium* mitochondrion. *FEMS Microbiol. Rev.*, **30**, 596–630.

2 Mather, M.W., Henry, K.W., and Vaidya, A.B. (2007) Mitochondrial drug targets in apicomplexan parasites. *Curr. Drug Targets*, **8**, 49–60.

3 Seeber, F., Limenitakis, J., and Soldati-Favre, D. (2008) Apicomplexan mitochondrial metabolism: a story of gains, losses and retentions. *Trends Parasitol.*, **24**, 468–478.

4 Vaidya, A.B. and Mather, M.W. (2009) Mitochondrial evolution and functions in malaria parasites. *Annu. Rev. Microbiol.*, **63**, 249–267.

5 Torrentino-Madamet, M., Desplans, J., Travaillé, C., James, Y., and Parzy, D. (2010) Microaerophilic respiratory metabolism of *Plasmodium falciparum* mitochondrion as a drug target. *Curr. Mol. Med.*, **10**, 29–46.

6 Rodrigues, T., Lopes, F., and Moreira, R. (2010) Inhibitors of the mitochondrial electron transport chain and de novo pyrimidine biosynthesis as antimalarials: The present status. *Curr. Med. Chem.*, **17**, 929–956.

7 Vander Jagt, D.L., Hunsaker, L.A., Kibirige, M., and Campos, N.M. (1989) NADPH production by the malarial parasite *Plasmodium falciparum*. *Blood*, **74**, 471–474.

8 Hunt, N.H. and Stocker, R. (1990) Oxidative stress and the redox status of malaria-infected erythrocytes. *Blood Cells*, **16**, 499–526; discussion 527–530.

9 Saleh, A., Friesen, J., Baumeister, S., Gross, U., and Bohne, W. (2007) Growth inhibition of *Toxoplasma gondii* and *Plasmodium falciparum* by nanomolar concentrations of 1-hydroxy-2-dodecyl-4(1*H*)quinolone, a high-affinity inhibitor of alternative (type II) NADH dehydrogenases. *Antimicrob. Agents Chemother.*, **51**, 1217–1222.

10 Melo, A.M., Bandeiras, T.M., and Teixeira, M. (2004) New insights into type II NAD(P)H:quinone oxidoreductases. *Microbiol. Mol. Biol. Rev.*, **68**, 603–616.

11 Biagini, G.A., Viriyavejakul, P., O'Neill, P.M., Bray, P.G., and Ward, S.A. (2006) Functional characterization and target validation of alternative complex I of *Plasmodium falciparum* mitochondria. *Antimicrob. Agents Chemother.*, **50**, 1841–1851.

12 Lin, S.S., Kerscher, S., Saleh, A., Brandt, U., Gross, U., and Bohne, W. (2008) The *Toxoplasma gondii* type-II

NADH dehydrogenase TgNDH2-I is inhibited by 1-hydroxy-2-alkyl-4(1*H*) quinolones. *Biochim. Biophys. Acta*, **1777**, 1455–1462.

13 Eschemann, A., Galkin, A., Oettmeier, W., Brandt, U., and Kerscher, S. (2005) HDQ (1-hydroxy-2-dodecyl-4(1*H*)quinolone), a high affinity inhibitor for mitochondrial alternative NADH dehydrogenase: evidence for a ping-pong mechanism. *J. Biol. Chem.*, **280**, 3138–3142.

14 Takeo, S., Kokaze, A., Ng, C.S., Mizuchi, D., Watanabe, J.I., Tanabe, K. *et al.* (2000) Succinate dehydrogenase in *Plasmodium falciparum* mitochondria: molecular characterization of the SDHA and SDHB genes for the catalytic subunits, the flavoprotein (Fp) and iron-sulfur (Ip) subunits. *Mol. Biochem. Parasitol.*, **107**, 191–205.

15 Mogi, T. and Kita, K. (2009) Identification of mitochondrial Complex II subunits SDH3 and SDH4 and ATP synthase subunits a and b in *Plasmodium* spp. *Mitochondrion*, **9**, 443–453.

16 Suraveratum, N., Krungkrai, S.R., Leangaramgul, P., Prapunwattana, P., and Krungkrai, J. (2000) Purification and characterization of *Plasmodium falciparum* succinate dehydrogenase. *Mol. Biochem. Parasitol.*, **105**, 215–222.

17 Kawahara, K., Mogi, T., Tanaka, T.Q., Hata, M., Miyoshi, H., and Kita, K. (2009) Mitochondrial dehydrogenases in the aerobic respiratory chain of the rodent malaria parasite *Plasmodium yoelii yoelii*. *J. Biochem.*, **145**, 229–237.

18 Denton, H., Roberts, C.W., Alexander, J., Thong, K.W., and Coombs, G.H. (1996) Enzymes of energy metabolism in the bradyzoites and tachyzoites of *Toxoplasma gondii*. *FEMS Microbiol. Lett.*, **137**, 103–108.

19 Fry, M. and Pudney, M. (1992) Site of action of the antimalarial hydroxynaphthoquinone, 2-[trans-4-(4′-chlorophenyl) cyclohexyl]-3-hydroxy-1,4-naphthoquinone (566C80). *Biochem. Pharmacol.*, **43**, 1545–1553.

20 Srivastava, I.K., Morrisey, J.M., Darrouzet, E., Daldal, F., and Vaidya, A.B. (1999) Resistance mutations reveal the atovaquone-binding domain of

cytochrome b in malaria parasites. *Mol. Microbiol.*, **33**, 704–711.

21 Kessl, J.J., Lange, B.B., Merbitz-Zahradnik, T., Zwicker, K., Hill, P., Meunier, B. *et al.* (2003) Molecular basis for atovaquone binding to the cytochrome bc₁ complex. *J. Biol. Chem.*, **278**, 31312–31318.

22 Mather, M.W., Darrouzet, E., Valkova-Valchanova, M., Cooley, J.W., McIntosh, M.T., Daldal, F., and Vaidya, A.B. (2005) Uncovering the molecular mode of action of the antimalarial drug atovaquone using a bacterial system. *J. Biol. Chem.*, **280**, 27458–27465.

23 Krause, P.J., Lepore, T., Sikand, V.K., GadbawJr, G., Burke, G., and Telford, S.R. 3rd *et al.* (2000) Atovaquone and azithromycin for the treatment of babesiosis. *N. Engl. J. Med.*, **343**, 1454–1458.

24 Fisher, N., Bray, P.G., Ward, S.A., and Biagini, G.A. (2008) Malaria-parasite mitochondrial dehydrogenases as drug targets: too early to write the obituary. *Trends Parasitol.*, **24**, 9–10.

25 Kessl, J.J., Meshnick, S.R., and Trumpower, B.L. (2007) Modeling the molecular basis of atovaquone resistance in parasites and pathogenic fungi. *Trends Parasitol.*, **23**, 494–501.

26 Rodrigues, T., Guedes, R.C., dos Santos, D.J., Carrasco, M., Gut, J., Rosenthal, P.J. *et al.* (2009) Design, synthesis and structure-activity relationships of (1*H*-pyridin-4-ylidene)amines as potential antimalarials. *Bioorg. Med. Chem. Lett.*, **19**, 3476–3480.

27 Winter, R.W., Kelly, J.X., Smilkstein, M.J., Dodean, R., Bagby, G.C., Rathbun, R.K. *et al.* (2006) Evaluation and lead optimization of anti-malarial acridones. *Exp. Parasitol.*, **114**, 47–56.

28 Xiang, H., McSurdy-Freed, J., Moorthy, G.S., Hugger, E., Bambal, R., Han, C. *et al.* (2006) Preclinical drug metabolism and pharmacokinetic evaluation of GW844520, a novel anti-malarial mitochondrial electron transport inhibitor. *J. Pharm. Sci.*, **95**, 2657–2672.

29 Yeates, C.L., Batchelor, J.F., Capon, E.C., Cheesman, N.J., Fry, M., Hudson, A.T.

et al. (2008) Synthesis and structure-activity relationships of 4-pyridones as potential antimalarials. *J. Med. Chem.*, **51**, 2845–2852.

30 Painter, H.J., Morrisey, J.M., Mather, M.W., and Vaidya, A.B. (2007) Specific role of mitochondrial electron transport in blood-stage *Plasmodium falciparum*. *Nature*, **446**, 88–91.

31 Malmquist, N.A., Baldwin, J., and Phillips, M.A. (2007) Detergent-dependent kinetics of truncated *Plasmodium falciparum* dihydroorotate dehydrogenase. *J. Biol. Chem.*, **282**, 12678–12686.

32 Sherman, I.W. (1979) Biochemistry of *Plasmodium* (malarial parasites). *Microbiol. Rev.*, **43**, 453–495. Review.

33 Gutteridge, W.E., Dave, D., and Richards, W.H. (1979) Conversion of dihydroorotate to orotate in parasitic protozoa. *Biochim. Biophys. Acta*, **582**, 390–401.

34 Rathod, P.K. and Reyes, P. (1983) Orotidylate-metabolizing enzymes of the human malarial parasite, *Plasmodium falciparum*, differ from host cell enzymes. *J. Biol. Chem.*, **258**, 2852–2855.

35 Xu, P., Widmer, G., Wang, Y., Ozaki, L.S., Alves, J.M., Serrano, M.G. *et al.* (2004) The genome of *Cryptosporidium hominis*. *Nature*, **431**, 1107–1112. Erratum in: *Nature* (2004) **432**, 415.

36 Hyde, J.E. (2007) Targeting purine and pyrimidine metabolism in human apicomplexan parasites. *Curr. Drug Targets*, **8**, 31–47.

37 Krungkrai, J. (1995) Purification, characterization and localization of mitochondrial dihydroorotate dehydrogenase in *Plasmodium falciparum*, human malaria parasite. *Biochim. Biophys. Acta*, **1243**, 351–360.

38 Hurt, D.E., Widom, J., and Clardy, J. (2006) Structure of *Plasmodium falciparum* dihydroorotate dehydrogenase with a bound inhibitor. *Acta Crystallogr. D Biol. Crystallogr.*, **62**, 312–323.

39 Nagy, M., Lacroute, F., and Thomas, D. (1992) Divergent evolution of pyrimidine biosynthesis between anaerobic and aerobic yeast. *Proc. Natl Acad. Sci. USA*, **89**, 8966–8970.

40 Baldwin, J., Michnoff, C.H., Malmquist, N.A., White, J., Roth, M.G., Rathod, P.K., and Phillips, M.A. (2005) High-throughput screening for potent and selective inhibitors of *Plasmodium falciparum* dihydroorotate dehydrogenase. *J. Biol. Chem.*, **280**, 21847–21853. Erratum in: *J. Biol. Chem.* (2005) **280**, 32048.

41 Deng, X., Gujjar, R., El Mazouni, F., Kaminsky, W., Malmquist, N.A., Goldsmith, E.J. *et al.* (2009) Structural plasticity of malaria dihydroorotate dehydrogenase allows selective binding of diverse chemical scaffolds. *J. Biol. Chem.*, **284**, 26999–27009.

42 Patel, V., Booker, M., Kramer, M., Ross, L., Celatka, C.A., Kennedy, L.M. *et al.* (2008) Identification and characterization of small molecule inhibitors of *Plasmodium falciparum* dihydroorotate dehydrogenase. *J. Biol. Chem.*, **283**, 35078–35085.

43 Boa, A.N., Canavan, S.P., Hirst, P.R., Ramsey, C., Stead, A.M., and McConkey, G.A. (2005) Synthesis of brequinar analogue inhibitors of malaria parasite dihydroorotate dehydrogenase. *Bioorg. Med. Chem.*, **13**, 1945–1967.

44 Krungkrai, J., Krungkrai, S.R., and Bhumiratana, A. (1993) *Plasmodium berghei*: partial purification and characterization of the mitochondrial cytochrome c oxidase. *Exp. Parasitol.*, **77**, 136–146.

45 Srivastava, I.K., Rottenberg, H., and Vaidya, A.B. (1997) Atovaquone, a broad spectrum antiparasitic drug, collapses mitochondrial membrane potential in a malarial parasite. *J. Biol. Chem.*, **272**, 3961–3966.

46 Uyemura, S.A., Luo, S., Moreno, S.N., and Docampo, R. (2000) Oxidative phosphorylation, Ca(27 +) transport, and fatty acid-induced uncoupling in malaria parasites mitochondria. *J. Biol. Chem.*, **275**, 9709–9715.

47 Foth, B.J., Stimmler, L.M., Handman, E., Crabb, B.S., Hodder, A.N., and McFadden, G.I. (2005) The malaria parasite *Plasmodium falciparum* has only one pyruvate dehydrogenase complex, which is located in the apicoplast. *Mol. Microbiol.*, **55**, 39–53.

48 Chan, M. and Sim, T.S. (2003) Recombinant *Plasmodium falciparum* NADP-dependent isocitrate dehydrogenase is active and harbours a unique 26 amino acid tail. *Exp. Parasitol.*, **103**, 120–126.

49 Wrenger, C. and Müller, S. (2003) Isocitrate dehydrogenase of *Plasmodium falciparum*. *Eur. J. Biochem.*, **270**, 1775–1783.

50 Sahni, S.K., Saxena, N., Puri, S.K., Dutta, G.P., and Pandey, V.C. (1992) NADP-specific isocitrate dehydrogenase from the simian malaria parasite *Plasmodium knowlesi*: partial purification and characterization. *J. Protozool.*, **39**, 338–342.

51 Wagner, J.T., Lüdemann, H., Färber, P.M., Lottspeich, F., and Krauth-Siegel, R.L. (1998) Glutamate dehydrogenase, the marker protein of *Plasmodium falciparum* – cloning, expression and characterization of the malarial enzyme. *Eur. J. Biochem.*, **258** (2), 813–819.

52 Werner, C., Stubbs, M.T., Krauth-Siegel, R.L., and Klebe, G. (2005) The crystal structure of *Plasmodium falciparum* glutamate dehydrogenase, a putative target for novel antimalarial drugs. *J. Mol. Biol.*, **349**, 597–607.

53 Boles, E., Lehnert, W., and Zimmermann, F.K. (1993) The role of the NAD-dependent glutamate dehydrogenase in restoring growth on glucose of a *Saccharomyces cerevisiae* phophoglucose isomerase mutant. *Eur. J. Biochem.*, **217**, 469–477.

54 Surolia, N. and Padmanaban, G. (1991) Chloroquine inhibits heme-dependent protein synthesis in *Plasmodium falciparum*. *Proc. Natl Acad. Sci. USA*, **88**, 4786–4790.

55 Lang-Unnasch, N. (1992) Purification and properties of *Plasmodium falciparum* malate dehydrogenase. *Mol. Biochem. Parasitol.*, **50**, 17–25.

56 Goward, C.R. and Nicholls, D.J. (1994) Malate dehydrogenase: a model for structure, evolution, and catalysis. *Protein Sci.*, **3**, 1883–1888.

57 Zhu, G. and Keithly, J.S. (2002) Alpha-proteobacterial relationship of apicomplexan lactate and malate dehydrogenases. *J. Eukaryot. Microbiol.*, **49**, 255–261.

58 Tripathi, A.K., Desai, P.V., Pradhan, A., Khan, S.I., Avery, M.A., Walker, L.A., and Tekwani, B.L. (2004) An alpha-proteobacterial type malate dehydrogenase may complement LDH function in *Plasmodium falciparum*. Cloning and biochemical characterization of the enzyme. *Eur. J. Biochem.*, **271**, 3488–3502.

17
Calcium-Dependent Protein Kinases as Drug Targets in Apicomplexan Parasites

Dominik Kugelstadt, Bianca Derrer, and Barbara Kappes[*]

Abstract

Calcium-dependent protein kinases (CDPKs) are the major transducers of calcium signals into protein phosphorylation events in apicomplexan parasites. CDPKs are abundant in plants and Alveolata to which the Apicomplexa belong, yet are absent from humans, which makes them attractive targets for anticoccidial drug development. Seven CDPK isoforms were identified in *Plasmodium* and *Cryptosporidium parvum*, and 12 in *Toxoplasma gondii*. Apicomplexan CDPKs are grouped into five classes, depending on their domain organization. However, the prototype of a classical CDPK is characterized by an N-terminal serine/threonine protein kinase domain fused to a regulatory calmodulin-like calcium-binding domain through a linker or junction region that can exhibit an autoinhibitory function. CDPK1 to 5 of the plasmodial CDPKs display this archetypical organization. Of the seven plasmodial CDPKs: (i) CDPK1 has been shown to be essential for blood-stage development, the phase of infection causing the symptoms of malaria; (ii) CDPK3, 4, and 6 are not vital for blood stages; and (iii) CDPK2, 5, and 7 have not been studied in enough detail to provide sufficient information on their essentiality. However, it is quite likely that one or more additional CDPKs are required for erythrocytic schizogony. Since PfCDPK1 is the only known plasmodial CDPK crucial for blood-stage development, efforts to identify selective inhibitors have concentrated on this particular enzyme. In total, three independent high-throughput screenings were carried out by different groups, resulting in the identification of highly specific inhibitors falling into at least three distinct chemical classes: the indol-2-ons; the 2,6,9-trisubstituted purines; and the imidazo[1,2-β]pyridazines. IC_{50} values for the inhibition of PfCDPK1 activity were in the low micromolar range for the indol-2-ons and in the low nanomolar range for the 2,6,9-trisubstituted purines and the imidazo[1,2-β]pyridazines. Their effectiveness on the parasite (EC_{50}) was in the low micromolar range for the indol-2-ons and imidazo[1,2-β]pyridazines, and in the middle to high nanomolar range for the 2,6,9-trisubstituted purines.

[*]Corresponding author

Apicomplexan Parasites. Edited by Katja Becker
Copyright © 2011 WILEY-VCH Verlag GmbH & Co. KGaA, Weinheim
ISBN: 978-3-527-32731-7

Introduction

Calcium is a major signaling molecule of the Apicomplexa, and controls processes such as cell motility, cell invasion, cell egress, development, differentiation, and protein secretion during the quite complex life cycles of the apicomplexan parasites [1–3]. In contrast to humans, where the vast majority of calcium-induced phosphorylation events occur through calmodulin-dependent protein kinases (CaMKs) and protein kinase C isoforms, the calcium-dependent protein kinases (CDPKs) are the primary source of calcium-induced phosphorylation events in the Apicomplexa. Until now, only a single classical CaMK has been described in *Plasmodium falciparum*, *Toxoplasma gondii*, and *Cryptosporidium parvum* [1–3]. The CaMKs, which are typical for animal cells but rare in plants, bind calmodulin through a C-terminal association domain. In contrast to the CaMKs, CDPKs contain a regulatory calmodulin-like Ca^{2+}-binding domain as an integral part of the protein. CDPKs are abundant in plants and Alveolata, including the Apicomplexa, but are absent from animal organisms, including humans [4]. This leads to the CDPKs becoming attractive chemotherapeutic targets for anticoccidial drug development.

The CDPKs form multigene families, emphasizing their importance for the respective organisms. In the plant organisms *Arabidopsis thaliana* and *Oryza sativa* 34 and 21, CDPK isoforms have been described, respectively [5]. In apicomplexan parasites, 12 CDPK isoforms have been indentified in *Toxoplasma gondii*, and seven in *Plasmodium* and in *Cryptosporidium parvum* (Figure 17.1) [1]. Based on their domain organization, the apicomplexan CDPKs have been grouped into five classes [1]. The archetypical domain structure of a classical CDPK is characterized by four C-terminal EF-hand calcium-binding domains, which are fused to a serine/threonine protein kinase domain through a linker or junction region that can have an autoinhibitory function [5]. However, other arrangements deviating from this model organization are also found. In the following, attention will be focused on the CDPK classes present in *Plasmodium*, since all inhibitor searches have been performed with PfCDPK1 of *P. falciparum*. The first two classes consist of CDPKs with the above-described classical organization. A distinction between classes 1 and 2 is that members of class 1 have a relatively short N terminus with acylation motifs, whereas the N termini of members of class 2 are extended by roughly 50 amino acids and do not possess N-terminal acylation motifs. PfCDPK1 and 4 fall into class 1, and PfCDPK1 has been shown to be myristoylated and palmitoylated [6]. Typical members of class 2 are PfCDPK3 and 5, whereas PfCDPK2 is an atypical member of this class. Its N terminus is roughly the same length as that of PfCDPK1 and PfCDPK5, and possesses a similar N-terminal myristoylation signal [7]. An example for the next class is PfCDPK6, consisting of a protein kinase domain framed N-terminally by two and C-terminally three EF-hand motifs. The N-terminal and C-terminal EF-hand motifs may vary in number from two to three in front of the protein kinase domain, and from three to four after the catalytic domain, depending on the individual CDPK [1]. The key feature of the fourth class of CDPKs, to which PfCDPK7 belongs, is a plekstrin homology domain separating two or more N-terminal EF-hands (two in the case of PfCDPK7) from the C-terminal protein

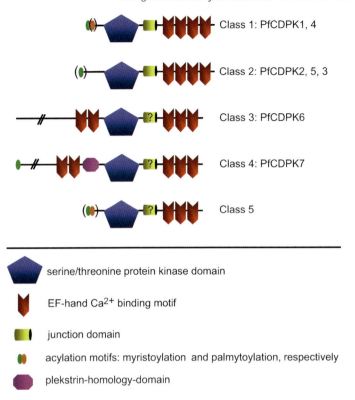

Class 1: PfCDPK1, 4

Class 2: PfCDPK2, 5, 3

Class 3: PfCDPK6

Class 4: PfCDPK7

Class 5

serine/threonine protein kinase domain

EF-hand Ca^{2+} binding motif

junction domain

acylation motifs: myristoylation and palmytoylation, respectively

plekstrin-homology-domain

Figure 17.1 Schematic representation of the CDPK classes identified in apicomplexan parasites. The parentheses indicate that the respective motif might be present or absent. The question marks indicate that the presence of a junction domain exhibiting autoinhibitory function has not been established.

kinase domain. The members of the fifth class are characterized by the classical organization described above, but possess three instead of four C-terminal EF-hands. *Plasmodium* has no members in this group.

Biological Functions of the Individual CDPKs in the Plasmodial Life Cycle

Calcium-dependent signaling is important for development and differentiation of the malarial parasite, and for its progression through the life cycle [8]. Accordingly, individual CDPKs have distinct key regulatory functions at defined phases of the malaria life cycle [3]. Chemical and genetic approaches have been used to clarify the function of individual CDPKs in *P. falciparum* and in the model rodent parasite *Plasmodium berghei*. As a review of the biological functions of the individual CDPKs is available [1], these details will be summarized only briefly at this point.

Gene knockouts of *cdpk*3, *cdpk*4, and *cdpk*6 have been successful in *P. berghei*, indicating that the activity of these kinases is not essential for blood-stage development [1]. These data also imply that the respective kinases are not suitable targets for the development of new antimalarial agents. In contrast, *pfcdpk1* is essential for blood-stage development, as its gene disruption is not possible in either *P. falciparum* or *P. berghei* [2]. As *pfcdpk2*, *pfcdpk5*, and *pfcdpk7* have not been studied in detail, no information is available on either their biological functions or the phase(s) of the malaria life cycle at which their activity is required. However, since PfCDPK1 during blood-stage development is primarily found in the parasitophorous vacuolar compartment and membranous systems derived thereof [6], this kinase is not able to transduce calcium signals within the parasite. Thus, it is most likely that one or more of the latter CDPKs are also required for erythrocytic schizogony, and may serve as potential targets for antimalarial drug development.

The *Plasmodium* infection is initiated when sporozoites are injected by the mosquito vector into the skin of a human host. Further development requires the sporozoites to travel to the liver, invade and develop into exoerythrocytic forms within the hepatocytes, and subsequently give rise to several thousand merozoites. The sporozoites contact and traverse many cells as they migrate from the skin to the liver. However in order to proceed in the life cycle, and to establish a schizogony in the liver, they need to switch from a migratory to an invasive mode. *Pb*CDPK6 has been shown to be one of the regulators controlling this switch [9]. Merozoites released from the liver schizonts enter the bloodstream, where they infect erythrocytes and proliferate, thereby causing the symptoms and pathology of malaria. PfCDPK1 is essential for erythrocytic schizogony, and most likely has multiple functions during blood-stage development such as invasion, membrane biogenesis, signaling tasks in the parasitophorous vacuolar compartment, and egress [2, 6, 10]. Phosphorylation of proteins of the parasite's motor complex, such as GAP45 and MTIP [10, 11], further indicates that PfCDPK1 activity is required for all of the parasite's invasive stages. This assumption is strengthened by data showing that sporozoite gliding motility is inhibited by a PfCDPK1-specific inhibitor (D. Kugelstadt, J. Baade, F. Frischknecht, B. Kappes, unpublished observations). Some of the blood-stage parasites differentiate into male and female gametocytes, which require passage into the mosquito's midgut to complete their development into male and female gametes. Gametocytogenesis is induced by three environmental signals: a drop in temperature; a small rise in pH; and the presence of xanthurenic acid, the mosquito's major catabolite of tryptophan [1, 12]. These factors lead to a sharp rise in cytosolic levels of Ca^{2+}. *Pb*CDPK4, a male-expressed CDPK, couples the increase in intracellular calcium to cell cycle progression and, as a consequence, its gene knockout arrests male gametocytes before S-phase and thus prevents exflagellation of the male gametes [13]. Following fertilization, the zygote transforms into a mobile ookinete that crosses the midgut epithelium and develops into an oocyst. *Pb*CDPK3 is required for ookinete motility, and ookinetes with a knockout of the *pbcdpk3* gene fail to establish sporogony in the mosquito vector [14, 15].

Regulatory Controls of CDPKs

Regulation Through Calcium

The regulatory calmodulin like calcium-binding apparatus of a classical CDPK consists of two structural domains which have been termed the N- and C-lobes, each containing two EF-hand helix–loop–helix calcium-binding motifs (Figure 17.2) [16]. A biophysical analysis of the *Arabidopsis thaliana* CDPK isoform CPK-1 revealed that the two lobes differed significantly in their affinity to calcium, and that the calcium affinity of the N-lobe was significantly lower than that of the C-lobe, which implied that the two lobes have differential roles in the activation of CDPKs [16]. The authors proposed a model according to which the C-lobe interacts with the junction domain at ordinary low cytosolic calcium concentrations, keeping the kinase in an autoinhibited state (Figure 17.2). Activation of the enzyme occurs when the calcium concentrations rise to a level that allows filling of the two weaker

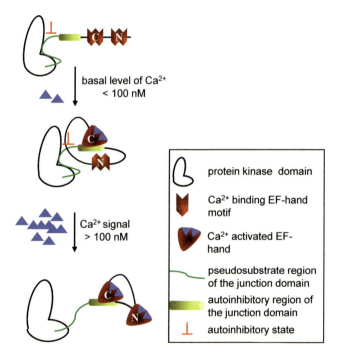

basal level of Ca^{2+}
< 100 nM

Ca^{2+} signal
> 100 nM

protein kinase domain

Ca^{2+} binding EF-hand motif

Ca^{2+} activated EF-hand

pseudosubstrate region of the junction domain

autoinhibitory region of the junction domain

autoinhibitory state

Figure 17.2 Activation model for CDPKs according to Christodoulou *et al.* [16]. At basal Ca^{2+} levels, the Ca^{2+}-binding sites of the C-loop (C) are likely to be occupied. The C-loop is interacting with the autoinhibitory domain of the junction region. The pseudosubstrate motif of the junction domain occupies the catalytic site, keeping the kinase in an inactive state. Activation of the kinase occurs when a Ca^{2+} signal reaches a certain threshold. Upon calcium binding to the N-lobe (N), the pseudosubstrate region is released from the catalytic cleft, allowing for kinase activation.

affinity binding sites in the N-lobe, thereby inducing the conformational change required for release of the autoinhibitory domain (Figure 17.2). The assignment of a Ca^{2+} sensor to the N-lobe is in line with the mutational analysis of the individual calcium-binding sites of PfCDPK1. Both, Ca^{2+}-dependent activation of the kinase as well as Ca^{2+}-induced conformational changes were prevented by mutations of the EF-hands I and II in the N-lobe, whereas mutations of the EF-hands III and IV had only minor effects [17].

The number and variations of the EF-hand motifs among the various CDPKs indicate that the activity of CDPKs may be fine-tuned for distinct functions. Harper *et al.* (2004) suggested that, in plants, multiple CDPK isoforms are needed in order to decode the different calcium signals, in response either to the amplitude or the frequencies of a calcium signal [18]. For example, soybean CDPK alpha and gamma isoforms differ dramatically in their thresholds required for their activation through calcium. Accordingly, the alpha isoform is tenfold more sensitive, and is activated by much lower calcium concentrations than is the gamma isoform, whereas at high calcium concentrations both isoforms are active. The most plausible explanation for how the different CDPKs sense different calcium thresholds, is that they have divergent EF-hands with differing calcium-binding affinities; this would result in calcium sensor systems responding to deviating calcium peaks and variable frequencies of calcium signals. PfCDPK1 might be an excellent example for the fine-tuning of enzymatic activity through the affinity of its calcium binding sites in dependence of the calcium amplitude. PfCDPK1 resides in the parasitophorous vacuolar compartment of the parasite and compartments derived thereof, in which an extracellular calcium milieu is mimicked. Gazarini and coworkers have shown that the calcium concentration in the parasitophorous vacuole is in the order of $40 \, \mu M$, and is thus about 100- to 1000-fold higher than that in the infected erythrocyte or the parasite cytoplasm [19]. The mean calcium dissociation constant of PfCDPK1 is $80 \, \mu M$, which is 20-fold higher than that of human calmodulin [17]. The lower affinity of PfCDPK1 to Ca^{2+} can easily be explained by its localization in the parasitophorous vacuolar compartment. In exerting its function in an environment of a high Ca^{2+} concentration, the calcium affinity of PfCDPK1 must be appropriately adjusted and fine-tuned to the environmental conditions coming across in this specific compartment.

However, modulation of the calcium affinities through the affinity of the individual calcium-binding sites is surely not the sole mechanism involved in threshold variations.

The Junction Domain

Plant CDPKs, which have been grouped as class 1 CDPKs [1], are regulated by an autoinhibitory region – the so-called "junction domain" – that is found immediately C-terminal to the kinase domain and N-terminal of the calmodulin-like Ca^{2+}-binding regulatory domain. Autoinhibition is relieved through a Ca^{2+}-dependent interaction with the N-lobe [16]. The junction region is approximately 30 amino acids in size, and comprises an N-terminal pseudosubstrate autoinhibitor domain and a

C-terminal region interacting with the C-lobe of the calmodulin-like domain [16]. In the presence of normal low cytosolic Ca^{2+} concentrations, the pseudosubstrate motif occupies the catalytic site, and the C-lobe of the calmodulin-like domain is bound to the junction region, whereas the N-lobe is free to respond to intracellular Ca^{2+} signals [20].

The junction domains of class 1 and 2 plasmodial CDPKs seem to have an organization akin to that of plant CDPKs. By using peptides directed against distinct parts of the junction domain, Ranjan and coworkers were able to confirm a similar organization of the junction domain for PfCDPK4 [21]. Once the catalytic cleft of PfCDPK4 is freed, autophosphorylation on T^{234} can occur, allowing for substrate interaction and phosphorylation [21]. Peptides used to inhibit PfCDPK4 were also effective against PfCDPK1, suggesting a similar mechanism of regulation [21].

Phosphorylation and Autophosphorylation

Although calcium-dependent autophosphorylation has been observed in almost all CDPKs investigated to date [22], its physiological role is only poorly understood, despite the fact that its importance for functional regulation has been discussed [22]. For example, tobacco CDPK2 undergoes an intra-molecular autophosphorylation within 2 min after exposure to stress, suggesting a biological function of autophosphorylation in immediate stress response [23].

All plasmodial CDPKs investigated to date – namely, PfCDPK1, PfCDPK2, PfCDPK4, and PfCDPK5 – are capable of a calcium-dependent autophosphorylation *in vitro* ([7, 21, 24, 25]; also D. Kugelstadt, B. Derrer, B. Kappes, unpublished observations). Furthermore, the autophosphorylation of PfCDPK1 can also occur in a calcium-independent manner, albeit to a minor extent. Interestingly, it has been shown that TgCDPK3, the *T. gondii* ortholog of PfCDPK1, is insensitive to EGTA and not dependent on exogenous calcium [26]. Autophosphorylation of PfCDPK1 occurs inter-molecularly [25].

The first evidence that autophosphorylation regulates CDPK kinase activity came from studies on the plasmodial raf kinase inhibitory protein (RKIP). This protein affected the autophosphorylation activity of PfCDPK1, which in turn led to an altered substrate phosphorylation activity of the kinase [27]. The first mechanistic clue, however, as to how autophosphorylation could be involved in the regulation of CDPK activities on the molecular level derived from the studies of Ranja and coworkers on PfCDPK4 regulation [21], who showed that the autophosphorylation of T^{234} in the activation loop, which resides between subdomains VII and VIII of PfCDPK4, results in the activation of kinase. PfCDPK1 has also been shown to be autophosporylated on a threonine residue in the activation loop, namely T^{231}, a site complementary to T^{234} of PfCDPK4 [22]. However, its significance for catalytic activation has not yet been demonstrated. Furthermore, several additional autophosphorylation sites of PfCDPK1 have been identified (about 20 in total), some of which have also been shown to be phosphorylated *in vivo*, suggesting a functional role of PfCDPK1 autophosphorylation in the living parasite (D. Kugelstadt, B. Derrer, B. Kappes, unpublished observations). The high number of these sites does, however, aggravate their functional assignment.

Besides autophosphorylation, phosphorylation by upstream kinases is most likely a regulatory tool to control the activation of the CDPKs, as has been demonstrated for CDPK2 and CDPK3 of tobacco [23]. However, despite their importance the signaling cascades involved in the regulation of CDPKs are poorly understood.

Regulatory Proteins

Besides calcium, regulatory proteins might be crucial for the functional regulation of individual CDPKs. The 14-3-3 proteins are small, highly conserved eukaryotic proteins that have been implicated in the regulation of multiple cellular enzymes, including protein kinases, in part by binding to specific phosphorylated residues [28]. Three different 14-3-3 isoforms have been shown to stimulate the activity of the *A. thaliana* CDPK isoform CPK-1 approximately twofold [29]. Furthermore, the 14-3-3 consensus binding site R-S/T-X-*S*-X-P (where the italic S is phosphorylated), has also been identified in AtCDPK24 and AtCDPK28, which suggests that these CDPKs might also be regulated via 14-3-3 proteins. The putative binding domain for the 14-3-3 proteins is located in the N-terminal variable domain of these kinases, suggesting that 14-3-3 proteins may be regulators of at least a CDPK subset [30]. While it is likely that some plant CDPKs are regulated by 14-3-3 proteins, their relevance for the regulation of plasmodial CDPKs has not been investigated. Another protein that serves as a modulator of signaling pathways by either promoting or inhibiting the formation of productive signaling complexes through protein–protein interactions is RKIP. Plasmodial RKIP increased the autophosphorylation activity of PfCDPK1 by about 5.5-fold, and reduced its substrate phosphorylation activity roughly in the same range; this suggests that plasmodial RKIP might indeed be involved in the regulation of PfCDPK1 in the parasite [27].

Protein Kinase Inhibitors Against CDPKs

There are many potential options to inhibit the phosphotransferase activity of a kinase. Some of these include: (i) antagonism to ATP binding; (ii) preventing the access of a kinase to its substrate; and (iii) abolishing kinase activation by interfering with regulatory subunits, or through stabilizing interactions that maintain a given kinase in an inactive mode. With respect to the inhibition of CDPKs, several scenarios of enzymatic inhibition are imaginable:

- The classical approach based on the identification of highly competitive ATP analogs for a given CDPK or a CDPK class. Current data suggest that it is quite unlikely that a selective inhibitor affecting the entire CDPK family will be identified, as the catalytic cleft of individual CDPKs is too divergent.
- The identification of antagonists interfering with Ca^{2+} binding and, as a consequence, kinase activation. This strategy may be more easily applicable to those CDPKs having a higher threshold in calcium activation.

- The identification of peptides or compounds keeping a given CDPK in an autoinhibited state.
- The discovery of compounds interfering with modifications states of a CDPK (e.g., acylation or phosphorylation).
- The identification of compounds abolishing the activation of a given CDPK through regulatory proteins.

Despite this multitude of possibilities, the majority of investigations on protein kinase inhibitors against plasmodial CDPKs have focused on the development of ATP competitive inhibitors, as is the case for other protein kinase inhibitors [31].

ATP Analogs

Until recent times it was believed that, due to the large size of the protein kinase family and the similarity of the ATP binding site, inhibitors would be nonspecific and thus ill-suited as good therapeutics. However, it transpired that such skepticism was unsubstantiated, and that it is possible and relatively easy to generate high-affinity inhibitors to the well-defined hydrophobic ATP-binding cleft [31]. This approach has also been chosen to identify inhibitors against PfCDPK1.

Activity of Known Protein Kinase Inhibitors Against PfCDPK1

Green *et al.* reported that the bisindolocarbazole K252a, a staurosporine analog, inhibited PfCDPK1 activity with an IC_{50} of 45 nM [10], which is a concentration known to inhibit serine/threonine protein kinase in general [32]. Similarly, staurosporine itself, which is a potent nonselective inhibitor of a multitude of protein kinases, inhibits PfCDPK1 kinase activity with an IC_{50} value of 60 nM (D. Kugelstadt, B. Derrer, B. Kappes, unpublished observations).

High-Throughput Screening for PfCDPK1 Specific Inhibitors

To date, three high-throughput screening (HTS) schemes have been conducted with PfCDPK1 as the target protein (Table 17.1). The first screen was initiated by the World Health Organization (WHO) and carried out by Discovery Technologies (Basle, Switzerland), with approximately 20 000 compounds from their compound library. The majority of the compounds were selected on the basis of high diversity, with the exception of 279 that were chosen on the basis of containing kinase inhibitor pharmacophores. The second screen was carried out by Kato *et al.* with a kinase-directed heterocyclic library consisting of 20 000 compounds [2, 33], and the third by Lemercier and coworkers in a collaborative project with the WHO with a kinase-directed proprietary compound library of 54 733 compounds from Merck-Serono [34].

The first screen resulted in the identification of seven compounds with IC_{50} values on PfCDPK1 activity ranging from 1.7 to 10.7 μM. Three of these were from the kinase inhibitor pharmacophore-biased selection (D. Kugelstadt, B. Derrer, B. Kappes, unpublished observations). All seven hits were assayed for their inhibitory

Table 17.1 Compounds identified through HTS against PfCDPK1.

Compound	IC_{50}[a]	EC_{50}[b]	Reference
[c]	2.9 µM	12 µM	Unpublished observations
	17 nM	230 nM	[2]
	10 nM	5.7 µM	[34]

a) Half-maximal inhibitory concentration (IC_{50}) against PfCDPK1.
b) Half-maximal response values in parasite proliferations assays (EC_{50}).
c) Only the chemical series is shown.

activity on parasite growth and were found to be active on *P. falciparum* strain HB3, with IC_{50} values ranging from 12 to 65 µM. The IC_{50} values on the parasite were in general higher than those on recombinant PfCDPK1, which most likely was due to a limited permeability of the compounds (D. Kugelstadt, B. Derrer, B. Kappes, unpublished observations). A total of 749 derivatives of the originally identified hits was screened to discover more active compounds that would inhibit both PfCDPK1 activity and *P. falciparum* growth. Seven of these derivates had IC_{50} values on PfCDPK1 kinase activity in the very low micromolar range, and were active on the parasite roughly in the same order of magnitude as on the enzyme. Amazingly, all active compounds among the tested derivatives were found to be indol-2 ones (Table 17.1). However, none of the tested derivatives was more effective than the originally identified hit compounds, which led to the project being stopped at this point.

In the second screening, conducted by *Kato et al.*, approximately 20 000 compounds from a kinase-directed heterocyclic library were examined [2]. This resulted in the identification of about 50 compounds which inhibited PfCDPK1 kinase activity by more than 80% at a concentration of 1 µM, with the 2,6,9-trisubstitued purines being identified as the most active compound class [2]. Thirteen analogs belonging to this class were tested for the inhibition of PfCDPK1 activity and of parasite growth, in order to obtain information regarding the structure–activity relationship. These analyses revealed that: (i) there was a strong preference of the purine C6-position R_1 for para substitutions; (ii) 1,4-*trans*-cyclohexane diamine as R_2 group at the position C2 resulted, in principle, in an optimal activity of the purine derivatives; (iii) the primary NH_2 group at position C2 is required for cellular activity; and (iv) the 9-phenyl ring tolerated substitutions with small R_3 groups. {4-[2-*trans*-(4-amino-cyclohexylamino)-9-(3-fluoro-phenyl)-9*H*-purin-6-ylamino]-phenyl}-piperidin-1-yl-methanone (known as purfalcamine) was chosen by the authors as the compound with the best combined properties on PfCDPK1 and the parasite, with IC_{50} and EC_{50} values of 17 and 230 nM, respectively (Table 17.1) [2]. The treated parasites progressed through the cell cycle and arrested before segmentation [2]. Purfalcamine proved to be effective against various parasite strains, including drug-resistant parasites, thus confirming its effectiveness as a lead compound for antimalarial drug development [2]. However, the compound failed to clear the parasite in the malaria model system *P. yoelii*, where it only delayed the increase in parasitemia. The authors suggested that the inability to eliminate parasites from the blood was most likely a combination of the compound being more effective on the *P. falciparum* enzyme than on the *P. yoelii* ortholog, and of its pharmacokinetic properties (low exposure and high clearance) in mice [2]. Although the data acquired by Kato *et al.* suggested that purfalcamine kills *P. falciparum* through the inhibition of PfCDPK1, the possibility that other intracellular targets were affected by the compound could not be ruled out, as structurally related analogs – for example {4-[2-*trans*-(4-amino-cyclohexylamino)-9-(3-benzyloxy-phenyl)-9*H*-purin-6-ylamino]-phenyl}-piperidin-1-yl-methanone – differed in their IC_{50} values for PfCDPK1 inhibition by 237.6-fold, and thus were much less active toward the enzyme but maintained their cellular activity. The inhibition of PfCDPK5 (the closest relative of PfCDPK1) with purfalcamine resulted in a 200-fold increase in IC_{50} value compared to PfCDPK1 (3.5 µM versus 17 nM), which suggested that this compound is not a CDPK class 1 specific inhibitor [2].

A third screen was performed with 54 733 molecules of a proprietary compound library from Merck-Serono, by Lemercier *et al.* [34]. The majority of the molecules were selected on the basis of bearing protein kinase inhibitor scaffolds. In total, 70 compounds were identified with submicromolar IC_{50} values on PfCDPK1, and these fell into two distinct chemical series. The most potent member of each series were 2-amino-3-[(4-methylphenyl)carbonyl]indolizine-1-carboxamide (Compound 1) and a substituted imidazo[1,2-β]pyridazine (Compound 2) (Table 17.1). These were further characterized through enzymatic and biophysical analyses. While exhibiting the same mechanism of action as competitive ATP analogs, Compounds 1 and 2 differed in the fact that the latter had an extended time of interaction with the kinase [34]. In order to understand the structural requirements governing the interaction between Compound 2 and PfCDPK1, the potency of close analogs was determined; the results led to the conclusion that the hydroxyl group could be replaced by other H-bond acceptor groups [34]. Unfortunately, Compounds 1 and 2 were far less effective on the parasite than on the kinase; whereas, Compound 2 inhibited PfCDPK1 with an IC_{50} of 10 nM, its EC_{50} on parasite growth was 5.7 µM, and thus 5700-fold higher (Table 17.1) [34].

In search of a potent and selective inhibitor of coccidial cGMP-dependent protein kinase (PKG), which should be particularly effective against *Eimeria* and *Toxoplasma*, medicinal chemistry efforts centered on an imidazopyridine scaffold, which is quite similar in structure to the imidazopyridazine scaffold of Compound 2 identified by Lemercier *et al.* [26, 34]. Interestingly, 4-[7-(dimethylamino)methyl] -2-(4-fluorophe-nyl)imidazo[1,2-*a*]pyridine-3-yl]pyrimidin-2-amine (also named Compound 2) recognized TgCDPK1 as a secondary intracellular target with an IC_{50} of 0.7 nM [26]; the IC_{50} against *Eimeria tenella* CDPK1 was 163 nM. Other coccidial CDPKs tested were TgCDPK3, the equivalent of PfCDPK1, and EtCDPK2; however, the IC_{50} values here were above 10 000 nM [26].

Calmodulin Antagonist

The present knowledge of calmodulin antagonists affecting the activity of plasmodial CDPKs is rather limited. Trifluoperazine, calmidazolium, W-7, and ophiobolin A each inhibited the activation of PfCDPK1 in the micromolar range, with IC_{50} values of 37, 47, 127, and 236 μM, respectively [25]. Except for trifluoperazine, the other compounds required much higher concentrations for inhibiting PfCDPK1 than for inhibiting calmodulin-mediated effects [25]. The result with calmidazole was most striking, as this drug is a very potent inhibitor of calmodulin-mediated effects, and is several hundred-fold more potent than trifluoperazine [35, 36]. Based on their data, these authors hypothesized that the conformational structure in the calcium binding domain of PfCDPK1 may be different from that of calmodulin, which was later verified by determination of the individual Ca^{2+}-binding constants of the N and C-lobe in the calmodulin-like domain of a CPK-1 (a CDPK from *A. thaliana*), and by structural resolution [16, 20]. The structure of the regulatory calmodulin-like calcium binding apparatus allows a computational searching for compounds, which may either abolish sufficient calcium binding or a calcium-induced conformational change of the domain, thus impeding kinase activation.

Conclusion

Apicomplexan calcium-dependent protein kinases, the function of which is crucial for parasite survival in those life cycle phases responsible for the symptoms of the disease, are proven targets for anticoccidial drug development. Until now, both public and private efforts have focused on malaria as a disease and PfCDPK1 as a drug target. Future efforts should concentrate on the identification of other essential CDPKs of *Plasmodium* (the essential role of CDPK2, 5 and 7 must be ruled out) and of other Apicomplexa. Activities aimed at developing PfCDPK1 inhibitors have utilized primarily the traditional approach of identifying highly competitive adenosine triphosphate (ATP) analogs. Other options (*vide infra*) have not been pursued and may represent promising alternatives. Three independent HTS approaches have resulted in the discovery of three chemical series of ATP-competitive inhibitors being active on PfCDPK1 and the parasite. Unfortunately, however, two of the projects were stopped, as the activities on the targets (PfCDPK1 and/or parasite) could not be improved. The optimization of a lead compound is achieved by:

- Co-crystallization, which allows a determination of the precise interaction of an inhibitor with its target, and thereby enables a structure-driven optimization of the binding interface.
- Molecular docking experiments, either with the structure of the target protein (if available) or with a closely related one.
- If the first two approaches are not possible, optimization of a lead through structure–activity-based drug design, if the two first approaches are not possible.

The lack of a CDPK structure is a major drawback in the process of continued optimization, and makes computational approaches more difficult to conduct. Docking experiments have been carried out with the structures of a plasmodial cyclin-dependent protein kinase and a CaMK of *C. parva*, although it is questionable whether these analyses are sufficiently precise for targeted optimization. The third HTS resulted in the discovery of a compound which was highly effective against PfCDPK1 and the parasite *in vitro*. However, when tested in mouse malaria this compound failed to clear the parasitemia, thus highlighting another significant hurdle that must be overcome, namely a suitable pharmacokinetic profile given that the potency of the compound is adequate.

References

1 Billker, O., Lourido, S., and Sibley, L.D. (2009) Calcium-dependent signaling and kinases in apicomplexan parasites. *Cell Host Microbe.*, **5**, 612–622.

2 Kato, N., Sakata, T., Breton, G., Le Roch, K.G., Nagle, A., Andersen, C., Bursulaya, B., Henson, K. *et al.* (2008) Gene expression signatures and small-molecule compounds link a protein kinase to *Plasmodium falciparum* motility. *Nat. Chem. Biol.*, **4**, 347–356.

3 Nagamune, K., Moreno, S.N., Chini, E.N., and Sibley, L.D. (2008) Calcium regulation and signaling in apicomplexan parasites. *Subcell. Biochem.*, **47**, 70–81.

4 Hook, S.S. and Means, A.R. (2001) Ca (2+)/CaM-dependent kinases: from activation to function. *Annu. Rev. Pharmacol. Toxicol.*, **41**, 471–505.

5 Harper, J.F. and Harmon, A. (2005) Plants, symbiosis and parasites: a calcium signalling connection. *Nat. Rev. Mol. Cell Biol.*, **6**, 555–566.

6 Moskes, C., Burghaus, P.A., Wernli, B., Sauder, U., Durrenberger, M., and

Kappes, B. (2004) Export of *Plasmodium falciparum* calcium-dependent protein kinase 1 to the parasitophorous vacuole is dependent on three N-terminal membrane anchor motifs. *Mol. Microbiol.*, **54**, 676–691.

7 Farber, P.M., Graeser, R., Franklin, R.M., and Kappes, B. (1997) Molecular cloning and characterization of a second calcium-dependent protein kinase of *Plasmodium falciparum*. *Mol. Biochem. Parasitol.*, **87**, 211–216.

8 Garcia, C.R. (1999) Calcium homeostasis and signaling in the blood-stage malaria parasite. *Parasitol. Today*, **15**, 488–491.

9 Coppi, A., Tewari, R., Bishop, J.R., Bennett, B.L., Lawrence, R., Esko, J.D., Billker, O. *et al.* (2007) Heparan sulfate proteoglycans provide a signal to *Plasmodium* sporozoites to stop migrating and productively invade host cells. *Cell Host Microbe.*, **2**, 316–327.

10 Green, J.L., Rees-Channer, R.R., Howell, S.A., Martin, S.R., Knuepfer, E., Taylor, H.M., Grainger, M., and Holder, A.A.

(2008) The motor complex of *Plasmodium falciparum*: phosphorylation by a calcium-dependent protein kinase. *J. Biol. Chem.*, **283**, 30980–30989.

11 Winter, D., Kugelstadt, D., Seidler, J., Kappes, B., and Lehmann, W.D. (2009) Protein phosphorylation influences proteolytic cleavage and kinase substrate properties exemplified by analysis of in vitro phosphorylated *Plasmodium falciparum* glideosome-associated protein 45 by nano-ultra performance liquid chromatography-tandem mass spectrometry. *Anal. Biochem.*, **393**, 41–47.

12 Garcia, G.E., Wirtz, R.A., Barr, J.R., Woolfitt, A., and Rosenberg, R. (1998) Xanthurenic acid induces gametogenesis in *Plasmodium*, the malaria parasite. *J. Biol. Chem.*, **273**, 12003–12005.

13 Billker, O., Dechamps, S., Tewari, R., Wenig, G., Franke-Fayard, B., and Brinkmann, V. (2004) Calcium and a calcium-dependent protein kinase regulate gamete formation and mosquito transmission in a malaria parasite. *Cell*, **117**, 503–514.

14 Ishino, T., Orito, Y., Chinzei, Y., and Yuda, M. (2006) A calcium-dependent protein kinase regulates *Plasmodium* ookinete access to the midgut epithelial cell. *Mol. Microbiol.*, **59**, 1175–1184.

15 Siden-Kiamos, I., Ecker, A., Nyback, S., Louis, C., Sinden, R.E., and Billker, O. (2006) *Plasmodium berghei* calcium-dependent protein kinase 3 is required for ookinete gliding motility and mosquito midgut invasion. *Mol. Microbiol.*, **60**, 1355–1363.

16 Christodoulou, J., Malmendal, A., Harper, J.F., and Chazin, W.J. (2004) Evidence for differing roles for each lobe of the calmodulin-like domain in a calcium-dependent protein kinase. *J. Biol. Chem.*, **279**, 29092–29100.

17 Zhao, Y., Pokutta, S., Maurer, P., Lindt, M., Franklin, R.M., and Kappes, B. (1994) Calcium-binding properties of a calcium-dependent protein kinase from *Plasmodium falciparum* and the significance of individual calcium-binding sites for kinase activation. *Biochemistry*, **33**, 3714–3721.

18 Harper, J.F., Breton, G., and Harmon, A. (2004) Decoding Ca(2+) signals through plant protein kinases. *Annu. Rev. Plant. Biol.*, **55**, 263–288.

19 Gazarini, M.L., Thomas, A.P., Pozzan, T., and Garcia, C.R. (2003) Calcium signaling in a low calcium environment: how the intracellular malaria parasite solves the problem. *J. Cell Biol.*, **161**, 103–110.

20 Chandran, V., Stollar, E.J., Lindorff-Larsen, K., Harper, J.F., Chazin, W.J., Dobson, C.M., Luisi, B.F. *et al.* (2006) Structure of the regulatory apparatus of a calcium-dependent protein kinase (CDPK): a novel mode of calmodulin-target recognition. *J. Mol. Biol.*, **357**, 400–410.

21 Ranjan, R., Ahmed, A., Gourinath, S., and Sharma, P. (2009) Dissection of mechanisms involved in the regulation of *Plasmodium falciparum* calcium-dependent protein kinase 4. *J. Biol. Chem.*, **284**, 15267–15276.

22 Hegeman, A.D., Rodriguez, M., Han, B.W., Uno, Y., Phillips, G.N. Jr, Hrabak, E.M., Cushman, J.C. *et al.* (2006) A phyloproteomic characterization of in vitro autophosphorylation in calcium-dependent protein kinases. *Proteomics*, **6**, 3649–3664.

23 Witte, C.P., Keinath, N., Dubiella, U., Demouliere, R., Seal, A., and Romeis, T. (2010) Tobacco calcium-dependent protein kinases are differentially phosphorylated in vivo as part of a kinase cascade that regulates stress response. *J. Biol. Chem.*, **285**, 9740–9748.

24 Kato, K., Sudo, A., Kobayashi, K., Sugi, T., Tohya, Y., and Akashi, H. (2009) Characterization of *Plasmodium falciparum* calcium-dependent protein kinase 4. *Parasitol. Int.*, **58**, 394–400.

25 Zhao, Y., Franklin, R.M., and Kappes, B. (1994) *Plasmodium falciparum* calcium-dependent protein kinase phosphorylates proteins of the host erythrocytic membrane. *Mol. Biochem. Parasitol.*, **66**, 329–343.

26 Donald, R.G., Zhong, T., Wiersma, H., Nare, B., Yao, D., Lee, A., Allocco, J., Liberator, P.A. *et al.* (2006) Anticoccidial kinase inhibitors: identification of protein

kinase targets secondary to cGMP-dependent protein kinase. *Mol. Biochem. Parasitol.*, **149**, 86–98.

27 Kugelstadt, D., Winter, D., Pluckhahn, K., Lehmann, W.D., and Kappes, B. (2007) Raf kinase inhibitor protein affects activity of *Plasmodium falciparum* calcium-dependent protein kinase 1. *Mol. Biochem. Parasitol.*, **151**, 111–117.

28 Obsilova, V., Silhan, J., Boura, E., Teisinger, J., and Obsil, T. (2008) 14-3-3 proteins: a family of versatile molecular regulators. *Physiol. Res.*, **57** (Suppl 3), S11–S21.

29 Camoni, L., Harper, J.F., and Palmgren, M.G. (1998) 14-3-3 proteins activate a plant calcium-dependent protein kinase (CDPK). *FEBS Lett.*, **430**, 381–384.

30 Cheng, S.H., Willmann, M.R., Chen, H.C., and Sheen, J. (2002) Calcium signaling through protein kinases. The *Arabidopsis* calcium-dependent protein kinase gene family. *Plant Physiol.*, **129**, 469–485.

31 McInnes, C. and Fischer, P.M. (2005) Strategies for the design of potent and selective kinase inhibitors. *Curr. Pharm. Des.*, **11**, 1845–1863.

32 Ruegg, U.T. and Burgess, G.M. (1989) Staurosporine. K-252 and UCN-01: potent but nonspecific inhibitors of protein kinases. *Trends Pharmacol. Sci.*, **10**, 218–220.

33 Ding, S., Gray, N.S., Wu, X., Ding, Q., and Schultz, P.G. (2002) A combinatorial scaffold approach toward kinase-directed heterocycle libraries. *J. Am. Chem. Soc.*, **124**, 1594–1596.

34 Lemercier, G., Fernandez-Montalvan, A., Shaw, J.P., Kugelstadt, D., Bomke, J., Domostoj, M., Schwarz, M.K. *et al.* (2009) Identification and characterization of novel small molecules as potent inhibitors of the plasmodial calcium-dependent protein kinase 1. *Biochemistry*, **48**, 6379–6389.

35 Gietzen, K., Sadorf, I., and Bader, H. (1982) A model for the regulation of the calmodulin-dependent enzymes erythrocyte Ca^{2+}-transport ATPase and brain phosphodiesterase by activators and inhibitors. *Biochem. J.*, **207**, 541–548.

36 Van Belle, H. (1981) R24 571: A potent inhibitor of calmodulin-activated enzymes. *Cell Calcium*, **2**, 483–494.

18
Protein Acylation: New Potential Targets for Intervention Against the Apicomplexa

*Joana M. Santos, Christian Hedberg, and Dominique Soldati-Favre**

Abstract

Apicomplexan parasites are obligate intracellular and treacherous pathogens responsible for lethal and debilitating diseases in animals and humans. These parasites undergo a complex lytic cycle, involving active host cell entry, a phase of fast intracellular replication, egress from the invaded host cells, and the invasion of neighboring host cells. Lipidation–delipidation cycles are of fundamental importance to maintain a correct function and localization of membrane-associated proteins. These processes are mediated by *N*-myristoyltransferases (NMTs), which catalyze myristoylation, and palmitoyl transferases (PATs) and acyl/palmitoyl protein thioesterases (APT/PPTs), that mediate palmitoylation or depalmitoylation, respectively. The apicomplexans possess a single NMT and up to a dozen genes coding for putative PATs, which exhibit the characteristic DHHC motif. These parasites also possess a couple of putative candidate genes resembling human APT1, which was found to depalmitoylate the H-Ras protein and the subunits of heterotrimeric G-protein. While *N*-myristoylation is considered to be static, *S*-palmitoylation has a rapid cycle turnover, offering therefore more occasions for interference. For this reason, the PATs and PPTs could be considered prime drug targets against the apicomplexan parasites. In this chapter, the current knowledge regarding acylation as a post-translational modification is described, and the potential for using currently available or new screened compounds targeting the apicomplexan NMT, PATs, and PPTs as drugs against these parasites is discussed.

Introduction

Translation of the genetic code into chains of amino acids is only the first of a series of steps leading to protein diversity. All proteins undergo a wide variety of co- and post-translational modifications that generate different isoforms, and which might have different steady-state levels, binding affinity, hydrophobicity, localization, and

*Corresponding author

Apicomplexan Parasites. Edited by Katja Becker
Copyright © 2011 WILEY-VCH Verlag GmbH & Co. KGaA, Weinheim
ISBN: 978-3-527-32731-7

regulation that ultimately fine-tune their function. In recent years, genomics and proteomics approaches, coupled with improved mass spectrometry technologies, have highlighted the vast underestimation of the importance of post-translational modifications (PTMs) for the diversity and regulation of protein function.

The term "PTM" refers to any modification that occurs on a protein after translation. Such modifications can involve proteolytic processing and/or the addition of functional groups or a peptide, a change in the chemical structure and mass of the amino acid residues, or a change on the protein's structure or conformation. Additions include phosphorylation, acetylation, methylation, glycosylation, ubiquitination, or lipidation.

PTMs remain a challenge to proteomics analysis because they are diverse, and in many cases difficult to predict by bioinformatics, due to the lack of signature motifs. They can also be difficult to demonstrate by experimental approaches. Nevertheless, great interest in the enzymes responsible for those modifications as potential drug targets has emerged due to the considerable impact of PTMs on protein function.

The importance of PTMs in the context of infectious diseases is only now materializing. The most common and well-described PTMs undergone by apicomplexan and kinetoplastid parasites proteins are the addition of glycosylphosphatidylinositols (GPIs) and phosphate groups (phosphorylation).

The GPIs, a family of glycolipids, are added to proteins by transamidases, their function being to anchor proteins primarily to the plasma membrane. The surface coating proteins of several protozoan parasites are predominantly anchored by this lipid moiety to the plasma membrane. Known examples of GPI-anchored proteins are the variant surface glycoproteins (VSGs) of *Trypanosoma*, the promastigote surface protease (PSP) of *Leishmania*, the merozoite surface proteins (MSPs) of *Plasmodium* merozoites, and the surface antigens (SAGs) of *Toxoplasma* tachyzoites. Furthermore, in *Toxoplasma gondii*, these GPI motifs are known to induce the host's inflammatory response by increasing tumor necrosis factor-alpha (TNF-α) production and the constitutive and interferon-gamma (IFN-γ)-induced expression of MHC classes I and II antigens at the surface of macrophages [1]. It has, therefore, been hypothesized that drugs mimicking this effect would improve immune response against infection. In fact, anti-GPI immunization has been shown to protect against *Trypanosoma* and to prevent systemic and cerebral malaria of mice infected with *Plasmodium berghei* [1].

Parasite kinases are also considered as attractive drug targets [2]. The kinome of *P. falciparum* has been unveiled and shown to encode approximately 100 protein kinases, half of which appear to be distinct from eukaryotic kinases [3]. Similar unique kinase families have been identified in other parasites, suggesting that the design of selective inhibitors against these enzymes is a feasible goal (for a review, see Ref. [4]).

These and other results uncovered the essential roles of PTMs in the regulation of protein function in apicomplexans and other pathogens, and indicate that the enzymes involved in these processes are potential targets for effective and specific therapeutics. *N*-myristoylation and *S*-palmitoylation, as lipid modifications, are discussed in detail in the following sections.

Protein Acylation: *N*-Myristoylation and *S*-Palmitoylation

The addition of lipids to proteins – which is referred to as lipidation or lipid modification – adds a hydrophobic anchor that promotes membrane association. The most common lipid modifications include the addition of a GPI lipid anchor, myristoylation, palmitoylation, and isoprenylation. In this section, attention will be focused on the biochemistry, biological roles and potential as drug targets of N-myristoylation and S-palmitoylation.

Biochemistry and Enzymology of Myristoylation and Palmitoylation

Myristoylation

Among the many possible lipidations, N-terminal myristoylation is one of the most widely studied. Found exclusively in eukaryotes and viral proteins, N-terminal myristoylation corresponds to the covalent addition – via a stable amide linkage – of a myristoyl group (14-carbon saturated fatty acids) to the N-terminal glycine residue of a protein, which became N-terminal after removal of the initial methionine residue and sometimes additional amino acids. The myristoyl group is catalytically transferred from myristoyl-CoA by the enzyme myristoyl-CoA : protein N-myristoyl-transferase (NMT; EC 2.3.1.97), a member of the Gcn5-related acetyltransferase (GNAT) superfamily of proteins. NMT is essential to several organisms, including mammals [5, 6] (Figure 18.1), and although it was initially considered to be cytosolic in nature, some studies have suggested that 10–50% of the protein may be membrane-associated [7].

The NMT from *Saccharomyces cerevisiae* has been extensively characterized and purified alone and also in binary (NMT: myristoyl-CoA) or tertiary (NMT: myristoyl-CoA : peptide) complexes, offering some insight into its mechanism of action. The binding of myristoyl-CoA occurs in two steps, and induces conformational changes that result in the opening of the peptide-binding site. In most cases, the proteins to be modified bind co-translationally to this site in the enzyme, following removal of the initiator methionine residue (for a review, see Ref. [8]). This binding leads to the formation of the tertiary complex myristoyl-CoA : NMT: protein and, consequently, to catalysis and product release [8].

Studies on NMTs from different species revealed that, while the myristoyl-CoA binding site is similar in all enzymes, the protein binding site is specific to each one [8]. In fact, the N-terminal six residues of a protein must fit deep into the binding pocket of the enzyme, involving specific van der Waals and/or hydrogen bond contacts in order to be recognized as a substrate [9]. Nevertheless, a single NMT can bind to different protein substrates, which is a unique characteristic among GNAT family members. Although many aspects of substrate recognition and specificity remain unclear, the results of some studies have indicated a role for the N-terminal domain in substrate recognition and protein localization [8].

Recently, considerable progress has been made with the availability of reliable bioinformatics prediction methods, resulting in a comprehensive collection of data in databases such as MYRbase [10]. However, it is still difficult to predict the *in silico*

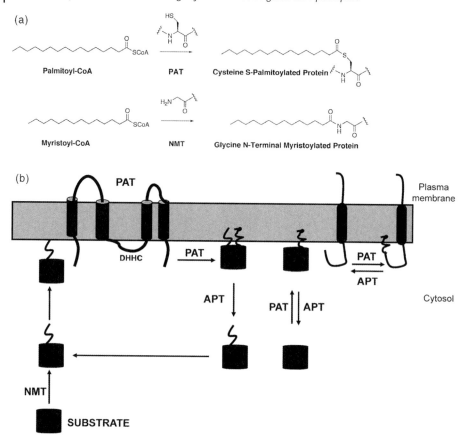

Figure 18.1 (a) Palmitoylation and myristoylation reactions catalyzed by PAT and NMT; (b) Substrate proteins (in black) can be weakly associated to membranes by myristoylation, mediated by NMT enzymes (in dark gray), at the cytoplasm. Once at the membrane, these myristoylated proteins can be further modified by palmitoylation, catalyzed by PAT enzymes (in light gray), located at the plasma membrane. Palmitoylation is a reversible mechanism and the palmitoylated proteins can return to a soluble or weakly membrane-associated status, when unmodified by APT/PAT enzymes (in white). Some proteins do not require a previous modification by myristoylation, in order to be palmitoylated (pathway represented on the right-hand side).

N-myristoylome of whole genomes because the myristoyl motif (MGxxxSxxx) is quite flexible. It is known, nevertheless, that there is a significant restriction in sequence variability up to position 17 [9], and that fusion of only the first ten amino acid residues of a modified protein (p60^{src}) to non-myristoylated proteins is sufficient for myristoylation [11]. This suggests that, at least for this protein, the complete signal for myristoylation resides in a short N-terminal segment.

Palmitoylation

S-acylation or thioacylation occurs when a protein undergoes a fatty acid modification involving a reversible thioester (*S*-acyl) linkage. The fatty acid added can vary (C14 or longer), but because the 16-carbon palmitate acyl group is the most common, this lipid modification is commonly described as *S*-palmitoylation [12]. At this point, only the addition of palmitate will be described.

The addition of a palmitate via a thioester linkage to a cysteine residue can occur at the N-terminal or C-terminal region of proteins, or at the juxtamembrane domains of transmembrane proteins. The two-signal hypothesis of palmitoylation suggests that a protein requires two modifications in order to become stably membrane-anchored. Consequently, most palmitoylated proteins are initially lipid-modified by *N*-myristoylation or *S*-farnesylation, which weakly increases their affinity to membranes, and only subsequently are palmitoylated [12].

The *kinetic membrane trapping model*, a revised version of the two-signal hypothesis, proposes that the first lipid modification, or another mechanism that provides a weak affinity to membranes, functions as membrane-targeting rather than as a specific recognition determinant for the palmitoylation catalytic enzyme. According to this model, a protein which is singly lipid-modified cycles on and off the membranes until it encounters one with an appropriate membrane-targeting receptor. This receptor is possibly the palmitoylation catalytic enzyme which, once it recognizes the protein substrate, modifies it and anchors it to the membrane [12]. Proteins that do not follow this model and are exclusively palmitoylated, have been reported. For SNAP-25, a neuronal protein, it was shown that the modified cysteine residues could be found within the 25 most N-terminal amino acids, and that palmitoylation also required a stretch of five amino acids located 25 amino acids downstream of the modified cysteine, as well as the presence of at least three cysteine residues [12].

No consensus amino acid sequence for palmitoylation has been reported, but specific cysteines are modified, and artificial constructs expressing known palmitoylation motifs are recognized by the modifying enzyme, indicating that specificity may exist [13, 14]. *N*-myristoylated proteins are usually palmitoylated at cysteine residues adjacent to the *N*-myristoylated glycine, but palmitoylation can also occur at cysteine residues up to 20 amino acids away. A common amino acid composition on residues surrounding palmitoylated cysteines has been reported [15], but the sequence context surrounding the palmitoylated cysteine was shown not to be important [12]; it is thus difficult to predict which adjacent amino acids provide a favorable environment for palmitoylation. Taking this discrepancy into account, the available prediction software tools for palmitoylation are not highly reliable.

The palmitoyl groups, donated by acyl-CoA, can be either spontaneously acquired or catalytically added to proteins by enzymes named protein *S*-acyl transferases (PATs). These enzymes are encoded in all eukaryotic genomes examined to date, and are classified into class 1 or 2 depending on if they modify previously *S*-farnesylated or *N*-myristoylated proteins [16]. The PATs can also be divided into three categories, depending on their structure: ankyrin-repeat containing; heterodimeric; or monomeric [17, 18]. All known PATs, excepting the yeast Yn1155W [19], are integral membrane proteins, and thus substrate recognition and catalysis can only occur if the protein substrates have been previously associated to the membrane via another lipidation. In some cases, this is assured by a transmembrane domain in the substrate (Figure 18.1).

The catalytic PAT domain was found to comprise the DHHC motif (Asp-His-His-Cys), a variant of the C2H2 zinc finger domain, which is found exclusively in eukaryotes [20]. Mutations of the cysteine residue abolish activity [19, 21], and this is the most homologous region of the enzyme among different species [19]. This motif is usually located within a cysteine-rich domain (CRD), between two transmembrane regions, facing the cytosol, which may aid in bringing closer the weakly membrane adherent substrates to the catalytic site, as predicted by the membrane-trapping model [19] (Figure 18.1). The other known PAT regions – the SH3 domain, the ankyrin repeat, and the PDZ-binding motif – may recruit specific substrates or regulators. Although most PATs act alone, in yeast the Erf2p needs to be associated to Erf4p [19].

Most organisms encode in their genome more than one PAT, and these enzymes are differently localized. To date, the mechanisms that dictate their specific subcellular distribution, substrate specificity, and mode of regulation are unknown. Extracellular factors – as well as phosphorylation and *S*-nitrosylation – have been suggested to influence PAT activity [18] and, indeed, the results of a recent large-scale study proposed that DHHC proteins are themselves palmitoylated [22]. An *in silico* analysis of the human PATs encoding mRNAs [21] also indicated that most, if not all, of the PAT genes are alternatively spliced at various sites, both on the translated and untranslated regions.

Deletion [15] and overexpression [23] studies with the yeast DHHC protein family revealed that, although deletion of all the DHHC-containing proteins is lethal, there is a considerable redundancy in PAT function.

S-palmitoylation has two unique aspects when compared to other lipidations: (i) it strongly affects the membrane affinity of a protein (a palmitoyl moiety alone provides a 100-fold stronger membrane association than a single *N*-myristoyl moiety); and (ii) it is reversible. Depalmitoylation returns the protein to a state where it is rapidly cycling on and off membranes and trafficking through a vesicle-independent mechanism. This reversibility offers a flexible, rapid and precise mode of regulation of protein activity that is comparable to phosphorylation [24]. Consequently, the *S*-acylation status of a protein will be changed several times during its lifetime [12].

Three soluble human thioesterases have been shown to perform depalmitoylation. These include: the palmitoyl protein thioesterases, PPT1 and PPT2, which function in the lysosomes during protein degradation; and the acyl protein thioesterase, APT1, a small cytosolic protein that is widely conserved from yeast to humans, and which preferentially depalmitoylates activated proteins [12, 25, 26]. Given that, to date, 23 different human PATs have been described [19], additional enzymes involved in depalmitoylation processes are expected to be identified.

The crystal structures of bovine PPT1 alone, in complex with palmitate, or with an inhibitor have been resolved [26]. However, no dramatic structural changes could be identified among the different situations, and consequently the mechanism of substrate recognition and catalysis remain unclear. The structure of APT1 has also been resolved, whereby the free status is characterized by dimerization and occlusion of the active site; this suggests that, upon substrate binding, dimer dissociation may occur [26]. APT1 does not recognize a specific protein sequence, and it has been suggested that a conformational change or dissociation of a binding partner in the vicinity of the palmitoyl group must occur in order for the substrate to become more accessible to cleavage [12].

Experimental Approaches to the Study of *N*-Myristoylation and *S*-Palmitoylation

The discovery and validation of drugs in the areas of protein myristoylation and palmitoylation requires the establishment of efficient assays that, on the one hand can allow the study of the mechanisms and enzymology of the reactions, and on the other hand make them amenable to the high-throughput screening (HTS) of putative inhibitors.

The validation of proteins as new myristoylation substrates commonly starts by analyzing the N terminus of the protein sequences in a prediction program; this is followed by a validation of the predicted modified glycine by mutation to an alanine residue. Specific monoclonal antibodies against the myristoylated tetrapeptide can also be used [9, 27]. The discovery of new palmitoylated proteins is much more cumbersome, because most modified proteins are low-abundant, the *S*-palmitate turns over rapidly, there is no general consensus palmitoylation motif, and no convenient method to detect and purify the native *S*-acylated proteins [22].

In the past, palmitoylation inhibitors – such as the palmitate analog 2-bromopalmitate (2-bromohexadecanoic acid) – were used to indicate if the membrane anchorage of a protein was palmitoyl-dependent [28]. Historically, however, the most frequently used myristoylation and palmitoylation assays involved the metabolic labeling of cultured cells with radioactive forms of palmitate (^{3}H-palmitate) or myristoyl-CoA (^{3}H-myristoyl-CoA or ^{14}C-myristoyl-CoA). After labeling, the modified proteins could be separated by either high-performance liquid chromatography (HPLC) or by selective adsorption onto acidic alumina. These methods can be used in live cells and with full-length proteins, and the reaction kinetics can be determined using pulse–chase assays. However, the extent of incorporation and the detection are highly variable, and these methods are not only time-consuming and expensive but also potentially dangerous [29]. Consequently, biotin switch methodologies – which are faster and less expensive and can still be used in live cells – have been replacing the radioactive labeling approach [29].

Most modern palmitoylation assays utilize the same principles as the acyl exchange method developed by Schmidt *et al.* in 1988 [30]. In this case, the palmitate of the cysteine residues of extracted proteins are removed by cleavage of the thioester bond with hydroxylamine, after which the available free thiols – which were all formerly palmitoylated – are modified using a thiol-specific reagent. As this strategy is specific to thioester bonds and not to *S*-acylation, it cannot be used to distinguish which acyl group was attached to the protein, and hence there is a high probability of obtaining false positives. It is, nonetheless, specific to palmitoylation, as it does not report the modification of cysteines by prenyl groups [21]. In order to increase the sensitivity of the test, some assays have combined acyl-exchange with the tagging of palmitoyl cysteines with radiolabeled *N*-ethylmaleimide (NEM) [21].

The yeast palmitome has been defined using a new acyl exchange-based strategy, referred to as acyl biotin exchange (ABE). For this, the total protein extracts were first prepared; following ABE, a thiol-reactive crosslinker was then used so that all of the palmitoylated proteins would have their modified cysteine residues exchanged by biotin. These tagged proteins were then trypsinized and analyzed using a multi-dimensional protein identification technology (MudPIT). The final repertoire of proteins contained 35 new palmitoyl proteins, and 12 of the 15 previously known modified proteins. By using the same methodology on yeast strains in which the different DHHC genes had been deleted, it was possible to identify the putative PAT enzyme–substrate pairs [15]. More recently, an updated method which had been adapted from ABE and was referred to as PalmPISC (palmitoyl protein identification and site characterization), was described for the isolation and purification of the *S*-acylated proteins and peptides of human lipid raft-enriched and non-raft membrane fractions [22].

The ABE method requires not only a complete blockade of all reduced cysteines, but also a very efficient thioester hydrolysis and disulfide-exchange reaction in order to obtain the efficient labeling of the palmitoylated proteins and to eliminate false positives. Coupled with the inherent background observed with streptavidin bead enrichments, almost one-third of the hits were estimated as false positives in the yeast palmitome. To avoid these problems, the myristoyl or palmitoyl proteins can be metabolically labeled with bio-orthogonal probes, such as an ω-azido- or alkyne-fatty acid analog, substituting the naturally occurring fatty acids of a cell at the sites of *N*-myristoylation or *S*-palmitoylation. Once bound to cysteines, the azido group on the fatty acid reacts (with high selectivity) with the (triaryl)phosphines that are coupled to a tag, to a fluorophore or to others, and in this way the azide/alkyne moiety can be detected and/or purified with minimal contamination [29] using matrix-assisted laser desorption/ionization–time of flight (MALDI-TOF) mass spectrometry [21]. By applying these principles, a global identification of the palmitoylated proteins from human Jurkat T-cells was carried out using a commercially available compound, 17-octadecynoic acid (17-ODYA), which can be used as a metabolically incorporated biorthogonal probe. Detection was achieved via the Cu(I)-catalyzed azide-alkyne cycloaddition reaction ("click chemistry"), while validation of the data generated was conducted with gel-based readouts, which are not only simpler to use but also have a higher throughput than large-scale mass spectrometry [31]. Although this strategy cannot be applied to tissue samples, the bio-orthogonal probes present a series of advantages over radioactive labels; notably, they are more sensitive, nontoxic, and extremely stable under physiological conditions. These features – and especially their ability to effectively substitute for endogenous fatty acids – make the bio-orthogonal probes ideal for labeling palmitoyl proteins in live cells, providing also a significantly more direct measure of protein palmitoylation [21].

The validation of putative new NMTs is routinely performed by expressing the recombinant protein in *Escherichia coli*, as this organism does not encode any endogenous enzyme [32]. It is also possible to validate putative new NMTs or PATs by cotransfecting the complementary DNA (cDNA) of the studied enzyme along with the one of the candidate substrates into heterologous cells; alternatively, the purified enzyme and substrate proteins can be mixed in a tube, after which metabolic labeling with either radioactive myristoyl or palmitic acid can be carried out. Unfortunately, the detection of the labeled proteins requires that the films are exposed for long periods of time [19, 21]. At present, the use of nonradioactive assays is becoming increasingly common.

The *in vitro* palmitoylation (IVP) assay is a highly sensitive fluorescence-based HPLC method, in which fluorescently labeled peptides mimicking the specific palmitoylation motifs of known modified proteins via C-terminal farnesylation and palmitoylation, or N-terminal myristoylation and palmitoylation, are used. To date, however, the application of this method to the screening of drugs has been of limited use, because it only tests the catalytic activity of the PAT in study against the motifs represented, and does not provide any information regarding the natural substrates [29]. However, performing the assay in live cells allows testing to be carried out if the PAT can use endogenous palmitoyl-CoA [29, 33].

The allocation of an enzyme to a substrate can also be achieved by transfecting small interfering RNA (siRNA) against an NMT or PAT into cells, and checking if there is a decrease in the myristoylation or palmitoylation of the protein substrate. Other studies opt for overexpressing the known substrates and PATs, and performing metabolic labeling with radiolabeled palmitate. An increase in the incorporation of the radiolabeled palmitate on the overexpressed substrate is then used to claim specificity (for a review, see Ref. [34]). However, such an artificial situation does not provide definitive proof that a PAT is responsible for the modification of a specific substrate. A newly developed and more specific assay, termed palmitoyl-cysteine isolation capture and analysis (PICA), involves a combination of the ABE method with siRNA-mediated knockdown. In this case, the DHHC protein under investigation is first transfected into HeLa cells; half of the culture is then subjected to siRNA-mediated PAT knockdown while the other half is mock-treated. On completion of the ABE treatment each of the isolated peptides, when analyzed by mass spectrometry, is shown to contain a cysteine which was palmitoylated in the original protein but is now attached to either a heavy or a light biotin group, so as to yield two peptides that can be separated using liquid column chromatography. Whereas, the palmitoylated proteins that are not substrates of the knockdown PAT are represented by the same amount of peptides from both the treated and mock cultures, the enzyme substrates are represented predominantly by peptides that have originated from the mock culture [13, 19].

Biological roles for *N*-Myristoylation and *S*-Palmitoylation

Both, myristoylation and palmitoylation function in concert with other lipid modifications and protein motifs to facilitate targeting to membranes, and also appropriate cellular destination through mechanisms that are just now beginning to be defined [12].

The spectrum of proteins modified by these two PTMs is very broad. Palmitoylation has been shown to modulate protein stability by functioning as a quality control checkpoint, indicating proper folding, and to regulate protein activity. Although the structural basis of this regulation is not well understood, *in vitro* assays are now revealing the importance of palmitoylation in promoting or inhibiting protein interactions [12].

Many proteins only function in defined membrane environments (known as membrane microdomains), and palmitoylation is known to target proteins to lipid rafts or caveolae in the plasma membrane. These microdomains are enriched in signaling molecules, and are thought to play a role in signal transduction by effectively increasing the local concentrations of effector molecules [16, 17].

In plants, palmitoylation is notably involved in GTPase and calcium signaling, pathogen defense, and microtubule function [17].

The current human palmitome includes Ras isoforms, many members of the Src family of protein tyrosine kinases, subunits of G-proteins and G-protein-coupled receptors, rhodopsin and several neuron-specific proteins [20]. In neurons, protein palmitoylation plays a key role in targeting proteins for transport to nerve terminals, and for regulating trafficking at the synapse [35]. The results of a recent study revealed that a mitochondrial RNA (miRNA) binds to the 3′-untreated region (UTR) of the human *APT1* in neurons, thus downregulating its activity and inhibiting the growth of human dendritic spines. This inhibition was found to be associated, at least in part, to an increase in membrane localization of a G-protein α-subunit [36].

S-palmitoylation is reversible, and hence induces cycles of modification on its substrates. The addition of a palmitate by a specific PAT at the endoplasmic reticulum (ER), Golgi complex or plasma membrane results in a firm anchoring of the protein to a specific membrane. Modified proteins can then leave these organelles and, depending on their sorting information, accumulate at the plasma membrane or be further trafficked to other compartments, via secretory or endocytic vesicles. Depalmitoylation can subsequently return a protein to a state where it samples between different membranes.

The small GTPases N- and H-Ras, which are key regulators of the mitogen-activated protein kinase (MAPK) cascade, cycle on and off different intracellular membranes following their initial modification by farnesylation. When at the Golgi complex, they become palmitoylated by a PAT localized in this compartment, and this results in an increased membrane affinity. Trafficking to the plasma membrane can then occur through the Golgi complex and via sorting to the exocytic vesicular pathway. Depalmitoylation then cycles the proteins back to the Golgi compartment, where they can undergo a new cycle of palmitoylation and targeting to the plasma membrane.

The smaller isoform of the gamma-aminobutyric acid (GABA)-synthesizing enzyme glutamic acid decarboxylase, GAD65, is another example of a protein that undergoes a cycle of palmitoylation/depalmitoylation. This cycle shuttles the protein between the ER/Golgi complex and the synaptic vesicle membranes. GAD65 is initially targeted to the ER/Golgi membranes by its own targeting signal, and cycles on and off these membranes until it is palmitoylated in the Golgi complex. Further trafficking via a post-Golgi vesicular pathway brings it to the membrane of the synaptic vesicles of the axon termini [19]. In summary, palmitoylation/depalmitoylation cycles are largely exploited for regulating the distribution of proteins between ER/Golgi and post-Golgi membrane compartments.

Role of Protein Lipidation for Pathogens

The results of recent studies have highlighted the importance of lipidation for protein function in pathogens, some examples of which are discussed below.

Lipid A, the hydrophobic anchor of lipopolysaccharide (LPS), which forms the outer leaflet of the outer membrane of Gram-negative bacteria, is modified in *Pseudomonas aeruginosa* by palmitoylation, resulting in a 10- to 100-fold attenuation of the immune response to lipid A. This occurs in response to activation of the PhoP/PhoQ system, when the bacteria are isolated from the lungs of cystic fibrosis patients, but not when isolated from chronic non-cystic fibrosis lung infections. The responsible PAT has not yet been identified [37].

Protozoan parasite surface antigens primarily anchor their plasma membrane via lipid modifications. In *Trypanosoma brucei*, the causative agent of African sleeping sickness, the VSGs – which cover the parasite's entire surface and participate in protection against the host immune system – are membrane-anchored by double myristoylation in the bloodstream stages, but not the procyclic stages. As VSG is such an abundant protein, *T. brucei* relies on an enormous supply of myristate, which is secured via multiple sources [38]. Moreover, as many as 60 additional *Trypanosoma* proteins are predicted to be *N*-myristoylated [39], among which are included the ADP-ribosylation factors TbARF and TbARL (ARF-like). These are a highly conserved family of small GTPases that function in membrane dynamics and trafficking. Human ADP-ribosylation factor-1 (ARF1) function is also regulated by *N*-myristoylation, whereby the GDP-bound inactive isoform is stabilized via binding of the *N*-myristoylated N-terminus to a pocket on its structure. During the exchange of GDP for GTP, changes in conformation cause the myristate group to be exposed, allowing the protein and its effectors to become associated with membranes [40].

Metabolic labeling studies have identified at least ten ^{3}H-myristate-labeled proteins in *Leishmania major*. Hydrophilic acylated surface protein B (HASPB), a lipoprotein that is exported to the extracellular space via myristoylation and palmitoylation, is one of these proteins [41, 42].

Giardia trophozoites are coated on their entire surface by palmitoylated variant-specific surface proteins (VSPs). This modification, which occurs on a cysteine residue at the C terminus, is not important for localizing the protein but rather controls parasite survival. The modification is performed by the *Giardia* PAT (gPAT), one of the three identified PATs in the parasite, localized to the plasma membrane [43].

N-Myristoylation and S-Palmitoylation in Apicomplexans

At present, although very little is known regarding myristoylation and palmitoylation in apicomplexans, a few studies support an essential function for these PTMs, thus validating an interest in the development of new therapeutics.

Blast analyses have indicated that all apicomplexan and kinetoplastid parasites possess a unique NMT gene. The identified *T. gondii* gene (TGGT1_022200) is 50% identical to the PfNMT and HsNMT, and 40% identical to the TbNMT and LmNMT enzymes (Figure 18.2a). The *Plasmodium falciparum* NMT is a single-copy gene (PF14_0127), expressed in the asexual blood-stage forms, and is approximately 50% identical to the human enzyme and 40% to the fungal form. With the exception of Ile[180] and His[46], all of the residues known to participate directly in the enzymatic activity, or to play critical regulatory roles in the yeast NMT, are conserved. An analysis of PfNMT activity by the expression of a recombinant form in *E. coli* showed that this enzyme is active against a peptide substrate based on the N-terminal sequence of PfARF1, a known modified protein [44].

A search for the presence of DHHC-CRD-containing proteins in apicomplexan genomes has revealed the presence of a series of potential PATs coding genes. The *T. gondii* genome codes for 18 DHHC-containing proteins, the *P. falciparum* for 12, the *Neospora caninum* for 15, and the *Theileria annulata* for only seven. This is a similar number to that found in the *T. brucei* and *L. major* genomes, which contain 12 and 21 genes, respectively (K. Frenal and D. Soldati-Favre, unpublished results). Not all of these proteins are predicted to function as PATs, because some of the DHHC motifs identified might correspond to zinc-binding motifs implicated in protein–protein or protein–DNA interactions (Figure 18.2b).

The PPT enzymes are also encoded in these parasites' genome. A blast search with the PPT1 and APT1 human sequences revealed the presence of four putative genes in *T. gondii* and three in *P. falciparum* (L. Kemp and D. Soldati-Favre, unpublished results) (Figure 18.2c).

Apicomplexan Proteins that Undergo N-Myristoylation and/or S-Palmitoylation

Although only a handful of proteins have been shown to date to be acylated in apicomplexans, their critical role suggests that drugs capable of interfering with this PTM might be considered for therapeutic interventions. Recently, bioinformatics and biochemical predictions have uncovered more than 40 *P. falciparum* proteins as potential PfNMT substrates. Among this list are included ARFs, calcium-dependent protein kinases (CDPKs), adenylate kinase 2 (PfAK2), Golgi reassembly stacking protein 1 (PfGRASP1), gliding associated protein 45 (PfGAP45) and several erythrocyte membrane proteins (EMPs) [45].

AK2 contains a myristoylation motif at its N-terminus (MGSCYSRKNK) exclusively in the human malarial parasite *P. falciparum*, and was shown to be myristoylated *in vitro* when expressed in *E. coli*, along PfNMT [46].

The GRASPs are peripheral membrane proteins involved in the stacking of Golgi cisternae, and *P. falciparum* possesses two genes coding for GRASPs. While

(a)

```
DmNMT: MPNENAEDLSGQELKQKAKEVADASEAMLEKVVAGLNIQDTASTNAAGNEDAEQPDGAKNEASVSANASKQALLQAVSD
TaNMT: ............MTEIPDTNTLNHDKDHLSSSKTGESEEDLSEKLTNVKLSVNNHSDPTPNSTTDSSSIRERILRKLS.
TgNMT: .......MAETHTVGADLSKDGVAAPALEVETKTEGNSSSVSEN.AEKKEGTGDKTGK.DRGALTSQGQLISLMREMRV
NcNMT: .......MADPNHSATDLPKDG.APSSLAPEQQKEGNAAQAANGPGSLDKKDAEKAGKGDRGALTSQGQLISLMREMRL
PfNMT: ...................................................................MNDDKKDFVGRDLYQ
ChNMT: .................................MSGETPEKSESSISNTKKITNLLKEMSL

DmNMT: .AMASTRQMAK.KFAFWSTQPVTKLDEQVTTNEC..IEPNKEISEIRALPYTLPGGFKWVTLDLND..ANDLKELYTLL
TaNMT: ..ISHTSSYIP.EHKFWDTQLVTKLTDVVNSNDYGPIDSNEDISKVKKNPIALPNGFEWVSLDIND..EEDRNQVYKLL
TgNMT: .SNKGVSPPFA.PHTFWDTQPVPKLNEPKEGKE.GPIET.KTVDQVRSEPYKLPDGFVWCECDVRD..PEELKEVYDLL
NcNMT: .ANKGVPPPFA.PHAFWDTQPVPKLNEPQEGRE.GPIET.KTVEQVRTEPYKLPDGFVWCECDVRN..PEELKEVYELL
PfNMT: .LIRNAKDKIKIDYKFWYTQPVPKINDEFDENVNEPFISDNKVEDVRKEEYKLPSGYAWCVCDITK..ENDRSDIYNLL
ChNMT: GSFMNTAANAIKPHKFWNTQPVVQNDDSSTEYSFGPIEI..EPDSFRKEIYKLPDGFSWFDCNLWDIESQDFEDTYQLL

DmNMT: NENYVEDDDAMFRFDYQPEFLKWSLQPPGWKRDWHVGVRVEKSGKLVGFISAIPSKLKSYDKVLKVVDINFLCVHKKLR
TaNMT: SENYVEDGDALFRFDYKPEFLIWALTVPNYNKEWQIGVQVSSCKTLIGYITAVPVNVNVIGNTLKLAEVNFLCIHKKFR
TgNMT: SQHYVEDDDNLFRFNYSADFLDWALTAPGCHRDWVIGVRVSSTNKLVGFITATPSQIRVFSDSVPMAEVNFLCVHKKLR
NcNMT: SQHYVEDDDNLFRFNYSADFLDWALTAPGCHRDWVIGVRVSSTKKLVGFITATPSQIHVFSNSVPMAEVNFLCVHKKLR
PfNMT: TDNYVEDDDNVFRFNYSSEFLLWALSSPNYVKNWHIGVKYESTNKLVGFISAIPIDMCVNKNIIKMAEVNFLCVHKSLR
ChNMT: KDHYVEDDDSQFRFNYSKEFLRWALCVPGQKRNWLVGVRVNETKKMVGFISAIPIKVRIHNCIMNTSVVNFLCVHKKLR

DmNMT: SKRVAPVLIREITRRVNLTGIFQAAYTAGVVLPTPVATCRYWHRSLNPKKLVDVRFSHLARNMTMQRTMKLYKLPDQPK
TaNMT: SKRLAPVLIKEITRRVNLSGIWQAIYTAGIVIPKPIAKCRYWHRPLDIKRLISARFSGVGRRMTISRAIRIYKVNDIPN
TgNMT: SKRLAPVLIKEITRRVNLRSIWQAVYTAGVVLPTPVAQCRYWHRSLNPKKLIEVGFSGLSERMTISRSIKLYRVKESPS
NcNMT: SKRLAPVLIKEITRRVNRSVWQAVYTAGVVLPTPVAQCRYWHRSLNPKKLIEVGFSGLSERMTISRSIKLYRVKESPS
PfNMT: SKRLAPVLIKEITRRINLESIWQAIYTAGVYLPKPISTARYFHRSINVKLIEIGFSCLNTRLTMSRAIKLYRIDDTLN
ChNMT: SKRLAPVLIKEITRRIRCEKIFQSIYTCGKNITKPFTIGTYWHRIINVKKLLETGFIGIPRNMTMSSLIKYHRIPADKR

DmNMT: TKGYRRITAKDMDKAHKLLEDYLKRFQ...............................LSPVFSKEEFR
TaNMT: VE.MRPMEQKDVLSVHKLLRKYLQKYK.................................LYQEFDVHEVE
TgNMT: TPGLRPAKPEDVPHIHKLLSNYLRNFK................................LHCEFTQEEVA
NcNMT: TAGLRPATKEDVPHIHKLLTNYLKNFK................................LYCEFTQDEVA
PfNMT: IKNLRLMKKKDIDGLQKLLNEHLKQYN...............................LHAIFSKEDVA
ChNMT: IEGFRPSVDSDAEQICKLFENYFMKYKDISNETMNNKINYDEINHSKELGKQAYMKLDKIEDLQDKIIIHQCFNVEDVK

DmNMT: HWFTPKEGIIDCFVVADEKGNITDLTSYYCLPSSVMHHPVHKTVRAAYSFYNVSTKTPWLDLMNDALISARNVQMDVYN
TaNMT: HQFMPREDIIQTFVKTNED.EVTDMVSYYSLPSTVINNRKVHTIRAAYSFYNIATTMPFKSLMEHAIFFAKSQGYDVYN
TgNMT: HWLLPREGVVHVYVRTSTKGTVTDLISFYELPSSVIGNQKYKEIKAAYSFYNVATTVPLKQLIEDALCLAKQLDFDVFN
NcNMT: HWLLPRDGVIHVYVRTSSKGTVTDLISFYELPSSVIGNQKYKEIKAAYSFYNVATTVSLRELIDDALCLAKQLGFDVFN
PfNMT: HWFTPIDQVIYTYV.NEENGEIKDLISFYSLPSKVLGNNKYNILNAAFSFYNITTTTTFKNLIQDAICLAKRNNFDVFN
ChNMT: HYFTNIDKVIVTYVRENKNKEITDLFSFFIIESTVINNERFPTINIAYSFYNIANTCSLKELFNEMIITAKNNNCDAFN

DmNMT: ALDLMENKKYFAPLKFGAGDGNLQYYLYNWRCPSMQPEEIALILM
TaNMT: ALDLMENSLVFKDLKFGMGDGDLHYYMFNYRVPDLKSTDVGMVLL
TgNMT: ALDVMENKSFVEVSRLA........FVPRQGFPEREGK.......
NcNMT: ALDVMENRSFVEDLKFGIGDGFLRYYIYNWRCPEMKHSDVGLVLL
PfNMT: ALEVMDNYSVFQDLKFGEGDGSLKYYLYNWKCASCHPSKIGIVLL
ChNMT: TLDLMQNLQVIQDNKFIIGTGKLRYYVFNWKIPQISPSNVGIILF
```

Figure 18.2 (a) Alignment of the protein sequence of the NMTs encoded in the genomes of *Drosophila melanogaster* (DmNMT: FBpp0076451), *Theileria annulata* (TaNMT: TA13465), *Toxoplasma gondii* (TgNMT: TGME49_009160), *Neospora caninum* (NcNMT: NCLIV_003570), *Plasmodium falciparum* (PfNMT: PF14_0127) and *Cryptosporodium hominis* (ChNMT: Chro.30047). The conserved residues are boxed in yellow; (b) Alignment of the DHHC-CRC domains (represented on the top) of the PAT Erf2 encoded by *Saccharomyces cerevisiae* (Q06551) along with the putative PATs encoded in *Plasmodium falciparum* genome (PfPAT1: MAL13P1.117, PfPAT2: PFI1580c, PfPAT3: PFE1415w, PfPAT4: PFB0725c, PfPAT5: PFF0485c, PfPAT6: PF10_0273, and PfPAT7:

PF11_0167). The conserved residues are boxed in cyan; (c) Alignment of the protein sequence of the APT1 of *Rattus norvegicus* (RnAPT1: D63885) and the putative PPT1 encoded in the genomes of *T. gondii* (TgPPT1: TGME49_028290), *N. caninum* (NcPPT1: NCLIV_045170) and *C. hominis* (ChPPT1: Chro.10422). The conserved residues are boxed in yellow, and the catalytic residues in cyan; (d) Alignment of the protein sequence of the APT1 of *Rattus norvegicus* (RnAPT1) and the putative PPTs encoded in the genome of *P. falciparum* (PfPPT2: MAL8P1.66, PfPPT3: PF11_0211 and PfPPT4: MAL8P1.138). The conserved residues are boxed in yellow, and the catalytic residues in cyan.

(b)

```
Cx₂Cx₃[R/K]PxRx₂HCx₂Cx₂Cx₄DHHCxW[V/I]xNC[I/V]Gx₂Nx₃F
Erf2:    CPSC.....R.IWRPPRSSHCSTCNVCVMVHDHHCIWVNNCIGKRNYRFFLI
PfPAT1:  CVNC.....N.HFKEPRSKHCYTCNNCVTKFDHHCVWIGNCVGNRNYRRFFF
PfPAT2:  CKIC.....D.VYQILRSKHCQMCKRCVRTFDHHCPWINNCVAENNRSFFLL
PfPAT3:  CKWC.....C.KYKPDRTHHCRVCKSCILKMDHHCPWIYNCVGYNNHKYFML
PfPAT4:  CKKC.....N.LLKIKRSHHCSVCDKCIMKMDHHCFWINSCVGLYNQKYFIL
PfPAT5:  CKIC.....N.VWKPDRTHHCSACNRCVLNMDHHCPWINNCVGFFNRRFFIQ
PfPAT6:  CDKC.....DFLVRPERAHHCRTCNKCILKMDHHCPWIGTCVGEKNLKFFFL
PfPAT7:  CKTC.....N.IIKPARSKHCSYCSSCISRYDHHCFLLNNCIGGYNNMYYLV
```

(c)

```
RnAPT1: ........MCGNNMSAPMPAVVPAARKATAAVIFLHGLGDTGHGWAEAFAGIKS....SHIKYICPHAPVMPVTLNMS
TgPPT1: MASLQPGDGYGGEGFHRFPGVCT...PQTASLVFMHGLGDTAAGWADLVSLLSSLSCFPALRVILPTAPVRPVTLNGG
NcPPT1: MASLQPGDGYGGDGFHRFPTVSVDGAPRPATIIFLHGLGDTAAGWADLISLLSSLPCFPSLRVILPTAPVRPVTLNGG
ChPPT1: ..MIQEGDGNNGQGFYYEP......KDYDSVLIWLHGKGDNANSYLDFIHTAQNYPELKKTKIILPTADIITFK.RFG

RnAPT1: MMMPSWFDIIGLSPDSQEDESGIKQAAETVKALIDQEVK.NGIPSNRIILGGFSQGGALSLYTALTT.QQKLAGVTAL
TgPPT1: FPAPAWTDIFSLSKDAPEDKPGFLASKQRIDAILAGELA.AGVAPERIILAGFSQGGALAYFTGLQA.SVRLGGIVAL
NcPPT1: FPAPAWTDIFSLSKDTPEDREGFLESKRRIDAILRGEIEDAHIPPERIVLAGFSQGGALAYFVGLQA.PYRLGGIVAL
ChPPT1: FSDNAWFDMEDLRPYALEDLDDINNSVSRITRLISLEIE.KGIDPKKISLGGFSQGSAIVFLISMASRKYTLGSCIVV

RnAPT1: SCWLPLRASFSQGPINSANRDIS...................................VLQCHGDCDPLVPLMF
TgPPT1: STWTPLAQELRVSAGCLGKRDTQSRKEALQTREEEKTEEEKEEEKKEEKKEEKEKRVEGPTPVLHCHGEQDELVLIEF
NcPPT1: STWTPLAQELRASDACLGKKDKEGQGQTTAEGETQETQGPR................GPTPVLHCHGEQDELVLFEF
ChPPT1: GGWLPLTERGFKEGKESKIATEELTFDVRESVKEH...................VDFIVLHGEADHVVLYQW

RnAPT1: GSLTVERLKGLVNPA........NVTFKVYEGMMHSSCQQEMMDVKYFIDKLLPPID
TgPPT1: GQESAAIVRRQYAEAWGEDVAKKAVKFLSFQGLGHSANAQELDQVRRFIENVLTTN.
NcPPT1: GEEESAALVKQQYAAACGEEVAKEAVKFRPFRGLGHSANPQELAEVRLFVESVLKPQ.
ChPPT1: SLMNKDFVLEFIKPK........KFIYKSYPGVVHTITSQMMVDIFNFLSKRN....
```

(d)

```
RnAPT1: ...............................................................
PfPPT2: ...................................................MRIIKYIFF
PfPPT3: MASYSKDSDLGECIKYVHLKEDKKEEDNLYSNDKDVEYFENLFDYEYDEDDEESTGGRRFGLTFFILGCMTICGFRGR
PfPPT4: ...............................................................

RnAPT1: .............................MCGNNMSAPMPAVVPAARKATAAVIFLHGLGDTGHGWAEAFAGI
PfPPT2: AFIILALFVCALNTYIYLKQDSFVFSNEF.PTVEEKNQTLGENYEIVTLTTKDNHKFTCWYIKTKDSENKPIMLYFQG
PfPPT3: MVKKMAFVPPIIKGYNIENDNKFIFHNSHHEEIKELMQINNIDINYKKLKRGSTEVSVIMLYKKPLDLNKQTILYSHG
PfPPT4: ...............MGNVLNRIIFNGPTEGYYEKF.....DLDFIYIETENNEKVAAHFINRNAPL...TILFCHG

RnAPT1: KSSHIKYICPHAPVMPVTLNMSMMMPSWFDIIGLSPDSQEDESGIKQAAETVKALIDQEVKNGIPSNRIILGGFSQGG
PfPPT2: NGGK..YMNLFNL.IIERVDVNIFSCSNRGC.GSNIAKPSEEYFYKDAHVYIEYVKT...K...NPKHLFIFGSSMGA
PfPPT3: NTTDIGYMTPFLLNLVTSNNVNVFSYDYSGY.GLSNKDPSEKNCYKSIKMSYDYLTK...DLNIKPENIIVYGHSLGS
PfPPT4: NGENVYMLYDYFYETSKIWNVNVFLYDYLGY.GESTGTASEKNMYLSGNAVYDYMVN...TLKINPNSIVLYGKSIGS

RnAPT1: ALSLYTALTTQQKLAGVTALSCWLPL..............RASFSQGPINSANRDISVLQCHGDCDPLVPLMFGSLT
PfPPT2: AVAIDTA....LKQHDHELSRYSHPFLNYFL.FDY..DMIIRSKMDNETKIKNITVPTLFTL.SEMDEKVPTSHTRTL
PfPPT3: ATSCYLINLKNVKVGGCILQSPLYSGLRLLLPLDYKKEMPWFDVFKNDKRLKNIPLLPLFIMHGKNDRDIPYQHSEYL
PfPPT4: CAAVDIAI..KRKVKGLILQSAILSLLNICFKTRF...IFPFDSFCNIKRIKLIPCF.VFFIHGTDDKIVPFYHGMCL

RnAPT1: VE.RLKGLVNPANVTFKVYEGMMHSSCQQEMMDVKYFIDKLLPPID..........................
PfPPT2: FQLSASTNKQLYLSKGGTHPNILKNDDGSYHKAMKKFIVTAISIREKNIQKAVERNPNTPN............
PfPPT3: LKIVKKNFAKKVQKNKSQFLRNKKINHLDVHDSILRFWGVENADHNDIDEKNPELFYHKLGEFLSYCSKFNMNE
PfPPT4: YE.KCKFKVHPYWVVDGKHNDIELIENERFNENVKSFLNFLYNSDL........................
```

Figure 18.2 (Continued).

PfGRASP1 was shown to be membrane-attached by myristoylation [47], PfGRASP2 is anchored to the membrane via N-terminal acetylation and an amphipathic helix [48]. The myristoyl motif present at the N-terminus of PfGRASP1 is conserved in all apicomplexan homologues. Myristoylation is essential for sorting to the Golgi apparatus, because the expression of a GRASP1–GFP (green fluorescent protein) fusion in which the predicted modified glycine was replaced by an alanine residue was mislocalized to the cytosol [47].

The myristoylated proteins PfAK2 and PfGRASP are also predicted to be palmitoylated, although to date this has not been validated experimentally [46, 47].

Plasmodium falciparum encodes a family of five classical CDPKs; these are found in several subcellular localizations, with some isoforms located in more than one compartment. Apart from their calcium requirement, very little is known concerning the mechanisms that regulate their activities and biological roles. In plants, the CDPKs are predicted to be *N*-myristoylated and *N*-palmitoylated [49], and several studies have shown that although these modifications aid in anchoring the kinases to membranes, other factors are involved that determine target-membrane specificity [24]. Three of the CDPK parasite kinases – PfCDPK1, PfCDPK2, and PfCDPK4 – contain motifs for myristoylation, while PfCDPK1 also contains a palmitoylation motif. So far, the contribution of these PTMs to protein targeting and function has only been studied for PfCDPK1 [50]. The metabolic labeling of parasites with ^{3}H-myristate, followed by immunoprecipitation, indicates that PfCDPK1 is myristoylated *in vivo*. Site-specific mutagenesis revealed that a myristoylation motif, a palmitoylated cysteine at position 3 and a cluster of polybasic residues, contribute equally to PfCDPK1 membrane attachment and parasitophorous vacuole localization. The *Plasmodium* genome also codes for other putative kinases containing motifs for myristoylation and palmitoylation, which suggests that this is a widespread mechanism used to regulate parasite kinases [50].

PfGAP45, a member of the conserved glideosome in apicomplexans, is both phosphorylated and acylated [51]. Staurosporine treatment and metabolic labeling has established that this is a dually modified protein by phosphorylation and acylation [52]. Palmitoylation was shown to increase at the same time as binding to PfGAP50, which is an integral membrane protein of the inner membrane complex. In the light of these results, the dual acylation of GAP45 may complement the role of GAP50 in capturing the motor complex to the membrane.

Finally, in all *Plasmodium* species, calpain – a calcium-dependent cysteine protease – is also known to be palmitoylated. This protein possesses an N-terminal extension, which spans into half of the protein length, and has no significant homology to any other known protein domain. This extension can be divided into four highly conserved subdomains, in the first of which can be found a myristoylation consensus sequence and a palmitoylated cysteine at position 3. The proposed targeting model indicates that membrane localization is conferred by the two lipid motifs, in particular by the palmitoylation motif at position 3, and a bipartite nuclear localization signal at the third subdomain is responsible for the nucleolar localization. This localization is possibly regulated by the reversible palmitoylation. However, as the localization of the active calpain is not known it is not possible to distinguish whether the protease is released from the membrane by depalmitoylation and then traffics to the nucleolus, or if it is initially trapped in the nucleolus and then re-routed by palmitoylation [53].

Potential of *N*-Myristoylation and *S*-Palmitoylation as Targets for Intervention

N-Myristoyltransferases as Drug Targets

NMT is ubiquitously present in eukaryotes, ranging from mammalians to parasitic protozoa, and is crucial for the survival of several human pathogens, including fungus, viruses, and eukaryotic parasites. Different classes of inhibitors have been identified, including analogs of myristoyl-CoA; however, as the myristoyl-CoA binding sites of NMT from different species are very similar, peptide derivatives are considered as potentially more specific drugs. In fact, by exploiting the differences in the peptide binding sites of the different NMTs, peptide and peptidomimetic inhibitors with selectivity for fungal NMTs, developed through the depeptidization of an octapeptide representing the N-terminal sequence of a known NMT substrate, have been designed [54–56]. Subsequent HTS led to the identification of low-molecular-weight bicyclic compounds that are selective inhibitors of the fungal enzyme at nanomolar concentrations [57]. These compounds have successfully controlled the opportunistic fungal pathogens *Candida albicans* and *Candida neoformans* [58].

A recent modeling study compared the NMT sequences of different species and investigated the three-dimensional (3-D) structural models of these NMTs in complex with substrates. The results of the analysis suggested that inhibitors selective for *C. albicans* could serve as a template for the construction of selective NMT inhibitors against other pathogenic organisms [9]. In fact, fungal and protozoan NMTs share some characteristics in terms of binding pockets, and this has allowed for the accelerated development of drugs targeting the apicomplexan NMTs [59]. Furthermore, a comparison of the sequences of PfNMT with the corresponding human enzyme, showed the potential for selective interference (Figure 18.2a), while a genome-wide search for similarities in other parasites (*Entamoeba histolytica, Giardia intestinalis, Leishmania major, T. brucei, Trypanosoma cruzi*) identified NMTs that could possibly be targeted [60].

The NMT of protozoan parasites (*L. major, T. brucei*), is coded by a single essential copy gene [61]. The establishment of robust expression and purification protocols for the recombinant forms of TbNMT and LmNMT in *E. coli* has enabled the testing of inhibitory compounds, via a "piggyback" approach. The incubation of *L. major* procyclic forms with myristate analogs led to morphological differences and a reduction of more than 50% in parasite numbers at 48 h, in a dose-dependent manner. When antifungal compounds were also tested, they were shown to inhibit LmNMT only weakly, but TbNMT was efficiently inhibited *in vitro* and *in vivo* by two of the four inhibitors tested [61]. When the effects of NMT depletion were tested *in vivo* in a rodent model, it was shown that *T. brucei* parasites knocked down for the enzyme were unable to establish themselves in the mammalian host, most likely due to defects in VSG trafficking [62]. NMT was also reported to be essential for the viability of *L. major* [61].

An exciting new study demonstrated that specific inhibitors targeting TbNMT could be generated [63]. The screening of a library comprising more than 60 000 compounds, with further optimization of the screening hit, identified highly potent and specific inhibitors of TbNMT. These compounds were shown to inhibit the proliferation of bloodstream stages in culture and to cure acute trypanosomiasis in an animal model.

In order to accelerate the discovery of novel parasite NMT inhibitors, it is important to solve the structure of the protozoan parasites NMTs. Until then, homology modeling might be of assistance in the design of structure-based inhibitors. A recent study reported the modeling of 3-D structures of PfNMT, LmNMT and TbNMT, using as a template the known structures of CaNMT and ScNMT in complex with a small-molecule inhibitor. Although the overall structure and active sites showed a similar 3-D folding pattern to the crystal structures of the templates, gaps were found near the non-peptide inhibitor binding site, which may affect the conformation of the active site and thus influence ligand specificity [64].

Most peptides and peptidomimetika reported to inhibit NMT are still far from being drug-like. The concept of the "depeptidization" of peptides and peptidomimetika is seldom successful, as typically too many peptide-like elements are left in the final structure. The *de novo* development of small-molecule inhibitors, without the inclusion of peptide-like fragments, is often based on the initial screening at "scaffold level," thus yielding low micromolar affinities, followed by optimization with traditional medicinal chemistry. The outcome of such a process is often dependent on the quality of the assay system. It is essential that the assay displays a high similarity with the native biochemistry for the evaluation of small-molecule inhibitors of lipidation processes. Often, the supply of the native protein is limiting, while the recombinant expression of eukaryotic proteins may also be problematic, and this applies equally to the NMTs. Screening against protozoan NMT also calls for parallel screening against human NMT (HsNMT), in order to avoid cross-affinity.

An interesting case is the design of small-molecule inhibitors around the 2-amino-benzothiazole scaffold [65]. An initial biochemical screening on PfNMT, employing five validated fungal NMT inhibitors, yielded two compounds showing IC_{50}-values of 360 and 280 nM, respectively. An additional set of 40 substances centered on the same amino-benzothiazole scaffold was investigated for inhibitory activity on PfNMT. With the acquired data, the authors were able to deduce a structural–activity relationship and to establish a pharmacophore model. Subsequently, the dimethyl amide substituent was found to be crucial for high potency, and further improvements in terms of selectivity and affinity could be addressed on the linker moiety and aromatic substituent. Further investigations in cell culture yielded inhibitors with an IC_{50} well below 50 μM. The combination of a well-working assay with a prevalidated (fungal NMT) inhibitor scaffold enabled the rapid development of a compound class which showed inhibitory activity in *P. falciparum* culture. In combination with information from structural biology and bioinformatics methods, apicomplexan NMT has an interesting future as a drug target.

Palmitoyl Transferases as Drug Targets

In eukaryotes, PAT-mediated *S*-palmitoylation is by far more abundant than spontaneous palmitoylation, and the DHHC enzyme family has been seen as an ideal therapeutic target for human diseases. In fact, palmitoylation has been shown to modify a variety of key proteins involved in cell growth, cell signaling, and synaptic transmission and many of its targets, such as the oncogenic Ras proteins, huntingtin and the anthrax-toxin receptors, are directly involved in disease processes [35]. Furthermore, the overexpression of some PATs has also been shown to alter cancer-related signaling [21], and mutations in DHHC genes have been associated with human diseases [18, 21, 66]. Limitations in terms of inhibitor design have, however, been raised because all eukaryotic PAT enzymes are highly conserved and dependent on palmitoyl-CoA as donor substrate, so the binding pocket of the peptide/protein substrate is the only targetable region of the protein in a competitive fashion.

At present, both lipid-based and non-lipid-based inhibitors of palmitoylation have been identified. In the first category can be found 2-bromopalmitate, tunicamycin and cerulenin, none of which is particularly selective and for which the mechanism of action is not known. The exception here is cerulenin, which inhibits the biosynthesis of fatty acids. The non-lipid compounds are more specific and were identified by a medium-throughput screening of a chemical library of diverse synthetic compounds [29]. This screen identified five low-molecular-weight drug-like molecules that could function in intact cells and have antitumor activity *in vivo*. Compounds I–IV were shown to be selective for type 1 PATs, and V to type 2 PATs; such specificity would suggest that they are peptide substrate competitors [16].

The global inhibition of lipidation at the level of fatty acid biosynthesis typically leads to lethal phenotypes in eukaryotic cells. A good example of this is 2-bromopalmitate, which is a highly efficient palmitoylation inhibitor but displays a high toxicity, and thus is of limited use in cell-based experimentation. In comparison to 2-bromopalmitate, small-molecule PAT inhibitors should be less toxic, but cannot be expected to be very specific. It is likely that a *vide spectra* inhibition of apicomplexan PATs would lead to various side effects due to interference with the PATs of the host cells, result of the relatively high level of conservation of PATs in eukaryotic organisms.

Palmitoyl Protein Thioesterases as Drug Targets

In sharp contrast to the inhibition of palmitoylation, interference with the depalmitoylation enzymes should allow for viable phenotypes by only downregulating biological responses (e.g., signaling) by disturbing the intracellular localization of the target protein. A good example is the human *S*-depalmitoylation enzyme APT1 which, upon inhibition, leads to a reduced MAPK-mediated signaling via

Ras-delocalization [67]. In this case, the inhibition of thioesterase activity responsible for Ras depalmitoylation increases the pool of palmitoylated Ras proteins and this counterintuitevly reduces oncogenic Ras signaling. In fact, since palmitoylated H- and N-Ras remain membrane-bound, a permanent unspecific cellular membrane exchange will lead to a loss of normal localization by entropy-driven redistribution to all cellular membranes. Thereby, the inhibition of depalmitoylation will lead to a downregulation rather than an inhibition of signaling, allowing cell viability. This observation verifies the strategy that disturbing dynamic lipidation cycles at the stage of delipidation leads to an entropy-driven loss of precise localization, thus affecting the signaling network. This concept can be translated and applied in a broader sense to a plethora of acylated proteins, which depend on a precise localization for their biological role, thus allowing for the small-molecule regulation on hard-to-affect targets that rely on dynamic acylation. When considering the relative number of turnovers during the lifetime of a eukaryotic organism, *S*-acylation is most likely to be effected at the delipidation stage by an inhibition of the corresponding hydrolases, whereas *N*- and *O*-lipidations are best targeted at the transferase stage of the cycle, due to a persistence of the modification. Several inhibitors designed around the β-lactone core to block APT1 have led to the creation of highly active compounds. Some of these inhibitors have proved to be very effective at low micromolar ranges against *P. falciparum* and *T. gondii*, not only by blocking invasion but also by impairing intracellular replication. Based on these observations, it appears relevant to investigate the potential of palmitoyl protein thioesterases as a novel drug target against the apicomplexans.

Conclusion

Recent studies conducted in yeast and neuronal tissues have underlined the importance of acylation as a reversible mode of regulating protein function. The repertoire of N-myristoylated and S-palmitoylated proteins is vast, and the enzymes responsible for protein acylation have lately been subject of intensive investigation. Although, in the Apicomplexa, very few studies have been reported on protein acylation, it is becoming increasingly clear that acylation plays an essential role in regulating protein function in numerous parasite pathways, and hence represents a plausible drug target. Furthermore, recent studies have shown that it is possible to design compounds targeting protein myristoylation in the trypanosomatids.

Acknowledgments

The authors are especially grateful to Prof. Steinnun Baekkeskov, who kindly agreed to review this chapter. J.S. is a recipient of the EU-funded Marie Curie Action MalParTraining (MEST-CT-2005-020492), "The challenge of Malaria in the Post-Genomic Era". D.S. is recipient of an International Scholar HHMI award on Infectious diseases. The authors are also grateful to Dr Karine Frenal and Louise

Kemp for providing experimentally annotated sequences information for the apicomplexan PATs and APTs, respectively. This project is financed with a grant from the Swiss SystemsX.ch initiative, grant LipidX-2008/011 to D.S. C.H also acknowledges AstraZeneca AB (Sweden) for financial support.

References

1 Debierre-Grockiego, F., Molitor, N., Schwarz, R.T., and Luder, C.G. (2009) *Toxoplasma gondii* glycosylphosphatidylinositols up-regulate major histocompatibility complex (MHC) molecule expression on primary murine macrophages. *Innate Immun.*, **15**, 25–32.

2 Leroy, D. and Doerig, C. (2008) Drugging the *Plasmodium* kinome: the benefits of academia-industry synergy. *Trends Pharmacol. Sci.*, **29**, 241–249.

3 Canduri, F., Perez, P.C., Caceres, R.A., and de Azevedo, W.F. Jr (2007) Protein kinases as targets for antiparasitic chemotherapy drugs. *Curr. Drug Targets*, **8**, 389–398.

4 Liotta, F. and Siekierka, J.J. (2010) Apicomplexa, trypanosoma and parasitic nematode protein kinases as antiparasitic therapeutic targets. *Curr. Opin. Investig. Drugs*, **11**, 147–156.

5 Yang, S.H., Shrivastav, A., Kosinski, C., Sharma, R.K., Chen, M.H., Berthiaume, L.G., Peters, L.L., Chuang, P.T., Young, S.G., and Bergo, M.O. (2005) N-myristoyltransferase 1 is essential in early mouse development. *J. Biol. Chem.*, **280**, 18990–18995.

6 Ducker, C.E., Upson, J.J., French, K.J., and Smith, C.D. (2005) Two N-myristoyltransferase isozymes play unique roles in protein myristoylation, proliferation, and apoptosis. *Mol. Cancer Res.*, **3**, 463–476.

7 Shrivastav, A., Selvakumar, P., Bajaj, G., Lu, Y., Dimmock, J.R., and Sharma, R.K. (2005) Regulation of N-myristoyltransferase by novel inhibitor proteins. *Cell Biochem. Biophys.*, **43**, 189–202.

8 Farazi, T.A., Waksman, G., and Gordon, J.I. (2001) The biology and enzymology of protein N-myristoylation. *J. Biol. Chem.*, **276**, 39501–39504.

9 Maurer-Stroh, S., Eisenhaber, B., and Eisenhaber, F. (2002) N-terminal N-myristoylation of proteins: prediction of substrate proteins from amino acid sequence. *J. Mol. Biol.*, **317**, 541–557.

10 Maurer-Stroh, S., Gouda, M., Novatchkova, M., Schleiffer, A., Schneider, G., Sirota, F.L., Wildpaner, M., Hayashi, N., and Eisenhaber, F. (2004) MYRbase: analysis of genome-wide glycine myristoylation enlarges the functional spectrum of eukaryotic myristoylated proteins. *Genome Biol.*, **5**, R21.

11 Pellman, D., Garber, E.A., Cross, F.R., and Hanafusa, H. (1985) Fine structural mapping of a critical NH2-terminal region of p60src. *Proc. Natl Acad. Sci. USA*, **82**, 1623–1627.

12 Smotrys, J.E. and Linder, M.E. (2004) Palmitoylation of intracellular signaling proteins: regulation and function. *Annu. Rev. Biochem.*, **73**, 559–587.

13 Huang, K., Sanders, S., Singaraja, R., Orban, P., Cijsouw, T., Arstikaitis, P., Yanai, A., Hayden, M.R., and El-Husseini, A. (2009) Neuronal palmitoyl acyl transferases exhibit distinct substrate specificity. *FASEB J.*, **23**, 2605–2615.

14 Navarro-Lerida, I., Alvarez-Barrientos, A., Gavilanes, F., and Rodriguez-Crespo, I. (2002) Distance-dependent cellular palmitoylation of de-novo-designed sequences and their translocation to plasma membrane subdomains. *J. Cell Sci.*, **115**, 3119–3130.

15 Roth, A.F., Wan, J., Bailey, A.O., Sun, B., Kuchar, J.A., Green, W.N., Phinney, B.S., Yates, J.R., III, and Davis, N.G. (2006) Global analysis of protein palmitoylation in yeast. *Cell*, **125**, 1003–1013.

16 Ducker, C.E., Griffel, L.K., Smith, R.A., Keller, S.N., Zhuang, Y., Xia, Z., Diller,

J.D., and Smith, C.D. (2006) Discovery and characterization of inhibitors of human palmitoyl acyltransferases. *Mol. Cancer Ther.*, **5**, 1647–1659.

17 Hemsley, P.A. and Grierson, C.S. (2008) Multiple roles for protein palmitoylation in plants. *Trends Plant Sci.*, **13**, 295–302.

18 Tsutsumi, R., Fukata, Y., and Fukata, M. (2008) Discovery of protein-palmitoylating enzymes. *Pflügers Arch.*, **456**, 1199–1206.

19 Baekkeskov, S. and Kanaani, J. (2009) Palmitoylation cycles and regulation of protein function (Review). *Mol. Membr. Biol.*, **26**, 42–54.

20 Mitchell, D.A., Vasudevan, A., Linder, M.E., and Deschenes, R.J. (2006) Protein palmitoylation by a family of DHHC protein *S*-acyltransferases. *J. Lipid Res.*, **47**, 1118–1127.

21 Planey, S.L. and Zacharias, D.A. (2009) Palmitoyl acyltransferases, their substrates, and novel assays to connect them (Review). *Mol. Membr. Biol.*, **26**, 14–31.

22 Yang, W., Di Vizio, D., Kirchner, M., Steen, H., and Freeman, M.R., Proteome scale characterization of human *S*-acylated proteins in lipid raft-enriched and non-raft membranes. *Mol. Cell Proteomics*, **9**, 54–70.

23 Hou, H., John Peter, A.T., Meiringer, C., Subramanian, K., and Ungermann, C. (2009) Analysis of DHHC acyltransferases implies overlapping substrate specificity and a two-step reaction mechanism. *Traffic*, **10**, 1061–1073.

24 Hemsley, P.A., Taylor, L., and Grierson, C.S. (2008) Assaying protein palmitoylation in plants. *Plant Methods*, **4**, 2.

25 Resh, M.D. (2006) Palmitoylation of ligands, receptors, and intracellular signaling molecules. *Sci. STKE*, **2006**, re14.

26 Linder, M.E. and Deschenes, R.J. (2003) New insights into the mechanisms of protein palmitoylation. *Biochemistry*, **42**, 4311–4320.

27 Podell, S. and Gribskov, M. (2004) Predicting N-terminal myristoylation sites in plant proteins. *BMC Genomics*, **5**, 37.

28 Webb, Y., Hermida-Matsumoto, L., and Resh, M.D. (2000) Inhibition of protein palmitoylation, raft localization, and T-cell signaling by 2-bromopalmitate and polyunsaturated fatty acids. *J. Biol. Chem.*, **275**, 261–270.

29 Draper, J.M. and Smith, C.D. (2009) Palmitoyl acyltransferase assays and inhibitors (Review). *Mol. Membr. Biol.*, **26**, 5–13.

30 Schmidt, M., Schmidt, M.F., and Rott, R. (1988) Chemical identification of cysteine as palmitoylation site in a transmembrane protein (Semliki Forest virus E1). *J. Biol. Chem.*, **263**, 18635–18639.

31 Martin, B.R. and Cravatt, B.F. (2009) Large-scale profiling of protein palmitoylation in mammalian cells. *Nat. Methods*, **6**, 135–138.

32 Boutin, J.A. (1997) Myristoylation. *Cell. Signal.*, **9**, 15–35.

33 Varner, A.S., De Vos, M.L., Creaser, S.P., Peterson, B.R., and Smith, C.D. (2002) A fluorescence-based high performance liquid chromatographic method for the characterization of palmitoyl acyl transferase activity. *Anal. Biochem.*, **308**, 160–167.

34 Fukata, Y., Iwanaga, T., and Fukata, M. (2006) Systematic screening for palmitoyl transferase activity of the DHHC protein family in mammalian cells. *Methods*, **40**, 177–182.

35 Linder, M.E. and Deschenes, R.J. (2007) Palmitoylation: policing protein stability and traffic. *Nat. Rev. Mol. Cell Biol.*, **8**, 74–84.

36 Siegel, G., Obernosterer, G., Fiore, R., Oehmen, M., Bicker, S., Christensen, M., Khudayberdiev, S., Leuschner, P.F., Busch, C.J., Kane, C., Hubel, K., Dekker, F., Hedberg, C., Rengarajan, B., Drepper, C., Waldmann, H., Kauppinen, S., Greenberg, M.E., Draguhn, A., Rehmsmeier, M., Martinez, J., and Schratt, G.M. (2009) A functional screen implicates microRNA-138-dependent regulation of the depalmitoylation enzyme APT1 in dendritic spine morphogenesis. *Nat. Cell Biol.*, **11**, 705–716.

37 Bishop, R.E., Kim, S.H., and El Zoeiby, A. (2005) Role of lipid A palmitoylation in bacterial pathogenesis. *J. Endotoxin Res.*, **11**, 174–180.

38 Paul, K.S., Jiang, D., Morita, Y.S., and Englund, P.T. (2001) Fatty acid synthesis in African trypanosomes: a solution to the myristate mystery. *Trends Parasitol.*, **17**, 381–387.

39 Mills, E., Price, H.P., Johner, A., Emerson, J.E., and Smith, D.F. (2007) Kinetoplastid PPEF phosphatases: dual acylated proteins expressed in the endomembrane system of *Leishmania*. *Mol. Biochem. Parasitol.*, **152**, 22–34.

40 Paris, S., Beraud-Dufour, S., Robineau, S., Bigay, J., Antonny, B., Chabre, M., and Chardin, P. (1997) Role of protein-phospholipid interactions in the activation of ARF1 by the guanine nucleotide exchange factor Arno. *J. Biol. Chem.*, **272**, 22221–22226.

41 Flinn, H.M., Rangarajan, D., and Smith, D.F. (1994) Expression of a hydrophilic surface protein in infective stages of *Leishmania major*. *Mol. Biochem. Parasitol.*, **65**, 259–270.

42 Pimenta, P.F., Pinto da Silva, P., Rangarajan, D., Smith, D.F., and Sacks, D.L. (1994) *Leishmania major*: association of the differentially expressed gene B protein and the surface lipophosphoglycan as revealed by membrane capping. *Exp. Parasitol.*, **79**, 468–479.

43 Touz, M.C., Conrad, J.T., and Nash, T.E. (2005) A novel palmitoyl acyl transferase controls surface protein palmitoylation and cytotoxicity in *Giardia lamblia*. *Mol. Microbiol.*, **58**, 999–1011.

44 Gunaratne, R.S., Sajid, M., Ling, I.T., Tripathi, R., Pachebat, J.A., and Holder, A.A. (2000) Characterization of N-myristoyltransferase from *Plasmodium falciparum*. *Biochem. J.*, **348** (Pt 2), 459–463.

45 Bowyer, P.W., Gunaratne, R.S., Grainger, M., Withers-Martinez, C., Wickramsinghe, S.R., Tate, E.W., Leatherbarrow, R.J., Brown, K.A., Holder, A.A., and Smith, D.F. (2007) Molecules incorporating a benzothiazole core scaffold inhibit the N-myristoyltransferase of *Plasmodium falciparum*. *Biochem. J.*, **408**, 173–180.

46 Rahlfs, S., Koncarevic, S., Iozef, R., Mailu, B.M., Savvides, S.N., Schirmer, R.H., and Becker, K. (2009) Myristoylated adenylate kinase-2 of *Plasmodium falciparum* forms a heterodimer with myristoyltransferase. *Mol. Biochem. Parasitol.*, **163**, 77–84.

47 Struck, N.S., de Souza Dias, S., Langer, C., Marti, M., Pearce, J.A., Cowman, A.F., and Gilberger, T.W. (2005) Re-defining the Golgi complex in *Plasmodium falciparum* using the novel Golgi marker PfGRASP. *J. Cell Sci.*, **118**, 5603–5613.

48 Struck, N.S., Herrmann, S., Langer, C., Krueger, A., Foth, B.J., Engelberg, K., Cabrera, A.L., Haase, S., Treeck, M., Marti, M., Cowman, A.F., Spielmann, T., and Gilberger, T.W. (2008) *Plasmodium falciparum* possesses two GRASP proteins that are differentially targeted to the Golgi complex via a higher- and lower-eukaryote-like mechanism. *J. Cell Sci.*, **121**, 2123–2129.

49 Ludwig, A.A., Romeis, T., and Jones, J.D. (2004) CDPK-mediated signalling pathways: specificity and cross-talk. *J. Exp. Bot.*, **55**, 181–188.

50 Moskes, C., Burghaus, P.A., Wernli, B., Sauder, U., Durrenberger, M., and Kappes, B. (2004) Export of *Plasmodium falciparum* calcium-dependent protein kinase 1 to the parasitophorous vacuole is dependent on three N-terminal membrane anchor motifs. *Mol. Microbiol.*, **54**, 676–691.

51 Jones, M.L., Cottingham, C., and Rayner, J.C. (2009) Effects of calcium signaling on *Plasmodium falciparum* erythrocyte invasion and post-translational modification of gliding-associated protein 45 (PfGAP45). *Mol. Biochem. Parasitol.*, **168**, 55–62.

52 Rees-Channer, R.R., Martin, S.R., Green, J.L., Bowyer, P.W., Grainger, M., Molloy, J.E., and Holder, A.A. (2006) Dual acylation of the 45kDa gliding-associated protein (GAP45) in *Plasmodium falciparum* merozoites. *Mol. Biochem. Parasitol.*, **149**, 113–116.

53 Russo, I., Oksman, A., and Goldberg, D.E. (2009) Fatty acid acylation regulates trafficking of the unusual *Plasmodium falciparum* calpain to the nucleolus. *Mol. Microbiol.*, **72**, 229–245.

54 Lodge, J.K., Jackson-Machelski, E., Devadas, B., Zupec, M.E., Getman, D.P., Kishore, N., Freeman, S.K., McWherter, C.A., Sikorski, J.A., and Gordon, J.I. (1997) N-myristoylation of Arf proteins in *Candida albicans*: an in vivo assay for evaluating antifungal inhibitors of myristoyl-CoA: protein N-myristoyltransferase. *Microbiology*, **143** (Pt 2), 357–366.

55 Lodge, J.K., Jackson-Machelski, E., Higgins, M., McWherter, C.A., Sikorski, J.A., Devadas, B., and Gordon, J.I. (1998) Genetic and biochemical studies establish that the fungicidal effect of a fully depeptidized inhibitor of *Cryptococcus neoformans* myristoyl-CoA:protein N-myristoyltransferase (Nmt) is Nmt-dependent. *J. Biol. Chem.*, **273**, 12482–12491.

56 Devadas, B., Freeman, S.K., McWherter, C.A., Kishore, N.S., Lodge, J.K., Jackson-Machelski, E., Gordon, J.I., and Sikorski, J.A. (1998) Novel biologically active nonpeptidic inhibitors of myristoylCoA: protein N-myristoyltransferase. *J. Med. Chem.*, **41**, 996–1000.

57 Georgopapadakou, N.H. (2002) Antifungals targeted to protein modification: focus on protein N-myristoyltransferase. *Expert Opin. Investig. Drugs*, **11**, 1117–1125.

58 Langner, C.A., Lodge, J.K., Travis, S.J., Caldwell, J.E., Lu, T., Li, Q., Bryant, M.L., Devadas, B., Gokel, G.W., Kobayashi, G.S. et al. (1992) 4-oxatetradecanoic acid is fungicidal for *Cryptococcus neoformans* and inhibits replication of human immunodeficiency virus I. *J. Biol. Chem.*, **267**, 17159–17169.

59 Bowyer, P.W., Tate, E.W., Leatherbarrow, R.J., Holder, A.A., Smith, D.F., and Brown, K.A. (2008) N-myristoyltransferase: a prospective drug target for protozoan parasites. *ChemMedChem*, **3**, 402–408.

60 Yamazaki, K., Kaneko, Y., Suwa, K., Ebara, S., Nakazawa, K., and Yasuno, K. (2005) Synthesis of potent and selective inhibitors of *Candida albicans* N-myristoyltransferase based on the benzothiazole structure. *Bioorg. Med. Chem.*, **13**, 2509–2522.

61 Price, H.P., Menon, M.R., Panethymitaki, C., Goulding, D., McKean, P.G., and Smith, D.F. (2003) Myristoyl-CoA:protein N-myristoyltransferase, an essential enzyme and potential drug target in kinetoplastid parasites. *J. Biol. Chem.*, **278**, 7206–7214.

62 Price, H.P., Guther, M.L., Ferguson, M.A., and Smith, D.F. (2010) Myristoyl-CoA: protein N-myristoyltransferase depletion in trypanosomes causes avirulence and endocytic defects. *Mol. Biochem. Parasitol.*, **169**, 55–58.

63 Frearson, J.A., Brand, S., McElroy, S.P., Cleghorn, L.A., Smid, O., Stojanovski, L., Price, H.P., Guther, M.L., Torrie, L.S., Robinson, D.A., Hallyburton, I., Mpamhanga, C.P., Brannigan, J.A., Wilkinson, A.J., Hodgkinson, M., Hui, R., Qiu, W., Raimi, O.G., van Aalten, D.M., Brenk, R., Gilbert, I.H., Read, K.D., Fairlamb, A.H., Ferguson, M.A., Smith, D.F., and Wyatt, P.G. (2010) N-myristoyltransferase inhibitors as new leads to treat sleeping sickness. *Nature*, **464**, 728–732.

64 Sheng, C., Ji, H., Miao, Z., Che, X., Yao, J., Wang, W., Dong, G., Guo, W., Lu, J., and Zhang, W. (2009) Homology modeling and molecular dynamics simulation of N-myristoyltransferase from protozoan parasites: active site characterization and insights into rational inhibitor design. *J. Comput. Aided Mol. Des.*, **23**, 375–389.

65 Panethymitaki, C., Bowyer, P.W., Price, H.P., Leatherbarrow, R.J., Brown, K.A., and Smith, D.F. (2006) Characterization and selective inhibition of myristoyl-CoA: protein N-myristoyltransferase from *Trypanosoma brucei* and *Leishmania major*. *Biochem. J.*, **396**, 277–285.

66 Nadolski, M.J. and Linder, M.E. (2007) Protein lipidation. *FEBS J.*, **274**, 5202–5210.

67 Dekker, F.J., Rocks, O., Vartak, N., Menninger, S., Hedberg, C., Balamurugan, R., Wetzel, S., Renner, S., Gerauer, M., Schölermann, B., Rusch, M., Kramer, J.W., Rauh, D., Coates, G.J., Brunsveld, L., Bastiaens, P., and Waldmann, H. (2010) Small molecule inhibition of depalmitoylation reverts unregulated Ras signaling. *Nature Chem. Biol.*, **6**, 449–456.

19
Drugs and Drug Targets in *Neospora caninum* and Related Apicomplexans

Joachim Müller, Norbert Müller, and Andrew Hemphill[*]

Abstract

The currently known arsenal of active drugs against *Neospora caninum* and related apicomplexans, such as *Toxoplasma*, *Besnoitia*, *Sarcocystis* and, to a certain extent, also *Cryptosporidium*, relies on a small number of compounds. The most effective drugs against these parasites are derived from antibacterial chemotherapy. With the exception of *Cryptosporidium*, they most likely interfere with essential functions of a unique organelle of apicomplexans, the apicoplast. During the past decade, several compounds with interesting *in vitro* properties have been identified, but few have been granted FDA approval and found their way to market. One compound, the nitro-thiazole nitazoxanide, exhibits a broad range of efficacy against protozoan parasites, helminthes, viruses, and also cancer cells, and is currently marketed for the treatment of persistent diarrhea caused by *Cryptosporidium parvum* and the intestinal protozoan *Giardia lamblia*. Nitazoxanide is also marketed for the treatment of equine myeloencephalitis, caused by another member of the Apicomplexa, *Sarcocystis neurona*. In this chapter, the modes of actions of established drugs and novel compounds potentially suitable for anti-apicomplexan chemotherapy will be discussed.

Introduction

Apicomplexan parasites are responsible for a variety of diseases in humans, pet animals, and livestock, and are thus of considerable medical and economic importance. Treatment options include: (i) the culling of infected livestock [1]; (ii) vaccination [2]; and (iii) chemotherapy. Of these approaches, chemotherapy would be the method of choice, although the number of available drugs is at present very limited (see Table 19.1). The most important Apicomplexon, *Plasmodium falciparum*, which is the causative agent of malaria, will not be discussed in this chapter, as attention is focused on chemotherapeutical treatment options against

[*]Corresponding author

Apicomplexan Parasites. Edited by Katja Becker
Copyright © 2011 WILEY-VCH Verlag GmbH & Co. KGaA, Weinheim
ISBN: 978-3-527-32731-7

Table 19.1 Overview of anti-apicomplexan drug families used as chemotherapeutics in humans or livestock.

Class	Drug	Apicomplexon
Aminoglycosides	Paromomycin	*Cryptosporidium*
Diamines	Pyrimethamine	*Cryptosporidium*
		Toxoplasma
Macrolides	Spiromycin	*Toxoplasma*
Sulfonamides		*Cryptosporidium*
		Toxoplasma
Thiazolides	Nitazoxanide	*Cryptosporidium*
		Sarcocystis
Triazines	Toltrazuril	*Cryptosporidium*
		Neospora
		Sarcocystis
		Toxoplasma

other Apicomplexa such as *Neospora*, *Toxoplasma*, *Sarcocystis*, and also *Cryptosporidium*. Nevertheless, a major research effort is currently under way with regards to antimalarial chemotherapy, and many "novel" compounds that have been tested against one of the coccidian Apicomplexa have been described previously as antimalarial drugs. In this chapter, attention will be focused on the mode of action of established drugs, and the current status of research for novel compounds summarized.

Neospora caninum and *Toxoplasma gondii* are phylogenetically closely related Apicomplexa that share many biological, immunological, and structural features. They can both infect a wide variety of hosts and tissues, and they exist in two distinct forms: (i) rapidly proliferating (and thus pathology-inducing) tachyzoites; and (ii) slowly dividing and cyst-forming bradyzoites, which can persist within their hosts for long periods of time. There are, however, distinct differences in terms of their host spectrum and transmission. The importance of *Neospora* lies in the fact that it can infect and cross the placental tissue during pregnancy, so as to cause stillbirth and abortion in not only cattle but also other species. Canids such as dogs, coyotes, and Australian dingoes are definitive hosts, which shed infective oocysts. *Neospora* thus represents an important economic burden to the cattle industry worldwide. To date, no human *Neospora* infections have been reported. In contrast, *Toxoplasma* is responsible for abortion, neonatal mortality, and severe encephalitis in newborn humans through infections to previously unexposed mothers, and represents an important opportunistic disease in immune-compromised individuals. *T. gondii* also causes abortion in many animal species, most importantly in sheep, but not in cattle.

Sarcocystis represents a large genus comprising about 130 recognized species. Infection in animals is common in cattle, sheep, pigs, and also horses, with the parasites often being located in muscular tissue cysts. Treatment is rarely required, except in cases where, for example, the central nervous system is involved. For instance, *Sarcocystis neurona* infection in horses causes equine protozoal myeloencephalitis with severe neurological symptoms, and is widespread in the US. The human infection is most likely severely underestimated, and occurs through consuming undercooked meat. Cryptosporidiosis in humans, which is characterized by persistent diarrhea, is caused by *Cryptosporidium parvum*; this is considered a distantly related lineage but is in fact not coccidian, although it occupies many of the same ecological niches.

Mode of Action and Drug Targets

General Remarks

From a theoretical standpoint, the mode of action of antiparasitic drugs consists of binding to a target protein (enzyme or receptor), which in turn results in the inhibition of essential cellular functions. Such targets are identified by analyzing structural and metabolic alterations to the parasite due to the drug, followed by the isolation of drug-binding proteins or the analysis of resistant versus sensitive parasites. The isolated targets are then validated by inhibition studies *in vitro*, the protein–ligand complexes are structurally analyzed and, if possible, knockout and overexpression studies of the corresponding gene in the target organism are conducted [3].

In an ideal situation, the drug–target interaction depends on distinct structural features in the target. In this case, point mutations which cause the presence or absence of a single amino acid or nucleotide (in the case of rRNA as a target) can discriminate between resistance and susceptibility. In reality, however, the situation is more complex. In resistant organisms, the "ideal" target may still be present, but the drug cannot access it due to either limitations in drug uptake or detoxification mechanisms. A search for drug targets (e.g., by protein binding studies) would, therefore, yield detoxification enzymes, transporters, or even irrelevant drug-binding proteins besides the "true" target. Conversely, susceptibility to a drug may be triggered by an ability to convert the (ineffective) prodrug to the (effective) drug or active compound. In this case, target searches must be performed with the active compound rather than with the prodrug; otherwise, the metabolizing or transforming enzyme would be identified rather than the drug target itself.

The most suitable targets are proteins associated with biosynthetic pathways that do not exist in the host cell. In the case of the Apicomplexa, there is a good opportunity for this approach due to the presence of a unique organelle, a plastid of probable green algal origin, the apicoplast [4], as a suitable drug target [5].

Methods for Determining Drug Efficacy

In order to evaluate the *in vitro* effectiveness of novel compounds against apicomplexan parasites such as *Neospora caninum*, host cell monolayers are infected with tachyzoites and cultured in the presence of the compounds for given period of time. Then, classically, the growth of intracellular parasites is evaluated by staining (e.g., with cresyl violet) and counting the parasitophorous vacuoles. A more accurate method is to determine the total and viable parasites by quantitative real-time polymerase chain reaction (PCR) [6]. A suitable alternative is the use of a reporter strain for *Toxoplasma gondii* expressing β-galactosidase [7], as shown in a recent application [8]. Moreover, the effects of the compounds on parasite morphology and ultrastructure are investigated at this stage by employing light and electron micros-copy analyses.

Since the drug acts in the host cell interior, the mode of action of a given compound may not be entirely dependent on the susceptibility of the parasite itself, but can also be modulated by – or even depend on – the susceptibility of the host cell, or the biochemical changes occurring therein. Thus, toxicity of the compounds to host cells and potential targets within the host cell must be evaluated and defined in order to distinguish between the direct and indirect killing of the parasite. Host cell toxicity can be evaluated by using standard methods such as Alamar blue, MTT(3-(4,5-Dimethylthiazol-2-yl)-2,5-diphenyltetrazolium bromide), or thymidine-incorpo-ration assays.

In the next step, compounds with promising results in cell cultures are tested in suitable animal models, which predominantly include normal or immunosup-pressed rodents. The animals are experimentally challenged with a defined number of tachyzoites which, if left untreated, lead to readily detectable clinical signs of disease. The animals are first treated with single or multiple doses of the compound, either by injection, orally by gavage, or simply by addition to their diet. The outputs of this type of experiments are: (i) the parasite load; (ii) the severity of symptoms; (iii) death or survival; (iv) serological parameters; and (v) , in the case of the pregnant mouse model, the transmission of parasites to the offspring and disease occurring in the offspring. Since not all parameters are tested in all studies, a comparison of different compounds is – at this stage – very difficult to perform, especially in metastudies and reviews.

Nucleic Acids and Gene Expression as Targets

DNA and DNA-Modifying Enzymes

Acridine and derivatives such as quinacrine were among the first-identified effica-cious drugs against *Plasmodium* and other parasites. Quinacrine is an intercalating agent that binds to DNA with a preference for (A+T)-rich regions, and thereby blocks DNA replication and RNA biosynthesis [9, 10]. More recently studied antiparasitic drugs with DNA-binding properties have included diamidines, such as pentamidine and its derivatives. In studies using pentamidine derivatives [11] and bis-benzimi-dazoles [12], very good correlations between the binding affinities of these com-pounds to calf thymus DNA and their efficacies against *Giardia lamblia* have been identified. Some diamidines are highly effective against *N. caninum* and *T. gondii* tachyzoites, with IC_{50}-values as low as 160 nM [13]. More recently, novel derivatives have been found to exhibit IC_{50}-values of 30 nM (C. Kropf and A. Hemphill, unpublished observations). Since DNA binding is by far not species-specific, and therefore prone to side effects on the host cells, the relevance of a DNA-binding agent as an antimicrobial drug resides in other, more species-specific, properties. These can include differential uptake or metabolism, or the presence of other targets besides DNA. For instance, diamidines (e.g., pentamidine and a variety of other DNA-binding compounds, such as etoposide) induce the cleavage of *Trypanosoma* DNA minicircles in a pattern that resembles the action of topoisomerase II inhibitors [14]. Since the DNA of the apicoplast is also circular, this may explain the effectiveness of these compounds in apicomplexan parasites. DNA-modifying enzymes may thus constitute more selective targets for antimicrobial agents. Ciprofloxacine and other fluoroquinolones inhibit DNA-gyrases and topoisomerases from prokaryotes [15], and are effective against *T. gondii in vitro*, most likely by also interfering with the apicoplast [16].

Histone Acetylation

Currently, there is increasing interest in drugs that interfere with epigenetic codes, especially histone methylation and acetylation, as these play a crucial role in developmental processes such as stage conversion [17]. The cyclic tetrapeptide apicidin, first isolated from *Fusarium*, is a well-established inhibitor of histone deacetylase, induces histone hyperacetylation, and is active against a number of apicomplexan parasites besides *Plasmodium*, with IC_{50}-values in the nanomolar range [18]. More recently, derivatives of apicidin with enhanced anti-protozoal properties have even been synthesized and characterized with respect to binding and inhibition properties on histone deacetylases from *Eimeria* and host cells. Since the inhibition of host cell histone deacetylases occurs in the same range as for parasite enzymes, it remains unclear how the inhibition of histone deacetylase can interfere with parasite survival [19]. The following hypotheses might be posed in order to explain the observed effects: (i) histone hyperacetylation in the parasite and host cell leads to the apoptosis of infected host cells, whereas noninfected cells remain nonapoptotic; (ii) histone hyperacetylation itself is lethal for the parasite, due to a generalized deregulation of gene expression; and (iii) histone deacetylase is not the only target, and another parasite-specific enzyme system is inhibited by the same class of compounds. Here, recent investigations using a member of another class of cyclic tetrapeptides derived from *Acremonium* species have provided some new insights. This tetrapeptide induces hyperacetylation and inhibits the proliferation of *T. gondii*, *N. caninum*, and *P. falciparum*, with an IC_{50} of approximately 10 nM, while host cells such as human foreskin fibroblasts (HFF) and a resistant *T. gondii* strain are affected at an IC_{50} of one magnitude higher. The histone deacetylase HDAC3 appears to be a likely target, as the recombinant enzyme is inhibited by the tetrapeptide (but with a K_i which is orders of magnitude higher than the *in vivo* IC_{50}), and since resistant mutants show distinct point mutations at a single locus in a conserved region of this gene [20].

Histone acetylation can also be influenced by inhibiting histone acetylation via histone acetyltransferase. A quinoline derivative inhibiting the histone acetyltransferase GCN5 from yeast does not inhibit the homologous enzyme from *T. gondii*, nor does it interfere with the histone acetylation of this parasite; however, it does inhibit the proliferation of *T. gondii* tachyzoites, with a moderate IC_{50} of approximately 100 μM. Thus, another target must be present in this case [21].

Protein Biosynthesis

Due to its prokaryotic origin, the apicomplexan plastid is a suitable target for antibiotics that interfere with its prokaryote-type protein biosynthesis. Tetracycline and other related aminoglycosides bind to the small subunit of ribosomes, more exactly to distinct features of the 16S-rRNA secondary structure, and thereby inhibit protein biosynthesis (see e.g., Ref. [22] for a detailed structural analysis). The aminoglycoside paromomycin is a well-established drug against intestinal infections, and also shows a limited efficacy against *C. parvum*. In this case, the large subunit of the ribosomes is targeted by some antibiotics which belong to the class of macrolides. One of these, spiromycin, may be used as a therapy against *T. gondii* [23].

Drugs Interfering with Intermediary Metabolism

The intermediary metabolism of pathogens provides suitable target options in all cases where a pathway is essential for the pathogen, and is absent or of minor importance for the host cell. In the case of apicomplexan parasites, the apicoplast again provides good examples. Apicoplasts have a type of fatty acid biosynthesis that resembles the biosynthetic pathways of prokaryotes and plastids from higher plants, and this therefore provides a suitable drug target [24]. Similar to higher plants, this type II pathway is encoded in the nucleus and targeted to the plastid. It can be inhibited, for instance, by thiolactomycin, a potent inhibitor of bacteria/plastid fatty acid biosynthesis, which also inhibits the growth of *P. falciparum* [25]. Conversely, *C. parvum*, which lacks an apicoplast, has a conventional type I fatty acid biosynthesis pathway and is resistant to thiolactomycin [26].

Folic acid biosynthesis represents another example of a prokaryote-like pathway in the Apicomplexa. The sulfonamides are inhibitors of H_2-pteroate synthase, while the compounds pyrimethamine, clindamycin, or trimethoprim inhibit dihydrofolate reductase. Either alone, or in combination with sulfonamides, these drugs constitute the main current therapy against *T. gondii*. Unfortunately, however, they are not specific for toxoplasmosis, they do not affect the tissue cyst stage, and they may have adverse side effects, mainly in pregnant women, where spiromycin is used instead [27].

Toltrazuril, a triazinone derivative with a broad spectrum of efficacy against cyst-forming and non-cyst-forming Apicomplexa, is used as a drug against various coccidioses in livestock [28]. Its mode of action is to inhibit the enzyme dihydroorotate dehydrogenase, and thereby pyrimidine biosynthesis. There is also some evidence that it affects the respiratory chain [29]. In mice artificially infected with *N. caninum*, toltrazuril and its main metabolite ponazuril have been shown to reduce both clinical signs and cerebral lesions [30], and also to interfere with the diaplacental transmission of *N. caninum* to newborns [31]. Moreover, toltrazuril is suitable for the treatment of congenitally infected newborn mice [32]. With regards to the economically more relevant situation in cattle, toltrazuril treatment does not reduce seropositivity in congenitally infected calves [33].

All of the above-mentioned drugs have been in use for at least three decades, and reports on novel drugs are indeed rare. Consequently, the release of a novel class of

Table 19.2 *In vivo* studies on the effects of nitazoxanide (NTZ) against cryptosporidiosis in cattle or human patients. P, positive effects; N, no effects; A, adverse effects.

Organism	Effects	Remarks	References
Calf			
Neonatal	A	No reduction of oocyst shedding. NTZ enhances diarrhea in the control group	[34]
Neonatal	P	NTZ reduces oocyst shedding and diarrhea with respect to a placebo treated group. No comparison to other drugs. No data concerning effects on uninfected animals	[35]
Goat (neonatal)	N–A	No reduction of oocyst shedding or improvement in weight gain. Early mortality increased in treated kids. Uninfected control groups are lacking	[36]
Mouse			
Neonatal	P	Efficacy depends on formulation. In no case better than paromomycin	[37]
Adult (immunosuppressed)	N	No reduction of oocyst shedding	[38]
Pig (neonatal)	P–N	Reduction of oocyst shedding, but less effective than paromomycin. No effect on reduction of diarrhea	[38]
Rat (immunosuppressed)	P	Reduction of oocyst shedding similar to paromomycin but no relapse after NTZ treatment	[39]
Human patients (immunocompromised)			
Meta-analysis of clinical studies until 2005	P–N	No reduction of diarrhea and oocyst shedding in HIV-seropositive patients. Tendency to better parasite clearance in seronegative patients. Overall similar to paromomycin	[40]
Clinical studies with HIV-seropositive children	N	No difference to placebo-control group with respect to parasite loss and mortality despite high-dose, prolonged treatment	[41, 42]

compounds – the thiazolides – has been followed with great interest. Nitazoxanide, the mother-compound of this class, has been approved by the FDA in the USA for the treatment of persistent diarrhea due to cryptosporidiosis and giardiasis in children aged over 12 years, and in adults. Several studies have also been reported describing the treatment of experimentally induced cryptosporidiosis in animals (Table 19.2). Overall, nitazoxanide does not provide a better treatment than paromomycin, with reported toxicity in two cases, namely in young goats [36] and calves [34]. Upon oral uptake, nitazoxanide is rapidly deacetylated to tizoxanide and further metabolized to tizoxanide glucuronide [43]. Tizoxanide has been reported to display antimicrobial activity similar to nitazoxanide, while tizoxanide-glucuronide is largely inactive against a number of pathogens. In anaerobic or microaerophilic pathogens, the mode of action is purportedly similar to metronidazole, with pyruvate oxidore-ductase as a potential target [44], although experimental evidence for this mode of action in the Apicomplexa is lacking. Recent studies have shown that not only nitazoxanide, but also some derivatives lacking the nitro group [45], exhibit *in vitro* activity against *Neospora caninum* [46], *Cryptosporidium parvum* [47], and *Besnoitia besnoiti* [48]. Moreover, nitazoxanide and some non-nitro-derivatives are toxic to proliferating mammalian (human) cells *in vitro*. By using affinity chromatography followed by mass spectroscopy, glutathione-*S*-transferase P1 has been identified in human colon carcinoma (Caco2) cells, and the inhibition of this enzyme by nitazoxanide and some non-nitro-thiazolides correlates well with their efficacy to induce apoptosis. The overexpression and downregulation of glutathione-*S*-trans-ferase P1 is correlated with an increase or decrease of sensitivity, respectively [49]. Therefore, it cannot be excluded that the efficacy of nitazoxanide against cryptospo-ridiosis in patients is partly due to apoptosis-inducing effects on the intestinal cells. The identification of thiazolide-binding proteins in the Apicomplexa has been unsuccessful so far, with the exception of *Neospora caninum*, where the protein disulfide isomerase (NcPDI) has been identified as a thiazolide-binding protein. As with the protein disulfide isomerases from *G. lamblia* [50], the *N. caninum* enzyme is inhibited by various thiazolides [51], which might account for the more indirect effect of these drugs, for example by interfering with protein targeting to subcellular organelles such as the apicoplast and with protein secretion. At the time of writing (Spring 2010), no information on clinical studies using non-nitro-thiazolides has yet been reported.

Compounds of Plant Origin with Miscellaneous Targets

In order to improve the situation in anti-parasitic chemotherapy, new potential pharmaceuticals from natural sources have been evaluated during the past few decades. In general, crude extracts from traditional medicinal herbs are tested and then fractionated in order to identify the active principles [52].

Although compounds of plant origin have been intensively investigated in the case of *Plasmodium*, with quinine and artemisinin and their derivatives as only two examples for effective drugs [53], reports concerning plant compounds with anti-coccidial effects are scarce. The mode of action of artemisinin and its derivatives allegedly relies on the cleavage of the endoperoxide bridge, resulting in the gener-ation of an unstable oxygen-centered radical and subsequent alkylation of various enzymes and membrane damage [54, 55]. Artemisinin has been shown to inhibit *N. caninum* proliferation in Vero cells, with little or no host cell toxicity [56]; however, the concentrations used were beyond reason for being active *in vivo*.

Another group of plant compounds with parasiticidal effects are the isoflavones, the targets of which are, in most cases, unknown. The exception is genistein, which has been well-characterized as a tyrosine kinase inhibitor (EGF receptor kinase inhibitor) in mammalian cells. This effect depends on the presence of hydroxyl groups in positions 5, 7, and 4′ of the isoflavone ring. These hydroxyl groups are crucial, as the closely related daidzein, with hydroxyl groups in positions 7 and 4′, is ineffective as a tyrosine kinase inhibitor [57]. With regards to anti-apicomplexan activities, a panel of 52 genistein derivatives, with a receptor kinase as an alleged target, has been described with good *in vitro* activities against *S. neurona*, *N. caninum*, and *C. parvum*, while two of them were also effective against *C. parvum* in immunosuppressed gerbils [58]. The authors, however, did not clarify whether these compounds really target receptor kinases, as the panel of compounds did not contain derivatives that differed only in the presence of an unsubstituted hydroxyl group in position 5. Moreover, simple controls with the mother compounds genistein and daidzein were lacking. Besides receptor kinases, genistein and related isoflavones with hydroxyl groups in positions 7 and 4' bind to estrogen receptors, which may lead to potential adverse side effects [59].

Another well-explored source of antimicrobial compounds are essential oils, with a broad spectrum of efficacy and, therefore, a potential toxicity toward host cells. One good example was a study in which active compounds against *G. lamblia* were identified from *Allium* extracts [60], when the compound with the lowest IC_{50} was allyl alcohol, a compound with an unspecific toxicity and a potential carcinogen. Allicin, another molecule which is effective against *Giardia* and other parasites such as *Entamoeba*, binds to the SH-groups from cysteine proteases and other SH-proteins including thioredoxin reductase [61, 62], a mode of action recently postulated for metronidazole [63]. One of the reports concerning the anti-coccidial effects of garlic extracts stated only a minor effectiveness against *C. parvum* in chickens [64].

Conclusion

This overview indicates that few new options exist for the treatment of diseases caused by the Apicomplexa. The only compound to have acquired clinical approval for the treatment of cryptosporidiosis within the past decade, nitazoxanide, does not perform any better than established therapies, and in fact leaves certain questions open with regards to its mode(s) of action, which could well be based on its effects on the host cells rather than on the parasites. Taken together, the most effective compounds target DNA (e.g., pentamidines) or the protein synthesis machinery (such as macrolides and other antibiotics) and interfere, in all likelihood, with the essential functions of the apicoplast. This organelle provides unique targets (with the exception of *Cryptosporidium*), and will undoubtedly become a focus of interest for future anti-apicomplexan drug development [65]. Accomplished and upcoming sequencing efforts on the Apicomplexa mentioned in this chapter will surely fuel these developmental efforts with novel target candidates [66, 67].

References

1 Häsler, B., Regula, G., Stärk, K.D., Sager, H., Gottstein, B., and Reist, M. (2006) Financial analysis of various strategies for the control of *Neospora caninum* in dairy cattle in Switzerland. *Prev. Vet. Med.*, **77**, 230–253.

2 Innes, E.A. and Vermeulen, A.N. (2007) Vaccination as a control strategy against the coccidial parasites *Eimeria*, *Toxoplasma* and *Neospora*. *Parasitology*, **133**, S145–S168.

3 Wang, C.C. (1997) Validating targets for antiparasite chemotherapy. *Parasitology*, **114** (Suppl.), S31–S44.

4 Köhler, S., Delwiche, D.F., Denny, P.W., Tilney, L.G., Webster, P. *et al.* (1997) A plastid of probable green algal origin in apicomplexan parasites. *Science*, **275**, 1485–1489.

5 Wiesner, J., Reichenberg, A., Heinrich, S., Schlitzer, M., and Jomaa, H. (2008) The plastid-like organelle of apicomplexan parasites as drug target. *Curr. Pharm. Des.*, **14**, 855–971.

6 Strohbusch, M., Müller, N., Hemphill, A., Greif, G., and Gottstein, B. (2008) NcGRA2 as a molecular target to assess the parasiticidal activity of toltrazuril against *Neospora caninum*. *Parasitology*, **135**, 1065–1073.

7 McFadden, D.C., Seeber, F., and Boothroyd, J.C. (1997) Use of *Toxoplasma gondii* expressing β-galactosidase for colorimetric assessment of drug activity in vitro. *Antimicrob. Agents Chemother.*, **41**, 1849–1853.

8 Müller, J., Limban, C., Stadelmann, B., Missir, A.V., Chirita, I.C. *et al.* (2009) Thioureides of 2-(phenoxymethyl)benzoic acid 4-R substituted: A novel class of anti-parasitic compounds. *Parasitol. Int.*, **58**, 128–135.

9 Wilson, W.D., Mizan, S., Tanious, F.A., Yao, S., and Zon, G. (1994) The interaction of intercalators and groove-binding agents with DNA triple-helical structures: the influence of ligand structure, DNA backbone modifications and sequence. *J. Mol. Recognit.*, **7**, 89–98.

10 Ciak, J. and Hahn, F.E. (1967) Quinacrine (atebrin): mode of action. *Science*, **156**, 655–656.

11 Bell, C.A., Hall, J.E., Cory, M., Fairley, T., and Tidwell, R.R. (1991) Structure-activity relationships of pentamidine analogs against *Giardia lamblia* and correlation of antigiardial activity with DNA binding affinity. *Antimicrob. Agents Chemother.*, **35**, 1099–1107.

12 Bell, C.A., Dykstra, C.C., Naiman, N.A., Cory, M., Fairley, T.A., and Tidwell, R.R. (1993) Structure-activity studies of dicationically substituted bis-benzimidazoles against *Giardia lamblia*: correlation of antigiardial activity with DNA binding affinity and giardial topoisomerase II inhibition. *Antimicrob. Agents Chemother.*, **37**, 2668–2673.

13 Leepin, A., Stüdli, A., Brun, R., Stephens, C.E., Boykin, D.W., and Hemphill, A. (2008) Host cells participate in the in vitro effects of novel diamidine analogues against tachyzoites of the intracellular apicomplexan parasites *Neospora caninum* and *Toxoplasma gondii*. *Antimicrob. Agents Chemother.*, **52**, 1999–2008.

14 Shapiro, T.A. and Englund, P.T. (1990) Selective cleavage of kinetoplast DNA minicircles promoted by antitrypanosomal drugs. *Proc. Natl Acad. Sci. USA*, **87**, 950–954.

15 Kidwai, M., Misra, P., and Kumar, P. (1998) The fluorinated quinolones. *Curr. Pharmaceut. Des.*, **4**, 101–118.

16 Fichera, M. and Roos, D.S. (1997) A plastid organelle as a drug target in apicomplexan parasites. *Nature*, **390**, 407–409.

17 Bougdour, A., Braun, L., Cannella, D., and Hakimi, M.A. (2010) Chromatin modifications: Implications in the regulation of gene expression in *Toxoplasma gondii*. *Cell. Microbiol.*, **12**, 413–423.

18 Darkin-Rattray, S.J., Gurnett, A.M., Myers, R.W., Dulski, P.M., Crumley, T.M. *et al.* (1996) Apicidin: a novel antiprotozoal agent that inhibits parasite histone deacetylase. *Proc. Natl Acad. Sci. USA*, **93**, 13143–13147.

19 Singh, S.B., Zink, D.L., Liesch, J.M., Mosley, R.T., Dombrowski, A.W. *et al.* (2002) Structure and chemistry of apicidins, a class of novel cyclic tetrapeptides without a terminal alpha-keto epoxide as inhibitors of histone deacetylase with potent antiprotozoal activities. *J. Org. Chem.*, **67**, 815–825.

20 Bougdour, A., Maubon, D., Baldacci, P., Ortet, P., Bastien, O. *et al.* (2009) Drug inhibition of HDAC3 and epigenetic control of differentiation in Apicomplexa parasites. *J. Exp. Med.*, **206**, 953–966.

21 Smith, A.T., Livingston, M.R., Mai, A., Filetici, P., Queener, S.F., and Sullivan, W.J., Jr (2007) Quinoline derivative MC1626, a putative GCN5 histone acetyltransferase (HAT) inhibitor, exhibits HAT-independent activity against *Toxoplasma gondii. Antimicrob. Agents Chemother.*, **51**, 1109–1111.

22 Brodersen, D.E., Clemons, W.M., Jr, Carter, A.P., Morgan-Warren, R.J., Wimberly, B.T., and Ramakrishnan, V. (2000) The structural basis for the action of the antibiotics tetracycline, pactamycin, and hygromycin B on the 30S ribosomal subunit. *Cell*, **103**, 1143–1154.

23 Greif, G., Harder, A., and Haberkorn, A. (2001) Chemotherapeutic approaches to protozoa: Coccidiae – current level of knowledge and outlook. *Parasitol. Res.*, **87**, 973–975.

24 Goodman, C.D. and McFadden, G.I. (2007) Fatty acid biosynthesis as a drug target in apicomplexan parasites. *Curr. Drug Targets*, **8**, 15–30.

25 Waller, R.F., Keeling, P.J., Donald, R.G., Striepen, B., Handman, E. *et al.* (1998) Nuclear-encoded proteins target to the plastid in *Toxoplasma gondii* and *Plasmodium falciparum. Proc. Natl Acad. Sci. USA*, **95**, 12352–12357.

26 Zhu, G., Marchewka, M.J., Woods, K.M., Upton, S.J., and Keithly, J.S. (2005) Molecular analysis of a Type I fatty acid synthase in *Cryptosporidium parvum. Mol. Biochem. Parasitol.*, **105**, 253–260.

27 Petersen, E. (2007) Prevention and treatment of congenital toxoplasmosis. *Expert Rev. Anti-Infect. Ther.*, **5**, 285–293.

28 Haberkorn, A. (1996) Chemotherapy of human and animal coccidioses: state and perspectives. *Parasitol. Res.*, **82**, 193–199.

29 Harder, A. and Haberkorn, A. (1989) Possible mode of action of toltrazuril: studies on two *Eimeria* species and mammalian and *Ascaris suum* enzymes. *Parasitol. Res.*, **76**, 8–12.

30 Gottstein, B., Eperon, S., Dai, W.J., Cannas, A., Hemphill, A., and Greif, G. (2001) Efficacy of toltrazuril and ponazuril against experimental *Neospora caninum* infection in mice. *Parasitol. Res.*, **87**, 43–48.

31 Gottstein, B., Razmi, G.R., Ammann, P., Sager, H., and Müller, N. (2005) Toltrazuril treatment to control diaplacental *Neospora caninum* transmission in experimentally infected pregnant mice. *Parasitology*, **130**, 41–48.

32 Strohbusch, M., Müller, N., Hemphill, A., Krebber, R., Greif, G., and Gottstein, B. (2009) Toltrazuril treatment of congenitally acquired *Neospora caninum* infection in newborn mice. *Parasitol. Res.*, **104**, 1335–1343.

33 Härdi, C., Hässig, M., Sager, H., Greif, G., Staubli, D., and Gottstein, B. (2006) Humoral immune reaction of newborn calves congenitally infected with *Neospora caninum* and experimentally treated with toltrazuril. *Parasitol. Res.*, **99**, 534–540.

34 Schnyder, M., Kohler, L., Hemphill, A., and Deplazes, P. (2009) Prophylactic and therapeutic efficacy of nitazoxanide against *Cryptosporidium parvum* in experimentally challenged neonatal calves. *Vet. Parasitol.*, **160**, 149–154.

35 Ollivett, T.L., Nydam, D.V., Bowman, D.D., Zambriski, J.A., Bellosa, M.L., *et al.* (2009) Effect of nitazoxanide on cryptosporidiosis in experimentally infected neonatal dairy calves. *J. Dairy Sci.*, **92**, 1643–1648.

36 Viel, H., Rocques, H., Martin, J., and Chartier, C. (2007) Efficacy of nitazoxanide against experimental cryptosporidiosis in goat neonates. *Parasitol. Res.*, **102**, 163–166.

37 Blagburn, B.L., Drain, K.L., Land, T.M., Kinard, R.G., Moore, P.H., *et al.* (1998) Comparative efficacy evaluation of dicationic carbazole compounds, nitazoxanide, and paromomycin against *Cryptosporidium parvum* infections in a neonatal mouse model. *Antimicrob. Agents Chemother.*, **42**, 2877–2882.

38 Theodos, C.M., Griffiths, J.K., D'Onfro, J., Fairfield, A., and Tzipori, S. (1998) Efficacy of nitazoxanide against *Cryptosporidium parvum* in cell culture and in animal models. *Antimicrob. Agents Chemother.*, **42**, 1959–1965.

39 Li, X., Brasseur, P., Agnamey, P., Lemeteil, D., Favennec, L., *et al.* (2003) Long-lasting anticryptosporidial activity of nitazoxanide in an immunosuppressed rat model. *Folia Parasitol.*, **50**, 19–22.

40 Abubakar, I., Aliyu, S.H., Arumugam, C., Usman, N.K., and Hunter, P.R. (2007) Treatment of cryptosporidiosis in immunocompromised individuals: systematic review and meta-analysis. *Br. J. Clin. Pharmacol.*, **63**, 387–393.

41 Amadi, B., Mwiya, M., Musuku, J., Watuka, A., Sianongo, S., Ayoub, A., and Kelly, P. (2002) Effect of nitazoxanide on morbidity and mortality in Zambian children with cryptosporidiosis: a randomised controlled trial. *Lancet*, **360**, 1375–1380.

42 Amadi, B., Mwiya, M., Sianongo, S., Payne, L., Watuka, A., Katubulushi, M., and Kelly, P. (2009) High dose prolonged treatment with nitazoxanide is not effective for cryptosporidiosis in HIV-positive Zambian children: a randomised controlled trial. *BMC Infect. Dis.*, **9**, 195.

43 Fox, L.M. and Saravolatz, L.D. (2005) Nitazoxanide: a new thiazolide antiparasitic agent. *Rev. Anti-Infect. Agents*, **40**, 1173–1180.

44 Hoffman, P.S., Sisson, G., Croxen, M.A., Welch, K., Harman, W.D. *et al.* (2007) Antiparasitic drug nitazoxanide inhibits the pyruvate oxidoreductases of *Helicobacter pylori* and selected anaerobic bacteria and parasites, and *Campylobacter jejuni. Antimicrob. Agents Chemother.*, **51**, 867–876.

45 Hemphill, A., Müller, N., and Müller, J. (2007) Structure-function relationship of thiazolides, a novel class of anti-parasitic drugs, investigated in intracellular and extracellular protozoan parasites and larval-stage cestodes. *Anti-Infect. Agents Med. Chem.*, **6**, 273–282.

46 Hemphill, A., Müller, J., and Esposito, M. (2006) Nitazoxanide, a broad-spectrum thiazolide anti-infective agent for the treatment of gastrointestinal infections. *Expert Opin. Pharmacother.*, **7**, 953–964.

47 Gargala, G., Le Goff, L., Ballet, J.J., Favennec, L., Stachulski, A.V., and Rossignol, J.F. (2010) Evaluation of new thiazolide/thiadiazolide derivatives reveals nitro group-independent efficacy against *Cryptosporidium parvum* in vitro development. *Antimicrob. Agents Chemother.*, **54**, 1315–1318.

48 Cortes, H.C.E., Müller, N., Esposito, M., Leitão, A., Naguleswaran, A., and Hemphill, A. (2007) In vitro efficacy of nitro- and bromo-thiazolyl-salicylamide compounds (thiazolides) against *Besnoitia besnoiti* infection in Vero cells. *Parasitology*, **134**, 975–985.

49 Müller, J., Sidler, D., Nachbur, U., Wastling, J., Brunner, T., and Hemphill, A. (2008) Thiazolides inhibit growth and induce glutathione-*S*-transferase pi (GSTP1)-dependent cell death in human colon cancer cells. *Int. J. Cancer*, **123**, 1797–1806.

50 Müller, J., Sterck, M., Hemphill, A., and Müller, N. (2007) Characterization of *Giardia lamblia* WB C6 clones resistant to nitazoxanide and to metronidazole. *J. Antimicrob. Chemother.*, **60**, 280–287.

51 Müller, J., Naguleswaran, A., Müller, N., and Hemphill, A. (2008) *Neospora caninum*: Functional inhibition of protein disulfide isomerase by the broad-spectrum anti-parasitic drug nitazoxanide and other thiazolides. *Exp. Parasitol.*, **118**, 80–88.

52 Anthony, J.-P., Fyfe, L., and Smith, H. (2005) Plant active components – a resource for antiparasitic agents? *Trends Parasitol.*, **21**, 462–468.

53 Jones, K.L., Donegan, S., and Lalloo, D.G. (2007) Artesunate versus quinine for treating severe malaria. *Cochrane Database Syst. Rev.*, **17**, article no. CD005967.

54 Asawamahasakda, W., Ittarat, I., Pu, Y.M., Ziffer, H., and Meshnick, S.R. (1994) Reaction of antimalarial endoperoxides with specific parasite proteins. *Antimicrob. Agents Chemother.*, **38**, 1854–1858.

55 Cui, L. and Su, X.Z. (2009) Discovery, mechanisms of action and combination therapy of artemisinin. *Expert Rev. Anti-Infect. Ther.*, **7**, 999–1013.

56 Kim, J.T., Park, J.Y., Seo, H.S., Oh, H.G., Noh, J.W. *et al.* (2002) *In vitro* antiprotozoal effects of artemisinin on

Neospora caninum. Vet. Parasitol., **103**, 53–63.

57 Akiyama, T., Ishida, J., Nakagawa, S., Ogawara, H., Watanabe, S. *et al.* (1987) Genistein, a specific inhibitor of tyrosine-specific protein kinases. *J. Biol. Chem.*, **262**, 5592–5595.

58 Gargala, G., Baishanbo, A., Favennec, L., François, A., Ballet, J.J., and Rossignol, J.F. (2005) Inhibitory activities of epidermal growth factor receptor tyrosine kinase-targeted dihydroxyisoflavone and trihydroxydeoxybenzoin derivatives on *Sarcocystis neurona, Neospora caninum,* and *Cryptosporidium parvum* development. *Antimicrob. Agents Chemother.*, **49**, 4628–4634.

59 Morito, K., Aomori, T., and Hirose, T. (2002) Interaction of phytoestrogens with estrogen receptors alpha and beta (II). *Biol. Pharm. Bull.*, **25**, 48–52.

60 Harris, J.C., Plummer, S., Turner, M.P., and Lloyd, D. (2000) The microaerophilic flagellate *Giardia intestinalis: Allium sativum* (garlic) is an effective antigiardial. *Microbiology*, **146**, 3119–3127.

61 Ankri, S. and Mirelman, D. (1999) Antimicrobial properties of allicin from garlic. *Microbes Infect.*, **1**, 125–129.

62 Ankri, S., Miron, T., Rabinkov, A., Wilchek, M., and Mirelman, D. (1997) Allicin from garlic strongly inhibits cysteine proteinases and cytopathic effects of *Entamoeba histolytica. Antimicrob. Agents Chemother.*, **41**, 2286–2288.

63 Leitsch, D., Kolarich, D., Binder, M., Stadlmann, J., Altmann, F., and Duchêne, M. (2009) *Trichomonas vaginalis*: metronidazole and other nitroimidazole drugs are reduced by the flavin enzyme thioredoxin reductase and disrupt the cellular redox system. Implications for nitroimidazole toxicity and resistance. *Mol. Microbiol.*, **72**, 518–536.

64 Sréter, T., Széll, Z., and Varga, I. (1999) Attempted chemoprophylaxis of cryptosporidiosis in chickens, using diclazuril, toltrazuril, or garlic extract. *J. Parasitol.*, **85**, 989–991.

65 Fleige, T. and Soldati-Favre, D. (2008) Targeting the transcriptional and translational machinery of the endosymbiotic organelle in apicomplexans. *Curr. Drug Targets*, **9**, 948–956.

66 Timmers, L.F., Pauli, I., Barcellos, G.B., Rocha, K.B., Caceres, R.A. *et al.* (2009) Genomic databases and the search of protein targets for protozoan parasites. *Curr. Drug Targets*, **10**, 240–245.

67 Unoki, M., Brunet, J., and Mousli, M. (2009) Drug discovery targeting epigenetic codes: the great potential of UHRF1, which links DNA methylation and histone modifications, as a drug target in cancers and toxoplasmosis. *Biochem. Pharmacol.*, **78**, 1279–1288.

Part Four
Compounds

Apicomplexan Parasites. Edited by Katja Becker
Copyright © 2011 WILEY-VCH Verlag GmbH & Co. KGaA, Weinheim
ISBN: 978-3-527-32731-7

20

Subversive Substrates of Glutathione Reductases from *Plasmodium falciparum*-Infected Red Blood Cells as Antimalarial Agents

Elisabeth Davioud-Charvet[*] *and Don Antoine Lanfranchi*

Abstract

Glutathione reductase (GR), which catalyzes NADPH-dependent glutathione disulfide reduction, is an important housekeeping enzyme for redox homeostasis both in human cells and in the causative agent of tropical malaria, *Plasmodium falciparum*. The aim of this interdisciplinary research was to substantiate glutathione reductase inhibitors as antimalarial agents. The strategy was based on the synthesis of subversive substrates or catalytic inhibitors of the selected targets, namely the GRs from human erythrocytes and from the malarial parasite *P. falciparum*. Both enzymes are essential proteins for survival of the parasite while infecting red blood cells, as they maintain the redox equilibrium in the cytosol by catalyzing the physiological reaction: $NADPH + H^+ + GSSG \rightarrow NADP^+ + 2\,GSH$, particularly in the course of the pro-oxidant process of hemoglobin digestion. The parasite evades the toxicity of the released heme by expressing two major detoxification pathways: hemozoin formation in the food vacuole, and an efficient thiol network in the cytosol. Different types of inhibitor (reversible, irreversible) were designed in the 1,4-naphthoquinone (NQ) series to evaluate the impact of each inhibition mode on the growth of the parasites: uncompetitive or catalytic inhibitors, fluorine-based suicide substrates, and hybrid dual drugs combining a GR inhibitor and a 4-aminoquinoline. The most potent antimalarial effects were observed for oxidant NQs acting as subversive substrates for both GRs. The findings suggest that antimalarial lead NQs act as redox-cyclers, being biometabolized by a heme-catalyzed oxidation reaction under the specific conditions found in the food vacuole of the parasites, and then by a GR-catalyzed reduction in the cytosol. In the oxidized form, the biometabolites acted as the most efficient subversive substrates of the malarial GR described to date while, in the reduced form, they were shown to reduce methemoglobin to oxyhemoglobin. Ultimately, the antimalarial NQs are suggested to affect the redox equilibrium, resulting in arrested trophozoite development by drowning the parasite in its own metabolic products. The results from this interdisciplinary approach should provide incisive insight into the understanding of redox-cyclers as small molecules inter-

[*]Corresponding author

Apicomplexan Parasites. Edited by Katja Becker
Copyright © 2011 WILEY-VCH Verlag GmbH & Co. KGaA, Weinheim
ISBN: 978-3-527-32731-7

fering with the half-site reactivity of malarial GR toward NADPH *in vivo*, and show how this property for disulfide reductase-catalyzed drug bioactivation might be exploited as a general concept to open new directions in medicinal chemistry.

Introduction

Malaria is a major world health tropical parasitic disease that, each year, affects 500 million people worldwide and kills two million people, including one million children in Africa. The most dangerous species of parasite involved is *Plasmodium falciparum*, which is responsible for cerebral malaria and transmitted to humans through the bite of the *Anopheles* mosquito. In its sporozoite stage, the parasite reaches the liver where it develops in the hepatocytes, which later release thousands of merozoites into the bloodstream. These merozoites invade the red blood cells and multiply, thus starting the so-called "intraerythrocytic cycle." It is at this stage of the disease that the symptoms appear. After intraerythrocytic development, the new merozoites are released via lysis of the red blood cell membranes every 48 h – the time that *P. falciparum* takes to complete the cycle and reinvade new erythrocytes. The red blood cells have no nucleus, they do not possess intracellular organelles or any endocytosis machinery, and they are incapable of *de novo* biosynthesis. Hence, in order to survive within erythrocytes, the *Plasmodium* parasites must emplace different membrane and subcellular compartments such that they can import different nutrients and export proteins/antigens at the surfaces of the host cells [1].

During their intracellular life cycle in the red blood cells, *Plasmodium* parasites are exposed to high fluxes of reactive oxygen species (ROS) due to both the host immune response to infection and to hemoglobin digestion. The parasite detoxifies free heme as a side product of hemoglobin digestion via biomineralization into the malarial pigment called *hemozoin* [2]. Glutathione participates in the degradation of non-polymerized heme [3, 4] through a cascade of redox reactions via the Fenton reaction (Equations 20.1–20.5; see also Figure 20.1):

$$2PFIX(Fe^{3+}) + 2GSH \rightarrow 2PFIX(Fe^{2+}) + GSSG \tag{20.1}$$

$$PFIX(Fe^{2+}) + O_2 \rightarrow PFIX(Fe^{3+}) + O_2^{\bullet-} \tag{20.2}$$

$$2O_2^{\bullet-} + 2H^+ \rightarrow O_2 + H_2O_2 \tag{20.3}$$

$$PFIX(Fe^{2+}) + H_2O_2 \rightarrow PFIX(Fe^{3+}) + HO^- + HO^{\bullet} \tag{20.4}$$

$$OH^{\bullet} + PFIX(Fe^{3+}) \rightarrow Fe^{3+} \text{ release and destruction of PFIX} \tag{20.5}$$

Methylene blue (MB), chloroquine (CQ), and related 4-aminoquinolines such as mefloquine or amodiaquine, quinine, and xanthones such as C5 (see structures in Scheme 20.1) are known inhibitors of hemozoin formation [5–10].

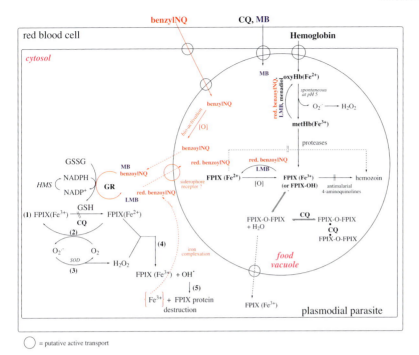

Figure 20.1 Hemoglobin catabolism, redox equilibrium, and heme detoxification in a *Plasmodium*-infected erythrocyte, and postulated mechanism of action of antimalarial redox-active methylene blue and 3-benzyl menadione (benzylNQ) derivatives.

However, nonpolymerized heme (Fe^{3+}) can also generate ROS, and is responsible for numerous heme-catalyzed oxidation reactions. For this reason, *Plasmodium* parasites are exposed to higher fluxes of ROS during their intraerythrocytic life, and require high activities of intracellular antioxidant systems that will provide a steady glutathione flux. The most important antioxidative systems are based on the glutathione reductases (GRs; E.C. 1.8.1.7) of the malarial parasite *Plasmodium falciparum* and the host erythrocyte, which catalyze the reduction of the disulfide bridge of oxidized glutathione (GSSG) into the thiol form, glutathione (GSH), according to the following equation:

$$NADPH + H^+ + GSSG \rightarrow NADP^+ + 2GSH \tag{20.6}$$

The GR enzymes from both the human erythrocyte and the malarial parasite have been identified as targets of antimalarial drugs [11, 12]. In particular, the parasites do not develop well in glucose-6-phosphate dehydrogenase (G6PDH)-deficient red blood cells [13], nor in erythrocytes depleted of GR activity [14, 15]. Notably, the prevalence of G6PDH deficiency coincides with that of malaria, with a very similar global distribution of 400 million people affected, supporting the so-called "malaria protection hypothesis" [16]. Evidence of G6PDH deficiency conferring resistance against malaria can be explained by both biochemical and clinical observations.

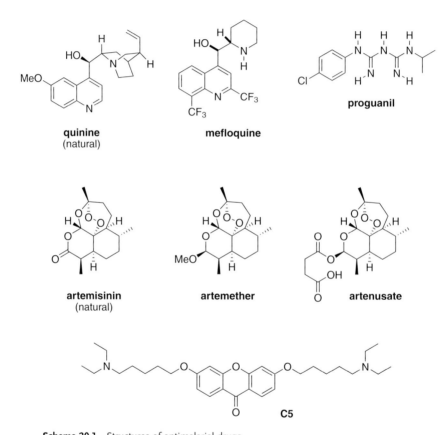

4-aminoquinolines

Scheme 20.1 Structures of antimalarial drugs.

At the biochemical level, G6PDH is the first enzyme of the pentose phosphate pathway (hexose monophosphate shunt, HMS in Figure 20.1) that is responsible for providing the reducing power of the cells in NADPH form. As NADPH is the main reductant provider for GR in erythrocytes, G6PDH deficiency in red blood cells can be regarded as a natural biomimetic knockout version of human GR. The deficiency

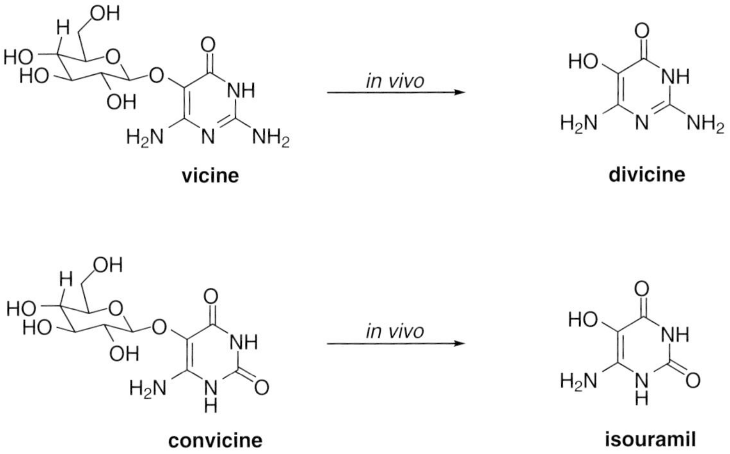
in vivo
vicine
divicine
in vivo
convicine
isouramil

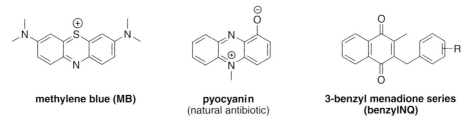

Scheme 20.2 Structures of pyrimidine glycosides and their generated redox-active biometabolites.

is not lethal for humans, but prevents a severe attack of malaria, as the oxidative stress released in the red blood cells makes the milieu hostile for *Plasmodium*. Consequently, a rapid elimination of these cells from the circulation occurs by enhanced phagocytosis via complement activation [17]. Despite the apparent absence of symptoms in G6PDH-deficient populations, a portion of G6PDH-deficient individuals are very sensitive to oxidative stressors (e.g., drugs and fava beans), resulting in most common manifestations, such as neonatal jaundice and acute hemolytic anemia episodes. Fava beans were shown to contain pyrimidine glycosides, vicine and convicine, which are known to be hydrolyzed *in vivo* into divicine and isouramil (Scheme 20.2). Both pyrimidine aglycones act as oxidative stressors, causing the depletion of reduced GSH in the host erythrocytes [18, 19]. Mimicking the G6PDH deficiency by developing GR inhibitors that function as redox-cyclers can pose a challenge when designing new antimalarial drugs. The essential condition is that these drugs not trigger hemolysis in any individuals, whether G6PDH-deficient or otherwise. This requirement could be reached via the prodrug approach, or by an enzyme–drug bioactivation process. The most relevant model of this strategy was exemplified in the past by the first-discovered synthetic antimalarial drug, methylene blue (MB) [20, 21], and more recently by its natural

methylene blue (MB)
pyocyanin
(natural antibiotic)
3-benzyl menadione series
(benzylNQ)

Scheme 20.3 Structures of antimalarial redox-cyclers acting as glutathione reductase-subversive substrates.

aza-analog, pyocyanin [22; see also Chapter 7], or by 1,4-naphthoquinones (NQs) [23]. Each of these compounds (Scheme 20.3) is capable of killing malarial parasites and acting as a redox-cycler and GR-subversive substrate.

Antimalarial Drugs

Most of the antimalarial drugs target the intraerythrocytic stages of the plasmodia (see Scheme 20.1). The first synthetic antimalarial drug, MB (Scheme 20.3), was discovered by Paul Ehrlich, following the observation that malarial parasites were specifically colored with MB, coupled with Ehrlich's intuition that this dye could destroy parasites without harming the human organism. Subsequently, Ehrlich applied MB in practice as the first synthetic antimalarial drug to treat malaria patients in 1891, and it was used in human medicine in Africa. However, despite extensive efforts to optimize the phenothiazine dye, a lack of efficacy in humans justified the withdrawal of MB from therapy once the 4-aminoquinoline CQ had been introduced to the market. The 4-aminoquinoline core was inspired by the structure of the first-discovered antimalarial natural product, a quinoline derivative known as *quinine* (Scheme 20.1), an alkaloid extracted from the bark of the tropical Cinchona tree. Since then, other antimalarial drugs have been designed by targeting various

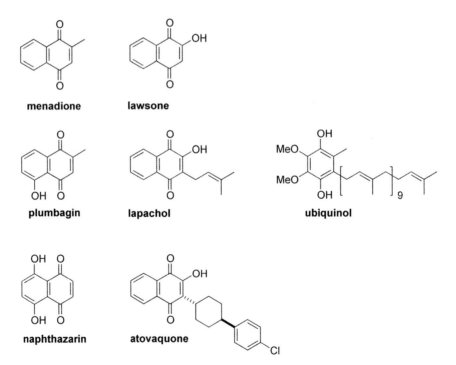

Scheme 20.4 Structures of quinones and related derivatives.

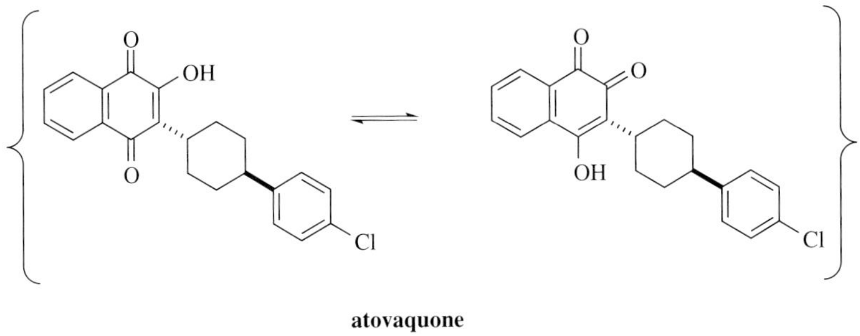

atovaquone

Scheme 20.5 Keto–enol tautomerization of atovaquone.

pathways from antimetabolites to compounds attacking validated protein targets. As an example, atovaquone (Scheme 20.4), a substituted 2-hydroxy-naphthoquinone, was developed as an inhibitor of the mitochondrial cytochrome bc_1 complex (E.C. 1.10.2.2) via interactions with the Rieske iron–sulfur protein and cytochrome b in the ubiquinol oxidation pocket [24]. Similarities between atovaquone and ubiquinol (Scheme 20.4) have been clearly evidenced, suggesting that the bc_1 complex of the mitochondrial respiratory chain could be the target [25]. While the bc_1 complex catalyzes electron transfer from ubiquinol to cytochrome c, and concomitantly translocates protons across membranes, atovaquone binds to the ubiquinol oxidation site between cytochrome b and the iron–sulfur protein. Due to the keto–enol tautomerization equilibrium (Scheme 20.5), atovaquone presumably mimics a transition state in the ubiquinol oxidation reaction that is catalyzed by the enzyme during this electron transfer. In the presence of atovaquone, the midpoint potential of the Rieske iron–sulfur cluster was increased from 285 to 385 mV, indicating that atovaquone binds approximately 50-fold more tightly when the Rieske center is reduced [24, 26]. Unfortunately, the malarial parasites rapidly developed a resistance to atovaquone, such that its monotherapy resulted in treatment failure in 30% of the patients [27]. The situation was rescued, however, by combining another antimalarial drug, proguanil (Scheme 20.1), with atovaquone to produce a very effective combination known as Malarone®, that could be used both as a prophylaxis and for the direct treatment of *P. falciparum* malaria. Although, CQ and other related 4-amino-quinolines have been the leaders in the antimalarial market for more than 50 years, new active, available, and cheap drugs are urgently required due to an increasing resistance of the parasites against standard drugs. Artemisinin (Scheme 20.1), a natural sesquiterpene endoperoxide isolated from the plant *Artemisia annua*, which has been used in Chinese medicine for more than 2000 years, has emerged as an important treatment alternative against drug-resistant strains of *P. falciparum*, even in severe cases of cerebral malaria. Semi-synthetic derivatives such as arte-mether and artesunate have also been developed to improve the relatively poor bioavailability of artemisinin (Scheme 20.1). Currently, as recommended by the World Health Organization (WHO), millions of people are being (or will be) treated for malaria with an artemisinin-based combination therapy (ACT). Unfortunately, as artemether-resistant parasites have recently been identified in French Guiana and

Senegal [28, 29], the treatment of malaria might reach crisis state if these strains were to develop rapidly.

More than 100 years after the discovery of MB, Schirmer, Becker and coworkers became pioneers in the analysis of the antioxidative flavoproteins in *P. falciparum* and in their identification as targets of known drugs, including MB [11, 12, 30, 31]. Thereby, MB has been shown to be an efficient substrate of GRs from both human and *P. falciparum* [20]. Likewise, MB was found to be active not only against the disease-causing asexual schizonts, but also against the disease-transmitting gametocytes of *P. falciparum* [32]. As MB lacks efficacy *in vivo* as a monotherapy, a drug combination with amodiaquine and artesunate was found to be very effective [33].

Oxidant 1,4-Naphthoquinones as Redox-Cyclers

The molecular basis of 1,4-naphthoquinone toxicity is the enzyme-catalyzed reduction to semiquinone radicals, which then reduce oxygen to superoxide anion radicals, thereby regenerating the quinone (Equation 20.7). This futile redox cycling and oxygen activation lead to increased levels of hydrogen peroxide (Equation 20.8) and GSSG:

$$NADPH + H^+ + 2NQ(or\ NQ) \rightarrow NADP^+ + 2H^+ + 2NQ^{\bullet-}(or\ NQ(H)_2)$$
$$(20.7)$$

$$NQ^{\bullet-} + O_2 \rightarrow NQ + O_2^{\bullet-}\ or\ NQ(H)_2 + O_2 \rightarrow NQ + H_2O_2 \qquad (20.8)$$

A well-studied example of this is menadione (2-methyl-1,4-naphthoquinone; see Scheme 20.4), which is an unspecific acceptor of electrons from numerous flavoproteins acting through a one-electron mechanism (e.g., NADPH cytochrome *c* reductase, NADH dehydrogenase, and ferredoxin NADP$^+$ reductase), or through a mixed mechanism (one- and two-electron) such as both NADPH-dependent GRs and trypanothione reductase [34, 35], NAD(P)H dehydrogenase (or DT diaphorase), or lipoamide dehydrogenase (LipDH) [36, 37]. Many naturally occurring naphthoquinones (NQs; see Scheme 20.4) such as menadione (Table 20.1), plumbagin, naphthazarin [38], lawsone, and lapachol, have displayed notable antimalarial activities on different *Plasmodium* strains responsible for the disease. Malarial parasites infecting red blood cells are particularly sensitive to oxidative stress, and have developed specific antioxidant systems [[11, 12, 39]; see also Chapter 16]

When studied as subversive substrates of GR by following the oxidation of NADPH in the presence of 25–100 μM NQ, the NQ reductase activity of GR (Equation 20.7) can be compared to the intrinsic NADPH oxidase activity of the enzyme in the absence of NQ (Equation 20.9):

$$2O_2 + NAD(P)H + H^+ \rightarrow 2O_2^{\bullet-} + NAD(P)^+ + 2H^+ \qquad (20.9)$$

These reactions (Equations 20.7–20.9) contribute to the production of a continuous flux of toxic reduced NQ species or of ROS [40].

Table 20.1 *In vitro* antiparasitic activity and cytotoxicity of menadione and its aza-analogs [41].

Quinone	IC$_{50}$ (μM) *P. falciparum*[a]		TC$_{50}$ (μM) cytotoxicity[a]
	3D7	K1	KB
menadione	9.6	12.0	34.3
quinoline-5,8-dione	19.45	2.79	5.36
6-methyl-quinoline-5,8-dione	13.97	12.09	73.16
6,7-dimethyl-quinoline-5,8-dione	2.18	2.27	5.8
quinoxaline-5,8-dione	21.6	1.70	9.34
2,3-dimethyl-quinoxaline-5,8-dione	>30	28.28	5.9

a) Chloroquine showed IC$_{50}$ values of 5.0 nM and of 550.0 nM against the CQ-sensitive *P. falciparum* 3D7 strain and the CQ-resistant *P. falciparum* K1 strain, respectively. Podophyllotoxin exhibited a TC$_{50}$ value of 4.8–72.4 nM against the human KB cell line (from Ref. [44]).

Six aza-analogs of 1,4-naphthoquinone and menadione were recently synthesized and evaluated as efficient subversive substrates of the NADPH-dependent disulfide reductases, namely the GR and the thioredoxin reductase [41]. The replacement of one to two carbons at the phenyl ring by one to two nitrogen atoms led to an increased oxidant character of the molecules, in accordance with redox potential values and enzyme kinetic studies. Based on the antimalarial effects against two strains of *P. falciparum* (Table 20.1), and the toxicity against one human cell line (V. Yardley, personal communication), it has been suggested that modulation of the antimalarial activities might be achieved by varying the number of nitrogen atoms and the substitution pattern of the aza-naphthoquinone core.

However, the substituted 2-hydroxy-naphthoquinone, atovaquone, is notably not a substrate of GR [41]. Also, the redox potential values measured by cyclic voltammetry are very distinct; that is, $-508 \pm 2\,\mathrm{mV}$ for atovaquone versus $-141 \pm 12\,\mathrm{mV}$ for menadione [42], evidencing the distinct mechanisms of action of redox-active oxidant NQ compared to atovaquone.

The Impact of GR Inhibition on the Growth of Malarial Parasites

The development of menadione chemistry [34, 35] has led to the selection of the carboxylic acid 6-[2′-(3′-methyl)-1′,4′-naphthoquinolyl] hexanoic acid $\mathbf{M_5}$ and its phenyl analog **1** (Scheme 20.6) as inhibitors of the parasitic GR [40, 44]. Studies of the $\mathbf{M_5}$ mechanism revealed an uncompetitive type of inhibition with respect to both NADPH and glutathione disulfide. Masking the polarity of the acidic function of $\mathbf{M_5}$ with ester [43] or amide [45] bonds improved the antiplasmodial activity. Bioisosteric replacement of the carboxylic function with tetrazole **2** (Scheme 20.6) to increase bioavailability and maintain comparable acidity led to improved antimalarial properties as well, but only with the cyanoethyl-protected tetrazole **3** [38].

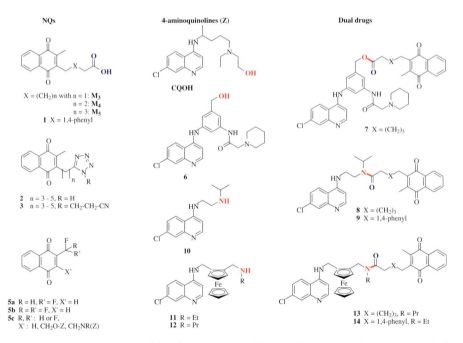

Scheme 20.6 Structures of glutathione reductase inhibitors, of 4-aminoquinolines used in the dual drug approach, and of dual drugs with high antimalarial activity.

Since the oxidant 1,4-naphthoquinones are substrates of GR, the methyl group of the parent menadione and of the carboxylic acid **M$_5$** was the target for fluorine introduction in order to transform the compounds into suicide-substrates **fluoroM$_5$** and its phenyl analog **4** (Scheme 20.7). A new route to prepare the 2-fluoromethyl menadione derivatives has been established, and proved to be convenient for synthesizing a large array of diverse 3-substituted-2-fluoromethyl menadione derivatives, as illustrated by the synthesis of **fluoroM$_5$** and its phenyl analog **4** [44]. Synthesis of the related difluoro- and trifluoro-menadione compounds **5a** and **5b** (Scheme 20.6) followed different routes (E. Davioud-Charvet, unpublished data). For the 2-difluoromethyl-1,4-naphthoquinone **5a**, the 1,4-dimethoxy-naphthalene-2-carbaldehyde was directly fluorinated by using the commercially available diethylaminosulfur trifluoride. For the introduction of the trifluoromethyl group, the method of Hünig was used [45]. The fluorine introduction at the methyl groups of menadione and **M$_5$** led to the most potent suicide-substrates of GR described to date, in accordance with the expected reactivity of the quinone methylene toward a nucleophile (Scheme 20.7). The kinetic parameters for the time-dependent inactivation of GR by **fluoroM$_5$** were determined, and both GR enzymes could be completely inactivated in GSSG reduction. When crystals of the alkylated enzyme were grown over a five-day period, the 1.7 Å resolution crystal structure of the alkylated GR showed that the inhibitor binds covalently to the active site Cys58, and interacts noncovalently with His467', Arg347, Arg37, and Tyr14. Interestingly, both fluoroM$_5$-inactivated GRs were active in the naphthoquinone reductase assay, revealing that the most exposed Cys of the active site was not responsible for naphthoquinone reduction. Subsequent stopped-flow kinetics studies were conducted to confirm that the naphthoquinone reduction occurred at the flavin of the reduced enzyme (a four-electron-reduced GR species EH$_4$ was prepared), but not at the active Cys pair of the two-electron-reduced GR species, EH$_2$. Based on the crystal structure of the inactivated human enzyme and on stopped-flow kinetics studies with two- and four-electron-reduced forms of unreacted *P. falciparum* enzymes, a mechanism has

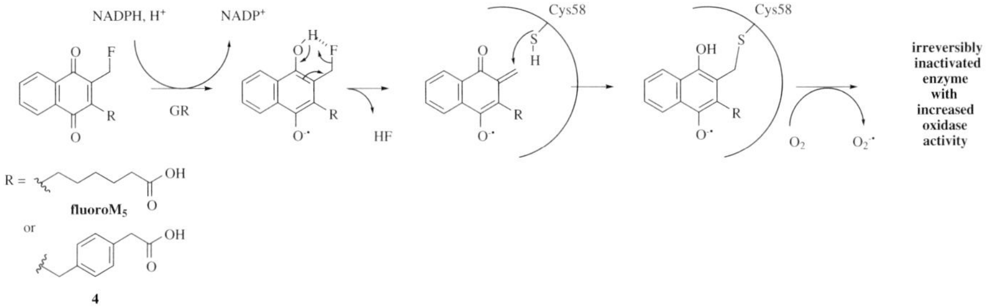

Scheme 20.7 Proposed mechanism of the irreversible inactivation of glutathione reductase by **FluoroM$_5$**. The naphthoquinone is activated via GR-catalyzed one-electron reduction (two-electron transfer is also possible). Elimination of HF promoted by a newly formed intramolecular hydrogen bond results in the generation of a quinone methide. Nucleophilic attack of the catalytic Cys58 leads to covalent modification of the enzyme. The radical can further react with oxygen, leading to the formation of superoxide.

been proposed which explains naphthoquinone reduction at the flavin of GR [44]. Thus, the 1,4-naphthoquinones are uncompetitive inhibitors, and act by diverting the electron pathway at the flavin of these disulfide reductases. The site where the naphthoquinone is reduced remains undefined. Besides monofluoromenadione derivatives such as **fluoroM₅**, the di- and tri-fluoromenadione **5a** and **5b** were synthesized and the kinetics of these oxidant NQs studied in GR assays when used as inhibitors or substrates (E. Davioud-Charvet, unpublished results). Both compounds are potent inhibitors of GSSG reduction, efficient subversive substrates, and time-dependent inhibitors of disulfide reductases [apparent $t_{1/2}$ (min) at 10 μM: 0.75 (CHF₂) and 3.5 (CF₃) versus no inactivation (CH₃)]. The high k_{cat}/K_m values of the compounds for *P. falciparum* GR (k_{cat}/K_m (mM⁻¹s⁻¹): 155.9 (CHF₂) and 46.7 (CF₃) versus 1.9 (CH₃)) correlated with their redox potential values (E_1^0 (V): −0.385 (CHF₂) and −0.308 (CF₃) versus −0.65 (CH₃)) [see Ref. [44], and H. Vezin, personal communication]. In these assays, both derivatives showed sigmoidal curves when tested as subversive substrates of PfGR. While PfGR inactivation was time-dependent, an electron spray ionization-mass spectrometry (ESI-MS) analysis of the reaction mixtures showed no mass peaks accounting for covalent inhibitor–enzyme complexes, but it instead showed mass peaks of monomers and dimers of the free enzyme, in addition to a high background that attested to enzyme crosslinking alkylation. Taken together, the kinetic studies with highly oxidant naphthoquinones as substrates (as conducted by the present authors) and the observed inactivation of PfGR at high NADPH concentration [21], indeed suggested a half-site reactivity mechanism for PfGR toward NADPH binding *in vivo*. The half-site reactivity implied that the NADPH binding at one site of the enzyme homodimer would inhibit the binding of a second NADPH molecule at the other site of the enzyme homodimer. This hypothesis was in accordance with the uncompetitive type of inhibition observed for NQ toward NADPH or GSSG in GR assays; that is, when one subunit of PfGR binds NADPH, the second subunit is thought to reduce GSSG in the catalytic cycle *in vivo*. The binding and reduction of NQ (or O₂) at the reduced flavin of the homodimer might shift this negative cooperativity behavior toward NADPH binding, leading to the formation of inactive enzyme dead-end GR dimers.

Finally, when studying the *in vitro* antiplasmodial activity of reversible redox-active NQs such as the carboxylic acid **1**, compared to the monofluoromenadione analog **4** acting as a time-dependent GR inhibitor, the data clearly showed that the most potent antimalarial activity was reached with redox-active oxidant low-weight 1,4-naphthoquinones rather than with the corresponding monofluoromethyl analogs acting as suicide-substrates, leading to irreversible inhibition of both GRs. In the latter case, unspecific cytotoxicity due to the broader distribution of the lipophilic fluorine-based molecules was observed in human cell assays [44].

The Dual Prodrug Approach

Glutathione, which is known to guard *P. falciparum* against oxidative damage, may have an additional protective role by promoting heme catabolism. An elevation of the GSH content in parasites leads to an increased resistance to CQ, whereas GSH depletion in resistant *P. falciparum* strains is expected to restore their sensitivity to CQ [46]. High intracellular GSH levels depend *inter alia* on the efficient reduction of GSSG by GR. On the basis of this hypothesis, a new strategy has been developed for overcoming GSH-dependent 4-aminoquinoline resistance. In an attempt to direct both a GR inhibitor and a 4-aminoquinoline to the parasite, dual prodrugs acting as "Trojan horses" were designed and synthesized (see Scheme 20.6). For this, quinoline-based alcohols (CQOH and **6**) with known antimalarial activity were combined with a GR inhibitor (M_3 to M_5) via a metabolically labile ester bond to create double-headed prodrugs, such as compound **7** (Scheme 20.6) [43, 47]. The biochemically most active dual prodrug **7** of this series was then evaluated as a growth inhibitor against six *P. falciparum* strains, which differed in their degree of resistance to CQ; typically, the IC_{50} values for CQ ranged from 14 to 183 nM. While the inhibitory activity of the original 4-aminoquinoline-based alcohol **6** followed that of CQ in these tests, the dual prodrug exhibited similar efficiency against all strains, the ED_{50} being as low as 28 nM. For the ester **7**, a dose-dependent decrease in GSH content and GR activity, and an increase in glutathione-*S*-transferase activity, were determined in treated parasites. The drug was subsequently tested for its antimalarial action *in vivo* by using murine malaria models infected with *Plasmodium berghei*. A 178% excess mean survival time was determined for the animals treated with 40 mg kg^{-1} compound **7** for four days, and no compound-related cytotoxicity was observed.

Following this first proof-of-concept to target the *P. falciparum* GR and the heme metabolism in one hybrid dual prodrug, the aim has therefore been to: (i) increase the oxidant character of the naphthoquinone; (ii) use a more stable bond between the two active moieties [48]; and (iii) improve the pharmacokinetic profile of the final molecule (i.e., decrease the molecular weight, introduce metabolically resistant groups such as –CF$_3$). These studies have been further extended to validate the strategy outlined here. As a future generation of dual drugs, a series of 18 naphthoquinones (or their related synthetic 1,4-dimethoxynaphthalene precursors from menadione and trifluoromenadione derivatives; Scheme 20.6) containing three different linkers – an amide, a 2-alkoxymethyl, or a 2-aminomethyl, as in **5c** – between the 4-aminoquinoline core and the redox active component was synthesized [45]. The antimalarial effects have been characterized in parasite assays using CQ-sensitive and -resistant strains of *Plasmodium*, either alone or in drug combination, and also in a *Plasmodium berghei* rodent model. In particular, two tertiary amides **8** and **9** showed potent antimalarial activity in the low nanomolar range against CQ-resistant parasites. The ability of these hybrid molecules to compete both for the (Fe^{3+}) protoporphyrin and for the CQ transporter was determined. The data were consistent with the presence of a carrier for uptake of the parent short CQ analog **10**, but not for the potent antimalarial amide **8**, which suggested a mode of action that was distinct from that of CQ. Some dual drugs also were designed by functionalizing various ferrocenyl analogs of ferroquine (FQ) (e.g., the ferrocenic 4-aminoquinolines **11** and **12**) with the GR inhibitor M_5 or its phenyl analog **1**, or glutathione depletors through a cleavable amide bond (Scheme 20.6) [49]. The two most potent antimalarial dual drugs **13** and **14** (Scheme 20.6), incorporating the GR inhibitors M_5 or its phenyl analog **1**, also showed a high inhibition of hematin polymerization in the same range as FQ, and even higher by taking into account the very low drug:hematin ratio ($IC_{50} = 0.2$ at 0.25:1 for **13**, and $IC_{50} = 0.5$ at 0.75:1 for **14** versus $IC_{50} = 0.4$ at 0.75:1 for FQ), necessary to reach the maximal inhibition. While various iron porphyrins as models of cytochrome P_{450} enzymes are used to mimic the oxidative conditions found in the liver cells, the hematin was used in the present studies to understand the drug catabolism of these amides under the conditions found in the parasitic food vacuole. At vacuolar pH (pH 5.2), the reaction of 5 mM dual drug **13** in the presence of 1 mM hematin, and 15 mM hydrogen peroxide resulted in a total disappearance of the starting dual molecule **13**, accompanied by the release of the GR inhibitor M_5. The half-life of the compound was estimated at between 0.5 and 1 h [49]. In the case of the nonferrocenyl dual drug **8**, only a small amount of the carboxylic acid M_5 was observed (E. Davioud-Charvet, unpublished results), suggesting a lower oxidative amide *N*-dealkylation rate upon reaction with hematin and hydrogen peroxide.

Low-Molecular-Weight Oxidant 1,4-Naphthoquinone as Antimalarial Agents

As seen earlier, the most potent antimalarial effects were observed for subversive substrates. Based both on published data and enzyme kinetic data using naphtho-quinones as substrates of *P. falciparum* GR, a putative half-site reactivity mechanism toward NADPH binding is suggested. The half-site reactivity toward NADPH binding, an extreme case of negative cooperativity, is proposed to be involved in *P. falciparum* GR catalysis, and to render the enzyme insensitive to environmental changes *in vivo*. Taking this assumption into account, the present strategy was guided by improving the catalytic efficiencies of subversive substrates in order to steadily generate GR-catalyzed reaction products so as to poison the entire redox metabolism in the parasite. The preparation of more than 150 new menadione derivatives, and the screening for antimalarial activity, allowed the selection of a series of potent antimalarial compounds that were active both *in vitro* against various multiresistant strains of *P. falciparum* (in the low nanomolar range) and *in vivo* in *P. berghei*-infected mice (intraperitoneally and *per os*) [23]. The compounds were devoid of toxicity and did not trigger erythrocyte hemolysis. The selected lead 3-benzyl menadione (benzylNQ; Scheme 20.3) are proposed to be biometabolized – via a cascade of redox reactions – by a heme-catalyzed oxidation reaction under the specific conditions found in the food vacuole of the parasites, leading to the benzoylNQ series, and then by a GR-catalyzed reduction in the cytosol (Scheme 20.8). In the PfGR assay in the absence of GSSG, the catalytic efficiencies expressed as the k_{cat}/K_m values of the benzoylNQ were determined to be as high as 12.5 $mM^{-1}s^{-1}$ (versus 13.7 $mM^{-1}s^{-1}$ for MB). The reduced biometabolites were shown to reduce the methemoglobin (Fe^{3+}) (MetHb) to the oxyhemoglobin (Fe^{2+}) (OxyHb), and to act as the most efficient subversive substrates of GR described so far (Figure 20.1; see also Figure 20.2). In order to evidence this reduction activity of MetHb into OxyHb and using MB as a positive control, a coupled assay was set up with the GR/NADPH-based system to continuously reduce the human MetHb (Figure 20.2). In this assay, MB was

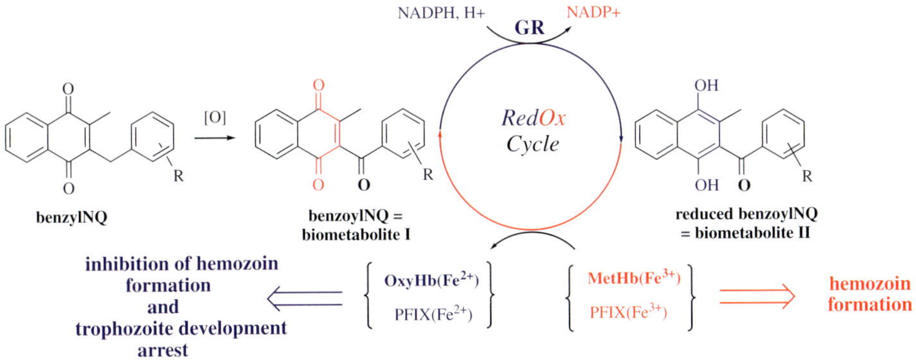

Scheme 20.8 Proposed cascade of redox reactions from 3-benzyl menadione derivatives (benzylNQ) leading to inhibition of *P. falciparum* trophozoite development and of hemozoin formation.

used as positive control (with a k_{cat}/K_m value of 13.7 mM^{-1} s^{-1} in the PfGR assay): the redox-cycler was first reduced by the NADPH-reduced GR, and the generated leucomethylene blue (LMB) was observed to reduce MetHb completely in 10 min (Figure 20.2). Under the same conditions, a benzoylNQ representative, which previously was shown to be an efficient GR subversive substrate with a k_{cat}/K_m value of 5.26 mM^{-1} s^{-1} in the PfGR assay, revealed the same complete effect, as observed with MB, after 15 min. As negative controls, CQ and FQ did not induce MetHb redox-cycling (Figure 20.2). Ferroquine, a 4-aminoquinoline with a ferrocenyl group in the amino side chain, which is not a substrate of both GRs, in accordance with the redox properties of the ferrocene(Fe^{2+})/ferricinium(Fe^{3+}) system of FQ ($E^{1/2} = +0.4615$ V versus ECS) [50], is detailed fully in Chapter 21 of this book.

Thus, the antimalarial 1,4-naphthoquinones might act as redox-active mediators as they cycle in and out the food vacuole in the parasites, thereby oxidizing major intracellular reductants (e.g., GR) and subsequently reducing the oxidants (heme, MetHb). Ultimately, the antimalarial 1,4-naphthoquinones are suggested to affect the redox equilibrium in the parasites, "drowning" the parasite in its own metabolic products.

Are *Plasmodium falciparum*-Infected Erythrocytes Electrochemically Active in the Presence of Redox-Cyclers?

As previously mentioned, the parasite has built different membrane and subcellular compartments in order to survive in the host milieu. While the inert noninfected red blood cells have no machinery/organelles to fulfill nutrient uptake, metabolite removal, volume regulation, and/or modification of cytosolic ion concentrations, *P. falciparum*-infected erythrocytes exhibit an increased permeability to a wide range of structurally unrelated solutes in order to fuel the vigorous parasite metabolism, and to remove waste products [51–53]. This is made possible because the infected red blood cell is complemented by a battery of transporters on the parasite plasma

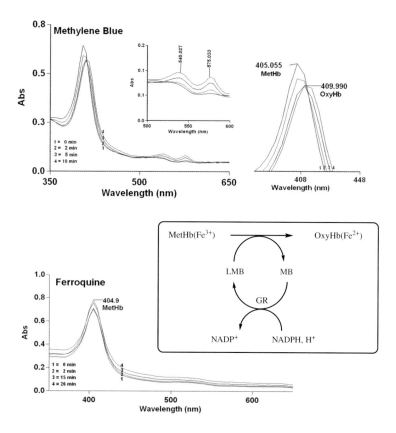

Figure 20.2 Coupled assay evidencing the reduction of MetHb(Fe^{3+}) into OxyHb (Fe^{2+}) in the presence of the subversive substrate methylene blue, and the GR/NADPH system. This figure shows the reduction effect of 100 μM MetHb in the presence of 200 μM MB, 400 μM NADPH and 1 μmol human GR. In this experiment, the UV spectrum of MetHb between 300 and 700 nm is characterized by a maximal absorbance at 405 nm and a broad band centered at around 630 nm. Following reduction, as a function of the time the spectrum of MetHb is shifted to the spectrum of OxyHb, displaying a maximal absorbance at 410 nm and two weak bands at 536 and 576 nm. MB with a k_{cat}/K_m value of 13.7 mM^{-1} s^{-1} was used as a positive control of this reduce activity; the shift of the maximal absorbance from 405 to 410 nm is already visible from 50 μM MB, or from 100 μM menadione or from 100 μM of a benzoylINQ representative with a k_{cat}/K_m of 5.26 mM^{-1} s^{-1} (data not shown). As expected, no shift was observed in the presence of CQ (data not shown), or FQ.

membrane which, in concert with the host cell membrane transporters, control the intracellular composition and the uptake and release of nutrient and waste compounds. This allows the parasite to control its ionic environment and use it for volume regulation. Unusual parasite-induced ion channels have recently been identified as enhanced activity versions of anion channels already present in uninfected red blood cells [54, 55]. As a source of amino acids to ensure its growth, the parasite imports hemoglobin from the host and digests it in its food vacuole. The hemoglobin digestion leads to a release of equivalent amounts of free heme (known as an "oxidation catalyst"), which is responsible for making the food vacuole the locus of intense oxidation reactions. In the cytosol, the disulfide reductases maintain the reducing milieu, with a rate dependent on NADPH regeneration from $NADP^+$. This particular situation might be naively compared to that of an electrochemical cell with two compartments – one for oxidation reactions (food vacuole or anode), and one for reduction reactions (cytosol or cathode) – with charged species being transported from one compartment to the other in the electrolytic milieu. In an electrolyzer, when the electrolytic milieu contains Fe^{2+} and when potentials values at the electrodes are in the appropriate scale, the Fe^{2+} species can be oxidized at the anode, generating Fe^{3+} ions which can be reduced back to Fe^{2+} at the cathode, rendering the cell submitted to redox-cycling. In the food vacuole of the malarial parasites, a similar situation could occur. Starting with *P. falciparum*-infected red blood cells (as organism fuel cells) and glucose as an energy source, MB, pyocyanin, and/or benzoylNQ might be used as cell-permeable mediators as the source of electrons to power the cells (and enable an electrical current to be generated). In the oxidized form, these "cathode" oxidants will accept the electrons produced by NADPH-dependent GR-catalyzed reduction in the cytosol, along with a concomitant NADPH oxidation to $NADP^+$. The now reduced form of the mediator (LMB, reduced pyocyanin, reduced benzoylNQ) might also be membrane-permeable and diffuse away from the cytosol – possibly via iron complexation – to the food vacuole, where the reduced species is then reoxidized during the course of the reduction of methemoglobin(Fe^{3+}) into oxyhemoglobin (Fe^{2+}). The oxidized mediator is then ready to repeat the cycle. This process continually drains off metabolic reducing power from *P. falciparum*-infected erythrocytes in order to provide "electrical power at the electrodes" (see Figure 20.1).

Malarial parasites use MetHb as a nutrient and digest it faster than OxyHb [56]; the reduction of MetHb can then used to slow down the parasites' MetHb digestion rate. Since FPIX(Fe^{2+}) is an inhibitor of hematin polymerization [57], compounds with GR redox-cycling activity displaying the ability to reduce FPIX(Fe^{3+}) into FPIX (Fe^{2+}) might synergistically contribute to increased oxidative stress in infected red blood cells and decrease hemozoin formation, with both processes participating in trophozoite development arrest. The high sensitivity of the parasite to oxidative stress, and the inhibitory capability of unpolymerized FPIX to inhibit different parasite enzymes and functional proteins, is also noteworthy. In addition, an expected positive side effect of the redox-cyclers is the prevention of methemoglobinemia, a frequent complication of anemia in childhood malaria and cerebral malaria [58].

Conclusion

The results from this interdisciplinary approach should provide incisive insight into understanding redox-cyclers as small molecules interfering with the putative half-site reactivity of *P. falciparum* GR toward NADPH *in vivo*, and also show how this property might be exploited for disulfide reductase-catalyzed drug bioactivation as a general concept to open new directions in medicinal chemistry. Despite the absence of MB-resistant parasites induced by drug pressure, it is essential to continue efforts in developing redox-cyclers, since the benzylNQ series – as new drugs targeting malarial parasites – can affect the thiol equilibrium of parasites more effectively/broadly than ACT.

Acknowledgments

The authors are indebted to the DFG through the collaborative center SFB 544 "Control of tropical infectious diseases," subproject B14, for continued generous support. The authors wish to thank Vanessa Yardley for providing residual unpublished data concerning the *in vitro* antiparasitic activity and cytotoxicity of menadione and its aza-analogs, and Hervé Vezin for the residual unpublished data regarding the redox potential values of di- and tri-fluoromenadiones, performed in the same set of cyclic voltammetry experiments previously published [44].

References

1 Baunaure, F. and Langsley, G. (2005) Protein traffic in *Plasmodium* infected-red blood cells. *Med. Sci.*, **21**, 523–529.

2 Egan, T.J. (2008) Recent advances in understanding the mechanism of hemozoin (malaria pigment) formation. *J. Inorg. Biochem.*, **102**, 1288–1299.

3 Atamna, H. and Ginsburg, H. (1995) Heme degradation in the presence of glutathione. A proposed mechanism to account for the high levels of non-heme iron found in the membranes of hemoglobinopathic red blood cells. *J. Biol. Chem.*, **270**, 24876–24883.

4 Ginsburg, H., Famin, O., Zhang, J., and Krugliak, M. (1998) Inhibition of glutathione-dependent degradation of heme by chloroquine and amodiaquine as a possible basis for their antimalarial mode of action. *Biochem. Pharmacol.*, **56**, 1305–1313.

5 Atamna, H., Krugliak, M., Shalmiev, G., Deharo, E., Pescarmona, G., and Ginsburg, H. (1996) Mode of antimalarial effect of methylene blue and some of its analogues on *Plasmodium falciparum* in culture and their inhibition of *P. vinckei petteri* and *P. yoelii nigeriensis* in vivo. *Biochem. Pharmacol.*, **51**, 693–700.

6 Ncokazi, K.K. and Egan, T.J. (2005) A colorimetric high-throughput beta-hematin inhibition screening assay for use in the search for antimalarial compounds. *Anal. Biochem.*, **338**, 306–319.

7 Riscoe, M., Kelly, J.X., and Winter, R. (2005) Xanthones as antimalarial agents: discovery, mode of action, and optimization. *Curr. Med. Chem.*, **12**, 2539–2549.

8 de Villiers, K.A., Marques, H.M., and Egan, T.J. (2008) The crystal structure of halofantrine-ferriprotoporphyrin IX and the mechanism of action of arylmethanol antimalarials. *J. Inorg. Biochem.*, **102**, 1660–1667.

9 Weissbuch, I. and Leiserowitz, L. (2008) Interplay between malaria, crystalline hemozoin formation, and antimalarial drug action and design. *Chem. Rev.*, **108**, 4899–4914.

10 Asher, C., de Villiers, K.A., and Egan, T.J. (2009) Speciation of ferriprotoporphyrin

IX in aqueous and mixed aqueous solution is controlled by solvent identity, pH, and salt concentration. *Inorg. Chem.*, **48**, 7994–8003.

11 Schirmer, R.H., Müller, J.G., and Krauth-Siegel, R.L. (1995) Disulfide reductase inhibitors as chemotherapeutic agents: the design of drugs for trypanosomiasis and malaria. *Angew. Chem. Int. Ed.*, **34**, 141–154.

12 Krauth-Siegel, R.L., Bauer, H., and Schirmer, R.H. (2005) Dithiol proteins as guardians of the intracellular redox milieu in parasites: old and new drug targets in trypanosomes and malaria-causing plasmodia. *Angew. Chem. Int. Ed.*, **44**, 690–715.

13 Ginsburg, H., Atamna, H., Shalmiev, G., Kanaani, J., and Krugliak, M. (1996) Resistance of glucose-6-phosphate dehydrogenase deficiency to malaria: effects of fava bean hydroxypyrimidine glucosides on *Plasmodium falciparum* growth in culture and on the phagocytosis of infected cells. *Parasitology*, **113**, 7–18.

14 Zhang, Y.A., Hempelmann, E., and Schirmer, R.H. (1988) Glutathione reductase inhibitors as potential antimalarial drugs. Effects of nitrosoureas on *Plasmodium falciparum* in vitro. *Biochem. Pharmacol.*, **37**, 855–860.

15 Zhang, Y., König, I., and Schirmer, R.H. (1988) Glutathione reductase-deficient erythrocytes as host cells of malarial parasites. *Biochem. Pharmacol.*, **37**, 861–865.

16 Cappellini, M.D. and Fiorelli, G. (2008) Glucose-6-phosphate dehydrogenase deficiency. *Lancet*, **371**, 64–74.

17 Cappadoro, M., Giribaldi, G., O'Brien, E., Turrini, F., Mannu, F., Ulliers, D., Simula, G. *et al.* (1998) Early phagocytosis of glucose-6-phosphate dehydrogenase (G6PD)-deficient erythrocytes parasitized by *Plasmodium falciparum* may explain malaria protection in G6PD deficiency. *Blood*, **92**, 2527–2534.

18 Chevion, M., Navok, T., Glaser, G., and Mager, J. (1982) The chemistry of favism-inducing compounds. The properties of isouramil and divicine and their reaction with glutathione. *Eur. J. Biochem.*, **127**, 405–409.

19 Clark, I.A., Cowden, W.B., Hunt, N.H., Maxwell, L.E., and Mackie, E.J. (1984) Activity of divicine in *Plasmodium vinckei*-infected mice has implications for treatment of favism and epidemiology of G-6-PD deficiency. *Br. J. Haematol.*, **57**, 479–487.

20 Färber, P.M., Arscott, L.D., Williams, C.H., Jr, Becker, K., and Schirmer, R.H. (1998) Recombinant *Plasmodium falciparum* glutathione reductase is inhibited by the antimalarial dye methylene blue. *FEBS Lett.*, **422**, 311–314.

21 Schirmer, M., Scheiwein, M., Gromer, S., Becker, K., and Schirmer, R.H. (1999) Disulfide reductases are destabilized by physiologic concentrations of NADPH, in (eds S. Ghisla, P.M., Kroneck, P., Macheroux and H. Sund), *Flavins and Flavoproteins*, Agency for Scientific Publications, Berlin, pp. 857–862.

22 Schirmer, R.H., Adler, H., Zappe, H.A., Gromer, S., Becker, K., Coulibaly, B., and Meissner, P. (2008) Disulfide reductases as drug targets: Methylene blue combination therapies for falciparum malaria in African children, in *Flavins and Flavoproteins* (eds S. Frago, C., Gómez-Moreno and M. Medina), Prensas Universitarias Zaragoza, pp. 481–486.

23 Davioud-Charvet, E., Müller, T., Bauer, H., and Schirmer, H.,1,4-naphthoquinones as inhibitors of glutathione reductases and antimalarial agents. European Patent EP 08290278.4 (26 March, 2008). PCT/EP2009/053483 (25 March, 2009).

24 Kessl, J.J., Lange, B.B., Merbitz-Zahradnik, T., Zwicker, K., Hill, P., Meunier, B., Pálsdóttir, H. *et al.* (2003) Molecular basis for atovaquone binding to the cytochrome bc1 complex. *J. Biol. Chem.*, **278**, 31312–31318.

25 Fry, M. and Pudney, M. (1992) Site of action of the antimalarial hydroxynaphthoquinone, 2-[trans-4-(4′-chlorophenyl) cyclohexyl]-3-hydroxy-1,4-naphthoquinone (566C80). *Biochem. Pharmacol.*, **43**, 1545–1553.

26 Kessl, J.J., Moskalev, N.V., Gribble, G.W., Nasr, M., Meshnick, S.R., and Trumpower, B.L. (2007) Parameters determining the relative efficacy of hydroxy-

naphthoquinone inhibitors of the cytochrome bc1 complex. *Biochim. Biophys. Acta*, **1767**, 319–326.

27 Looareesuwan, S., Viravan, C., Webster, H.K., Kyle, D.E., Hutchinson, D.B., and Canfield, C.J. (1996) Clinical studies of atovaquone, alone or in combination with other antimalarial drugs for treatment of acute uncomplicated malaria in Thailand. *Am. J. Trop. Med. Hyg.*, **54**, 62–66.

28 Jambou, R., Legrand, E., Niang, M., Khim, N., Lim, P., Volney, B., Ekala, M.T. *et al.* (2005) Resistance of *Plasmodium falciparum* field isolates to in-vitro artemether and point mutations of the SERCA-type PfATPase6. *Lancet*, **366**, 1960–1963.

29 Noranate, N., Durand, R., Tall, A., Marrama, L., Spiegel, A., Sokhna, C., Pradines, B. *et al.* (2007) Rapid dissemination of *Plasmodium falciparum* drug resistance despite strictly controlled antimalarial use. *PLoS ONE*, **2**, e139.

30 Buchholz, K., Schirmer, R.H., Eubel, J.K., Akoachère, M.B., Dandekar, T., Becker, K. *et al.* (2008) Interactions of methylene blue with human disulfide reductases and their orthologues from *Plasmodium falciparum*. *Antimicrob. Agents Chemother.*, **52**, 183–191.

31 Buchholz, K., Comini, M.A., Wissenbach, D., Schirmer, R.H., Krauth-Siegel, R.L., and Gromer, S. (2008) Cytotoxic interactions of methylene blue with trypanosomatid-specific disulfide reductases and their dithiol products. *Mol. Biochem. Parasitol.*, **160**, 65–69.

32 Coulibaly, B., Zoungrana, A., Mockenhaupt, F.P., Schirmer, R.H., Klose, C., Mansmann, U., Meissner, P. *et al.* (2009) Strong gametocytocidal effect of methylene blue-based combination therapy against falciparum malaria: a randomised controlled trial. *PLoS ONE*, **4**, e5318.

33 Zoungrana, A., Coulibaly, B., Sié, A., Walter-Sack, I., Mockenhaupt, F.P., Kouyaté, B., Schirmer, R.H. *et al.* (2008) Safety and efficacy of methylene blue combined with artesunate or amodiaquine for uncomplicated falciparum malaria: a randomized

controlled trial from Burkina Faso. *PLoS ONE*, **3**, e1630.

34 Salmon-Chemin, L., Lemaire, A., De Freitas, S., Deprez, B., Sergheraert, C., and Davioud-Charvet, E. (2000) Parallel synthesis of a library of 1,4-naphthoquinones and automated screening of potential inhibitors of trypanothione reductase from *Trypanosoma cruzi*. *Bioorg. Med. Chem. Lett.*, **10**, 631–635.

35 Salmon-Chemin, L., Buisine, E., Yardley, V., Kohler, S., Debreu, M.A., Landry, V., Sergheraert, C. *et al.* (2001) 2- and 3-substituted-1,4-naphthoquinone derivatives as subversive substrates of trypanothione reductase and lipoamide dehydrogenase from *Trypanosoma cruzi*: synthesis and correlation between redox-cycling activities and in vitro cytotoxicity. *J. Med. Chem*, **44** 548–565.

36 Iyanagi, T. and Yamazaki, I. (1970) One-electron-transfer reactions in biochemical systems V. Difference in the mechanism of quinone reduction by the NADH dehydrogenase and the NAD(P)H dehydrogenase (DT-diaphorase). *Biochim. Biophys. Acta*, **216**, 282–294.

37 Nakamura, M. and Yamazaki, I. (1972) One-electron-transfer reactions in biochemical systems VI. Changes in electron-transfer mechanism of lipoamide dehydrogenase by modification of sulfhydryl groups. *Biochim. Biophys. Acta*, **267**, 249–257.

38 Biot, C., Dessolin, J., Grellier, P., and Davioud-Charvet, E. (2003) Double-drug development against antioxidant enzymes from *Plasmodium falciparum*. *Redox Report*, **8**, 280–283.

39 Krauth-Siegel, R.L., and Coombs, G.H. (1999) Enzymes of parasite thiol metabolism as drug targets. *Parasitol. Today*, **15**, 404–409.

40 Biot, C., Bauer, H., Schirmer, R.H., and Davioud-Charvet, E. (2004) 5-Substituted tetrazoles as bioisosteres of carboxylic acids. Bioisosterism and mechanistic studies on glutathione reductase inhibitors as antimalarials. *J. Med. Chem.*, **47**, 5972–5983.

41 Morin, C., Besset, T., Moutet, J.C., Fayolle, M., Brückner, M., Limosin, D., Becker, K.,

and Davioud-Charvet, E. (2008) The aza-analogues of 1,4-naphthoquinones are potent substrates and inhibitors of plasmodial thioredoxin and glutathione reductases and of human erythrocyte glutathione reductase. *Org. Biomol. Chem.*, **6**, 2731–2742.

42 Lopez-Shirley, K., Zhang, F., Gosser, D., Scott, M., and Meshnick, S.R. (1994) Antimalarial quinones: redox potential dependence of methemoglobin formation and heme release in erythrocytes. *J. Lab. Clin. Med.*, **123**, 126–130.

43 Davioud-Charvet, E., Delarue, S., Biot, C., Schwöbel, B., Böhme, C.C., Müssigbrodt, A., Maes, L. *et al.* (2001) A prodrug form of a *Plasmodium falciparum* glutathione reductase inhibitor conjugated with a 4-anilinoquinoline. *J. Med. Chem.*, **44**, 4268–4276.

44 Bauer, H., Fritz-Wolf, K., Winzer, A., Kühner, S., Little, S., Yardley, V., Vezin, H. *et al.* (2006) A fluoro analogue of the menadione derivative 6-[2′-(3′-Methyl)-1′,4′-naphthoquinolyl]hexanoic acid is a suicide substrate of glutathione reductase. Crystal structure of the alkylated human enzyme. *J. Am. Chem. Soc.*, **128**, 10784–10794.

45 Friebolin, W., Jannack, B., Wenzel, N., Furrer, J., Oeser, T., Sanchez, C.P., Lanzer, M. *et al.* (2008) Antimalarial dual drugs based on potent inhibitors of glutathione reductase from *Plasmodium falciparum*. *J. Med. Chem.*, **51**, 1260–1277.

46 Meierjohan, S., Walter, R.D., and Müller, S. (2002) Regulation of intracellular glutathione levels in erythrocytes infected with chloroquine-sensitive and chloroquine-resistant *Plasmodium falciparum*. *Biochem. J.*, **368**, 761–768.

47 Ginsburg, H. (2002) A double-headed pro-drug that overcomes chloroquine resistance. *Trends Parasitol.*, **18**, 103.

48 Bernadou, J. and Meunier, B. (2004) Biomimetic chemical catalysts in the oxidative activation of drugs. *Adv. Synth. Catal.*, **346**, 171–184.

49 Chavain, N., Davioud-Charvet, E., Trivelli, X., Mbeki, L., Rottmann, M., Brun, R., and Biot, C. (2009) Antimalarial activities of ferroquine conjugates with either glutathione reductase inhibitors or glutathione depletors via a hydrolyzable amide linker. *Bioorg. Med. Chem.*, **17**, 8048–8059.

50 Chavain, N., Vezin, H., Dive, D., Touati, N., Paul, J.F., Buisine, E., and Biot, C. (2008) Investigation of the redox behavior of ferroquine, a new antimalarial. *Mol. Pharm.*, **5**, 710–716.

51 Kutner, S., Ginsburg, H., and Cabantchik, Z.I. (1983) Permselectivity changes in malaria (*Plasmodium falciparum*) infected human red blood cell membranes. *J. Cell. Physiol.*, **114**, 245–251.

52 Ginsburg, H., Kutner, S., Krugliak, M., and Cabantchik, Z.I. (1985) Characterization of permeation pathways appearing in the host membrane of *Plasmodium falciparum*-infected red blood cells. *Mol. Biochem. Parasitol.*, **14**, 313–322.

53 Kutner, S., Breuer, W.V., Ginsburg, H., Aley, S.B., and Cabantchik, Z.I. (1985) Characterization of permeation pathways in the plasma membrane of human erythrocytes infected with early stages of *Plasmodium falciparum*: association with parasite development. *J. Cell. Physiol.*, **125**, 521–527.

54 Bouyer, G., Egée, S., and Thomas, S.L. (2007) Toward a unifying model of malaria-induced channel activity. *Proc. Natl Acad. Sci. USA*, **104**, 11044–11049.

55 Staines, H.M., Alkhalil, A., Allen, R.J., De Jonge, H.R., Derbyshire, E., Egée, S., Ginsburg, H. *et al.* (2007) Electrophysiological studies of malaria parasite-infected erythrocytes: current status. *Int. J. Parasitol.*, **37**, 475–482.

56 Hogg, T., Nagarajan, K., Herzberg, S., Chen, L., Shen, X., Jiang, H., Wecke, M. *et al.* (2006) Structural and functional characterization of Falcipain-2, a hemoglobinase from the malarial parasite *Plasmodium falciparum*. *J. Biol. Chem.*, **281**, 25425–25437.

57 Monti, D., Vodopivec, B., Basilico, N., Olliaro, P., and Taramelli, D. (1999) A novel endogenous antimalarial: Fe(II)-protoporphyrin IX alpha (heme) inhibits hematin polymerization to beta-hematin (malaria pigment) and kills

malaria parasites. *Biochemistry*, **38**, 8858–8863.

58 Anstey, N.M., Hassanali, M.Y., Mlalasi, J., Manyenga, D., and Mwaikambo, E.D. (1996) Elevated levels of methaemoglobin in Tanzanian children with severe and uncomplicated malaria. *Trans. R. Soc. Trop. Med. Hyg.*, **90**, 147–151.

21
Ferroquine: A Concealed Weapon

Christophe Biot*, Bruno Pradines, and Daniel Dive

Abstract

The development of ferroquine (FQ; SR97193) is the result of a successful strategy consisting of the incorporation of a ferrocene moiety into the structure of chloro-quine (CQ) to overcome the resistance of *Plasmodium falciparum* to the parent drug, or to improve its antimalarial action. This approach has been applied to several classes of known antimalarials (arylaminoalcohols, artemisinin derivatives, naphtho-quinones), without any real gain of activity. Nevertheless, the modification of amino-4-quinolines and, more recently, of fluoroquinolones has provided highly significant results. The efficacy of ferrocene-quinoline antimalarials is directly dependent not only on the position of ferrocene in the molecule, but also on several other properties of the drugs. Among numerous molecules synthesized, FQ appears so far to be the most promising among antimalarials currently in development. The potent activity of FQ and the absence of cross-resistance with other antimalarials were demonstrated both *in vitro* on a large number of clones and field isolates of *P. falciparum*, and *in vivo* on rodent models that demonstrated the high bioavailability of the product. Excellent properties of FQ were observed during a pharmacology survey. The drug has been developed by Sanofi-Aventis, and is currently in Phase IIb clinical trials.

Introduction

The synthesis of organometallic molecules as drugs can follow two main strategies: (i) the modification of already known molecules in order to improve their activity or to overcome resistance problems encountered with the drugs; or (ii) the *de novo* synthesis of entirely new molecules which mimic known structure–activity relation-ships that have already been described for other antimalarial agents.

In the search for new antimalarials, several metals – including Au, Rh, Fe, and Ga – have been used in the design of active compounds [1], with the developed products having IC_{50}-values ranging from hundreds of nanomolar to the micromolar level.

*Corresponding author

Apicomplexan Parasites. Edited by Katja Becker
Copyright © 2011 WILEY-VCH Verlag GmbH & Co. KGaA, Weinheim
ISBN: 978-3-527-32731-7

Unfortunately, two major problems have arisen when these metals are used: (i) a potentially unexpected deleterious reactivity might occur within the molecule; and (ii) if the metal were to be liberated within the organism, it might prove to be toxic. Indeed, the latter property has been established clinically for some of these materials [2]. In contrast, ferrocene, with its "sandwich" structure [3] and small, rigid, and lipophilic nature, does not display such disadvantages. Moreover, the stability of ferrocene, either in aqueous media or in aerobic conditions, in conjunction with its electrochemical properties, have proven to be highly advantageous for potential biological applications [4]. The minimal toxicity of ferrocene is also in keeping with its use in drug development, as shown previously when ferrocerone was used to treat iron-deficiency anemia [5], or as ferrocifens and ferrociphenols to treat cancer [6, 7].

From A Ferrocene Strategy to Ferroquine

The Ferrocene Strategy: Deception and Success

The concept of grafting a ferrocene moiety inside an established antimalarial molecule was first applied during the mid-1990s, and these investigations are ongoing. Surprisingly, ferrocenyl derivatives of mefloquine (MQ) and quinine (QN) [8], artemisinin (ART) [9, 10], or atovaquone [11] provided disappointing results, with the antimalarial activity of the new molecules being inferior or equal to that of the parent drugs. It was suggested that the results obtained with MQ and QN might be related to the instability of the molecules in the acidic conditions of the digestive vacuole [12].

Ferrocenyl derivatives of fluoroquinolones were recently obtained via a double strategy that involved an esterification of the carboxylic acid of ciprofloxacin (the prodrug approach) with the insertion of a ferrocenyl moiety into the molecule [13]. Whilst the compounds produced in this way had a much greater antimalarial activity than ciprofloxacin, the esterification appeared to enhance the activity, and the ferrocene core provided a complementary efficacy.

Finally, chloroquine (CQ) derivation enabled the development of very active antimalarial agents. As many ferrocene derivatives of CQ were synthesized, it was possible to define the main rules directing the antimalarial activity of this class of products. Notably, the position of the ferrocene moiety within the parent molecule was shown to determine the drug's efficacy (Figure 21.1).

The primary finding was that the simple association of basic CQ and ferrocene carboxylic acid via a salt bridge resulted in a poor antimalarial activity, due to an antagonistic effect between the two compounds [14]. The condensation of ferrocene at C-3, or on the endocyclic nitrogen, led to a decrease in activity compared to CQ [15]. Monoferrocenyl conjugates in which the ferrocenyl group was attached to the terminal nitrogen associated with a modulation of the lateral chain length did not provide any increase in activity compared to CQ [16]. Moreover, these results indicated a high risk of cross-resistance with CQ for these molecules. The diferrocenyl conjugates provided inconsistent results in antimalarial tests, most likely due to

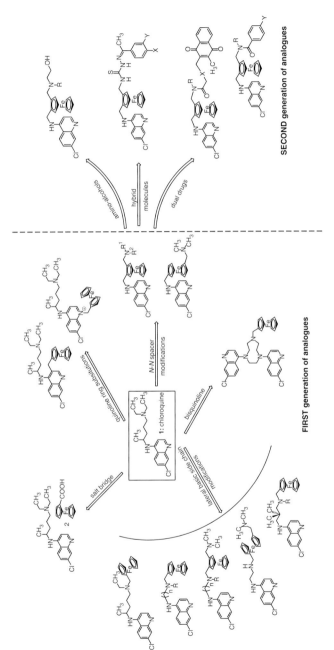

Figure 21.1 Strategy of development of ferrocene conjugates or derivatives of CQ.

their instability under the experimental conditions used [16]. The insertion of ferrocene into the lateral chain of CQ resulted in a set of products which were very active, against both CQ-susceptible and CQ-resistant clones of *Plasmodium falciparum*, whatever the substitution of the terminal nitrogen atom [16–19]. The location of the ferrocene moiety inside the lateral chain was one of the key determinants of the antimalarial activity of these compounds toward CQ-resistant parasites.

Among these compounds, ferroquine (FQ, SR97193) proved to be the most active, was quickly recognized as showing huge potential, and thus was protected by patents [20, 21]. Subsequent studies conducted to produce a molecule that was more active and of a more affordable cost for industrial production failed to result in a viable competitor for FQ [16], as did the dual drug strategy of an association of FQ with glutathione reductase (GR) inhibitors [22] or thiosemicarbazones [23]. Nonetheless, further *in vitro* and *in vivo* studies confirmed the value of FQ as a potential antimalarial agent, and this led to its entry into preclinical industrial development during 2003, and to its current status in Phase II clinical trials.

Antimalarial Activity of Ferroquine

Ferroquine Efficacy and Mechanisms of Resistance in P. falciparum

Ferroquine is highly active against CQ-resistant *P. falciparum* laboratory strains [14], and also against *P. falciparum* isolated from infected patients [24–29]. Notably, FQ has also been shown as highly active against parasites with a reduced susceptibility toward QN, monodesethylamodiaquine (MDAQ; the active metabolite of amodiaquine), or MQ. In addition, FQ was more active than CQ, QN, MDAQ, and even MQ. Encouragingly, no correlation was found between the activity of FQ and those of other quinoline drugs, such as CQ, QN, MDAQ, or MQ. These results suggested that no cross-resistance existed between FQ and CQ or quinoline antimalarial drugs, and these data were in accordance with the results of previous studies that had identified weak coefficients of determination that ranged from 0.096 to 0.127 between FQ and CQ [25, 27, 28]. The high activity of FQ on CQ-, QN-, MDAQ-, or MQ-resistant *P. falciparum* strains, and the absence of any cross-resistance, suggested that these drugs had different modes of action and/or mechanisms of resistance. The essential information relating to genes which encode transporters involved in drug transport and quinoline resistance (*pfcrt*, *pfmrp*, *pfmdr1* and *pfnhe-1*) is summarized in Table 21.1.

The *pfcrt* gene was first identified in 2000 [30]. Among 18 point mutations described [31–34], only one is considered to be the marker of the CQR phenotype, namely amino acid K76 which becomes T76 when mutated. PfCRT is involved in an essential membrane transport function in *P. falciparum*, indicated by its predicted channel-like structure. Mutations affecting transmembrane domains are associated with altered digestive vacuole pH regulation and changes in the accumulation of certain drugs [35–37]. Mutations of *pfcrt* may result in a modified CQ flux or a reduced drug binding to hematin, via an effect on the digestive vacuole [30]; moreover, PfCRT could interact directly with quinoline-based drugs [35, 38] and, indeed, it has been

Table 21.1 Genes involved in resistance to quinoline antimalarials.

Gene	Protein localization	Expected function	Polymorphism associated with resistance	Resistance or decreased susceptibility to
Pfcrt	Membrane of the digestive vacuole	Transport (channel-like structure).	18 known mutations, but one essential K76T.	Resistance to CQ.
		Digestive vacuole pH regulation	Different haplotypes according to geographic origin and CQ resistance	Demonstrated by genetic cross; field molecular studies and experimental transfections
Pfmrp	Membrane of the digestive vacuole	Putative GSSG transport	191Y, 437A	Decrease in CQ response
Pfmdr1	Membrane of the digestive vacuole	Transport (channel- like structure)	Multicopies of gene.	Resistance to MQ.
			Mutations of codons 86, 184, 1034, 1042, 1246	Modulation of response to MQ, HAL, ART. Demonstrated by field molecular monitoring and experimental transfections.
Pfnhe-1		Na/H$^+$ exchanger.	Polymorphism of microsatellite Pfnhe-1 ms4760	Decrease in QN response
		Cytoplasm pH modulator		

shown recently that PfCRT binds CQ at physiologically relevant concentrations [39]. Another hypothesis relates to the efflux of CQ by PfCRT. In this case, the mutation K76T removes the positive charge of the lysine side chain, possibly allowing the doubly protonated form of CQ to interact with PfCRT. Consequently, the greater negative mean charge, together with the increase in molecular hydrophobicity and a voluminous reduction of the side chains would facilitate the efflux of a hydrophilic, positively charged bulky drug such as CQ [40, 41].

The transporter *pfmdr1* gene encodes a P-glycoprotein homolog (Pgh1), located in the membrane of the parasite digestive vacuole [42]. No relationship (or, at best, only a weak relationship) was found in *P. falciparum* between CQ resistance and mutations in *pfmdr1* [43–46]. Through the heterologous expression of *pfmdr1* and transfection experiments, it has been shown that mutations at different codons can modulate the parasite's susceptibility to MQ, halofantrine, or ART derivatives [47–54].

The *Pfa0590w* gene encodes a protein termed PfMRP or CAB63558 [55], This transporter is also localized in the digestive vacuole membrane, and its structural similarities with human MRP2 suggest a function as a transporter of oxidized glutathione (GSSG) in *P. falciparum*. The results of previous studies identified two point mutations that were correlated with CQ resistance [56–58], while the *pfmrp* haplotypes 191Y and 437A were associated with a decrease in the CQ response.

Studies to identify markers of QN resistance in *P. falciparum* parasites have been very limited [59–62]. A quantitative trait locus (QTL) analysis of the progeny of HB3 $\times$ Dd2 clones recently highlighted the role of the *P. falciparum* Na^+/H^+ exchanger in this reduction of susceptibility toward QN. A variation in the coding sequence of a microsatellite DNA located in this gene (*Pfnhe-1* ms4760, a number of DNNND repeats) might be a predictor of decreased QN efficiency [59]. Clones with a decreased susceptibility to QN were associated with a greater activity of the exchanger, and with a higher cytosolic pH under physiological conditions. However, evidence for the involvement of PfNHE in resistance to quinine is limited, and remains the subject of debate.

The susceptibility *P. falciparum* to FQ was shown to be unrelated to mutations that occurred in the genes of transport proteins involved in quinoline antimalarial drug resistance, such as *pfcrt*, *pfmdr1*, *pfmrp*, or *pfnhe-1* [26, 56, 63–65]. These data suggest that FQ may not be expelled by transport proteins in quinoline-resistant parasites, possibly as a result of a strong avidity of *P. falciparum* for the iron moiety of the molecule [66]. In comparison to CQ, the presence of the ferrocene moiety with a different shape, volume, lipophilicity, effects on basicity, and electronic profile dramatically modifies the pharmacological behavior of the parent drug [67]. Therefore, FQ appears to present a reduced affinity for the transporters involved in CQ and quinoline drugs. This may partially explain the high FQ activity against multidrug-resistant *P. falciparum* parasites. In addition, resistance has not been obtained *in vitro* in laboratory-adapted strains submitted to drug pressure [68].

Ferroquine Efficacy and Characteristics of Resistance in Rodent Malaria

The efficacy of FQ was monitored successfully in several rodent malarial strains that are widely used for *in vivo* tests (e.g., *Plasmodium berghei* N, *Plasmodium yoelii* NS,

Plasmodium vinckei). Each of these strains showed a very wide range of curative doses for CQ, ranging from $50\,\mathrm{mg\,kg^{-1}}$ per day to more than $100\,\mathrm{mg\,kg^{-1}}$ per day in the standard four-day test. Subsequently, *P. yoelii* NS was tested in the absence and presence of CQ pressure (CQ ED_{90}: 2.78 and $8.33\,\mathrm{mg\,kg^{-1}}$ per day, respectively). The CQ-sensitive *P. vinckei* had a CQ curative dose of $70\,\mathrm{mg\,kg^{-1}}$ per day, while the curative dose for the CQ-resistant clone was up to $400\,\mathrm{mg\,kg^{-1}}$ per day. For all strains tested, ED_{90}-values were measured for FQ that ranged from 1.96 to $3.89\,\mathrm{mg\,kg^{-1}}$ per day. The curative dose observed was $10\,\mathrm{mg\,kg^{-1}}$ per day for all strains tested, irrespective of the route of administration (see Refs [18, 69]; also C. Biot *et al.*, unpublished results). These results demonstrated the powerful activity of FQ *in vivo*, and its excellent bioavailability when administered orally. Other strains which are known to have an affinity for reticulocytes, and to be more resistant to CQ in this type of host cell (*P. berghei*, *P. yoelii*) were cured as efficiently by FQ as those which invaded only mature red blood cells.

When selecting a rodent strain of *Plasmodium* that was resistant to FQ, *P. yoelii* NS proved to be the most suitable for establishing resistance to a number of antimalarial agents. By using the 2% relapse method of Peters with the PyNS strain, the PyCQ^R strain was obtained under chloroquine pressure. Although the CQ resistance of PyCQ^R could be reversed by verapamil, its susceptibility to FQ remained unchanged. However, following a 23-week period of FQ pressure ($60\,\mathrm{mg\,kg^{-1}}$) on PyCQ^R, a PyFQ^R strain was obtained that was multiresistant to CQ, MQ, and FQ, and only partially susceptible to verapamil action [12]. This unusual phenotype of resistance did not seem to be fixed genetically, because either the release of pressure or simply cryopreservation of the strain resulted in an immediate loss of FQ resistance, while the CQ resistance persisted. The molecular analysis of both the *pymdr1* and *pycrt* genes failed to show any mutations in critical amino-acid positions [12]. To date, no stable resistance to FQ has been established in any rodent or human *Plasmodium* species submitted to drug pressure.

Ferroquine as a New Antimalarial

Since FQ was wrongly considered to be a CQ-like derivative, the possibility that *P. falciparum* could rapidly become resistant to the drug could not be excluded. However, the absence of cross-resistance with other antimalarials in *P. falciparum*, the characteristics of efficacy on rodent malarial parasites, and the absence of genetically fixed resistance raised many questions regarding the mechanism of action of FQ, and the way in which it would overcome CQ resistance. Consequently, the chemical structure and properties of FQ were closely studied in an effort to determine the reasons for its potent activity.

Chemical Properties and Reactivity of Ferroquine

Properties Shared with Chloroquine
Ferroquine shares with CQ various important properties that are directly involved in its antimalarial activity. For example, CQ is believed to act by concentrating in the

parasite digestive vacuole and preventing the crystallization of toxic heme into hemozoin, leading to membrane damage and parasite death [69, 70]. CQ also forms complexes with hematin in solution, and is an inhibitor of β-hematin formation. Like CQ, FQ is a weak base that potentially is able to accumulate in the parasite digestive vacuole. The pK_a of FQ is lower than that of CQ, and it might be speculated that this property may be less marked than for CQ [67]. FQ is able to form complexes with hematin (ratio 1:1 for both FQ and CQ), and to inhibit hematin crystallization into β-hematin (IC$_{50}$ 0.8 versus 1.9 for CQ) [67, 71].

Chirality

Ferroquine is a chiral molecule, and is used as a racemic mixture. Studies on separate pure enantiomers showed no dramatic difference in activity between them, either on human or rodent parasites, and an equal cytotoxicity. This enabled the development of the racemic mixture as an antimalarial agent [72].

Lipophilicity and Protonation

A comparison of the apparent partition coefficient (log$D$) at postulated cytosolic and vacuolar pH (7.4 and 5.2, respectively) might result in a higher concentration of FQ in the digestive vacuole, despite its weaker base properties [71, 73]. The protonation of FQ is also different from that of CQ at these two pH-values; at equal concentrations of the two drugs in the vacuole, a higher concentration of FQ in its most active form toward hematin crystallization can be predicted compared to CQ [74].

An Apparently Crucial Hydrogen Bond

The solid-state structure of neutral FQ is stabilized by a strong intramolecular hydrogen bond (d = 2.17 Å) between the amine N(11) and the amine N(24) (Figure 21.2) [67]. This conformation is also found in solution in organic solvents such as chloroform [74]. Compared to the more flexible lateral side chain of CQ, the intramolecular hydrogen bond should be stronger in FQ. Therefore, the role of this noncovalent interaction on the antimalarial activity was studied. To this end, an analog FQ-Me bearing a methyl group instead of the hydrogen atom on the anilino N(11) was designed and synthesized (Figure 21.3).

FQ-Me shared similar physico-chemical properties with FQ, and was also able to inhibit β-hematin formation (Table 21.2). Nevertheless, this structural modification led to a significant reduction in activity against CQ-susceptible and CQ-resistant strains (Table 21.2). The presence of a hydrogen-bonding interaction in the lateral side chain of FQ might contribute to the antiplasmodial activity.

This important structural factor probably serves to maintain the molecule in a particular geometry; when comparing CQ to FQ, the part of the molecule modified by isosterism (the ferrocenyl moiety instead of the alkyl chain) should not be involved in the interaction with the receptor, but only serve to correctly position the other elements of the molecule. The interaction of the ferrocenic moiety with lipidic structures might enable the drug to escape the transport system involved in resistance, and to concentrate the drug at the site where its action is optimal for the inhibition of hemozoin formation [73].

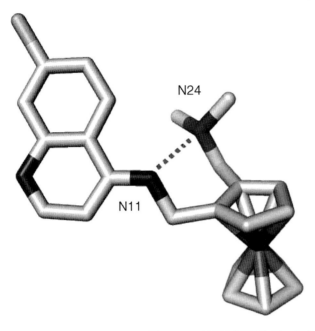

Figure 21.2 Crystal structure of ferroquine (CCDC 262108). The dashed line indicates the intramolecular hydrogen bond.

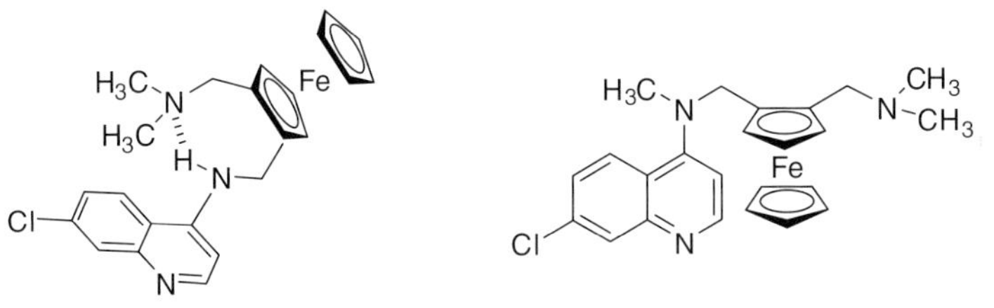

Figure 21.3 Comparative structures of ferroquine (FQ) and methyl-ferroquine (FQ-Me).

Table 21.2 Comparative properties of FQ and FQ-methyl.

	IC_{50} (nM)			β-Hematin inhibition	Lipophilicity		Basicity	
	3D7	**W2**	**FCM29**	**IC_{50}**	**$logD_{7.4}$**	**$logD_{5.2}$**	**pK_{a1}**	**pK_{a2}**
FQ	3.5	7.1	10.0	0.3	2.95	−1.20	7.00	8.45
FQ-Me	99.8	200.9	230.1	2.0	3.04	−0.56	7.05	8.80

Production of Hydroxyl Radicals

In conditions encountered in the digestive vacuole (acid pH and occurrence of H_2O_2), FQ is able to produce significant amounts of hydroxyl radicals, which leads to damage of the parasite. CQ is unable to produce such an effect, as the reaction is directly linked to the reactivity of the ferrocene core via the generation of ferricinium species [74].

Conclusion

An organometallic strategy for the development of new antimalarials led to the discovery of FQ, which is currently in Phase IIb clinical trials, in association with artesunate, in a new proposed artemisinin-based combination therapy (ACT) [75]. While many organometallic chemistry approaches to synthesize affordable antimalarial agents have been disappointing, the present authors' observations have shown this route to be open for new drug discovery, not only in the domain of malaria but also for other diseases. These examples of organometallic anticancer drugs, FQ and ferrocenic ciprofloxacin continue to provide much encouragement in this area.

Beyond its antimalarial activity, FQ may provide new strategies for the development of potent drugs. The elucidation of its mechanism of action, and how it can overcome the resistance mechanisms of *P. falciparum*, is an important area of research that may lead to the blockade of resistance caused by transport mechanisms preventing drugs from reaching their targets. The deciphering of the mechanism of action of FQ on resistant strains of *P. falciparum* clearly represents a promising future strategy for the development of new active drugs based on organometallic chemistry.

Acknowledgments

The authors are very grateful to Dr Jamal Khalife and Dr Ray Pierce for their critical reading of the manuscript, and for helpful comments and improvements.

References

1 Biot, C. and Dive, D. (2010) Metallocenic antimalarials and development of ferroquine, in *Medicinal Organometallic Chemistry* (eds G. Jaouen and N. Metzler-Nolte), Topics in Organometallic Chemistry, Vol. **32**, Springer-Verlag.

2 Desoize, B. (2004) Metals and metal compounds in cancer treatment. *Anticancer Res.*, **24**, 1529–1544.

3 Wilkinson, G., Rosenblum, M., Whiting, M.C., and Woodward, R.B. (1952) The structure of iron bis-cyclopentadienyl. *J. Am. Chem. Soc.*, **74**, 2125–2126.

4 van Staveren, D.R. and Metzler-Nolte, N. (2004) Bioorganometallic chemistry of ferrocene. *Chem. Rev.*, **104**, 5931–5985.

5 Nesmeyanov, A.N., Bogomolova, L.G., Viltcheskaya, V., Palitsyne, N., Andrianova, I., and Belozerova, O. (1971) US Patent 119 356.

6 Top, S., Vessières, A., Leclercq, G., Quivy, J., Tang, J., Vaissermann, J., Huché, M.,

and Jaouen, G. (2003) Synthesis, biochemical properties and molecular modelling studies of organometallic specific estrogen receptor modulators (SERMs), the ferrocifens and hydroxyferrocifens: evidence for an antiproliferative effect of hydroxyferrocifens on both hormone-dependent and hormone-independent breast cancer cell lines. *Chem. Eur. J.*, **9**, 5223–5236.

7 Vessières, A., Top, S., Pigeon, P., Hillard, E.A., Boubeker, L., Spera, D., and Jaouen, G. (2005) Modification of the estrogenic properties of diphenols by the incorporation of ferrocene. Generation of antiproliferative effects in vitro. *J. Med. Chem.*, **48**, 3937–3940.

8 Biot, C., Delhaës, L., Maciejewski, L.A., Mortuaire, M., Camus, D., Dive, D., and Brocard, J.S. (2000) Synthetic ferrocenic mefloquine and quinine analogues as potential antimalarials. *Eur. J. Med. Chem.*, **35**, 707–714.

9 Paitayatat, S., Tarnchompoo, B., Thebtaranonth, Y., and Yuthavong, Y. (1997) Correlation of antimalarial activity of artemisinin derivatives with binding affinity with ferroprotoporphyrin IX. *J. Med. Chem.*, **40**, 633–638.

10 Delhaës, L., Biot, C., Berry, L., Maciejewski, L.A., Camus, D., Brocard, J.S., and Dive, D. (2000) Novel Ferrocenic artemisinin derivatives: Synthesis, in vitro antimalarial activity and affinity of binding with Ferroprotoporphyrin IX. *Bioorg. Med. Chem.*, **8**, 2739–2745.

11 Baramee, A., Coppin, A., Mortuaire, M., Pelinski, L., Tomavo, S., and Brocard, J. (2006) Synthesis and in vitro activities of ferrocenic aminohydroxynaphthoquinones against *Toxoplasma gondii* and *Plasmodium falciparum*. *Bioorg. Med. Chem.*, **14**, 1294–1302.

12 Dive, D. and Biot, C. (2008) Ferrocene conjugates of chloroquine and other antimalarials: the development of ferroquine, a new antimalarial. *ChemMedChem*, **3**, 383–391.

13 Dubar, F., Anquetin, G., Pradines, B., Dive, D., Khalife, J., and Biot, C. (2009) Enhancement of the antimalarial activity of ciprofloxacin using a double prodrug/ bioorganometallic approach. *J. Med. Chem.*, **52**, 7954–7957.

14 Domarle, O., Blampain, G., Agnaniet, H., Nzadiyabi, T., Lebibi, J., Brocard, J., Maciejewski, L. *et al.* (1998) In vitro antimalarial activity of a new organometallic analog, ferrocene-chloroquine. *Antimicrob. Agents Chemother.*, **42**, 540–544.

15 Biot, C., Delhaës, L., Abessolo, H., Domarle, O., Maciejewski, L.A., Mortuaire, M., Delcourt, P. *et al.* (1999) Novel metallocenic compounds as antimalarials agents. Study of the position of ferrocene in chloroquine. *J. Organomet. Chem.*, **589**, 59–65.

16 Biot, C., Daher, W., Jarry, C., Ndiaye, C.H., Pelinski, L., Khalife, J., Fraisse, L. *et al.* (2006) Probing the role of the covalent linkage of ferrocene into a chloroquine template. *J. Med. Chem.*, **49**, 4707–4714.

17 Biot, C., Glorian, G., Maciejewski, L.A., Brocard, J.S., Millet, P., Georges, A.J., Abessolo, H. *et al.* (1997) Synthesis and antimalarial activity in vitro and in vivo of a new ferrocene-chloroquine analogue. *J. Med. Chem.*, **40**, 3715–3718.

18 Delhaës, L., Abessolo, H., Biot, C., Deloron, P., Karbwang, J., Mortuaire, M., Maciejewski, L.A. *et al.* (2001) Ferrochloroquine, a ferrocenyl analogue of chloroquine, retains a potent activity against resistant *Plasmodium falciparum* in vitro and *P. vinckei* in vivo. *Parasitol. Res.*, **87**, 239–244.

19 Biot, C., Daher, W., Chavain, N., Fandeur, T., Khalife, J., Dive, D., and De Clercq, E. (2006) Design and synthesis of hydroxyferroquine derivatives with antimalarial and antiviral activities. *J. Med. Chem.*, **49**, 2845–2849.

20 Brocard, J., Lebibi, J., and Maciejewski, L. (1995) Patent France 9505532.

21 Brocard, J., Lebibi, J., and Maciejewski, L. (1996) Antimalarial organometallic iron complexes. Patent international PCT/FR 96/00721.

22 Chavain, N., Davioud-Charvet, E., Trivelli, X., Mbeki, L., Rottmann, M., Brun, R., and Biot, C. (2009) Antimalarial activities of ferroquine conjugates with either glutathione reductase inhibitors or

glutathione depletors via a hydrolyzable amide linker. *Bioorg. Med. Chem.*, **17**, 8048–8059.

23 Biot, C., Pradines, B., Sergeant, M.H., Gut, J., Rosenthal, P.J., and Chibale, K. (2007) Design, synthesis, and antimalarial activity of structural chimeras of thiosemicarbazone and ferroquine analogues. *Bioorg. Med. Chem. Lett.*, **17**, 6434–6438.

24 Atteke, C., Ndong, J.N., Aubouy, A., Maciejewski, L., Brocard, J., Lebibi, J., and Deloron, P. (2001) In vitro susceptibility to a new antimalarial organometallic analogue, ferroquine, of *Plasmodium falciparum* isolates from the Haut-Ogooue region of Gabon. *J. Antimicrob. Chemother.*, **51**, 1021–1024.

25 Barends, M., Jaidee, A., Khaohirun, N., Singhasivanon, P., and Nosten, F. (2007) In vitro activity of ferroquine (SSR 97193) against *Plasmodium falciparum* isolates from the Thai-Burmese border. *Malar. J.*, **6**, 81.

26 Chim, P., Lim, P., Sem, R., Nhem, S., Maciejewski, L., and Fandeur, T. (2004) The *in vitro* antimalarial activity of ferrochloroquine measured against Cambodian isolates of *Plasmodium falciparum*. *Ann. Trop. Med. Parasitol.*, **98**, 419–424.

27 Kreidenweiss, A., Kremsner, P.G., Dietz, K., and Mordmuller, B. (2006) In vitro activity of ferroquine (SAR97193) is independent of chloroquine resistance in *Plasmodium falciparum*. *Am. J. Trop. Med. Hyg.*, **75**, 1178–1181.

28 Pradines, B., Fusai, T., Daries, W., Laloge, V., Rogier, C., Millet, P., Panconi, E. *et al.* (2001) Ferrocene-chloroquine analogues as antimalarial agents: in vitro activity of ferrochloroquine against 103 Gabonese isolates of *Plasmodium falciparum*. *J. Antimicrob. Chemother.*, **48**, 179–184.

29 Pradines, B., Tall, A., Rogier, C., Spiegel, A., Mosnier, J., Marrama, L., Fusai, T. *et al.* (2002) In vitro activities of ferrochloroquine against 55 Senegalese isolates of *Plasmodium falciparum* in comparison with those of standard antimalarial drugs. *Trop. Med. Int. Health*, **7**, 265–270.

30 Fidock, D.A., Nomura, T., Talley, A.K., Cooper, R.A., Dzekunov, S.M., Ferdig, M.T., Ursos, L.M.B., Sidhu, A.B.S. *et al.* (2000) Mutations in the *P. falciparum* digestive vacuole transmembrane protein PfCRT and evidence for their role in chloroquine resistance. *Mol. Cell*, **6**, 861–871.

31 Durrand, V., Berry, A., Sem, R., Glaziou, P., Beaudou, J., and Fandeur, T. (2004) Variations in the sequence and expression of the *Plasmodium falciparum* chloroquine resistance transporter (Pfcrt) and their relationship to chloroquine resistance in vitro. *Mol. Biochem. Parasitol.*, **136**, 273–285.

32 Chen, N., Kyle, D.E., Pasay, C., Fowler, E.V., Baker, J., Peters, J.M., and Cheng, Q. (2003) pfcrt Allelic types with two novel amino acid mutations in chloroquine-resistant *Plasmodium falciparum* isolates from the Philippines. *Antimicrob. Agents Chemother.*, **47**, 3500–3505.

33 Nagesha, H.S., Casey, G.J., Rieckmann, K.H., Fryauff, D.J., Laksana, B.S., Reeder, J.C., Maguire, J.D., and Baird, J.K. (2003) New haplotypes of the *Plasmodium falciparum* chloroquine resistance transporter (pfcrt) gene among chloroquine-resistant parasite isolates. *Am. J. Trop. Med. Hyg.*, **68**, 398–402.

34 Johnson, D.J., Fidock, D.A., Mungthin, M., Lakshmanan, V., Sidhu, A.B., Bray, P.G., and Ward, S.A. (2004) Evidence for a central role for PfCRT in conferring *Plasmodium falciparum* resistance to diverse antimalarial agents. *Mol. Cell*, **15**, 867–877.

35 Cooper, R.A., Ferdig, M.T., Su, X.Z., Ursos, L.M., Mu, J., Nomura, T., Fujioka, H. *et al.* (2002) Alternative mutations at position 76 of the vacuolar transmembrane protein PfCRT are associated with chloroquine resistance and unique stereospecific quinine and quinidine responses in *Plasmodium falciparum*. *Mol. Pharmacol.*, **61**, 35–42.

36 Sanchez, C.P., Stein, W., and Lanzer, M. (2003) Trans stimulation provides evidence for a drug efflux carrier as the mechanism of chloroquine resistance in *Plasmodium falciparum*. *Biochemistry*, **42**, 9383–9394.

37 Bennett, T.N., Kosar, A.D., Ursos, L.M., Dzekunov, S., Singh Sidhu, A.B., Fidock, D.A., and Roepe, P.D. (2004) Drug resistance-associated PfCRT mutations confer decreased *Plasmodium falciparum* digestive vacuolar pH. *Mol. Biochem. Parasitol.*, **133**, 99–114.

38 Sidhu, A.B., Verdier-Pinard, D., and Fidock, D.A. (2002) Chloroquine resistance in *Plasmodium falciparum* malaria parasites conferred by pfcrt mutations. *Science*, **298**, 210–213.

39 Zhang, H., Paguio, M., and Roepe, P.D. (2004) The antimalarial drug resistance protein *Plasmodium falciparum* chloroquine resistance transporter binds chloroquine. *Biochemistry*, **43**, 8290–8296.

40 Warhurst, D.C., Craig, J.C., and Adagu, I.S. (2002) Lysosomes and drug resistance in malaria. *Lancet*, **360**, 1527–1529.

41 Warhurst, D.C. (2003) Polymorphism in the *Plasmodium falciparum* chloroquine-resistance transporter protein links verapamil enhancement of chloroquine sensitivity with the clinical efficacy of amodiaquine. *Malar. J.*, **2**, 31.

42 Wilson, C.M., Serrano, A.E., Wasley, A., Bogenschutz, M.P., Shankar, A.H., and Wirth, D.F. (1989) Amplification of a gene related to mammalian mdr genes in drug-resistant *Plasmodium falciparum*. *Science*, **244**, 1184–1186.

43 Basco, L.K., de Pecoulas, P.E., Le Bras, J., and Wilson, C.M. (1996) *Plasmodium falciparum*: molecular characterization of multidrug-resistant Cambodian isolates. *Exp. Parasitol.*, **82**, 97–103.

44 Basco, L.K. and Ringwald, P. (1998) Molecular epidemiology of malaria in Yaoundé, Cameroon. III. Analysis of chloroquine resistance and point mutations in the multidrug resistance 1 (pfmdr1) gene of *Plasmodium falciparum*. *Am. J. Trop. Med. Hyg.*, **59**, 577–581.

45 Povoa, M.M., Adagu, I.S., Oliveira, S.G., Machado, R.L.D., Miles, M.A., and Warhurst, D.C. (1998) Pfmdr1 Asn1042Asp and Asp1246Tyr polymorphisms, thought to be associated with chloroquine resistance, are present in chloroquine-resistant and -sensitive Brazilian field isolates of *Plasmodium falciparum*. *Exp. Parasitol.*, **88**, 64–68.

46 Pickard, A.L., Wongsrichanalai, C., Purfield, A., Kamwendo, D., Emery, K., Zalewski, C., Kawamoto, F. *et al.* (2003) Resistance to antimalarials in Southeast Asia and genetic polymorphisms in pfmdr1. *Antimicrob. Agents Chemother.*, **47**, 2418–2423.

47 Price, R.N., Cassar, C., Brockman, A., Duraisingh, M., van Vugt, M., White, N.J., Nosten, F., and Krishna, N. (1999) The pfmdr1 gene is associated with a multidrug-resistant phenotype in *Plasmodium falciparum* from the western border of Thailand. *Antimicrob. Agents Chemother.*, **43**, 2943–2949.

48 Duraisingh, M.T., Jones, P., Sambou, I., von Seidlein, L., Pinder, M., and Warhurst, D.C. (2000) The tyrosine-86 allele of the pfmdr1 gene of *Plasmodium falciparum* is associated with increased sensitivity to the anti-malarials mefloquine and artemisinin. *Mol. Biochem. Parasitol.*, **108**, 13–23.

49 Reed, M.B., Saliba, K.J., Caruana, S.R., Kirk, K., and Cowman, A.F. (2000) Pgh1 modulates sensitivity and resistance to multiple antimalarials in *Plasmodium falciparum*. *Nature*, **403**, 906–909.

50 Basco, L.K. and Ringwald, P. (2002) Molecular epidemiology of malaria in Cameroon. X. Evaluation of pfmdr1 mutations as genetic markers for resistance to amino alcohols and artemisinin derivatives. *Am. J. Trop. Med. Hyg.*, **66**, 667–671.

51 Mawili-Mboumba, D.P., Kun, J.F.J., Lell, B., Kremsner, P.G., and Ntoumi, F. (2002) Pfmdr1 alleles and response to ultralow-dose mefloquine treatment in Gabonese patients. *Antimicrob. Agents Chemother.*, **46**, 166–170.

52 Pillai, D.R., Hijar, G., Montoya, I., Marquino, W., Ruebush, T.K., Wrongsrichanalai, C., and Kain, K.C. (2003) Lack of prediction of mefloquine and mefloquine-artesunate treatment outcome by mutations in the *Plasmodium falciparum* multidrug resistance 1 (pfmdr1) gene for *P. falciparum* malaria in Peru. *Am. J. Trop. Med. Hyg.*, **68**, 107–110.

53 Duraisingh, M.T., Roper, C., Walliker, D., and Warhurst, D.C. (2000) Increased sensitivity to the antimalarials mefloquine and artemisinin is conferred by mutations in the pfmdr1 gene of *Plasmodium falciparum*. *Mol. Microbiol.*, **36**, 955–961.

54 Ngo, T., Duraisingh, M., Reed, M., Hipgrave, D., Biggs, B., and Cowman, A.F. (2003) Analysis of pfcrt, pfmdr1, dhfr, and dhps mutations and drug sensitivities in *Plasmodium falciparum* isolates from patients in Vietnam before and after treatment with artemisinin. *Am. J. Trop. Med. Hyg.*, **68**, 350–356.

55 Klokouzas, A., Tiffert, T., van Schalkwyk, D., Wu, C.P., van Veen, H.W., Barrand, M.A., and Hladky, S.B. (2004) *Plasmodium falciparum* expresses a multidrug resistance-associated protein. *Biochem. Biophys. Res. Commun.*, **321**, 197–201.

56 Henry, M., Briolant, S., Fontaine, A., Mosnier, J., Baret, E., Amalvict, R., Fusai, T. *et al.* (2008) In vitro activity of ferroquine is independent of polymorphisms in transport proteins genes implicated in quinoline resistance in *Plasmodium falciparum*. *Antimicrob. Agents Chemother.*, **52**, 2755–2759.

57 Mu, J., Ferdig, M.T., Feng, X., Joy, D.A., Duan, J., Furuya, T., Subramanian, G. *et al.* (2003) Multiple transporters associated with malaria parasite responses to chloroquine and quinine. *Mol. Microbiol.*, **49**, 977–989.

58 Ursing, J., Zakeri, S., Gil, J.P., and Bjorkman, A. (2006) Quinoline resistance associated polymorphisms in the pfcrt, pfmdr1 and pfmrp genes of *Plasmodium falciparum* in Iran. *Acta Trop.*, **97**, 352–356.

59 Ferdig, M.T., Cooper, R.A., Mu, J., Deng, B., Joy, D.A., Su, X.Z., and Wellems, T.E. (2004) Dissecting the loci of low-level quinine resistance in malaria parasites. *Mol. Microbiol.*, **52**, 985–997.

60 Bennett, T.N., Patel, J., Ferdig, M.T., and Roepe, P.D. (2007) *Plasmodium falciparum* Na(+)/H(+) exchanger activity and quinine resistance. *Mol. Biochem. Parasitol.*, **153**, 48–58.

61 Henry, M., Briolant, S., Zettor, A., Pelleau, S., Baragatti, M., Baret, E., Mosnier, J. *et al.* (2009) *Plasmodium falciparum* Na^+/H^+ exchanger 1 transporter is involved in reduced susceptibility to quinine. *Antimicrob. Agents Chemother.*, **53**, 1926–1930.

62 Vinayak, S., Tauqeer Alam, M., Upadhyay, M., Das, M.K., Dev, V., Singh, N., Dash, A.P. *et al.* (2007) Extensive genetic diversity in the *Plasmodium falciparum* Na^+/H^+ exchanger 1 transporter protein implicated in quinine resistance. *Antimicrob. Agents Chemother.*, **51**, 4508–4511.

63 Pradines, B., Pagès, J.-M., and Barbe, J. (2005) Chemosensitizers in drug transport mechanisms involved in protozoan resistance. *Curr. Drug Targets - Infect. Disorders*, **5**, 411–431.

64 Henry, M., Alibert, S., Rogier, C., Barbe, J., and Pradines, B. (2008) Inhibition of efflux of quinolines as new therapeutic strategy in malaria. *Curr. Top. Med. Chem.*, **8**, 563–578.

65 Alibert Franco, S., Pradines, B., Mahamoud, A., Davin-Regli, A., and Pagès, J.-M. (2009) Efflux mechanism, a new target to combat multidrug resistant pathogen. *Curr. Med. Chem.*, **16**, 301–317.

66 Macreadie, I., Ginsburg, H., Siriwaraporn, W., and Tilley, L. (2000) Antimalarial drug development and new targets. *Parasitol. Today*, **16**, 438–444.

67 Biot, C., Taramelli, D., Forbar-Bares, I., Maciejewski, L.A., Boyce, M., Nowogrocki, G., Brocard, J.S. *et al.* (2005) Insights into the mechanism of action of ferroquine. Relationship between physicochemical properties and antiplasmodial activity. *Mol. Pharm.*, **2**, 185–193.

68 Daher, W., Biot, C., Fandeur, T., Jouin, H., Pelinski, L., Viscogliosi, E., Fraisse, L. *et al.* (2006) Assessment of *P. falciparum* resistance to ferroquine in field isolates and in W2 strain under pressure. *Malar. J.*, **5**, 11.

69 Yayon, A., Cabantchik, Z.I., and Ginsburg, H. (1985) Susceptibility of human malaria parasites to chloroquine is pH dependent. *Proc. Natl Acad. Sci. USA*, **82**, 2784–2787.

70 Dorn, A., Vippagunta, S.R., Matile, H., Jacquet, C., Vennerstrom, J.L., and Ridley, R.G. (1998) An assessment of drug-haematin binding as a mechanism for inhibition of haematin polymerisation by quinoline antimalarials. *Biochem. Pharmacol.*, **55**, 727–736.

71 Dubar, F., Khalife, J., Brocard, J., Dive, D., and Biot, C. (2008) Ferroquine, an ingenious antimalarial drug. Thoughts on the mechanism of action. *Molecules*, **13**, 2900–2907.

72 Delhaes, L., Biot, C., Berry, L., Delcourt, P., Maciejewski, L.A., Camus, D., Brocard, J.S. *et al.* (2002) Synthesis of ferroquine enantiomers. First investigation of metallocenic, chirality upon antimalarial activity and cytotoxicity. *ChemBioChem*, **3**, 101–106.

73 Biot, C., Chavain, N., Dubar, F., Pradines, B., Brocard, J., Forfar, I., and Dive, D. (2009) Structure-activity relationships of 4-N-substituted ferroquine analogues. Time to re-evaluate the mechanism of action of ferroquine. *J. Organomet. Chem.*, **694**, 845–854.

74 Chavain, N., Vezin, H., Dive, D., Touati, N., Paul, J.F., Buisine, E., and Biot, C. (2008) Investigation of the redox behavior of the new antimalarial, ferroquine. *Mol. Pharmaceutics*, **5**, 710–716.

75 Fraisse, L., and Ter Minassian, D. (2006) Association between ferroquine and an artemisinine derivative for treating malaria. International Patent PCT/FR2006/000842.

22
Current Aspects of Endoperoxides in Antiparasitic Chemotherapy

*Denis Matovu Kasozi, Stefan Rahlfs, and Katja Becker**

Abstract

Endoperoxides such as artemisinin derivatives are prospective, novel, broad-spectrum apicomplexicidal agents. However, a short plasma half-life, limited bioavailability, poor solubility either in oil or water, and recently reducing susceptibility remain significant challenges. Furthermore, the mechanism of action of endoperoxides remains incompletely understood. A multitarget mechanism of action is proposed, at the center of which is the endoperoxide moiety which, upon reductive cleavage, leads to the formation of highly reactive oxygen- or carbon-centered radicals that attack cellular targets such as the SERCA-type Ca^{2+}-ATPase. As a result of efforts to improve their properties, diverse endoperoxides have either been synthesized or extracted from natural sources; however, with the exception of *Plasmodium* species, they have not been systematically examined in other apicomplexan parasites. In this chapter, recent developments in the field of endoperoxide centered therapy are reviewed, and approaches for rational endoperoxide-based drug development discussed. The data suggest the further development and improvement of drug-hybrids, and advocate the implementation of endoperoxide combination therapies, as already exist for *Plasmodium* species, also against other apicomplexan parasites.

Introduction

Endoperoxides are sesquiterpene lactone compounds containing the active trioxane pharmacophore. Compounds with the 1,2,4-trioxane ring system, as exemplified by artemisinin and its semi-synthetic derivatives, represent a novel class of antiparasitic agents, including those of the phylum Apicomplexa. Numerous endoperoxides have either been synthesized or extracted from natural sources. Important derivatives have been obtained through functionalization at the C-3, C-9, C-10, and O-11 positions of artemisinin [1]. However, a short plasma half-life, limited bioavailability, and poor solubility in either oil or water remain major limitations. Although to

*Corresponding author

Apicomplexan Parasites. Edited by Katja Becker
Copyright © 2011 WILEY-VCH Verlag GmbH & Co. KGaA, Weinheim
ISBN: 978-3-527-32731-7

a large extent developed as antimalarial agents, artemisinin and its derivatives have been shown to exhibit activity against nonplasmodial apicomplexan parasites. Besides *Plasmodium* spp., endoperoxides have already been tested against *Toxoplasma* spp., *Babesia* spp., *Eimeria* spp., *Cryptosporidium* spp., *Haemoproteus* spp., and *Neospora* spp. Strategies to develop novel efficient endoperoxide-based chemotherapies include more systematic analyses on the different apicomplexa, an improvement of the physico-chemical properties of the endoperoxides, an enhanced knowledge of the mechanism of drug action, and the development of drug combination therapies.

First-Generation Endoperoxides

Artemisinin (Figure 22.1, **A**), also referred to as *qinghaosu*, is a natural sesquiterpene lactone, isolated from *Artemisia annua* [2]. However, it has a short plasma half-life, a limited bioavailability, and poor solubility in either oil or water. As a result, efforts to improve these properties led to the development of first-generation artemisinin derivatives (Figure 22.1, **B–E**). Dihydroartemisinin (**B**), which can be obtained from artemisinin via reduction with sodium borohydride, is almost sixfold more potent *in vitro* than the parent compound. Artemether (**C**) and arteether (**D**), both of which are ethers of dihydroartemisinin, are oil-soluble drugs and are well absorbed upon

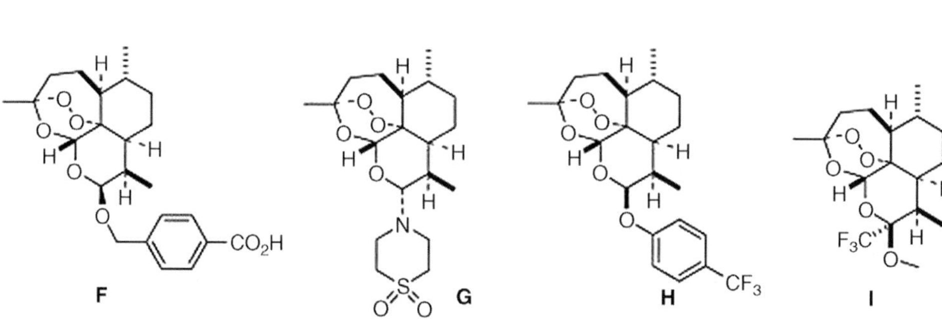

Figure 22.1 First-generation endoperoxides. Artemisinin (A), the natural product isolated from the Sweet Wormwood tree (*Artemisia annua*), from which the endoperoxides (**B–I**) are derived: Dihydroartemisinin (**B**), artemether (**C**), arteether (**D**), artesuanate (**E**), artelinic acid (**F**), artemisone (**G**), trifluoromethyl derivative (**H**), phenoxy derivative (**I**).

intramuscular administration [3]. Artesunate (**E**) is a water-soluble hemisuccinate derivative that usually is administered intravenously [4].

Due to their short plasma half-lives, the artemisinins are administered as combination therapies with longer-acting drugs. Artelinate (**F**), a water-soluble derivative that is equi-effective as artesunate and more stable in solution, was developed at the Walter Reed Army Institute of Research [5]. Additionally, artemisone (**G**) is a C-10-alkylamino artemisinin derivative with superior metabolic stabilities and oral bioavailability [6]. However, due to its short half-life it is currently considered a promising candidate for artemisinin-based combination therapy [7].

Magueur *et al.* [8] reported the synthesis of lead compounds (Figure 22.1, **H** and **I**) with a C-10-trifluoromethyl group, which resulted in an improved hydrolytic stability of acetal functionality. Several semisynthetic artemisinin derivatives as ethers or esters of dihydroartemisinin have been produced and tested on apicomplexan parasites.

Toxoplasma Species

Several artemisinin derivatives have been tested *in vitro* or in mouse models against *Toxoplasma gondii*. *In vitro*, *T. gondii* has been reported to be susceptible at micromolar concentrations to artemisinin, with IC_{50}-values of $8\,\mu M$ [9], $2\,\mu M$ [10], and $0.35\,\mu M$ [11]. Nagamune *et al.* [12] reported that the IC_{50}-values of artemisinin, dihydroartemisinin, artemether, artesunate, and artemisone against *T. gondii* were 0.8, 0.4, 0.1, 0.1, and $0.03\,\mu M$, respectively. In contrast, with the exception of artesunate, which has been reported to increase survival in mouse models [13], no curative effect was found with either artemisinin [14] or artemether [15].

Hencken *et al.*, [16] reported the synthesis of dehydroartemisinin (DART) trioxane derivatives (Figure 22.2) consisting of thiazoles (**A**), oxadiazoles (**B**),

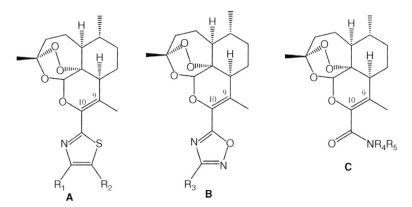

Figure 22.2 Dehydroartemisinin trioxane (DART) derivatives. DART-1,3 thiazoles (**A**), DART-1,2,4-oxadiazoles (**B**) and DART-carboxamides (**C**) with enhanced *in vitro* activity against the *Toxoplasma* lytic cycle.

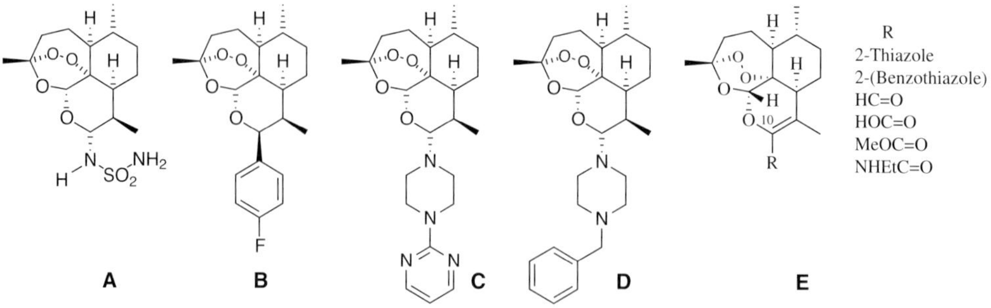

R
2-Thiazole
2-(Benzothiazole)
HC=O
HOC=O
MeOC=O
NHEtC=O

Figure 22.3 Artemisinin derivatives active against *T. gondii*. The sulfamide derivative (**A**), a 10,4′-fluorophenyl derivative (**B**), a pyrimidylpiperazine derivative (**C**) and a benzylpiperazine derivative (**D**) were more active than artesunate. The artemisinin carba-derivatives (**E**) effectively inhibited *T. gondii* growth better than artemisinin.

and 10-carboxamides (**C**), and tested them *in vitro* for activity in the *Toxoplasma* lytic cycle.

Eight thiazole derivatives and two carboxamide derivatives displayed effective inhibition of *Toxoplasma* growth (IC$_{50}$ = 0.25–0.42 μM), comparable in potency to artemether (IC$_{50}$ = 0.31 μM) and more than 100-fold more inhibitory than trimethoprim (IC$_{50}$ = 46 μM), the current standard treatment. Interestingly, two thiazole trioxanes were parasiticidal, inhibiting parasite replication irreversibly after exposure to 10 μM of drug for 24 h, whereas the standard trioxane drugs artemisinin and artemether were not parasiticidal. Furthermore, artemisone and artemiside were reported to control acute and reactivated toxoplasmosis in a murine model [17]. In this study, the *in vitro* activities of artemiside (IC$_{50}$ = 0.11 μM) and artemisone (IC$_{50}$ = 0.12 μM) were slightly better compared to artesunate (IC$_{50}$ = 0.21 μM), a sulfamide derivative (Figure 22.3, **A**) (IC$_{50}$ = 0.16 μM), a benzylpiperazine derivative (**B**) (IC$_{50}$ = 0.21 μM), a pyrimidylpiperazine derivative (**C**) (IC$_{50}$ = 0.16 μM), and a 10,4′-fluorophenyl derivative (**D**) (IC$_{50}$ = 0.18 μM) against the *T. gondii* strain RH (ATCC 50838). Furthermore, while most of the artemisinin-treated mice succumbed to infection, 60% of the artemiside-treated mice and >50% of the artemisone-treated mice survived the infection [17].

D'Angelo *et al.* [18] reported the synthesis of new, unsaturated carba derivatives, and then tested them for *in vitro* efficacy against three steps of the lytic cycle of *T. gondii* tachyzoites. In this study, the artemisinin carba derivatives (Figure 22.3, **E**) effectively inhibited *T. gondii* growth (IC$_{50}$ = 1.0–4.4 μM), tachyzoite replication, and attachment to and invasion of host cells as effectively as, or better than, the parent compound, artemisinin.

Babesia Species

The IC$_{50}$-values of artesunate against *Babesia equi* and *Babesia caballi* in *in vitro* cultures were 0.26 and 0.47 μM, respectively [19]. In a hamster model, artesunate had no effect, and artelinate reached 80% parasite suppression against *Babesia microti* [20]. In a donkey model, arteether had no effect, and artesunate increased survival against *Babesia equi* [21]. The IC_{50}-values of artesunate and dihydroartemisinine against *B. gibsoni* (Aomori strain) were 879 nM and 938 nM, respectively [22]. Artemisinin has been shown to accumulate in the food vacuoles of *Plasmodium* parasites that bind heme incorporated into hemozoin granules [23]. However, it is known that *Babesia* parasites do not contain food vacuoles or hemozoin granules [24]. Thus, the lack of a food vacuole, and hence heme, in *Babesia* spp. might reduce the activity of artemisinin derivatives against them compared to *Plasmodium* parasites.

Eimeria Species

The anticoccidial effect of artemisinin, when administered as a food supplement at 10 and 17 ppm in the diet, was tested in *Eimeria tenella* [25]. In chicken infected with *E. tenella* [25] *and E. acervalina* [27], artemisinin led to a reduction in oocyst output. Although gametogenesis was reported to be normal in artemisinin-treated chickens, the oocyst wall formation was significantly altered, resulting in both the death of developing oocysts and a reduced sporulation rate. Furthermore, Del Cacho *et al.* [25] reported an inhibition of the sarcoplasmic-endoplasmic reticulum calcium ATPase (SERCA) expression in macrogametes following artemisinin treatment. Thus, the deleterious effect of artemisinin on fertilized macrogametes (early zygotes) may be a result of SERCA inhibition.

Cryptosporidium Species

Artemisinin, artemether, and arteether had no curative effects against *Cryptosporidium parvum* in mice [28]. *In vitro*, artemisinin proved to be ineffective at 7 μM against *C. parvum* [29].

Neospora Species

The IC_{50}-value of artemisinin against *Neospora caninum* (KB-2 strain) in *in vitro* cultures was 4 μM [30]. Artemisinin, deoxyartemisinin, OZ277, and carbaOZ277 were each inactive against *N. caninum*, at concentrations of up to $1.0 \, \mu g \, ml^{-1}$ [31].

Yingzhaosu A and Arteflene

Yingzhaosu A (Figure 22.4, **A**) is a natural product isolated from *Artabotrys uncinatus* [32]. This second series of phytochemical endoperoxides is structurally unrelated to the artemisininoids. Hofheinz *et al.*, [33] reported the synthesis of arteflene (Ro 42-1611; Figure 22.4, **B**) one of the yingzhaosu derivatives which was effective both *in vitro* and *in vivo* against *Plasmodium* species. However, arteflene offered no comparative advantages over artemisinin, and thus its development was halted [23].

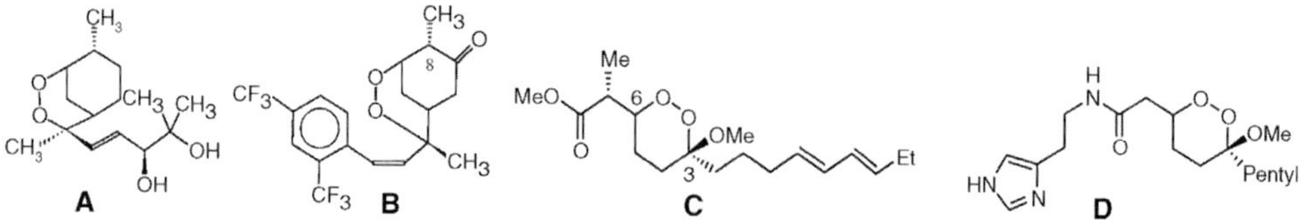

Figure 22.4 Endoperoxides from *Artabotrys uncinatus* and the marine sponge *Plakortis*. These are a different class of endoperoxides structurally unrelated to artemisinin including yingzhaosu A (**A**), arteflene (**B**) and methyl esters of peroxyplakoric acids (**C** and **D**).

Second-Generation Endoperoxides

Second-generation endoperoxides are fully synthetic endoperoxide bridges containing the products, not ethers or esters, of dihydroartemisinin [23]. With the understanding that only the endoperoxide bridge is required for activity, several analogs have been synthesized and tested both *in vitro* and *in vivo* as antimalarial drugs.

Trioxanes, Tetraoxanes and Spiro-1,2-Dioxolanes

This is a broad class of endoperoxide compounds (Figure 22.5) that comprises tricyclic trioxanes (**A** and **B**), cyclopentene-1,2,4-trioxanes, and spirocyclopentyl trioxane [34]. Developed as antimalarial agents, the trioxanes [36] were tested both *in vitro* against strains of *P. falciparum* and *in vivo* against *P. berghei* [36]. However despite being active, many of the trioxanes were not developed further because their synthesis was impractical.

Opsenica *et al.* [38] reported the synthesis of 11 new tetraoxanes possessing cholic acid-derived carrier and isopropylidene moieties. The tetraoxanes (Figure 22.5, **C**) were tested against a chloroquine (CQ)-resistant *P. falciparum* isolate from South America (RCS), and found to be more active against this strain than against the Southeast Asian strains W2 and TM91C235. The amides exhibited a better *in vitro* activity than the corresponding esters and acids.

Wang *et al.* [39] reported the synthesis of 14 spiro- and dispiro-1,2-dioxolanes via peroxycarbenium ion annulations with alkenes. The peroxycarbenium ion precursors included triethylsilyldiperoxyketals and -acetals derived from geminal dihydroperoxides, and from a new method which employed triethylsilylperoxyketals

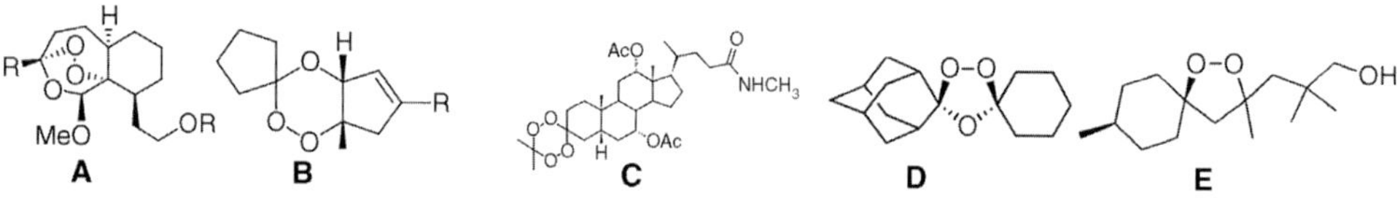

1 R=P(O)(OPh)₂   2 R=CH₂PH

Figure 22.5 Second-generation endoperoxides. Several fully synthetic endoperoxides including trioxanes (**A**) tetraoxanes (**B**) and spiro-1,2-dioxolanes (**C**) were active against *P. falciparum in vitro* and *P. berghei in vivo*.

and -acetals derived from the ozonolysis of alkenes. The 1,2,4-trioxolanes (Figure 22.5, **D**) were active against *P. falciparum in vitro* and *P. berghei in vivo*. Furthermore, Martyn *et al*. [40] reported the synthesis of spiro-1,2-dioxolanes (Figure 22.5, **E**) via the SnCl$_4$-mediated annulation of a bis-silylperoxide and an alkene. The reported eight analogs had EC$_{50}$-values of 50 to 150 nM against the 3D7 and Dd2 strains of *P. falciparum*. The main structural feature of these compounds was the presence of alkyl and cyclohexyl groups at the 3- and 5-positions of the dioxolane, respectively. A second series, synthesized by coupling a spiro-1,2-dioxolane carboxylic acid to four separate amines, afforded the most potent compound (EC$_{50}$ = 5 nM). The *in vitro* antimalarial activity of spiro- and dispiro-1,2,4-trioxolane antimalarials was found to correlate with the extent of heme alkylation [41].

Endoperoxide Dimers

Dimers, trimers and tetramers of artemisinin were synthesized by dimerizing artemisinin through reactive functionalities at the C-9 or C-10 positions. The increased peroxide content would enhance pro-oxidant activity, leading to greater antiparasitic action. Posner *et al*. [42] reported the synthesis of highly potent artemisinin dimers (Figure 22.6, **A**) when tested in *P. berghei*-infected mice.

Artemisinin dimers are known to possess not only significant antimalarial activity but also remarkable cytotoxicity, with inhibitory activity in the nanomolar to micromolar range [36]. Slade *et al*. [43] reported the synthesis of dihydroartemisinin acetal dimers (Figure 22.6, **B**), which were more active than artemisinin when tested against D6 (CQ-sensitive) and W2 (CQ-resistant) strains of *P. falciparum*. Compounds 9B and 9C and had a range of IC$_{50}$-values: 3.0–6.7 nM (D6) and 4.2–5.9 nM (W2) compared to 32.9 nM (D6) and 42.5 nM (W2) for artemisinin. Additionally, dihydroartemisinin acetal monomers and a trimer with moderate activity were synthesized as byproducts.

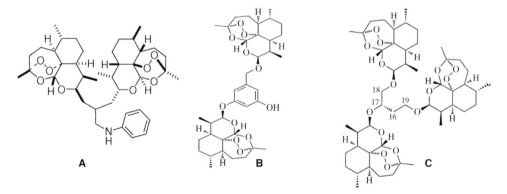

Figure 22.6 Endoperoxide dimers. Artemisinin dimers (**A**) and dihydroartemisinin acetal dimers (**B–E**) were more active than artemisinin and dihydroartemisinin against D6 (chloroquine-sensitive) and W2 (chloroquine-resistant) strains of *P. falciparum*.

Mannich Artemisinin Derivatives

Semisynthetic C-10 pyrrole Mannich artemisinin derivatives with nanomolar anti-malarial activity and a threefold greater potency than artemisinin have been reported by *in vitro* analysis against both 3D7 (CQ-sensitive) and K1 (CQ-resistant) strains [44]. When these water-soluble analogs of dihydroartemisinin (Figure 22.1, **B**) were synthesized by a modular approach, Peter's four-day test revealed that morpholine, *N*-methylpiperazine, and sulfonylmorpholine heterocyles in the Mannich side chains provided the best activity profiles, both *in vitro* and *in vivo* [44]. The best three compounds were significantly more active than sodium artesunate and artemether.

Epoxy Endoperoxide

Taylor *et al.* [45] reported the synthesis of epoxy endoperoxides, a novel class that contain not only an endoperoxide linkage but also a third oxygen atom positioned in an environment similar to that of oxygen atom 3 within artemisinin. These epoxy endoperoxides were active at submicromolar ranges (IC_{50} 300–500 nM) against the D10 (CQ-sensitive) strain of *P. falciparum*. Those compounds containing the epoxy group and two cyclic substituents showed the highest activity.

Artemisinin-Hybrid Molecules

Hybrid molecules in which two compounds are combined via a linker can offer an effective means of delivering these agents to the site of action within the parasite cell [46]. For example, Capela *et al.* [47] reported the synthesis of artemisinin-dipeptidyl vinyl sulfone hybrid molecules (Figure 22.7, **A**). These hybrid compounds, which consisted of peptidyl vinyl sulfones (which are potent, irreversible inhibitors of the falcipains) act as Michael acceptors of the catalytic cysteine residue [48]. This hybrid displayed activity in the nanomolar range, being more active than artemisinin and equipotent to artelinic acid. All conjugates were active against the *Plasmodium falciparum* W2 strain in the low nanomolar range, while those containing the Leu-hPhe core inhibited falcipain-2 in the low micromolar range [47].

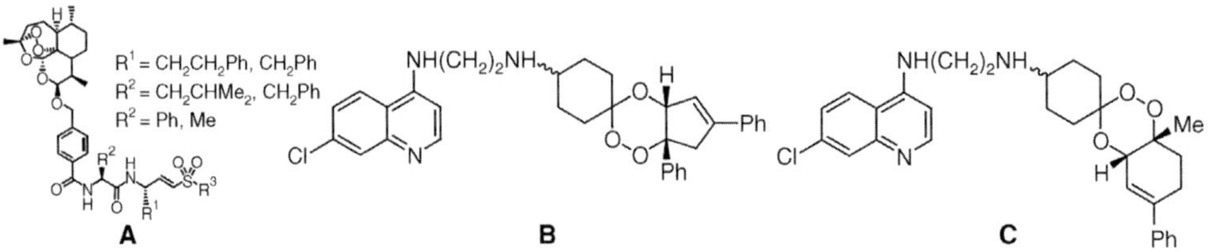

Figure 22.7 Hybrid compounds. The artemisinin-dipeptidyl vinyl sulfone hybrid (**A**) comprises a vinyl sulfone group linked to the endoperoxide moiety via the N terminus, using a 4-hydroxymethylbenzoic acid linker. Trioxaquines (**B**) comprising a trioxane reductively combined with an aminoquinoline.

Dechy-Cabaret *et al.* [49] reported the preparation of trioxaquines (Figure 22.7) that comprised a trioxane combined reductively with an aminoquinoline. The rationale for trioxaquines (Figure 22.7, **B** and **C**) has been questioned [50]. On the one hand, the quinoline part, upon binding with heme, would prevent its degradation to hemozoin, thus potentiating its toxic effect on the parasite. On the other hand, the trioxane part reacts differently with heme, converting it to hemozoin while producing a carbon radical as the lethal agent. Although these hybrids were active both *in vitro* and *in vivo* against *Plasmodium* species, they showed no major improvement over either the parent trioxane [51] or artemisinin.

Besides the lack of a significant cooperative effect on activity, these hybrids were unstable and poorly soluble in oil and water; hence their evaluation as citrates. The confirmation of a *synergistic in vitro* interaction, as opposed to addition or antagonism between artemisinin and potential partner compounds, would significantly enhance the hybrid approach.

Peroxyplakoric Endoperoxides

Kawanishi *et al.* [52] reported the isolation of methyl esters of peroxyplakoric acids with antimalarial activity (Figure 22.4, **C**) from the marine sponge *Plakortis*. The best of these was an imidazole derivative (Figure 22.4, **D**), which showed an ED_{50} value of $18 \, mg \, kg^{-1}$ against *P. berghei*, compared to a value of $5 \, mg \, kg^{-1}$ for artemisinin [53].

Mechanism of Endoperoxide Action

The mechanism of action of the endoperoxides, including artemisinin, its semi-synthetic derivatives and the recently synthesized peroxides, remains ambiguous – not only in *Plasmodium* species [54], where it has been intensively investigated, but also in *Toxoplasma* spp. [12], *Babesia* spp., *Eimeria* spp., *Cryptosporidium* spp., *Haemoproteus* spp., and *Neospora* spp.

By using a fluorescent dansyl trioxane derivative [55] and the pH-sensitive probe LysoSensor Blue [56], endoperoxides have been shown to accumulate initially to the digestive vacuole, which suggests that this is an important initial site of endoperoxide antimalarial activity. The fluorescent trioxanes were shown to localize to the neutral lipids of the digestive vacuole [55] and induce oxidative damage. These cellular accumulation patterns and effects on lipids were entirely endoperoxide-dependent, as inactive dioxolane analogs lacking the endoperoxide moiety failed to label neutral lipid bodies or to induce any oxidative membrane damage. Furthermore, the neutral lipids were reported to be closely associated with heme [55] that was generated as a result of hemoglobin degradation and which, besides iron Fe(II), may promote reductive cleavage of the peroxide bond. The key steps involved in this process are shown schematically in Figure 22.8 [1]. The oxygen- or carbon-centered radicals, high-valent iron-oxo species, or electrophilic agents such as epoxides (e.g., species **A–F**) potentially may alkylate vital target cellular components, leading to parasitic death.

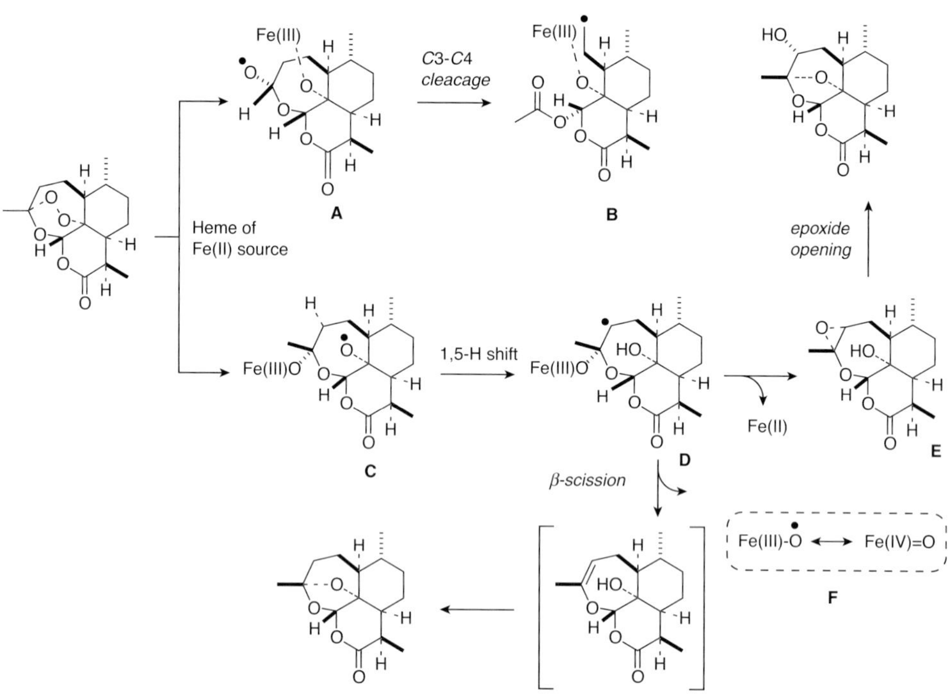

Figure 22.8 Reaction scheme for activation of endoperoxides. Endoperoxides may react with heme or iron leading to the formation of reactive oxygen species, high-valent iron-oxo species (**A–D**) or electrophilic agents such as epoxides (**E**) that alkylate cellular targets.

Putative targets of alkylation such as the PfATP6 (a SERCA-type Ca^{2+} ATPase [57]), translationally controlled tumor protein (TCTP) [58], heme [59], reduced glutathione [60], as well as the cysteine proteases [61], proteins of the electron transport chain [62], and mitochondrial membranes [56], have been suggested. Indeed, the identification of artemisinin-heme adducts in the spleen of infected mice by Robert *et al.* [63] supports this view. Furthermore, when parasite lysates were incubated with radiolabeled artemisinin, several interacting proteins were identified, suggesting that parasite death might result from the alkylation and inactivation of parasitic proteins [64].

Eckstein-Ludwig *et al.* [57] showed that the PfATP6 activity in transfected *Xenopus laevis* oocytes is abolished by artemisinin, but not by the inactive compound desoxyartemisinin, which lacks the endoperoxide bridge and is unaffected by other antimalarials. Furthermore, Uhlemann *et al.* [65] reported that a L263E point mutation in PfATP6 in *X. laevis* abolished inhibition by artemisinins. Another report by Jambou *et al.* [66], using field isolates, demonstrate a correlation between S769N PfATPase-6 mutation and their reduced susceptibility to artemether. In contrast, neither artemisinin nor the novel endoperoxides, at 10-fold their IC_{50}-value, caused any alterations in the morphology of the endoplasmic reticulum, the site of PfATPase-

6 [56]. Nonetheless the effects of thapsigargin, a sesquiterpene lactone that specifically inhibits SERCA, differed from that of artemisinin.

However, by using a yeast model Wang *et al.* [67] showed that artemisinin would directly target the malarial mitochondria by producing reactive oxygen species (ROS), resulting in a depolarization of their membrane potential [62]. Furthermore, the deletion of genes encoding the mitochondrial NADH dehydrogenases *NDE1* or *NDI1* (both enzymes of the electron transport chain) conferred resistance to artemisinin, while overexpression increased the sensitivity to artemisinin. Nonetheless, by employing rhodamine 123 uptake studies, del Pilar Crespo *et al.* [56] also reported a disruption of the mitochondrial membrane potential, albeit more as a downstream effect than as an initiator of parasite killing.

Despite a decade of debate, a multiple target mechanism of endoperoxide action is emerging, which could explain the broad activity of this class of compounds. Thus, neither *P. falciparum* ring-stage parasites that do not seem to have high concentrations of heme [68], nor *Babesia* parasites that do not contain food vacuoles or hemozoin granules [24], remain susceptible to endoperoxides.

Endoperoxide-Based Combinations

These combinations comprise an endoperoxide compound paired with another drug from a distinct chemical class, and with a different mechanism of action. To date, most of the endoperoxide combinations are artemisinin-based combination therapy (ACTs), developed as treatments against malaria. These combinations increase the rates of clinical and parasitological cures, and decrease the selection pressure for the emergence of antimalarial resistance. Artesunate in combination with mefloquine (AS + MQ), with amodiaquine (AS + AQ), with sulfadoxine pyrimethamine (AS+SP), with artemether–lumefantrine (AL), and with piperaquine–dihdyroartemimin (PD) include the ACTs that have now been adopted across all malaria-endemic regions. The 28-day clinical and parasitological failure rate adjusted after genotyping for the combination artesunate + amodiaquine (AM), was 1.2% in Angola, 3.5% in Senegal, 4% in Uganda, 6.8% in the Democratic Republic of the Congo, 9.2% in Zanzibar (United Republic of Tanzania), 7.2% in Sudan, and 19.5% in Rwanda. Similarly, with artesunate + sulfadoxine–pyrimethamine (ASP), the failure rate was 1.2% in Angola, 5% in South Africa, 5.5% in Ghana, 8.8% in Sudan, 16.5% in Zambia, 19.7% in the Democratic Republic of the Congo, and 30.3% in Rwanda. In contrast to Thailand, the efficacy of artesunate + mefloquine rarely exceeded 90% in the African countries of Benin, Cameroon, Côte d'Ivoire, and Senegal [69]. The efficacy of both AL (Coartem®, Novartis) and PD (Artekin®) in Africa was over 95% in Uganda and Tanzania, although in Uganda the 42-day rate of parasitological failure when adjusted for genotyping was 16% for AL and 6.9% for PD. Pyronaride–artesunate (PA; Pyramax®) has proved to be a promising combination in Phase II and III clinical trials in Africa [69]. Comparatively, endoperoxide combinations are likely to have a remarkable efficacy against *Toxoplasma* spp., *Babesia* spp., *Eimeria* spp., *Cryptosporidium* spp., *Haemoproteus* spp., and *Neospora* spp. Thus, a rational development of endoperoxide combination therapies is also advocated against other apicomplexan parasites.

Endoperoxide Resistance

Although no endoperoxide resistance has been confirmed, recent reports of *P. falciparum* strains with a reduced susceptibility to artemisinin [70], in addition to the generation of artemisinin-resistant mutants of *T. gondii* [12], may indicate the onset of endoperoxide resistance. Notably, evidence of four early clinical failure cases – two in India and two in Thailand – has been reported. In India, an adult still had parasitemia following a five-day parenteral treatment with artemether (total dose 480 mg, administered by intramuscular injection), while another adult reported recrudescence on day 14 after a seven-day treatment with artesunate at a dose of 13.3 mg kg^{-1} [71]. In Thailand, two children aged 2 and 5 years had positive blood smears on day 7 after a dose of 12 mg kg^{-1}, and one had persistent parasitemia throughout treatment [72]. Correspondingly, evidence of reduced *in vitro* susceptibility was reported from Vietnam between 1998 and 2001, where the 90% inhibitory concentration (IC$_{90}$) and IC$_{99}$ of artemisinin were reported to have doubled and quadrupled, respectively [73]. Additionally, in China between 1988 and 1999, the artesunate IC$_{50}$ was tripled and the IC$_{99}$ doubled [74]. Furthermore, a rise in the IC$_{50}$ for artemether was reported in 2002 in French Guyana [75], as well as a resistance to dihydroartemisinin [76]. Noedl *et al.* [77], in a study conducted in Battambang Province, Cambodia, reported *P. falciparum* strains with a decreased *in vitro* sensitivity to artesunate and delayed parasite clearance times, despite adequate plasma drug concentrations. Much slower clearance rates were reported in a study from Pailin, Western Cambodia, and Wang Pha, Western Thailand [78]. Furthermore, artemisinin-resistant mutants of *T. gondii* were reported to differ *in vitro* by three- to fivefold in their sensitivities compared to that of the wild-type parasites [12].

Although delayed clearance may be due to parasite, host, or other factors specific to this population, a high heritability is an indication of a genetic basis to resistance [79]. However, the mechanism of resistance to artemisinin and to endoperoxides remains unknown. Indeed, neither in *P. falciparum* nor in *T. gondii* have the genetic determinants of resistance been identified. Similarly, in *Babesia* spp., *Eimeria* spp., *Cryptosporidium* spp., *Haemoproteus* spp., and *Neospora* spp., neither action nor resistance to endoperoxides has received attention. Nonetheless, the difference in clearance rates was not explained by genetic polymorphisms in previously resistance-associated genes. In *Plasmodium* species, for example, an amplification in *pfmdr1* and mutations in the *P. falciparum* genes *pfcrt* (chloroquine resistance transporter gene), *pfmdr1* (multidrug resistance gene 1), or *pfserca* (a sarcoplasmic reticulum Ca^{2+} ATPase), all of which had been suggested previously as the targets of artemisinins, were not associated with clearance rates. *In vitro*, the W2, D6, and TM91C235 strains of *P. falciparum* have been reported to develop resistance to artelinic acid and artemisinin [80]. A stable and transmissible artemisinin resistance was also induced in a related species, *Plasmodium chabaudi chabaudi*, although there was no associ-ation with single nucleotide polymorphisms (SNPs) or copy numbers in the candidate genes *atp6* (encoding the SERCA-type Ca^{2+}-ATPase), *tctp* (encoding translationally controlled tumor protein, TCTP), *mdr1*, or *crt* [81]. Interestingly, the ubiquitin-specific protease-1 (*pfubp-1*) was found in *P. chabaudi chabaudi* to be associated with artemisinin resistance [82]. It is still unclear whether *pfubpp-1* is the only factor involved, or whether this extends to other *Plasmodium* species or to the entire Apicomplexa phylum.

Uhlemann *et al.* [65] had reported that, in *X. laevis*, an L263E point mutation in PfATPase could abolish inhibition by artemisinins, though this mutation has never been identified in field studies. Alternatively, the PfATPase 6S769N mutation was found to be strongly associated with artemether's high IC_{50}-values in French Guyana [75], and hence was proposed as a molecular marker of resistance against artemisinin. Indeed, the molecular characterization of several *P. falciparum* strains has revealed only the genetic diversity of pfATPase. Over 30 SNPs have been identified, of which only three codons – H431K, N569K, and A630S – were above 5% prevalent [83]. The diversity of PfATPase would provide the selection capacity for a differential susceptibility of *P. falciparum* to endoperoxides. Similarly, the artemisi-nin-resistant mutants of *T. gondii* failed to display any differences in either the mutation or expression of *SERCA*, *mdr*, or *tctp*, but rather were resistant to the induction of protein secretion from the micronemes, a calcium-dependent process triggered by artemisinin. A reduced susceptibility or resistance to endoperoxides has not been identified in other apicomplexan species. It remains to be seen if any reduced susceptibility or resistance is due to single or multiple genes involving known and unknown genes, and the contribution of each to the resistance phenotype must also be determined.

Conclusion

Endoperoxides currently comprise potentially broad, specific anti-apicomplexan agents. Efforts to improve the pharmacological limitations of this important class of agents should yield compounds with proven activity not only against *Plasmodium*, but also against *Toxoplasma*, *Babesia*, *Eimeria*, *Cryptosporidium*, *Haemoproteus*, and *Neospora* species. In particular, a short plasma half-life, a limited bioavailability, poor solubility in either oil or water, and recently reducing susceptibility are challenges that require urgent attention. As already evident, the first-generation endoperoxides are substantially more active against *Plasmodium* species than other apicomplexan parasites. Yet, as shown by D'Angelo *et al.* [18], carba derivatives that retain the artemisinin core showed remarkable activity against *T. gondii* tachyzoites. Therefore, such structure–optimization strategies that retain the artemisinin core remain to be explored for other apicomplexan parasites. Besides the artemisinoids, several endo-peroxides containing compounds from plant and marine sources hold much promise. Already, an understanding that the endoperoxide bridge is central to the mechanism of action has resulted in the synthesis of several second-generation endoperoxides. Both, natural and synthetic compounds should be improved through structure–activity relationships to increase their oxidative potential.

Notably, by increasing the oxidative content, Posner *et al.* [36] found that the acetal dimers of both artemisinin and dihydroartemisinin were significantly more active than artemisinin itself. In the future, endoperoxide derivatives with diverse cores, endoperoxide hybrids, and combinations will form the center of development for anti-apicomplexan agents.

References

1 Muraleedharan, K.M. and Avery, M.A. (2009) Progress in the development of peroxide based anti-parasitic agents. *Drug Discov. Today*, **14**, 793–803.

2 Klayman, D.L. (1985) Qinghaosu (artemisinin): an antimalarial drug from China. *Science*, **228**, 1049–1055.

3 Brossi, A., Venugopalan, B., Dominguez-Gerpe, L., Yeh, H.J., Flippen-Anderson, J.L., Buchs, P. *et al.* (1988) Arteether, a new antimalarial drug: synthesis and antimalarial properties. *J. Med. Chem.*, **31**, 645–650.

4 White, N.J. (1994) Artemisinin: current status. *Trans. R. Soc. Trop. Med. Hyg.*, **88**, S3–S4.

5 Lin, A.J., Klayman, D.L., and Milhous, W.K. (1987) Antimalarial activity of new water-soluble dihydroartemisinin derivatives. *J. Med. Chem.*, **30**, 2147–2150.

6 Haynes, R.K., Fugmann, B., Stetter, J., Rieckmann, K., Heilmann, H.D., Chan, H.W. *et al.* (2006) Artemisone – a highly active antimalarial drug of the artemisinin class. *Angew. Chem.*, **45**, 2082–2088.

7 Nagelschmitz, J., Voith, B., Wensing, G., Roemer, A., Fugmann, B., Haynes, R.K. *et al.* (2008) First assessment in humans of the safety, tolerability, pharmacokinetics, and ex vivo pharmacodynamic antimalarial activity of the new artemisinin derivative artemisone. *Antimicrob. Agents Chemother.*, **52**, 3085–3091.

8 Magueur, G., Crousse, B., Charneau, S., Grellier, P., Bégué, J.P., and Bonnet-Delpon, D. (2004) Fluoroartemisinin: trifluoromethyl analogues of artemether and artesunate. *J. Med. Chem.*, **47**, 2694–2699.

9 Jones-Brando, L., D'Angelo, J., Posner, G.H., and Yolken, R. (2006) In vitro inhibition of *Toxoplasma gondii* by four new derivatives of artemisinin. *Antimicrob. Agents Chemother.*, **50**, 4206–4208.

10 Holfels, E., McAuley, J., Mack, D., Milhous, W.K., and McLeod, R. (1994) In vitro effects of artemisinin ether, cycloguanil hydrochloride (alone and in combination with sulfadiazine), quinine sulfate, mefloquine, primaquine phosphate, trifluoperazine hydrochloride, and verapamil on *Toxoplasma gondii*. *Antimicrob. Agents Chemother.*, **38**, 1392–1396.

11 Berens, R.L., Krug, E.C., Nash, P.B., and Curiel, T.J. (1998) Selection and characterization of *Toxoplasma gondii* mutants resistant to artemisinin. *J. Infect. Dis.*, **177**, 1128–1131.

12 Nagamune, K., Moreno, S.N., and Sibley, L.D. (2007) Artemisinin-resistant mutants of *Toxoplasma gondii* have altered calcium homeostasis. *Antimicrob. Agents Chemother.*, **51**, 3816–3823.

13 Yao, J.M., Cai, Y., and Shi, X.H. (2003) Early treatment of *Toxoplasma gondii* infections with artemether in mice. *Zhongguo Ji Sheng Chong Xue Yu Ji Sheng Chong Bing Za Zhi*, **21**, 371.

14 Amato-Neto, V., Braz, L.M., Campos, R., Pinto, P.L., Moreira, A.A., Boulos, M., and Nahkle, M.C. (1991) Eventual effect of artemisinine on the experimental infection of mice by *Toxoplasma gondii*. *Rev. Soc. Bras. Med. Trop.*, **24**, 141–143.

15 Brun-Pascaud, M., Chau, F., Derouin, F., and Girard, P.M. (1996) Lack of activity of artemether for prophylaxis and treatment of *Toxoplasma gondii* and *Pneumocystis carinii* infections in rat. *Parasite*, **3**, 187–189.

16 Hencken, C.P., Jones-Brando, L., Bordón, C., Stohler, R., Mott, B.T., Yolken, R. *et al.* (2010) Thiazole, oxadiazole, and carboxamide derivatives of artemisinin are highly selective and potent inhibitors of *Toxoplasma gondii*. *J. Med. Chem.*, **53**, 3594–3601.

17 Dunay, I.R., Chan, W.C., Haynes, R.K., and Sibley, L.D. (2009) Artemisone and artemiside control acute and reactivated toxoplasmosis in a murine model. *Antimicrob. Agents Chemother.*, **53**, 4450–4456.

18 D'Angelo, J.G., Bordón, C., Posner, G.H., Yolken, R., and Jones-Brando, L. (2009) Artemisinin derivatives inhibit *Toxoplasma gondii* in vitro at multiple steps in the lytic cycle. *J. Antimicrob. Chemother.*, **63**, 146–150.

19 Nagai, A., Yokoyama, N., Matsuo, T., Bork, S., Hirata, H., Xuan, X. *et al.* (2003) Growth-inhibitory effects of artesunate, pyrimethamine, and pamaquine against *Babesia equi* and *Babesia caballi* in in vitro cultures. *Antimicrob. Agents Chemother.*, **47**, 800–803.

20 Marley, S.E., Eberhard, M.L., Steurer, F.J., Ellis, W.L., McGreevy, P.B., and Ruebush, T.K. 2nd (1997) Evaluation of selected antiprotozoal drugs in the *Babesia microti*-hamster model. *Antimicrob. Agents Chemother.*, **41**, 91–94.

21 Kumar, S., Gupta, A.K., Pal, Y., and Dwivedi, S.K. (2003) In vivo therapeutic efficacy trial with artemisinin derivative, buparvaquone and imidocarb dipropionate against *Babesia equi* infection in donkeys. *J. Vet. Med. Sci.*, **65**, 1171–1177.

22 Matsuu, A., Yamasaki, M., Xuan, X., Ikadai, H., and Hikasa, Y. (2008) In vitro evaluation of the growth inhibitory activities of 15 drugs against *Babesia gibsoni* (Aomori strain). *Vet. Parasitol.*, **157**, 1–8.

23 Meshnick, S.R., Taylor, T.E., and Kamchonwongpaisan, S. (1996) Artemisinin and the antimalarial endoperoxides: from herbal remedy to targeted chemotherapy. *Microbiol. Rev.*, **60**, 301–315. Review.

24 Kawai, S., Igarashi, I., Abgaandorjiin, A., Miyazawa, K., Ikadai, H., Nagasawa, H. *et al.* (1999) Ultrastructural characteristics of *Babesia caballi* in equine erythrocytes in vitro. *Parasitol. Res.*, **85**, 794–799.

25 Del Cacho, E., Gallego, M., Francesch, M., Quílez, J., and Sánchez-Acedo, C. (2010) Effect of artemisinin on oocyst wall formation and sporulation during *Eimeria tenella* infection. *Parasitol. Int.*, **59**, 506–511.

26 Allen, P.C., Lydon, J., and Danforth, H.D. (1997) Effects of components of *Artemisia annua* on coccidia infections in chickens. *Poult. Sci.*, **76**, 1156–1163.

27 Arab, H.A., Rahbari, S., Rassouli, A., Moslemi, M.H., and Khosravirad, F. (2006) Determination of artemisinin in *Artemisia sieberi* and anticoccidial effects of the plant extract in broiler chickens. *Trop. Anim. Health Prod.*, **38**, 497–503.

28 Fayer, R. and Ellis, W. (1994) Qinghaosu (artemisinin) and derivatives fail to protect neonatal BALB/c mice against *Cryptosporidium parvum* (Cp) infection. *J. Eukaryot. Microbiol.*, **41**, 14S.

29 Giacometti, A., Cirioni, O., and Scalise, G. (1996) In vitro activity of macrolides alone and in combination with artemisin, atovaquone, dapsone, minocycline or pyrimethamine against *Cryptosporidium parvum*. *J. Antimicrob. Chemother.*, **38**, 399–408.

30 Kim, J.T., Park, J.Y., Seo, H.S., Oh, H.G., Noh, J.W., Kim, J.H. *et al.* (2002) In vitro antiprotozoal effects of artemisinin on *Neospora caninum*. *Vet. Parasitol.*, **103**, 53–63.

31 Kaiser, M., Wittlin, S., Nehrbass-Stuedli, A., Dong, Y., Wang, X., Hemphill, A. *et al.* (2007) Peroxide bond-dependent antiplasmodial specificity of artemisinin and OZ277 (RBx11160). *Antimicrob. Agents Chemother.*, **51**, 2991–2993.

32 Liang, X.-T., Yu, D.-Q., Wu, W.L., and Deng, H.-C. (1979) The structure of yingzhousu A. *Huaxue Xuebao (Acta Chim. Sinica)*, **37**, 215–230.

33 Hofheinz, W., Bürgin, H., Gocke, E., Jaquet, C., Masciadri, R., Schmid, G. *et al.* (1994) Ro 42-1611 (arteflene), a new effective antimalarial: chemical structure and biological activity. *Trop. Med. Parasitol.*, **45**, 261–265.

34 Ploypradith, P. (2004) Development of artemisinin and its structurally simplified trioxane derivatives as antimalarial drugs. *Acta Trop.*, **89**, 329–42.

35 Tang, Y., Dong, Y., and Vennerstrom, J.L. (2004) Synthetic peroxides as antimalarials. *Med. Res. Rev.*, **24**, 425–48.

36 Jefford, C.W. (2001) Why artemisinin and certain synthetic peroxides are potent antimalarials. Implications for the mode of action. *Curr. Med. Chem.*, **8**, 1803–1826.

37 Posner, G.H., D'Angelo, J., O'Neill, P.M., and Mercer, A. (2006) Anticancer activity of artemisinin-derived trioxanes. *Exp. Opin. Ther. Patents*, **16**, 1665.

38 Opsenica, I., Opsenica, D., Lanteri, C.A., Anova, L., Milhous, W.K., Smith, K.S., and Solaja, B.A. (2008) New chimeric antimalarials with 4-aminoquinoline moiety linked to a tetraoxane skeleton. *J. Med. Chem.*, **51**, 6216–6219.

39 Wang, X., Dong, Y., Wittlin, S., Creek, D., Chollet, J., Charman, S.A. *et al.* (2007) Spiro- and dispiro-1,2-dioxolanes: contribution of iron(II)-mediated one-electron vs two-electron reduction to the activity of antimalarial peroxides. *J. Med. Chem.*, **50**, 5840–5847.

40 Martyn, D.C., Ramirez, A.P., Beattie, M.J., Cortese, J.F., Patel, V., Rush, M.A. *et al.* (2008) Synthesis of spiro-1,2-dioxolanes and their activity against *Plasmodium falciparum*. *Bioorg. Med. Chem. Lett.*, **18**, 6521–6524.

41 Creek, D.J., Charman, W.N., Chiu, F.C., Prankerd, R.J., Dong, Y., Vennerstrom, J.L., and Charman, S.A. (2008) Relationship between antimalarial activity and heme alkylation for spiro- and dispiro-1,2,4-trioxolane antimalarials. *Antimicrob. Agents Chemother.*, **52**, 1291–1296.

42 Posner, G.H., Jeon, H.B., Ploypradith, P., Paik, I.H., Borstnik, K., Xie, S., and Shapiro, T.A. (2002) Orally active, water-soluble antimalarial 3-aryltrioxanes: short synthesis and preclinical efficacy testing in rodents. *J. Med. Chem.*, **45**, 3824–3828.

43 Slade, D., Galal, A.M., Gul, W., Radwan, M.M., Ahmed, S.A., Khan, S.I. *et al.* (2009) Antiprotozoal, anticancer and antimicrobial activities of dihydroartemisinin acetal dimers and monomers. *Bioorg. Med. Chem.*, **17**, 7949–7957.

44 Pacorel, B., Leung, S.C., Stachulski, A.V., Davies, J., Vivas, L., Lander, H. *et al.* (2010) Modular synthesis and in vitro and in vivo antimalarial assessment of C-10 pyrrole Mannich base derivatives of artemisinin. *J. Med. Chem.*, **53**, 633–640.

45 Taylor, D.K., Avery, T.D., Greatrex, B.W., Tiekink, E.R., Macreadie, I.G., Macreadie, P.I. *et al.* (2004) Novel endoperoxide antimalarials: synthesis, heme binding, and antimalarial activity. *J. Med. Chem.*, **47**, 1833–1839.

46 Walsh, J.J., Coughlan, D., Heneghan, N., Gaynor, C., and Bell, A. (2007) A novel artemisinin-quinine hybrid with potent antimalarial activity. *Bioorg. Med. Chem. Lett*, **17**, 3599–3602.

47 Capela, R., Oliveira, R., Gonçalves, L.M., Domingos, A., Gut, J., Rosenthal, P.J. *et al.* (2009) Artemisinin-dipeptidyl vinyl sulfone hybrid molecules: design, synthesis and preliminary SAR for antiplasmodial activity and falcipain-2 inhibition. *Bioorg. Med. Chem. Lett*, **19**, 3229–3232.

48 Santos, M.M. and Moreira, R. (2007) Michael acceptors as cysteine protease inhibitors. *Mini Rev. Med. Chem.*, **7**, 1040–1050. Review.

49 Dechy-Cabaret, O., Benoit-Vical, F., Robert, A., and Meunier, B. (2000) Preparation and antimalarial activities of "trioxaquines," new modular molecules with a trioxane skeleton linked to a 4-aminoquinoline. *Chembiochem.*, **1**, 281–283.

50 Jefford, C.W. (2007) New developments in synthetic peroxidic drugs as artemisinin mimics. *Drug Discov. Today*, **12**, 487–495. Review.

51 Singh, C., Malik, H., and Puri, S.K. (2004) Synthesis and antimalarial activity of

a new series of trioxaquines. *Bioorg. Med. Chem.*, **12**, 1177–1182.

52 Kawanishi, M., Kotoku, N., Itagaki, S., Horii, T., and Kobayashi, M. (2004) Structure-activity relationship of anti-malarial spongean peroxides having a 3-methoxy-1,2-dioxane structure. *Bioorg. Med. Chem.*, **12**, 5297–5307.

53 Murakami, N., Kawanishi, M., Itagaki, S., Horii, T., and Kobayashi, M. (2002) New readily accessible peroxides with high anti-malarial potency. *Bioorg. Med. Chem. Lett.*, **12**, 69–72.

54 O'Neill, P.M., Barton, V.E., and Ward, S.A. (2010) The molecular mechanism of action of artemisinin – the debate continues. *Molecules*, **15**, 1705–1721.

55 Hartwig, C.L., Rosenthal, A.S., D'Angelo, J., Griffin, C.E., Posner, G.H., and Cooper, R.A. (2009) Accumulation of artemisinin trioxane derivatives within neutral lipids of *Plasmodium falciparum* malaria parasites is endoperoxide-dependent. *Biochem. Pharmacol.*, **77**, 322–336.

56 del Pilar Crespo, M., Avery, T.D., Hanssen, E., Fox, E., Robinson, T.V., Valente, P. *et al.* (2008) Artemisinin and a series of novel endoperoxide antimalarials exert early effects on digestive vacuole morphology. *Antimicrob. Agents Chemother.*, **52**, 98–109.

57 Eckstein-Ludwig, U., Webb, R.J., Van Goethem, I.D., East, J.M., Lee, A.G., Kimura, M. *et al.* (2003) Artemisinins target the SERCA of *Plasmodium falciparum*. *Nature*, **424**, 957–961.

58 Bhisutthibhan, J., Pan, X.Q., Hossler, P.A., Walker, D.J., Yowell, C.A., Carlton, J. *et al.* (1998) The *Plasmodium falciparum* translationally controlled tumor protein homolog and its reaction with the antimalarial drug artemisinin. *J. Biol. Chem.*, **273**, 16192–16198.

59 Kannan, R., Sahal, D., and Chauhan, V.S. (2002) Heme-artemisinin adducts are crucial mediators of the ability of artemisinin to inhibit heme polymerization. *Chem. Biol.*, **9**, 321–332.

60 Wang, D.-Y. and Wu, Y.L. (2000) A possible antimalarial action mode of qinghaosu (artemisinin) series compounds.

Alkylation of reduced glutathione by C-centered primary radicals produced from antimalarial compound qinghaosu and 12-(2,4-dimethoxyphenyl)-12-deoxoqinghaosu. *Chem. Commun.*, 2193–2194.

61 Pandey, A.V., Tekwani, B.L., Singh, R.L., and Chauhan, V.S. (1999) Artemisinin, an endoperoxide antimalarial, disrupts the hemoglobin catabolism and heme detoxification systems in malarial parasite. *J. Biol. Chem.*, **274**, 19383–19288.

62 Li, W., Mo, W., Shen, D., Sun, L., Wang, J., Lu, S. *et al.* (2005) Yeast model uncovers dual roles of mitochondria in action of artemisinin. *PLoS Genet.*, **1**, e36.

63 Robert, A., Benoit-Vical, F., Claparols, C., and Meunier, B. (2005) The antimalarial drug artemisinin alkylates heme in infected mice. *Proc. Natl Acad. Sci. USA*, **102**, 13676–13680. Erratum in: *Proc. Natl Acad. Sci. USA*, **103**, 3943.

64 Bhisutthibhan, J. and Meshnick, S.R. (2001) Immunoprecipitation of [(3)H] dihydroartemisinin translationally controlled tumor protein (TCTP) adducts from *Plasmodium falciparum*-infected erythrocytes by using anti-TCTP antibodies. *Antimicrob. Agents Chemother.*, **45**, 2397–2399.

65 Uhlemann, A.C., Cameron, A., Eckstein-Ludwig, U., Fischbarg, J., Iserovich, P., Zuniga, F.A. *et al.* (2005) A single amino acid residue can determine the sensitivity of SERCAs to artemisinins. *Nat. Struct. Mol. Biol.*, **12**, 628–629.

66 Jambou, R., Legrand, E., Niang, M., Khim, N., Lim, P., Volney, B. *et al.* (2005) Resistance of *Plasmodium falciparum* field isolates to in-vitro artemether and point mutations of the SERCA-type PfATPase6. *Lancet*, **366**, 1960–1963.

67 Wang, J., Huang, L., Li, J., Fan, Q., Long, Y., Li, Y., and Zhou, B. (2010) Artemisinin directly targets malarial mitochondria through its specific mitochondrial activation. *PLoS One*, **5**, e9582.

68 Olliaro, P.L., Haynes, R.K., Meunier, B., and Yuthavong, Y. (2001) Possible modes of action of the artemisinin-type compounds. *Trends Parasitol.*, **17**,

122–126. Review. Erratum in *Trends Parasitol.*, **17**, 268.

69 World Health Organization (2005) Susceptibility of *Plasmodium falciparum* to antimalarial drugs: report on global monitoring: 1996–2004.

70 Dondorp, A.M., Yeung, S., White, L., Nguon, C., Day, N.P., Socheat, D., and von Seidlein, L. (2010) Artemisinin resistance: current status and scenarios for containment. *Nat. Rev. Microbiol.*, **8**, 272–280.

71 Gogtay, N.J. (2000) Probable resistance to parenteral arthemether in Plasmodium falciparum: case reports from Mumbai (Bombay), India. *Ann. Trop. Med. Parasitol.*, **94**, 519–520.

72 Luxemburger, C. (1998) Two patients with falciparum malaria and poor *in vivo* responses to artesunate. *Trans. R. Soc. Trop. Med. Hyg.*, **92**, 668–669.

73 Huong, N.M., Hewitt, S., Davis, T.M.E., Dao, L.D., Toan, T.Q. *et al.* (2001) Resistance of *Plasmodium falciparum* to antimalarial drugs in highly endemic area of southern Vietnam: a study in vivo and in vitro. *Trans. R. Soc. Trop. Med. Hyg.*, **95**, 325–329.

74 Henglin, Y., Deguan, L., Yaming, Y., Bo, F., Pinfang, Y. *et al.* (2003) Changes in susceptibility of *Plasmodium falciparum* to artesunate in vitro in Yunnan province, China. *Trans. R. Soc. Trop. Med. Hyg.*, **97**, 226–228.

75 Legrand, E., Volney, B., Meynard, J.B., Mercereau-Puijalon, O., and Esterre, P. (2008) In vitro monitoring of *Plasmodium falciparum* drug resistance in French Guiana: a synopsis of continuous assessment from 1994 to 2005. *Antimicrob. Agents Chemother*, **52**, 288–298.

76 Legrand, E., Volney, B., Meynard, J.B., Esterre, P., and Mercereau-Puijalon, O. (2007) Resistance to dihydroartemisinin. *Emerg. Infect. Diseases*, **13**, 808–809.

77 Noedl, H., Socheat, D., and Satimai, W. (2009) Artemisinin-resistant malaria in Asia. *N. Engl. J. Med.*, **361**, 540–541.

78 Dondorp, A.M., Nosten, F., Yi, P., Das, D., Phyo, A.P., Tarning, J. *et al.* (2009) Artemisinin resistance in *Plasmodium falciparum* malaria. *N. Engl. J. Med.*, **361**, 455–467. Erratum in: *N. Engl. J. Med.*, **361**, 1714.

79 Anderson, T.J., Nair, S., Nkhoma, S., Williams, J.T., Imwong, M., Yi, P., Socheat, D. *et al.* (2010) High heritability of malaria parasite clearance rate indicates a genetic basis for artemisinin resistance in western Cambodia. *J. Infect. Dis.*, **201**, 1326–1330.

80 Chavchich, M., Gerena, L., Peters, J., Chen, N., Cheng, Q., and Kyle, D.E. (2010) Role of *pfmdr1* amplification and expression in induction of resistance to artemisinin derivatives in *Plasmodium falciparum*. *Antimicrob. Agents Chemother.*, **54**, 2455–2464.

81 Afonso, A., Hunt, P., Cheesman, S., Alves, AC., Cunha, C.V., do Rosário, V., and Cravo, P. (2006) Malaria parasites can develop stable resistance to artemisinin but lack mutations in candidate genes atp6 (encoding the sarcoplasmic and endoplasmic reticulum Ca^{2+} ATPase), tctp, mdr1, and cg10. *Antimicrob. Agents Chemother.*, **50**, 480–489.

82 Hunt, P., Afonso, A., Creasey, A., Culleton, R., Sidhu, A.B., Logan, J., Valderramos, S.G. *et al.* (2007) Gene encoding a deubiquitinating enzyme is mutated in artesunate- and chloroquine-resistant rodent malaria parasites. *Mol. Microbiol.*, **65**, 27–40.

83 Dahlström, S., Veiga, M.I., Ferreira, P., Mårtensson, A., Kaneko, A., Andersson, B., Björkman, A., and Gil, J.P. (2008) Diversity of the sarco/endoplasmic reticulum Ca(2 +)-ATPase orthologue of *Plasmodium falciparum* (PfATP6). *Infect. Genet. Evol.*, **8**, 340–345.

23
Plasmodium Hsp90 as an Antimalarial Target

G. Sridhar Prasad, Nicholas D.P. Cosford, and Sailen Barik[*]

Abstract

Antimalarial drug resistance has emerged as one of the greatest challenges facing malaria control today, creating an urgent need for new drugs. Accumulating evidence suggests that *Plasmodium falciparum* (Pf), the causative agent of pernicious malaria, relies heavily on its chaperones and cochaperones for growth and survival in human red blood cells. Underscoring their functional importance, the chaperone genes in fact constitute as much as 2% of the Pf genome. The Hsp90 chaperone holds a special place in all living cells, due to its multiple clients. Given the high degree of sequence homology between the plasmodial and human Hsp90, and the important role of Hsp90 in both species, it is crucial to develop inhibitors that are specific toward *Plasmodium* in order to avoid human toxicity. To this end, a highly specific drug-design approach – differential fragment-based screening (DFS) – has been adopted which can be combined with biological studies and lead optimization to probe the active sites that are different between the *Plasmodium* and human orthologs. In this chapter, proof-of-concept examples of novel drug-like inhibitors of PfHsp90 that evolved from crystal structure, synthetic chemistry and structure-based compound–Hsp90 interaction analysis, are presented. The results should form the foundation for the discovery and optimization of highly specific and potent PfHsp90 inhibitors with improved antimalarial efficacy and minimal side effects.

Introduction

The heat shock response, first described in 1962 [1, 2], triggers the induction of heat shock proteins (HSPs). This family of proteins is modulated also by other forms of stress, such as nutrient deprivation, oxidative stress, and exposure to certain

---

[*]Corresponding author

Apicomplexan Parasites. Edited by Katja Becker
Copyright © 2011 WILEY-VCH Verlag GmbH & Co. KGaA, Weinheim
ISBN: 978-3-527-32731-7

chemicals – conditions that can promote protein denaturation [3–7]. The HSPs generally form multisubunit chaperone complexes, which help their client protein fold properly, thus affecting global cellular pathways and essentially all organelles. Mammalian HSPs, including those of humans, have been classified into several families according to their molecular size, with Hsp90 being the largest and by far the most extensively studied [3–5]. In this chapter, the status of Hsp90 in *Plasmodium falciparum*, the causative agent of lethal malaria, is reviewed. Attention is also focused on how HSPs can serve as potential targets for antimalarial drug design.

Molecular Chaperones and *P. falciparum* Hsp90

Recent years have seen significant emphasis placed on chaperone function in human disease. Whilst chaperones have been implicated in various human ailments, such as cancer, neurodegenerative diseases, and infections [4, 5, 8, 9] they are, more importantly, also being viewed as lucrative drug targets [4, 8, 9]. Of all the molecular chaperones, Hsp90 has been the focus of multiple drug discovery programs by pharmaceutical agencies and academia–industry collaborative efforts [8–14]. This situation has occurred in part because Hsp90 is the most prevalent chaperone present in the cell, representing 1–2% of all protein in the unstressed cell, and as much as 6% in the stressed cell [3–5]. It is also because a number of Hsp90's client proteins are essential signaling proteins [4, 15, 16]; these include ErbB2/Her-2, EGFR, Hif1a, c-Met, Akt/PKB, Raf-1, Cdk-1 and 4, Aurora B, CHK1, CKII, and p53, many of which are also known targets for cancer chemotherapy [17, 18].

Biochemistry and Structure of Hsp90

Detailed studies have revealed four structural domains of Hsp90 in the following order:

- The N-terminal domain (NTD), which binds ATP ($\sim$25 kDa) and shows similarity to members of the ATPase/kinase GHKL (Gyrase, Hsp90, Histidine Kinase, MutL) superfamily [3, 4, 19–21]. Seven amino acid residues of NTD are directly involved in the interaction with ATP. In addition, Mg^{2+} and several water molecules form bridging electrostatic and hydrogen-bonding interactions, respectively, between Hsp90 and ATP. In addition, Glu33 is required for ATP hydrolysis.
- The charged linker region that links the N terminus to the middle domain.
- The middle domain (MD).
- The C-terminal domain (CTD), which serves as the second, cryptic ATP-binding domain.

As detailed later, these functional domains and motifs also appear to be conserved in the *P. falciparum* Hsp90.

Hsp90 Inhibitors

In view of the important biological role of Hsp90 in maintaining the structural integrity of a number of client proteins – which, by themselves, are involved in cellular and signal transduction processes – it is fair to postulate that the inhibition of Hsp90 would circumvent the need for inhibiting multiple drug targets associated with many pathological conditions, such as ischemia and reperfusion, infections, neurodegenerative disease, and cancer [4–13]. The classic inhibitor of Hsp90 is geldanamycin (GA) [22], a benzoquinone ansamycin antibiotic that occupies the N-terminal ATP-binding domain of Hsp90 [5, 9, 20–22]. In support of the promising anticancer role of Hsp90 [23], GA induces the degradation of oncoproteins such as v-Src, Bcr-Abl, and p53 [24] preferentially over their normal cellular counterparts. Despite its potent antitumor potential, GA presents several major drawbacks as a drug candidate (e.g., hepatotoxicity), and this has led to the development of its analogs, in particular those containing derivations at the 17 position, namely 17-*N*-allylamino-17-demethoxygeldanamycin (17AAG), and 17-desmethoxy-17-*N*,*N*-dimethoxyaminoethylaminogeldanamycin (17DMAG) [25].

The highest developmental stages of the various Hsp90 compounds are listed in Table 23.1; of the 18 Hsp90 inhibitors currently undergoing clinical trials, eight belong to the GA class and the other 10 are synthetic. Interestingly, all but one of the compounds are being evaluated for oncology indications. In spite of the growing evidence regarding the critical role of Hsp90 in the life-cycle of *P. falciparum*, very little effort has been exerted in pursuing this important drug target as a novel antimalarial therapeutic. Given the fact that a growing number of Hsp90 inhibitors are synthetic, a fragment-based screening approach, combined with rational structure-based drug discovery, would provide an excellent opportunity for the rapid identification of novel hits. Moreover, these compounds could be optimized by synthetic routes, ultimately to discover and develop potent compounds capable of targeting this novel mechanism of action to inhibit Hsp90.

Structure-Based Drug Discovery

Advances in both biology and crystallography have made it possible to generate high-resolution three-dimensional (3-D) structures of protein–ligand complexes, significantly impacting on the drug discovery process [26]. Notably, the drug candidates discovered and developed using this structural information have been predominantly inhibitors of either the active or allosteric site of the enzyme. Thus, fragment-based screening has been effectively combined with structure-based methods for the discovery and development of multiple drug candidates, and these have entered clinical trials for the treatment of type 2 diabetes mellitus [27] and HIV-1 AIDS [28–31]. Given the recent success of multiple human Hsp90 inhibitors entering clinical trials, and which have been discovered and developed by using a combination of fragment and structure-based approaches [32, 33], this method may also be applied successfully to the discovery and development of highly specific PfHsp90 inhibitors for the treatment of malaria.

Table 23.1 Current status of Hsp90 inhibitors.[a]

Highest development status	Discontinued	No development reported	Suspended	Discovery	Phase I clinical	Phase II clinical	Phase III clinical	Pre-registration	Total
Cancer	1	6	2	14	7	3	2	1	36
Hematological neoplasm	–	–	–	–	3	2	–	–	5
Solid tumor	1	–	2	1	3	3	–	–	10
Breast tumor	–	1	–	–	–	2	2	1	6
Gastrointestinal stromal tumor	–	–	–	–	–	2	1	–	3
Non-small-cell lung cancer	–	–	–	–	–	2	1	–	3
Multiple myeloma	–	–	–	–	–	1	2	–	3
Prostate tumor	–	2	–	–	–	–	1	–	3
Melanoma	–	–	–	–	–	–	2	–	2
Fungal infection	–	1	–	–	–	–	–	1	2
Total	2	12	3	15	7	3	2	1	

a) 2010 Thompson Reuter.

Fragment-Based Drug Discovery

In drug development, a "fragment" is the term typically applied to a low-molecular-weight chemical (ca. <200 Da) which is eventually developed into a "lead compound" to be tested as a drug. Fragment-based drug discovery (FBDD), which represents an efficient means of identifying novel leads for targets, begins with low-molecular-weight chemicals (100–200 Da) which, due to their low complexity, will have high hit rates (3–4%). These hits are then expanded into novel leads that possess the desired absorption–distribution–metabolism–excretion (ADMET) properties. Such small molecular fragments will have a weaker affinity, because fewer atoms necessarily means a lower free energy of binding. However, they are typically highly efficient – that is, they have a high free energy of binding per heavy atom. Because of their modest affinity, fragments typically cannot be screened using standard high-through-put screening (HTS) techniques, as the required high concentration of ligands will interfere with the assay and lead to false-positive results. Consequently, alternative screening methods, such as nuclear magnetic resonance (NMR) [29], surface plasmon resonance (SPR) [30, 31], biochemical assays [32], and X-ray crystallography [33] are currently used to detect the binding of the small fragments to protein targets. Of these methods, crystallography has provided the most valuable information regarding fragment binding and key interactions [34]. Over the past decade, leads developed through fragment-based screening have led to almost 48 programs reaching various stages of development [35]. These examples demonstrate that clinically useful compounds can be generated using a fragment-based approach, by increasing the success rate and reducing the time and cost of modern drug discovery. However, given the generally high degree of sequence homology between the *Plasmodium* and human Hsp90 (Figure 23.1), and the important role of Hsp90 in both species, it is crucial to develop inhibitors that are specific toward *Plasmodium*, in order to avoid human toxicity. In order to address this critical issue of selectivity and off-species interaction during the process of drug discovery, a novel and unique approach has been developed, termed differential fragment-based screening (DFS).

Differential Fragment-Based Screening

In general, a significant proportion of compounds that exhibit excellent *in vitro* properties will fail during *in vivo* studies and clinical trials, due to a lack of specificity. One approach to overcoming this problem would be to build such specificity into the compound during the very early stages of lead identification. This possibility is offered by DFS, in which the same set of fragments is screened against two or more drug targets, and against closely related off-targets, such that *specificity* can be achieved [27]. Likewise, DFS can also be used to develop inhibitors that are capable of targeting multiple drug targets simultaneously, which may result in an increased clinical *efficacy*.

Due to high sequence homology, an overall similarity in the 3-D structures of both species of Hsp90 is to be expected (Figure 23.1). Consequently, in order to achieve

```
hHsp90-A1   --------------------------------------------------------------------
PfHsp90-1   --------------------------------------------------------------------
PfHsp90-2   ---MKLNNIYSFFFLFFVLCVIQENVRRVLCDSSVEGDKGPSDDVSDSSGEKKEV----------  52
PfHsp90-3   MQNVYVGNKIKFIILYFFCVLFLKDYERSEAFNLARTTEKLNYILNYKTPNRYDLNNNVNKLFFE  65
PfHsp90-4   ---MKCSTQLSRRLSNFEGKGTFNKSAFYNCTREKCSIVCLRKKMN-------------------  43


hHsp90-A1   ----------------MPEETQTQDQPMEEEEVETFAFQAEIAQLMSLIINTFYSNKEIFLREL  48
PfHsp90-1   ----------------------------MSTETFAFNADIRQLMSLIINTFYSNKEIFLREL  34
PfHsp90-2   --------------KRDRDTLEEIEEGEKPTESMESHQYQTEVTRLMDIIVNSLYTQKEVFLREL  103
PfHsp90-3   KQKKKIEFSRKPLNSFNEDVKTIREDISSDSSPVEKYNFKAEVNKVMDIIVNSLYTDKDVFLREL  130
PfHsp90-4   ------------VELKKICEISKMNKRNYSSECENYEFKAETKKLLQIVAHSLYTDKEVFIREL  95
                           *.. ::::  ::::.:: :::*::*::*:***


hHsp90-A1   ISNSSDALDKIRYESLTD----------------------------------PSKLDSGKELH  77
PfHsp90-1   ISNASDALDKIRYESITD----------------------------------TQKLSAEPEFF  63
PfHsp90-2   ISNAADALEKIRFLSLSD----------------------------------ESVLGEEKKLE  132
PfHsp90-3   ISNASDACDKKRIILENNKLIKDAEVVTNEEIKNETEKEKTENVNESTDKKENVEEEKNDIKKLI  195
PfHsp90-4   ISNSSDAIEKLRFLLQSG----------------------------------NIKASENITFH  124
            ***::**  :* *          ..                             .  :


hHsp90-A1   INLIPNKQDRTLTIVDTGIGMTKADLINNLGTIAKSGTKAFMEALQAG-ADISMIGQFGVG----  137
PfHsp90-1   IRIIPDKTNNTLTIEDSGIGMTKNDLINNLGTIARSGTKAFMEAIQAS-GDISMIGQFGVG----  123
PfHsp90-2   IRISANKEKNILSITDTGIGMTKVDLINNLGTIAKSGTSNFLEAISKSGGDMSLIGQFGVNDKKL  197
PfHsp90-3   IKIKPDKEKKTLTITDNGIGMDKSELINNLGTIAQSGTAKFLKQIEEGKADSNLIGQFGVG----  256
PfHsp90-4   IKVSTDENNNLFIIEDSGVGMNKEEIIDNLGTIAKSGSLNFLKKLKEQKESINRNNNINKNEIDN  189
            *.: ..:: .. :  *  *.*:** *  ::*:******;**:  *:: :.    . .  .::. .


hHsp90-A1   --------------------------FYSAYLVAEKVTVITKHNDDEQ---YAWESSAGGS  169
PfHsp90-1   --------------------------FYSAYLVADHVVVISKNNDDEQ---YVWESAAGGS  155
PfHsp90-2   MDLISKYSQFIQFPIYLLHENVYTEEVLAGFYSAFLVADKVIVYTKNNDDEQ---YIWESTADAK  259
PfHsp90-3   --------------------------FYSSFLVSNRVEVYTK-KEDQI---YRWSSDLKGS  288
PfHsp90-4   KINNNNNMDVPIEGNEKSQEGDIIGQFGVGFYSSFVVSNKVEVFTRSYDNNSSKGYHWVSYGNGT  254
                              ***::*:::* *:  *: :::      * * *   ..


hHsp90-A1   FTV-----RTDTG-EPMGRGTKVILHLKEDQTEYLEERRIKEIVKKHSQFIGYPITLFVEKERDK  228
PfHsp90-1   FTV-----TKDETNEKLGRGTKIILHLKEDQLEYLEEKRIKDLVKKHSEFISFPIKLYCERQNEK  215
PfHsp90-2   FTI-----YKDPRGATLKRGTRISLHLKEDATNLLD--IAKDMVND------------------  297
PfHsp90-3   FSVNEIKKYDQEYDDIKGSGTKIILHLKEECDEYLEDYKLKELIKKYSEFIKFPIEIWSEK----  348
PfHsp90-4   FTL--------KEVDNIPKGTKIICHLKDSCKEFSNIQNVQKIVEKFSSFINFPVYVLKKKKILQ  311
            *::            **::  ***:. :


hHsp90-A1   EVSDDEAEE------------------------KEDKEEEKEKEEKESEDK----PEIEDV  261
PfHsp90-1   EITASEEEEGEGEGEREGEEEEEKKKKTGEDKNADESKEENEDEEKKEDNEEDDNKTDHPKVEDV  280
PfHsp90-2   ----------------------------------------------------------PNYDSV  304
PfHsp90-3   ----------------------------------------------------------IDYERV  354
PfHsp90-4   TRKDIEQNELN---------DHTQQNELNHHNDQNDHTQQNELNHHNDQNDHTQQNELNHHNDQN  367
            :    :.::::.                                              . :


hHsp90-A1   GSDEEEEKKDGDKKKKKKIKEKYIDQEE-------------------------------------  289
PfHsp90-1   TEELENAEKKKKEKRKKKIHTVEHEWEE-------------------------------------  308
PfHsp90-2   KVEETDDPN----KKTRTVEKKVKKWTL-------------------------------------  328
PfHsp90-3   P-DDSVSLKDGDKMKMKTITKRYHEWEK-------------------------------------  381
PfHsp90-4   DHTQQNELNHHVEHHKNVVENKNDEYEKNMDNRTSDNIKPESLSKQNDYENMNQCNNEETVLNRT  432
                    :      :  . :


hHsp90-A1   -------------LNKTKPIWTRNPDDITNEEYGEFYKSLTN------DWEDHLAVKHFSVEGQL  335
PfHsp90-1   -------------LNKQKPLWMRKPEEVTNEEYASFYKSLTN------DWEDHLAVKHFSVEGQL  354
PfHsp90-2   -------------MNEQRPIWLRSPKELKDEDYKQFFSVLSG------YNDQPLYHIHFFAEGEI  374
PfHsp90-3   -------------INVQLPIWKQDEKSLTENDYYSFYKNTFK------AYDDPLAYVHFNVEGQI  427
PfHsp90-4   KELNEEHIVEEILVNKQKPLWCKDN--VTEEEHRHFFHFLNKNKSYNEDNKSYLYNMLYKTDAPL  495
                        :*  *: * :.   :.:::: *:          .. *   : .:. :
```

Figure 23.1 Similarities and differences among *P. falciparum* and the representative human Hsp90 sequences. For the gray highlighted residues, please refer to the legend of Figure 23.6.

```
hHsp90-A1   EFRALLFVPRRAPFDLFENRKKKNN--IKLYVRRVFIMDNCEELIPEYLNFIRGVVDSEDLPLNI 398
PfHsp90-1   EFKALLFIPKRAPFDMFENRKKRNN--IKLYVRRVFIMDDCEEIIPEWLNFVKGVVDSEDLPLNI 417
PfHsp90-2   EFKCLIYIPSKAPSMNDQLYSKQNS--LKLYVRRVLVADEFVEFLPRYMSFVKGVVDSDDLPLNV 437
PfHsp90-3   SFNSILYIPGSLPWELSKNMFDEESRGIRLYVKRVFINDKFSESIPRWLTFLRGIVDSENLPLNV 492
PfHsp90-4   SIKSVFYIPEEAPSRLFQ---QSNDIEISLYCKKVLVKKNADNIIPKWLYFVKGVIDCEDMPLNI 557
            .:..::::*    *      :     : :** ::*:: ..  : :*.:: *::*::*.:::***:

hHsp90-A1   SREMLQQSKILKVIRKNLVKKCLELFTELAED-KEN----------------------------- 433
PfHsp90-1   SRESLQQNKILKVIKKNLIKKCLDMFSELAEN-KEN----------------------------- 452
PfHsp90-2   SREQLQQNKILKAVSKRIVRKILDTFHKLYKEGKKNKETLRSELENETDEEKKKEITKKLSEPST 502
PfHsp90-3   GREILQKSKMLSIINKRIVLKSISMMKGLKETGGDK---------------------------- 528
PfHsp90-4   SRENMQDSSLINKLSRVVVSKILKTLEREADINEEK---------------------------- 593
            .** :*...::. : : :: * :. :         .  .:

hHsp90-A1   YKKFYEQFSKNIKLGIHEDSQN---RKKLSELLRYYTSASGDEMVSLKDYCTRMKENQKHIYYIT 495
PfHsp90-1   YKKFYEQFSKNLKLGIHEDNAN---RTKITELLRFQTSKSGDEMIGLKEYVDRMKENQKDIYYIT 514
PfHsp90-2   YKLIYKEYRKFLKSGCYEDDIN---RNKIAKLLLFKTMQYP-KSISLDTYIEHMKPDQKFIYYAS 563
PfHsp90-3   WTKFLNTFGKYLKIGVVEDKEN---QEEIASLVEFYSINSGDKKTDLDSYIENMKEDQKCIYYIS 590
PfHsp90-4   YLKFYKNYNYNLKEGVLEDSNKNHYKNSLMNLLRFYSINQN-KFISLKQYVNNFRNNQKNIYYFS 657
            : : : :  :* *  **. :   : .: .*: : :    : .*. *  .:: :** *** :

hHsp90-A1   GETKDQVANSAFVERLRKHGLEVIYMIEPIDEYCVQ--QLKEFEGKTLVSVTK-------EGLEL 551
PfHsp90-1   GESINAVSNSPFLEALTKKGFEVIYMVDPIDEYAVQ--QLKDFDGKKLKCCTK-------EGLDI 570
PfHsp90-2   GDSYEYLAKIPQLQIFKKKNIDVLFLTESVDESCIQ--RVQEYEGKKFKSIQK-------GEISF 619
PfHsp90-3   GENKKTAQNSPSLEKLKALNYDVLFSLEPIDEFCLSSLTVNKYKGYEVLDVNK-------ADLKL 648
PfHsp90-4   ANDKNVALNSPYMEPFKKQNIDVLLLLEEIDEFVLMN--LQTYKDAKFVSIDTSQNEDFDEAVLN 720
            .:  .  : .:: .  .:*: : :**  :  ::.. .   .    :

hHsp90-A1   PEDEEEKKKQ--EEKKTKFENLCKIMKDILEKKVEKVVVSNRLVTSPCCIVTSTYG-WTANMERI 613
PfHsp90-1   DDSEEAKKDF--ETLKAEYEGLCKVIKDVLHEKVEKVVVGQRITDSPCVLVTSEFG-WSANMERI 632
PfHsp90-2   ELTEEEEKKKE--QQMQKMYKALIDVISDTLKNQIFKVEISRRLVDAPCAVVSTEWG-LSGQMEKL 681
PfHsp90-3   KKENDQNKSDSLDKQKMEYEILCRWLHNKFSHKVHEVRISDRLINSPALLVQGEMG-MSPSMQKY 712
PfHsp90-4   TNKNDNEKKQSIFFNDEQKKELQAYFKQVLGSKCSDVKFSERLTTSPAVVTGFLSPTLRKVMKAT 785
            ::  :*.    .   :  * ::: : .*  ..  *: :*. :.       *:

hHsp90-A1   MKAQALRDNSTMGYMAA-----KKHLEINPDHSIIETLRQKAEADKNDKSVKDLVILLYETALLS 673
PfHsp90-1   MKAQALRDNSMTSYMLS-----KKIMEINARHPIISALKQKADADKSDKTVKDLIWLLFDTSLLT 692
PfHsp90-2   MKMNVS-NSDQIKAMSG-----QKILEINPNHPIMIDLLKRSVTNPKDLELTNSIKIMYQSAKLA 740
PfHsp90-3   MKQQATAQGISENEMFGGQSANQPVLEINPNHFIIKQLNHLIQIDKMNLQNSEIAEQIFDVASMQ 777
PfHsp90-4   MKNSDFNDNTNNSNNMNMFQNLPATLELNPSHTIVTSIYHLKNTN--QEVAKLLVQQLYDNACIA 848
            **  .  :.            :*:*. * *: : :    : :   .      ::: : :

hHsp90-A1   SGFSLEDPQTHANRIYRMIKLGLGID-------------------EDDPTADDTSAAVTEEMPP 718
PfHsp90-1   SGFALEEPTTFSKRIHRMIKLGLSID-------------------EEE--------NNDIDLPP 721
PfHsp90-2   SGFDLEDTADLAQIVYDHINQKLGVDHINLKIDDLDPSIFETKKIEDENDSSKFEEEINIDDEIQK 805
PfHsp90-3   GGYTIDDTGLFAKRVIGMMEKNAEQYLMNVQSNISNNTLNNNTSGSEMPQNNSPNELQSEMKSTN 842
PfHsp90-4   AG--ILEDPRSLLSKLNELLLLTARYAYHYEKKDIQEN-------TKDQPIDVKDNTMAIDNIVNT 905
            .*  :::.    . :    :
```

```
hHsp90-A1   LEG--DDDTS-RMEEVD 732 (End)
PfHsp90-1   LEETVDATDS-KMEEVD 745(End)
PfHsp90-2   KDNNVDNESNDKSDEL 821(End)
PfHsp90-3   GIDDNSNISENKINESSSNQNNIGENSIAEENNIKNIAESDVNKINLGENDVSQNTMHKQDSGLF 907
PfHsp90-4   KENNTNDDSVMKEAQSSNL 924(End)
            .     : :
```

```
PfHsp90-3   NLDPSILNSNMLSGSDKTLL 927(End)
```

Figure 23.1 (*Continued*).

specificity it would be critical to take advantage of the subtle but important differences in amino acid sequence that exist between the inhibitor binding pockets of these closely related enzymes. In summary, the DFS approach provides an excellent avenue to exploit such differences and introduce specificity during the process hit identification and optimization, which should eventually lead to compounds that are highly specific toward PfHsp90.

Hsp90 in the Apicomplexa

In the apicomplexan family, essentially all members have Hsp90 homologs that appear to be essential for parasite viability. The genome of *P. falciparum* (Pf) – the deadly malarial parasite – contains four highly conserved Hsp90 paralogs, of which one has been biochemically characterized and shown to have properties character-istic of Hsp90, such as ATP-binding and inhibition by GA [36–38]. The essentiality of PfHsp90 was evidenced by the inhibitory effect of GA on *P. falciparum* growth [37, 38]. Essential roles of Hsp90 were also demonstrated in at least two other members of this family, namely *Toxoplasma gondii* and *Eimeria tenella* [39–42]. Based on these facts, it has been speculated [43] that plasmodial Hsp90 may associate with the following Pf orthologs, which include other chaperones and co-chaperones, many protein kinases and phosphatases, and various tRNA aminoacyl tRNA synthetases (ARS): Hsp70, CyP40, FKBP52, p23, HOP1, ERK2, CSNK, Akt, CHIP, PP1A, PP5, PP2B, HDAC1, RBMX, CALM1, DARS, EPRS, LARS, MARS, QARS, and RARS. Although an association has been demonstrated experimentally for a few, such as PfPP2B [44], overall there is little doubt that the complete clientele of PfHsp90 includes many other *Plasmodium* proteins, which makes it an important signaling node in the parasite and, hence, a promising antimalarial drug target [43, 45].

Interestingly, the *P. falciparum* (Pf) genome encodes four Hsp90 paralogs, which have been tentatively named PfHsp90-1 to PfHsp90-4 (Figure 23.1). All four paralogs are highly similar in sequence, and contain conserved residues found in all Hsp90s, including the human ortholog; however, each sequence – and PfHsp90-3 and PfHsp90-4 in particular – contains unique insertions (Figure 23.1). Although, at present, the relative role and importance of the four malarial Hsp90s remain unexplored, it is interesting to note that humans have five Hsp90 paralogs. Three of these paralogs belong to the Hsp90A family (Hsp90-α_1, Hsp90-α_2, and Hsp90-β; all of which are cytosolic), one paralog is an endoplasmic reticular Hsp90B (also called GRP94), while the fifth, unique mitochondrial Hsp90 is known as TRAP1 (TNF-receptor-associated protein 1). It is currently unknown whether any of the PfHsp90 paralogs reside exclusively in the parasitic organelles.

Recently, PfHsp90-1 – the smallest Pf ortholog – has been cloned and expressed in *Escherichia coli*, with the recombinant PfHsp90-1 being confirmed as capable of encoding all biochemical hallmarks of the Hsp90 family, including an ability to bind ATP; the inhibition of this binding by 17AAG was also confirmed [38]. In the following text, this best-studied paralog is referred to simply as PfHsp90. In addition to the full-length PfHsp90, multiple constructs of its N-terminal domain have been designed, cloned, expressed and purified for fragment- and structure-based drug discovery, as well as for the crystallization experiments that currently are under way.

Designing and Building a Successful Fragment Library

In the following, the present authors' preliminary results are detailed as a proof-of-concept that PfHsp90 may indeed represent a viable drug target. It is important to remember at the outset that, in order for fragment-based drug discovery to be successful, it is necessary to design a fragment library such that the hits will be amenable to efficient synthetic routes. Additionally, the fragments created during optimization should possess chemical characteristics that will allow the introduction of suitable functional groups to achieve the desired pharmacokinetic properties. The present library design followed the general "rule of three," which was developed based on the properties of hits that have been identified to bind a range of drug targets, using X-ray crystallography [38]. The molecular weight of the fragments was less than 300 Da; there were three or fewer H-bond donors, three of fewer H-bond acceptors, and a cLogP-value of about 3. A database search based on these criteria and additional filtering resulted in almost 40 000 fragments that were subsequently sorted by shape and size into groups of 50. One fragment from each group, representing the mid-point of this distribution, was chosen as a representative. An experienced chemist then examined the fragments and discarded them if they were likely to be unstable or reactive; in such a case, another compound was chosen for the group. The physical compounds were then purchased and dissolved in either dimethylsulfoxide (DMSO) or water (if insoluble in DMSO); any compound that proved to be insoluble in both water and DMSO was discarded. These search criteria resulted in a total of 800 fragments, which constituted the library; the latter was further formatted by grouping bins of four compounds into 1 ml-deep well blocks. The groups of four and the original unmixed solutions were plated into 96-well plates, with 5 μl per well (which allowed only the amount required for each experiment to be thawed); the blocks and plates were then stored at $-20\,^{\circ}\mathrm{C}$.

Fragment Screening Using Human Hsp90 N-Terminal Domain

Since a significant sequence homology exists between the *Plasmodium* and human Hsp90, and crystallization experiments are under way for PfHsp90, fragment screening was performed using available human Hsp90 crystals, not only to demonstrate its feasibility but also to use this information for advancing the design and synthesis of PfHsp90 inhibitors. After screening 36 fragments for the collection of the proprietary library, three positive hits were successfully identified (Figure 23.2), which translates to a success rate of 8%. Although the typical success rate using this approach has been in the range of 3–4%, an efficient design of the fragment library might have resulted in a greater number of positive hits.

The fragments identified using crystallography screening experiments occupy the same region where inhibitors of Hsp90 are known to bind, including GA [20] and radicicol [46]. All three fragments formed key interactions with the main chain atoms and side-chain residues of Hsp90, which included H-bonded and hydrophobic interactions. Most of these interactions were novel compared to known inhibitors.

CPF000013 **CPF000017** **CPF000079**

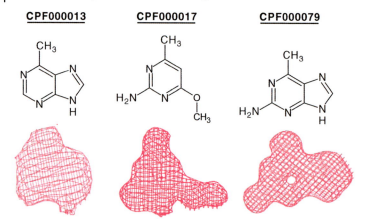

Figure 23.2 Positive fragment hits identified using crystallographic screening experiments known to bind to the ATP pocket of the N-terminal domain of Hsp90. The upper row indicates the molecular structure of the fragments; the lower row represents the corresponding 2|Fo |- |Fc| electron density of the fragments contoured at 1.0 s. CPF: CalAsia Pharmaceuticals Fragment.

A 3-D homology model of the N-terminal domain of PfHsp90 was then generated, using a combination of sequence alignment and the deposited coordinates of the N-terminal domain of PfHsp90-3 (Figure 23.3), in a complex with adenosine monophosphate phosphoramidate (AMPPN) deposited in the Protein Data Bank (PDB: 3IED).

Three-Dimensional Model Structure of *Pf*Hsp90 and its Comparison to the Human Ortholog

The overall fold of the model PfHsp90 is similar to the N-terminal domain of human Hsp90 and other related enzymes, as shown in Figure 23.3 [20, 46–48]. Briefly, the structure includes eight α-helices and seven strands, forming a α + β sandwich fold. One face of the β sheet, which is hydrophobic, packs against a set of α-helices; these are arranged into two sets, where one set packs onto the second at the periphery of the sandwich.

At its center, the helical face of the sandwich has a wide opening extending into the hydrophobic core, resulting in a deep pocket, about 15 Å deep (Figure 23.3). This region contains residues that are highly conserved across the species, including human Hsp90 (Figure 23.1), and forms the active site of the enzyme. Consistent with this observation, AMPPN binds into this pocket (Figure 23.4). The small molecule is sandwiched between the two set of α-helices, with the adenine ring forming a hydrophobic interaction with a number of conserved residues, while a number of H-bonded interactions further stabilize binding. Additionally GA, the known inhibitor of Hsp90, could be modeled into this hydrophobic pocket (Figure 23.5).

A comparison of the PfHsp90 (model) and human Hsp90 structures further shows that they are very closely related in overall fold, and can be superimposed with a root

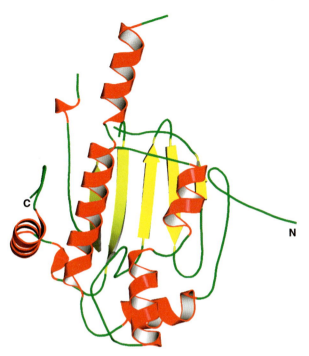

Figure 23.3 Overall three-dimensional structure of the N-terminal domain of PfHsp90-3. The α-helices are shown in red, β-strands in yellow, and loops in green, generated by using Pymol [51].

mean square (RMS) deviation of 1.5 Å for equivalent residues (Figure 23.6). In particular, the β-sheet that forms the core and backbone of the binding pocket superimposes within 0.5 Å RMS between the two structures, but the deviation is significant between the α-helices. In all, based on co-crystal structures, there are about 15 residues that are known to interact with inhibitors and substrates of Hsp90 (Figure 23.6). It is worth noting that, of the 15 residues, only three were different between the two Hsp90s, and were mapped onto the 3-D structure in the context of inhibitor binding.

Molecular Modeling of the Fragments Bound to Human Hsp90 in the Active Site of *Pf*Hsp90-3 Crystal Structure

There is currently no biological evidence for the functionality or essentiality of PfHsp90-3, although the crystal structure of the recombinant protein has been solved (PDB:3IED). Thus, whilst PfHsp90-1 crystallography efforts are currently under way, the aim was to capitalize on the available PfHsp90-3 structure. It was reasoned that a comparison of fragment-binding interactions between human Hsp90 and PfHsp90-3 structures would facilitate a rational homology modeling of these

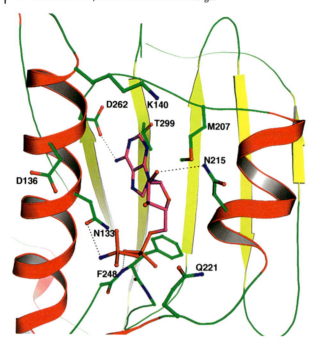

Figure 23.4 Binding of AMPNN into the ATP pocket of Hsp90; residues involved in hydrophobic and H-bonded interactions are shown in ball-and-stick fashion. The PfHsp90-1 model is shown in red, and the human-Hsp90 crystal structure in green.

fragments with PfHsp90-3, which would expedite the lead discovery efforts. The three positive hits which resulted from an ongoing fragment screening against the human Hsp90, and which subsequently were used for PfHsp90-3, are shown in Figure 23.2.

The docking of CPF000013 fragment into the PfHsp90-3 active site (Figure 23.7a) placed the fragment in the same position as the adenine of AMPNN, as observed in the crystal structure (PDB:3IED), thereby providing confidence to the modeling results. The fragment oriented similarly to the adenine, forming hydrophobic interactions with M207, I200, and F248; moreover, there were no direct H-bonding interactions between the fragment and enzyme residues. All polar interactions were mediated via water molecules. The NH_2 group of the fragment appeared not to participate in the interaction, but may be used to introduce appropriate substituents. Finally, the binding of the fragment created four pockets that were suitable for evolving it into lead compounds.

The binding of fragment CPF000017 was similarly modeled (Figure 23.7b). Similar to CPF000013, this fragment also occupied the same position and orientation, but the amino group was shown to form a direct H-bonded interaction with the carboxylate O of D202. This group was also shown to be involved in a water-mediated interaction involving the backbone carbonyl O atom of L130. In this case, the OMe group extended into one of the hydrophobic pockets, offering new ideas for the design of inhibitors.

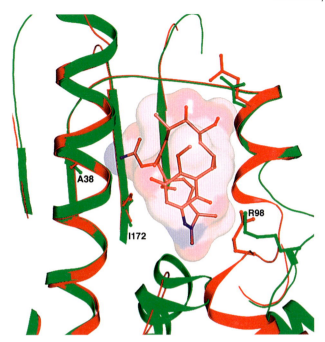

Figure 23.5 Superposition showing the close-up view of the inhibitor binding pockets of *Plasmodium* and human Hsp90. Shown is the most widely studied natural inhibitor, geldanamycin; also mapped are the three nonconserved residues. Color coding as Figure 23.4.

Lastly, CPF000079 docking (Figure 23.7c) resembled that of CPF000013 (where $NH_2 = H$), whereby all hydrophobic and H-bonded interactions were conserved. However, the presence of an additional amino group resulted in direct H-bonding with the side chain of D202. This was identical to that of the CPF00017 amino group, and suggested that an amino functional group would be well tolerated in this position and would stabilize the bound inhibitor.

Structure-Based Lead Optimization

Analyses of small molecules modeled into the PfHsp90-3 ATP-binding site have revealed the presence of four regions that are of significance to the structure-based development of small-molecule inhibitors (Figure 23.8). There exists a purine-binding region containing multiple moieties capable of making favorable H-bonding interactions with the heteroaromatic nitrogen atoms. The purine-binding region is occupied by the adenine-derived fragments STF000013 and CPF000079, but it also accommodates the pyrimidine fragment CPF000017 (Figure 23.9).

Adjacent to this site is a small lipophilic (L1) binding pocket which is proximal to allow for a small molecule to feasibly occupy this region in addition to the ATP site (Figure 23.8). It was envisioned that an exploration of the structure–activity relationship (SAR) around this lipophilic region would allow the development of potent

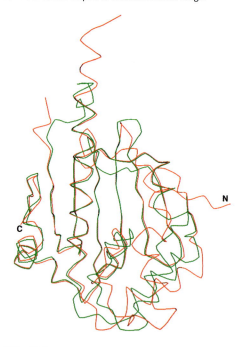

***Pf*Hsp90-1**	N	A	D	A	K	M	D	L	L	T	R	G	F Y I
*Hs*Hsp90	N	S	D	A	K	M	D	L	L	T	K	G	F Y V

Figure 23.6 Upper panel: Superposition of Cα-atoms of PfHsp90-1 (model) (red) and human (green) Hsp90. Lower panel: Key variant and invariant residues involved in interaction with the inhibitors in the N-terminal domain of Pf and human Hsp90. The variant residues are shown in red. The residue numbers, corresponding to PfHsp90-1 (see Figure 23.1) are: N37, A38, D40, A41, K44, M84, D88, L89, L93, T95, R98, G121, F124, Y125, and I173. Note that these residues are based on 3-D structure superposition, and therefore do not represent linear sequences and will not directly align with the sequences in Figure 23.1. To prevent any confusion, these residues have all been highlighted in Figure 23.1.

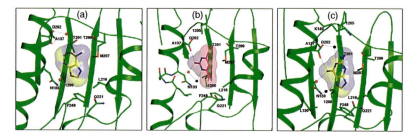

Figure 23.7 (a) Molecular modeling of CPF000013 into the active site of PfHsp90-3 crystal structure. For this and subsequent modeling studies, all solvent molecules were included. The fragment is shown as yellow ball-and-stick and also as a surface representation; (b) Binding of CPF000017 (in magenta) modeled into the active site of PfHsp90-3; (c) Binding of CPF000079 (yellow) into the active site, using molecular modeling.

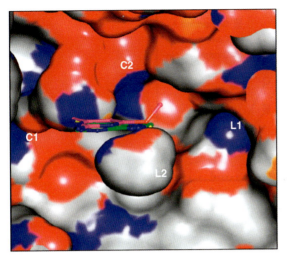

Figure 23.8 Accessible surface area showing lipophilic (L1 and L2) and charged (C1 and C2) pockets in the inhibitor/ATP pocket of PfHsp90-3 from modeling studies of the fragments.

and selective inhibitors. The strategy here was to utilize novel, highly efficient synthetic chemistry to provide a series of lead-like molecules that contained two components: (i) purine- or pyrimidine- and pyrimidinone-like moieties [49]; and (ii) a lipophilic group consisting of aromatic or heteroaromatic moieties capable of exploiting nonpolar interactions to afford selectivity (Figure 23.9). After establishing and validating the activity of these small molecule leads, a crystallographic structure-based design will subsequently be used to optimize these inhibitors for potency, selectivity, and physico-chemical properties.

Design and Synthesis of Analogs Using the Structural Information

By using the structural information of the fragments bound to the active site of PfHsp90, two compounds (A and B) were designed, synthesized, and characterized.

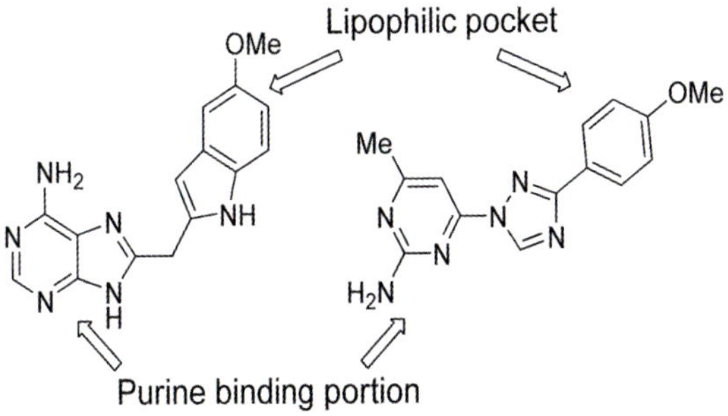

Figure 23.9 Schematic representation showing the introduction of lipophilic groups with drug-like properties to the bound fragments.

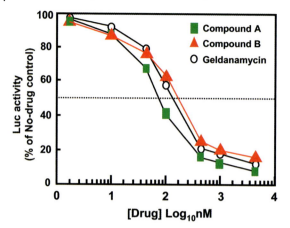

Figure 23.10 Luciferase refolding assay for the chaperone activity of Hsp90 to measure the IC_{50} of compounds A and B. The assay was performed using full-length Hsp90. Renaturation of heat-denatured luciferase (Luc) was performed in the presence of 0 (no inhibitor), 2, 10, 50, 100, 500, 1000, and 5000 nM inhibitors and geldanamycin, a well-known Hsp90 inhibitor [22–24, 38].

These were evaluated for their inhibition potency in a functional enzymatic assay, using recombinant full-length Hsp90 [50]. Briefly, the renaturation of heat-denatured luciferase (Luc) was measured in the presence of 0 (no inhibitor), 2, 10, 50, 100, 500, 1000, and 5000 nM inhibitor. Subsequently, GA – a well-known Hsp90 inhibitor – was used as a control. The IC_{50}-value of GA in this assay was determined as 150 nM, in close agreement with the published value of $\sim$0.2 μM [38], thus validating the experimental method.

The *in vitro* enzyme assay data and the Luc activity (as % of no-drug control) as a function of inhibitor concentration (nM) is shown in Figure 23.10. Compounds A and B were both found to be highly potent, and were comparable to the well-characterized inhibitor of Hsp90, GA (150 nM). The fact that these compounds had evolved from the very first round of fragment optimization, and had low ($\sim$300 Da) molecular weights and excellent potency, provided a significant opportunity to further optimize them to achieve the desired efficacy and pharmacological properties.

Co-Crystal Structures of Compounds A and B in Complex with N-Terminal Domain of Hsp90

In an attempt to understand the SAR and mode of binding of the compounds, and to guide further lead optimization, the structures of the compounds in complex with Hsp90 have been resolved. The binding of A and B into the ATP-binding pocket of the N-terminal domain of Hsp90 is shown in Figure 23.11. Both compounds were designed using the structural information of the fragment CPF000017 (Figure 23.2) bound to Hsp90 (Figure 23.7b), identified by crystallography screening experiments.

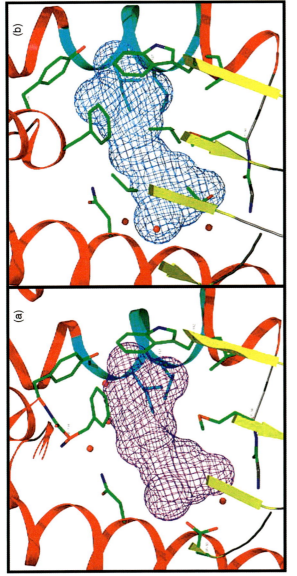

Figure 23.11 (a) Binding of compound A into the ATP binding pocket of the N-terminal domain of Hsp90: the enzyme residues are shown as green sticks, the secondary structure elements in ribbon representation, the ligand as mesh (purple), and the solvent molecules in red spheres; (b) Binding of compound B into the same pocket: the coloring scheme is the same for the enzyme and the ligand is shown as mesh (cyan).

Compounds of A and B occupied the same position as that observed in the fragment-bound structures (Figure 23.7). The added substituents of compounds A and B respectively occupied a newly identified lipophilic region L1, as defined in the section, *Structure-based lead optimization*. This region of the ligand is surrounded by six hydrophobic residues, three being aromatic and three aliphatic. The observed twofold potency of A (80 nM) compared to B (200 nM) could be explained due to the presence of two additional H-bonding interactions that were absent from compound B.

Conclusion

Based on finer sequence comparisons and structure-based homology modeling between human and *Plasmodium* orthologs, it was postulated that the parasitic Hsp90 would be a promising drug target. This suggestion was supported by preliminary results obtained when using differential fragment-based screening, a new and improved approach for specific fragment-based drug discovery. The nanomolar potencies of the two novel compounds described here were highly encouraging. It was concluded that this discovery approach could be used very effectively to identify hits that, in turn, would evolve into potent compounds, and eventually be optimized as "drug-like" candidates for novel antimalarial therapies.

References

1 Ritossa, F. (1962) A new puffing pattern induced by temperature shock and DNP in *Drosophila*. *Cell. Mol. Life Sci.*, **18**, 571–573.

2 Ritossa, F. (1996) Discovery of the heat shock response. *Cell Stress Chaperones*, **1**, 97–98.

3 Pearl, L.H. and Prodromou, C. (2006) Structure and mechanism of the Hsp90 molecular chaperone machinery. *Annu. Rev. Biochem.*, **75**, 271–294.

4 Goetz, M.P., Toft, D.O., Ames, M.M., and Erlichman, C. (2003) The Hsp90 chaperone complex as a novel target for cancer therapy. *Ann. Oncol.*, **14**, 1169–1176.

5 Neckers, L. and Ivy, S.P. (2003) Heat shock protein 90. *Curr. Opin. Oncol.*, **15**, 419–424.

6 Kalmar, B. and Greensmith, L. (2009) Induction of heat shock proteins for protection against oxidative stress. *Adv. Drug Deliv. Rev.*, **61**, 310–318.

7 Akerfelt, M., Trouillet, D., Mezger, V., and Sistonen, L. (2007) Heat shock factors at a crossroad between stress and development. *Ann. N. Y. Acad. Sci.*, **1113**, 15–27.

8 Neckers, L. (2007) Heat shock protein 90: the cancer chaperone. *J. Biosci.*, **32**, 517–530.

9 Neckers, L. (2002) Hsp90 inhibitors as novel cancer chemotherapeutics agents. *Trends Mol. Med.*, **8**, S55–S61.

10 Kim, Y.S., Alarcon, S.V., Lee, S., Lee, M.J., Giaccone, G., Neckers, L., and Trepel, J.B. (2009) Update on Hsp90 inhibitors in clinical trial. *Curr. Top. Med. Chem.*, **9**, 1479–1492.

11 Verkhivker, G.M., Dixit, A., Morra, G., and Colombo, G. (2009) Structural and computational biology of the molecular chaperone Hsp90: from understanding molecular mechanisms to computer-based inhibitor design. *Curr. Top. Med. Chem.*, **9**, 1369–1385.

12 Kang, B.H., Plescia, J., Song, H.Y., Meli, M., Colombo, G., Beebe, K., Scroggins, B., Neckers, L., and Altieri, D.C. (2009) Combinatorial drug design targeting multiple cancer signaling networks controlled by mitochondrial Hsp90. *J. Clin. Invest.*, **119**, 445–448.

13 Porter, J.R., Ge, J., Lee, J., Normant, E., and West, K. (2009) Ansamycin inhibitors of Hsp90: Nature's prototype for anti-chaperone therapy. *Curr. Top. Med. Chem.*, **9**, 1386–1418.

14 Taldone, T. and Chiosis, G. (2009) Purine-scaffold Hsp90 inhibitors. *Curr. Top. Med. Chem.*, **9**, 1436–1446.

15 Young, J.C., Moarefi, I., and Hartl, F.U. (2001) Hsp90: a specialized but essential protein folding tool. *J. Cell. Biol.*, **154**, 267–273.

16 Richter, K. and Buchner, J. (2001) Hsp90: chaperoning signal transduction. *J. Cell. Physiol.*, **188**, 281–290.

17 Giamas, G., Man, Y.L., Hirner, H., Bischof, J., Kramer, K., Khan, K., Lavina Ahmed, S.S., Stebbing, J., and Knippschild, U. (2010) Kinases as targets in the treatment of solid tumors. *Cell. Signal.*, **22**, 984–1002.

18 Faivre, S., Djelloul, S., and Raymond, E. (2006) New paradigms in anticancer therapy: targeting multiple signaling pathways with kinase inhibitors. *Semin. Oncol.*, **33**, 407–420.

19 Grenert, J.P., Sullivan, W.P., Fadden, P., Haystead, T.A., Clark, J., Mimnaugh, E., Krutzsch, H., Ochel, H.J., Schulte, T.W., Sausville, E., Neckers, L.M., and Toft, D.O. (1997) The amino-terminal domain of heat shock protein 90 (hsp90) that binds geldanamycin is an ATP/ADP switch domain that regulates hsp90 conformation. *J. Biol. Chem.*, **272**, 23843–23850.

20 Stebbins, C.E., Russo, A.A., Schneider, C., Rosen, N., Hartl, F.U., and Pavletich, N.P. (1997) Crystal structure of an hsp90–geldanamycin complex: targeting of a protein chaperone by an anti-tumor agent. *Cell*, **89**, 239–250.

21 Prodromou, C., Roe, S.M., O'Brien, R., Ladbury, J.E., Piper, P.W., and Pearl, L.H. (1997) Identification and structural characterization of the ATP/ADP-binding site in the Hsp90 molecular chaperone. *Cell*, **90**, 65–75.

22 DeBoer, C., Meulman, P.A., Wnuk, R.J., and Peterson, D.H. (1970) Geldanamycin, a new antibiotic. *J. Antibiot. (Tokyo)*, **23**, 442–447.

23 Whitesell, L., Mimnaugh, E.G., De Costa, B., Myers, C.E., and Neckers, L.M. (1994) Inhibition of heat shock protein HSP90-pp60v-src heteroprotein complex formation by benzoquinone ansamycins: essential role for stress proteins in oncogenic transformation. *Proc. Natl Acad. Sci. USA*, **91**, 8324–8328.

24 Blagosklonny, MV., Toretsky, J., and Neckers, L. (1995) Geldanamycin selectively destabilizes and conformationally alters mutated p53. *Oncogene*, **11**, 933–939.

25 Schulte, T.W. and Neckers, L.M. (1998) The benzoquinone ansamycin 17-allylamino-17-demethoxygeldanamycin binds to HSP90 and shares important biologic activities with geldanamycin. *Cancer Chemother. Pharmacol.*, **42**, 273–279.

26 Scapin, G. (2006) Structural biology and drug discovery. *Curr. Pharm. Des.*, **12**, 2087–2097.

27 Wallace, M.B., Feng, J., Zhang, Z., Skene, R.J., Shi, L., Caster, C.L., Kassel, D.B., Xu, R., and Gwaltney, S.L. 2nd (2008) Structure-based design and synthesis of benzimidazole derivatives as dipeptidyl peptidase IV inhibitors. *Bioorg. Med. Chem. Lett.*, **8** 2362–2367.

28 Zhao, Z., Wolkenberg, S.E., Lu, M., Munshi, V., Moyer, G., Feng, M., Carella, A.V., Ecto, L.T., Gabryelski, L.J., Lai, M.T., Prasad, S.G., Yan, Y., McGaughey, G.B., Miller, M.D., Lindsley, C.W., Hartman, G.D., Vacca, J.P., and Williams, T.M. (2008) Novel indole-3-sulfonamides as potent HIV non-nucleoside reverse transcriptase inhibitors (NNRTIs). *Bioorg. Med. Chem. Lett.*, **18** 554–559.

29 Shuker, S.B., Hajduk, P.J., Meadows, R.P., and Fesik, S.W. (1996) Discovering high-affinity ligands for proteins: SAR by NMR. *Science*, **274**, 1531–1534.

30 Huber, W. (2005) A new strategy for improved secondary screening and lead

optimization using high-resolution SPR characterization of compound-target Interactions. *J. Mol. Recognit.*, **18**, 273–281.

31 Neumann, T., Junker, H.D., Schmidt, K., and Sekul, R. (2007) SPR-based fragment screening: advantages and applications. *Curr. Top. Med. Chem.*, **7**, 1630–1642.

32 Mercier, K.A., Germer, K., and Powers, R. (2006) Design and characterization of a functional library for NMR screening against novel protein targets. *Comb. Chem. High-Throughput Screen.*, **9**, 515–534.

33 Eitner, K. and Kochu, U. (2009) From fragment screening to potent binders: Strategies for fragment-to-lead evolution. *Mini Rev. Med. Chem.*, **9**, 956–961.

34 Hartshorn, M.J., Murray, C.W., Cleasby, A., Frederickson, M., Tickle, I.J., and Jhoti, H. (2005) Fragment-based lead discovery using X-ray crystallography. *J. Med. Chem.*, **48**, 403–413.

35 Hajduk, P.J. and Greer, J. (2007) A decade of fragment-based drug design: strategic advances and lessons learned. *Nat. Rev. Drug Discov.*, **6**, 211–219.

36 Bonnefoy, S., Attal, G., Langsley, G., Tekaia, F., and Mercereau-Puijalon, O. (1994) Molecular characterization of the heat shock protein 90 gene of the human malaria parasite *Plasmodium falciparum*. *Mol. Biochem. Parasitol.*, **67**, 157–170.

37 Banumathy, G., Singh, V., Pavithra, S.R., and Tatu, U. (2003) Heat shock protein 90 function is essential for *Plasmodium falciparum* growth in human erythrocytes. *J. Biol. Chem.*, **278**, 18336–18345.

38 Kumar, R., Musiyenko, A., and Barik, S. (2003) The heat shock protein 90 of *Plasmodium falciparum* and antimalarial activity of its inhibitor, geldanamycin. *Malar. J.*, **2**, 30.

39 Rojas, P.A., Martin, V., Nigro, M., Echeverria, P.C., Guarnera, E.A., Pszenny, V., and Angel, S.O. (2000) Expression of a cDNA encoding a *Toxoplasma gondii* protein belonging to the heat-shock 90 family and analysis of its antigenicity. *FEMS Microbiol. Lett.*, **190**, 209–213.

40 Ahn, H.J., Kim, S., and Nam, H.W. (2003) Molecular cloning of the 82-kDa heat shock protein (HSP90) of *Toxoplasma gondii* associated with the entry into and

growth in host cells. *Biochem. Biophys. Res. Commun.*, **311**, 654–659.

41 Echeverria, P.C., Matrajt, M., Harb, O.S., Zappia, M.P., Costas, M.A., Roos, D.S., Dubremetz, J.F., and Angel, S.O. (2005) *Toxoplasma gondii* Hsp90 is a potential drug target whose expression and subcellular localization are developmentally regulated. *J. Mol. Biol.*, **350**, 723–734.

42 Péroval, M., Péry, P., and Labbé, M. (2006) The heat shock protein 90 of *Eimeria tenella* is essential for invasion of host cell and schizont growth. *Int. J. Parasitol.*, **36**, 1205–1215.

43 Acharya, P., Kumar, R., and Tatu, U. (2007) Chaperoning a cellular upheaval in malaria: heat shock proteins in *Plasmodium falciparum. Mol. Biochem. Parasitol.*, **153**, 85–94.

44 Kumar, R., Musiyenko, A., and Barik, S. (2005) *Plasmodium falciparum* calcineurin and its association with heat shock protein 90: mechanisms for the antimalarial activity of cyclosporin A and synergism with geldanamycin. *Mol. Biochem. Parasitol.*, **141**, 29–37.

45 Shonhai, A. (2010) Plasmodial heat shock proteins: targets for chemotherapy. *FEMS Immunol. Med. Microbiol.*, **58**, 61–74.

46 Roe, S.M., Prodromou, C., O'Brien, R., Ladbury, J.E., Piper, P.W., and Pearl, L.H. (1999) Structural basis for inhibition of the Hsp90 molecular chaperone by the antitumor antibiotics radicicol and geldanamycin. *J. Med. Chem.*, **42**, 260–266.

47 Prodromou, C., Roe, S.M., Piper, P.W., and Pearl, L.H. (1997) A molecular clamp in the crystal structure of the N-terminal domain of the yeast Hsp90 chaperone. *Nat. Struct. Biol.*, **4**, 477–482.

48 Cheung, K.M., Matthews, T.P., James, K., Rowlands, M.G., Boxall, K.J., Sharp, S.Y., Maloney, A., Roe, S.M., Prodromou, C., Pearl, L.H., Aherne, G.W., McDonald, E., and Workman, P. (2005) The identification, synthesis, protein crystal structure and *in vitro* biochemical evaluation of a new 3,4-diarylpyrazole class of Hsp90

inhibitors. *Bioorg. Med. Chem. Lett*, **15**, 3338–3343.

49 Chiang, A.N., Valderramos, J.C., Balachandran, R., Chovatiya, R.J., Mead, B.P., Schneider, C., Bell, S.L., Klein, M.G., Huryn, D.M., Chen, X.S., Day, B.W., Fidock, D.A., Wipf, P., and Brodsky, J.L. (2009) Select pyrimidinones inhibit the propagation of the malarial parasite, *Plasmodium falciparum*. *Bioorg. Med. Chem.*, **17**, 1527–1533.

50 Galam, L., Hadden, M.K., Ma, Z., Ye, Q.Z., Yun, B.G., Blagg, B.S., and Matts, R.L. (2007) High-throughput assay for the identification of Hsp90 inhibitors based on Hsp90-dependent refolding of firefly luciferase. *Bioorg. Med. Chem.*, **15**, 1939–1946.

51 DeLano, W.L. (2002) The PyMOL Molecular Graphics System. DeLano Scientific, San Carlos, CA, USA. Available at: http://www.pymol.org.

24
Drug Discovery Against *Babesia* and *Toxoplasma*

Mohamad Alaa Terkawi and Ikuo Igarashi*

Abstract

The Apicomplexa, which consist of numerous genera of pathogenic protozoa, constitute a global medical and economic problem. *Toxoplasma gondii* and certain *Babesia* spp. have recently emerged as lethal opportunistic pathogens in immuno-compromised patients, and as systemic veterinary pathogens associated with significant economic losses in the animal industry. The widespread parasitic infections and the increase of drug resistance in many areas of the world have exacerbated the problem and demonstrated the need for the development of safe, effective, and affordable agents. The genomics and proteomics of apicomplexan parasites provide comprehensive insights into their biochemistry and cell processes, and highlight possible drug targets. On the other hand, a recombinant protein technique combined with three-dimensional structure elucidation using X-ray crystallography are critical for identifying the structural information of target molecules and obtaining a model of the target binding site, leading to the best design of antiparasitic drugs. However, the efficacy, toxic side effects, pharmacokinetic compatibility, and potential for developing resistance are major parameters in the design of a successful future drug. In this chapter, the biochemical features of *Babesia* spp. and *Toxoplasma gondii* are discussed and updated, and possible molecular targets available in the field of drug discovery are highlighted.

Introduction

The Impact of *Babesia* and *Toxoplasma* Infection

The phylum Apicomplexa is a group of obligate intracellular parasites responsible for a wide range of important diseases of both human and livestock hosts. *Babesia* organisms are obligate intraerythrocytic protozoa capable of invading a broad range of vertebrate hosts, subsequently leading to serious economic problems within the

*Corresponding author

Apicomplexan Parasites. Edited by Katja Becker
Copyright © 2011 WILEY-VCH Verlag GmbH & Co. KGaA, Weinheim
ISBN: 978-3-527-32731-7

livestock industry throughout tropical and subtropical regions of the world. Human babesiosis is caused by certain babesial species that have distinct geographical distributions on the basis of the presence of competent hosts. In North America, human babesiosis is caused by *Babesia microti*, while in Europe the condition is considerably more rare, but more lethal, and is caused by *Babesia divergens*. The parasite normally causes clinical manifestations that range from an asymptomatic carrier state to potentially life-threatening malaria-like episodes. *Babesia bovis* is the most economically important species to the cattle industry due to its ability to sequester in the microcapillaries of the kidneys, lungs, and brain, with consequent organ failure and systemic shock leading to death [1, 2]. *Toxoplasma gondii*, another member of this phylum, can invade and multiply in the nucleated cells of virtually all warm-blooded animals, causing significant morbidity and mortality in patients with acquired immunodeficiency syndrome (AIDS), and also serious congenital birth defects in both humans and livestock [3]. Despite the medical and veterinary importance of babesiosis and toxoplasmosis, no effective vaccine exists for these infections, and chemotherapy remains the mainstay for treatment and control. However, the emergence of drug resistance and the prevalence of such opportunistic pathogens in immunocompromised patients emphasize the need to identify a new chemotherapeutic agent that can counter the spread of parasitic infection.

Current Status of Chemotherapy

At present, imidocarb dipropionate is the only babesiacide which is documented consistently to clear the parasites, but it is often associated with solid sterile immunity in the case of bovine babesiosis. Diminazene diaceturate is effective for the treatment of bovine and canine babesiosis, but not of human babesiosis caused by *B. divergens*. The combination of azithromycin with atovaquone or quinine has been effective in both experimental animals and humans. However, the current therapy for symptomatic human babesiosis is quinine and clindamycin, sometimes supplemented with exchange transfusion [4, 5]. On the other hand, the most effective therapy for congenital or cerebral toxoplasmosis is the synergistic combination of sulfadiazine and pyrimethamine. Azithromycin is also efficient *in vivo*, and can be administered synergistically with pyrimethamine. In those patients who are intolerant to sulfonamides or pyrimethamine, atovaquone can be used as an alternative. In immunocompromised patients (e.g., AIDS), pyrimethamine in combination with either sulfadiazine or clindamycin is the first choice of regimen, while pregnant patients usually receive spiramycin [6].

Although these components are often effective, the emergence of drug resistance in endemic areas, as well as the toxicity and cost of some of these compounds, has hampered their wide use and provoked the need for exploring new effective chemotherapies. Recently, natural products have attracted increasing attention in an effort to develop new antiparasitic drugs. Indeed, many well-known drugs listed in the modern pharmacopoeia have their origins in Nature, and some have been documented as being active against a broad range of parasitic infections. The most common components used for the clinical treatment of parasitic infections are quinine, from the bark of the *Cinchona* tree, and artemisinin, from the herb *Artemisia annua* [5–7].

Chemotherapeutic Targets for *Babesia* and *Toxoplasma*

Genomic research on apicomplexan parasites has generated a wealth of information related to their biochemistry that will undoubtedly lead to new therapies for these pathogens. In the same regard, the proteins involved in metabolism, signaling, protein trafficking pathways, cell division, and extracellular communication mechanisms could logically be attractive drug targets [8–11]. Targets for therapy will be considered based on their locations and function within the parasites.

Drug Targets Associated with Membrane Transport

Membrane transport proteins play an integral role in the biochemistry of cells, mediating ion homeostasis, the uptake of metabolic and biosynthetic substrates, and the expulsion of metabolic wastes. Hexose transport is a potential drug target for apicomplexan parasites that depend on glucose for their energy source [12]. Cytochalasin B and β-glucogallin, both of which are known potent inhibitors of glucose transport, completely inhibited the *in vitro* growth of *B. bovis* [13]. In addition, a long-chain O-3 hexose derivative (compound 3361), which killed malarial parasites *in vitro* and *in vivo* by selectively inhibiting hexose transport, has blocked the invasion of *B. bovis* into erythrocytes [12, 13].

Cytosolic Drug Targets

The cytosol of apicomplexan parasites contains numerous enzymes involved in metabolic pathways that could be potential drug targets. Details of the drugs proposed for targeting cytosolic metabolic pathways in *T. gondii* and *Babesia* spp. are listed in Table 24.1.

Folate Metabolic Pathway
The folate metabolic pathway, which includes the enzymes dihydrofolate reductase (DHFR) and dihydropteroate synthase (DHPS), is essential for DNA synthesis and the metabolism of certain amino acids. Generally, pyrimethamine and cycloguanil are known to target DHFR activity, while sulfa drugs inhibit DHPS. Antifolates have been identified as effective components for both bacterial and protozoan infections, with minimal toxicity to the host [14, 15]. Indeed, for many decades, they have represented the only inexpensive regime for combating clinical malaria and toxoplasmosis. However, clinical malaria and toxoplasmosis are widely treated by a combination of sulfadoxine/pyrimethamine or chlorproguanil/dapsone. Although the extreme toxicity of antifolates against *Babesia* spp. has been reported *in vitro*, they

Table 24.1 Drugs and herbicides proposed as a cytosolic target in *T. gondii* and *Babesia* spp.

Pathway/Mechanism	Putative target	Therapies
Folate	DHFR	Pyrimethamine, cycloguanil, and methotrexate
	DHPS	Sulfadoxine
Purine salvage	AK	Adenosine analogs (NBMPR)
	IMPDH	Ribavirin and mycophenolic acid and adenine dinucleotide analogs (TAD)
Pyrimidine salvage	OPRT	Pyrazofurin
	DOD	Hydroxynaphtoquinone BW58C
	UPRT[a]	Pyrimidine analogs (5-FU)
Shikimate	EPSPS[a]	Glyphosate and fluorinated analogs of shikimate
Glycolysis	LDH	Gossypol and gossypol derivatives
Antioxidant	Unknown	Artemisinin and NSC3852

Abbreviation: DHFR, dihydrofolate reductase; DHPS, dihydropteroate synthase; AK, adenosine kinase; IMPDH, inosine 5-monophosphate dehydrogenase; OPRT, orotate phosphoribosyl transferase; DOD, dihydroorotate dehydrogenase; UPRT, uracil phosphoribosyltransferase; EPSP, 5-enolpyruvylshikimate-3-phosphate; LDH, lactate dehydrogenase.
a) The target is absent in *Babesia* spp.

are not used for the clinical treatment of babesiosis [4, 5, 14, 16]. Recently, resistance to these components has emerged in many areas of the world, and their side effects have also been reported in immunocompromised patients, suggesting the need to seek new inhibitors [15].

Nucleic Acid Metabolic Pathways

Salvage pathways represent a major source of nucleotides for the synthesis of DNA, RNA, and enzyme cofactors. The purine salvage pathway is known to be crucial in the metabolism of protozoan parasites, and is believed to be an attractive target for drug intervention. Unlike their mammalian host cells, most of the parasites that have been studied lack the ability to synthesize purine nucleotides *de novo*, and are dependent on the host for their purine demands. The biochemical and physiological differences between parasites and host cells found in purine metabolism provide potential targets for the design of safe and effective antiparasitic drugs. The therapeutic application within the pathway includes the inhibition of nucleoside or nucleobase transport and enzymes [8]. Apicomplexan parasites are quite diverse in their complement of purine salvage enzymes. For example, *Babesia* spp., *Cryptosporidium* spp., and *Theileria* spp. rely on adenosine kinase (AK), while *Plasmodium* spp. utilize the nucleoside phosphorylase pathway for purine salvage [15–17]. Both, AK and hypoxanthine–xanthine–guanine phosphoribosyltransferase (HXGPRT) provide redundant routes of purine salvage in *T. gondii*, but adenosine is the major purine source [17–19]. Inosine 5-monophosphate dehydrogenase (IMPDH) has been suggested as a drug target due to its critical role in converting inosine monophosphate (IMP) to guanosine monophosphate (GMP). Previously, IMPDH inhibitors – including ribavirin and mycophenolic acid – have been used clinically as immunosuppressive antimicrobial, antiviral, and anticryptosporidial agents, without side effects to the host cells [19–21]. The apicomplexan parasites can uptake purines either by specific carrier-mediated nucleoside and nucleobase transporters, or by passive diffusion, but not via a sodium-dependent transporter. This distinction from mammalian cells might provide a novel foundation for alternative antiparasitic drug development [21, 22]. Purine analogs that can selectively block or disrupt the parasite purine uptake and salvage represent potential chemotherapeutic agents with a high degree of selectivity [19]. For instance, nitrobenzylthioinosine (NBMPR) and 6-benzylthioinosine were shown to be toxic against parasites that contain adenosine kinase, but showed a low toxicity against the host cells [23]. Likewise, the adenosine analogs tubercidin (7-deaza-adenosine), 2-bromo-adenosine, 8-bromo-3-ribosyl adenine, and 6-phenylamino-deoxyadenosine were highly effective against the *in vitro* growth of *B. bovis* [24]. The adenine dinucleotide analog thiazole-4-carboxamide adenine dinucleotide (TAD) and the nonhydrolyzable β-methylene derivatives were shown to be selective inhibitors of *T. gondii* IMPD (TgIMPD), and were highly toxic toward the parasites but not toward the host cells [25].

Another salvage pathway of interest is that of the pyrimidines. The presence of all six enzymes of the *de novo* pyrimidine pathway, leading to the biosynthesis of uridine monophosphate (UMP), suggests the existence of a functional pyrimidine synthesis pathway in *T. gondii* and *Babesia* spp. [26]. However, *T. gondii* tachyzoites are also capable of using pyrimidines from their host cell – a process that seems to be vital for their virulence [26, 27]. The uracil phosphoribosyltransferase (UPRT) of *T. gondii* plays a central role in the salvage of pyrimidine nucleobases and, as a consequence, serves as a potential therapeutic target. TgUPRT has been shown to bind to a variety of pyrimidine analogs (e.g., 5-fluorouracil; 5-FU), all of which were very toxic to the parasites [28]. Moreover, pyrimidine biosynthesis inhibitors such as pyrazofurin (specific for orotate phosphoribosyl transferase and orotidine 5-phosphate decarboxylase) and the hydroxynaphthoquinone BW58C (an inhibitor of dihydroorotate dehydrogenase) have shown strong toxicity against the growth of *Babesia* spp. *in vitro* and *in vivo* in a mouse model [29, 30].

Shikimate Pathway

The shikimate pathway is essential for the production of a plethora of aromatic compounds in plants, bacteria, and fungi. This pathway has also been reported in apicomplexan parasites, including *T. gondii*, *Plasmodium falciparum*, *Cryptosporidium parvum*, and *Eimeria bovis*, as an attractive target for the development of a new antiparasitic drug [31, 32]. This pathway possesses seven enzymes that catalyze the sequential conversion of erythrose 4-phosphate and phosphoenol pyruvate to chorismate, which is used as a substrate for other pathways that culminate in the production of folates, ubiquinone, naphthoquinones, and the aromatic amino acids tryptophan, phenylalanine, and tyrosine. The herbicide glyphosate showed significant inhibition against the *in vitro* growth of *T. gondii*, *P. falciparum*, and *C. parvum* by inhibiting 5-enolpyruvylshikimate-3-phosphate (EPSP) synthase, the sixth enzyme in the pathway [31].

Glycolysis Pathway

Glycolysis is believed to be the main energy source for many organisms, including the apicomplexan parasites. Lactate dehydrogenase (LDH) is the terminal enzyme of anaerobic Embden–Meyerhoff glycolysis, and plays an indispensable role within this pathway, catalyzing the interconversion of pyruvate to lactate with concomitant interconversion of NADH and NAD$^+$ [33]. The LDH knockdown *T. gondii* parasites resulted in an impairment of the differentiation processes, such that the parasites were unable to form significant numbers of tissue cysts, or to establish a chronic infection [34]. Additionally, the unique structural and kinetic properties of LDH, which differ from those of mammalian host cells, have suggested the enzyme's potential as a parasitic drug target [35]. Gossypol, a natural product isolated from cotton plants, and its derivatives, have been suggested as a selective inhibitor of LDH. These components significantly inhibited the *in vitro* growth of *P. falciparum*, *T. gondii*, and *B. bovis*, as well as the *in vivo* growth of *B. microti*, without any signs of toxicity in mice [36, 37].

Redox System

Whilst oxidative stress is an important mechanism that is used by the host to destroy intracellular parasites, the parasite has developed defense tactics to prevent such attack, by producing antioxidant enzymes. The parasite's antioxidant defense might, therefore, represent a potential target for chemotherapeutics [38]. The antioxidant enzymes superoxide dismutase (SOD), glutathione peroxidase (GPx) and catalase are each known to be present in most apicomplexan parasites. Within this context, Ding et al. [39] documented that *T. gondii* catalase conferred protection against H_2O_2 exposure and contributed to virulence in mice. It was also noted that peroxidases can function as a defense mechanism against endogenously produced reactive oxygen intermediates and the oxidative stress imposed by the host. Artemisinins, which are thought to increase oxidative stress, were shown to be toxic against *Plasmodium* spp., *T. gondii*, and *Babesia* spp. [14, 40]. Likewise, (−)-epigallocatechin-3-gallate (EGCG), which was derived from green tea and thought to act as an endoperoxide, proved to be highly toxic against the growth of *Babesia*, both *in vitro* and *in vivo* [41]. Recently, a redox active antiproliferative and tumor cell differentiation agent, NSC3852, was demonstrated to be a potent inhibitor for *T. gondii* and *P. falciparum* [42].

Mitochondrial Drug Targets

The mitochondria in apicomplexan parasites contain the smallest known mitochondrial genome, a 6–7 kb that encodes only three proteins – cytochrome *b*, cytochrome oxidase I, and cytochrome oxidase III – and fragmented rRNA genes. The mitochondria have two main functions, namely electron transport and protein synthesis, both of which appear to be essential for the parasite's energy metabolism and survival. Consequently, the mitochondria constitute a potential target for antiparasitic drugs [43]. Atovaquone, a licensed antiparasitic agent for the treatment of clinical human malaria, toxoplasmosis, and babesiosis, inhibits electron transport at the *bc*1 complex (complex III) by interfering with the ubiquinol oxidation site of cytochrome *b*; this results in a collapse of the mitochondrial membrane and death of the parasite [44]. An alternative NADH (type II) dehydrogenase, which is located at the inner mitochondrial membrane but is absent from mammalian cells, is therefore considered to be an antiparasitic drug target [45]. The alternative type II NADH dehydrogenases (NDH2s) do not transport protons across the membrane and, unlike complex I, are not sensitive to rotenone [46]. A quinolone-like compound, 1-hydroxy-2-dodecyl-4 (1*H*) quinolone (HDQ) was shown to effectively reduce the replication rates of *P. falciparum* and *T. gondii in vitro* by inhibiting the electron transport chain at the level of ubiquinone reduction [45, 47].

Apicoplast Drug Targets

The apicoplast of apicomplexan parasites (notably, *Cryptosporidium* spp. lack apicoplasts) is a plastid of cyanobacterial origin, which bears similarities to both plant and algal chloroplasts and contains a circular genome of about 35 kb. Like plant chloroplasts, the apicoplasts are semi-autonomous, having their own genome and expression machinery. The apicoplast contains a range of metabolic pathways and housekeeping processes that differ radically from those of the host, thereby presenting ideal strategies for drug therapy. Cellular processes, including DNA replication, transcription, translation, post-translational modification, catabolism, and anabolism, all of which are known to be vital for survival of these parasites, are present in apicoplasts [48, 49]. A summary of drugs proposed for the targeting of apicoplast metabolic pathways in apicomplexan parasites is provided in Table 24.2.

DNA Replication, Transaction, and Protein Translation
Plastid genomes contain genes that encode elements of their own transcriptional and translational machinery, which are highly similar to those found in bacteria. Several antibacterials that target this machinery are also active against apicomplexan parasites. Antibiotics, including tetracyclines, clindamycin, and azithromycin, caused delayed death in *P. falciparum* and *T. gondii* [50]. These drugs are more potent against the progeny of treated parasites that inherit nonfunctional apicoplasts, leading to the death of the parasites. Antibacterial quinolones and their derivative (ciprofloxacin), which specifically disrupt two essential bacterial-type II topoisomerases, DNA gyrase and topoisomerase IV, were toxic to the growth of *T. gondii*, *Cryptosporidium*, *Theileria*, and *Plasmodium* parasites [51–54]. The presence of the β and β′ subunits of the polymerase (encoded by *rpoB*, *rpoC1*, and *rpoC2* genes) in the apicoplast genome suggests a transcription system [55]. The bacterial α2, β, and β′ polymerases were highly sensitive to rifampicin, which suggests that this drug blocks apicoplast transcription. Likewise, rifampicin was toxic toward parasites, including *Theileria* spp., *P. falciparum*, and *T. gondii* [53, 56, 57]. The organellar protein-synthesizing system was also suggested to occur within the apicoplast. The plastid genome encodes a comprehensive set of tRNAs and protein synthesis as an elongation factor (*tuf*A). However, numerous antibacterial agents, including doxycycline, clindamycin, and spiramycin, all of which block protein synthesis, have shown potent parasiticidal properties and have been used clinically for the treatment of malaria and toxoplasmosis [51, 58]. Similarly, amythiamicin, which prevents the release of activated *TufA*

Table 24.2 Drugs and herbicides proposed as targets of apicoplast metabolic pathways in apicomplexan parasites.

Metabolic pathway	Putative target	Component	IC$_{50}$
DNA replication	DNA gyrase	Ciprofloxacin	*Pf* 8–38 μM
			Tg 30 μM
			Cp 80 μg ml^{-1}
			Tp 123 μM
RNA transcription	RNA polymerase	Rifampicin	*Pf* 3 μM
			Tg 26.5 μM
			Tp 96 μM
Protein translation	23S rRNA	Clindamycin	*Pf* 20 nM
			Tg 10 nM
			Cp 20 μM
			Tp >400 μM
		Azithromycin	*Pf* 2 μM
			Tg 2 μM
			Cp 90 μM
		Spiramycin	*Tg* 40 ng ml^{-1}
		Thiostrepton	*Pf* 2 μM
			Tg NA
		Micrococcin	*Pf* 35 nM
		Chloramphenicol	*Pf* 10 μM
			Tg 5 μM
			Cp 300 μM
	16S rRNA	Doxycycline	*Pf* 100 μM
			Cp 200 μM
		Tetracycline	*Pf* 10 μM
			Tg 20 μM
			Cp 100 μM
	TufA	Amythiamicin	*Pf* 10 nM
Fatty acid biosynthesis	FabF and FabH	Thiolactomycin	*Pf* 50 μM
			Tg 100 μM
	FabI	Triclosan	*Pf* 1 μM
			Bb 50–100 μg ml^{-1}
			Tp 20 μM
	Acetyl-CoA carboxylase	Clodinafop	*Tg* 10 μM
		Quizlofop	*Tg* 100 μM
		Haloxypop	*Tg* 100 μM
		Fenoxaprop	*Pf* 144 μM
		Tralkoxydim	*Pf* 181 μM
		Diclofop	*Pf* 210 μM
Isoprenoid biosynthesis	DOXP reductoisomerase	Fosmidomycin	*Pf* 290–370 nM
			Bb 0.88 μg ml^{-1}
			Bbg 0.63 μg ml^{-1}
			Tg NA
			Tp NA
	FPPS	bisphosphonates	*Pf, Tg, Cp* < 200 μM

Abbreviation: *Tg, Toxoplasma gondii*; *Pf, Plasmodium falciparum*; *Cp, Cryptosporidium parvum*; *Bb, Babesia bovis*; *Bbg, Babesia bigemina*; *Tp, Theileria parva*.
NA: not active.

from the ribosome, lincosamides (lincomycin and clindamycin), and macrolides (erythromycin and azithromycin), which interact with the peptidyl transferase domain of 23S rRNA, proved to be toxic against the growth of many apicomplexan parasites [51, 54, 59–63]. Both, thiostrepton and micrococcin inhibited the growth of *P. falciparum in vitro* by interfering with the guanosine triphosphatase (GTPase) binding domain of the large subunit rRNA, which provides energy for translation [64]. Surprisingly, the same antibiotics were ineffective against *T. gondii*, which suggests that this parasite does not share the complete bacterial genotype [59]. In contrast, tetracyclines, which inhibit translation by binding to a small subunit of the ribosome, have been shown to decrease the growth of *Plasmodium* spp., *Cryptosporidium* spp., and *T. gondii* [54, 59, 65]. However, tetracyclines appear to inhibit protein translation not only in the apicoplast but also in the mitochondrion [50, 59].

Fatty Acid Synthesis

The apicoplast-localized fatty acid synthesis (FAS II) pathway is a metabolic process that is fundamentally divergent from the cytosolic type I pathway of the mammalian host cells and, therefore, represents a rational drug target in parasites. Fatty acids are the most abundant components in cells, and are essential components of membrane biogenesis, cell signaling, and the survival and pathogenesis of the parasites [66, 67]. The enzymes involved in fatty acid biosynthesis are present in *T. gondii* and *Plasmodium* spp., but not in *Babesia* and *Theileria*, suggesting that fatty acid synthesis is lacking in these parasites; consequently, the parasites are entirely dependent on the lipid from the host cells [11]. This adaptation might be due to an absence of a parasitophorous vacuole, which allows a more direct access to host metabolites. Thiolactomycin, which inhibits the type II pathway for fatty acid synthesis in bacteria by binding specifically to β-ketoacyl-ACP synthases, caused a reduction in the growth of both *T. gondii* and *P. falciparum* [66–68]. Likewise, triclosan was toxic to both *T. gondii* and *P. falciparum* by inhibiting enoyl-ACP-CoA reductase, which mediates the final step in the fatty acid synthesis cycle [61]. Unexpectedly, triclosan inhibited the *in vitro* growth of both *Babesia* spp. and *Theileria parva*, despite the absence of any fatty acid synthesis enzymes in these parasites [53, 69]. Currently, plastid acetyl-CoA carboxylase (ACC) is also being investigated as a drug target, and can be inhibited by aryloxyphenoxypropionate herbicides [59]. A number of herbicides, including clodinafop, quizalofop, and haloxyfop, have also been tested; of these, clodinafop was shown to be the most inhibitory, reducing the growth of *T. gondii in vitro* [70].

Isoprenoid Biosynthesis

Isoprenoid synthesis is an apicoplast lipid pathway that has also been suggested as a chemotherapeutic target for apicomplexan parasites. The isoprenoid precursor isopentenyl diphosphate can be synthesized via two independent pathways: (i) the standard acetate-mevalonic acid (MVA) pathway; and (ii) the alternative pathway, which utilizes pyruvate and glyceraldehyde 3-phosphate as metabolic precursors. 1-Deoxy-D-xylulose 5-phosphate (DOXP) reductoisomerase, the key enzyme of the non-mevalonate DOXP pathway that is absent from mammalian cells, is inhibited by fosmidomycin, a natural antibiotic originally developed for the treatment of bacterial infections [49, 59], and which has been shown to be toxic to the growth of *P. falciparum*, *B. bovis*, and *B. bigemina* [71, 72]. Whilst clinical trials with fosmidomycin verified its efficacy and safety in therapeutic applications against human malaria [71], the drug was found to be inactive against *T. gondii*, despite a database analysis of the *T. gondii* genome revealing sequences that were highly homologous to the DOXP pathway enzymes. The bisphosphonates, which target farnesyl pyrophosphate synthesis (FPPS), a key intermediate of the isoprenoid biosynthesis pathway, have also demonstrated potent effects against many protozoan parasites [73].

Invasion of the Host Cell as a Drug Target

The invasion process is an essential step in the life cycle of all apicomplexan parasites, and includes a concerted action of secretory adhesins, a myosin motor, factors regulating actin dynamics, and proteases. A current model of invasion proposes that the initial interaction occurs between host and parasite surface molecules, and is followed by reorientation and junction formation mediated by the release of higher-affinity transmembrane adhesins from a set of micronemes at the apical end of the parasite. These adhesins bind to the host cell-surface receptors through their ectodomains, and engage with a cortical actinomyosin motor through their cytoplasmic domains. The transmembrane microneme proteins and rhoptry proteins that serve in attachment and moving junction formation are believed to be the key steps in the invasion, and have in recent years been the major focus of vaccine research [74, 75]. Apicomplexan parasites rely on gliding motility for migration across biological barriers and host cell invasion. Such movement is dependent on an intact parasite actomyosin system. The glideosome complex in *T. gondii* comprises the motor complex (the major myosin heavy-chain TgMyoA, light-chain TgMLC1, and glideosome-associated proteins TgGAP45 and TgGAP50), which is anchored in the outer membrane of the inner membrane complex (IMC) and connected via filamentous actin (F-actin) and aldolase to the hexameric microneme protein complex that interacts with host-cell receptors. Interestingly, these homologs are present in all apicomplexan parasites, which suggests a conserved function throughout the entire phylum [76]. Drugs targeted at the apicomplexan class XIV myosin function, including butanedione monoxime (BDM), reduced the invasion of *T. gondii*, *B. bovis*, *C. parvum*, and *P. falciparum* into host cells [77–79]. Moreover, the compounds that disrupt the actin filament polymerization dynamics, including cytochalasin D, latrunculin, and jasplakinolide, had potent inhibitory effects on parasite replication [77, 78]. Proteins involved in calcium control and signaling represent another potential option for drug targets, due to the fact that calcium [Ca^{2+}] plays an important role in regulating the microneme secretion and the interaction between the protozoan ligands and the corresponding host-cell receptors [74, 75]. Recent investigations have suggested that the mechanism behind the efficacy of artemisinins against *T. gondii* involves the disruption of calcium homeostasis in the parasites, by inhibiting the sarcoplasmic-endoplasmic reticulum calcium ATPase (SERCA) protein [80]. Moreover, proteases have recently attracted much attention as drug targets, due to their indispensable role in the invasion of apicomplexan parasites, mediating the cleavage and shedding of secreted proteins. Protease inhibitors, including *N*-acetylleucylleucylnorleucinal (ALLN), chymostatin, phenylmethylsulfo-nyl fluoride (PMSF) (which inhibit the maturation of microneme proteins), 3,4-dichloro-2-benzopyran-1-one (3,4-DCI) and *N*-tosyl-L-lysine chloromethylketone (TLCK) (which inhibit the shedding of microneme proteins) each blocked the invasion of *T. gondii*, *Plasmodium* spp., and *B. divergens* [81–83]. A cysteine protease has also been noted to be involved in the invasion process, as evidenced by the effects of their inhibitors, which blocked the invasion and gliding motility in *T. gondii* and *Babesia* spp. by selectively impairing the release of microneme contents [84, 85].

Conclusion

The development of new drugs targeted at parasite pathogens represents an important step toward decreasing the morbidity and mortality associated with infection. Most of the drugs used today were discovered decades ago, and are often ineffective against resistant organisms and/or possess toxic side effects. Currently, the genomic databases of certain parasites provide a comprehensive insight into the biochemical aspects and metabolic map of these organisms, and into the transport processes that operate in the infected cell. Whilst such knowledge offers many possibilities for selectively attacking the parasites rather than their hosts, many of these targets have, unfortunately, not been fully studied. There is, however, a need to validate these target molecules, including their biochemical and structural characteristics. Likewise, a better understanding of the mechanisms that underlie drug resistance is also required in order to circumvent the emergence of resistance to new generations of parasite therapy in the near future. Although such studies will be time-consuming, they should not be seen as an obstacle to achieving the ultimate goal of developing a successful drug.

References

1 Homer, M.J., Aguilar-Delfin, S.R., Telford, P.J., Krause, P., and Persing, D.H. (2000) Babesiosis. *Clin. Microbiol. Rev.*, **13**, 451–469.

2 Kuttler, K.L. (1988) Worldwide impact of babesiosis, in *Babesiosis of Domestic Animals* (Ed. M. Ristic), CRC Press, Inc., Boca Raton, Florida, pp. 1–22.

3 Dubey, J.P. and Beattie, C.P. (1988) *Toxoplasmosis of Animals and Man*, CRC Press, Boca Raton, Florida, pp. 1–220.

4 Vial, H.J. and Gorenflot, A. (2006) Chemotherapy against babesiosis. *Vet. Parasitol.*, **138** (1–2), 147–160.

5 Weiss, L.M. (2002) Babesiosis in humans: A treatment review. *Expert Opin. Pharmacother.*, **3**, 1109–1115.

6 Khaw, M. and Panosian, C.B. (1995) Human antiprotozoal therapy: Past, present, and future. *Clin. Microbiol. Rev.*, **8** (3), 427–439.

7 Tagboto, S. and Townson, S. (2001) Antiparasitic properties of medicinal plants and other naturally occurring products. *Adv. Parasitol.*, **50**, 199–295.

8 Kim, K. and Weiss, L.W. (2004) Toxoplasma *gondii*: The model apicomplexan. *Int. J. Parasitol.*, **34**, 423–432.

9 Ajioka, J.W., Fitzpatrick, J.M., and Reitter, C.P. (2001) Toxoplasma *gondii* genomics: Shedding light on pathogenesis and chemotherapy. *Expert Rev. Mol. Med.*, **6**, 1–16.

10 Brayton, K.A., Lau, A.O., Herndon, D.R., Hannick, L., Kappmeyer, L.S., Berens, S.J., Bidwell, S.L. *et al.* (2007) Genome sequence of *Babesia bovis* and comparative analysis of apicomplexan hemoprotozoa. *PLoS Pathog.*, **3** (10), 1401–1413.

11 Fleige, T., Limenitakis, J., and Soldati-Favre, D. (2010) Apicoplast: keep it or leave it. *Microbes Infect.*, **12** (4), 253–262.

12 Joet, T., Holterman, L., Stedman, T.T., Kocken, C.H., Van Der Wel, A., Thomas, A.W., and Krishna, S. (2002) Comparative characterization of hexose transporters of *Plasmodium knowlesi*, *Plasmodium yoelii* and *Toxoplasma gondii* highlights functional differences within the apicomplexan family. *Biochem. J.*, **368**, 923–929.

13 Derbyshire, E.T., Franssen, F.J., de Vries, E., Morin, C., Woodrow, C.J., Krishna, S., and Staines, H.M. (2008) Identification, expression and characterization of a *Babesia bovis* hexose transporter. *Mol. Biochem. Parasitol.*, **161** (2), 124–129.

14 Nagai, A., Yokoyama, N., Matsuo, T., Bork, S., Hirata, H., Xuan, X., Zhu, Y., Claveria, F.G., Fujisaki, K., and Igarashi, I. (2003) Growth-inhibitory effects of artesunate, pyrimethamine, and pamaquine against *Babesia equi* and *Babesia caballi* in in vitro cultures. *Antimicrob. Agents Chemother.*, **47**, 800–803.

15 Hyde, J.E. (2007) Targeting purine and pyrimidine metabolism in human apicomplexan parasites. *Curr. Drug Targets*, **8** (1), 31–47.

16 Aboge, G.O., Jia, H., Terkawi, M.A., Goo, Y.K., Nishikawa, Y., Sunaga, F., Namikawa, K., Tsuji, N., Igarashi, I., Suzuki, H., Fujisaki, K., and Xuan, X. (2008) Cloning, expression, and characterization of *Babesia gibsoni* dihydrofolate reductase-thymidylate synthase: Inhibitory effect of antifolates on its catalytic activity and parasite proliferation. *Antimicrob. Agents Chemother.*, **52** (11), 4072–4080.

17 Gherardi, A. and Sarciron, ME. (2007) Molecules targeting the purine salvage pathway in Apicomplexan parasites. *Trends Parasitol.*, **23** (8), 384–389.

18 Hyde, J.E. (2008) Fine targeting of purine salvage in *Cryptosporidium* parasites. *Trends Parasitol.*, **24** (8), 336–339.

19 Chaudhary, K., Darling, J.A., Fohl, L.M., Sullivan, W.J. Jr, Donald, R.G., Pfefferkorn, E.R., Ullman, B., and Roos, D.S. (2004) Purine salvage pathways in the apicomplexan parasite *Toxoplasma gondii*. *J. Biol. Chem.*, **279** (30), 31221–31227.

20 Galazka, J., Striepen, B., and Ullman, B. (2006) Adenosine kinase from *Cryptosporidium parvum*. *Mol. Biochem. Parasitol.*, **149** (2), 223–230.

21 El Kouni, M.H. (2003) Potential chemotherapeutic targets in the purine metabolism of parasites. *Pharmacol. Ther.*, **99** (3), 283–309.

22 De Koning, H.P., Bridges, D.J., and Burchmore, R.J. (2005) Purine and pyrimidine transport in pathogenic protozoa: From biology to therapy. *FEMS Microbiol. Rev.*, **29** (5), 987–1020.

23 El Kouni, M.H., Guarcello, V., Al Safarjalani, O.N., and Naguib, F.N.M. (1999) Metabolism and selective toxicity of 6-nitrobenzylthioinosine in *Toxoplasma gondii*. *Antimicrob. Agents Chemother.*, **43**, 2437–2443.

24 Kerr, E.A. and Gero, A.M. (1991) The toxicity of adenosine analogues against *Babesia bovis in vitro*. *Int. J. Parasitol.*, **21**, 747–751.

25 Sullivan, W.J. Jr., Dixon, S.E., Li, C., Striepen, B., and Queener, S.F. (2005) IMP dehydrogenase from the protozoan parasite *Toxoplasma gondii*. *Antimicrob. Agents Chemother.*, **49** (6), 2172–2179.

26 Iltzsch, M.H. (1993) Pyrimidine salvage pathways in *Toxoplasma gondii*. *J. Euk. Micro.*, **40**, 24–28.

27 Fox, B.A. and Bzik, D.J. (2002) De novo pyrimidine biosynthesis is required for virulence of *T. gondii*. *Nature*, **415** (6874), 926–929.

28 Schumacher, M.A., Carter, D., Scott, D.M., Roos, D.S., Ullman, B., and Brennan, R.G. (1998) Crystal structures of *Toxoplasma gondii* uracil phosphoribosyltransferase reveal the atomic basis of pyrimidine

discrimination and prodrug binding. *EMBO J.*, **17** (12), 3219–3232.

29 Gero, A.M., O'Sullivan, W.J., Wright, I.G., and Mahoney, D.F. (1983) The enzymes of pyrimidine biosynthesis in *Babesia bovis* and *Babesia bigemina*. *Aust. J. Exp. Biol. Med. Sci.*, **61**, 239–243.

30 Holland, J.W., Gero, A.M., and O'Sullivan, W.J. (1983) Enzymes of de novo pyrimidine biosynthesis in *Babesia rodhaini*. *J. Protozool.*, **30** (1), 36–40.

31 Bentley, R. (1990) The shikimate pathway – a metabolic tree with many branches. *Crit. Rev. Biochem. Mol. Biol.*, **25**, 307–384.

32 Roberts, F., Roberts, C.W., Johnson, J.J., Kyle, D.E., Krell, T., Coggins, J.R., Coombs, G.H. *et al.* (1989) Evidence for the shikimate pathway in apicomplexan parasites. *Nature*, **393**, 801–805.

33 Wiwanitkit, V. (2007) Plasmodium and host lactate dehydrogenase molecular function and biological pathways: Implication for antimalarial drug discovery. *Chem. Biol. Drug Des.*, **69** (4), 280–283.

34 Al-Anouti, F., Tomavo, S., Parmley, S., and Ananvoranich, S. (2004) The expression of lactate dehydrogenase is important for the cell cycle of *Toxoplasma gondii*. *J. Biol. Chem.*, **279** (50), 52300–52311.

35 Kavanagh, K.L., Elling, R.A., and Wilson, D.K. (2004) Structure of *Toxoplasma gondii* LDH1: Active-site differences from human lactate dehydrogenases and the structural basis for efficient $APAD^+$ use. *Biochemistry*, **43** (4), 879–889.

36 Conners, R., Schambach, F., Read, J., Cameron, A., Sessions, R.B., Vivas, L., Easton, A., Croft, S.L., and Brady, R.L. (2005) Mapping the binding site for gossypol-like inhibitors of *Plasmodium falciparum* lactate dehydrogenase. *Mol. Biochem. Parasitol.*, **142**, 137–148.

37 Bork, S., Okamura, M., Boonchit, S., Hirata, H., Yokoyama, N., and Igarashi, I. (2004) Identification of *Babesia bovis* L-lactate dehydrogenase as a potential chemotherapeutical target against bovine babesiosis. *Mol. Biochem. Parasitol.*, **136** (2), 165–172.

38 Muller, S. (2004) Redox and antioxidant systems of the malaria parasite

Plasmodium falciparum. Mol. Microbiol., **53** (5), 1291–1305.

39 Ding, M., Kwok, L.Y., Schlüter, D., Clayton, C., and Soldati, D. (2004) The antioxidant systems in *Toxoplasma gondii* and the role of cytosolic catalase in defense against oxidative injury. *Mol. Microbiol.*, **51** (1), 47–61.

40 Cumming, J.N., Ploypradith, P., and Posner, G.H. (1997) Antimalarial activity of artemisinin (qinghaosu) and related trioxanes: Mechanism (s) of action. *Adv. Pharmacol.*, **37**, 253–297.

41 Aboulaila, M., Yokoyama, N., and Igarashi, I. (2010) Inhibitory effects of (−)-epigallocatechin-3-gallate from green tea on the growth of *Babesia* parasites. *Parasitology*, **137** (5), 785–791.

42 Strobl, J.S., Seibert, C.W., Li, Y., Nagarkatti, R., Mitchell, S.M., Rosypal, A.C., Rathore, D., and Lindsay, D.S. (2009) Inhibition of *Toxoplasma gondii* and *Plasmodium falciparum* infections in vitro by NSC3852, a redox active antiproliferative and tumor cell differentiation agent. *J. Parasitol.*, **95** (1), 215–223.

43 Mather, M.W., Henry, K.W., and Vaidya, A.B. (2007) Mitochondrial drug targets in apicomplexan parasites. *Curr. Drug Targets*, **8** (1), 49–60.

44 Mather, M.W. and Vaidya, A.B. (2008) Mitochondria in malaria and related parasites: Ancient, diverse and streamlined. *J. Bioenerg. Biomembr.*, **40** (5), 425–443.

45 Saleh, A., Friesen, J., Baumeister, S., Gross, U., and Bohne, W. (2007) Growth inhibition of *Toxoplasma gondii* and *Plasmodium falciparum* by nanomolar concentrations of 1-hydroxy-2-dodecyl-4 (1*H*)quinolone, a high-affinity inhibitor of alternative (type II) NADH dehydrogenases. *Antimicrob. Agents Chemother.*, **51** (4), 1217–1222.

46 Kerscher, S., Zickermann, V., and Brandt, U. (2008) The three families of respiratory NADH dehydrogenases. *Res. Probl. Cell Differ.*, **45**, 185–222.

47 Lin, S.S., Gross, U., and Bohne, W. (2009) Type II NADH dehydrogenase inhibitor 1-hydroxy-2-dodecyl-4(1*H*)quinolone leads to collapse of mitochondrial inner-

membrane potential and ATP depletion in *Toxoplasma gondii. Eukaryot. Cell*, **8** (6), 877–887.

48 Roos, D.S. (1999) The apicoplast as a potential therapeutic target in *Toxoplasma* and other apicomplexan parasites: Some additional thoughts. *Parasitol. Today*, **15** (1), 5–7.

49 Ralph, S.A., D'Ombrain, M.C., and McFadden, G.I. (2001) The apicoplast as an antimalarial drug target. *Drug Resist. Update*, **4** (3), 145–151.

50 Fleige, T. and Soldati-Favre, D. (2008) Targeting the transcriptional and translational machinery of the endosymbiotic organelle in apicomplexans. *Curr. Drug Targets*, **9** (11), 948–956.

51 Fichera, ME. and Roos, DS. (1997) A plastid organelle as a drug target in apicomplexan parasites. *Nature*, **390**, 407–409.

52 Anquetin, G., Greiner, J., and Vierling, P. (2005) Quinolone-based drugs against *Toxoplasma gondii* and *Plasmodium* spp. *Curr. Drug Targets Infect. Disord.*, **5** (3), 227–245.

53 Lizundia, R., Werling, D., Langsley, G., and Ralph, S.A. (2009) Theileria apicoplast as a target for chemotherapy. *Antimicrob. Agents Chemother.*, **53** (3), 1213–1217.

54 Woods, K.M., Nesterenko, M.V., and Upton, S.J. (1996) Efficacy of 101 antimicrobials and other agents on the development of *Cryptosporidium parvum* in vitro. *Ann. Trop. Med. Parasitol.*, **90** (6), 603–615.

55 Gray, M.W. and Lang, B.F. (1998) Transcription in chloroplasts and mitochondria: A tale of two polymerases. *Trends Microbiol.*, **6**, 1–3.

56 Pukrittayakamee, S., Viravan, C., Charoenlarp, P., Yeamput, C., Wilson, R.J., and White, N.J. (1994) Antimalarial effects of rifampin in *Plasmodium vivax* malaria. *Antimicrob. Agents Chemother.*, **38** (3), 511–514.

57 Olliaro, P., Gorini, G., Jabes, D., Regazzetti, A., Rossi, R., Marchetti, A., Tinelli, C., and Della Bruna, C. (1994) In-vitro and in-vivo activity of rifabutin against *Toxoplasma gondii. J. Antimicrob. Chemother.*, **34** (5), 649–657.

58 Fichera, M.E., Bhophale, M.K., and Roos, D.S. (1995) In vitro assays elucidate peculiar kinetics of clindamycin action against *Toxoplasma gondii. Antimicrob. Agents Chemother.*, **39**, 1530–1537.

59 Waller, R.F. and McFadden, G.I. (2005) The apicoplast: A review of the derived plastid of apicomplexan parasites. *Curr. Issues Mol. Biol.*, **7** (1), 57–79.

60 Derouin, F. (1995) New pathogens and mode of action of azithromycin: *Toxoplasma gondii. Pathol. Biol.*, **43** (6), 561–564.

61 McConkey, G.A., Rogers, M.J., and McCutchan, T.F. (1997) Inhibition of *Plasmodium falciparum* protein synthesis: Targeting the plastid-like organelle with thiostrepton. *J. Biol. Chem.*, **272**, 2046–2049.

62 Beckers, C.J., Roos, D.S., Donald, R.G., Luft, B.J., Schwab, J.C., Cao, Y., and Joiner, K.A. (1995) Inhibition of cytoplasmic and organellar protein synthesis in *Toxoplasma gondii*. Implications for the target of macrolide antibiotics. *J. Clin. Invest.*, **95** (1), 367–376.

63 Pfefferkorn, E.R. and Borotz, S.E. (1994) Comparison of mutants of *Toxoplasma gondii* selected for resistance to azithromycin, spiramycin, or clindamycin. *Antimicrob. Agents Chemother.*, **338**, 31–37.

64 Rogers, M.J., Cundliffe, E., and McCutchan, T.F. (1998) The antibiotic micrococcin is a potent inhibitor of growth and protein synthesis in the malaria parasite. *Antimicrob. Agents Chemother.*, **42**, 715–716.

65 Tabbara, K.F., Sakuragi, S., and O'Connor, G.R. (1982) Minocycline in the chemotherapy of murine toxoplasmosis. *Parasitology*, **84**, 297–302.

66 Sonda, S. and Hehl, A.B. (2006) Lipid biology of Apicomplexa: Perspectives for new drug targets, particularly for *Toxoplasma gondii. Trends Parasitol.*, **22** (1), 41–47.

67 Goodman, C.D. and McFadden, G.I. (2007) Fatty acid biosynthesis as a drug target in apicomplexan parasites. *Curr. Drug Targets*, **8** (1), 15–30.

68 Gornicki, P. (2003) Apicoplast fatty acid biosynthesis as a target for medical intervention in apicomplexan parasites. *Int. J. Parasitol.*, **33**, 885–896.

69 Bork, S., Yokoyama, N., Matsuo, T., Claveria, F.G., Fujisaki, K., and Igarashi, I. (2003) Growth inhibitory effect of triclosan on equine and bovine *Babesia* parasites. *Am. J. Trop. Med. Hyg.*, **68**, 334–340.

70 Zuther, E., Johnson, J.J., Haselkorn, R., McLeod, R., and Gornicki, P. (1999) Growth of *Toxoplasma gondii* is inhibited by aryloxyphenoxypropionate herbicides targeting acetyl-CoA carboxylase. *Proc. Natl Acad. Sci. USA*, **96** (23), 13387–13392.

71 Wiesner, J. and Jomaa, H. (2007) Isoprenoid biosynthesis of the apicoplast as drug target. *Curr. Drug Targets*, **8** (1), 3–13.

72 Sivakumar, T., Aboulaila, M.R.A., Khukhuu, A., Iseki, H., Alhassan, A., Yokoyama, N., and Igarashi, I. (2008) In vitro inhibitory effect of fosmidomycin on the asexual growth of *Babesia bovis* and *Babesia bigemina*. *J. Protozool. Res.*, **18**, 71–78.

73 Srivastava, A., Mukherjee, P., Desai, P.V., Avery, M.A., and Tekwani, B.L. (2008) Structural analysis of farnesyl pyrophosphate synthase from parasitic protozoa, a potential chemotherapeutic target. *Infect. Disord. Drug Targets*, **8** (1), 16–30.

74 Sibley, L.D. (2004) Intracellular parasite invasion strategies. *Science*, **304**, 248–253.

75 Morgan, R.E., Evans, K.M., Patterson, S., Catti, F., Ward, G.E., and Westwood, N.J. (2007) Targeting invasion and egress: From tools to drugs? *Curr. Drug Targets*, **8** (1), 61–74.

76 Soldati-Favre, D. (2008) Molecular dissection of host cell invasion by the apicomplexans: The glideosome. *Parasite.*, **15** (3), 197–205.

77 Keeley, A. and Soldati, D. (2004) The glideosome: A molecular machine powering motility and host-cell invasion by Apicomplexa. *Trends Cell Biol.*, **14** (10), 528–532.

78 Lew, A.E., Dluzewski, A.R., Johnson, A.M., and Pinder, J.C. (2002) Myosins of *Babesia bovis*: Molecular characterization, erythrocyte invasion, and phylogeny. *Cell Motil. Cytoskeleton*, **52** (4), 202–220.

79 Wetzel, D.M., Schmidt, J., Kuhlenschmidt, M.S., Dubey, J.P., and Sibley, L.D. (2005) Gliding motility leads to active cellular invasion by *Cryptosporidium parvum* sporozoites. *Infect Immun.*, **73** (9), 5379–5387.

80 Nagamune, K., Beatty, W.L., and Sibley, L.D. (2007) Artemisinin induces calcium-dependent protein secretion in the protozoan parasite *Toxoplasma gondii*. *Eukaryot. Cell*, **6** (11), 2147–2156.

81 Dowse, T.J. and Soldati, D. (2005) Rhomboid-like proteins in Apicomplexa: Phylogeny and nomenclature. *Trends Parasitol.*, **21** (6), 254–258.

82 Carruthers, V.B. (2006) Proteolysis and *Toxoplasma* invasion. *Int. J. Parasitol.*, **36** (5), 595–600.

83 Montero, E., Rafiq, S., Heck, S., and Lobo, C.A. (2007) Inhibition of human erythrocyte invasion by *Babesia divergens* using serine protease inhibitors. *Mol. Biochem. Parasitol.*, **153** (1), 80–84.

84 Teo, C.F., Zhou, X.W., Bogyo, M., and Carruthers, V.B. (2007) Cysteine protease inhibitors block *Toxoplasma gondii* microneme secretion and cell invasion. *Antimicrob. Agents Chemother.*, **51** (2), 679–688.

85 Okubo, K., Yokoyama, N., Govind, Y., Alhassan, A., and Igarashi, I. (2007) *Babesia bovis*: effects of cysteine protease inhibitors on in vitro growth. *Exp. Parasitol.*, **117** (2), 214–217.

25

Search for Drugs and Drug Targets against *Babesia bovis*, *Babesia bigemina*, *Babesia caballi*, and *Babesia* (*Theileria*) *equi*

Sabine Bork-Mimm

Abstract

Babesiosis is one of the most important protozoan diseases of livestock from an economic viewpoint. Moreover, due to its zoonotic character and widespread occurrence, it is also gaining attention as an emerging malaria-like disease. Whilst several drugs have been used to treat babesiosis, and most exert adequate babesia-cidal activities, some have demonstrated severe side effects, toxicity, and potent resistance phenomena. Thus, the quest for new and effective drugs and drug combinations, as well as chemotherapeutic targets, is a continuing endeavor. In this chapter, for a better understanding of the organism, the parasite will be introduced in terms of its etiology and life cycle, followed by details of compounds recently and successfully tested in *in vitro* cultured bovine and equine *Babesia* spp. The elucidation of possible drug targets, and their future application *in vivo*, is also discussed. Notably, the focus of the chapter relates to the development or detection of possible compounds and drug targets, rather than to current standard treatments and established drug targets.

Introduction

Characterization of the *Babesia* Parasites

History

In 1888, the Hungarian pathologist, Dr Babes, reported the "intra-erythrocytic bacterium *Hematococcus bovis*" as being the cause of death in more than 40 000 cattle in Romania [1]. Five years later, the causative agent was identified as *Babesia bigemina*, and the role of the tick clarified in its transmission, thus establishing the life cycle of the *Babesia* species [2]. Both, *Babesia caballi* and *Babesia equi* are the causative agents of equine babesiosis, which was first described in 1883 as anthrax fever [3] and later as African horse sickness [4]. In 1901, intraerythrocytic parasites were identified in the blood of African horses, and classified as *Piroplasma equi* [5]; however, when a few years later two alternative species were identified, they were initially named

Apicomplexan Parasites. Edited by Katja Becker
Copyright © 2011 WILEY-VCH Verlag GmbH & Co. KGaA, Weinheim
ISBN: 978-3-527-32731-7

Nuttallia equi and *Piroplasma caballi* [6, 7], but subsequently were reclassified as *B. equi* and *B. caballi* [8, 9]. Based on several characteristics of the *Theileria* spp., attempts were made to re-describe *B. equi* as *Theileria equi* [10].

Etiology and Life Cycle

For their maintenance and developmental cycles, the apicomplexan *Babesia* spp. [11] require an invertebrate host (e.g., tick) and a vertebrate host (e.g., mammals and avian species) [12, 13]. Within the tick vector, the ingested babesial organisms infect the intestinal tissues and move to the salivary glands, from which they are introduced into a new vertebrate host [14]. In the vertebrate host, the bovine *Babesia* spp. and *B. caballi* develop exclusively in the erythrocytes [15, 16]; this contrasts with *B. equi*, which has a stage of schizogony in the lymphocytes before the erythrocyte invasion occurs [17]. The intraerythrocytic stages include trophozoites, dividing forms, and merozoites; the latter are released from the erythrocyte and subsequently invade new erythrocytes.

Epidemiology and Impact

Bovine *Babesia* spp. are widely distributed and of major importance in most tropical and subtropical parts of Africa and Asia, as well as in Australia and Central and South America [18, 19]. In Europe, only *B. bigemina* has been reported in Sicily [20]. The equine *Babesia* spp. occur in Portugal, France, Italy, Hungary, in many parts of the former Union of Soviet Socialist Republics [21], Switzerland [22], African countries, Madagascar [23], South and Central America [24], and in some Asian countries [25]. Currently, the United States of America [26], the United Kingdom, Germany, Austria, Australia, and Japan are free from equine babesiosis [27, 28], although due to the horse trade and equestrian events equine babesiosis may become a major constraint for these countries in the future [29].

Clinical Signs

Babesia bovis and Babesia bigemina
Although the clinical symptoms are similar and caused by both parasites, *B. bovis* is more virulent, displaying a high fever ($>40\,°C$), weakness, depression, hemoglobinuria, anemia, icterus, diarrhea, abortion, tremors, and recumbency [30]. Cerebral babesiosis in *B. bovis* infection shows hyperesthesia, nystagmus, circling, head pressing, convulsions, paralysis, and coma that inevitably leads to death [19]. The clinical signs in subacute cases of *B. bovis* infection are similar to those in acute cases, but less pronounced [30].

Babesia equi and Babesia caballi
Clinical cases are caused by the more widespread and virulent *B. equi* (10–50% mortality), while *B. caballi* infection (10% mortality) is usually clinically unapparent [11], occasionally leading to mild symptoms [23]. In the peracute case of *B. equi*

infection, horses are either found dead or moribund [31]. The acute cases are characterized by high fever, anorexia, malaise, elevated respiratory and pulse rates, congestion of the mucous membranes, and tachycardia [32]. Subacute cases additionally show weight loss, increased pulse and respiratory rates, intermittent fever, hemoglobinuria, and splenomegaly [31]. Chronic cases present nonspecific clinical signs [23]. Foals and neonates display weakness at birth or a rapid onset of symptoms such as listlessness, anemia, and severe icterus [3].

Diagnosis

It is not possible to distinguish infections caused by *B. equi* and *B. caballi* in equidae [23] and *B. bovis* and *B. bigemina* in bovidae [33] on the basis of clinical manifestations alone. Romanowsky-type staining methods, such as thin or thick blood-smear techniques stained with Giemsa, are useful for the identification of clinically infected animals, but are unsuitable for the detection of carriers [34]. Serological tests, notably the complement fixation (CF) test [35] and the indirect fluorescent antibody (IFA) test [36], both of which are prescribed for the international horse trade, and the enzyme-linked immunosorbent assay (ELISA) are preferred [37]. DNA probes for the detection of equine *Babesia* DNA are highly sensitive [38]. Latent infections in carrier animals can also be detected using xenodiagnosis, *in vitro* cultivation [39], or the polymerase chain reaction (PCR) [40].

Control and Treatment

Diamidine derivatives, such as diminazene diaceturate, amicarbalide diisethionate, phenamidine, and imidocarb dipropionate are effective [41, 42]. Several compounds such as acridine dyes, and bisazo dyes (trypan blue) have been tested over the years, but not recommended due to the occurrence of severe dermatological irritations [43]. The development of drug resistance, as reported in *Plasmodium* spp. infections [44], has not yet become a major menace to babesial chemotherapy. However, one group has reported that drug-resistant *Babesia* spp. can be produced experimentally, and may also occur naturally [45]. During the past few decades, the ultimate breakthrough in the development of highly effective drugs and the detection of important drug targets have been lacking, and this has resulted in a need to develop more specific, fast-acting new treatments for babesiosis [46]. Supportive therapies include the administration of antibiotics, essential phospholipids, and blood-, fluid-, and electrolyte-infusions [19]. Moreover, in contrast to equine babesiosis, several different live vaccines are available to control *B. bovis* and *B. bigemina* infections [47].

The First Approaches Start at the Bench

In most countries, appropriate *in vivo* experimental systems are unavailable because the inoculation of *Babesia* parasites into equidae and bovidae is outlawed [48]. Thus, two different models have been established: (i) the *in vitro* culture system, and (ii) a mouse model using severe combined immunodeficient (SCID) mice. Both models represent well-established, economical, and reliable systems for the investigation of drug efficacy in protozoan parasites [49]. The *in vitro* culture system is easily applicable to the bovine and equine *Babesia* spp., whereas the SCID model requires a greater degree of sophistication and only covers the two bovine *Babesia* spp. and *B. caballi* [50]. To date, a successful and applicable SCID mouse model for the *B. equi* is not available.

The *in vitro* culture technique, using a well-established continuous microaerophilous stationary phase culture system [51], was developed as an *in vitro* test protocol to assess parasite susceptibility to babesiacidal compounds. Thus, it represents a very important first approach for the possible application of the tested compound *in vivo*.

Recently, the important roles of calcium ions and sialic acids in the invasion of equine and/or bovine *Babesia* spp. into erythrocytes were elucidated, which represent significant contributions to the development of appropriate culture systems for the discovery of babesiacidal compounds in future [52–54]. Moreover, a hexose-transporter in the genome of *B. bovis* that provides new insights into the glucose metabolism of erythrocytic-stage *Babesia* infections – and thus may contribute to the development of suitable compounds – was identified, expressed, and characterized [55].

During the past few years, a plethora of compounds has been successfully tested against *in vitro*-cultured equine and bovine *Babesia* spp. [56, 57]. Some of these are well-known drugs that have already proved their efficacy in protozoa, such as *Plasmodium* spp. The most promising approaches are introduced in the following subsections.

Compounds Successfully Tested *In Vitro*

Among the strategies applied to the identification and application of new drugs, some have been designed to directly target essential enzymes in the parasitic pathways. Typical examples include fatty acid synthesis (e.g., triclosan, clodinafop-propargyl) or the microsomal cytochrome P450-dependent system (e.g., ketoconazole). Other drugs are either based on: (i) the detection of specific drug targets; (ii) a indirect intervention of the drug with the parasite's metabolism (i.e., to disturb the parasitic heme catabolism (e.g., clotrimazole); or (iii) to disturb the parasite's invasion into host erythrocytes (e.g., glycosaminoglycans).

Triclosan

Triclosan, a synthetic 2-hydroxydiphenyl ether, has been used for more than 35 years as a widely accepted and safe antiseptic ingredient in healthcare products and many household fabrics and plastics [58]. It targets the *trans*-2-enoyl-acyl-carrier-protein-(ACP)-reductase (ENR, inhA or FabI), an enzyme located in the plastid-like organelle in *B. bovis* that catalyzes the final, regulatory step in the parasitic type II fatty acid synthesis [59, 60]. The growth of *in vitro*-cultured equine and bovine *Babesia* spp. was inevitably inhibited at concentrations as low as $50\,\mu g\,ml^{-1}$, without showing any toxicity to the host cells (i.e., erythrocytes) over time [61]. To date, the exact mode of drug action in bovine and equine *Babesia* spp. remains unknown, but clearly requires clarification.

Clotrimazole, Ketoconazole, and Clodinafop-Propargyl

The azole derivative ketoconazole (KC) was released as the first available oral treatment of non-life-threatening endemic mycoses, while the imidazole clotrimazole (CLT) was the first antifungal azole for clinical applicability. Clodinafop-propargyl (CP) is an official registered herbicide for use in spring wheat. Both imidazole derivatives and CP have been successfully tested for their inhibitory efficacy in several protozoa [62, 63]. Despite the antimalarial effects of imidazole compounds having been studied since the early 1980s [64], their modes of action are still not fully understood. In several *Plasmodium* spp., KC inhibits the sterol-14-α-dimethylase, a microsomal cytochrome P450-dependent enzyme [65], whereas CLT binds to the erythrocytic heme to disturb the parasitic hemoglobin catabolism [66]. The herbicide, CP, interacts with the fatty acid/isoprenyl pyrophosphate biosynthesis by targeting the essential enzyme, acetyl-CoA-carboxylase (ACC) [67].

In vitro-cultured equine and bovine *Babesia* spp. were exposed to the compounds, which were applied either independently or as various combinations [68, 69]. In this case, the IC_{50}-values ranged between 2 and $23.5\,\mu M$ for CLT, between 6 and $50\,\mu M$ for KC, and between 265 and $450\,\mu M$ for CP. Whereas, in *B. bovis*, all combinations showed a synergistic effect, the KC/CP combination failed to have any effect on the efficacy in *B. bigemina*, and even acted antagonistically in *B. caballi*. No beneficial synergistic effects were observed in *B. equi*, for any of the combinations used.

Glycosaminoglycans (GAGs)

During the asexual growth cycle in a natural host, *Babesia* merozoites first invade the host erythrocytes via multiple interactions of several protozoan adhesive ligands with unknown receptors on the host cell surface; subsequently, they divide within the infected erythrocytes, which are finally destroyed. Thus, an understanding of the basic molecular interaction(s) that occur in erythrocyte invasion by *Babesia* merozoites may accelerate the successful development of new therapeutic and preventive measures for babesiosis [70]. Most of the GAGs defined below successfully inhibited the invasion of erythrocytes and hepatocytes by several *Plasmodium* and *Theileria* spp. [71, 72]. Various studies have implicated a range of molecular interactions between sulfated GAGs on the surface of host cells and the hemoprotozoan adhesive ligands. Thus, several sulfated and nonsulfated GAGs were tested against bovine and equine *Babesia* spp. [73]. Despite extensive studies, the precise inhibitory mechanisms against these hemoprotozoa remain poorly understood. Notably, the multiplication of *B. bovis*, *B. bigemina*, *B. equi*, and *B. caballi* was significantly inhibited *in vitro* in the presence of several heparin preparations, showing a complete clearance of the intracellular parasites. Additionally, in *B. bovis*, dextran sulfate, fucoidan, and chondroitin sulfate B strongly inhibited parasitic growth, while all but chondroitin sulfate B induced a significant accumulation of extracellular merozoites in the culture. In contrast, chondroitin sulfate A, keratan sulfate, and protamine sulfate, as well as nonsulfated dextran and hyaluronic acid, had no influence on merozoite growth. The indication that the asexual growth of *B. bovis* was inhibited by specific sulfated GAGs provided important insights into the molecular interactions occurring during erythrocyte invasion by *B. bovis* merozoites. Both, equine and bovine *Babesia* spp. appeared to express GAG-affinity molecules on the surfaces of the extracellular merozoites, and these might serve as host receptors for merozoite attachment [74].

Recently Discovered or Proposed Drug Targets

Phosphatidylcholine Formation in *Babesia bovis* [75]

Erythrocytes infected with bovine *Babesia* spp. accumulate lipids and show significant increases in phosphatidylcholine (PC; the most abundant phospholipid), phosphatidic acid, diacylglycerol, and cholesteryl esters. Notably, radiolabeled choline was incorporated into complex lipids, especially PC. Moreover, the incorporation of [^{3}H]myoinositol into phosphatidylinositol by the parasite strongly indicated that the lipid changes in infected erythrocytes could be explained on the basis of the lipid biosynthetic activities of *B. bovis*. Infected erythrocytes had higher ratios of saturated to unsaturated fatty acids, and *B. bovis* cultures did not desaturate [^{14}C]stearate. However, the erythrocyte membrane phospholipid composition remained unchanged by the parasite. Subsequently, PC formation was shown to be the main biosynthetic process in infected erythrocytes. The striking differences in the contents of PC between host erythrocytes and the parasite may represent an interesting target for chemotherapy and immunoprophylaxis against bovine babesiosis.

Cyclin-Dependent Kinase Inhibitors in *Babesia bovis* [57]

Cyclin-dependent kinases (CDKs) are essential for regulation of the eukaryotic cell cycle, and the CDK inhibitors roscovitine, purvalanol A, CDK2 Inhibitor II, and CGP74514A each significantly suppressed the growth of *in vitro*-cultured *B. bovis*. Whereas, the first three of these inhibitors exerted their effects on the early stages of the intraerythrocytic stages of *B. bovis*, CGP74514A (a CDK1-specific inhibitor) hampered the invasion of merozoites into the erythrocyte. These findings supported the chemotherapeutic potential of the CDK inhibitors to treat babesiosis.

Transfection Systems for *Babesia bovis* [76]

When, recently, the *B. bovis* genome was sequenced, many genes were identified with unknown functions. The ability to supplement and knockout unknown and previously identified genes in the parasite represents a valuable tool not only to better understand gene function, but also to design specific drug targets. Initially, transient transfection constructs were generated by use of the promoter and the 3' region of the rap-1 genes of *B. bovis* controlling the expression of luciferase as a reporter. Thereafter, stronger promoters (i.e., the ef-1α promoter) were identified and further characterized. It was essential to possess a transient transfection system in order to develop a stable transfection technique using a plasmid designed to target the integration of a *GFP-BSD* gene into the *B. bovis* ef-1α locus. After transfection, parasites which were resistant to the anti-babesial antibiotic drug blasticidin (BSD), and constantly expressed the *GFP-BSD* gene, were generated. Thus, the introduction, integration, and expression of exogenous genes in *B. bovis* were realized.

Babesia bovis ʟ-Lactate Dehydrogenase [77]

The successful isolation and characterization of the enzyme kinetics of a novel cDNA clone encoding the *B. bovis* ʟ-lactate dehydrogenase (LDH; ʟ-lactate:NAD$^+$ oxidore-ductase, E.C. 1.1.1.27; BbLDH) was subsequently realized. This enzyme is essential for the anaerobic phase of parasite growth, as it catalyzes the reversible dehydration of ʟ-lactate to pyruvic acid, using nicotinamide adenine dinucleotide (NAD$^+$) or its reduced form (NADH) as a cofactor. The amino acid sequence of BbLDH showed high identities to other protozoan LDHs, and conserved four LDH motifs. A typical pentapeptide insertion, which is detectable in the mobile loop of all *Plasmodia* species and *T. gondii* [78, 79], was also found especially in BbLDH. This suggested that the enzyme kinetics and biological roles of the protozoan LDHs were not the same as those of the mammalian LDHs as the insertion, located in the active loop site of the enzyme, had never been observed in mammalian LDHs. It has since been suggested that the BbLDHs spontaneously escape the parasitic body and partially imitate the function of the host LDH. Such a role would be essential for the anaerobic glycolysis pathway of the host erythrocytes, when exchanging to a suitable environment for parasitic survival.

The polyphenolic binaphthyl disesquiterpene, gossypol, which competitively inhibits the binding of NADH to LDH, occurs in the seed, stem, and roots of the cotton plant (*Gossypium* sp.), where it plays a role in the plant's natural defense mechanism by causing infertility in insects that feed on it [80]. The strong inhibitory effect of gossypol in *in vitro* cultures of *B. bovis*, in reasonable concentrations, suggested a possible role for BbLDH as a drug target when developing a structure-based design for anti-babesial chemotherapeutics.

Conclusion

The widespread and unrestricted application of currently existing babesiacidal compounds may lead to the possible development of side effects and resistance phenomena. The search for new compounds and novel drug targets is, therefore, highly desirable. Effective drugs used to treat various infectious diseases commonly function by targeting the specific pathway or activity of the pathogens. For example, many antibiotics inhibit protein synthesis in bacteria, but not in mammals, due to differences in the bacterial and mammalian pathways. Hence, the goal of sustained drug research should be based on the finding and exploitation of differences between parasites and their hosts, based on their life cycle and metabolism. In order to prevent adverse side effects and possible drug-resistance phenomena, however, it will become necessary to evaluate parasite-specific targets and subsequently to develop target-directed compounds.

An understanding of the biology of the organism may be significant when developing novel strategies to prevent bovine babesiosis. Subsequent structure-based drug design studies may, therefore, lead to the production of new compounds as anti-babesial drugs.

References

1 Babes, V. (1888) Sur l'hémoglobinurie bactérienne du boeuf. *Compt. Rend. Acad. Sci. (Paris)*, **107**, 692–694.

2 Smith, T. and Kilbourne, F.L. (1893) Investigations into the nature, causation, and prevention of southern cattle fever. *USDA Bureau Anim. Indust. Bull.*, **1**, 1–301.

3 Henning, M.W. (ed.) (1956) *Animal Diseases in South Africa*, 3rd edn, Central News Agency Ltd, Pretoria.

4 Roberts, E.D., Morehouse, L.G., Gainer, J.H., and McDaniel, H.A. (1962) Equine piroplasmosis. *J. Vet. Med. Assoc.*, **141**, 1323–1329.

5 Laveran, A. (1901) Contribution à l'étude du *Piroplasm equi*. *C. R. Seances Soc. Biol.*, **12**, 385–388.

6 Nuttall, G.H.F. and Strickland, C. (1910) Die Parasiten der Pferdepiroplasmose resp. der "Biliary Fever". *Zentralbl. Bakteriol. Parasitenkd. Infkrankh. Abtlg. I*, **56**, 524–525.

7 Nuttall, G.H.F. and Strickland, C. (1912) On the occurrence of two species of parasites in equine piroplasmosis or biliary fever. *Parasitology*, **5**, 65–96.

8 Franca, C. (1910) Sur la classification des piroplasmes et description de deux formes de ces parasites. *Inst. Roy. Bact. Camera Pestana*, **3**, 11–18.

9 Neitz, W.O. (1956) Classification, transmission and biology of piroplasms of domestic animals. *Ann. N. Y. Acad. Sci.*, **64**, 56–111.

10 Mehlhorn, H. and Schein, E. (1998) Redescription of *Babesia equi* Laveran 1901 as *Theileria equi*. *Parasitol. Res.*, **84**, 467–475.

11 Levine, N.D. (1985) *Veterinary Protozoology*, Iowa State University Press, Ames, Iowa, p. 414.

12 Schein, E., Rehbein, G., Voigt, W.P., and Zweygarth, E. (1981) Babesia *equi* (Laveran 1901). Development in horses and in lymphocyte culture. *Tropenmed. Parasitol.*, **32**, 223–227.

13 Spielman, A. (1976) Human babesiosis on Nantucket Island: Transmission by nymphal *Ixodes* ticks. *Am. J. Trop. Med. Hyg.*, **25**, 784–787.

14 Spielman, A., Wilson, M.L., Levine, J.F., and Piesman, J. (1985) Ecology of *Ixodes dammini*-borne human babesiosis and Lyme disease. *Annu. Rev. Entomol.*, **30**, 439–460.

15 Holbrook, A.A., Johnson, A.J., and Madden, P.A. (1968) Equine piroplasmosis: Intraerythrocytic development of *Babesia caballi* (Nutall) and *Babesia equi*. *Am. J. Vet. Res.*, **29**, 297–303.

16 Friedhoff, K.T. (1981) Morphologic aspects of *Babesia* in the tick, in *Babesiosis* (eds M. Ristic and J.P. Kreier), Academic Press, New York.

17 Moltmann, U.G., Mehlhorn, H., Schein, E., Rehbein, G., Voigt, W.P., and Zweygarth, E. (1983) Fine structure of *Babesia equi* (Leveran, 1901) within lymphocytes and erythrocytes of horses:

An *in vitro* and *in vivo* study. *J. Parasitol.*, **68**, 111–120.

18 Mahoney, D.F., Wright, I.G., Goodgerm, B.V., Mirre, G.B., Sutherst, R.W., and Utech, K.B. (1981) The transmission of *Babesia bovis* in herds or European and Zebu×European cattle infested with the tick, *Boophilus microplus*. *Aust. Vet. J.*, **57**, 461–469.

19 De Vos, A.J. and Potgieter, F.T. (1994) Bovine babesiosis, in *Infectious Diseases of Livestock*, vol. 1 (eds J.A.W. Coetzer, G.R. Thomson, and R.C. Tustin), Oxford University Press, pp. 278–294.

20 De Vico, G., Macri, B., Sammartino, C., and Loria, G.R. (1999) Bovine babesiosis in Sicily: Preliminary study on pathology. *Parassitologia*, **41**, 37–38.

21 Friedhoff, K.T. and Soule, C. (1996) An account on equine babesiosis. *Rev. Sci. Tech. Office Int. Epizootics*, **15**, 1191–1201.

22 Meier, H., Hauser, R., and Hentrich, B. (1997) Equinella 1996. Schweizerische Vereinigung fuer Pferdewesen (SVPM) und Bundesamt fuer Veterinaerwesen (BVET). Available at: http://www.sciquest.org.nz/node/62597.

23 de Waal, D.T. and van Herden, J. (1994) Equine babesiosis, in *Infectious Diseases of Livestock*, vol. 1 (eds J.A.W. Coetzer, G.R. Thomson, and R.C. Tustin), Oxford University Press, pp. 295–304.

24 Knowles, R.C., Mathis, R.M., Bryant, J.E., and Willers, K.H. (1966) Equine piroplasmosis. *J. Am. Vet. Med. Assoc.*, **148**, 407–410.

25 Xuan, X., Chahan, B., Huang, X., Yokoyama, N., Makala, L.H., Igarashi, I., Fujisaki, K. *et al.* (2002) Diagnosis of equine piroplasmosis in Xinjiang province of China by the enzyme-linked immunosorbent assays using recombinant antigens. *Vet. Parasitol.*, **108**, 179–182.

26 Holman, P.J., Hietala, S.K., Kayashima, L.R., Olson, D., Waghela, S.D., and Wagner, G.G. (1997) Case report: Field-acquired subclinical *Babesia equi* infection confirmed by *in vitro* culture. *J. Clin. Microbiol.*, **35**, 474–476.

27 Joyner, L.P., Donnelly, J., and Huck, R.A. (1981) Complement fixation tests for equine piroplasmosis (*Babesia equi* and *B. caballi*) performed in the UK during 1976 to 1979. *Equine Vet. J.*, **13**, 103–106.

28 Friedhoff, K.T. (1982) Die Piroplasmen der Equiden -Bedeutung für den Internationalen Pferdeverkehr. *Berl. Muench. Tieraerztl. Wochenschr.*, **95**, 368–374.

29 Friedhoff, K.T., Tenter, A.M., and Mueller, I. (1990) Haemoparasites of equines: Impact on international trade of horses. *Rev. Sci. Tech. Office Int. Epizooties*, **9**, 1187–1194.

30 Callow, L.L. (1984) *Protozoal and Rickettsial Diseases. Animal Health in Australia*, vol. V, Australian Bureau of Animal Health, Australian Government Public Services, Canberra, p. 264.

31 Littlejohn, A. (1963) Babesiosis, in *Equine Medicine and Surgery* (eds J.F. Bone, E.J. Catcott, A.A. Gabel, L.E. Johnson, and W.F. Riley), American Veterinary Publications, Santa Barbara, CA.

32 Schein, E. (1988) Equine babesiosis, in *Babesiosis of Domestic Animals and Man* (ed. M. Ristic), CRC Press, Boca Raton, Florida USA, pp. 197–208.

33 Boese R., Jorgensen, W.K., Dalgliesh, R.J., Friedhoff, K.T., and de Vos, A.J. (1995) Current state and future trends in the diagnosis of babesiosis. *Vet. Parasitol.*, **57**, 61–74.

34 Mahoney, D.F. and Saal, J.R. (1961) Bovine babesiosis: Thick blood films for the detection of parasitaemia. *Aust. Vet. J.*, **37**, 44–47.

35 USDA (United States Department of Agriculture) Animal and Plant Health Inspection Service, Veterinary Services (1997) Complement fixation test for the detection of antibodies to *Babesia caballi* and *Babesia equi* – microtitration test. USDA, National Veterinary Services Laboratories, Ames, Iowa, USA. Available at: http://www.aphis.usda.gov/about_aphis/.

36 Madden, P.A. and Holbrook, A.A. (1968) Equine piroplasmosis: Indirect fluorescent antibody test for *Babesia caballi*. *Am. J. Vet. Res.*, **29**, 117–123.

37 Goetz, F. (1982) Untersuchungen ueber die Brauchbarkeit von ELISA,

IFAT, IHA, und KBR zum Nachweis von *Babesia equi* Infektionen. DVM Thesis, University of Munich, Germany.

38 Posnett, E.S. and Ambrosio, R.E. (1989) DNA probes for the detection of *Babesia equi*. *Mol. Biochem. Parasitol.*, **34**, 75–78.

39 Zweygarth, E., Just, M.C., and de Waal, D.T. (1997) *In vitro* cultivation of *Babesia equi*: Detection of carrier animals and isolation of parasites. *Onderstepoort J. Vet. Res.*, **64**, 51–56.

40 Figueroa, J.V., Chieves, L.P., Johnson, G.S., and Buening, G.M. (1992) Detection of *Babesia bigemina*-infected carriers by polymerase chain reaction amplification. *J. Clin. Microbiol.*, **30**, 2576–2582.

41 Anonymous (1984) Tick and tick-borne disease control. A practical field manual, in *Tick-borne disease control*. Vol. 11, Food and Agriculture Organisation of the United Nations (FAO), Rome, Italy.

42 Rashid, H.B., Chaudhry, M., Rashid, H., Pevez, K., Khan, M.A., and Mahmood, A.K. (2008) Comparative efficacy of diminazene diaceturate and diminazene aceturate for the treatment of babesiosis in horses. *Trop. Anim. Health Prod.*, **40**, 463–467.

43 Jansen, B.C. (1953) The parasiticidal effect of aureomycin (Lederle) on *Babesia equi* (Laveran 1899) in splenectomised donkeys. *Onderstepoort J. Vet. Res.*, **26**, 175–182.

44 Hyde, J.E. (2002) Mechanism of resistance of *Plasmodium falciparum* to antimalarial drugs. *Microbes Infect.*, **4**, 165–174.

45 Dalgliesh, R.J. and Stewart, N.P. (1977) Tolerance to imidocarb induced experimentally in tick-transmitted *Babesia argentina*. *Aust. Vet. J.*, **53**, 176–180.

46 Vial, H.J. and Gorenflot, A. (2006) Chemotherapy against babesiosis. *Vet. Parasitol.*, **138**, 147–160.

47 de Waal, D.T. and Combrink, M.P. (2006) Live vaccines against bovine babesiosis. *Vet. Parasitol.*, **138**, 88–96.

48 Tsuji, M., Yutaka, T., Arai, S., Okada, H., and Ishihara, C. (1995) Use of Bo-RBC-

SCID mouse model for isolation of a parasite from grazing calves in Japan. *Exp. Parasitol.*, **81**, 512–518.

49 McColm, A.A. and McHardy, N. (1984) Evaluation of a range of antimicrobial agents against the parasitic protozoa, *Plasmodium falciparum*, *Babesia rodhaini* and *Theileria parva in vitro*. *Ann. Trop. Med. Parasitol.*, **78**, 345–354.

50 Rodríguez Bautista, J.L., Ikadai, H., You, M., Battsetseg, B., Igarashi, I., Nagasawa, H., and Fujisaki, K. (2001) Molecular evidence of *Babesia caballi* (Nuttall and Strickland, 1910) parasite transmission from experimentally-infected SCID mice to the ixodid tick, *Haemaphysalis longicornis* (Neuman, 1901). *Vet. Parasitol.*, **102**, 185–191.

51 Avarzed, A., Igarashi, I., Kanemura, T., Hirumi, K., Omata, Y., Saito, A., Oyamada, T. *et al.* (1997) Improved *in vitro* cultivation of *Babesia caballi*. *J. Vet. Med. Sci.*, **59**, 479–481.

52 Okamura, M., Yokoyama, N., Wickramathilaka, N.P., Takabatake, N., Ikehara, Y., and Igarashi. I. (2005) *Babesia caballi* and *Babesia equi*: implications of host sialic acids in erythrocytes infection. *Exp Parasitol.*, **110**, 406–411.

53 Okubo, K., Wilawan, P., Bork, S., Okamura, M., Yokoyama, N., and Igarashi, I. (2006) Calcium-ions are involved in erythrocyte invasion by equine *Babesia* parasites. *Parasitology*, **133** (Pt 3), 289–294.

54 Takabatake, N., Okamura, M., Yokoyama, N., Okubo, K., Ikehara, Y., and Igarashi, I. (2007) Involvement of a host erythrocyte sialic acid content in *Babesia bovis* infection. *J. Vet. Med. Sci.*, **69**, 999–1004.

55 Derbyshire, E.T., Franssen, F.J., de Vries, E., Morin, C., Woodrow, C.J., Krishna, S., and Staines, H.M. (2008) Identification, expression and characterisation of a *Babesia bovis* hexose-transporter. *Mol. Biochem. Parasitol.*, **161**, 124–129.

56 Nagai, A., Yokoyama, N., Matsuo, T., Bork, S., Hirata, H., Xuan, X., Zhu, Y. *et al.* (2003) Growth-inhibitory effects of artesunate, pyrimethamine, and pamaquine against *Babesia equi* and

Babesia caballi in *in vitro* cultures. *Antimicrob. Agents Chemother.*, **47**, 800–803.

57 Nakamura, K., Yokoyama, N., and Igarashi, I. (2007) Cyclin-dependent kinase inhibitors block erythrocyte invasion and intraerythrocytic development of *Babesia bovis in vitro*. *Parasitology*, **134** (Pt 10), 1347–1353.

58 Levy, C.W., Roujeinikova, A., Sedelnikova, S., Baker, P.J., Stuitje, A.R., Slabas, A.R., Rice, W.R. *et al.* (1999) Molecular basis of triclosan activity. *Nature*, **398**, 383–384.

59 McLeod, R., Muench, S.P., Rafferty, J.B., Kyle, D.E., Mui, E.J., Kirisits, M.J., Mack, D.G. *et al.* (2001) Triclosan inhibits the growth of *Plasmodium falciparum* and *Toxoplasma gondii* by inhibition of apicomplexan Fab I. *Int. J. Parasitol.*, **31**, 109–113.

60 Heath, R.J., White, S.W., and Rock, C.O. (2002) Inhibitors of fatty acid synthesis as antimicrobial chemotherapeutics. *Appl. Microbiol. Biotechnol.*, **58**, 695–703.

61 Bork, S., Yokoyama, N., Matsuo, T., Claveria, F.G., Fujisaki, K., and Igarashi, I. (2003) Growth inhibitory effect of triclosan on equine and bovine *Babesia* parasites. *Am. J. Trop. Med.*, **68**, 334–340.

62 El-Sayed, M. and Anwar, A. (2010) Intralesional sodium stibogluconate alone or its combination with either intramuscular sodium stibogluconate or oral ketoconazole in the treatment of localized cutaneous leishmaniasis: a comparative study. *Eur. Acad. Dermatol. Venereol.*, **24** (3), 335–340.

63 Tiffert, T., Ginsburg, H., Krugliak, M., Elford, B.C., and Lew, V.L. (2000) Potent antimalarial activity of clotrimazole in *in vitro* cultures of *Plasmodium falciparum*. *Proc. Natl Acad. Sci. USA*, **97**, 331–336.

64 Pfaller, M.A. and Krogstad, D.J. (1981) Imidazole and polyene activity against chloroquine-resistant *Plasmodium falciparum*. *J. Infect. Dis.*, **4**, 372–375.

65 Srivastava, P. and Pandey, V.C. (2000) Studies on hepatic mitochondrial cytochrome P-450 during *Plasmodium yoelii* infection and pyrimethamine

treatment in mice. *Ecotoxicol. Environ. Saf.*, **46**, 19–22.

66 Huy, N.T., Kamei, K., Kondo, Y., Serada, S., Kanaori, K., Takano, R., Tajima, K. *et al.* (2001) Effect of antifungal azoles on the heme detoxification system of malarial parasite. *J. Biochem.*, **131**, 437–444.

67 Wilson, R.J. (2002) Progress with parasite plastids. *J. Mol. Biol.*, **31**, 257–274.

68 Bork, S., Yokoyama, N., Matsuo, T., Claveria, F.G., Fujisaki, K., and Igarashi, I. (2003) Clotrimazole, ketoconazole, and clodinafop-propargyl inhibit the *in vitro* growth of *Babesia bigemina* and *Babesia bovis* (Phylum Apicomplexa). *Parasitology*, **127** (Pt 4), 311–315.

69 Bork, S., Yokoyama, N., Matsuo, T., Claveria, F.G., Fujisaki, K., and Igarashi, I. (2003) Clotrimazole, ketoconazole, and clodinafop-propargyl as potent growth inhibitors of equine *Babesia* parasites during *in vitro* culture. *J. Parasitol.*, **89**, 604–606.

70 Yokoyama, N., Suthisak, B., Hirata, H., Matsuo, T., Inoue, N., Sugimoto, C., and Igarashi, I. (2002) Cellular localization of *Babesia bovis* merozoite rhoptry-associated protein 1 and its erythrocyte-binding activity. *Infect. Immun.*, **70**, 5822–5826.

71 Xiao, L., Yang, C., Patterson, P.S., Udhayakumar, V., and Lal, A.A. (1996) Sulfated polyanions inhibit invasion of erythrocytes by plasmodial merozoites and cytoadherence of endothelial cells to parasitized erythrocytes. *Infect. Immun.*, **4**, 1373–1378.

72 Hagiwara, K., Takahashi, M., Ichikawa, T., Tsuji, M., Ikuta, K., and Ishihara, C. (1997) Inhibitory effect of heparin on red blood cell invasion by *Theileria sergenti* merozoites. *Int. J. Parasitol.*, **27**, 535–539.

73 Bork, S., Yokoyama, N., Ikehara, Y., Kumar, S., Sugimoto, C., and Igarashi, I. (2004) Growth-inhibitory effect of heparin on *Babesia* parasites. *Antimicrob. Agents Chemother.*, **48**, 236–241.

74 Bork, S., Yokoyama, N., Hashiba, S., Nakamura, K., Takabatake, N., Okamura,

M., lkehara, Y., and lgarashi, I. (2007) Asexual growth of *Babesia bovis* is inhibited by specific sulfated glycoconjugates. *J. Parasitol.*, **93**, 1501–1504.

75 Florin-Christensen, J., Suarez, C.E., Florin-Christensen, M., Hines, S.A., McElwain, T.F., and Palmer, G.H. (2000) Phosphatidylcholine formation is the predominant lipid biosynthetic event in the hemoparasite *Babesia bovis*. *Mol. Biochem. Parasitol.*, **106**, 147–156.

76 Suarez, C.E., and McElwain, T.F. (2010) Transfection systems for *Babesia bovis*: A review of methods for the transient and stable expression of exogenous genes. *Vet. Parasitol.*, **167** (2–4), 205–215.

77 Bork, S., Okamura, M., Boonchit, S., Hirata, H., Yokoyama, N., and Igarashi, I. (2004) Identification of *Babesia bovis* L-lactate dehydrogenase as a potential chemotherapeutical target against bovine babesiosis. *Mol. Biochem. Parasitol.*, **136**, 165–172.

78 Winter, V.J., Cameron, A., Tranter, R., Sessions, R.B., and Brady, R.L. (2003) Crystal structure of *Plasmodium berghei* lactate dehydrogenase indicates the unique structural differences of these enzymes are shared across the *Plasmodium* genus. *Mol. Biochem. Parasitol.*, **131**, 1–10.

79 Yang, S., and Parmley, S.F. (1997) Toxoplasma *gondii* expresses two distinct lactate dehydrogenase homologous genes during its life cycle in intermediate hosts. *Gene*, **184**, 1–12.

80 Coutinho, E.M. (2002) Gossypol: A contraceptive for men. *Contraception*, **65**, 259–263.

26
Orlistat: A Repositioning Opportunity as a Growth Inhibitor of Apicomplexan Parasites?

*Christian Miculka, Hon Q. Tran, Thorsten Meyer, Anja R. Heckeroth, Stefan Baumeister, Frank Seeber, and Paul M. Selzer**

Abstract

Orlistat, a saturated derivative of naturally occurring lipstatin, is administered orally as an anti-obesity agent in humans. It acts as a covalent inhibitor of lipases, thereby blocking the hydrolysis of triglycerides from nutrition into free fatty acids. Orlistat also inhibits the C-terminal enzymatic domain of fatty acid synthetase (FAS). In both cases, binding to and inhibition of the serine hydrolase domain is the presumed mode of action. As the fatty acid synthesis of protozoan parasites has been extensively discussed as a valid target for pharmaceutical intervention, an investigation was made into orlistat's effect on *Plasmodium falciparum* and *Toxoplasma gondii* in cell cultures. The *in vitro* replication of blood-stage *P. falciparum* in human erythrocytes and intracellular growth of the tachyzoite form of *T. gondii* was found to be inhibited at submicromolar concentrations. Blast searches for a serine hydrolase consensus pattern in the proteomes of *P. falciparum, T. gondii* and *Eimeria tenella* showed that a number of proteins contain the serine hydrolase pattern. The amino acid sequences surrounding the serine active sites, including the human enzyme, displayed three conserved amino acid residues, $G \times S \times G$. While these proteins are serine hydrolases with high probability, their cellular functions are unknown. These studies provide the foundation to conduct efficacy studies on *E. tenella* in cell culture, and to explore orlistat's potential for the prophylaxis or treatment of coccidiosis in chicken and diseases caused by other apicomplexans.

Repositioning of Actives: New, But Not Really New

Approved drugs are the pharmaceutical industry's greatest assets; they already have distinguished themselves from hundreds of thousands of other compounds by passing the myriad of tests related to disease relevancy, safety and efficacy. In the human pharmaceuticals field, finding a new application and indication for an

*Corresponding author

Apicomplexan Parasites. Edited by Katja Becker
Copyright © 2011 WILEY-VCH Verlag GmbH & Co. KGaA, Weinheim
ISBN: 978-3-527-32731-7

approved drug has become an increasingly important approach in recent years. Repositioning success stories and start-up companies betting on repositioning as a business model are increasing significantly in number [1]. Drug candidates from a repositioning – also termed redirecting, repurposing, reprofiling – approach lead to significantly lower risks in development [2]. Toxicology data, manufacturing processes and, in part, formulation approaches, may have been generated already. The triad of higher success probability, lower cash requirements and shorter duration drive the value of such R&D projects up [3]. The annual conference "Drug Repositioning Summit," held by the Cambridge Healthtech Institute (http://www.healthtech.com/rjv/09) with participants from both industry and academia, also proves the increasing interest in repositioning actives.

Interestingly, the repositioning of drug actives from either the human to the veterinary field (or originally, to a lesser extent *vice versa*) occurred earlier than within the human field [4]. For example, diethylcarbamazine, an inhibitor of arachidonic acid metabolism, and halofuginone, a specific inhibitor of collagen type 1 synthesis, which originally were intended for the treatment of filariasis and malaria in humans, respectively, were later introduced into the veterinary field for heartworm prophylaxis (diethylcarbamazine) and avian coccidiosis or bovine cryptosporidiosis (halofuginone), respectively [5–7]. Most interestingly, halofuginone is currently under evaluation for use against pulmonary fibrosis, pancreatic or liver fibrosis, and cirrhosis [8–10]. Due to the inhibition of collagen synthesis in connective tissue, halofuginone is used also as a coating material for medical devices, such as sutures, catheters, and stents [11, 12]. In some cases, the human pharma companies' R&D programs around particular modes of action or structural classes have yielded dedicated actives for the veterinary field. Examples of this include Hoechst's fourth-generation cephalosporin cefquinome [13] for the treatment of bovine pneumonia and mastitis, and Merck's cyclooxygenase-2 (COX-2) inhibitor firocoxib [14], an anti-inflammatory used in dogs and horses. Allopurinol, approved for the treatment of hyperuricemia in humans, is also used in the treatment of canine leishmaniasis [15]. Miltefosine, originally intended as an antineoplastic agent, is applied in the treatment of canine leishmaniasis caused by *Leishmania infantum* [16]. The antibacterial agent tetracycline has been shown to lead to defects in apicoplast function in *Plasmodium falciparum* [17] and to affect the bacterium *Wolbachia*, which is endosymbiotic to *Brugia malayi* [18]; elimination of the endosymbiont is known to lead to sterilization of the adult female *Brugia* worm. Entacapone, used for treatment against Parkinson's disease (Comtan®)[1]), was discovered as lead compound against multi-drug-resistant tuberculosis [19]. Recently, a set of more than 800 approved human drugs has been used in a phenotypic screen for the identification of antischistosomal compounds [20]. A similar screen with 2160 FDA-approved drugs, bioactive compounds and natural products has been performed for *Trypanosoma brucei* and *P. falciparum* [21, 22]. Systematic text mining and literature searches aimed at the identification of new drug uses, or the correlation of drug action and disease

---

1) Comtan is a registered trademark of Orion Corporation in the United States of America and elsewhere.

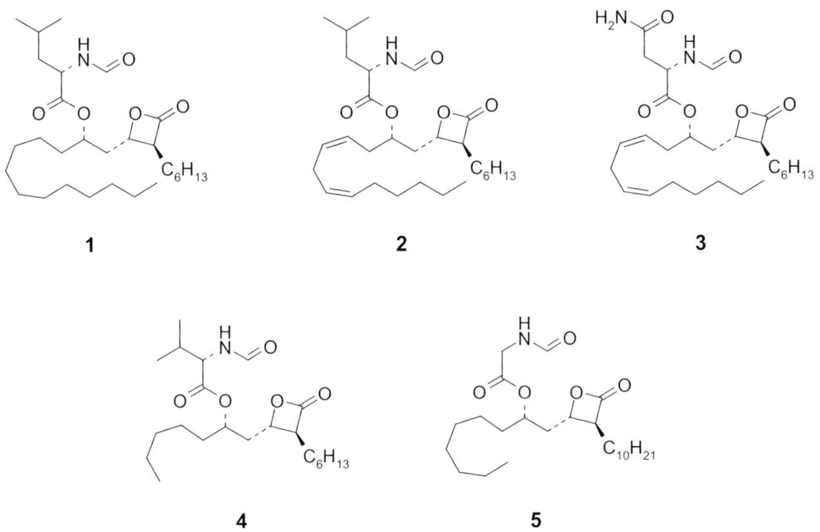

Figure 26.1 Examples of β-lactone pancreatic lipase inhibitors. Orlistat (**1**), lipstatin (**2**), esterastin (**3**), valilactone (**4**), and panclicin D (**5**).

mechanisms by comparing label and off-label use of FDA-approved drugs, have been undertaken [23, 24].

Another potentially very interesting molecule for the treatment of apicomplexan parasites of veterinary and human importance could be orlistat. This is marketed under the trade names Xenical®[2] or Alli®[3] as an orally administered anti-obesity agent and has a described mode of action toward lipases and fatty acid synthetase (FAS) [25, 26]. Both enzymatic activities could also be essential for parasite growth. Here, the current status of orlistat as a repositioning opportunity is summarized.

Chemistry and Synthesis of Orlistat

Orlistat (**1**, tetrahydrolipstatin; see Figure 26.1) [[(1′S)-1-[(2″S,3″S)-3-hexyl-4-oxo-oxetan-2-yl]methyl]dodecyl]-(2S)-2-formamido-4-methyl-pentanoate (CAS 96829-58-2)] is the saturated derivative of the naturally occurring lipstatin (**2**) from which it is obtained as a white stable crystalline solid following chemical reduction [27]. Lipstatin itself was isolated as a pale yellow oil from *Streptomyces toxytricini* by a group from Hoffmann-La Roche [28, 29], and it belongs to a class of β-lactone pancreatic lipase inhibitors of microbial origin. Members of this class also include structurally related esterastin (**3**) [30], valilactone (**4**) [31], and panclicin D (**5**) [32], all of which have an unusual δ-hydroxy-β-lactone moiety in common. This core feature is derived from

2) Xenical is a registered trademark of Hoffmann-La Roche in the United States of America and elsewhere.
3) Alli is a registered trademark of Glaxo Smith Kline in the United States of America and elsewhere.

either substituted 3,5-dihydroxy-2-hexyl- or decylpentanoic acids, and the natural products distinguish themselves only by the nature of the amino acid residue attached. Extensive spectroscopic and chemical studies on **2** have revealed absolute configuration [33] (Figure 26.1).

Due to its intriguing structure and use for prevention and treatment of obesity and as a cholesterol-lowering agent, orlistat has been at the center of interest for synthetic chemists in both academia and in industry. Until 2009, four formal, seven enantiospecific and 16 asymmetric total syntheses have been reported in scientific journals [34], and numerous patents have been filed claiming improved chemical processes [35] and new methods for producing crystalline forms of **1** [36]. Synthetic approaches toward orlistat include, among others, a stereoselective Prins cyclization [37], aldol [38], asymmetric allylboronation [39], [2 + 2], and [2 + 3] cycloaddition reactions [40, 41].

The first syntheses of orlistat and lipstatin stereoisomers were reported from the research group of Hoffmann-La Roche [33], who aimed at establishing the absolute configuration of the natural product and its saturated derivative. The first example of a stereoselective total synthesis of **1** was reported soon after, and its disconnection strategy comprised an up-front establishment of the desired hydroxy acid stereochemistry prior to β-lactone formation with late-stage introduction of the *N*-formylleucine residue [42]. The key transformation of the forward synthesis, setting the desired (2*S*,3*S*) configuration, is a chiral auxiliary controlled aldol-type reaction between ketene silyl acetal **7** and the pre-functionalized aldehyde **6**. During the course of the reaction, two new chiral centers are formed, thereby yielding two hydroxy esters as a 3:1 mixture of diastereoisomers. The major isomer **8** that comprises the desired *threo* configuration at C-2″,C-3″ was separated from the reaction mixture and further converted to the corresponding δ-hydroxy-β-lactone **9**. After final insertion of the (*S*)-*N*-formylleucine moiety under Mitsunobu conditions proceeding with concurrent inversion of configuration at C-1′, desired (−)-tetrahydrolipstatin (**1**) was obtained (Figure 26.2).

Orlistat's Modes of Action

Orlistat reversibly inhibits gastric and pancreatic lipases by covalently binding to the active serine residue [25]. With only low bioavailability, the site of action of orlistat is the lumen of the stomach and small intestine, where it blocks the hydrolysis of nutritionally derived triglycerides into free fatty acids. Thus, because only the latter form can be absorbed, the caloric intake from fat is reduced. Another target of orlistat is fatty acid synthetase (FAS). FAS (type I) catalyzes the reaction of acetyl-CoA, malonyl-CoA, and NADPH through multiple cycles of fatty acid elongation to form palmitate. The enzyme belongs to the class of megasynthetases and acts exclusively as a homodimer, with each monomer containing 2511 amino acid residues with a size of 272 kDa [43]. Orlistat inhibits the C-terminal enzymatic domain possessing thioesterase activity, and elicits cytostatic and cytotoxic effects at concentrations close to

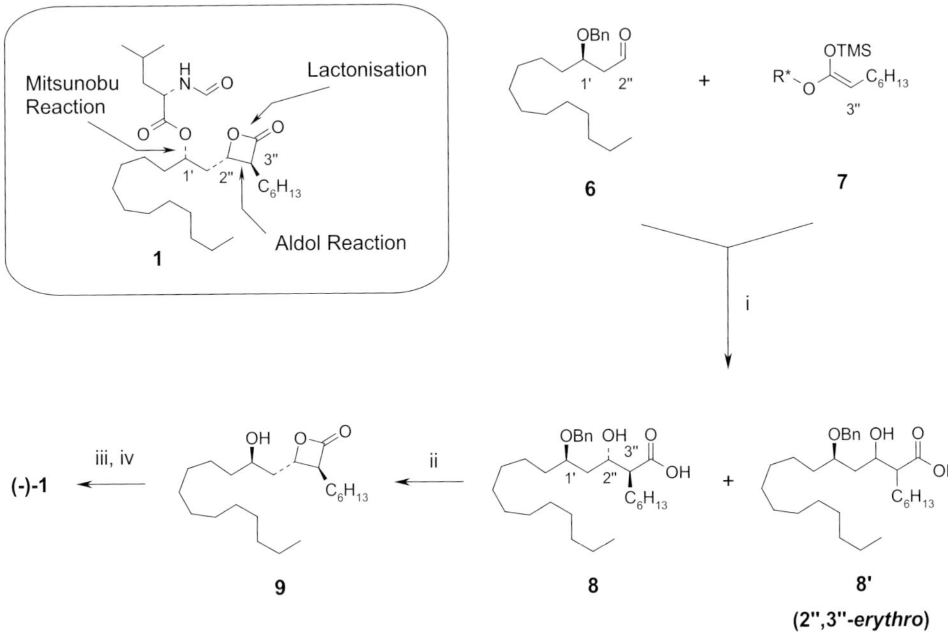

Figure 26.2 Barbier's retrosynthetic analysis and forward synthesis of (−)-tetrahydrolipstatin (1) (R* = (−)-N-Methylephedrine, Bn = benzyl, TMS = trimethylsilyl) [42]. Reagents and conditions: (i) Aldol reaction and saponification: TiCl$_4$, dichloromethane, −78 °C, then separation of diastereoisomers, then KOH, methanol, reflux, 26% overall; (ii) Lactonization: PhSO$_2$Cl, pyridine, 0 °C, 61%; (iii) Debenzylation: Pd/C, hydrogen, tetrahydrofuran, 70%; (iv) Mitsunobu reaction: (S)-N-formylleucine, triphenylphosphine, diethylazodicarboxylate, tetrahydrofuran, 0 °C, 77%.

cellular IC$_{50}$-values of the inhibition of FAS [26]. The thioesterase domain is responsible for releasing palmitate from the acyl carrier protein of the enzyme. The covalent binding of orlistat to human FAS has been demonstrated, based on crystal structure analysis [44] (Figure 26.3). A phenotype identical to that induced by orlistat treatment – that is, the suppression of tumor growth – was obtained by knocking down the expression of FAS with small interfering RNAs (siRNAs) [45]. The inhibition of FAS by orlistat was shown to induce caspase-8-mediated tumor cell apoptosis by upregulating DDIT4 [46]. Whereas, in mammals all functions are comprised in one large, multifunctional FAS enzyme (type I FAS), in prokaryotes they are distributed across multiple distinct enzymes (type II FAS); the same applies to apicomplexan parasites such as *P. falciparum* and *Toxoplasma gondii* [47, 48]. Notably, coccidia such as *T. gondii* and *Eimeria tenella*, as well as Cryptosporidia, contain also a large type-1 FAS enzyme in their genomes [48]. The fatty acid synthesis of protozoan parasites has been extensively reviewed recently [48, 49], and represents a valid target for pharmaceutical intervention. Because fatty acid metabolism also is an important area of research for new antibacterials, anticancer agents, and for the reduction of obesity, the cross-fertilization of these fields can be anticipated. Orlistat

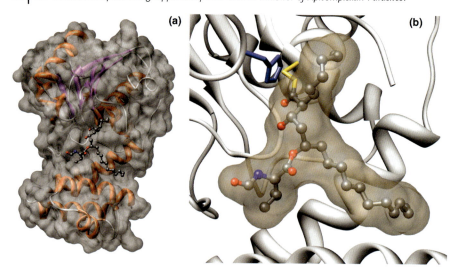

(a)　　　　　　　　　　　　　　　　**(b)**

Figure 26.3 Orlistat covalently bound to the active serine of the human FAS thioesterase domain. Crystal structure analysis revealed that orlistat was bound in the active site of two thioesterase molecules as a stable acyl–enzyme intermediate (as shown), and as the hydrolyzed product (not shown here) [44]. (a) Overall structure of one thioesterase domain with bound orlistat; (b) Close-up of orlistat covalently bound to the active site serine (yellow), histidine (blue). The asparagine of the catalytic triad has not been visualized.

had previously been shown to inhibit the growth of *T. brucei in vitro* [22]. Based on the fact that the polyketide antibiotic cerulenin inhibits both, FAS I and the bacterial ketoacyl synthase of type II FAS [50], orlistat also was evaluated as a potential type II FAS inhibitor in the apicoplast.

Orlistat's Effect on *Plasmodium falciparum* and *Toxoplasma gondii*

The effect of orlistat was determined on cell cultures of *P. falciparum* and *T. gondii*. The *in vitro* replication of blood-stage *P. falciparum* in human erythrocytes was inhibited at submicromolar concentrations, and intracellular growth of the tachyzoite form of *T. gondii* was inhibited at a similar concentration ($IC_{50} \sim 5 \times 10^2$ nM) (Figure 26.4).

In the case of *Plasmodium*, it was recently shown that type II FAS is not essential in blood stages, whereas several fatty acid biosynthesis enzymes (FabI, FabB/F, FabZ) are essential for the development of late liver-stage malarial parasites [55–57]. Therefore, the observed effect of orlistat on blood-stage *P. falciparum* can hardly be mediated by type II FAS inhibition. Currently, it can only be speculated that it is orlistat's inhibition of either a parasite or a host lipase that mediates its effect, thereby restricting the parasites' ability to convert host cell lipids into fatty acids. In the case of *T. gondii*, the situation is more complex due to the above-mentioned presence of a FAS I enzyme in the parasite, and the proven essentiality of the apicoplast-resident FAS II

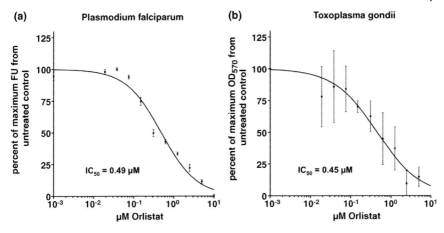

Figure 26.4 IC_{50}-values for the effects of orlistat on *P. falciparum* blood stages (a) and *T. gondii* tachyzoites (b). Orlistat was extracted from Xenical® capsules as described [45] and serially diluted into the growth medium. Parasites were cultivated under standard conditions [51, 52] in triplicates in the presence of drug or diluent only (=control) for 48 h in the case of *P. falciparum*, and 74 h for *T. gondii*. Growth inhibition was recorded as described previously [53, 54]. The values of the untreated controls were set to 100% in each case, and those for the treated cultures were calculated accordingly. IC_{50}-values were determined using PRISM™[4) 4 software (GraphPad).

machinery for parasite survival *in vitro* and *in vivo* [58]. Evidently, further experiments are required to define the molecular target(s) of orlistat in these two different apicomplexans.

Bioinformatic Analysis for Potential Orlistat Targets in Apicomplexan Parasites

The observed killing effects on *T. gondii* and *P. falciparum* in *in vitro* cultures raise the question about orlistat's mode of action in apicomplexan parasites. Therefore, bioinformatic analyses using knowledge of orlistat's mode of action in human was undertaken to identify putative molecular targets in these parasites.

Both, FAS-thioesterase and lipases (gastric and pancreatic) hydrolyze the ester bond of triglycerides, and share a highly conserved region around the serine active site. The Prosite protein domain database (www.expasy.ch/prosite/) presents the following consensus pattern for serine hydrolases (PS00120): [LIV] - {KG} - [LIVFY] - [LIVMST] - G - [HYWV] - S - {YAG} - G - [GSTAC]. The apicomplexan parasite proteomes were searched for this consensus pattern using the *Emboss* software tool Fuzzpro (www.emboss.org). The organisms included were *P. falciparum* (www.plasmodb.org), *T. gondii* (www.toxodb.org), and *E. tenella* (www.sanger.ac.uk).

---

4) PRISM is a trademark of GraphPad Software, Inc.

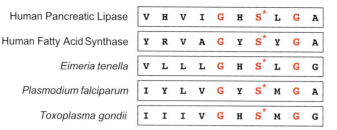

Human Pancreatic Lipase	V	H	V	I	G	H	S*	L	G	A
Human Fatty Acid Synthase	Y	R	V	A	G	Y	S*	Y	G	A
Eimeria tenella	V	L	L	L	G	H	S*	L	G	G
Plasmodium falciparum	I	Y	L	V	G	Y	S*	M	G	A
Toxoplasma gondii	I	I	I	V	G	H	S*	M	G	G

Figure 26.5 Amino acid sequence surrounding the serine active site of putative serine hydrolases in human and apicomplexan parasites. The fixed pattern G × S × G (red) is including the active serine residue (asterisk) defined in the Prosite motif (PS00120) is conserved throughout all identified putative serine hydrolases.

BLAST$^{SM5)}$ searches confirmed that the FAS type I is absent in *P. falciparum* and *E. tenella*. However, a number of proteins containing the serine hydrolase pattern can be found: 15 in *P. falciparum*, 14 in *T. gondii*, and six in *E. tenella*. Figure 26.5 shows the amino acid sequence surrounding the serine active site from the human pancreatic lipase and FAS and one representative from each of the three parasites. All five sequences differ slightly from each other, and from the consensus pattern. Nevertheless, the three fixed amino acid residues G × S × G defined in the consensus pattern are conserved throughout all identified parasite proteins. Therefore, these proteins are highly probable serine hydrolases: however, except for the coccidian FAS I, their cellular functions are currently unknown. It should be noted that other cellular targets for orlistat besides serine hydrolases have been recently proposed [59].

Conclusion

Based on knowledge of orlistat's targets in human, bioinformatic analyses have identified a number of putative targets in the apicomplexan parasites *P. falciparum*, *T. gondii*, and *E. tenella*. Orlistat could address one – or perhaps more than one – of these targets within the same parasite. Target classes other than serine hydrolases are as yet unknown for orlistat; however, provided that this is the only target class, at least one of the identified putative serine hydrolases within the same parasite must be essential for survival. This hypothesis would explain orlistat's activity on the apicomplexan parasites *P. falciparum* and *T. gondii*. Coccidiosis in chickens is an area of veterinary therapeutic interest and, therefore, preliminary investigations were performed in that area. To date, efficacy data on *E. tenella* cell cultures are pending, however. *In vivo* studies in chickens treated with orlistat have revealed no adverse effects (e.g., diarrhea), indicating its safe use. Nonetheless,

---

5) BLAST is a registered service mark of The National Library of Medicine in the United States of America.

further *in vitro* and *in vivo* studies are required to investigate the activity of orlistat against *Eimeria* spp. and other apicomplexa, and to evaluate this drug as a potential medication.

Acknowledgment

The authors thank Richard Marhöfer for his help in generating Figure 26.3. F.S. is supported by a grant from the Deutsche Forschungsgemeinschaft (Se 642/3).

References

1 Ashburn, T.T. and Thor, K.B. (2004) Drug repositioning: identifying and developing new uses for existing drugs. *Nat. Rev. Drug Disc.*, **3**, 673–683.

2 Stuart, M. (2004) Rediscovering existing drugs. *Start-Up*, **9**, 23–30.

3 Tobinick, E.L. (2009) The value of drug repositioning in the current pharmaceutical market. *Drug News Perspect.*, **22**, 119–125.

4 Geary, T.G., Woods, D.J., Williams, T., and Nwaka, S. (2009) Target identification and mechanism-based screening for anthelmintics: application of veterinary antiparasitic research programs to search for new antiparasitic drugs for human indications, in *Drug Discovery in Infectious Diseases. Antiparasitic and Antibacterial Drug Discovery: From Molecular Targets to Drug Candidates* (ed. P.M. Selzer), Wiley-VCH, Weinheim.

5 Pailet, F.A., Abadie, S.H., Smith, M.W., and Gonzalez, R.R. (1968) Chemotherapeutic heartworm control - the use of diethylcarbamine in the control of *Dirofilaria immitis* infection in dogs in clinical trials. *Vet. Med. Sm. Anim. Clin.*, **63**, 691–693.

6 McQuistion, T.E. and McDougald, L.R. (1981) Anticoccidial activity of arprinocid and halofuginone. *Vet. Parasitol.*, **9**, 27–33.

7 Villacorta, I., Peeters, J.E., Vanopdenbosch, E., Ares-Mazas, E., and Theys, H. (1991) Efficacy of halofuginone lactate against *Cryptosporidium parvum* in calves. *Antimicrob. Agents Chemother.*, **35**, 283–287.

8 Pines, M. and Nagler, A. (1998) Halofuginone: a novel antifibrotic therapy. *Gen. Pharmacol.*, **30**, 445–450.

9 Pines, M., Vlodavsky, I., and Nagler, A. (2000) Halofuginone: from veterinary use to human therapy. *Drug Dev. Res.*, **50**, 371–378.

10 Popov, Y. and Schuppan, D. (2009) Targeting liver fibrosis: strategies for development and validation of antifibrotic therapies. *Hepatology*, **50**, 1294–1306.

11 Hodges, S.J. (2008) Urologic devices incorporating collagen inhibitors. WO2008069961.

12 Sullivan, C.A. and Hodges, S.J. (2009) Medical devices incorporating collagen inhibitors. US20090028914.

13 Boettner, A., Schmid, P., and Humke, R. (1995) *In vitro* efficacy of cefquinome (INN) and other anti-infective drugs against bovine bacterial isolates from Belgium, France, Germany, The Netherlands, and the United Kingdom. *J. Vet. Med., Ser. B*, **42**, 377–383.

14 McCann, M.E., Andersen, D.R., Zhang, D., Brideau, C., Black, W.C., Hanson, P.D., and Hickey, G.J. (2004) In vitro effects and in vivo efficacy of a novel cyclooxygenase-2 inhibitor in dogs with experimentally induced synovitis. *Am. J. Vet. Res.*, **65**, 503–512.

15 Cavaliero, T., Arnold, P., Mathis, A., Glaus, T., Hofmann-Lehmann, R., and Deplazes, P. (1999) Clinical, serologic, and parasitologic follow-up after long-term allopurinol therapy of dogs naturally

infected with *Leishmania infantum. J. Vet. Intern. Med.*, **13**, 330–334.

16 Mateo, M., Maynard, L., Vischer, C., Bianciardi, P., and Miró, G. (2009) Comparative study on the short term efficacy and adverse effects of miltefosine and meglumine antimoniate in dogs with natural leishmaniasis. *Parasitol. Res.*, **105**, 155–162.

17 Goodman, C.D., Su, V., and McFadden, G.I. (2007) The effects of anti-bacterials on the malaria parasite *Plasmodium falciparum. Mol. Biochem. Parasitol.*, **152**, 181–191.

18 Ghedin, E., Hailemariam, T., DePasse, J.V., Zhang, X., Oksov, Y., Unnasch, T.R., and Lustigman, S. (2009) Brugia *malayi* gene expression in response to the targeting of the *Wolbachia* endosymbiont by tetracycline treatment. *PLoS Negl. Trop. Dis.*, **3**, e525.

19 Kinnings, S.L., Liu, N., Buchmeier, N., Tonge, P.J., Xie, L., and Bourne, P.E. (2009) Drug discovery using chemical systems biology: repositioning the safe medicine Comtan to treat multi-drug and extensively drug resistant tuberculosis. *PLoS Comput. Biol.*, **5**, e1000423.

20 Abdulla, M.-H., Ruelas, D.S., Wolff, B., Snedecor, J., Lim, K.-Ch., Xu, F., Renslo, A.R., Williams, J., McKerrow, J.H., and Caffrey, C.R. (2009) Drug discovery for schistosomiasis: hit and lead compounds identified in a library of known drugs by medium-throughput phenotypic screening. *PLoS Negl. Trop. Dis.*, **3**, e478.

21 Weisman, J.L., Liou, A.P., Shelat, A.A., Cohen, F.E., Guy, R.K., and DeRisi, J.L. (2006) Searching for new antimalarial therapeutics amongst known drugs. *Chem. Biol. Drug Des.*, **67**, 409–416.

22 Mackey, Z.B., Baca, A.M., Mallari, J.P., Apsel, B., Shelat, A., Hansell, E.J., Chiang, P.K., Wolff, B., Guy, K.R., Williams, J., and McKerrow, J.H. (2006) Discovery of trypanocidal compounds by whole cell HTS of *Trypanosoma brucei. Chem. Biol. Drug Des.*, **67**, 355–363.

23 Chiang, A.P. and Butte, A.J. (2009) Systematic evaluation of drug-disease relationships to identify leads for novel drug uses. *Clin. Pharm. Ther.*, **86**, 507–510.

24 Qu, X.A., Gudivada, R.C., Jegga, A.G., Neumann, E.K., and Aronow, B.J. (2009) Inferring novel disease indication for known drugs by semantically linking drug action and disease mechanism relationships. *BMC Bioinform.*, **10** (Suppl. 5), S4.

25 Hadvary, P., Sidler, W., Meister, W., Vetter, W., and Wolfer, H. (1991) The lipase inhibitor tetrahydrolipstatin binds covalently to the putative active site serine of pancreatic lipase. *J. Biol. Chem.*, **266**, 2021–2027.

26 Kridel, S.J., Axelrod, F., Rozenkrantz, N., and Smith, J.W. (2004) Orlistat is a novel inhibitor of fatty acid synthase with antitumor activity. *Cancer Res.*, **64**, 2070–2075.

27 Hochuli, E., Kupfer, E., Maurer, R., Meister, W., Mercadal, Y., and Schmidt, K. (1987) Lipstatin, an inhibitor of pancreatic lipase produced by *Streptomyces toxytricini*, I. Chemistry and structure elucidation. *J. Antibiot.*, **40**, 1086–1091.

28 Hadvary, P., Hochuli, E., Kupfer, E., Lengsfeld, H., and Weibel, E.K. (1985) Hexadecanoic acid and hexadecadienoic acid derivatives. EP129748.

29 Weibel, E.K., Hadvary, P., Hochuli, E., Kupfer, E., and Lengsfeld, H. (1987) Lipstatin, an inhibitor of pancreatic lipase produced by *Streptomyces toxytricini*, II. Producing organism, fermentation, isolation and biological activity. *J. Antibiot.*, **40**, 1081–1085.

30 Umezawa, H., Aoyagi, T., Hazato, T., Uotani, K., Kojima, F., Hamada, M., and Takeuchi, T. (1978) Esterastin, an inhibitor of esterase, produced by Actinomycetes. *J. Antibiot.*, **31**, 639–641.

31 Kitahara, M., Asano, M., Naganawara, H., Maeda, K., Hamada, M., Aoyagi, T., Umezawa, H., Itiaka, H.Y., and Nakamura, H. (1987) Valilactone, produced by Actinomycetes. *J. Antibiot.*, **40**, 1647–1650.

32 Mutoh, M., Nakada, N., Matsukuma, S., Ohshima, S., Yoshinari, K., Watanabe, J., and Arisawa, M. (1994) Panclicins, novel pancreatic lipase inhibitors. I. Taxonomy, fermentation, isolation and biological activity. *J. Antibiot.*, **47**, 1369–1375.

33 Barbier, P. and Schneider, F. (1987) Syntheses of tetrahydrolipstatin and absolute configuration of tetrahydrolipstatin and lipstatin. *Helv. Chim. Acta*, **70**, 196–202.

34 Case-Green, S.C., Davies, S.G., Roberts, P.M., Russel, A.J., and Thomson, J.E. (2008) Asymmetric synthesis of tetrahydrolipstatin and valilactone. *Tetrahedron: Asymm.*, **19**, 2620–2631.

35 Srinivas, P.V., Aswathanarayanappa, C., Sannachikkanna, U., and Guptha, R. (2009) Process for preparation of tetrahydrolipstatin via hydrogenation of lipstatin in polyethylene glycol. WO2009040827.

36 Simonic, I., Kramar, A., Benkic, P., Simac, A., and Vajs, A. (2007) Process for preparing crystalline forms of orlistat. EP1803714.

37 Yadav, J.S., Reddy, M.S., and Prasad, A.R. (2006) Stereoselective synthesis of (−)-tetrahydrolipstatin via Prins cyclisations. *Tetrahedron Lett.*, **47**, 4995–4998.

38 Ghosh, A.K. and Fidanze, S. (2000) Asymmetric synthesis of (−)-tetrahydrolipstatin: An anti-aldol-based strategy. *Org. Lett.*, **2**, 2405–2407.

39 Hanessian, S., Tehim, A., and Chen, P. (1993) Total synthesis of (−)-tetrahydrolipstatin. *J. Org. Chem.*, **58**, 7768–7781.

40 Pons, J.-M. and Kocienski, P. (1989) A synthesis of (−)-tetrahydrolipstatin. *Tetrahedron Lett.*, **30**, 1833–1836.

41 Dirat, O., Kouklovsky, C., and Langlois, Y. (1999) Oxazoline N-oxide-mediated [2 + 3] cycloadditions. Application to a synthesis of (−)-tetrahydrolipstatin. *Org. Lett.*, **1**, 753–755.

42 Barbier, P., Schneider, F., and Widmer, U. (1987) Stereoselective syntheses of tetrahydrolipstatin and of an analogue, potent pancreatic-lipase inhibitor containing a β-lactone moiety. *Helv. Chim. Acta*, **70**, 1412–1418.

43 Leibundgut, M., Maier, T., Jenni, S., and Ban, N. (2008) The multienzyme architecture of eukaryotic fatty acid synthases. *Curr. Opin. Struct. Biol.*, **18**, 714–725.

44 Pemble, C.W., Johnson, L.C., Kridel, S.J., and Lowther, W.T. (2007) Crystal structure of the thioesterase domain of human fatty acid synthase inhibited by orlistat. *Nat. Struct. Mol. Biol.*, **14**, 704–709.

45 Knowles, L.M., Axelrod, F., Browne, C.D., and Smith, J.W. (2004) A fatty acid synthase blockade induces tumor cell-cycle arrest by down-regulating Skp2. *J. Biol. Chem.*, **279**, 30540–30545.

46 Knowles, L.M., Yang, C., Osterman, A., and Smith, J.W. (2008) Inhibition of fatty-acid synthase induces caspase-8-mediated tumor cell apoptosis by up-regulating DDIT4. *J. Biol. Chem.*, **283**, 31378–31384.

47 Surolia, A., Ramya, T.N., Ramya, V., and Surolia, N. (2004) 'FAS't inhibition of malaria. *Biochem. J.*, **383**, 401–412.

48 Mazumdar, J. and Striepen, B. (2007) Make it or take it: fatty acid metabolism of apicomplexan parasites. *Eukaryot. Cell*, **6**, 1727–1735.

49 Goodman, C.D. and McFadden, G.I. (2008) Fatty acid synthesis in protozoan parasites: Unusual pathways and novel drug targets. *Curr. Pharm. Des.*, **14**, 901–916.

50 Omura, S. (1976) The antibiotic cerulenin, a novel tool for biochemistry as an inhibitor of fatty acid synthesis. *Microbiol. Mol. Biol. Rev.*, **40**, 681–697.

51 Trager, W. and Jensen, J.B. (1976) Human malaria parasites in continuous culture. *Science*, **193**, 673–675.

52 Roos, D.S., Donald, R.G., Morrissette, N.S., and Moulton, A.L. (1994) Molecular tools for genetic dissection of the protozoan parasite *Toxoplasma gondii*. *Meth. Cell Biol.*, **45**, 27–63.

53 Smilkstein, M., Sriwilaijaroen, N., Kelly, J.X., Wilairat, P., and Riscoe, M. (2004) Simple and inexpensive fluorescence-based technique for high-throughput antimalarial drug screening. *Antimicrob. Agents Chemother.*, **48**, 1803–1806.

54 Crawford, M.J., Thomsen-Zieger, N., Ray, M., Schachtner, J., Roos, D.S., and Seeber, F. (2006) Toxoplasma *gondii* scavenges host-derived lipoic acid despite its de novo synthesis in the apicoplast. *EMBO J.*, **25**, 3214–3222.

55 Yu, M., Kumar, T.R., Nkrumah, L.J., Coppi, A., Retzlaff, S., Li, C.D., Kelly, B.J., Moura, P.A., Lakshmanan, V., Freundlich, J.S., Valderramos, J.C., Vilcheze, C., Siedner, M., Tsai, J.H., Falkard, B., Sidhu, A.B., Purcell, L.A., Gratraud, P., Kremer, L., Waters, A.P., Schiehser, G., Jacobus, D.P., Janse, C.J., Ager, A., Jacobs, W.R. Jr, Sacchettini, J.C., Heussler, V., Sinnis, P., and Fidock, D.A. (2008) The fatty acid biosynthesis enzyme FabI plays a key role in the development of liver-stage malarial parasites. *Cell Host Microbe.*, **11**, 567–578.

56 Vaughan, A.M., O'Neill, M.T., Tarun, A.S., Camargo, N., Phuong, T.M., Aly, A.S., Cowman, A.F., and Kappe, S.H. (2009) Type II fatty acid synthesis is essential only for malaria parasite late liver stage development. *Cell Microbiol.*, **11**, 506–520.

57 Tarun, A., Vaughan, A., and Kappe, S. (2009) Redefining the role of de novo fatty acid synthesis in *Plasmodium* parasites. *Trends Parasitol.*, **25**, 545–550.

58 Mazumdar, J., Wilson, E.H., Masek, K., Hunter, C.A., and Striepen, B. (2006) Apicoplast fatty acid synthesis is essential for organelle biogenesis and parasite survival in *Toxoplasma gondii*. *Proc. Natl Acad. Sci. USA*, **103**, 13192–13197.

59 Yang, P.Y., Liu, K., Ngai, M.H., Lear, M.J., Wenk, M.R., and Yao, S.Q. (2010) Activity-based proteome profiling of potential cellular targets of Orlistat–an FDA-approved drug with anti-tumor activities. *J. Am. Chem. Soc.*, **132**, 656–666.

27
Recent Drug Discovery Against *Cryptosporidium*

*Jean-François Rossignol, J. Edward Semple, Andrew V. Stachulski, and Gilles Gargala**

Abstract

Cryptosporidium parvum, an apicomplexan protozoan, is regarded as one of the most common enteric pathogens affecting humans worldwide, and is the cause of cryptosporidiosis. *C. parvum* primarily targets the small intestine, gall bladder, lungs and conjunctiva, causing acute, watery, and non-bloody diarrhea, enteritis, anorexia, nausea/vomiting, and abdominal pain. The infection is generally self-limiting in immunocompetent people; however, its emergence as an opportunistic infection entered the forefront of public attention during the 1980s in HIV patients with CD4 counts of <50 cell mm^{-3}, leading to severe dehydration, malnutrition, and death. In 2002, nitazoxanide (NTZ) was licensed as the first FDA-approved drug in the United States for treatment of cryptosporidiosis in non-immunodeficient children and adults. Although paromomycin and azithromycin show limited efficacy, alternate treatments for cryptosporidiosis in immunodeficient patients are still urgently needed. Subsequent to the commercialization of NTZ, focused medicinal chemistry efforts at Romark led to the identification of two novel classes of compounds, the thiazolides and the isoflavones. The elucidation of structure–activity relationships (SARs) was followed by the identification of several potent inhibitors. The activities of NTZ and top lead compounds were studied *in vivo* using immunosuppressed gerbils, *Merioneas unguiculatus*, which were infected with *C. parvum*. Subsequently, two compounds emerged as potential candidates for development in humans: RM6427, an isoflavone; and RM4865, a thiazolide. The historical background and early treatments of *C. parvum*, and the development and SARs of the novel isoflavone and thiazolide drug classes, are reviewed in this chapter.

Introduction

First described by Tizzer in 1907 [1], *Cryptosporidium parvum* was regarded for many years as a rare apicomplexan protozoa that caused only a very limited number of

*Corresponding author

Apicomplexan Parasites. Edited by Katja Becker
Copyright © 2011 WILEY-VCH Verlag GmbH & Co. KGaA, Weinheim
ISBN: 978-3-527-32731-7

illnesses. Although long believed to be an occupational hazard that affected only those people exposed to infected animals, cryptosporidiosis is now probably the most common – and certainly the most devastating – gastrointestinal infection in people with AIDS. In 1993, a dramatic reemergence occurred during a large outbreak of diarrheal disease in the city of Milwaukee in the United States, where more than 400 000 people were affected; this resulted in the hospitalization of 4400 patients and led to about 100 deaths. The infectious agent was quickly identified as *C. parvum*, which had contaminated the city's water distribution system. In most of the affected individuals the symptoms were relatively mild and self-limiting, but in a significant proportion the symptoms were severe and persisted such that these patients required hospitalization. A subsequent analysis of this specific subgroup showed them to be immunocompromised, with the vast majority being infected by HIV [2]. Consequently, the quest for a treatment for cryptosporidiosis was initiated with a great degree of urgency.

The treatment of anaerobic parasitic protozoa such as *Trichomonas vaginalis*, *Entamoeba histolytica*, and *Giardia intestinalis* was revolutionized during the 1960s when the nitro-imidazole compounds such as metronidazole, tinidazole, or ornidazole were introduced [3]. Whilst the search for an antimicrobial agent to treat *C. parvum* became an urgent matter, it quickly became apparent that the development of an effective drug would present a major challenge. Fortuitously, several test systems of the infection were available at the time, including a simple cell-line infection of *C. parvum* that could be used as primary screening. In addition, various animal models of *C. parvum* experimental infection were available, using suckling mice, severe combined immune deficiency (SCID) mice, gerbils, rats, or piglets. Of these models, only the piglet elicited a robust diarrheal disease state; the rodents failed to exhibit many symptoms of cryptosporidial infection, and did not experience diarrhea. Such availability of other systems to investigate not only cryptosporidiosis but also other emerging infections (e.g., microsporidiosis) led to the discovery of new effective antiparasitic drugs, using rationally designed methodologies.

Although, during the early 1990s, the search for a drug to treat cryptosporidiosis was initiated worldwide, the process proved to be highly disappointing. with only one drug – paromomycin – being shown as partially effective. At the time, two groups – one group in the United States, using a cell-line infection with *C. parvum*, and one group in France, using an immunosuppressed rat model infected with *C. parvum* – reported their results. In total, 50 existing antimicrobial, antiviral, or antifungal agents were tested either in cell culture or *in vivo*, but with little success [4, 5].

The Discovery of the First Treatments

Paromomycin sulfate (also known as monomycin A and aminosidine) is an aminoglycoside antibiotic that was first isolated from *Streptomyces rimosus* forma *paromomycinus* by Frohardt *et al.* in 1959 [6]. In humans, the drug is almost entirely contained within the intestinal tract, and demonstrates a very low level of systemic absorption. For many years, paromomycin was primarily used as a safe and effective luminal amoebicide when combined with a systemic amoebicide (e.g., metronida-zole) for the treatment of potentially invasive intestinal amoebiasis caused by *Entamoeba histolytica*. Although the discovery of paromomycin's anticryptosporidial effects was not surprising (as it was an antiprotozoal agent), the drug's main problems were its poor biodisposition and a complete lack of absorption. Yet, despite these problems, paromomycin still demonstrated a degree of anticryptosporidial activity, even though a series of double-blind, placebo-controlled studies with the drug had failed to show any treatment benefit compared to placebo [7].

The search for a new small-molecule drug for cryptosporidiosis began by a perusal of the chemical structures of existing antiprotozoal agents, especially metronidazole. Whilst nitro-imidazole derivatives had first been reported to be effective against *T. vaginalis* by Cosar and Julou in 1959 [3], a second generation of long-acting derivatives (e.g., tinidazole and ornidazole) demonstrated much improved pharma-cokinetics, including longer plasma half-lives. At the time, the spectrum of activity of these drugs included *E. histolytica* and, a little later, also *Giardia intestinalis*. The biodisposition of these compounds was ideal, as they were almost entirely absorbed from the gastrointestinal tract and extensively distributed in various tissues, which made them very effective for treating the tissue stages of *Trichomonas* and *Entamoeba* species-based infections. As their mechanism of action was to target anaerobic protozoa, and later bacteria, it was hypothesized that they might also be effective against *Cryptosporidium* species; however, this proved not to be the case.

Hence, attention was turned to a new class of drugs that had been synthesized during the early 1970s and were related to the nitro-imidazoles [8]. These drugs – the salicylamide derivatives of 2-amino-5-nitrothiazole (a heteroaromatic ring system closely related to the nitro-imidazole ring) – proved to be highly effective against anaerobic protozoa, and also had potential efficacy against the emerging apicom-plexan parasites. Subsequently, they were shown to be effective not only in cell line assays but also in animal models of cryptosporidiosis in mice, gerbils, rats, or piglets [9–13].

NTZ was first shown to be effective against *C. parvum* in a cell-line experimental infection, with an IC_{90} of $10 \, \mu g \, ml^{-1}$ [9]. Subsequently, the abilities of NTZ (the parent compound), tizoxanide (TIZ; the active circulating metabolite), and tizox-anide-glucuronide (TIZ-*glu*; the metabolite of detoxification) to inhibit parasite development were confirmed, with an IC_{50}-value of $1.2 \, \mu g \, ml^{-1}$, for NTZ. This study showed that NTZ and its circulating metabolites were effective against the various stages of the parasite life cycle, whether sexual or asexual, intracellular or extracellular [10]. A review of the structure–activity relationship (SAR) trends for the currently available thiazolide analogs led to a retesting of TIZ, using an indirect fluorescence assay (IFA), with IC_{50}- and IC_{90}-values of 0.6 and $3.1 \, \mu g \, ml^{-1}$, respectively for TIZ (G. Gargala, unpublished results). These new data, which were more consistent with those acquired for the other analogs, are listed in Table 27.1.

When various animal models of cryptosporidial infections were employed to test orally administered NTZ, different degrees of antiparasitic activity were recorded in suckling mice [11], SCID mice [9], gerbils [13], rats [12], and piglets [9].

Table 27.1 Chemical structures, physical properties and *in vitro* inhibitory activities of nitazoxanide (NTZ), tizoxanide (TIZ) and 39 new thiazolide/thiadiazolide derivatives against *Cryptosporidium parvum* development.

Agent	MW (Da)	Thiazole ring X	Thiazole ring Y	Benzene R1	Benzene R2	Benzene R3	Benzene R4	IC50 (mg l⁻¹)	IC90 (mg l⁻¹)	IC50 (µM)	IC90 (µM)	m.p. (°C)	ClogP	TPSA	N_{rot} (ACD)	pKa (ACD. amide)	Total MR (cm3/mol)	Mass Intrinsic Solubility (ACD, 25 °C, mg l⁻¹)	Mass Solubility (ACD, unbuffered water, 25 °C, mg/L)	π	σ_m	σ_p	MR (cm3/mol)	F	R
NTZ	307	NO$_2$	H	OAc	H	H	H	1.2	10	3.9	32.6	202	1.24	119.6	4	6.18 ± 0.50	72.9	52.0	55.0	−0.28	0.71	0.78	7.36	0.67	0.16
TIZ	265	NO$_2$	H	OH	H	H	H	0.6[a]	3.1[a]	2.3	11.7	254	2.42	113.5	3	6.70 ± 0.50	63.6	24.0	25.0	−0.28	0.71	0.78	7.36	0.67	0.16
RM4815	279	NO$_2$	CH$_3$	OH	H	H	H	0.6	1[b]	2.2	3.6	240–242	2.84	113.5	3		69.3			−0.28	0.71	0.78	7.36	0.67	0.16
RM4814	279	NO$_2$	H	OMe	H	H	H	5.5	NA	19.7	NA	230–232	1.81	102.5	3	6.60 ± 0.50	69.0	39.0	42.0	−0.28	0.71	0.78	7.36	0.67	0.16
RM4802	321	NO$_2$	H	OAc	CH$_3$	H	H	0.5	1[b]	1.6	3.1		1.74	119.6	4	6.25 ± 0.50	78.8	28.0	28.0	−0.28	0.71	0.78	7.36	0.67	0.16
RM4801	279	NO$_2$	H	OH	CH$_3$	H	H	0.7	5.8[b]	2.5	20.8	208–210	2.87	113.5	3	6.77 ± 0.50	69.5	14.0	14.0	−0.28	0.71	0.78	7.36	0.67	0.16
RM4805	295	NO$_2$	H	OH	OCH$_3$	H	H	0.4	1[b]	1.4	3.4		0.85	130.6	4	6.58 ± 0.50	70.8	24.0	24.0	−0.28	0.71	0.78	7.36	0.67	0.16
RM4809	279	NO$_2$	H	OH	H	CH$_3$	H	1.4	5.0[b]	5.0	17.9	246–248	2.92	113.5	3	6.87 ± 0.50	69.5	14.0	14.0	−0.28	0.71	0.78	7.36	0.67	0.16
RM4807	279	NO$_2$	H	OH	H	H	CH$_3$	0.5	0.9[b]	1.8	3.2	234–236	2.92	113.5	3	6.77 ± 0.50	69.5	14.0	14.0	−0.28	0.71	0.78	7.36	0.67	0.16
RM4848	255	Cl	H	OH	H	H	H	2.6	6[b]	10.2	23.5	217.8–218.6	3.26	61.7	2	6.95 ± 0.50	61.9	25.0	25.0	0.71	0.37	0.23	6.03	0.41	−0.15
RM4804	311	Cl	H	OAc	CH$_3$	H	H	0.4	1[b]	1.3	3.2	146.5–147.5	2.58	67.8	3	6.50 ± 0.50	77.2	28.0	30.0	0.71	0.37	0.23	6.03	0.41	−0.15
RM4851	269	Cl	H	OH	CH$_3$	H	H	2.3	4.6[b]	8.6	17.1	195.5–197.5	3.71	61.7	2	7.02 ± 0.50	67.8	14.0	15.0	0.71	0.37	0.23	6.03	0.41	−0.15
RM4852	269	Cl	H	OH	H	CH$_3$	H	0.8	5.5[b]	3.0	20.5	205.6–206.5	3.76	61.7	2	7.12 ± 0.50	67.8	14.0	15.0	0.71	0.37	0.23	6.03	0.41	−0.15

RM4865	311	Cl	H	H	CH_3	OAc	H	0.1	3.8[b]	0.3	12.2	165–166	2.58	67.8	3		77.2	14.0	15.0	0.71	0.37	0.23	6.03	0.41	−0.15
RM4850	269	Cl	H	H	CH_3	OH	H	0.6	6.7[b]		24.9	212	3.76	61.7	2	7.02 ± 0.50	67.8	61.0	65.0	0.71	0.37	0.23	6.03	0.41	−0.15
RM4820	341	Br	H	H	H	OAc	H	0.6	4[b]	1.8	11.7	110–112	2.23	67.8	2	6.43 ± 0.50	74.4	61.0	65.0	0.86	0.39	0.23	8.9	0.44	−0.17
RM4832	299	Br	H	H	H	OH	H	2.5	8	8.4	26.8	198–200	3.41	67.8	2	6.95 ± 0.50	65.1	30.0	30.0	0.86	0.39	0.23	8.9	0.44	−0.17
RM4803	355	Br	H	H	H	OAc	CH_3	0.6	5.6	1.7	15.8	180–181	2.73	67.8	3	6.50 ± 0.50	80.3	33.0	35.0	0.86	0.39	0.23	8.9	0.44	−0.17
RM4819	313	Br	H	H	H	OH	CH_3	1.4	4.8[b]	4.5	15.4	187–188	3.86	61.7	2	7.02 ± 0.50	71.0	17.0	18.0	0.86	0.39	0.23	8.9	0.44	−0.17
RM4806	371	Br	H	H	H	OAc	OCH_3	4.6	NA	12.4	NA	NA	0.65	84.8	4	6.31 ± 0.50	81.7	48.0	52.0	0.86	0.39	0.23	8.9	0.44	−0.17
RM4822	355	Br	H	H	CH_3	OAc	H	0.7	3.7[b]	2.0	10.4	145–146	2.73	67.8	2	6.60 ± 0.50	80.3	17.0	18.0	0.86	0.39	0.23	8.9	0.44	−0.17
RM4847	313	Br	H	CH_3	H	OH	H	5.2	9.3	16.6	29.7	206–207	3.91	61.7	2	7.12 ± 0.50	71.0	33.0	33.0	0.86	0.39	0.23	8.9	0.44	−0.17
RM4821	355	Br	H	CH_3	H	OAc	H	1.7	4.4[b]	4.8	12.4	163–164	2.73	67.8	2	6.50 ± 0.50	80.3	33.0	35.0	0.86	0.39	0.23	8.9	0.44	−0.17
RM4823	355	Br	CH_3	H	H	OAc	H	2.1	4.8[b]	5.9	13.5		2.73	67.8	3	6.43 ± 0.50	80.2	33.0	35.0	0.86	0.39	0.23	8.88	0.44	−0.13
RM4858	313	Br	CH_3	H	H	OH	H	2.9	9.1	9.3	29.1	215–216	3.91	61.7	2		70.8			0.86 / +0.56	0.39 / −0.07	0.23 / −0.17	8.88 / +5.65	0.44 / −0.04	−0.17 / −0.13
RM4859	327	Br	CH_3	H	CH_3	OH	H	0.1	1[b]	0.3	3.1	177–178	4.36	61.7	2		76.7			0.86 / +0.56	0.39 / −0.07	0.23 / −0.17	8.88 / +5.65	0.44 / −0.04	−0.17 / −0.13
RM4860	389	Br	C_6H_5	H	CH_3	OH	H	0.1	1[b]	0.3	2.6	194.3–195.4	5.65	61.7	3		96.5			0.86 / +0.56	0.39 / −0.07	0.23 / −0.17	8.88 / +5.65	0.44 / −0.04	−0.17 / −0.13
RM4861	375	Br	C_6H_5	H	H	OH	H	0.4	1	1.1	2.7	198.6–199.6	5.21	61.7	3	6.77 ± 0.50	90.6	1.7	2.0	0.86 / +1.96	0.39 / +0.06	0.23 / −0.01	8.88 / +25.63	0.44 / +0.08	−0.17 / −0.08
RM4857	220	H	H	H	H	OH	H	>10	NA	>45.3	NA	250.3–251.6	2.51	61.7	2	7.62 ± 0.50	57.2	81.0	81.0	1.96	0.00	0.05	1.0	0.00	−0.08
RM4813[c]	263	H	H	H	H	OAc	H	5	NA	19.0	NA	130–132	0.61	80.1	3		64.5			0.00	0.00	0.00	1.0	0.00	0.00
RM4854	238	F	H	H	H	OH	H	1.8	8.6	7.6	36.1	201.3–203.1	2.69	61.7	2	7.34 ± 0.50	57.6	60.0	60.0	0.14	0.34	0.15	0.92	0.43	−0.34
RM4816[c]	331	CF_3	H	H	H	OAc	H	0.7	6.6	2.1	19.9	166–168	1.50	80.1	3	7.66 ± 0.50	70.1	60.0	60.0	0.88	0.43	0.54	5.0	0.38	0.19
RM4853	234	CH_3	H	H	H	OH	H	>10	NA	>47.7	NA	240	3.01	61.7	3		63.0	45.0	47.0	0.56	−0.07	−0.17	5.7	−0.04	−0.13
RM4855	262	C_3H_7	H	H	H	OH	H	6.7	NA	25.6	NA	194–195	3.94	61.7	3		72.4			1.53	−0.07	−0.15	14.98	−0.05	−0.10
RM4856	302	C_6H_{11}	H	H	H	OH	H	>10	NA	>33.1	NA	220–221	5.13	61.7	3		83.9			2.51	−0.22		26.69	−0.05	
RM4862	331	C_6H_4Cl	H	H	H	OH	H	6.1	NA	18.4	NA	277–279	5.33	61.7	3		87.3			2.67[e]	0.06[e]	−0.01[a]	31.39[e]	0.08[e]	−0.08[e]
RM4817[c]	309	SMe	H	H	H	OAc	H	6.4	9.2	20.7	29.8	142–144	1.45	80.1	4	7.44 ± 0.50	76.4	74.0	74.0	0.61	0.15	0.00	13.8	0.20	−0.18
RM4863	298	SO_2Me	H	H	H	OH	H	6.9	NA	23.2	NA	282–283	1.51	95.8	3	6.81 ± 0.50	71.2	260	260	−1.63	0.60	0.72	13.5	0.54	0.22
RM4864	340	SO_2Me	H	H	H	OAc	H	5	NA	14.7	NA	173–175	0.34	101.9	4	6.29 ± 0.50	80.6	480.0	480.0	−1.63	0.60	0.72	13.5	0.54	0.22
RM4818[d]	511	H	CH_2CO_2Et	H	H	OAc	H	>10	NA	>19.6	NA	130–132	2.70	128.6	9	NA	128.1	NA	NA	−0.16[e]	NA	0.05[e]	16.5[e]	NA	NA
RM4824	348	H	CH_2CO_2Et	H	H	OAc	H	8.4	NA	24.1	NA	NA	1.55	94.1	7	6.93 ± 0.50	87.8	56.0	59.0	−0.16[e]	NA	0.05[e]	16.5[e]	NA	NA

a) New data determined via IFA (unpublished, see text).
b) Maximal inhibitory activity >99% (complete growth inhibition).
c) 1,3,4-thiadiazole analog.
d) Bis-acetylsalicyloyl thiazolide derivative.
e) Estimated value for this parameter.
MW, molecular weight; NA, not available; TPSA, topological polar surface area.

Following these initial investigations, the clinical and commercial development of NTZ was vigorously pursued. In 2002, this led to its regulatory approval in the United States, where it currently remains the only approved drug for the treatment of cryptosporidiosis.

Nitazoxanide

The biodisposition of NTZ was the subject of several studies carried out *in vitro*, in animals and in human volunteers, including patients with AIDS and cryptosporidial diarrhea. Of these studies, the most interesting were conducted in rats, dogs, and human volunteers, using orally administered [^{14}C]-NTZ. The results showed that the drug was only partially absorbed from the gastrointestinal tract, and excreted mainly via the urine and feces. Only two metabolites were identified: TIZ, the active circulating metabolite; and TIZ-*glu*, its detoxification metabolite [14]. In human volunteers, 32% of an oral dose of [^{14}C]-NTZ was excreted in the urine as TIZ and TIZ-*glu*, and 68% in the feces as TIZ. Both, NTZ and TIZ were highly bound to the plasma proteins (>99%). In rats, when tissue concentrations were studied over time, the non-absorbed fraction of the drug was concentrated in the gut, whereas the absorbed portion of the molecule was located in the liver and biliary tract [15].

NTZ was tested in both immunocompetent and AIDS patients infected with *C. parvum*. The studies in immunocompetent patients were conducted in the Middle East and in Africa, where three double-blind, placebo-controlled studies included a total of 240 adults and children aged >12 months. When compared to a placebo, NTZ given twice daily for three consecutive days was shown to be safe and effective in reducing the duration of diarrhea, and also eliminating the parasite from the gastrointestinal tract [16–18]. In the case of immunosuppressed patients (such as those with AIDS), two double-blind, placebo-controlled studies were carried out in Mexico and Thailand in 66 and 50 adult patients, respectively. In patients with CD_4 counts >50 cells mm^{-3}, a 14-day treatment with NTZ was sufficient to reduce the duration of diarrhea and to eradicate the parasite when compared to placebo [19]. However, a 60-day treatment was required to achieve a similar result in patients with a CD_4 count <50 mm^{-3} (J.F. Rossignol, personal communication).

The subsequent clinical introduction of highly active antiretroviral therapy (HAART) rendered difficult – and somewhat less urgent – the development of NTZ in AIDS patients. Notably, cryptosporidiosis was becoming more rare, as patients were able to reconstitute their immune system under HAART treatment. Nevertheless, results obtained from the compassionate use program in the United States provided important data for NTZ from a cohort that included 365 patients. Although not controlled, this study confirmed the results of the two double-blind, placebo-controlled studies, and showed NTZ to be safe and effective in AIDS patients with cryptosporidial diarrhea, with both a reduction in the duration of diarrhea and an elimination of parasites from the feces [20].

Further research conducted in the United Kingdom and in France to identify antiparasitic agents effective against *C. parvum* led to the synthesis of two novel classes of potentially effective compounds, and their subsequent testing. Based on the results of these studies, 146 compounds were identified in the thiazolide class, while 54 were isoflavones and their deoxybenzoin precursors. From a total of 200 compounds, two emerged as potential candidates for development in humans, namely RM6427 (an isoflavone) and RM4865 (a thiazolide).

Cryptosporidium parvum Testing System

Cell Culture Assay

The test compounds were solubilized in dimethylsulfoxide (DMSO) at a maximum concentration of $5\,g\,l^{-1}$ and stored at $-20\,°C$ until used. The final concentrations in cultures ranged from 0.1 to $10\,mg\,l^{-1}$, which corresponding to a maximum DMSO concentration of 0.2% (v/v). HCT-8 cells (ATCC CRL244, MD, USA) were cultured as previously described [10, 21]. Cell toxicity induced by the test compounds was monitored via light microscopy examinations and a tetrazolium assay (Cell Titer 96 Aqueous non-radioactive cell proliferation assay; Promega, WI, USA). Absent, moderate and severe toxicities were rated as previously described [10, 21]. The *C. parvum* oocysts were obtained from Dr Naciri at the National Institute of Agronomical Research (INRA) in Nouzilly, France. They were separated from calf feces and left to excystate as previously described [10]. After sieving through a $5\,\mu m$ nucleopore filter, $2.5–5 \times 10^5$ sporozoites were added in each confluent HCT-8 monolayer well. After 2 h, the supernatants were removed and replaced by the test or control solutions. After a further 46 h, all cryptosporidial developmental stages of methanol-fixed cultures were counted in 20 microscopic fields ($\times 1250$ magnification), using an indirect immunofluorescence technique as previously described [10, 22]. Each set of experiments was conducted at least in duplicate, and the inhibitory activity (in %) calculated as:

$$[(P_{UC} - P_{TC})/(P_{UC})] \times 100$$

where P_{TC} is the mean number of parasite forms in treated cultures and P_{UC} is the mean number of parasite forms in untreated cultures.

The results were expressed in 50% inhibitory concentrations (IC_{50}), defined as the concentration (w/v) of agent that would result in a mean 50% inhibitory activity. For agents exhibiting a $\geq 90\%$ inhibitory activity, an IC_{90} was similarly calculated. The significance of differences between end-point values of experimental and control cultures was determined using Student's *t*-tests, thus assuming a normal-like distribution of values. A *P*-value <0.05 was considered to be statistically significant. High-yield, complete (asexual and sexual stages) parasite development was obtained [21, 22]. Assays were performed at host cell confluence to limit the apoptosis-inducing activity recently reported for NTZ and RM4819, for instance in human enterocytic Caco-2 cells [23]. At concentrations of 1 and $5\,mg\,l^{-1}$ of all agents (i.e., 1.96 to $22.70\,\mu M$), no alteration of HCT-8 cells was observed after a 48 h period of contact.

In Vivo Assay

The activities of the top lead compounds from the thiazolide and isoflavone classes were studied *in vivo* using immunosuppressed gerbils, *Merioneas unguiculatus*, which were infected with *C. parvum* (see Table 27.3) under two experimental conditions that differed in their duration of infection prior to treatment with drug. In the first study with RM6427, NTZ and paromomycin (control), one-month-old male and female gerbils were immunosuppressed by injections of dexamethasone (0.8 mg at two-day intervals for at least 10 consecutive days) before infection. Prior to the start of treatment, each gerbil was inoculated orally with 10^5 *C. parvum* oocysts by gavage. The test compounds were suspended in 5% carboxymethylcellulose and administered orally twice daily, at a constant dosing volume of 20 ml kg^{-1}, commencing 4 h after infection. The test drugs were generally administered for 8–12 consecutive days. The *Cryptosporidium* infection was assessed by measuring oocyst shedding in the feces, which were collected for 24 h from each animal on post-infection days 4, 8, and 12. The feces were suspended in 10% (w/v) formalin solution and homogenized. The oocysts were counted using a phase-contrast microscopy examination of smears prepared by mixing the fecal suspension with carbol fuchsine solution. Oocysts numbers were expressed per microscopic field, at a magnification of $\times 400$. All animals were killed at 12 days after infection, using deep pentobarbitone anesthesia, and necropsied. The distal ileal segments and biliary tracts were fixed in 10% formalin, cut, embedded in paraffin for blind histological examination of 4 μm sections stained with hematoxylin and eosin–saffron and Giemsa, and considered to be infected if at least one cryptosporidial developmental form was microscopically observed within an epithelial cell, by two independent investigators [13].

In the second study with RM4865 and NTZ, the gerbils were immunosuppressed prior to oocyst ingestion as described above, but were also administered dexamethasone until the end of the experiment (i.e., day 32 post-infection). On day 0, each gerbil was administered orally (by gavage) 6×10^5 *C. parvum* oocysts, and allowed to develop chronic cryptosporidiosis for 21 days before any drug treatment. On day 21 after oocyst ingestion, the animals were allocated to three treatment groups that received NTZ, RM4865, or DMSO (as control). The test compounds were dissolved in pure DMSO, and each treatment was administered at 200 mg kg^{-1} twice daily by oral gavage for eight consecutive days. On day 32 post-infection (i.e., at three days after the treatment had ceased), the gerbils were euthanized and the *Cryptosporidium* infection was assessed by using flow cytometry to count oocyst numbers in the cecum from each animal. For this, the cecal contents (stored in formalin) were vortexed for 30 s and allowed to settle for an additional 30 s; a 100 μl aliquot of the suspension was then removed and added to 900 μl of phosphate-buffered saline (PBS). After washing, the pellet containing oocysts was stained with an oocyst-specific monoclonal antibody conjugated with fluorescein isothiocyanate (Crypt-a-Glo; Waterborne, New Orleans, LA, USA). The samples were analyzed using flow cytometry (FACScalibur; Becton Dickinson Immunocytometry Systems, San Jose, CA, USA). Oocyst gating was obtained using a suspension of purified oocysts. In order to quantify the oocyst numbers, calibrated fluorescent beads (Flow-Count fluorospheres; Beckman Coulter, Fullerton, CA, USA) were added to each sample-containing tube before cytometric analysis.

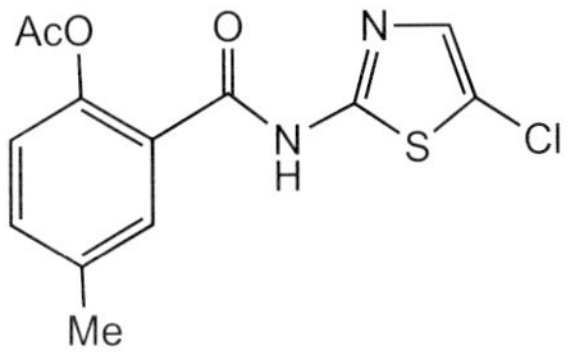

Figure 27.1 RM4865, a thiazolide and new drug candidate.

New Thiazolides

Based on the interesting activity of the prototypes NTZ and TIZ, a representative series of 144 novel thiazolide analogs was designed and synthesized which embraced a wide structural variety (Table 27.1). In addition to NTZ and TIZ, 39 new analogs were screened which featured the standard functional moieties encountered in small drug-like molecules. The amide group linking the substituted phenyl and thiazole fragments was held constant. The structural features, along with calculated physical properties (as further detailed in the following subsections) are summarized in Table 27.1.

In the thiazole moiety, substituents at the 5-position included nitro, hydrogen, halogen (F, Cl, Br), trifluoromethyl (CF_3–), alkyl (methyl, isopropyl, cyclohexyl), aryl (*p*-chlorophenyl), methylthio (CH_3S–), and methylsulfonyl (CH_3SO_2–). Although, at the thiazole 4-position, fewer analogs were prepared, attention was focused primarily on the substituents hydrogen, methyl, phenyl, and ethoxycarbonylmethylene (--CH_2CO_2Et). In the phenyl (salicyloyl) moiety, the *ortho*-hydroxy and -acetoxy groups were deemed essential pharmacophores (cf. methyl ether RM4814), and retained throughout the series. Other phenyl substituents included hydrogen, methyl, and methoxy. All 39 thiazolides were tested using the cell-line assay and, for the most effective compound, an experimental *in vivo* infection in gerbils was conducted (as described above), with the exception that gerbils were infected with $6 \times 10^5 C.$ *parvum* oocysts, allowed to develop chronic cryptosporidiosis for 21 days before treatment, and euthanized at three days after the treatment had ceased. RM4865 (see Figure 27.1 and Table 27.3) proved to be equally effective as NTZ when given at a daily dose of 400 mg kg^{-1} for eight consecutive days, and showed 89% and 96% reductions in oocyst excretion, respectively.

New Thiazolides: Structure–Activity Relationships

In this section, the SARs of 41 thiazolides and some related isosteric 1,3,4-thiadiazole analogs against *C. parvum*, are reviewed. Several potent compounds were identified, 16 of which proved to be more potent than NTZ, while 21 were more potent than TIZ. The chemical structures, physical properties and *in vitro* inhibitory activity of NTZ, TIZ, and 39 new thiazolide/thiadiazolide derivatives on *C. parvum* development, are summarized in Table 27.1.

With regards to Lipinski's [24] and Weber's [25] rules, all of these analogs can be classified as drug-like, with low to moderate molecular weights (MWs) ranging from 220 to 389 Da (except for RM4818, an inactive bis-acylated analog, with an MW of 510 Da), and with only one or two hydrogen bond donors (HBD), two to four hydrogen bond acceptors (HBA), and two to seven rotatable bonds (N_{rot}). During the course of this analysis, a wide variety of standard biological and physico-chemical parameters was perused [26]. In addition to the IC_{50}, IC_{90}, melting point, aqueous solubility, and pK_a, the $LogP$, $ClogP$, topological polar surface area (TPSA, in $\mathring{A}^2$), number of N_{rot}, molar refractivity (MR) of both the individual heterocyclic moiety 4- and/or 5-substituents and of the total molecule, were examined [27]. For fragment SAR of the core thiazole and 1,3,4-thiadiazole moieties, the classic aromatic Hansch and Lupton parameters were adopted [28–30]; these included lipophilicity (π), the electronic components sigma *meta* and *para* (σ_m, σ_p), and field ($\mathcal{F}$) and resonance (R) values, the latter two being derived from the sigma values.

As summarized in Table 27.1, in the 5-nitrothiazolide series, where $X = NO_2$, $R_2–R_4$, and $Y = H$, the IC_{50} potency decreased in the order $R_1 = OH$ (TIZ) > OAc (NTZ) $\gg$ OMe (RM4814); this activity trend was also observed in other pathogens, including viruses [31, 32]. All three groups possess HBA properties, while the acetate moiety of NTZ is readily hydrolyzed to a free hydroxyl group (TIZ) by ubiquitous esterases in the cells and biological fluids. The SAR data acquired indicated that a free hydroxyl group is required in the thiazolides for high *in vivo* potency. In addition to energetically favorable hydrogen bonding interactions with receptor surface proteins, the *ortho*-hydroxy group may also confer potency via an intramolecular hydrogen bond with the adjacent carbonyl group, inducing further degrees of conformational rigidity in addition to that imposed by the *trans*-amide bond.

In the core unsubstituted *ortho*-hydroxybenzoyl analogs, where $R_1 = OH$, $R_2–R_4$, and $Y = H$, the IC_{50} potency decreased in the order where $X = NO_2 > F > Br \sim Cl > p$-Cl-Ph > $SO_2Me \sim i$-Pr > H $\sim$ Me $\sim$ cyclo-C_6H_{11}. Thus, the greatest potency was expressed by the fluoro (RM4854, $IC_{50} = 1.8$ mg l^{-1}, 7.6 μM) and nitro (TIZ, $IC_{50} = 0.6$ mg l^{-1}, 2.3 μM) analogs, both of which were regarded as electron-withdrawing and weak HBA groups. However, the larger (MR), hydrophobic (π), polarizable bromo and chloro groups demonstrated similar, albeit somewhat reduced, activity. Hydrogen, methyl sulfone and, increasingly, hydrophobic, nonpolarizable groups including alkyl, cycloalkyl, and aryl moieties, showed a low activity, which indicates that the core SAR is governed by a balance of several factors ($LogP$, MR, π, σ_m, σ_p), whereas a high activity mandates the presence of strongly to moderately electronegative, electron-withdrawing, polarizable groups, with the halo substituents also possessing hydrophobic characteristics.

Furthermore, substitution at the Y position is well tolerated in these core analogs. When $X = $ bromo, the potency decreases in the order $Y = Ph > H > Me$. With $ClogP = 5.2$, the phenyl analog RM4861 was relatively hydrophobic, but demonstrated the highest inhibitory potency in the prototype series, with $IC_{50} = 0.4$ mg l^{-1} (1.1 μM). Similarly, when $X = NO_2$, the IC_{50} potency decreased in the order $Y = Me > H$. RM4815 ($Y = Me$), with an IC_{50} of 0.6 mg l^{-1} (2.2 μM), was fourfold more potent than TIZ ($Y = H$, IC_{50} 2.2 mg l^{-1}, 8.3 μM). Although currently limited in scope, this

SAR trend clearly indicates that further studies with a range of Y substituents is warranted.

In analog pairs where R_2–R_4 and Y substituents are held constant, acetate ($R_1 =$ OAc) prodrugs are generally more potent than their corresponding parent phenols ($R_1 =$ OH). For example, in the case where X = Br and R_2–R_4 and Y = H, RM4820 ($R_1 =$ OAc) with an $IC_{50} = 0.6\,mg\,l^{-1}$ (1.8 µM) was more potent than RM4832 ($R_1 =$ OH) with an $IC_{50} = 2.5\,mg\,l^{-1}$ (8.4 µM). This result was consistent with differences in their physical properties. Although acetates have a slightly higher TPSA, they are at least one $\log_{10}$ ($\geq$10-fold) more hydrophilic and thus more water-soluble than the parent phenols. In small molecules such as the thiazolides, the increased polarity (lower Clog$P$) of the acetates should confer not only improved solubility in biological matrices, but also improved cell permeability.

Some of the preceding core analogs were not available as the acetate prodrugs; however, a survey of 10 acetate prodrugs in the unsubstituted *ortho*-acetoxybenzoyl thiazole and thiadiazole series, where $R_1 =$ AcO, R_2–R_4 and Y = H, showed that potency decreased in the order X = Br $\geq$ CF$_3$ > NO$_2$ > SO$_2$Me > H > SMe. Thus, a trend similar to that observed in the parent phenols ($R_1 =$ OH) also holds in the acetate series, with hydrophobic, polarizable (Br, CF$_3$) and nitro moieties demonstrating the greatest potencies. Substitution at Y with a small methyl group (cf. RM4823 versus RM4820) retains activity, but in the case where X = H, substitution at Y with the larger CH_2CO_2Et moiety (RM4818 and RM4824) led to essentially inactive compounds, indicative of an unfavorable steric effect at the Y position.

Turning to thiazolides bearing substitution at the R_2–R_4 positions, the IC_{50} and IC_{90} potency trends of acetates ($R_1 =$ OAc) were up to eightfold higher than the corresponding phenols ($R_1 =$ OH). In the 5-nitrothiazolide series (X = NO$_2$, Y = H), the introduction of methyl or methoxy groups at R_2–R_4 led to a consistently increased potency versus the parent compounds NTZ and TIZ, in the order Me $\sim$ MeO > H. When $R_1 =$ OH, the methyl-substituted analogs showed potency at $R_4 > R_2 > R_3$. From this group, several highly potent analogs have emerged, including RM4801, RM4802, RM4805, RM4807, and RM4809. The most potent nitro analog in this series was RM4805 ($R_2 =$ MeO), with $IC_{50} = 0.4\,mg\,l^{-1}$ (1.4 µM), $IC_{90} = 1.0\,mg\,l^{-1}$ (3.4 µM), and this was also the most hydrophilic (Clog$P = 2.2$) with the highest TPSA (131 Å^2).

The SAR trend in the 5-chlorothiazolide series (X = Cl, Y = H) was slightly different. The introduction of methyl groups at R_2–R_4 led to an increased potency compared to the parent compound RM4848, although IC_{50} potency was at $R_4 > R_3 > R_2$. From this group, several highly potent acetate analogs were identified, which included RM4804 with $IC_{50} = 0.4\,mg\,l^{-1}$ (1.3 µM), $IC_{90} = 1.0\,mg\,l^{-1}$ (3.2 µM), and RM4865 with $IC_{50} = 0.1\,mg\,l^{-1}$ (0.32 µM), $IC_{90} = 3.8\,mg\,l^{-1}$ (12.2 µM).

In the 5-bromothiazolide series (X = Br, Y = H), the introduction of methyl or methoxy groups at R_2–R_4 led to an equal or slightly decreased potency relative to the parent compounds RM4820 and RM4832. At R_2, potency decreased in the order H > Me $\gg$ OMe. As discussed previously, when X = Br and R_2–$R_4 =$ H, the potency decreased in the order Y = Ph > H > Me. However, in contrast to the preceding analogs where Y = H, when Y = Ph or Me the introduction of methyl groups at R_2

led to dramatic increases in potency. In these investigations, other substitutions at R_2–R_4 were not studied. RM4859, with $IC_{50} = 0.1$ mg l^{-1} (0.31 µM), $IC_{90} = 1.0$ mg l^{-1} (3.1 µM) and RM4860, with $IC_{50} = 0.1$ mg l^{-1} (0.26 µM), $IC_{90} = 1.0$ mg l^{-1} (2.6 µM) were identified as two of the most potent new compounds. While RM4859 has Clog$P = 4.4$ and MR $= 77$, RM4860 with the additional orthogonal phenyl moiety is a slightly larger, more hydrophobic compound with Clog$P = 5.7$ and MR $= 97$; moreover, both had a low TPSA (62 Å^2). Less hydrophobic analogs, which would retain the general molecular topology of the parent compounds, are envisioned for both leads.

Two *in vivo* studies (see Table 27.3) demonstrated that the thiazolides NTZ and RM4865 had similar levels of efficacy in immunosuppressed gerbil cryptosporidiosis models. In the first study, where drugs were administered at 4 h post-infection, NTZ produced a 55% reduction in mean oocyst shedding at eight days post-infection ($P < 0.001$), while the standard paromomycin resulted in an 18% reduction in oocyst shedding ($P < 0.05$). In the second study, drugs were administered at 21 days post-infection, at a point where the *Cryptosporidium* infection was well established. At three days after cessation of therapy, both RM4865 and NTZ demonstrated comparable efficacies (89% and 96%, respectively), with statistically significant reductions of oocyst shedding compared to infected controls (P-values of <0.05 and <0.05, respectively). Thus, the novel non-nitro thiazolide RM4865 was shown to be as effective as the nitro-thiazolide prototype NTZ in reducing oocyst shedding.

In conclusion, SAR studies were conducted on a group of 41 thiazolides and isosteric 1,3,4-thiadiazole analogs that were tested against the parasite *C. parvum*. Several potent compounds were identified, 16 of which were more potent than NTZ, and 21 of which were more potent than TIZ. From this group, RM4805, a 5-nitrothiazolide featuring a methoxy group at R_2, RM4865, a 5-chlorothiazolide acetate prodrug featuring methyl at R_4, and the 5-bromothiazolides RM4860 (Y $=$ Ph) and RM4859 (Y $=$ Me), both featuring methyl groups at R_2, were identified as representative prototypes from this group. The thiazolides were tested using the cell-line assay and one of the most potent analogs, RM4865, demonstrated efficacy *in vivo* similar to NTZ against oocyst shedding ($P < 0.05$), when each was given at a daily dose of 400 mg kg^{-1} for eight consecutive days in the gerbil model. Only highly effective drugs are readily identified in this model, where gerbils are rendered very susceptible to infection by treatment with the powerful immunosuppressive drug, dexamethasone. Although several physico-chemical parameters were utilized for establishing useful SAR trends, high potency resulted from a careful balance between several of these parameters, including steric, electronic, hydrophilic, and hydrophobic factors. Further studies including quantitative SAR (QSAR) trends for thiazolides employing several of these parameters in multiple regression analysis are currently under way, the results of which will be reported in a forthcoming publication.

New Isoflavones

A series of isoflavone derivatives was also synthesized, starting with genistein (Table 27.2, R_1–R_3, R_5–$R_6 =$ H, $R_4 =$ OH), a naturally occurring compound that has been widely studied as a potential therapeutic agent. In particular, genistein is known to be

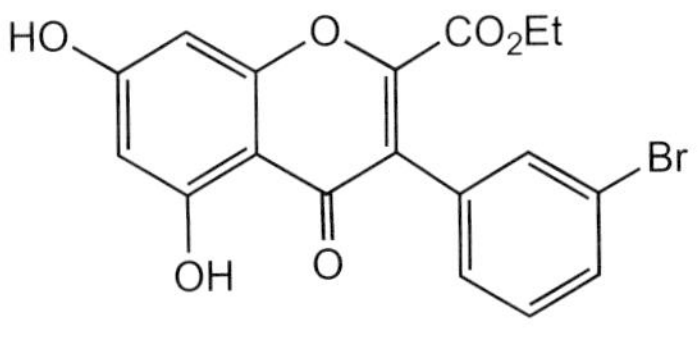

Figure 27.2 RM6427, an isoflavone and new drug candidate.

an inhibitor of protein tyrosine kinases found in the epidermal growth factor receptors (EGRFs). As EGRF peptides can mediate host–parasite infections, they appear to be potential targets in antiparasitic chemotherapy, and consequently genistein was identified as a logical starting point. Subsequently, 42 isoflavones and 12 of their precursors deoxybenzoins were synthesized. All of these generally retained the 5,7-dihydroxyisoflavone core of genistein, and were direct genistein analogs (2-*H* isoflavones), 2-carboxy isoflavones and the precursor deoxybenzoins (Figure 27.2). Their structural features and calculated physical properties are summarized in Table 27.2, and further detailed in the following section. The genistein analogs were substituted on the benzene ring with NO_2, NH_2, $NH–CO–CH_3$, Br, Cl and F in all positions on the benzene rings in lieu of the hydroxyl group (OH) in the *para* position of the genistein. Among this group, 15 isoflavones and five deoxybenzoin analogs were tested using the cell-line assay, and for the most effective compounds an experimental infection in gerbils was investigated (as described above). Typically, RM6427 (Table 27.3, First Study) was found to be more effective than both NTZ and paromomycin when administered 4 h post-infection at a daily dose of 200 mg kg^{-1} for eight consecutive days. The three compounds showed 91%, 55% and 18% reductions of the parasite infection, respectively [33].

New Isoflavones: Structure–Activity Relationships
In this section, the SARs of 15 isoflavones and five related deoxybenzoin analogs against *C. parvum* are reviewed. As shown in Table 27.2, four potent compounds were identified from this group. Similar to the thiazolides, each of these analogs may be classified as drug-like and obeys Lipinski's [24] and Weber's [25] rules. Moreover, they have low to moderate MWs ranging from 262 to 419 Da, two HBDs, four to six HBAs, and one to three N_{rot}. As detailed above, for the SAR analysis a wide variety of standard biological and physico-chemical parameters was perused [26, 27]. In the isoflavone series (see Table 27.2), eight analogs were synthesized where R_1 was hydrogen and $R_2–R_4$ were selected from either halo (F, Cl, Br) or nitro groups. The preliminary SAR of these phenyl-substituted analogs indicated a preference for the relatively larger, electronegative, polarizable nitro or bromo groups at R_4 over R_3 (cf. total MR), while the trend with the smaller fluoro atom was opposite, with R_3 preferred over R_4. At the *para*-position (R_4), the IC$_{50}$ potency decreased in the order Br > NO_2 > F; at the *meta*-position (R_3), IC$_{50}$ potency decreased in the order F > Br ~ NO_2; and at the

Table 27.2 Chemical structures, physical properties and *in vitro* inhibitory activities of dihydroxyisoflavone and trihydroxydeoxybenzoin derivatives on *Cryptosporidium parvum* development.

Agent	MW	R_1	R_2	R_3	R_4	R_5	R_6	IC$_{50}$ (mg l^{-1})	MI (%)[a]	IC$_{50}$ (µM)	ClogP	LogP	TPSA	N$_{rot}$	Total MR (cm³ mol^{-1})	Physico-chemical parameters of phenyl substituents					
																π	σ_m	σ_p	MR (cm³ mol^{-1})	F	R
Isoflavones:																					
RM-6403	299.2	H	H	H	NO2	H	H	1.25	98 ± 1.4	4.2	2.81	2.16	118.6	2	76.3	−0.28	0.71	0.78	7.36	0.67	0.16
RM-6411	299.2	H	H	NO2	H	H	H	2.3	99.5 ± 0.3	7.7	2.81	2.16	118.6	2	76.3	−0.28	0.71	0.78	7.36	0.67	0.16
RM-6424	333.1	H	Br	H	H	H	H	2.4	98.8 ± 1.7	7.2	3.93	2.96	66.8	1	77.6	0.86	0.39	0.23	8.9	0.44	−0.17
RM-6425	333.1	H	H	H	Br	H	H	0.9	99.4 ± 0.8	2.7	3.93	2.96	66.8	1	77.6	0.86	0.39	0.23	8.9	0.44	−0.17
RM-6436	288.7	H	Cl	H	H	H	H	1.8	96.9 ± 1.6	6.2	3.53	2.69	66.8	1	74.5	0.71	0.37	0.23	6.03	0.41	−0.15
RM-6439	272.2	H	F	H	H	H	H	5.75	77.6 ± 2.6	21.1	3.21	2.29	66.8	1	70.3	0.14	0.34	0.15	0.92	0.43	−0.34
RM-6440	272.2	H	H	F	H	H	H	0.8	98.1 ± 2.6	2.9	3.21	2.29	66.8	1	70.3	0.14	0.34	0.15	0.92	0.43	−0.34
RM-6441	272.2	H	H	H	F	H	H	3.4	97.1 ± 3.5	12.5	3.21	2.29	66.8	1	70.3	0.14	0.34	0.15	0.92	0.43	−0.34
RM-6426	405.2	CO2Et	Br	H	H	H	H	3.75	96.5 ± 2.2	9.3	4.32	2.47	93.1	3	94.5	0.86	0.39	0.23	8.9	0.44	−0.17

	MW							MI												
RM-6427	405.2	CO$_2$Et	H	Br	H	H	0.75	97 ± 2.6	1.85	4.32	2.47	93.1	3	94.5	0.86	0.39	0.23	8.9	0.44	−0.17
RM-6448[b]	419.2	CO$_2$Et	H	Br	H	H	2.85	83.4 ± 4.5	6.8	5.14	2.73	82.1	3	100.0	0.86	0.39	0.23	8.9	0.44	−0.17
RM-6428	405.2	CO$_2$Et	H	H	Br	H	0.9	99 ± 1.4	2.2	4.32	2.47	93.1	3	94.5	0.86	0.39	0.23	8.9	0.44	−0.17
RM-6442	360.8	CO$_2$Et	Cl	H	H	H	2.4	100 ± 0	6.7	4.22	2.19	93.1	3	91.4	0.71	0.37	0.23	6.03	0.41	−0.15
RM-6443	360.8	CO$_2$Et	H	Cl	H	H	0.85	90.1 ± 6.2	2.4	4.22	2.19	93.1	3	91.4	0.71	0.37	0.23	6.03	0.41	−0.15
RM-6446	344.3	CO$_2$Et	H	F	H	H	3	91.2 ± 3.3	8.7	3.90	1.79	93.1	3	87.2	0.14	0.34	0.15	0.92	0.43	−0.34
Deoxybenzoins:																				
RM-6430	278.7	—	Cl	H	H	H	3.9	68.1 ± 3.9	14.0	3.27	2.58	77.8	3	71.0	0.71	0.37	0.23	6.03	0.41	−0.15
RM-6431	278.7	—	H	Cl	H	H	6.9	52 ± 9.5	24.8	3.27	2.58	77.8	3	71.0	0.71	0.37	0.23	6.03	0.41	−0.15
RM-6433	262.2	—	F	H	H	H	7.75	66.6 ± 7.8	29.6	2.7	2.18	77.8	3	66.8	0.14	0.34	0.15	0.92	0.43	−0.34
RM-6434	262.2	—	H	F	H	H	7.75	69 ± 10.2	29.6	2.7	2.18	77.8	3	66.8	0.14	0.34	0.15	0.92	0.43	−0.34
RM-6435	262.2	—	H	H	F	H	4.75	63.9 ± 6.7	18.1	2.7	2.18	77.8	3	66.8	0.14	0.34	0.15	0.92	0.43	−0.34

a) MI, maximum inhibition of *Cryptosporidium parvum* form development. MI results are expressed as mean ± standard deviation.
b) Methoxy group at *.
MW, molecular weight.

Table 27.3 In vivo efficacy of RM6427, RM4865, nitazoxanide, and paromomycin against *Cryptosporidium parvum* in immunosuppressed Mongolian gerbils.

Agent	No. of animals	Regimen	No. of animals with no oocyst shedding (day post-infection)	% Reduction in mean oocyst shedding[a] (day post-infection)	No. of animals (day 12 post-infection) with no intracellu-lar parasites in:	
					Ileum	Biliary tract
First Study:						
None (5% CMC)	10	None	0		2	4
RM6427	10	200 mg kg^{-1} day^{-1} × 8 days*	6(8)[b]	90.5(8)[e]	5[f]	8[f]
RM6427	10	400 mg kg^{-1} day^{-1} × 8 days*	6(8)[b]	92.0(8)[e]	7[i]	10[k]
Nitazoxanide	14	200 mg kg^{-1} day^{-1} × 12 days*	2(8)[c]	55.0(8)[b]	11[j]	9[l]
Paromomycin	14	100 mg kg^{-1} day^{-1} × 12 days*	2(8)[c]	17.6(8)[f]	9[f]	9[f]
Second Study:						
None (DMSO)	5	None	0		ND	ND
RM4865	5	400 mg kg^{-1} day^{-1} × 8 days**	0(32)	89(32)[f]	ND	ND
Nitazoxanide	5	400 mg kg^{-1} day^{-1} × 8 days**	1(32)[c]	96(32)[f]	ND	ND

a) [(Mean oocyst count in treated animals)/(mean oocyst count in untreated animals)] × 100.
b) $P < 0.001$.
c) $P = 0.0001$.
d) $P < 0.001$.
e) $P < 0.0001$.
f) $P < 0.05$.
g) $P < 0.01$.
h) $P < 0.02$.
i) $P < 0.067$.
j) $P < 0.011$.
k) $P < 0.002$.
l) $P < 0.005$.
* Commencing 4 h after infection with 10^5 oocysts.
** Commencing 21 days after infection with 6×10^5 oocysts.
DMSO, dimethylsulfoxide; ND, not determined.

ortho-position (R$_2$), IC$_{50}$ potency decreased in the order Cl ≫ F. From this group, the *para*-bromo analog RM6425 (IC$_{50}$ = 0.9 mg l; 2.7 μM) and the *meta*-fluoro analog RM6440 (IC$_{50}$ = 0.8 mg l^{-1}; 2.9 μM) emerged as potent analogs.

The replacement of hydrogen at R$_1$ with carboethoxy (–CO$_2$Et, seven analogs) led to increases in potency, although the inhibitory activity was seen to depend on the substitution pattern at R_2–R_4. When $R_2 = Cl$, the potency was decreased at R_1 in the order $H > CO_2Et$, most likely due to unfavorable steric interactions between the spatially close *ortho*-chloro and ester moieties. However, when R_3 or $R_4 = Br$, the potency was decreased at R_1 in the order $CO_2Et > H$. From this group of 15 analogs, RM6427 ($R_1 = CO_2Et$, $R_3 = Br$) with an $IC_{50} = 0.75$ mg l^{-1} (1.85 µM) and RM6428 ($R_1 = CO_2Et$, $R_4 = Br$) with an $IC_{50} = 0.9$ mg l^{-1} (2.2 µM) were identified as the top lead compounds. Methylation of the 7-hydroxyl group of RM6427 gave RM6448, with an $IC_{50} = 2.85$ mg l^{-1} (6.8 µM), a 3.7-fold reduction in inhibitory potency relative to the phenolic parent compound RM6427. As discussed in the preceding section, when the isoflavones were tested using the cell-line assay, the most potent analog RM6427 (Table 27.3, First Study) was more effective than NTZ or paramomycin when given at a daily dose of 200 mg kg^{-1} for eight consecutive days in the gerbil model.

Screening results for the five deoxybenzoin intermediates RM6430, RM6431 and RM6433-6435 indicated only low to modest levels of inhibitory activity against *C. parvum*, with RM6430 ($R_2 = Cl$) showing the highest potency ($IC_{50} = 3.9$ mg l^{-1}; 14.0 µM).

In conclusion, SAR studies were conducted on a group of 15 isoflavones and five related deoxybenzoin analogs, which were tested against the parasite *C. parvum*. From this group, RM6425, a *p*-bromophenyl isoflavone, RM6440, a *m*-fluorophenyl isoflavone, and the 2-carboethoxy isoflavones RM6427 ($R_3 = Br$) and RM6428 ($R_4 = Br$) were identified as potent lead prototypes from this group. The most potent isoflavone analog, RM6427, proved to be more effective than NTZ or paramomycin *in vivo* in a gerbil model of cryptosporidiosis. In addition to geometric constraints imposed by the substitution pattern on the phenyl moiety, the relative potency in this series appears to be determined by a combination of steric (MR), hydrogen bond donation (cf. isoflavone moiety 7-OH versus 7-OMe), electronic (σ_m, σ_p, *F* and R), and hydrophobic (π, Clog*P*) effects.

Conclusion

The discovery of the anticryptosporidial activity of paramomycin was clearly the result of a massive screening effort of existing antimicrobial agents, which at that point in time were considered as attractive and potentially effective against the new emerging protozoa. To some extent, the same could be said for NTZ, although the drug was not commercially available and had to undergo a complete development program in order to obtain regulatory approval. Yet, it did so, and was approved for the treatment of *C. parvum* in the United States in 2002. The follow-up program undertaken by Romark Laboratories targeted two novel classes of potentially effective antimicrobials: (i) the thiazolides, which represented the second generation of NTZ analogs; and (ii) a completely new class, the isoflavones. This research was greatly facilitated by the availability of an efficient and rapid screening system that utilized a cell culture infection with *C. parvum*, followed by a robust animal model – the gerbil – that was capable of simultaneously producing both intestinal and biliary cryptosporidiosis. These tools were essential in order to identify test compounds that were efficacious against disseminated cryptosporidiosis, a condition often seen in advanced immunosuppression, such as that seen in AIDS patients.

Subsequently, the program identified two novel drug candidates – RM4865 (a thiazolide) and RM6427 (an isoflavone) – both of which were more effective than NTZ in cell assays, and could be shown to be at least as potent in reducing oocyst shedding. In contrast, RM6427 demonstrated a significantly greater potential than NTZ and paromomycin for the treatment of biliary cryptosporidiosis. These results were obtained after synthesizing 200 drug candidates, only 61 of which were screened – a remarkably small number of compounds.

References

1 Tyzzer, E.E. (1907) A sporozoan fund in the peptic glands of the common mouse. *Proc. Soc. Experiment. Biol. Méd.*, **5**, 12–13.

2 MacKenzie, W.R., Hoxie, N.J., Proctor, M.E., Gradus, M.S., Blair, K.A., Peterson, D.E., Kamierczak, J.J., Addis, D.G., Fox, K.R., Rose, J.D., and Davis, J.P. (1994) A massive outbreak in Milwaukee of *Cryptosporidium* infection transmitted through the public water supply. *N. Engl. J. Med.*, **331**, 161–167.

3 Cosar, C. and Julou, L. (1959) Activité de l'1-(hydroxymethyl)-2-methyl-5-nitoimidazole (8823RP) vis à vis des infections expérimentales à *Trichomonas vaginalis. Ann. Inst. Pasteur*, **96**, 238–241.

4 Woods, K.M., Nesterenko, M.V., and Upton, S.J. (1996) Efficacy of 101 antimicrobials and other agents on the development of *C. parvum* in vitro. *Ann. Trop. Med. Parasitol.*, **90**, 603–615.

5 Lemeteil, D., Roussel, F., Favennec, L., Ballet, J.J., and Brasseur, P. (1993) Assessment of candidate anticryptosporidial agents in an immunosuppressed rat model. *J. Infect. Dis.*, **167**, 766–769.

6 Frohardt, R.P., Haskell, T.H., and Ehrlich, J. (1959) Antibiotic and method for obtaining same. United States Patent No. 2,916,485.

7 White, A.C. Jr, Chappell, C.L., Hayat, C.S., Kimball, K.T., Flanigan, T.P., and Goodgame, R.W. (1994) Paromomycin for cryptosporidiasis in AIDS: a prospective double-blind trial. *J. Infect. Dis.*, **170**, 419–424.

8 Rossignol, J.F. and Cavier, R. (1975) 2-Benzamido-5-nitrothiazoles. *Chem. Abstract.*, **83**, 28216n.

9 Theodos, C.M., Griffith, J.K., D'Onfro, J., Fairfield, A., and Tzipori, S. (1998) Efficacy of nitazoxanide against *Cryptosporidium parvum* in cell culture and in animal models. *Antimicrob. Agents Chemother.*, **42**, 1959–1965.

10 Gargala, G., Delaunay, A., Li, X.D., Brasseur, P., Favennec, L., and Ballet, J.J. (2000) Efficacy of nitazoxanide, tizoxanide and tizoxanide-glucuronide against *Cryptosporidium parvum* development in sporozoite infected HCT-8 enterocytic cells. *J. Antimicrob. Chemother.*, **46**, 57–60.

11 Blagburn, B., Drain, K., Land, T., Kinard, R., Moore, P.H., Lindsay, D., Patrick, D., Boykin, D., and Tidwell, R. (1998) Comparative efficacy evaluation of dicationic carbazole compounds, nitazoxanide and paromomycin against *Cryptosporidium parvum* in neonatal mouse model. *Antimicrob. Agents Chemother.*, **42**, 2877–2882.

12 Li, X., Brasseur, P., Agnamey, P., Lemeteil, D., Favennec, L., Ballet, J.J., and Rossignol, J.F. (2003) Long-lasting anticryptosporidial activity of nitazoxanide in an immunosuppressed rat model. *Folia Parasitol.*, **50**, 19–22.

13 Baishanbo, A., Gargala, G., Duclos, C., François, A., Rossignol, J.F., Ballet, J.J., and Favennec, L. (2006) Efficacy of nitazoxanide and paromomycin on biliary tract cryptosporidiosis in an immunosuppressed gerbil model. *J. Antimicrob. Chemother.*, **57**, 353–355.

14 Rossignol, J.F. and Stachulski, A. (1999) Synthesis and antibacterial activities of tizoxanide and its *O*-aryl glucuronide. *J. Chem. Res. (S)*, 44–45.

15 Broekhuysen, J., Stockis, A., Lins, R.L., De Graeve, J., and Rossignol, J.F. (2000) Nitazoxanide: pharmacokinetics and metabolism in man. *Int. J. Clin. Pharm. Ther.*, **38**, 387–394.

16 Rossignol, J.F., Ayoub, A., and Ayers, M.S. (2001) Treatment of diarrhea caused by cryptosporidium parvum: a prospective randomized double-blind placebo-controlled study of nitazoxanide. *J. Infect. Dis.*, **184**, 103–106.

17 Amadi, B., Mwiya, M., Musuku, J., Watuka, A., Sianongo, S., Ayoub, A., and Kelly, P. (2002) Effect of nitazoxanide on morbidity and mortality in Zambian children with cryptosporidiasis: a randomized controlled trial. *Lancet*, **360**, 1375–1380.

18 Rossignol, J.F., Kabil, S.M., Younis, A.M., and El-Gohary, Y. (2006) Double-blind placebo controlled study of nitazoxanide in the treatment of cryptosporidial diarrhea in 90 immunocompetent adults and adolescents from the Nile Delta of Egypt. *Clin. Gastroenterol. Hepatol.*, **4**, 320–324.

19 Rossignol, J.F., Hidalgo, H., Feregrino, M., Higuera, F., Gomez, W.H., Romero, J.L., Padierna, J., Geyne, A., and Ayers, M.S. (1998) A double-blind placebo controlled study of nitazoxanide in the treatment of cryptosporidial diarrhea in AIDS patients. *Trans. R. Soc. Trop. Med. Hyg.*, **92**, 663–666.

20 Rossignol, J.F. (2006) Nitazoxanide in the treatment of acquired immune deficiency syndrome-related cryptosporidiosis: results of the united states compassionate use program in 365 patients. *Aliment. Pharmacol. Ther.*, **24**, 887–894.

21 Gargala, G., Baishanbo, A., Favennec, L., François, A., Ballet, J.J., and Rossignol, J.F. (2005) Inhibitory activities of epidermal growth factor receptor tyrosine kinase-targeted dihydroxyisoflavone and trihydroxydeoxybenzoin derivatives on *Sarcocystis neurona, Neospora caninum* and *Cryptosporidium parvum* development.

Antimicrob. Agents Chemother., **49**, 4628–4634.

22 Gargala, G., Delauney, A., Favennec, L., Brasseur, P., and Ballet, J.J. (1999) Enzyme immunoassay detection of *Cryptosporidium parvum* inhibition by sinefungin in sporozoite infected HCT-8 enterocytic cells. *Int. J. Parasitol.*, **29**, 703–709.

23 Müller, J., Sidier, D., Nachbur, U., Wastling, J., Brunner, T., and Hemphill, A. (2008) Thiazolides inhibit growth and induce glutathione-*S*-transferase Pi (GSTP1)-dependent cell death in human colon cancer cells. *Int. J. Cancer*, **123**, 1797–1806.

24 Lipinski, C.A., Lombardo, F., Dominy, B.W., and Feeney, P.J. (1997) Experimental and computational approaches to estimate solubility and permeability in drug discovery and development settings. *Adv. Drug Deliv. Rev.*, **23**, 3–25.

25 Veber, D.F., Johnson, S.R., Cheng, H.Y., Smith, B.R., Ward, K.W., and Kopple, K.D. (2002) Molecular properties that influence the oral bioavailability of drug candidates. *J. Med. Chem.*, **45**, 2615–2623.

26 In addition to references cited herein, physical property parameters were calculated using Advanced Chemistry Development (ACD/Labs) Software V9.04 for Solaris; Advanced Chemistry Development, Inc., Toronto, Ontario, Canada M5C 1T4 and ChemDraw Ultra v. 11.0, CambridgeSoft, Cambridge, MA 02140.

27 Molar refractivity (MR) is a steric effect parameter that describes the relative volume occupied by a functional group and is derived from an equation containing MW, density, and index of refraction terms. For a recent review, see Estrada, E. (2008) How the parts organize in the whole? A top-down view of molecular descriptors and properties for QSAR and drug design. *Mini Rev. Med. Chem.*, **8**, 213–221.

28 Hansch, C., Leo, A., and Taft, R.W. (1991) A survey of Hammett substituent constants and resonance and field parameters. *Chem. Rev.*, **91**, 165–195.

29 Hansch, C., Leo, A., Unger, S., Kim, K.H., Nikaitani, D., and Lien, E.J. (1973) "Aromatic" substituent constants for structure–activity correlations. *J. Med. Chem.*, **16**, 1207–1216.

30 Kubinyi, H. (2004) 2D QSAR models: Hansch and Free-Wilson analyses, in *Computational Medicinal Chemistry for Drug Discovery* (eds P. Bultinck, H.D. Winter, W. Langenaeker, and J.P. Tollenaere), Marcel Dekker, Inc., New York, pp. 539–570.

31 Esposito, M., Müller, N., and Hemphill, A. (2007) Structure–activity relationships from in vitro efficacies of the thiazolide series against intracellular apicomplexan protozoan *Neospora caninum*. *Int. J. Parasitol.*, **37**, 183–190.

32 Gargala, G., Le Goff, L., Ballet, J.J., Favennec, L., Stachulski, A.V., and Rossignol, J.F. (2010) Evaluation of new thiazolide/thiadiazolide derivatives reveals nitro-group-independent efficacy against Cryptosporidium parvum. *Antimicrob. Agents Chemother.*, **54**, 1315–1318.

33 Stachulski, A.V., Berry, N.G., Lilian Low, A.C., Moores, S.L., Row, E., Warhurst, D.C., Adagu, I.S., and Rossignol, J.F. (2006) Identification of isoflavone derivatives as effective anticryptosporidial agents in vitro and in vivo. *J. Med. Chem.*, **49**, 1450–1454.

Index

Apicomplexan Parasites. Edited by Katja Becker
Copyright © 2011 WILEY-VCH Verlag GmbH & Co. KGaA, Weinheim
ISBN: 978-3-527-32731-7